Urban Patterns

Urban Patterns:
STUDIES IN HUMAN ECOLOGY

Revised Edition

Edited by
GEORGE A. THEODORSON

The Pennsylvania State University Press
University Park and London

To my Parents

Library of Congress Cataloging in Publication Data

Main entry under title:

Urban patterns.

   Rev. ed. of: Studies in human ecology. [1961]
   Includes bibliography and index.
   1. Human ecology—Addresses, essays, lectures. I. Theodorson, George A.
HM206.U7   1982   304.2   81-83145
ISBN 0-271-00297-2   AACR2

Copyright © 1982 The Pennsylvania State University
All rights reserved
Designed by Dolly Carr
Printed in the United States of America

Contents

Preface	ix
Part I: The Development of the Ecological Framework	1
Introduction	3
Section A: *Early Spatial Studies and the Rise of "Classical" Human Ecology*	8
Century-Old Ecological Studies in France *M. C. Elmer*	8
English Ecology and Criminology of the Past Century *Yale Levin and Alfred Lindesmith*	11
An Early Study in Medical Ecology *L. D. Stamp*	19
Human Ecology *Robert E. Park*	20
The Scope of Human Ecology *R. D. McKenzie*	28
The Growth of the City: An Introduction to a Research Project *Ernest W. Burgess*	35
The Pattern of Movement of Residential Rental Neighborhoods *Homer Hoyt*	42
The Natural Areas of the City *Harvey W. Zorbaugh*	50
The Distribution of Commercialized Vice in the City: A Sociological Analysis *Walter C. Reckless*	55
Vicksburg: A Study in Urban Geography *Preston E. James*	61
The Gold Coast and the Slum *Harvey W. Zorbaugh*	67

Section B: *The Decline of Classical Ecology* — 73

"Community" and Ecological Studies — 73
Milla A. Alihan

The Concept of Natural Area — 78
Paul Hatt

A Re-examination of Ecological Theory — 82
A. B. Hollingshead

Ecological Correlation Reexamined: A Refutation of the Ecological Fallacy — 88
Donald J. Bogue and Elizabeth J. Bogue

Section C: *Reconciling Conflicting Conceptions of Social Ecology: Neo-Orthodox Approach vs. Sociocultural Approach* — 104

Ecology and Human Ecology — 104
Amos H. Hawley

Human Ecology, Space, Time, and Urbanization — 111
Amos H. Hawley

Social Morphology and Human Ecology — 115
Leo F. Schnore

From Social System to Ecosystem — 123
Otis Dudley Duncan

Sentiment and Symbolism as Ecological Variables — 129
Walter Firey

Communities in the Salt Lake Basin — 137
Albert L. Seeman

The Origin and Spread of the Grid-Pattern Town — 142
Dan Stanislawski

Issues in Sociocultural Ecology — 150
Walter Firey and Gideon Sjoberg

Sociocultural Versus Neoclassical Ecology: A Contribution to the Problem of Scope in Sociology — 165
Kenneth D. Bailey and Patrick Mulcahy

Part II: Human Ecology as a Framework for the Study of the City — 173

Introduction — 175

Section A: *Ethnic and Racial Groups* — 179

Negro Harlem: An Ecological Study — 179
E. Franklin Frazier

Black Harlem Revisited: Patterns of Ecological and Social Organizational Change, 1940–1970 — 188
Walter R. Allen and Walter C. Farrell, Jr.

Cultural Variables in the Ecology of an Ethnic Group — 194
Christen T. Jonassen

The Ecology of Norwegian–Americans in Metropolitan New York from 1940–1980 — 202
Knight E. Hoover

CONTENTS

Ethnic Boundaries: A Comparison of Two Urban Neighborhoods ... 207
   Leo Driedger

Section B: *Neighborhoods* ... 218

Beale Street, Memphis: A Study in Ecological Succession ... 218
   Robert W. O'Brien

Notes on Beale Street Forty Years Later, and "Beale Street" and Temporal Ecology ... 221
   Robert W. O'Brien and Robert M. O'Brien

The Contrived Community: 1970–1980 ... 224
   Gerald D. Suttles

Children of the Ghetto: The Evolution of an Urban Neighborhood ... 230
   Laurence S. Rosen

Changing Land-Use Patterns in SoHo: Residential Invasion of an Industrial Area ... 241
   James R. Hudson

Cognitive Dimensions of Space and Boundary in Urban Areas ... 249
   Roger M. Downs

Section C: *Mental Illness* ... 257

The Ecology of Mental Disorder ... 257
   Leo Levy and Louis Rowitz

Section D: *Commuting* ... 266

Urban Daytime Population: A Field for Demographic–Ecological Analysis ... 266
   Donald L. Foley

Nighttime and Daytime Populations of Large American Suburbs ... 273
   Avery M. Guest

The Journey to Work in Australia ... 287
   Ian Manning

Section E: *Social Area Analysis (Factorial Ecology) of the City* ... 297

Residential Area Characteristics: Research Methods for Identifying Urban Sub-areas—Social Area Analysis and Factorial Ecology ... 297
   R. J. Johnston

Social Area Analysis and Factorial Ecology: A Review of Substantive Findings ... 316
   Bernd Hamm

Testing the Theory of Social Area Analysis: The Ecology of Cairo, Egypt ... 338
   Janet L. Abu-Lughod

Part III: Urban Patterns in Different Cultural Settings ... 353

Introduction ... 355

The Social Ecology of Guatemala City ... 359
   Theodore Caplow

The Social Ecology of Latin-American Cities: Recent Evidence ... 374
   Bruce London

The Anatomy of Eleven Towns in Michoacán ... 379
   Dan Stanislawski

The Preindustrial City — 386
Gideon Sjoberg

A Society Without Wheels — 394
Richard W. Bulliet

Moscow in 1897 as a Preindustrial City: A Test of the Inverse Burgess Zonal Hypothesis — 398
Walter F. Abbott

The Social Environment of Rio de Janeiro in 1960 — 406
Fred B. Morris and Gerald F. Pyle

The Shantytowns of Montego Bay, Jamaica — 416
L. Alan Eyre

Urban Structure in France — 425
Theodore Caplow

Urban Structure in Europe: Recent Evidence — 429
Bruce London

Ways of Dwelling in the Communities of India — 433
Radhakamal Mukerjee

Comparative Ecologies of Large Indian Cities — 444
Brian J. L. Berry and Howard Spodek

Some Ecological Patterns of Community Disorganization in Honolulu — 454
Andrew W. Lind

Honolulu—Fifty Years Later — 462
Andrew W. Lind

Index — 465

Preface

While I had been thinking for some time of revising *Studies in Human Ecology,* originally published in 1961 by Row, Peterson (later reprinted by Harper and Row), much of the impetus to finally undertake the project came from encouragement received during my participation in the "Conference on Research Trends and Policy Implications in Sociological Human Ecology." The conference, involving twenty-five participants, was organized by Michael Micklin and Harvey M. Choldin and held at the Battelle Conference Center in Seattle in October 1977. Many of the conference participants expressed the view that a revised edition of *Studies in Human Ecology* would be a helpful and welcome addition to the field.

A number of writers have commented that the first edition of this book clarified the issues in the field. Human ecology was a sharply divided field at that time, but many of the divisions were clouded and poorly understood. Attacks were still appearing on aspects of classical ecological theory that had been abandoned for some years. It was my hope in publishing that edition to reveal the state of the subdiscipline at the time so that real differences might be seen and outdated issues discarded. I think the volume did contribute to that purpose.

Several goals motivated me to undertake a revision of that book twenty years later. One was to examine theoretical developments in the field and see to what extent divisions of the past have persisted and to what extent there has been a growth of consensus. It is my conclusion that, while differences certainly persist, sociological ecology is not the bitterly divided field it was a score of years ago. While surveying the present state of the field, significant early contributions are maintained in this edition, for a second purpose of this volume was to preserve a historical perspective so that present-day ecological theory may be understood in the context of its development.

A third purpose of this volume was to undertake something rarely done in sociological research; that is, to attempt to bring selected older studies up to date by examining ecological changes that have occurred since the original studies were published. The "updates" take a variety of forms, depending on the judgment of each author. In most cases it was not possible to have a complete replication of the original study, but available data have been marshaled in each case to show the degree to which earlier patterns have persisted or changed.

Finally, I felt it would be worthwhile to focus attention on the utility of the ecological framework in understanding urban phenomena. I think it is worth emphasizing that ecological principles and research findings are not merely an accumulation of esoteric knowledge of interest to a small group of specialists, but are applicable to a number of aspects of urban life.

Twenty-four articles in this volume have been retained from the first edition. Thirty-two articles are new to this edition. Of these, six are "updates" of earlier articles, six are articles published for the first time in this book, and twenty are articles previously published elsewhere. Because of the new additions, the nature of new material available, and economic pressures limiting the size of the volume, the sections on human geography and metropolitan regions found in the first edition have been omitted. There have been a number of changes in the format of the book, particularly the consolidation of the old Parts I and II into Part I, and the substitution of a new Part II: "Human Ecology as a Framework for the Study of the City."

I should like to express my deep appreciation, first of all, to the authors who wrote original papers for this volume, and to the authors and publishers who have given permission to reprint previously published articles. Thanks are due also to the director and staff of The Pennsylvania State University Press. In addition, I wish to thank the many colleagues and fellow ecologists who encouraged me to undertake this project. Finally, I thank my wife, Lucille, for her indispensable assistance in this project.

Part I **The Development of the Ecological Framework**

Introduction

Human ecology has had three main sources of development: plant and animal ecology, geography, and studies of the spatial distribution of social phenomena. Best known to American sociologists is the development which stemmed from biological writings of the late nineteenth century—the works of Darwin and his followers, of Haeckel, and of the plant and animal ecologists. The term "human ecology" was introduced by Park and Burgess in 1921 and represented an attempt to apply systematically the basic theoretical scheme of plant and animal ecology to the study of human communities. A body of ecological theory was developed in the twenties and thirties to form what has been called the classical position. Its derivation from the biological sciences had a strong influence on classical human ecology, giving it a naturalistic emphasis.

However, long before the writings of the classical human ecologists, indeed even before the development of plant and animal ecology, human ecology, although not so named, was emerging in social studies of spatial distribution. Elmer's article tells of studies made in France more than a century ago of the distribution of crime, suicide, and other social phenomena, classified by age, sex, and type of crime. Levin and Lindesmith tell of a similar development of spatial studies of crime in England from 1830 to 1860. An early venture in medical ecology may be seen in the exerpt from Stamp. Although these early studies tended to be forgotten and did not lead to the development of a continuous school of thought, they nevertheless represent a very early emergence of what has always been an important aspect of human ecology, the study of the spatial distribution of interrelated social variables. With the formal development of human ecology, many studies were published, and in fact continue to be published, which essentially have represented studies of spatial distribution with very little integration into any systematic ecological theory. Thus the spatial studies of a century ago represent a beginning of what was to become an important part of human ecology. However, a theoretical development of human ecology as a defined and self-conscious field of study awaited the development of the classical position.

The theoretical system of the classical ecologists is presented in Park's article. According to the classical position, the basic process in human relationships is competition, largely involving a struggle for space. However, because of the high degree of interdependence and division of labor among humans, competition must always involve an automatic and unplanned degree of cooperation, forming what is called competitive cooperation. As a result of competitive cooperation, men form relationships of unplanned interdependence called symbiotic relationships. Human society is seen as organized on two levels: the biotic and the cultural. The biotic level involves basic, nonthoughtful adjustments made in the struggle for existence. This level is regarded as subsocial and is based on the organization of symbiotic relationships. The struggle for existence, based on competitive cooperation, and resulting in the organization of the biotic level of society, also determines the spatial distribution of persons. The spatial distribution is therefore seen as reflecting the organization of the biotic level of society. The cultural level of society, based on communication and consensus, is seen as a superstructure resting upon the biotic level. The biotic level is referred to as community, and the cultural level as society. The biotic level of human organization is regarded as the proper field of investigation for human ecology, and therefore cultural factors are excluded from ecological investigations.

McKenzie was another influential classical ecologist, but his conception of human ecology appears to have had more of an economic and less of a biological orientation than Park's. The reader may also note that McKenzie did not adhere to a strict exclusion of cultural factors from ecological analysis. In fact, he includes cultural and technical factors as one of the four types of ecological factors. However, McKenzie did not criticize Park's exclusion of culture from ecological analysis, nor did he specifically challenge the basic tenets of classical ecological theory. He accepted the conception of competition as the basic ecological process, and agreed with the other classical ecologists that the organization of the ecological community is based upon the dominance of the central business district. Dominance, competition, and human mobility were regarded as resulting in the processes of centralization of services, concentration, segregation, invasion, and succession of populations.

Burgess emphasized the dominance of the central business district and the variation in naturally formed areas of homogeneous population and land use which occurs because of varying distance from and degree of influence of the central business district. He developed this conception in his famous theory of concentric zones of city growth.

Hoyt modified Burgess's concentric zone theory by suggesting that development, rather than occurring in rings around the central business district, tends to follow a pattern of sectors. There are, for example, high rental sectors and low rental sectors. Within each sector, however, the pattern of concentric zones tends to occur. Hoyt's sector theory, while proposing a modification to the theory of concentric zones, did not basically challenge the Burgess model.

In Zorbaugh's article on "The Natural Areas of the City," we find the characteristic emphasis of the classical school on the city as a natural phenomenon. Zorbaugh, accepting Burgess's conception of concentric zones, suggests a further division of the city into smaller units which he calls natural areas because they are the unplanned result of natural city growth.

Most of the empirical studies of the classical period were concerned with the distribution of particular variables in relation to the natural areas and concentric zones of the city. Thus Reckless discusses the distribution of commercialized vice in relation to the concentric zones of the city, the characteristics of the zones where vice predominates, and the forms it takes in different zones.

In James's study of Vicksburg, we come to a geographical analysis of urban structure. Geography, as was mentioned earlier, was one of three sources from which human ecology developed. While human geography, a development of the late nineteenth and early twentieth centuries, has been concerned primarily with regions and rural areas, geographers in the twentieth century also turned their attention to studies of the structure of cities. These studies were generally known as urban morphology or urban geography. It is interesting to compare James's study of Vicksburg with the writings of the classical school that dominated the sociologists' approach to human ecology at the time. The urban geographers did not develop an elaborate theoretical scheme comparable to that of the sociological ecologists. They did not distinguish between a biotic and a cultural level, and they never sought to exclude social or cultural data from their studies. At the same time, the picture James gives us of Vicksburg appears basically similar to the structure of other American cities as described by sociological ecologists of the classical school. The geographic and sociological studies of human ecology developed in relative isolation from each other, but in recent years there has been increasing mutual recognition and interdisciplinary cooperation.

The final selection in Section A is taken from Zorbaugh's famous study of one part of Chicago. Vividly describing an area of stark contrasts, Zorbaugh—like most researchers of the classical period—does not concern himself with theoretical distinctions between biotic and cultural levels, and thus his study is still very relevant.

Attacks on the classical school began in the late 1930s. In 1938 a book by Milla Alihan, entitled *Social Ecology,* appeared that devastatingly criticized basic theoretical elements of classical human ecology. Alihan particularly criticized the distinction between the biotic and cultural levels of human organization. The selection from Alihan's book which is included in this volume deals specifically with her criticisms of this distinction and the classical ecologists' assumption that competition is the basic process in human relationships. In reading this selection, it should be kept in mind that Alihan uses the terms "community" and "society" in accordance with the classical ecological definition of "community" as the biotic level and "society" as the cultural level of organization.

Criticisms of classical human ecology continued through the forties. Hatt, in a paper published in 1946, shows that the concept of natural areas does not apply to the section of Seattle he studied. He suggests that if the concept of natural areas is to be of value it should be defined by criteria relevant to a specific research project, rather than assuming the existence of fixed natural areas of a city. He accuses ecologists of a tendency toward reification of their concepts. Hollingshead's article, published in 1947, discusses the discrepancy between ecological theory and investigation, but above all stresses that man is a social animal with cultural factors surrounding and influencing all phases of his activities so that it is impossible to isolate a truly subsocial or noncultural level of behavior.

The final article in Section B deals with a major statistical criticism that challenged much ecological research. As early as 1934 Gehlke and Biehl discussed the effect of the grouping of data on the size of the correlation coefficient, thus questioning its validity.[1] Robinson, in an article published in 1950, carried this criticism further and attempted to demonstrate that ecological correlations cannot validly be substituted for individual correlations.[2] His analysis received a great deal of attention, and "ecological correlations" became the subject of a lengthy debate. While some writers felt that Robinson had exposed an "ecological fallacy" in demonstrating the invalid use of ecological correlation to infer individual correlation, others attempted to define circumstances under which valid inferences could be made. In the article reprinted here, Bogue and Bogue review the issues involved in this controversy and set forth guidelines under which ecological correlations may properly be used.

By 1950 the classical position had been subjected to more than a decade of sharp criticism and was seriously undermined. Although some lingering attacks continued into the fifties, most ecologists turned their attention to building a new foundation for their subdiscipline. The distinction between the biotic and cultural levels of human organization was no longer accepted, and it was agreed that cultural factors cannot be excluded, either theoretically or practically, from ecological investigations. Beyond these basic agreements, however, sharp divisions arose; and by the time the first edition of this volume was published in 1961, two major approaches, referred to as "neo-orthodox" and "sociocultural," vied with each other to define the field. As the names suggest, the neo-orthodox approach was closer to traditional ecology and, while not excluding culture, rejected the notion that it should have a central place in ecological theory. Sociocultural ecologists, on the other hand, regarded social and cultural factors, such as values, as the primary explanatory concepts in ecological analysis. While sociocultural ecologists sought to move ecology closer to the mainstream of sociological thought, neo-orthodox ecologists were concerned with differentiating ecology from other branches of sociology.

In the years that have passed since the publication of the first edition of this book, divisions among ecologists have become less acrimonious. While differences still exist, they have increasingly been seen as differences in emphasis. Many writers, seeing the two approaches as essentially compatible, do not feel a need to choose between them. Yet it would be incorrect to give the impression of total harmony and agreement: different emphases remain important to their proponents.

The most influential figure in what has been termed neo-orthodox ecology is Amos Hawley. The first selection in Section C, written by Hawley in 1944 and reprinted in his book in 1950, had a profound impact on ecological thought. In this article, Hawley rejects the classical distinction between a cultural and a biotic level of society and also rejects a distinction proposed by James A. Quinn between a social and a subsocial level of interaction.[3] He regards all human interrelationships as social. Hawley considers the main task of human ecology to be the analysis of community structure, which he conceives of primarily in terms of the division of labor. Community structure from the ecological point of view is seen as the organization of sustenance activities—the way a population organizes itself for survival in a particular habitat. This organization of sustenance activities results in a spatial distribution. Those activities least able to

---

[1] C. E. Gehlke and Katherine Biehl, "Certain Effects of Grouping upon the Size of the Correlation Coefficient in Census Tract Material," *Journal of the American Statistical Association,* 29 (March 1934, Supplement), pp. 169–70.

[2] W. S. Robinson, "Ecological Correlations and the Behavior of Individuals," *American Sociological Review,* 15 (June 1950), pp. 351–57.

[3] See James A. Quinn, "The Nature of Human Ecology: Reexamination and Redefinition," *Social Forces,* 18 (December 1939), pp. 161–68.

withstand the time and energy costs of more distant locations and with a maximum need for accessibility will tend toward a central location. However, Hawley emphasizes that nonspatial aspects of the organization of sustenance activities in community structure are also part of the proper scope of human ecology. Hawley's orientation to human ecology expressed here may be regarded as basically economic. He feels that economic data are readily available and are often indices of social phenomena. Hawley also feels that culture has become too inclusive a concept to be of primary explanatory value.

The second selection by Hawley represents a more recent statement of his position. While not basically different from his earlier conception, there is now a greater stress on demographic analysis, for the focus is on the organization of a population in a particular environment. This point of view will become clearer in the two following selections.

Those ecologists concerned with defining a distinctive domain for human ecology were much influenced by Leo Schnore's suggestion that the foundations of ecological theory may be traced back to Durkheim. In Durkheim's conception of social morphology, Schnore suggests, may be found key elements forming a cornerstone for ecological theory. These are: analysis on a macro (group rather than individual) level, concern with the relationship between the social organization of a population and its environment, and the study of population characteristics. These elements, traced by Schnore to Durkheim, have become central concepts in the theoretical framework of an important segment of ecologists.

In Duncan's article, the variables population (P), organization (O), environment (E), and technology (T) are set forth as comprising the basic categories in the ecological complex, and an illustration of analysis in terms of this framework is given. Duncan also proposed this framework in other writings and coined the acronym POET. This framework offers at least three advantages: (1) a distinguishable ecological perspective operating on the macro level, (2) an opportunity to use readily available demographic data, and (3) an interrelationship with the broader theoretical orientation known as functionalism. It should be noted, however, that while the POET framework is essentially an offshoot of functional theory, it does not suffer from the criticism leveled at functionalism in recent years that it tends to neglect change. Quite the contrary. In the POET framework, changing relationships of the basic elements are a major focus of analysis.

By no means do all ecologists embrace the POET schema. The sociocultural perspective remains a vigorous force among ecologists. Most influential in launching this perspective was Walter Firey, particularly through the article reprinted here and through his book *Land Use in Central Boston*,[4] which was published two years later. Firey maintains that space may have symbolic value and should not always be regarded as having only cost-imposing qualities. He has emphasized that space takes on meaning for man through cultural definition, and that at every point cultural values intervene between the physical environment and the human community. In his article Firey presents evidence in support of his position from a study of Boston. He shows that the maintenance of Beacon Hill as an upper-class residential district can only be explained by symbolic values. He also finds symbolic values the only adequate explanation of what he calls "space fetishes," such as the Common, three cemeteries, old churches, and old meetinghouses, all maintained in the central business district despite their serious economic dysfunctions. Patterns of land use and community structure in the North End are also explained in terms of cultural values.

Seeman, as a geographer, did not refer to himself as a sociocultural ecologist nor specifically discuss the assumptions of that position. Yet in his study of Mormon communities in Utah we can see essentially a presentation of evidence in support of the importance of the sociocultural variable in ecological analysis. Seeman shows the importance of religious values in shaping the form of Mormon communities, leading to a pattern of nucleated agricultural settlements, in contrast to the typical American pattern of scattered homesteads. Moreover, the physical form of the communities themselves, including the larger urban centers, has been affected by the Mormon emphasis on planning communities in accordance with their religious values.

Leaving the question of ethnic groups, we come to an interesting study by the geographer Stanislawski of a basic factor in the ecological structure of most American cities—the grid-pattern layout of streets. The grid pattern is so characteristic of American cities, with very few exceptions, that we generally assume its existence or consider it a natural, perhaps inevitable, development. How-

---

[4]Berkeley: University of California Press, 1947.

ever, Stanislawski seeks to demonstrate that it is a cultural conception, spread through culture contact from its probable original source in India. Thus we may say that Stanislawski has placed another important feature of ecological structure within a cultural context, giving further support to the sociocultural approach to ecological analysis.

There have been numerous other studies showing the importance of cultural factors in understanding ecological patterns. Some of these may be found in other parts of this book (for example, Part III: Urban Patterns in Different Cultural Settings). In the next paper, however, Firey and Sjoberg go beyond specific studies to address the basic theoretical differences between the sociocultural and neo-orthodox (or POET) approaches (or paradigms, as they are referred to here) and the issues facing sociocultural ecology today. First of all, they note an essential difference in the type of data employed by each. Neo-orthodox ecologists prefer quantitative data measuring overt demographic, physical, and behavioral characteristics, whereas to sociocultural ecologists interpretive data reflecting environmentally relevant attitudes, motivations, and meanings are most important. (It may be noted at this point that sociocultural ecologists do not consider it crucial to keep their analysis on a "macro" level, and they have been criticized for this by some POET ecologists.) Sociocultural ecologists are interested in the ways these meanings and symbols, built into norms and institutions, shape community structure. The authors particularly stress the importance of the concept of "power" in the sociocultural analysis of ecological structure. Power as a reflection of political and class structure shapes social planning and policies relating to land use. Finally, Firey and Sjoberg consider three basic forms which sociocultural analyses may take: descriptive, causal, and functional.

In the final paper in Part I, Bailey and Mulcahy suggest that the basic difference between sociocultural and neo-orthodox (referred to here as neoclassical—the POET approach) ecology is a matter of level of analysis. The authors maintain that the sociocultural approach is in reality microecology, dealing with the ecological consequences of the values, motivations, and decisions of individuals; whereas the neoclassical approach is macroecology, stressing the characteristics of populations. The two approaches thus are seen as complementary, both being required for a complete ecological analysis. One caveat should be noted here. As may be seen in the preceding paper by Firey and Sjoberg, sociocultural ecologists do not restrict themselves to the micro level. They extend their analysis to the macro level in studying the manifestation of values not only through individual decisions but also through institutionalized forms, social organizations, class structure, and the exercise of power. However, this does not contradict the accuracy of Bailey and Mulcahy's analysis. Neoclassical ecologists (POETs) deliberately limit their analysis to the macro level. They are consciously macroecologists. Sociocultural ecologists, on the other hand, are willing to use micro-level concepts and data in their research and analysis. Going beyond the question of levels, however, there can be little question that the two approaches are complementary. Complete ecological understanding requires both demographic-economic, quantitative knowledge and cultural-political, interpretive knowledge.

Section A *Early Spatial Studies and the Rise of "Classical" Human Ecology*

Century-Old Ecological Studies in France*

M. C. Elmer

There is a tendency for students of human relations to emphasize a particular method of study during each period of years. Following the time of Comte the analysis of social activities was largely of a non-statistical nature. The general impression was developed that statistical methods could not be applied to the study of social phenomena. However, during the past twenty-five years sociologists have placed an unusual amount of emphasis upon attempts to develop objective standards for the measurement of social phenomena, and to determine what relationship, if any, exists between human activities and measurable situations. In our enthusiasm to learn more about the "unknown" which is all about us we have neglected many important studies of earlier periods which should be of much value and interest to us.

Quetelet (1796–1874) was active for nearly half a century in the attempt to measure social phenomena statistically and hence much of his work is generally known. There were, however, other important studies which preceded the work of Quetelet, but which have not received much attention, perhaps because it has been only recently that the interests of sociologists have been directed toward certain fields.

Just about the time when Auguste Comte was beginning to formulate the social philosophy which played an important part in the development of sociology for the next seventy-five years, M. de Guerry de Champneuf completed a remarkable study which he called *Statistique morale de la France*.[1] In this study of one hundred years ago he revived with fresh contributions the principles from which Muenster, Sansovino, Bodin, Graunt, Petty, Süssmilch, and Achenwall had been approaching the study of society in previous centuries. M. de Guerry de Champneuf took particular phases of life concerning which erroneous ideas were afloat, and made careful studies of the data available. In his analysis of crime, suicides, illegitimacy, and similar phenomena, he was so far in advance of most sociologists who followed him that it has been only within the last twenty years that we have made specific investigations and regional studies and surveys along the lines he

*Reprinted from *The American Journal of Sociology*, 39 (July 1933), 63–65, 70, by permission of the author and The University of Chicago Press.

[1] M. de Guerry de Champneuf was Director of Affaires Criminelles in the Ministry of Justice (1821–35).

proposed. He lacked an *understanding and interested audience*. In this study de Champneuf brought forth the view that relatively constant factors and conditions are determinants in the study of social phenomena and that these environmental conditions may be modified with a consequent modification in the acts. Due to the lack of supporting data of history, economics, biology, and psychology the analyses made did not reach the degree of reliability now possible in similar studies. However, conclusions reached compare very favorably with the results obtained by students who have the advantage of the additional knowledge and machinery in their study of human ecology. In 1823 Quetelet spent some time in Paris and became acquainted with the studies being made there. From that time he began to interest himself in the scientific view of statistics. In April, 1825, he published his first statistical work—*Memoir sur les lois des naissances et de la mortalité à Bruxelles*.

M. de Champneuf divided France into five regions or districts. Each one was composed of seventeen departments. For six successive years he compiled data on various types of crime in each department and made interesting tables, maps, and comparisons based upon geographical location and considering age, sex, and instruction. He classified crimes as crimes against the person and crimes against property. The calculations were made on the number of persons accused of the crime rather than on the basis of convictions. It is rather significant that the study showed a consistent ratio during six successive years, 1825–30, for all crimes and including all departments of France. Concerning these data he stated,

> There is the influence of climate, and there is the influence of seasons, for whereas the crimes against persons are always more numerous in the summer, the crimes against property are more numerous in winter—so of the crimes committed in the south, the crimes against the person are far more numerous than those against property, while in the north the crimes against property are, in the same proportion, more numerous than those against the person.

M. de Champneuf did not personally collect all these data. From 1821–29 M. le Comte de Chabral, Prefet de la Seine, published *Recherches statistiques sur la ville de Paris et le Departement de la Seine*. These studies were made under the direction of Fourier and contained a judicial study of populations by him. They covered a very wide range of topics which, as stated in the reports of Dr. John Bowring in 1832,

> . . . contain the most curious and interesting and valuable information. . . . Here it will appear that there is hardly any subject which can interest the inhabitants of Paris,—which may be curious to the traveller, or interesting to the statesman, that the government has not found it possible to secure and to give, not with perfect accuracy perhaps, but still with sufficient accuracy to enable one, over a long series of years, to come to certain conclusions.

It was with such a mass of general data as a background that this most significant series of studies and charts . . . was made a few years later. Because of lack of space we shall only mention the remarkable studies made of the suicides, distributed according to age and means employed in the various divisions of France. Likewise the distribution of the seventeen most common crimes by age and sex groups. Space will permit only brief mention of data showing the average geographical distribution of crimes against property and the average geographical distribution of instruction in France from 1825 to 1830 inclusive.

In this study France was divided according to its 86 political departments. M. de Champneuf shows . . . the number of persons in each department for one person accused of a crime against persons . . . [and] the number of persons for one person accused of a crime against property. . . . The incidence of instruction based on the number of young men and of each 100 inscribed is shown [as well]. . . .

We are not attempting to justify the studies of M. de Guerry de Champneuf nor to present them as models. We wish, however, to call attention to some of the interesting contributions made by the forerunners of modern sociology and human ecology.

Table 1. Distribution of Crimes Against Persons

Region	Percent of Total Crimes Committed in France						Average	Percent of Total Population
	1825	1826	1827	1828	1829	1830		
North	25	24	23	26	25	24	25	27
South	28	26	22	23	25	23	24	15
East	17	21	19	20	19	19	19	18
West	18	16	21	17	17	16	18	22
Central	12	13	15	14	14	18	14	17

Table 2. Distribution of Crimes Against Property

Region	Percent of Total Crimes Committed in France						Average	Percent of Total Population
	1825	1826	1827	1828	1829	1830		
North	41	42	42	43	44	44	42	27
South	12	11	11	12	12	11	12	15
East	18	16	17	16	14	15	16	18
West	17	19	19	17	17	17	18	22
Central	12	12	11	12	13	13	12	17

English Ecology and Criminology of the Past Century*

Yale Levin
and Alfred Lindesmith

The emphasis that has been placed in recent years upon what is known as the "ecological approach" to the study of crime makes it appropriate to pay attention to a period in English history when ecological studies of this subject appear to have been as much in fashion as they are today. Roughly between the years 1830 and 1860, a considerable interest in territorial or regional studies of crime was manifested in England. Over a period of several decades there were accumulated a mass of data and a body of knowledge which were never really discredited or displaced by work of superior scientific merit along the same lines, but were simply relegated to the background in favor of the psychiatric, biological and other types of theories of the later nineteenth century, and eventually forgotten or disregarded. Although present day criminologists who adopt the ecological approach do not refer to their English predecessors for guidance and corroboration, it is surprising to find that the emphasis which is being placed upon social factors in the causation of crime is closely paralleled in these earlier studies of what

*Reprinted from the *Journal of Criminal Law and Criminology,* 27 (March 1937), 801–16, by permission of the authors and the *Journal of Criminal Law and Criminology.*

might be called the pre-Lombrosian era. The recent revival of some of these old points of view and techniques suggests the comparison of the older studies with contemporary ones in order to evaluate more precisely the progress criminology has made in the last hundred years. The enthusiasm of social scientists often leads them to attribute greater originality to contemporary studies and less value to the old than is actually warranted by the facts in the case. In the descriptions of some of these older studies which follow, we have attempted to keep in mind contemporary work along the same lines so as to facilitate comparison. Limitations of space prevent more than passing reference to many maps, tables, or discussion which deserve far more extended treatment than we shall attempt in the present article.

I

Before analyzing the statistical data in these earlier studies, it should be noticed that numerous general observations regarding the concentration of crime in "low" neighborhoods were made by writers and officials dealing with criminals. Thus, on the basis of his observations, and not with the aid of statistics, Walter Buchanan, one of Her Majesty's

Justices of the Peace for the County of Middlesex, writing in 1846, noted that

> The great recesses of juvenile crime in the metropolitan districts to the north of the Thames are Spitalfields, Bethnal-Green, Shoreditch, Hoxton, Wapping, Ratcliffe, White Chapel, Shaffron-Hill, Almonry, Tothill Fields, Gray's Inn Lane, St. Giles, Seven Dials, Drury Lane, Field Lane, and Lisson Grove; and although in some parts of Maryle-Bone, St. Pancras, Chelsea, Islington, Clerkenwell, Limehouse, Paddington, Kensington and elsewhere in and about the metropolis, young thieves resort, they are not to be compared in number to those who are to be found issuing from the above named places. In the densely crowded lanes and alleys of these areas, wretched tenements are found, containing in every cellar and on every floor, men and women, children both male and female, all huddled together, sometimes with strangers, and too frequently standing in very doubtful consanguinity to each other. In these abodes decency and shame have fled; depravity reigns in all its horrors.[1]

That juvenile delinquents and adult criminals were concentrated in the deteriorated areas of the large towns and cities was a matter of common observation. Not only those whose work brought them in direct contact with criminals and youthful delinquents, but others, notably writers on social and political economy, observed the effects of deteriorated housing conditions. Allison observes that

> If any person will walk thru St. Giles, the crowded alleys of Dublin, or the poorer quarters of Glasgow at night, he will no longer worry at the disorderly habits and profligate enjoyments of the lower order; his astonishment will be, that there is so little crime in the world.... The great cause of human corruption in these crowded situations is the contagious nature of bad example.... A family is compelled by circumstances or induced by interest to leave the country. The extravagant price of lodgings compels them to take refuge in one of the crowded districts of the town, in the midst of thousands in similar necessitous circumstances with themselves. Under the same roof they probably find a nest of prostitutes, in the next door a den of thieves. In the room which they occupy they hear incessantly the revel of intoxication or are compelled to witness the riot of licentiousness.[2]

The author relates of this family that one of the sons becomes a member of one of the numerous bands of thieves, commits a housebreaking, and is sentenced to be transported. The daughters become prostitutes and the children of a once happy and virtuous family are thrown upon the streets to pick up a precarious subsistence. He concludes that this unhappy history of a family proceeds not from any extraordinary depravity in their character, but from the almost irresistible nature of the temptations to which the poor are exposed.

Contemporary observers of what we now call the Industrial Revolution carefully noted the growth of large towns, which was one of the marked features of the transformation of England from an agricultural country to an industrial one. In the early decades of the nineteenth century, students of political economy began to assess the growth of the factory towns, a growth which was apparent to every one. In a volume published in 1843, a writer discusses the growth of manufacturing in England and its attendant good effects on the population, such as the growth of large and princely fortunes, the encouragement given to the arts, the enterprise and energy created by the establishment of factories. At the same time he declares that

> Among the numerous causes which appear inseparable from manufactories, producing crime and immorality, the following deserve particular notice. The crowding together of the working classes in narrow streets, filthy lanes, alleys and yards, is a serious evil and one which has hitherto increased in all manufacturing towns. The poor are not resident in these places from choice, but from necessity. Families are not huddled together into dark ill-ventilated rooms from any peculiar pleasure it affords. They may indeed have become insensible of the inconvenience and wretchedness of such situations, but slender and uncertain means do not enable them to command more comfortable abodes. They are fixed there by circumstances.[3]

In his evidence before a Select Committee of Crime in 1830, the Governor of Coldbath Prison stated that:

> In my opinion the crowning cause of crime in the metropolis is to be found in the shocking state of the habitations of the poor, their confined and fetid localities, the consequent necessity for consigning children to the streets for requisite air and exercise. These causes combine to produce a state of frightful demoralization. The absence of cleanliness, of decency, and of all decorum; the disregard of any

---

[1] Walter Buchanan, *Remarks on the Causes and State of Juvenile Crime in the Metropolis With Hints for Preventing Its Increase* (1846), pp. 6–7.

[2] A. Allison, *Principles of Population* (1840), p. 76.

[3] G. C. Holland, *Vital Statistics of Sheffield* (1843), p. 138.

heedful separation between the sexes; the polluting language; and the scenes of profligacy, hourly occurring, all tend to foster idleness and vicious abandonment.[4]

When the reformatories were established in the 1850 decade, the Chief Inspector, Sydney Turner, noted in his annual report for 1856 that the juvenile delinquents committed from the deteriorated districts of London presented a special problem because of their association in gangs. He advocated that these delinquents be committed to various reformatories instead of being permitted to concentrate in any one reformatory.

The following quotation from M. D. Hill, Recorder of Birmingham, will serve to illustrate how the effects of city life upon personal conduct were analyzed in the middle of the nineteenth century:

> A century and a half ago, as far as I have been able to ascertain, there was scarcely a large town in the island except London. When I use the term 'large town' I mean where an inhabitant of the humbler classes is unknown to the majority of the inhabitants of that town. By a small town, I mean a town where, 'a converso' every inhabitant is more or less known to the mass of people of that town. I think it will not require any long train of reflection to show that in small towns there must be a sort of natural police, of a very wholesome kind, operating upon the conduct of every individual, who lives, as it were, under the public eye. But in a large town, he lives, as it were, in absolute obscurity; and we know that large towns are sought by way of refuge, because of that obscurity, which, to a certain extent, gives impunity. Again, there is another cause which I have never seen much noticed, but which, having observed its operation for many years, I am disposed to consider it very important, and that is the gradual separation of classes which takes place in towns by a custom which has gradually grown up, that every person who can afford it lives out of town, and at a spot distant from his place of business. Now this was not formerly so; it is a habit which has, practically speaking, grown up within the last half century. The result of the old habit was that rich and poor lived in proximity; and the superior classes exercised that species of silent but very efficient control over their neighbors, to which I have already referred. They are now gone, and the consequence is, that large masses of the population are gathered together without those wholesome influences which operated upon them when their congregation was more mixed, when they were divided, so to speak, by having persons of a different class of life, better educated, amongst them. These two causes, namely, the magnitude of towns and the separation of classes, have acted so concurrently, and the effect has been that we find in very large towns, which I am acquainted with, that in some quarters there is a public opinion and a public standard of morals very different from what we should desire to see. Then the children who are born amongst these masses grow up under that opinion, and make that standard of morals their very own; and with them the best lad, or the best man, is he who can obtain subsistence, or satisfy the wants of life, with the least labour, by begging or by stealing, and who shows the greatest dexterity in accomplishing his object, and the greatest wariness in escaping the penalties of the law; and lastly the greatest power of endurance and defiance, when he comes under the lash of the law.[5]

II

We have selected for examination the works of two authors, Henry Mayhew[6] and Joseph Fletcher, who utilized official statistics in investigating the problems of crime in their wider aspects. Mayhew's volume, *The Criminal Prisons of London,* in addition to containing a detailed description of the prisons of London, as the title indicates, includes also a wealth of statistics and illuminating observations on such subjects as: juvenile delinquency, the evolution of the juvenile offender into the habitual criminal, recidivism, female crime, the concentration of various types of crime in certain localities within London and in certain counties of England and Wales, classifications of crime and criminals,

---

[4]Quoted in "The Causes of Crime in the Metropolis," *Taits Edinburgh Magazine,* Vol. 17 (1830), p. 332.

[5]Evidence before Select Committee on Criminal and Destitute Juveniles, *Report of the Select Committee on Criminal and Destitute Juveniles* (1852), p. 33.

[6]Henry Mayhew's . . . principal work, in which he was assisted by John Binny and others, was *London Labour and the London Poor,* a series of articles, anecdotic and statistical, on the petty trades of London, originally appearing in the "Morning Chronicle." Two volumes were published in 1851; but their circulation was interrupted by litigation in Chancery. In 1856 a continuation of it appeared in monthly parts as . . . "The Great World of London" which was ultimately completed and published as *The Criminal Prisons of London and Scenes of Prison Life* (1862). A portion of this volume was written by John Binny. Mayhew's *London Labour and the London Poor* (4 vols.) appeared in its final form in 1864 and again in 1865. The title page of each volume is as follows: *London Labour and the London Poor: a Cyclopedia of the Condition and Earnings of Those That Will Work, Those That Cannot Work, and Those That Will Not Work.* The fourth volume, acknowledging the assistance of John Binny and other contributors, is devoted to thieves, swindlers, beggars, and prostitutes. . . .

the evaluation of police statistics, the history of the "delinquency areas" of London, methods of prison administration and prison discipline, and the role of early family and community conditions in producing criminals. Many of the statistical tables cover all of England and Wales by counties; some give data by police districts within London; other tables compare cities and other territorial divisions. It is interesting in connection with the general problem of the relation of crime to the social life of the period that Mayhew introduces his study with a general topographical description of London and London streets, giving population and other general descriptive data for the city as a whole. He has a section entitled "Some Idea of the Size and Population of London" and a "Table Showing the Area, Number of Houses, and Proportion of Houses to Each Acre in London, 1851" by districts—36 of them. In a similar manner he notes the "Distribution and Density of the Population of London in 1851" in terms of the same 36 districts and, on the page facing this table, represents the same data on a shaded ecological map of the city. He does the same for the average income tax assessments and poor rate assessments per house in these districts, and gives us an idea of "mobility" by listing the number of vehicles passing through each of the principal London streets in 24 hours.

Perhaps of even greater interest in this volume is the ecological study of the residences of the members of the various branches of the legal profession in London, in which Mayhew shows the concentration in what he calls the "legal capital," Chancery Lane, which he describes in detail. He traces the ramifications radiating out from this legal capital and describes the legal "suburbs" of the city, listing more than a hundred "legal localities." In order to place this material in its proper setting, he precedes it with the statistics showing the proportion of the population included in the professional classes of each of the counties of England and Wales and in London.

The other volume, "Those Who Will Not Work," is remarkable in revealing the full extent and detailed character of Mayhew's ecological description of London crime. In it he classifies London "beggars, thieves, prostitutes, cheats and swindlers" into a total of more than one hundred specific groups. He and his collaborators discuss and describe the habitat and mode of making a living of each of these groups and specify quite exactly the districts in which they commit their depredations as well as the streets and localities where they live. This volume abounds in graphic descriptions of the various crime areas and in personal testimony obtained by interviews with persons in the walks of life and areas under consideration, in the manner of the "participant observer" of contemporary sociology. There are recorded in this volume more than a dozen narratives of professional criminals, written in the first person in the criminal's own words, telling his life history, describing the natural evolution of the professional from the juvenile criminal, and giving vivid descriptions of the modus operandi in the various "rackets," as we would call them.

In the appendix of Mayhew's "Those Who Will Not Work" there is a series of fifteen maps with accompanying tables showing the distribution (by rates whenever appropriate) of the following, in each of the counties of England and Wales:

Map No. 1— Density of Population
2— Intensity of Criminality
3— Intensity of Ignorance
4— Number of Illegitimate Children
5— Number of Early Marriages
6— Number of Females
7— Committals for Rape
8— Committals for Carnally Abusing Girls
9— Committals for Disorderly Houses
10— Concealment of Births
11— Attempts at Miscarriage
12— Assaults with Intent
13— Committals for Bigamy
14— Committals for Abduction
15— Criminality of Females

The tables usually cover a ten-year period (1841–1850). In arriving at correlations without the use of the coefficient, which was not known at that time, Mayhew lists the counties above and below the average according to their respective deviations from the average, and then juxtaposes two such series and analyzes their differences and similarities.

Perhaps one of the major points made in recent ecological studies of crime, and one that has received a great deal of attention and been heralded as a landmark in the scientific study of crime, is that crime rates, juvenile and adult, vary from one community to another within cities; and that crime is concentrated in certain areas and not distributed uniformly. This fact was well known to Mayhew, who, in addition to working out rates by counties and cities, also computed rates for police districts within London, and went a step farther in specifying what particular kinds of crimes were to be found in particular areas within the city. He calls attention to the fact that London's "rookeries" of crime have long histories, some of which extend back more than five hundred years. He made personal investigations of these areas, which have been "nests of London's beggars, prostitutes and thieves" continuously for centuries. His masterly descriptions of such districts as St. Giles, Spitalfields, Westminster, and the Borough are precise delimitations of characteristic areas of London vice and crime. The following excerpt is typical:

> There is no quarter of the Metropolis impressed with such strongly-marked features as the episcopal city of Westminster. We do not speak of that vague and straggling electoral Westminster, which stretches as far as Kensington and Chelsea to the west, and even Temple Bar to the east; but of that Westminster proper—that triangular snip of the Metropolis which is bounded by the Vauxhall Road on one side, St. James Park on another, and by the Thames on the third—that Westminster which can boast of some of the noblest and some of the meanest buildings to be found throughout London (the grand and picturesque old Abbey, and the filthy and squalid Duck Lane—the brand new and ornate Houses of Parliament, and the half-dilapidated and dingy old Almonry) which is the seat at once of the great mass of law makers and lawbreakers—where there are more almshouses, and more prisons and more schools—more old noblemen's mansions and more costermonger's hovels—more narrow lanes, and courts, and more broad unfinished highways—whose Hall is frequented by more lawyers, and whose purlieus are infested by more thieves—whose public houses are resorted to by more paviors—whose streets are thronged by more soldiers—on whose doorsteps sit more bareheaded wantons—and whose dry arches shelter more vagabond urchins than are to be noted in any other part of the Metropolis—ay, and perhaps in any other part of the world[7]

In his analysis of juvenile crime Mayhew compares rates in the various counties and notes that the rates of juvenile delinquency are highest in those counties which have large cities in them. He takes note of the difference in age distribution from one locality to another when he makes these comparisons. In the county containing London he shows that 41% of the juvenile offenders came from one of the seven police districts and 24%

[7]*The Criminal Prisons of London*, p. 353.

from another. The other districts contributed an average of between 5% and 8% and the country only 5½% of the total. He further splits up the rural returns to show that most of the rural offenders came from one district—Hammersmith. He lists areas and streets of London which particularly abound in gangs of juvenile delinquents. The following excerpts taken from "Those Who Will Not Work" show clearly an amazing ecological knowledge of London crime of that day:

> In order to find these houses it is necessary to journey eastwards, and leave the artificial glitter of the West-end, where vice is pampered and caressed. Whitechapel, Wapping, Ratcliff Highway, and analogous districts are prolific in the production of these infamies. St. Georges in-the-east abounds with them.... Whitechapel has always been looked upon as a suspicious unhealthy locality. To begin with, its population is a strange amalgamation of Jews, English, French, Germans and other antagonistic elements.... Ship alley is full of foreign lodging houses.... Tiger Bay like Frederick Street is full of brothels and thieves lodging houses.... The most of those engaged in this kind of robbery in Oxford Street come from the neighborhood of St. Giles and Lisson Grove.... The most accomplished pickpockets reside at Islington, Hoxton, Kingsland Road, St. Lukes, the Borough, Camberwell and Lambeth in quiet respectable streets, and occasionally change their lodging if watched by the police.... Some Londoners are in the habit of stealing horses. These often frequent the Old Kent Road and are dressed as grooms or stablemen.... Dog stealing is very prevalent, particularly in the West-end of the Metropolis, and is a rather profitable class of felony. These thieves reside at the Seven Dials, the neighborhood of Belgravia, Chelsea, Knightsbridge, and low neighborhoods, some of them men of mature age.... There are great numbers of expert cracksmen known to the police in the different parts of the Metropolis. Many of these reside on the Surrey side, about Waterloo Road and Kent Road, the Borough, Hackney and Kingsland Road and other localities.

It is no doubt true that many of the facts having to do with the concentration of crime in particular areas were noted long before the time of Mayhew, inasmuch as London's crime areas had acquired histories of several centuries when he wrote. What is particularly noteworthy about Mayhew, as well as the other students of his day, was that they used these facts definitely and consciously for the purposes of what was known as "moral" or "social science." Thus Mayhew remarks:

Surely even the weakest-minded must see that our theories of crime, to be other than mere visionary hypotheses, must explain roguery and vagabondage *all over the world,* and not merely be framed with reference to that little clique among human society which we happen to call our State.[8]

Students of today who are in the habit of considering Lombroso the first scientific student of crime will be surprised to find Mayhew anticipating in the middle of the nineteenth century the criticisms of the early Lombrosian viewpoint which were advanced near the end of the nineteenth and in the first part of the twentieth century. He states:

> But crime, we repeat, is an effect with which the shape of the head and the form of the features appear to have no connection whatever. . . . Again we say that the great mass of crime in this country is committed by those who have been bred and born to the business, and who make a regular trade of it, living as systematically by robbery or cheating as others do by commerce or the exercise of intellectual or manual labour.[9]

He thus definitely rejects the view of the criminal as a distinct physical type in favor of what might be called an environmental or sociological view. In fact, if we were to select the main theme of his books we should say that it was the point that habitual crime is the result of a natural evolution of juvenile crime in response to the impact of social factors. He even calculates the number of juvenile offenders who each year must have graduated to the ranks of the adult convicts to have maintained this latter group at a constant figure.

III

In 1847 and 1849 Joseph Fletcher read three papers before the British Association for the Advancement of Science and the Statistical Society of London, which he later incorporated in a book, *Summary of Moral Statistics of England and Wales,* which might, in many respects, be taken as a model ecological work of the period. It is not as encyclopedic as Mayhew's work but it is more minute and specialized. The entire book is centered around a series of 12 ecological maps in the appendix of the volume and an ecological map in the frontispiece colored to represent what we might call "natural areas" in England and Wales. These areas were determined on the basis of the prevailing economic organization, whether agricultural, mining, manufacturing, et cetera, in the various counties, grouping like ones together. He proceeds by means of a complex series of tables, coordinated with these maps to analyze what he calls "indices to moral influences" and "indices to moral results" in the various regions specified, listing 36 conclusions as a result of this analysis. Some 80 pages are then devoted to a more detailed tabular presentation of the data on which the conclusions are based, and this is followed by the series of 12 maps and accompanying tables giving the distribution of the above-mentioned "indices of moral influences and results" by counties and by districts in England and Wales. The maps are shaded in seven tints and those relating to crime are in terms of rates adjusted to the ages of the population. The following "indices" are graphically represented on these maps:

1. Dispersion of the population
2. Real property in proportion to the population
3. Persons of independent means in proportion to the population
4. Ignorance, as measured by the percentage of signatures by marks in the marriage registers
5. Crime as indicated by criminal commitments of males (allowance made for the ages of the population)
6. Commitments for the more serious offences against the person and malicious offences against property (allowance made for age distributions)
7. Commitments for all offences against property, excepting the malicious (allowance made for age distribution)
8. Commitments for assaults and miscellaneous offences for males in proportion to the total male population
9. Improvident marriages, or those entered into by males less than 21 years old
10. Bastardy as indicated by the registers of births
11. Pauperism
12. Deposits in savings banks in proportion to the population

Commenting on his method, Fletcher remarks:

> Rather than rush to one generalization upon aggregate results, it is better to retain the facts in manageable groups, by means of which to compare one class with another, one district with

---
[8]Ibid., p. 383.
[9]Ibid., p. 413.

another, and one period with another; and by the alternate use of analytical and synthetical methods, to bring the several elements into every possible combination, and detect the laws of their coincidence and relationship, or obtain new views as to the direction which should be given to more refined observation. . . . In framing the accompanying tables I have throughout adhered to one general division of the Kingdom into distinct industrial provinces, drawn with as much accuracy as was permitted by the large and varying size of the counties; the civil divisions which are the integral ones for nearly all my data. These provinces are portrayed in the accompanying map which will serve as a key to the whole of the following tables. A glance down the vertical columns of these tables will convey all that could be pictured forth by an expensive series of shaded maps showing the relative intensity of each element; at the same time that their horizontal lines will convey the collective results in a manner far more compendious than could be obtained by any pictorial means.[10]

His attempt to obtain an index to the crime of the various counties and districts of England and Wales which would not be affected by the migration of the "depraved" is interesting. He tries to make allowances for the "influence of the denser populations rather to assemble the demoralized than to breed an excess of demoralization." He is led by his reasoning along these lines to accept the rate of "more serious offences against the person and malicious offences against property" as a truer index of the moral state of a community less affected by mere migration, than other forms. He finds this form of crime to be highly correlated with ignorance, speaking of "its universal excess wherever ignorance is in excess." His principal concern is with the importance of education. Mayhew, however, refuted this view by contending that all education did was to increase the proportion of educated criminals without reducing the total number of them at all; but he, unlike Fletcher, regarded the habitual offences, or professional crime, as constituting the heart of the problem.

IV

In conclusion, we wish to call attention again to the fact that the scientific study of crime is usually said to have begun in the last quarter of the nineteenth century with the founding of the "Italian School" by Lombroso. Thus, George W. Kirchwey remarks, "It is incredible, but it is a fact, that, prior to the publication of Lombroso's *L'uomo delinquente* (The Criminal) which was given to the world in 1876, there had never been offered a serious, scientific approach to the study of the criminal."[11] Of Lombroso, Harry Elmer Barnes stated: "By taking the discussion of crime out of the realm of theology and metaphysics, and putting it on the positivistic basis of a consideration of the characteristics of the criminal, he may be said to have founded modern criminology."[12] All of the English studies to which we have referred were published at least a decade before the appearance of *L'uomo delinquente*. Although this early regional and sociological approach suffered an eclipse in the later decades of the nineteenth century (particularly in England), due perhaps to the philosophic pre-occupations of sociologists under the influence of Spencer and Comte, it has been resumed by American sociologists in the past two decades. In the evolution of criminological theory, it would appear that the work of Lombroso was in the nature of an interlude and interruption.

Contrary to widely circulated assertions which speak of the "ecological approach" to the study of crime as a twentieth-century development, we believe that it has been amply demonstrated that this approach was systematically employed in the early part of the nineteenth century by scholars in different countries who were aware of each other's work. The first systematic work in this field was apparently done in France by A. M. Guerry and in Belgium by A. Quetelet in the 1830s. The English writers to whom we have referred were influenced by both of these men and frequent references to their publications are found in their works. The work of Comte did not directly affect this movement at all. It was, in fact, too philosophical to have interested these students who were concerned with empirical research and were not interested in the broad speculative problems with which the Comtean tradition dealt. Mayhew, for example, heaped scorn upon the classical political economists of his day, speaking of them as a "sect of social philosophers" who "sat beside a snug sea-coal fire and tried to think out the several matters affecting the working classes, or else they have retired to some obscure corner, and there

---

[10] J. Fletcher, *Summary of Moral Statistics of England and Wales* (1850), p. 3.

[11] *Encyclopaedia Britannica*, 14th ed., article on Criminology.

[12] *Encyclopedia of the Social Sciences*, Vol. 4, article on Criminology.

remained, like big-bottomed spiders, spinning their cobweb theories among heaps of rubbish.[13]

The "moral statistics" of the eighteenth century were too defective to have made possible any such significant widespread development as occurred in the nineteenth century. When accurate governmental statistics became available, English and French students (and those in other countries) zealously employed these statistics in making regional studies of crime, suicide, insanity, illegitimacy, vagrancy, pauperism, and other social problems which interest the sociologist today. Individualistic theories which sought causal explanations of crime in terms of the characteristics of the criminal had not yet come into vogue, and in the works of these earlier writers there were elaborated many viewpoints which attributed primary causal significance to external environment or social factors in much the same manner as do present-day sociological theories.

[13]H. Mayhew, *Low Wages: Their Causes, Consequences and Remedies,* London (1851), p. 126. . . .

An Early Study in Medical Ecology*

L. D. Stamp

A major breakthrough in the battle against cholera was made by Dr. John Snow (1813–1858), then a general practitioner in the Golden Square area of London. Over 500 deaths occurred within the space of 10 days in August and early September 1848. On a large-scale map, he plotted exactly the house where each of the victims was attacked. It centered round a spot in Broad Street where there was a manual pump from which local residents obtained their drinking water. On 8 September the handle of this pump was removed at Snow's request and incidence of new cases ceased almost miraculously.

*Reprinted from L. Dudley Stamp, *Some Aspects of Medical Geography* (New York: Oxford University Press, 1964), pp. 15–16, by permission of Humanities Press, Inc., of New Jersey.

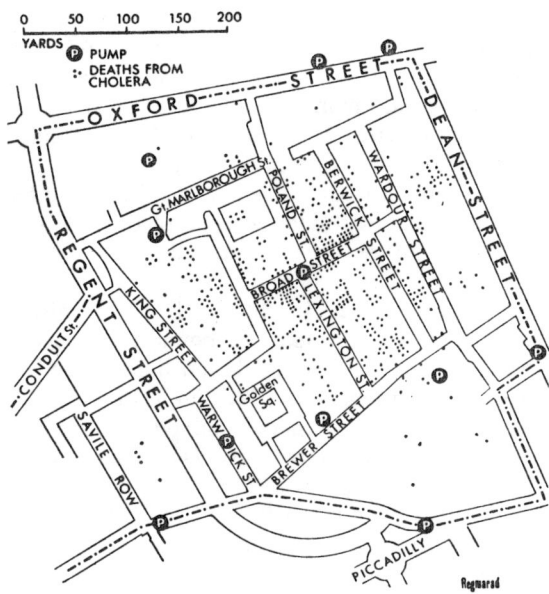

Figure 1—Dr. John Snow's map of cholera deaths in the Soho district of London, 1848. Notice the affected pump in Broad Street.

Human Ecology*

Robert E. Park

I. The web of life

Naturalists of the last century were greatly intrigued by their observation of the interrelations and co-ordinations, within the realm of animate nature, of the numerous, divergent, and widely scattered species. Their successors, the botanists and zoölogists of the present day, have turned their attention to more specific inquiries, and the "realm of nature," like the concept of evolution, has come to be for them a notion remote and speculative.

The "web of life," in which all living organisms, plants and animals alike, are bound together in a vast system of interlinked and interdependent lives, is nevertheless, as J. Arthur Thompson puts it, "one of the fundamental biological concepts" and is "as characteristically Darwinian as the struggle for existence."[1]

Darwin's famous instance of the cats and the clover is the classic illustration of this interdependence. He found, he explains, that humblebees were almost indispensable to the fertilization of the heartsease, since other bees do not visit this flower. The same thing is true with some kinds of clover. Humblebees alone visit red clover, as other bees cannot reach the nectar. The inference is that if the humblebees became extinct or very rare in England, the heartsease and red clover would become very rare, or wholly disappear. However, the number of humblebees in any district, depends in a great measure on the number of field mice, which destroy their combs and nests. It is estimated that more than two-thirds of them are thus destroyed all over England. Near villages and small towns the nests of humblebees are more numerous than elsewhere and this is attributed to the number of cats that destroy the mice. Thus next year's crop of purple clover in certain parts of England depends on the number of humblebees in the district, the number of humblebees depends upon the number of field mice, the number of field mice upon the number and the enterprise of the cats, and the number of cats—as someone has added—depends on the number of old maids and others in neighboring villages who keep cats.

These large food chains, as they are called, each link of which eats the other, have as their logical prototype the familiar nursery rhyme, "The House that Jack Built." You recall:

> The cow with the crumpled horn,
> That tossed the dog,
> That worried the cat,
> That killed the rat,
> That ate the malt
> That lay in the house that Jack built.

---

*Reprinted from *The American Journal of Sociology*, 42 (July 1936), 1–15, by permission of The University of Chicago Press.

[1] *The System of Animate Nature* (Gifford Lectures, 1915–16), II (New York, 1920), p. 58.

Darwin and the naturalists of his day were particularly interested in observing and recording these curious illustrations of the mutual adaptation and correlation of plants and animals because they seemed to throw light on the origin of the species. Both the species and their mutual interdependence, within a common habitat, seem to be a product of the same Darwinian struggle for existence.

It is interesting to note that it was the application to organic life of a sociological principle—the principle, namely, of "competitive co-operation"—that gave Darwin the first clue to the formulation of his theory of evolution.

"He projected on organic life," says Thompson, "a sociological idea," and "thus vindicated the relevancy and utility of a sociological idea within the biological realm."[2]

The active principle in the ordering and regulating of life within the realm of animate nature is, as Darwin described it, "the struggle for existence." By this means the numbers of living organisms are regulated, their distribution controlled, and the balance of nature maintained. Finally, it is by means of this elementary form of competition that the existing species, the survivors in the struggle, find their niches in the physical environment and in the existing correlation or division of labor between the different species. J. Arthur Thompson makes an impressive statement of the matter in his *System of Animate Nature*. He says:

> The hosts of living organisms are not . . . isolated creatures, for every thread of life is intertwined with others in a complex web. . . . Flowers and insects are fitted to one another as hand to glove. Cats have to do with the plague in India as well as with the clover crop at home. . . . *Just as there is a correlation of organs in the body, so there is a correlation of organisms in the world of life.* When we learn something of the intricate give and take, supply and demand, action and reaction between plants and animals, between flowers and insects, between herbivores and carnivores, and between other conflicting yet correlated interests, we begin to get a glimpse of a vast self-regulating organization.

These manifestations of a living, changing, but persistent order among competing organisms—organisms embodying "conflicting yet correlated interests"—seem to be the basis for the conception of a social order transcending the individual species, and of a society based on a biotic rather than a cultural basis, a conception later developed by the plant and animal ecologists.

In recent years the plant geographers have been the first to revive something of the earlier field naturalists' interest in the interrelations of species. Haeckel, in 1878, was the first to give to these studies a name, "ecology," and by so doing gave them the character of a distinct and separate science, a science which Thompson describes as "the new natural history."

The interrelation and interdependence of the species are naturally more obvious and more intimate within the common habitat than elsewhere. Furthermore, as correlations have multiplied and competition has decreased, in consequence of mutual adaptations of the competing species, the habitat and habitants have tended to assume the character of a more or less completely closed system.

Within the limits of this system the individual units of the population are involved in a process of competitive co-operation, which has given to their interrelations the character of a natural economy. To such a habitat and its inhabitants—whether plant, animal, or human—the ecologists have applied the term "community."

The essential characteristics of a community, so conceived, are those of: (1) a population, territorially organized, (2) more or less completely rooted in the soil it occupies, (3) its individual units living in a relationship of mutual interdependence that is symbiotic rather than societal, in the sense in which that term applies to human beings.

These symbiotic societies are not merely unorganized assemblages of plants and animals which happen to live together in the same habitat. On the contrary, they are interrelated in the most complex manner. Every community has something of the character of an organic unit. It has a more or less definite structure and it has "a life history in which juvenile, adult and senile phases can be observed."[3] If it is an organism, it is one of the organs which are other organisms. It is, to use Spencer's phrase, a superorganism.

What more than anything else gives the symbiotic community the character of an organism is the fact that it possesses a mechanism (competition) for (1) regulating the numbers, and (2) preserving the balance between the competing species of which it is composed. It is by maintaining this biotic balance that the community preserves its identity and integrity as an individual unit through

---

[2] J. Arthur Thompson, *Darwinism and Human Life* (New York, 1911), p. 72.

[3] Edward J. Salisbury, "Plants," *Encyclopaedia Britannica* (14th ed.).

the changes and the vicissitudes to which it is subject in the course of its progress from the earlier to the later phases of its existence.

II. The balance of nature

The balance of nature, as plant and animal ecologists have conceived it, seems to be largely a question of numbers. When the pressure of population upon the natural resources of the habitat reaches a certain degree of intensity, something invariably happens. In the one case the population may swarm and relieve the pressure of population by migration. In another, where the disequilibrium between population and natural resources is the result of some change, sudden or gradual, in the conditions of life, the pre-existing correlation of the species may be totally destroyed.

Change may be brought about by a famine, an epidemic, or an invasion of the habitat by some alien species. Such an invasion may result in a rapid increase of the invading population and a sudden decline in the numbers if not the destruction of the original population. Change of some sort is continuous, although the rate and pace of change sometimes vary greatly. Charles Elton says:

> The impression of anyone who has studied animal numbers in the field is that the "balance of nature" hardly exists, except in the minds of scientists. It seems that animal numbers are always tending to settle down into a smooth and harmonious working mechanism, but something always happens before this happy state is reached.[4]

Under ordinary circumstances, such minor fluctuations in the biotic balance as occur are mediated and absorbed without profoundly disturbing the existing equilibrium and routing of life. When, on the other hand, some sudden and catastrophic change occurs—it may be a war, a famine, or pestilence—it upsets the biotic balance, breaks "the cake of custom," and releases energies up to that time held in check. A series of rapid and even violent changes may ensue which profoundly alter the existing organization of communal life and give a new direction to the future course of events.

The advent of the boll weevil in the southern cotton fields is a minor instance but illustrates the principle. The boll weevil crossed the Rio Grande at Brownsville in the summer of 1892. By 1894 the pest had spread to a dozen counties in Texas, bringing destruction to the cotton and great losses to the planters. From that point it advanced, with every recurring season, until by 1928 it had covered practically all the cotton-producing area in the United States. Its progress took the form of a territorial succession. The consequences to agriculture were catastrophic but not wholly for the worse, since they served to give an impulse to changes in the organization of the industry long overdue. It also hastened the northward migration of the Negro tenant farmer.

The case of the boll weevil is typical. In this mobile modern world, where space and time have been measurably abolished, not men only but all the minor organisms (including the microbes) seem to be, as never before, in motion. Commerce, in progressively destroying the isolation upon which the ancient order of nature rested, has intensified the struggle for existence over an ever-widening area of the habitable world. Out of this struggle for a new equilibrium and a new system of animate nature, the new biotic basis of the new world-society is emerging.

It is, as Elton remarks, the "fluctuation of numbers" and "the failure" from time to time "of the regulatory mechanism of animal increase" which ordinarily interrupts the established routine, and in so doing releases a new cycle of change. In regard to these fluctuations in numbers Elton says:

> These failures of the regulating mechanism of animal increase—are they caused by internal changes, after the manner of an alarm clock which suddenly goes off, or the boilers of an engine blowing up, or are they caused by some factors in the outer environment—weather, vegetation, or something like that?

and he adds:

> It appears that they are due to both but that the latter (external factor) is the more important of the two, and usually plays the leading role.

The conditions which affect and control the movements and numbers of populations are more complex in human societies than in plant and animal communities, but they exhibit extraordinary similarities.

The boll weevil, moving out of its ancient habitat in the central Mexican plateau and into the virgin territory of the southern cotton plantations, incidentally multiplying its population to the limit of the territories and resources, is not unlike the Boers of Cape Colony, South Africa, trekking out

---

[4] "Animal Ecology," ibid.

into the high veldt of the central South African plateau and filling it, within a period of one hundred years, with a population of their own descendants.

Competition operates in the human (as it does in the plant and animal) community to bring about and restore the communal equilibrium, when, either by the advent of some intrusive factor from without or in the normal course of its life-history, that equilibrium is disturbed.

Thus every crisis that initiates a period of rapid change, during which competition is intensified, moves over finally into a period of more or less stable equilibrium and a new division of labor. In this manner competition brings about a condition in which competition is superseded by co-operation.

It is when, and to the extent that, competition declines that the kind of order which we call society may be said to exist. In short, society, from the ecological point of view, and in so far as it is a territorial unit, is just the area within which biotic competition has declined and the struggle for existence has assumed higher and more sublimated forms.

III. Competition, dominance and succession

There are other and less obvious ways in which competition exercises control over the relations of individuals and species within the communal habitat. The two ecological principles, dominance and succession, which operate to establish and maintain such communal order as here described are functions of, and dependent upon, competition.

In every life-community there is always one or more dominant species. In a plant community this dominance is ordinarily the result of struggle among the different species for light. In a climate which supports a forest the dominant species will invariably be trees. On the prairie and steppes they will be grasses.

> Light being the main necessity of plants, the dominant plant of a community is the tallest member, which can spread its green energy-trap above the heads of the others. What marginal exploitation there is to be done is an exploitation of the dimmer light below this canopy. So it comes about in every life-community on land, in the cornfield just as in the forest, that there are layers of vegetation, each adapted to exist in a lesser intensity of light than the one above. Usually there are but two or three such layers; in an oak-wood for example there will be a layer of moss, above this herbs or low bushes, and then nothing more to the leafy roof; in the wheat-field the dominating form is the wheat, with lower weeds among its stalks. But in tropical forests the whole space from floor to roof may be zoned and populated.[5]

But the principle of dominance operates in the human as well as in the plant and animal communities. The so-called natural or functional areas of a metropolitan community—for example, the slum, the rooming-house area, the central shopping section and the banking center—each and all owe their existence directly to the factor of dominance, and indirectly to competition.

The struggle of industries and commercial institutions for a strategic location determines in the long run the main outlines of the urban community. The distribution of population, as well as the location and limits of the residential areas which they occupy, are determined by another similar but subordinate system of forces.

The area of dominance in any community is usually the area of highest land values. Ordinarily there are in every large city two such positions of highest land value—one in the central shopping district, the other in the central banking area. From these points land values decline at first precipitately and then more gradually toward the periphery of the urban community. It is these land values that determine the location of social institutions and business enterprises. Both the one and the other are bound up in a kind of territorial complex within which they are at once competing and interdependent units.

As the metropolitan community expands into the suburbs the pressure of professions, business enterprises, and social institutions of various sorts destined to serve the whole metropolitan region steadily increases the demand for space at the center. Thus not merely the growth of the suburban area, but any change in the method of transportation which makes the central business area of the city more accessible, tends to increase the pressure at the center. From thence this pressure is transmitted and diffused, as the profile of land values discloses, to every other part of the city.

Thus the principle of dominance, operating within the limits imposed by the terrain and other natural features of the location, tends to determine the general ecological pattern of the city and the functional relation of each of the different areas of the city to all others.

[5] H. G. Wells, Julian S. Huxley, and G. P. Wells, *The Science of Life* (New York, 1934), pp. 968–69.

Dominance is, furthermore, in so far as it tends to stabilize either the biotic or the cultural community, indirectly responsible for the phenomenon of succession.

The term "succession" is used by ecologists to describe and designate that orderly sequence of changes through which a biotic community passes in the course of its development from a primary and relatively unstable to a relatively permanent or climax stage. The main point is that not merely do the individual plants and animals within the communal habitat grow but the community itself, i.e., the system of relations between the species, is likewise involved in an orderly process of change and development.

The fact that, in the course of this development, the community moves through a series of more or less clearly defined stages is the fact that gives this development the serial character which the term "succession" suggests.

The explanation of the serial character of the changes involved in succession is the fact that at every stage in the process a more or less stable equilibrium is achieved, which in due course, and as a result of progressive changes in life-conditions, possibly due to growth and decay, the equilibrium achieved in the earlier stages is eventually undermined. In such case the energies previously held in balance will be released, competition will be intensified, and change will continue at a relatively rapid rate until a new equilibrium is achieved.

The climax phase of community development corresponds with the adult phase of an individual's life.

> In the developing single organism, each phase is its own executioner, and itself brings a new phase into existence, as when the tadpole grows the thyroid gland which is destined to make the tadpole state pass away in favour of the miniature frog. And in the developing community of organisms, the same thing happens—each stage alters its own environment, for it changes and almost invariably enriches the soil in which it lives; and thus it eventually brings itself to an end, by making it possible for new kinds of plants with greater demands in the way of mineral salts or other riches of the soil to flourish there. Accordingly bigger and more exigent plants gradually supplant the early pioneers, until a final balance is reached, the ultimate possibility for that climate.[6]

The cultural community develops in comparable ways to that of the biotic, but the process is more complicated. Inventions, as well as sudden or catastrophic changes, seem to play a more important part in bringing about serial changes in the cultural than in the biotic community. But the principle involved seems to be substantially the same. In any case, all or most of the fundamental processes seem to be functionally related and dependent upon competition.

Competition, which on the biotic level functions to control and regulate the interrelations of organisms, tends to assume on the social level the form of conflict. The intimate relation between competition and conflict is indicated by the fact that wars frequently, if not always, have, or seem to have, their source and origin in economic competition which, in that case, assumes the more sublimated form of a struggle for power and prestige. The social function of war, on the other hand, seems to be to extend the area over which it is possible to maintain peace.

IV. Biological economics

If population pressure, on the one hand, co-operates with changes in local and environmental conditions to disturb at once the biotic balance and social equilibrium, it tends at the same time to intensify competition. In so doing it functions, indirectly, to bring about a new, more minute and, at the same time, territorially extensive division of labor.

Under the influence of an intensified competition, and the increased activity which competition involves, every individual and every species, each for itself, tends to discover the particular niche in the physical and living environment where it can survive and flourish with the greatest possible expansiveness consistent with its necessary dependence upon its neighbors.

It is in this way that a territorial organization and a biological division of labor, within the communal habitat, is established and maintained. This explains, in part at least, the fact that the biotic community has been conceived at one time as a kind of superorganism and at another as a kind of economic organization for the exploitation of the natural resources of its habitat.

In their interesting survey, *The Science of Life*, H. G. Wells and his collaborators, Julian Huxley and G. P. Wells, have described ecology as "biological economics," and as such very largely concerned with "the balances and mutual pressures of species living in the same habitat."

"Ecology," as they put it, is "an extension of Economics to the whole of life." On the other

---
[6]Ibid., pp. 977–78.

hand the science of economics as traditionally conceived, though it is a whole century older, is merely a branch of a more general science of ecology which includes man with all other living creatures. Under the circumstances what has been traditionally described as economics and conceived as restricted to human affairs, might very properly be described as Barrows some years ago described geography, namely as human ecology. It is in this sense that Wells and his collaborators would use the term.

> The science of economics—at first it was called Political Economy—is a whole century older than ecology. It was and is the science of social subsistence, of needs and their satisfactions, of work and wealth. It tries to elucidate the relations of producer, dealer, and consumer in the human community and show how the whole system carries on. Ecology broadens out this inquiry into a general study of the give and take, the effort, accumulation and consumption in every province of life. Economics, therefore, is merely Human Ecology, it is the narrow and special study of the ecology of the very extraordinary community in which we live. It might have been a better and brighter science if it had begun biologically.[7]

Since human ecology cannot be at the same time both geography and economics, one may adopt, as a working hypothesis, the notion that it is neither one nor the other but something independent of both. Even so the motives for identifying ecology with geography on the one hand, and economics on the other, are fairly obvious.

From the point of view of geography, the plant, animal, and human population, including their habitations and other evidence of man's occupation of the soil, are merely part of the landscape, of which the geographer is seeking a detailed description and picture.

On the other hand ecology (biologic economics), even when it involves some sort of unconscious co-operation and a natural, spontaneous, and non-rational division of labor, is something different from the economics of commerce; something quite apart from the bargaining of the market place. Commerce, as Simmel somewhere remarks, is one of the latest and most complicated of all the social relationships into which human beings have entered. Man is the only animal that trades and traffics.

Ecology, and human ecology, if it is not identical with economics on the distinctively human and cultural level is, nevertheless, something more than and different from the static order which the human geographer discovers when he surveys the cultural landscape.

The community of the geographer is not, for one thing, like that of the ecologist, a closed system, and the web of communication which man has spread over the earth is something different from the "web of life" which binds living creatures all over the world in a vital nexus.

V. Symbiosis and society

Human ecology, if it is neither economics on one hand nor geography on the other, but just ecology, differs, nevertheless, in important respects from plant and animal ecology. The interrelations of human beings and interactions of man and his habitat are comparable but not identical with interrelations of other forms of life that live together and carry on a kind of "biological economy" within the limits of a common habitat.

For one thing man is not so immediately dependent upon his physical environment as other animals. As a result of the existing world-wide division of labor, man's relation to his physical environment has been mediated through the intervention of other men. The exchange of goods and services has co-operated to emancipate him from dependence upon his local habitat.

Furthermore man has, by means of inventions and technical devices of the most diverse sorts, enormously increased his capacity for reacting upon and remaking, not only his habitat but his world. Finally, man has erected upon the basis of the biotic community an institutional structure rooted in custom and tradition.

Structure, where it exists, tends to resist change, at least change coming from without, while it possibly facilitates the cumulation of change within. In plant and animal communities structure is biologically determined, and so far as any division of labor exists at all it has a physiological and instinctive basis. The social insects afford a conspicuous example of this fact, and one interest in studying their habits, as Wheeler points out, is that they show the extent to which social organization can be developed on a purely physiological and instinctive basis, as is the case among human beings in the natural as distinguished from the institutional family.[8]

---

[7] H. H. Barrows, "Geography as Human Ecology," *Annals of the Association of American Geographers*, 13 (1923), 1–14. See H. G. Wells et al., op. cit., pp. 961–62.

[8] William Morton Wheeler, *Social Life among the Insects* (Lowell Institute Lectures, March 1922), pp. 3–18.

In a society of human beings, however, this communal structure is reinforced by custom and assumes an institutional character. In human as contrasted with animal societies, competition and the freedom of the individual is limited on every level above the biotic by custom and consensus.

The incidence of this more or less arbitrary control which custom and consensus impose upon the natural social order complicates the social process but does not fundamentally alter it—or, if it does, the effects of biotic competition will still be manifest in the succeeding social order and the subsequent course of events.

The fact seems to be, then, that human society, as distinguished from plant and animal society, is organized on two levels, the biotic and the cultural. There is a symbiotic society based on competition and a cultural society based on communication and consensus. As a matter of fact the two societies are merely different aspects of one society, which, in the vicissitudes and changes to which they are subject, remain, nevertheless, in some sort of mutual dependence each upon the other. The cultural superstructure rests on the basis of the symbiotic substructure, and the emergent energies that manifest themselves on the biotic level in movements and actions reveal themselves on the higher social level in more subtle and sublimated forms.

However, the interrelations of human beings are more diverse and complicated than this dichotomy, symbiotic and cultural, indicates. This fact is attested by the divergent systems of human interrelations which have been the subject of the special social sciences. Thus human society, certainly in its mature and more rational expression, exhibits not merely an ecological, but an economic, a political, and a moral order. The social sciences include not merely human geography and ecology, but economics, political science, and cultural anthropology.

It is interesting also that these divergent social orders seem to arrange themselves in a kind of hierarchy. In fact they may be said to form a pyramid of which the ecological order constitutes the base and the moral order the apex. Upon each succeeding one of these levels, the ecological, economic, political, and moral, the individual finds himself more completely incorporated into and subordinated to the social order of which he is a part than upon the preceding.

Society is everywhere a control organization. Its function is to organize, integrate, and direct the energies resident in the individuals of which it is composed. One might, perhaps, say that the function of society was everywhere to restrict competition and by so doing bring about a more effective co-operation of the organic units of which society is composed.

Competition, on the biotic level, as we observe it in the plant and animal communities, seems to be relatively unrestricted. Society, so far as it exists, is anarchic and free. On the cultural level, this freedom of the individual to compete is restricted by conventions, understandings, and law. The individual is more free upon the economic level than upon the political, more free on the political than the moral.

As society matures control is extended and intensified and free commerce of individuals restricted, if not by law then by what Gilbert Murray refers to as "the normal expectation of mankind." The mores are merely what men, in a situation that is defined, have come to expect.

Human ecology, in so far as it is concerned with a social order that is based on competition rather than consensus, is identical, in principle at least, with plant and animal ecology. The problems with which plant and animal ecology have been traditionally concerned are fundamentally population problems. Society, as ecologists have conceived it, is a population settled and limited to its habitat. The ties that unite its individual units are those of a free and natural economy, based on a natural division of labor. Such a society is territorially organized and the ties which hold it together are physical and vital rather than customary and moral.

Human ecology has, however, to reckon with the fact that in human society competition is limited by custom and culture. The cultural superstructure imposes itself as an instrument of direction and control upon the biotic substructure.

Reduced to its elements the human community, so conceived, may be said to consist of a population and a culture, including in the term culture (1) a body of customs and beliefs and (2) a corresponding body of artifacts and technological devices.

To these three elements or factors—(1) population, (2) artifacts (technicological culture), (3) custom and beliefs (non-material culture)—into which the social complex resolves itself, one should, perhaps, add a fourth, namely, the natural resources of the habitat.

It is the interaction of these four factors—(1) population, (2) artifacts (technicological culture), (3) custom and beliefs (non-material culture), and (4) the natural resources—that maintains at once the biotic balance and the social equilibrium, when and where they exist.

The changes in which ecology is interested are the movements of population and of artifacts (commodities) and changes in location and occupation—any sort of change, in fact, which affects an existing division of labor or the relation of the population to the soil.

Human ecology is, fundamentally, an attempt to investigate the processes by which the biotic balance and the social equilibrium (1) are maintained once they are achieved and (2) the processes by which, when the biotic balance and the social equilibrium are disturbed, the transition is made from one relatively stable order to another.

The Scope of Human Ecology*

R. D. McKenzie

In the struggle for existence in human groups social organization accommodates itself to the spatial and sustenance relationships existing among the occupants of any geographical area. All the more fixed aspects of human habitation, the buildings, roads, and centers of association, tend to become spatially distributed in accordance with forces operating in a particular area at a particular level of culture. In society physical structure and cultural characteristics are parts of one complex.

The spatial and sustenance relations in which human beings are organized are ever in process of change in response to the operation of a complex of environmental and cultural forces. It is the task of the human ecologist to study these processes of change in order to ascertain their principles of operation and the nature of the forces producing them.

It is perhaps necessary at the outset to indicate the relation of human ecology to the kindred sciences of geography and economics. It has been claimed that geography is human ecology.[1] There are doubtless many points in common between the two disciplines; but geography is concerned with place; ecology, with process. Location, as a geographical concept, signifies position on the earth's surface; location as an ecological concept signifies position in a spatial grouping of interacting human beings or of interrelated human institutions.

Research in economics and commercial geography on land value, marketing, transportation, commerce, factory and business location frequently has ecological significance. The difference between economics and ecology lies mainly in the direction of attention. Business economics, the division of economics having most ecological significance, is usually approached from the point of view of the business man who may want to know the best place to locate a factory or the best method of marketing a commodity. The ecologist studies the same economic problems, but in relation to the processes of human distribution. The chain-store system of marketing goods, for instance, might be studied by the economist as a system of retail marketing, whereas the ecologist might study it as an index of the process of decentralization.

ECOLOGICAL DISTRIBUTION. By this term is meant the spatial distribution of human beings and human activities resulting from the interplay of forces which effect a more or less conscious, or at any rate dynamic and vital, relationship among the units comprising the aggregation. An ecological distribution should be distinguished from a fortu-

---

*Reprinted from *Publications of the American Sociological Society*, 20 (1926), 141–54, by permission of The American Sociological Association.

[1] H. H. Barrows, "Geography as Human Ecology," *Annals of the Association of American Geographers*, 13 (March 1923), 1–14.

itous or accidental distribution, where spatial relationships are, or seem to be, largely a matter of chance rather than the resultant of competing forces. For example, the aggregation of people waiting for the door of a theater to open represents a fortuitous spatial distribution; but their distribution in the theater, according to the kind of tickets they present, is a temporary ecological distribution. Although less complex and exacting, this distribution is quite similar to that which takes place in the community at large under conditions of free competition and choice.

The spatial distribution of economic utilities, shops, factories, offices, is the product of the operation of ecological forces quite as much as is the distribution of residence. The business man who attempts to locate his factory or place of business with scientific exactness seeks the position of maximum advantage: that is, he seeks a point of equilibrium among competing forces. For this reason the value of location is always relative and changes as one or more of the co-operating forces gain or lose in relative significance. A community, then, is an ecological distribution of people and services in which the spatial location of each unit is determined by its relation to all other units. A network of interrelated communities is likewise an ecological distribution. In fact, civilization, with its vast galaxy of communities, each of which is more or less dependent upon some or all of the others, may be thought of as an ecological distribution or organization.[2]

ECOLOGICAL UNIT. Any ecological distribution—whether of residences, shops, offices, or industrial plants—which has a unitary character sufficient to differentiate it from surrounding distributions may be defined as an ecological unit. On the other hand, an interdependent grouping of ecological units around a common center may be called an "ecological constellation." The metropolitan area, with its various districts of residence, business, and industry integrated about a common center usually called the city is an ecological constellation. Such groupings may vary in degree of ecological interdependence from the conurbations which are found in each of the strategic areas of commerce and industry to the larger national or international communal federations linked financially and industrially with a metropolitan center such as London or New York.

MOBILITY AND FLUIDITY. An ecological organization is in process of constant change, the rate depending upon the dynamics of cultural, and particularly technical, advance. Mobility is a measure of this rate of change; it is represented in change of residence, change of employment, or change of location of any utility or service. Mobility must be distinguished from fluidity, which represents movement without change of ecological position. Modern means of transportation and communication have greatly increased the fluidity of both people and commodities. Increased fluidity, however, does not necessarily imply increased mobility. In fact, it frequently produces the opposite effect by making residence relatively independent of the place of work; also by extending the territorial zone in which the individual may seek the satisfaction of his wishes.

Fluidity tends to vary inversely with mobility. Slums are the most mobile but least fluid sections of a city. Their inhabitants come and go in continuous succession, but, while domiciled within a given area, have a smaller range of movement than the residents of any of the higher economic districts. The unequal fluidity of different districts of the city and of different individuals within the same district is an important factor in the processes of segregation and centralization. Youth tends to be more fluid than old age, or childhood, giving rise to characteristically different centers of interest and varying regions of experience for each age group.

DISTANCE. Ecological distance is a measure of fluidity. It is a time-cost concept rather than a unit of space. It is measured by minutes and cents rather than by yards and miles. By time-cost measurement the distance from A to B may be farther than from B to A, provided B is upgrade from A.

Communal growth and structure are largely functions of ecological distance as a time-cost concept. This basis of distance determines the currents of travel and traffic, which in turn determine the areas of concentration and the locations of cities. Likewise, communal structure is a response to distance in the local movements of commodities and people. The uneven expansion of cities along the routes of rapid and cheap transportation is but an obvious result of the time-cost measurement of distance. American cities, unlike European cities, are seldom circular in shape, owing to the fact that they have usually grown up without systematic planning, and therefore their intramural transportation is frequently less uniformly developed than is the case in most European cities. American cities—and this is particu-

---

[2] Ecological distribution, as here used, is synonymous with ecological organization.

larly true since the advent of the automobile—tend to spread out in starlike fashion along the lines of rapid communication. The maximum linear distance from the periphery to the center of the city is seldom over an hour's travel by the prevailing form of transportation.

ECOLOGICAL FACTORS. The changing spatial relations of human beings are the result of the interplay of a number of different forces, some of which have general significance throughout the entire cultural area in which they operate; others have limited reference, applying merely to a specific region or location. For instance, the shaft elevator, introduced in the seventies, and steel construction, introduced in the nineties, and the more recent advent of the automobile have acted as general factors in affecting the concentration of population and organization of communities. On the other hand, geographic factors, such as rivers, hills, lakes, and swamps, may have either general or limited significance with regard to ecological distribution, depending upon the peculiarities of local conditions. Certain factors, such as bridges, public buildings, cemeteries, parks, and other institutions or forces have only limited significance in attracting or repelling population.

Ecological factors may be classified under four general heads: (1) geographical, which includes climatic, topographic, and resource conditions; (2) economic, which comprises a wide range and variety of phenomena such as the nature and organization of local industries, occupational distribution, and standard of living of the population; (3) cultural and technical, which include, in addition to the prevailing condition of the arts, the moral attitudes and taboos that are effective in the distribution of population and services; (4) political and administrative measures, such as tariff, taxation, immigration laws, and rules governing public utilities.

Ecological factors are either positive or negative; they either attract or repel. It is part of the task of the ecologist to measure the dispersive and integrative influence of typical communal institutions upon different elements of the population. Such knowledge would be of great value in city-planning, as it would enable the community to control the direction of its growth and structure. Effort must always be made to isolate the determining or limiting factors in a specific ecological situation.

ECOLOGICAL PROCESSES. By ecological process is meant the tendency in time toward special forms of spatial and sustenance groupings of the units comprising an ecological distribution. There are five major ecological processes: concentration, centralization, segregation, invasion, succession. Each of these has an opposite or negative aspect, and each includes one or more subsidiary processes.

REGIONAL CONCENTRATION. This is the tendency of an increasing number of persons to settle in a given area or region. Density is a measure of population concentration in a given area at a given time. World-population density maps indicate in a general way the significance of geographical factors in the distribution of human beings. While formerly the limits of concentration were defined by the conditions of local food supply, modern industrialism has created new regions of concentration, the limits of which are defined not by the local food supply but by the strategic significance of location with reference to commerce and industry.

> The townward tendency is operating in every civilized country. As in other countries so in Japan the dominant characteristic of the new industrialism is the trend of population from the country to the city.... In the case of Tokyo, the capital, population during the last twenty-five years has increased from 857,780 to 2,500,000 while Osaka, the greatest industrial center of the Empire, during the same period has grown from 500,000 to over 1,500,000; Nagoya, from 200,000 to 450,000, Yokohama has increased fourfold, and Kobe, fivefold. The five greatest industrial centers above mentioned have thus increased 325 percent, or 300 percent more than the nation as a whole.... Great areas which ten years ago were taken up with rice fields or marshes are now reclaimed and covered with factories or labor tenements, and property values at the same time have gone up more than 1,000 percent.... These cities may be justly taken as focal points to reveal the metamorphosis of Japan from a feudal to an agricultural country, and now to the age of steam, electricity, and steel.[3]

The territorial concentration of population resulting from industrialism and modern forms of transportation and communication is more dynamic and unpredictable than were the older concentrations controlled by factors of the local environment. Modern territorial concentration is never the result of natural population increase alone. It always represents the shifting of population from one territory to another. Practically all food-producing areas of countries which have come under the influence of modern machine industry have decreased in population during the last few decades.

[3]*Present-Day Impressions of Japan* (1919), p. 539.

The limits of regional concentration of population in a world-economy of large-scale industry are determined by the relative competitive strength which the particular region possesses over other regions in the production and distribution of commodities. The degree of concentration attained by any locality is therefore a measure of its resource and location advantages as compared with those of its competitors. This strength is shown in the struggle for *hinterland*, raw materials, and markets, and depends upon the conditions of transportation and communication.

REGIONAL SPECIALIZATION. Regional specialization in production is the natural outcome of competition under prevailing conditions of transportation and communication. Territorial specialization has two points of special significance for the human ecologist. In the first place it produces an economic interdependence between different regions and communities which changes the sustenance relations not only of the individuals within the community but also of the different communities to one another. In the second place it makes for regional selection of population, by age, sex, race, and nationality in conformity with the occupational requirements of the particular form of specialized production.[4]

DISPERSION. The obverse of concentration is dispersion. Concentration in one region usually implies dispersion in another. Steam transportation, by increasing the fluidity of commodities, ushered in a new epoch in regional concentration; motor and electric transportation, by increasing the fluidity of people, is now producing a new era in dispersion. Whatever retards the movement of commodities limits concentration, and whatever facilitates the movement of people makes for dispersion. The forces at work during the past few years have been favorable to dispersion. High freight-rates, high taxes, and labor costs are forcing many industries to disperse or relocate. On the other hand, the automobile and rapid-transit lines are permitting the concentrated urban populations to spread out over adjacent territory.

CENTRALIZATION. Centralization as an ecological process should be distinguished from concentration, which is mere regional aggregation. Centralization is an effect of the tendency of human beings to come together at definite locations for the satisfaction of specific common interests, such as work, play, business, education. The satisfaction of each specific interest may be found in a different region. Centralization, therefore, is a temporary form of concentration, an alternate operation of centripetal and centrifugal forces. Centralization implies an area of participation with center and circumference. It is the process of community formation. The fact that people come together at specific locations for the satisfaction of common interests affords a territorial basis for group consciousness and social control. Every communal unit, the village, town, city, and metropolis, is a function of the process of centralization.

The focal point of centralization in the modern community is the retail shopping center. The market place, at which buyers and sellers meet, has always had a potent centralizing community-making significance. Since economic contacts are more abstract and impersonal than other kinds of contacts, the trade center has more general attractive significance, and therefore more community-making influence, than the school, the church, the theater, or any other type of interest center. It is retail shopping that creates the "Main Street" of the little town and the city of the metropolitan community.

The distance from the center to the periphery of any unit of centralization depends upon the degree of specialization which the center has attained and on the conditions of transportation and communication. In regions or districts where human energy is the chief motor power the units of centralization are seldom more than a few miles in radius as is illustrated by the village communities of the Orient. In the agricultural town of America, prior to the advent of the automobile, Warren H. Wilson found that the "team-haul" (the distance that a team could travel to the center and return on the same day) defined the outer limits of the trade area.

Focal points of centralization are invariably in competition with other points for the attention and patronage of the inhabitants of the surrounding area. Thus the present conditions of centralization always represent but a temporary stage of unstable equilibrium within a zone of competing centers. The degree of centralization at any particular center is, therefore, a measure of its relative drawing-power under existing cultural and eco-

---

[4]Few American cities at the present time have normal age and sex distribution of the population. The percentage of persons in the age group fifteen to forty-five is usually much higher for cities than for rural districts or for the country as a whole. Furthermore, industrial specialization tends to create single-sex cities. Textile cities such as Lowell, Paterson, New Bedford, have a predominance of women, while heavy-industry cities, such as Pittsburgh, Akron, Seattle, have a predominance of men.

nomic conditions. The introduction of a new form of transportation, such as the automobile, completely disturbs the ecological equilibrium and makes for a reaccommodation on a new scale of distance.

Centralization under any given conditions of transit and concentration takes place in cumulative fashion, increasing with its own momentum until it reaches the point of equilibrium or saturation. Then, unless relief is afforded by the introduction of new avenues of transit, a retrograde movement commences, giving rise to new units of centralization or new developments of old units. In this way new communities are born within the metropolitan area.

Centralization may take place in two ways: first, by an addition to the number and variety of interests at a common location, as, for instance, when the rural trade center becomes also the locus of the school, church, post-office, and dance hall; second, by an increase in the number of persons finding satisfaction of a single interest at the same location.

SPECIALIZATION AND CENTRALIZATION. As the regional concentration and fluidity of the population increases, territorial specialization of interest satisfaction follows. The urban area becomes studded with centers of various sizes and degrees of specialization, which is a magnet drawing to itself the appropriate age, sex, cultural, and economic groups. Time specialization takes place as well as place specialization. At different hours of the day and night the waves of selective centralization ebb and flow. As a New York bohemian facetiously remarked, the commuters' train carries to the city in the early morning the workers, an hour or so later the clerkers, and about midday the shirkers. A similar cycle is repeated by the night population of amusement-seekers.

TYPES OF CENTERS. Communal points of centralization may be classified according to (1) size and importance as indicated by land values and concentration; (2) the dominant interest producing the centralization, such as work, business, amusement; (3) the distance or area of the zone of participation.

Every community has its main center called the main street, the town, or the city, which is a constellation of specialized centers. The larger the community, the more specialized are the divisions of its center and the wider the zone of patronage. Civilization is a product of centralization. The evolution of economic organization from village and town to metropolitan economy is but the extension and specialization of centralization of each of the dominant interests of life.

LOCATION AND MOVEMENT OF CENTERS. Centralization is a function of transportation and communication. Centers are located where lines of traffic meet or intersect, and vary in importance, other things equal, with the number and variety of converging lines of transit. The "city" is the point of convergence of all the main avenues of transportation and communication, both local and intercommunal.

Most centers are responsive to the trends of distribution and segregation of the local population. The main retail shopping center, which is usually the point of highest land value, tends to move in the direction of the higher economic residential areas, but is held fairly close to the median center of population within the zone of participation. Local business centers are more mobile, they respond quite accurately to local trends of segregation and fluidity. Financial centers are less responsive to the currents of travel. Being centers of wide participation, they tend to become of great physical value, and therefore acquire great stability.[5] Work centers are controlled by forces which frequently transcend the bounds of community; those of the basic manufacturing type tend to move out to the fringe of the community, thus making for decentralization.

Leisure-time centers, not associated with trade centers, are comparatively unstable, as is indicated by the dynamic changes in land values.[6] Conditions of concentration and fluidity become determining factors in their distribution. The motion-picture theater, operating on the chain-store principle, is causing new centers to be established far from the downtown center, and new white-light areas are arising in different sections of the city.

DECENTRALIZATION AND RECENTRALIZATION. These are but phases of the centralization process. New units of centralization are constantly appearing and established units constantly changing in significance. By decentralization is meant the tendency for zone areas of centralization to decrease in size, which of course implies a multiplication of centers, each of relatively less importance. In this sense decentralization is taking place in all metropolitan areas with reference to some interests, while at the same time more extreme centralization is occurring in connection with other interests. In studying the process of centralization, therefore, it is important to find what particular

[5]Note the location and great stability of Wall Street.
[6]See Felix Isman, *Real Estate* (1924).

aspects of life are being organized on the basis of smaller centers, what on the larger centers, and what seem to be the factors involved.

General observation leads one to believe that the centralization of any interest varies directly with the element of choice involved in the satisfaction of the interest. Standardization of commodities, both in quality and in price, minimizes the element of choice, with the result that all primary standardized services, such as grocery stores, drug stores, soft-drink parlors, are very widely distributed. On the other hand, the more specialized services tend to become more and more highly centralized.[7]

SEGREGATION. Segregation is used here with reference to the concentration of population types within a community. Every area of segregation is the result of the operation of a combination of forces of selection. There is usually, however, one attribute of selection that is more dominant than the others, and which becomes the determining factor of the particular segregation. Economic segregation is the most primary and general form. It results from economic competition and determines the basic units of the ecological distribution. Other attributes of segregation, such as language, race, or culture, function within the spheres of appropriate economic levels.

Economic segregation decreases in degree of homogeneity as we ascend the economic scale; the lower the economic level of an area, the more uniform the economic status of the inhabitants, because the narrower the range of choice. But as we ascend the economic scale each level affords wider choice, and therefore more cultural homogeneity.

The slum is the area of minimum choice. It is the product of compulsion rather than design. The slum, therefore, represents a homogeneous collection as far as economic competency is concerned, but a most heterogeneous aggregation in all other respects. Being an area of minimum choice, the slum serves as the reservoir for the economic wastes of the city. It also becomes the hiding-place for many services which are forbidden by the mores but which cater to the wishes of residents scattered throughout the community.

[7] A study of the shopping habits of about two thousand families of a middle-class residential district in Seattle showed that about 90 percent bought their groceries in the neighborhood; 70 percent, their drugs; 50 percent, their hardware; and a smaller percentage, their furniture and clothes. In leisure-time activities, a much higher percentage attended local, rather than downtown churches, but the opposite was true of the attendance at the moving-picture theater.

INVASION. Invasion is a process of group displacement; it implies the encroachment of one area of segregation upon another, usually an adjoining, area. The term "invasion," in the historic sense, implies the displacement of a higher by a lower cultural group. While this is perhaps the more common process in the local community, it is not, however, the only form of invasion. Frequently a higher economic group drives out the lower-income inhabitants, thus enacting a new cycle of the succession.

Invasion should be distinguished from atomatization; the latter is a consequence of individual displacement without consciousness of displacement or change in cultural level.

SUCCESSION. In human and plant communities change seems to take place in cyclic fashion. Regions within a city pass through different stages of use and occupancy in a regularity of manner which may eventually be predictable and expressible in mathematical terms. The process of obsolescence and physical deterioration of buildings makes for a change in type of occupancy which operates in a downward tendency in rentals, selecting lower and lower income levels of population, until a new cycle is commenced, either by a complete change in use of the territory, such as a change from residence to business, or by a new development of the old use, the change, say, from an apartment to a hotel form of dwelling.

The thing that characterizes a succession is a complete change in population type between the first and last stages, or a complete change in use. While there is not the intimate connection between the different stages in a human succession that is found between the stages in a plant succession, nevertheless there is an economic continuity which makes the cycles in a human succession quite as pronounced and as inevitable as those in the plant succession. Real-estate investigators are beginning to plot the stages in use succession by mathematical formulas.

The entire community may pass through a series of successions, due to mutations of its economic base affecting its relative importance in the larger ecological constellation. The population type usually changes with the changing of the economic base, as, for instance, when an agricultural community changes to a mining or a manufacturing community.

STRUCTURE. Ecological processes always operate within a more or less rigid structural base. The relative spatial fixity of the road and the establishment furnishes the base in which the ecological processes function. The fact that the movements of men and commodities follow narrow channels or rather fixed spatial significance gives a structural foundation to human spatial relations which is absent in the case of plant and animal communities.

The history of civilization shows a gradually increasing flexibility of the structural skeleton in which ecological processes operate. Prior to the advent of the railroad the movements of people and commodities were largely controlled by the course of the water systems: rivers, lakes, and seas. The coming of the railroads in the early part of the nineteenth century marked the first great release with regard to population distribution. New regions of concentration immediately arose, while old regions either declined or commenced a new cycle of growth. The advent of motor transportation and the good-roads movement affords a freedom to human distribution which is unique in history, making for a redistribution of people and institutions on a much more flexible base than was ever known before.

The Growth of the City: An Introduction to a Research Project*

Ernest W. Burgess

The outstanding fact of modern society is the growth of great cities. Nowhere else have the enormous changes which the machine industry has made in our social life registered themselves with such obviousness as in the cities. In the United States the transition from a rural to an urban civilization, though beginning later than in Europe, has taken place, if not more rapidly and completely, at any rate more logically in its most characteristic forms.

All the manifestations of modern life which are peculiarly urban—the skyscraper, the subway, the department store, the daily newspaper, and social work—are characteristically American. The more subtle changes in our social life, which in their cruder manifestations are termed "social problems," problems that alarm and bewilder us, such as divorce, delinquency, and social unrest, are to be found in their most acute forms in our largest American cities. The profound and "subversive" forces which have wrought these changes are measured in the physical growth and expansion of cities. That is the significance of the comparative statistics of Weber, Bücher, and other students.

*Reprinted from *The City,* ed. Robert E. Park, Ernest W. Burgess, and R. D. McKenzie (Chicago: University of Chicago Press, 1925), pp. 47–62, by permission of The University of Chicago Press.

These statistical studies, although dealing mainly with the effects of urban growth, brought out into clear relief certain distinctive characteristics of urban as compared to rural populations. The larger proportion of women to men in the cities than in the open country, the greater percentage of youth and middle-aged, the higher ratio of the foreign-born, the increased heterogeneity of occupation increase with the growth of the city and profoundly alter its social structure. These variations in the composition of population are indicative of all the changes going on in the social organization of the community. In fact, these changes are a part of the growth of the city and suggest the nature of the processes of growth.

The only aspect of growth adequately described by Bücher and Weber was the rather obvious process of the *aggregation* of urban population. Almost as overt a process, that of *expansion,* has been investigated from a different and very practical point of view by groups interested in city planning, zoning, and regional surveys. Even more significant than the increasing density of urban population is its correlative tendency to overflow, and so to extend over wider areas, and to incorporate these areas into a larger communal life. This paper, therefore, will treat first of the expansion of the city, and then of the less-known processes of urban metabolism and mobility which are closely related to expansion.

Expansion as physical growth

The expansion of the city from the standpoint of the city plan, zoning and regional surveys is thought of almost wholly in terms of its physical growth. Traction studies have dealt with the development of transportation in its relation to the distribution of population throughout the city. The surveys made by the Bell Telephone Company and other public utilities have attempted to forecast the direction and the rate of growth of the city in order to anticipate the future demands for the extension of their services. In the city plan the location of parks and boulevards, the widening of traffic streets, the provision for a civic center, are all in the interest of the future control of the physical development of the city.

This expansion in area of our largest cities is now being brought forcibly to our attention by the Plan for the Study of New York and its Environs, and by the formation of the Chicago Regional Planning Association, which extends the metropolitan district of the city to a radius of 50 miles, embracing 4,000 square miles of territory. Both are attempting to measure expansion in order to deal with the changes that accompany city growth. In England, where more than one-half of the inhabitants live in cities having a population of 100,000 and over, the lively appreciation of the bearing of urban expansion on social organization is thus expressed by C. B. Fawcett:

> One of the most important and striking developments in the growth of the urban populations of the more advanced peoples of the world during the last few decades has been the appearance of a number of vast urban aggregates or conurbations, far larger and more numerous than the great cities of any preceding age. These have usually been formed by the simultaneous expansion of a number of neighboring towns, which have grown out toward each other until they have reached a practical coalescence in one continuous urban area. Each such conurbation still has within it many nuclei of denser town growth, most of which represent the central areas of the various towns from which it has grown, and these nuclear patches are connected by the less densely urbanized areas which began as suburbs of these towns. The latter are still usually rather less continuously occupied by buildings, and often have many open spaces.
>
> These great aggregates of town dwellers are a new feature in the distribution of man over the earth. At the present day there are from thirty to forty of them, each containing more than a million people, whereas only a hundred years ago there were, outside the great centers of population on the waterways of China, not more than two or three. Such aggregations of people are phenomena of great geographical and social importance; they give rise to new problems in the organization of the life and well-being of their inhabitants and in their varied activities. Few of them have yet developed a social consciousness at all proportionate to their magnitude, or fully realized themselves as definite groupings of people with many common interests, emotions and thoughts.[1]

In Europe and America the tendency of the great city to expand has been recognized in the term "the metropolitan area of the city," which far overruns its political limits, and in the case of New York and Chicago, even state lines. The metropolitan area may be taken to include urban territory that is physically contiguous, but it is coming to be defined by that facility of transportation that enables a business man to live in a suburb of Chicago and to work in the Loop, and his wife to shop at Marshall Field's and attend grand opera in the Auditorium.

Expansion as a process

No study of expansion as a process has yet been made, although the materials for such a study and intimations of different aspects of the process are contained in city planning, zoning, and regional surveys. The typical process of the expansion of the city can best be illustrated, perhaps, by a series of concentric circles, which may be numbered to designate both the successive zones of urban extension and the types of areas differentiated in the process of expansion.

This chart represents an ideal construction of the tendencies of any town or city to expand radially from its central business district—on the map "The Loop" (I). Encircling the downtown area there is normally an area of transition, which is being invaded by business and light manufacture (II). A third area (III) is inhabited by the workers in industries who have escaped from the area of deterioration (II) but who desire to live within easy access of their work. Beyond this zone is the "residential area" (IV) of high-class apartment buildings or of exclusive "restricted" districts of single family dwellings. Still farther, out beyond

---

[1] "British Conurbations in 1921," *Sociological Review*, 14 (April 1922), pp. 111–12.

the city limits, is the commuters' zone—suburban areas, or satellite cities—within a thirty- to sixty-minute ride of the central business district.

This chart brings out clearly the main fact of expansion, namely, the tendency of each inner zone to extend its area by the invasion of the next outer zone. This aspect of expansion may be called *succession*, a process which has been studied in detail in plant ecology. If this chart is applied to Chicago, all four of these zones were in its early history included in the circumference of the inner zone, the present business district. The present boundaries of the area of deterioration were not many years ago those of the zone now inhabited by independent wage-earners, and within the memories of thousands of Chicagoans contained the residences of the "best families." It hardly needs to be added that neither Chicago nor any other city fits perfectly into this ideal scheme. Complications are introduced by the lake front, the Chicago River, railroad lines, historical factors in the location of industry, the relative degree of the resistance of communities to invasion, etc.

Besides extension and succession, the general process of expansion in urban growth involves the antagonistic and yet complementary processes of concentration and decentralization. In all cities there is the natural tendency for local and outside transportation to converge in the central business district. In the downtown section of every large city we expect to find the department stores, the skyscraper office buildings, the railroad stations, the great hotels, the theaters, the art museum, and the city hall. Quite naturally, almost inevitably, the economic, cultural and political life centers here. The relation of centralization to the other processes of city life may be roughly gauged by the fact that over half a million people daily enter and leave Chicago's "Loop." More recently sub-business centers have grown up in outlying zones. These "satellite loops" do not, it seems, represent the "hoped for" revival of the neighborhood, but rather a telescoping of several local communities into a larger economic unity. The Chicago of yesterday, an agglomeration of country towns and immigrant colonies, is undergoing a process of reorganization into a centralized decentralized system of local communities coalescing into sub-business areas visibly or invisibly dominated by the central business district. . . .

Expansion, as we have seen, deals with the physical growth of the city, and with the extension of the technical services that have made city life not only livable, but comfortable, even luxurious.

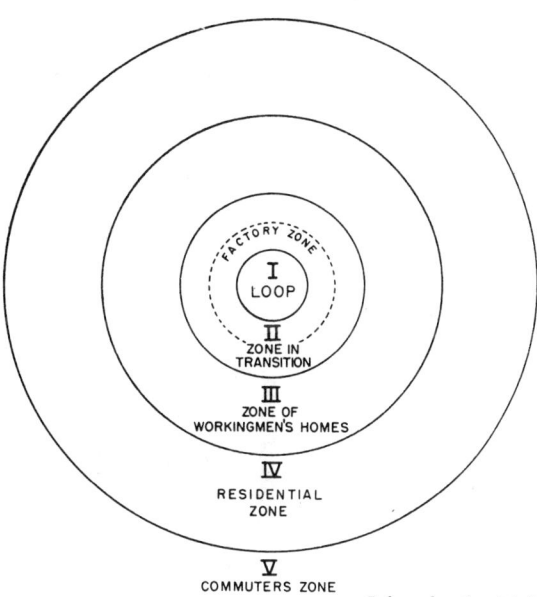

Chart 1—The growth of the city.

Certain of these basic necessities of urban life are possible only through a tremendous development of communal existence. Three millions of people in Chicago are dependent upon one unified water system, one giant gas company, and one huge electric light plant. Yet, like most of the other aspects of our communal urban life, this economic co-operation is an example of co-operation without a shred of what the "spirit of co-operation" is commonly thought to signify. The great public utilities are a part of the mechanization of life in great cities, and have little or no other meaning for social organization.

Yet the processes of expansion, and especially the rate of expansion, may be studied not only in the physical growth and business development, but also in the consequent changes in the social organization and in personality types. How far is the growth of the city, in its physical and technical aspects, matched by a natural but adequate readjustment in the social organization? What, for a city, is a normal rate of expansion, a rate of expansion with which controlled changes in the social organization might successfully keep pace?

Social organization and disorganization as processes of metabolism

These questions may best be answered, perhaps, by thinking of urban growth as a resultant of

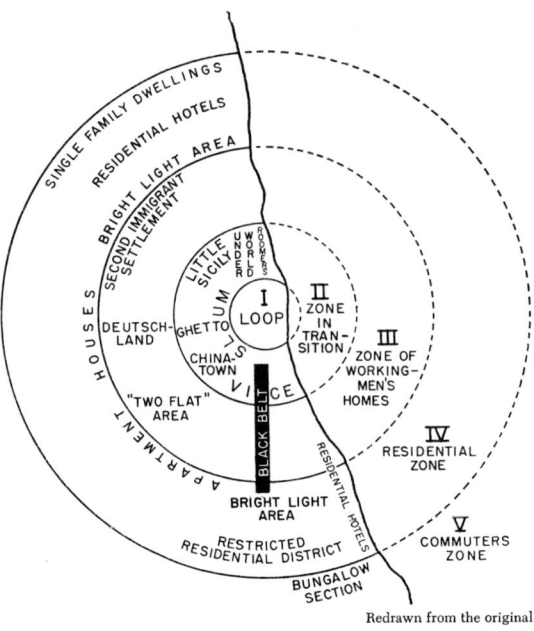

Chart 2—Urban areas.

organization and disorganization analogous to the anabolic and katabolic processes of metabolism in the body. In what way are individuals incorporated into the life of a city? By what process does a person become an organic part of his society? The natural process of acquiring culture is by birth. A person is born into a family already adjusted to a social environment—in this case the modern city. The natural rate of increase of population most favorable for assimilation may then be taken as the excess of the birth-rate over the death-rate, but is this the normal rate of city growth? Certainly, modern cities have increased and are increasing in population at a far higher rate. However, the natural rate of growth may be used to measure the disturbances of metabolism caused by any excessive increase, as those which followed the great influx of southern Negroes into northern cities since the war. In a similar way all cities show deviations in composition by age and sex from a standard population such as that of Sweden, unaffected in recent years by any great emigration or immigration. Here again, marked variations, as any great excess of males over females, or of females over males, or in the proportion of children, or of grown men or women, are symptomatic of abnormalities in social metabolism.

Normally the processes of disorganization and organization may be thought of as in reciprocal relationship to each other, and as co-operating in a moving equilibrium of social order toward an end vaguely or definitely regarded as progressive. So far as disorganization points to reorganization and makes for more efficient adjustment, disorganization must be conceived not as pathological, but as normal. Disorganization as preliminary to reorganization of attitudes and conduct is almost invariably the lot of the newcomer to the city, and the discarding of the habitual, and often of what has been to him the moral, is not infrequently accompanied by sharp mental conflict and sense of personal loss. Oftener, perhaps, the change gives sooner or later a feeling of emancipation and an urge toward new goals.

In the expansion of the city a process of distribution takes place which sifts and sorts and relocates individuals and groups by residence and occupation. The resulting differentiation of the cosmopolitan American city into areas is typically all from one pattern, with only interesting minor modifications. Within the central business district or on an adjoining street is the "main stem" of "hobohemia," the teeming Rialto of the homeless migratory man of the Middle West.[2] In the zone of deterioration encircling the central business section are always to be found the so-called "slums" and "bad lands" with their submerged regions of poverty, degradation, and disease, and their underworlds of crime and vice. Within a deteriorating area are rooming-house districts, the purgatory of "lost souls." Near by is the Latin Quarter, where creative and rebellious spirits resort. The slums are also crowded to overflowing with immigrant colonies—the Ghetto, Little Sicily, Greektown, Chinatown—fascinatingly combining old world heritages and American adaptations. Wedging out from here is the Black Belt, with its free and disorderly life. The area of deterioration, while essentially one of decay, of stationary or declining population, is also one of regeneration, as witness the mission, the settlement, the artists' colony, radical centers—all obsessed with the vision of a new and better world.

The next zone is also inhabited predominantly by factory and shop workers, but skilled and thrifty. This is an area of second immigrant settlement, generally of the second generation. It is the region of escape from the slum, the *Deutschland* of the aspiring Ghetto family. For *Deutschland* (literally "Germany") is the name

[2] For a study of this cultural area of city life see Nels Anderson, *The Hobo* (Chicago, 1923).

given half in envy, half in derision, to that region beyond the Ghetto where successful neighbors appear to be imitating German Jewish standards of living. But the inhabitant of this area in turn looks to the "Promised Land" beyond, to its residential hotels, its apartment-house region, its "satellite loops," and its "bright light" areas.

This differentiation into natural economic and cultural groupings gives form and character to the city. For segregation offers the group, and thereby the individuals who compose the group, a place and role in the total organization of city life. Segregation limits development in certain directions, but releases it in others. These areas tend to accentuate certain traits, to attract and develop their kind of individuals, and so to become further differentiated.

The division of labor in the city likewise illustrates disorganization, reorganization, and increasing differentiation. The immigrant from rural communities in Europe and America seldom brings with him economic skill of any great value in our industrial, commercial, or professional life. Yet interesting occupational selection has taken place by nationality, explainable more by racial temperament or circumstance than by old-world economic background, as Irish policemen, Greek ice-cream parlors, Chinese laundries, Negro porters, Belgian janitors, etc.

The facts that in Chicago one million (996,589) individuals gainfully employed reported 509 occupations, and that over 1,000 men and women in *Who's Who* gave 116 different vocations, give some notion of how in the city the minute differentiation of occupation "analyzes and sifts the population, separating and classifying the diverse elements."[3] These figures also afford some intimation of the complexity and complication of the modern industrial mechanism and the intricate segregation and isolation of divergent economic groups. Interrelated with this economic division of labor is a corresponding division into social classes and into cultural and recreational groups. From this multiplicity of groups, with their different patterns of life, the person finds his congenial social world and—what is not feasible in the narrow confines of a village—may move and live in widely separated, and perchance conflicting, worlds. Personal disorganization may be but the failure to harmonize the canons of conduct of two divergent groups.

If the phenomena of expansion and metabolism indicate that a moderate degree of disorganization may and does facilitate social organization, they indicate as well that rapid urban expansion is accompanied by excessive increases in disease, crime, disorder, vice, insanity, and suicide, rough indexes of social disorganization. But what are the indexes of the causes, rather than of the effects, of the disordered social metabolism of the city? The excess of the actual over the natural increase of population has already been suggested as a criterion. The significance of this increase consists in the immigration into a metropolitan city like New York and Chicago of tens of thousands of persons annually. Their invasion of the city has the effect of a tidal wave inundating first the immigrant colonies, the ports of first entry, dislodging thousands of inhabitants who overflow into the next zone, and so on and on until the momentum of the wave has spent its force on the last urban zone. The whole effect is to speed up expansion, to speed up industry, to speed up the "junking" process in the area of deterioration (II). These internal movements of the population become the more significant for study. What movement is going on in the city, and how may this movement be measured? It is easier, of course, to classify movement within the city than to measure it. There is the movement from residence to residence, change of occupation, labor turnover, movement to and from work, movement for recreation and adventure. This leads to the question: What is the significant aspect of movement for the study of the changes in city life? The answer to this question leads directly to the important distinction between movement and mobility.

Mobility as the pulse of the community

Movement, per se, is not an evidence of change or of growth. In fact, movement may be a fixed and unchanging order or motion, designed to control a constant situation, as in routine movement. Movement that is significant for growth implies a change of movement in response to a new stimulus or situation. Change of movement of this type is called *mobility*. Movement of the nature of routine finds its typical expression in work. Change of movement or mobility is characteristically expressed in adventure. The great city, with its "bright lights," its emporiums of novelties and bargains, its palaces of amusement, its underworld of vice and crime, its risks of life and property from accident, robbery, and homicide, has become the

---

[3] Weber, *The Growth of Cities*, p. 442.

region of the most intense degree of adventure and danger, excitement and thrill.

Mobility, it is evident, involves change, new experience, stimulation. Stimulation induces a response of the person to those objects in his environment which afford expression for his wishes. For the person, as for the physical organism, stimulation is essential to growth. Response to stimulation is wholesome so long as it is a correlated *integral* reaction of the entire personality. When the reaction is *segmental*, that is, detached from, and uncontrolled by, the organization of personality, it tends to become disorganizing or pathological. That is why stimulation for the sake of stimulation, as in the restless pursuit of pleasure, partakes of the nature of vice.

The mobility of city life, with its increase in the number and intensity of stimulations, tends inevitably to confuse and to demoralize the person. For an essential element in the mores and in personal morality is consistency, consistency of the type that is natural in the social control of the primary group. Where mobility is the greatest, and where in consequence primary controls break down completely, as in the zone of deterioration in the modern city, there develop areas of demoralization, of promiscuity, and of vice.

In our studies of the city it is found that areas of mobility are also the regions in which are found juvenile delinquency, boys' gangs, crime, poverty, wife desertion, divorce, abandoned infants, vice.

These concrete situations show why mobility is perhaps the best index of the state of metabolism of the city. Mobility may be thought of, in more than a fanciful sense, as the "pulse of the community." Like the pulse of the human body, it is a process which reflects and is indicative of all the changes that are taking place in the community, and which is susceptible of analysis into elements which may be stated numerically.

The elements entering into mobility may be classified under two main heads: (1) the state of mutability of the person, and (2) the number and kind of contacts or stimulations in his environment. The mutability of city populations varies with sex and age composition, the degree of detachment of the person from the family and from other groups. All these factors may be expressed numerically. The new stimulations to which a population responds can be measured in terms of change of movement or of increasing contacts. Statistics on the movement of urban population may only measure routine, but an increase at a higher ratio than the increase of population measures mobility. In 1860 the horse-car lines of New York City carried about 50,000,000 passengers; in 1890 the trolley-cars (and a few surviving horse-cars) transported about 500,000,000; in 1921, the elevated, subway, surface, and electric and steam suburban lines carried a total of more than 2,500,000,000 passengers.[4] In Chicago the total annual rides per capita on the surface and elevated lines were 164 in 1890; 215 in 1900; 320 in 1910; and 338 in 1921. In addition, the rides per capita on steam and electric suburban lines almost doubled between 1916 (23) and 1921 (41) and the increasing use of the automobile must not be overlooked.[5] For example, the number of automobiles in Illinois increased from 131,140 in 1915 to 833,920 in 1923.[6]

Mobility may be measured not only by these changes of movement, but also by increase of contacts. While the increase of population of Chicago in 1912–22 was less than 25 percent (23.6 percent), the increase of letters delivered to Chicagoans was double that (49.6 percent)—(from 693,084,196 to 1,038,007,854).[7] In 1912 New York had 8.8 telephones; in 1922, 16.9 per 100 inhabitants. Boston had, in 1912, 10.1 telephones; ten years later 19.5 telephones per 100 inhabitants. In the same decade the figures for Chicago increased from 12.3 to 21.6 per 100 population.[8] But increase of the use of the telephone is probably more significant than increase in the number of telephones. The number of telephone calls in Chicago increased from 606,131,928 in 1914 to 944,010,586 in 1922,[9] an increase of 55.7 percent, while the population increased only 13.4 percent.

Land values, since they reflect movement, afford one of the most sensitive indexes of mobility. The highest land values in Chicago are at the point of greatest mobility in the city, at the corner of State and Madison streets, in the Loop. A traffic count showed that at the rush period 31,000 people an hour, or 210,000 men and women in sixteen and one-half hours, passed the southwest corner. For over ten years land values in the Loop have been stationary, but in the same time they have dou-

---

[4] Adapted from W. B. Monro, *Municipal Government and Administration*, II, p. 377.

[5] *Report of the Chicago Subway and Traction Commission*, p. 81, and the *Report on a Physical Plan for a Unified Transportation System*, p. 391.

[6] Data compiled by automobile industries.

[7] Statistics of mailing division, Chicago Post Office.

[8] Determined from *Census Estimates for Intercensual Years*.

[9] From statistics furnished by Mr. R. Johnson, traffic supervisor, Illinois Bell Telephone Company.

bled, quadrupled, and even sextupled in the strategic corners of the "satellite loops,"[10] an accurate index of the changes which have occurred. Our investigations so far seem to indicate that variations in land values, especially where correlated with differences in rents, offer perhaps the best single measure of mobility, and so of all the changes taking place in the expansion and growth of the city.

In general outline, I have attempted to present the point of view and method of investigation which the department of sociology is employing in its studies in the growth of the city, namely, to describe urban expansion in terms of extension, succession, and concentration; to determine how expansion disturbs metabolism when disorganization is in excess of organization; and, finally, to define mobility and to propose it as a measure both of expansion and metabolism, susceptible to precise quantitative formulation, so that it may be regarded almost literally as the pulse of the community. In a way, this statement might serve as an introduction to any one of five or six research projects under way in the department.[11] The project, however, in which I am directly engaged is an attempt to apply these methods of investigation to a cross-section of the city—to put this area, as it were, under the microscope, and so to study in more detail and with greater control and precision the processes which have been described here in the large. For this purpose the West Side Jewish community has been selected. This community includes the so-called "Ghetto," or area of first settlement, and Lawndale, the so-called "Deutschland," or area of second settlement. This area has certain obvious advantages for this study, from the standpoint of expansion, metabolism, and mobility. It exemplifies the tendency to expansion radially from the business center of the city. It is now relatively a homogeneous cultural group. Lawndale is itself an area in flux, with the tide of migrants still flowing in from the Ghetto and a constant egress to more desirable regions of the residential zone. In this area, too, it is also possible to study how the expected outcome of this high rate of mobility in social and personal disorganization is counteracted in large measure by the efficient communal organization of the Jewish community.

[10]From 1912–23, land values per front foot increased in Bridgeport from $600 to $1,250; in Division-Ashland-Milwaukee district, from $2,000 to $4,500; in "Back of the Yards," from $1,000 to $3,000; in Englewood, from $2,500 to $8,000; in Wilson Avenue, from $1,000 to $6,000; but decreased in the Loop from $20,000 to $16,500.

[11]Nels Anderson, *The Slum: An Area of Deterioration in the Growth of the City;* Ernest R. Mowrer, *Family Disorganization in Chicago;* Walter C. Reckless, *The Natural History of Vice Areas in Chicago;* E. H. Shideler, *The Retail Business Organization as an Index of Business Organization;* F. M. Thrasher, *One Thousand Boys' Gangs in Chicago: A Study of Their Organization and Habitat;* H. W. Zorbaugh, *The Lower North Side: A Study in Community Organization.*

The Pattern of Movement of Residential Rental Neighborhoods*

Homer Hoyt

The high rent neighborhoods of a city do not skip about at random in the process of movement—they follow a definite path in one or more sectors of the city.

Apparently there is a tendency for neighborhoods within a city to shift in accordance with what may be called the sector theory of neighborhood change. The understanding of the framework within which this principle operates will be facilitated by considering the entire city as a circle and various neighborhoods as falling into sectors radiating out from the center of that circle. No city conforms exactly to this ideal pattern, of course, but the general figure is useful inasmuch as in our American cities the different types of residential areas tend to grow outward along rather distinct radii, and new growth on the arc of a given sector tends to take on the character of the initial growth in that sector.

Thus if one sector of a city first develops as a low rent residential area, it will tend to retain that character for long distances as the sector is extended through process of the city's growth. On the other hand, if a high rent area becomes established in another sector of the city, it will tend to grow or expand within that sector, and new high grade areas will tend to establish themselves in the sector's outward extension. This tendency is portrayed in Figure 1 by the shifts in the location of the fashionable residential areas in six American cities between 1900 and 1936. Generally speaking, different sectors of a city present different characters according to the original types of the neighborhoods within them.

In considering the growth of a city, the movement of the high rent area is in a certain sense the most important because it tends to pull the growth of the entire city in the same direction. The homes of the leaders of society are located at some point in the high rent area. This location is the point of highest rents or the high rent pole. Residential rents grade downward from this pole as lesser income groups seek to get as close to it as possible. This high rent pole tends to move outward from the center of the city along a certain avenue or lateral line. The new houses constructed for the occupancy of the higher rental groups are situated on the outward edges of the high rent area. As

*Reprinted from Homer Hoyt, *The Structure and Growth of Residential Neighborhoods in American Cities* (Washington, D.C.: Federal Housing Administration, 1939), pp. 114–22, by permission of the author.

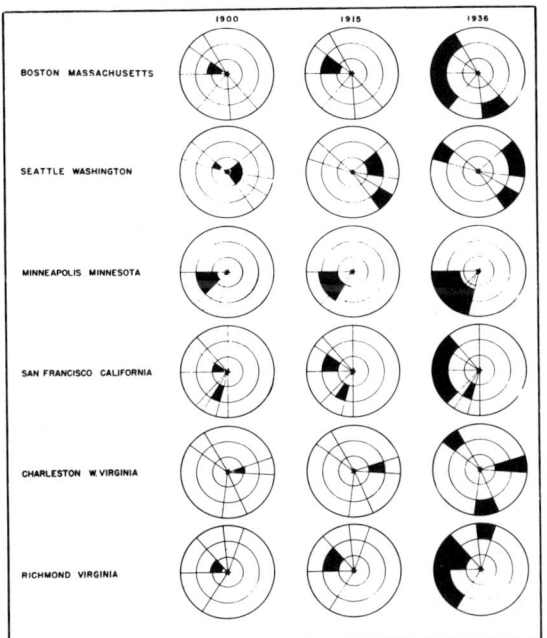

Figure 1—Shifts in location of fashionable residential areas in six American cities (fashionable residential areas indicated by solid black).

these areas grow outward, the lower and intermediate rental groups filter into the homes given up by the higher income groups. In New York City the movement was up Fifth Avenue, starting at Washington Square and proceeding finally to Ninety-sixth Street in the course of a century. In Chicago, there were three high rental areas, moving southward along Michigan and Wabash Avenues, westward in the band between Jackson and Washington Streets, and northward along La Salle and Dearborn Streets to the Lake Shore Drive.

Sometimes the high rent pole jumps to new areas on the periphery of the city, as in the case of the development of Shaker Heights in Cleveland, Ohio, and Coral Gables in Miami, Fla., but usually these new areas are in the line of growth of the high rent areas. In Charleston, W. Va., the high grade neighborhood moved from the center of the city along Kanawha Street until it reached the river, and then the new high grade area jumped to new locations in the hills in the south and north. In Seattle, Wash., the high grade neighborhood started near the center of the city and moved northeast in one sector of the city—the location along the lake on the periphery. At the same time the high grade development sprang up to the northwest, jumping intervening low grade areas.

In Minneapolis, Minn., there was a movement of the high grade neighborhood to the southwest, starting at the center of the city and repeating the same type of growth until it reached the outer edge of the city in a lake region. In Richmond, Va., the sector of the city containing Monument Avenue first developed as a high grade area. The movement of the high grade neighborhood continued out along the line of Monument Avenue until it reached the city limits and then it expanded fan shape in a sector to the north and west. At the same time a high grade development started to the north in a sector which was bisected by Chamberlayne Street.

In Detroit, Mich., the growth of the high grade neighborhood proceeded eastward along Jefferson Avenue out to Grosse Pointe along Lake St. Clair. There was another band of high grade development west of the axis of Woodward Avenue. In Miami, Fla., bands of high grade development followed Biscayne Bay to the north and south and also to Miami Beach.

As a result of the outward movement of the high rent neighborhoods in American cities, present fashionable areas are mostly located beyond the earlier settled areas of American cities. Thus, in Washington, D.C., practically all of the high rent area of today is located in a section that lies beyond the area occupied by houses in 1887. Similarly, in the 14 other illustrative cities referred to in this chapter, most of the high rent areas of today are located beyond the areas occupied by houses at a relatively recent period of time.

High rent or high grade residential neighborhoods must almost necessarily move outward toward the periphery of the city. The wealthy seldom reverse their steps and move backward into the obsolete houses which they are giving up. On each side of them is usually an intermediate rental area, so they cannot move sideways. As they represent the highest income group, there are no houses above them abandoned by another group. They must build new houses on vacant land. Usually this vacant land lies available just ahead of the line of march of the area because, anticipating the trend of fashionable growth, land promoters have either restricted it to high grade use or speculators have placed a value on the land that is too high for the low rent or intermediate rental groups. Hence the natural trend of the high rent area is outward, toward the periphery of the city in the very sector in which the high rent area started. The exception to this outward movement is the development of de luxe apartment areas in old residential areas. This will be treated more fully on a following page.

What determines the point of origin of the highest rental areas of the city and the direction and pattern of their future growth? The answer to this question is of vital importance to all students of urban growth, for the high rent sector is the pole or center of attraction that pulls the other residential areas with it.

In all of the cities studied, the high grade residential area had its point of origin near the retail and office center. This is where the higher income groups work, and is the point that is the farthest removed from the side of the city that has industries or warehouses. In each city, the direction and pattern of its future growth then tends to be governed by some combination of the following considerations:

(1) *High grade residential growth tends to proceed from the given point of origin, along established lines of travel or toward another existing nucleus of buildings or trading centers.*—This principle is illustrated by the movement of the high grade residential neighborhood of Chicago along the main axes of the roads like Cottage Grove Avenue, leading south around the bend of Lake Michigan to the east, of main roads like Madison Street leading westward, and of roads following the lake northward to Milwaukee. In Detroit, Mich., there was a trend of fashionable growth along the radial line of Woodward Avenue, the main thoroughfare to Flint and Pontiac, beginning within the Grand Boulevard Circuit and later extending to Highland Park, Palmer Woods, Ferndale, Royal Oak, and Birmingham.

(2) *The zone of high rent areas tends to progress toward high ground which is free from the risk of floods and to spread along lake, bay, river, and ocean fronts, where such water fronts are not used for industry.*—The movement of high grade residential neighborhoods away from river bottoms to higher ground or to wooded hills is illustrated by numerous examples. In San Francisco, Calif., the wealthy moved from the lowland along the bay to Knob Hill which was less subject to fogs and smoke. In Washington, D.C., the high grade neighborhoods moved from the mud flats along the Potomac in the southeast quadrant and from the lowland in the southwest quadrant, to the higher land in the northwest section. In Springfield, Mass., the best areas moved from the lowland along the Connecticut River to rising land and to Longmeadow. In Kansas City, Mo., St. Louis, Mo., and Cincinnati, Ohio, there has been a movement of settlement away from the river bottoms to the higher land.

In cities located on relatively flat land near rivers, bays, lakes, or oceans, the high grade residential neighborhood tends to expand in long lines along the water front that is not used for industrial purposes. Thus in Chicago, the lake front on the north side is the front yard of the city and is preempted for high grade residential use for a distance of nearly 30 miles north of the business center. In New York City, a high grade residential area grew northward along the Hudson River on Riverside Drive from 72d Street to Riverdale in the West Bronx. In Miami, Fla., the high rent areas extend along Biscayne Bay to the north and southeast and along the ocean front on Miami Beach. In Detroit, Mich., a high grade development extends along Lake St. Clair at Grosse Pointe. On the New Jersey coast, there is a long string of resorts along the ocean front with the highest paid residential use confined to the strip along the beach. In Charleston, W. Va., one high grade residential area extends along the high bank of the Kanawha River.

Thus, where such lakes, rivers, bays, or ocean fronts exist and offer the attractions of bathing, yachting, cool breezes in summer, and a wide expanse of water with its uninterrupted view, rent areas tend to follow the contour of the water front in long, narrow lines of growth.

(3) *High rent residential districts tend to grow toward the section of the city which has free, open country beyond the edges and away from "dead end" sections which are limited by natural or artificial barriers to expansion.*—The lure of open fields, golf courses, country clubs, and country estates acts as a magnet to pull high grade residential areas to sections that have free, open country beyond their borders and away from areas that run into "dead ends." Thus the high grade neighborhood of Washington, D.C., grows northwest toward expanding open country and estates. Thus, the expansion of high grade neighborhoods to the north of Baltimore, Md., to the south of Kansas City, Mo., and to the north of New York City in Westchester County is into areas with a wide expanse of country beyond them.

(4) *The higher priced residential neighborhood tends to grow toward the homes of the leaders of the community.*—In Washington, D.C., the White

House; in New York, the homes of the Astors and the Vanderbilts were the magnets that pulled the members of society in their direction. One fashionable home, an outpost on the prairie, standing near Sixteenth Street and Prairie Avenue in Chicago in 1836, gave prestige to the section and caused other leaders of fashion to locate near the same spot.[1]

(5) *Trends of movement of office buildings, banks, and stores, pull the higher priced residential neighborhoods in the same general direction.*—The stores, offices, and banks in the central business district usually move in the direction of the high rent area, but follow rather than lead the movement of the high rent neighborhood. Sometimes, however, when an office building center becomes established at a certain point, it facilitates the growth of a high rent area in sections that are conveniently accessible to it. Thus the office building center in the Grand Central District in New York City has aided the growth of the de luxe apartment area in Park Avenue and also the exclusive suburban towns in Westchester that are served by fast express trains entering the Grand Central Station. The establishment of an office building center at Grand Boulevard and Woodward Avenue in Detroit, Mich., aided the growth of the high grade area to the north and west of it. In Washington, D.C., the northwestward trend of the office buildings, while the result of the pull of the high grade areas to the northwest, also favored the further growth of the northwest area because it made those areas more accessible to offices. Similarly, the trend of office buildings on North Michigan Avenue in Chicago favored the northward growth of the de luxe apartment area.

(6) *High grade residential areas tend to develop along the fastest existing transportation lines.*—The high grade residential areas in Chicago grew along the main plank road, horse car, cable car, and suburban railroad routes. In New York City, the elevated lines and subways paralleled Fifth Avenue. Fast commuters' trains connect New York City with the high grade suburban homes in Montclair, the Oranges, and Maplewood in New Jersey, in Scarsdale, Pelham, and Bronxville in Westchester, and in Forest Hills, Kew Gardens, Flushing, and Hempstead in Long Island. In Detroit, Mich., the high grade areas are located close to main arteries leading directly to the center of the city—Jefferson, Woodward, and Grand River Avenues. In Washington, D.C., the best areas are on the main transportation arteries—Connecticut Avenue, Massachusetts Avenue, and Sixteenth Street leading directly to the White House.

(7) *The growth of high rent neighborhoods continues in the same direction for a long period of time.*—In New York City, the march of the fashionable areas continued up Fifth Avenue from Washington Square to Central Park for over a century. The high grade neighborhoods in Chicago moved south, west, and north from their starting points in or near the present "Loop" to present locations—7 to 20 miles distant—in the course of a century. In the century after the Revolutionary War, the high grade area of Washington, D.C., moved from the Capitol to the Naval Observatory. The high rent areas of Detroit, Mich., moved from points near the present business center to Grosse Pointe, Palmer Woods, and Birmingham, 6 to 10 miles away.

In Miami, Fla., Minneapolis, Minn., Seattle, Wash., Charleston, W. Va., Salt Lake City, Utah, and many other cities, this same continuous outward movement of high rent areas has been maintained for long periods of time. Except under the unusual conditions now to be described, there have been no reversals of this long continued trend.

(8) *De luxe high rent apartment areas tend to be established near the business center in old residential areas.*—One apparent exception to the rule that high rent neighborhoods do not reverse their trend of growth is found in the case of de luxe apartment areas like Streeterville in Chicago and Park Avenue in New York City. This exception is a very special case, however, and applies only to intensive high grade apartment developments in a few metropolitan centers. When the high rent single-family home areas have moved far out on the periphery of the city, some wealthy families desire to live in a colony of luxurious apartments close to the business center. Because of both the intensive use of the land by use of multiple-family structures and the high rents charged it pays to wreck existing improvements.

Such apartments can rise even in the midst of a poor area because the tall building itself, rising from humble surroundings like a feudal castle above the mud huts of the villeins, is a barrier against intrusion. Thus, when the railroad tracks

---

[1] See Robert S. and Helen M. Lynd, *Middletown in Transition* (New York: Harcourt, Brace & Co., 1937), pp. 81–82, for an interesting example of how the northwest section of Middletown became the outstanding residential section as a result of the movement of the most prominent family to that section.

were depressed under Park Avenue in New York City and the railroads were electrified, that street, originally lined with shanties, became the fashionable apartment avenue of New York City. In Chicago, the wall of apartments on the sands where Captain Streeter once had his shack is now occupied by the most exclusive social set. In both cases, there was a renaissance of an old neighborhood. It is only where intensive apartment uses occupy the land that such an apparent reversal of trend occurs.

(9) *Real estate promoters may bend the direction of high grade residential growth.*—While it is almost impossible for real estate developers to reverse the natural trend of growth of high grade neighborhoods, even by the expenditure of large sums of money and great promotional effort, it is possible for them to accelerate a natural trend or to bend a natural trend of growth.

Miami Beach, directly on the Gulf Stream in Florida, was favored by nature as the site for high grade resort homes. When it was a mangrove swamp, separated from the mainland by Biscayne Bay, it was almost inaccessible. Carl Fisher, by building a million dollar causeway and by pumping up 2,800 acres of land out of the bay and erecting thereon golf courses and hotels, made it possible for these natural advantages of Miami Beach to be utilized. Similarly, George Merrick acquired a great tract of land at Coral Gables, Fla., and, by spending millions of dollars in laying out streets, in planting flowering trees, and in establishing restrictions, gave the area a high grade character which it did not otherwise possess. So, likewise, did the developers of Roland Park in Baltimore, Shaker Heights near Cleveland, and the Country Club District of Kansas City take large areas in the line of growth and establish high grade communities by means of building restrictions, architectural control, community planning, and other barriers against invasion.

In all these cases, the high rent area was in the general path of growth; but which area of the many in the favored area became the fashionable center depended upon the promotional skill and the money expended by individual promoters.

As a result of some or all of these forces, high rent neighborhoods thus become established in one sector of the city, and they tend to move out in that sector to the periphery of the city. Even if the sector in which the high rent growth begins does not possess all of the advantages, it is difficult for the high rent neighborhood to change its direction suddenly or to move to a new quarter of the city. For as the high rent neighborhood grows and expands, the low and intermediate areas are likewise growing and expanding, and they are taking up and utilizing land alongside the high rent area as well as in other sectors of the city. When these other areas have acquired a low rent character, it is very difficult to change that character except for intensive apartment use. Hence, while in the beginning of the growth of the city, high rent neighborhoods may have a considerable choice of direction in which to move, that range of choice is narrowed as the city grows and begins to be filled up on one or more sides by low rent structures.

It is possible for high rent neighborhoods to take over sections which are marred by a few shacks. These are swept aside or submerged by the tide of growth. Negro houses have even been bought up and moved away in some southern cities to make way for a high grade development. This possibility exists where the houses are flimsy or scattered, where the land is cheap, where it is held by one owner, or where the residents are under the domination of others. It is extremely difficult otherwise. The cost of acquiring and tearing down substantial buildings and the practical impossibility of acquiring large areas from scattered owners, usually prevent high grade areas from taking over land once it has been fairly well occupied by middle or low grade residential uses.

Now that the radius of the settled area of cities has been greatly extended by the automobile, however, there is little difficulty in securing land for the expansion of high rent areas; for the high rent sector of the city expands with an ever widening arc as one proceeds from the business center.

The next vital question to be considered is how the various types of high rent areas are affected by the process of dynamic growth of the city and how the various types are related to each other in historical sequence.

The first type of high rent development was the axial type with high grade homes in a long avenue or avenues leading directly to the business center. The avenue was a social bourse, communication being maintained by a stream of fashionable carriages, the occupants of which nodded to their acquaintances in other passing carriages or to other friends on the porches of the fine residences along the way. Such avenues were lined with beautiful shade trees and led to a park or parks through a series of connecting boulevards. Examples of this type of development, in the decades from 1870 to 1900, are Prairie and South Michigan

Avenues, Washington and Jackson Streets, and the Lake Shore Drive in Chicago, Fifth Avenue in New York City, Monument Avenue in Richmond, Va., and Summit Street in St. Paul, Minn. The fashionable area in this type of development expanded in a long string in a radial line from the business center. There was usually an abrupt transition within a short distance on either side of the high grade street.

The axial type of high rent area rapidly became obsolete with the growth of the automobile. When the avenues became automobile speedways, dangerous to children, noisy, and filled with gasoline fumes, they ceased to be attractive as home sites for the well-to-do. No longer restricted to the upper classes, who alone could maintain prancing steeds and glittering broughams, but filled with *hoi polloi* jostling the limousines with their flivvers, the old avenues lost social caste. The rich then desired seclusion—away from the "madding crowd" whizzing by and honking their horns. Mansions were then built in wooded areas, screened by trees. The very height of privacy is now attained by some millionaires whose homes are so protected from the public view by trees that they can be seen from outside only from an airplane.

The well-to-do who occupy most of the houses in the high rent brackets have done likewise in segregated garden communities. The new type of high grade area was thus not in the form of a long axial line but in the form of a rectangular area, turning its back on the outside world, with winding streets, woods, and its own community centers. Such new square or rectangular areas are usually located along the line of the old axial high grade areas. The once proud mansions still serve as a favorable approach to the new secluded spots. As some of the old axial type high rent areas still maintain a waning prestige and may still be classed as high rent areas, the new high rent area takes a fan-shaped or funnel form expanding from a central stem as it reaches the periphery of the city.

The old stringlike development of high rent areas still asserts itself, however, in the cases of expansion of high rent areas along water fronts like Lake Michigan, Miami Beach, and the New Jersey coast. The automobile, however, has made accessible hilly and wooded tracts on which houses are built on the crest of hills along winding roads.

The fashionable suburban town, which had its origin even before the Civil War, has remained a continuous type of high grade area. Old fashionable towns like Evanston, Oak Park, and Lake Forest near Chicago, have maintained their original character and expanded their growth. Other new high grade suburban towns have been established. The de luxe apartment area has been a comparatively recent development, coming after 1900, when the wealthy ceased to desire to maintain elaborate town houses and when the high grade single-family home areas began to be located far from the business center. A group of wealthy people, desiring to live near the business center and to avoid the expense and trouble of maintaining a retinue of servants, sought the convenience of tall elevator apartments.

The high grade areas thus tend to preempt the most desirable residential land by supporting the highest values. Intermediate rental groups tend to occupy the sectors in each city that are adjacent to the high rent area. Those in the intermediate rental group have incomes sufficient to pay for new houses with modern sanitary facilities. Hence, the new growth of these middle-class areas takes place on the periphery of the city near high grade areas or sometimes at points beyond the edge of older middle-class areas.

Occupants of houses in the low rent categories tend to move out in bands from the center of the city mainly by filtering up into houses left behind by the high income groups, or by erecting shacks on the periphery of the city. They live in either second-hand houses in which the percentage needing major repairs is relatively high or in newly constructed shacks on the periphery of the city. These shacks frequently lack modern plumbing facilities and are on unpaved streets. The shack fringe of the city is usually in the extension of a low rent section.

Within the low rent area itself there are movements of racial and national groups. Until only comparatively recently, the immigrants poured from Europe into the oldest and cheapest quarters on the lower East Side of New York and on the West Side of Chicago. The earlier immigrants moved out toward the periphery of the city. These foreign groups moved in bands or straight lines out from the railroad stations near the central business district. The Italian colony of Chicago moved westward along the area in the point between Harrison Street and Roosevelt Road and northwestward along Grand Avenue. The Poles proceeded northwest along Milwaukee Avenue and expanded southwest along the stockyards. The Russian Jews moved west between Roosevelt Road and Sixteenth Street. The Czechoslovakians shifted southwest from Eighteenth and Loomis Streets to Twenty-second and thence westward to Cicero.

With the decline of immigration after the World War, new immigrants ceased to fill the old houses in the downtown area and this outward progression of foreign groups slackened. Many of the tenements in the lower east side were boarded up, and some of the oldest quarters near the central business district of Chicago were demolished.

During the World War and after, however, there was a great influx of Negroes into the northern cities to take the place of European immigration. The Negro neighborhood in Harlem, New York, expanded in concentric circles. In Chicago, the Negroes burst the bounds of their old area along State Street and the Rock Island tracks, Twenty-second and Thirty-ninth Streets, and spread eastward to Cottage Grove Avenue and south to Sixty-seventh Street. In this movement in Chicago, they spread into an area formerly occupied by middle-class and some high income families. The area, however, was becoming obsolete and did not offer vigorous resistance to the incoming of other racial groups.

Thus, in the framework of the city there is a constant dynamic shifting of rental areas. There is a constant outward movement of neighborhoods because as neighborhoods become older they tend to be less desirable.

Forces constantly and steadily at work are causing a deterioration in existing neighborhoods. A neighborhood composed of new houses in the latest modern style, all owned by young married couples with children, is at its apex. At this period of its vigorous youth, the neighborhood has the vitality to fight off the disease of blight. The owners will strenuously resist the encroachment of inharmonious forces because of their pride in their homes and their desire to maintain a favorable environment for their children. The houses, being in the newest and most popular style, do not suffer from the competition of any superior house in the same price range, and they are marketable at approximately their reproduction cost under normal conditions.

Both the buildings and the people are always growing older. Physical depreciation of structures and the aging of families constantly are lessening the vital powers of the neighborhood. Children grow up and move away. Houses with increasing age are faced with higher repair bills. This steady process of deterioration is hastened by obsolescence; a new and more modern type of structure relegates these structures to the second rank. The older residents do not fight so strenuously to keep out inharmonious forces. A lower income class succeeds the original occupants. Owner occupancy declines as the first owners sell out or move away or lose their homes by foreclosure. There is often a sudden decline in value due to sharp transition in the character of the neighborhood or to a period of depression in the real estate cycle.

These internal changes due to depreciation and obsolescence in themselves cause shifts in the locations of neighborhoods. When, in addition, there is poured into the center of the urban organism a stream of immigrants or members of other racial groups, these forces also cause dislocations in the existing neighborhood pattern.

The effects of these changes vary according to the type of neighborhood and can best be described by discussing each one in turn. The highest grade neighborhood, occupied by the mansions of the rich, is subject to an extraordinary rate of obsolescence. The large scale house, modeled after a feudal castle or a palace, has lost favor even with the rich. When the wealthy residents seek new locations, there is no class of a slightly lower income which will buy the huge structures because no one but wealthy persons can afford to furnish and maintain them. There is no class filtering up to occupy them for single-family use. Consequently, they can only be converted into boarding houses, offices, clubs, or light industrial plants, for which they were not designed. Their attraction of these types of uses causes a deterioration of the neighborhood and a further decline in value. These mansions frequently become white elephants like those on Arden Park and East Boston Boulevard in Detroit, Mich.

On the other hand, houses in intermediate rental neighborhoods designed for small families can be handed down to a slightly lower income group as they lose some of their original desirability because of age and obsolescence. There is a loss of value when a transition to a lower income group occurs, but the house is still used for the essential purpose for which it was designed; and the loss of value is not so great. There is always a class filtration to occupy the houses in the intermediate rental neighborhoods. Hence, a certain stability of value is assured.

Since the buildings in low rent areas are occupied by the poorest unskilled or casual workers, collection losses and vacancy ratios are highest. The worst buildings are condemned or removed by demolition to save taxes. Formerly these worst quarters in the old law tenements of New York or the West Side of Chicago were occupied by newly arrived immigrants. With the decline of immigration, this submarginal fringe of housing is being wrecked or boarded up as the residents filter up to better houses.

Thus, intermediate rental neighborhoods tend to preserve their stability better than either the highest or lowest rental areas.

The erection of new dwellings on the periphery of a city, made accessible by new circulatory systems, sets in motion forces tending to draw population from the older houses and to cause all groups to move up a step leaving the oldest and cheapest houses to be occupied by the poorest families or to be vacated. The constant competition of new areas is itself a cause of neighborhood shifts. Every building boom, with its new crop of structures equipped with the latest modern devices, pushes all existing structures a notch down in the scale of desirability.

The Natural Areas of the City*

Harvey W. Zorbaugh

The city as artifact and as natural phenomenon

To the philosophically minded the city has often seemed to be the most colossal artifact of man's creation. The towering skyscrapers of a New York or a Chicago, palatial banking houses, the frenzied stock exchange, a Fifth or a Michigan Avenue with its ceaseless stream of automobiles and busses, its smart shops, and its brilliant hotels, underground tubes with roaring trains, or elevated railroads clattering overhead, great belts of smoking industries, miles of canyon-like streets flanked with tall apartments, magnificent park and boulevard systems, water works beside which the Roman aqueducts fall into insignificance—all in all the city seems the most exotic and artificial flower of a man-made civilization, a product not alone of man's brawn, but of man's brain and man's will.

Yet the city is curiously resistant to the fiats of man. Like the Robot, created by man, it goes its own way indifferent to the will of its creator. Reformers have stormed, the avaricious have speculated, and thoughtful men have planned. But again and again their programs have met with obstacles. Human nature offers some opposition; traditions and institutions offer more; and—of

*Reprinted from *Publications of the American Sociological Society,* 20 (1926), 188–97, by permission of The American Sociological Association.

especial significance—the very physical configuration of the city is unyielding to change. It becomes apparent that the city has a natural organization that must be taken into account.

In the latter part of the past century and the early years of this present century a tidal wave of reform swept over the city, culminating in the "Man with the Muckrake" and the "Yellow Press." Jacob Riis painted the descent into the slum. Parkhurst crusaded against vice in New York; and Stead, in *If Christ Came to Chicago,* lashed the lords of Customs House Place. Ida M. Tarbell and Upton Sinclair took the muckrake into industry, while Lincoln Steffens laid bare the rotten spots in city government. There was a tremendous stir, public interest was aroused, reforms were proposed, but little happened. Practically all these movements for social reform met with unexpected obstacles: influential persons, "bosses," "union leaders," "local magnates," and powerful groups such as party organizations, "vested interests," "lobbies," unions, manufacturers' associations, and the like. Candid recognition of the role of these persons and groups led writers on social, political, and economic questions to give them the impersonal designation of "social forces."

The concept of social forces was a commonsense generalization. But implicit in Steffens' book, *The Shame of the Cities,* was a far more sophisticated insight. Steffens maintained that with his knowledge of New York he could go into any city and quickly gauge conditions; that conditions in New York were not due to a failure of institutions peculiar to itself, but to a condition incident to the growth of all cities. This was the first recognition of the fact that the city is a natural phenomenon and has a natural history.

Meantime, realtors, public utilities, city-planning and zoning commissions, and others interested in predicting the future of the city were discovering much about the way in which the city grows. Richard Hurd, in a small volume, *The Principles of City Land Values,* attempting to generalize fluctuations of city land values, formulated certain typical processes of the city's growth. Most instructive are the more recent statistical studies of the American Bell Telephone Company and other utilities for the purposes of extension in anticipation of future service. The city is discovered to be an organization displaying certain typical processes of growth. Knowledge of these processes makes possible prediction of the direction, rate, and nature of its growth. That is, the city is found to be not an artifact but a natural phenomenon.

A human ecology

In an address in 1922, before the meeting at which the Russell Sage Foundation's proposal for a regional plan for metropolitan New York was first outlined, Elihu Root recognized this fact of the natural organization of the city when he said: "A city is a growth. It is not the result of political decrees or control. You may draw all the lines you please between counties and states; a city is a growth responding to forces not at all political, quite disregarding political lines. It is a growth like that of a crystal responding to forces inherent in the atoms that make it up." In the three years that have elapsed since Elihu Root wrote these words, a mass of material about the city has been gathered and analyzed that enables us to describe these "atoms" to which he referred.

Studies of the expansion of the city have shown that all American cities exhibit certain typical processes in their growth.[1] To begin with, they segregate into broad zones as they expand radially from the center—a "loop," or central business district, a zone of transition between business and resident; an invasion by business and light manufacturing, involving physical deterioration and social disorganization; a zone of working men's homes, cut through by rooming-house districts along focal lines of transportation; a zone of apartments and "restricted" districts of single-family dwellings; and, farther out, beyond city limits, a commuters' zone of suburban areas. Ideally, this gross segregation may be represented by a series of concentric circles, and such tends to be the actual fact where there are no complicating geographical factors.

Such is a generalized description of the gross anatomy of the city—the typical structure of a modern American commercial and industrial city. Of course, no city quite conforms to this ideal scheme. Physical barriers such as rivers, lakes, and rises of land may modify the growth and structure of the individual city, as is strikingly demonstrated in the cases of New York, Pittsburgh, and Seattle. Railroads, with their belts of industry, cut through this generalized scheme, breaking the city up into sections; and lines of local transportation, along the more travelled of which grow up retail business streets, further modify the structure of the city.

The structure of the individual city, then, while always exhibiting the generalized zones described above, is built about this framework of transportation, business organization and industry, park and boulevard systems, and topographical features. All of these break the city up into numerous smaller areas, which we may call natural areas, in that they are the unplanned, natural product of the city's growth. Railroad and industrial belts, park and boulevard systems, rivers and rises of land acting as barriers to movements of population tend to fix the boundaries of these natural areas, while their centers are usually intersections of two or more business streets. By virtue of proximity to industry, business, transportation, or natural advantages, each area acquires a physical individuality accurately reflected in land values and rents.

Now, in the intimate economic relationships in which all people are in the city, everyone is, in a sense, in competition with everyone else. It is an impersonal competition—the individual does not know his competitors. It is a competition for other values in addition to those represented by money. One of the forms it takes is competition for position in the community. We do not know all the factors involved, but each individual influences the ultimate position of every other individual.

In this competition for position the population is segregated over the natural areas of the city. Land values, characterizing the various natural areas, tend to sift and sort the population. At the same

---

[1] E. W. Burgess, "The Growth of the City: An Introduction to a Research Project," in *The City,* ed. Robert E. Park et al., pp. 50ff.

time segregation re-emphasizes trends in values. Cultural factors also play a part in this segregation, creating repulsions and attractions. From the mobile competing stream of the city's population each natural area of the city tends to collect the particular individuals predestined to it. These individuals, in turn, give to the area a peculiar character. And as a result of this segregation, the natural areas of the city tend to become distinct cultural areas as well—a "black belt" or a Harlem, a Little Italy, a Chinatown, a "stem" of the "hobo," a rooming-house world, a "Towertown," or a "Greenwich Village," a "Gold Coast," and the like—each with its characteristic complex of institutions, customs, beliefs, standards of life, traditions, attitudes, sentiments, and interests. The physical individuality of the natural areas of the city is re-emphasized by the cultural individuality of the populations segregated over them. Natural areas and natural cultural groups tend to coincide.

A natural area is a geographical area characterized both by a physical individuality and by the cultural characteristics of the people who live in it. Studies in various cities have shown, to quote Robert E. Park, that "Every American city of a given size tends to reproduce all the typical areas of all the cities, and that the people in these areas exhibit, from city to city, the same cultural characteristics, the same types of institutions, the same social types, with the same opinions, interests, and outlook on life." That is, just as there is a plant ecology whereby, in the struggle for existence, like geographical regions become associated with like "communities" of plants, mutually adapted, and adapted to the area, so there is a human ecology whereby, in the competition of the city and according to definable processes, the population of the city is segregated over natural areas into natural groups. And these natural areas and natural groups are the "atoms" of city growth, the units we try to control in administering and planning for the city.

Administrative area and natural area

The distinction between the natural area and the administrative area is apparent. The city is broken up into administrative units, such as the ward, the school district, the police precinct, and the health district, for the purposes of administrative convenience. The object is usually to apportion either the population or area of the city into equal units. The natural area, on the other hand, is a unit in the physical structure of the city, typified by a physical individuality and the characteristic attitudes, sentiments, and interests of the people segregated within it. Administrative areas and natural areas may coincide. In practice they rarely do. Administrative lines cut across the boundaries of natural areas, ignoring their existence.

The contrast between administrative and natural areas is not new. Historians long ago pointed out the international complications that have arisen because state lines were not drawn with reference to natural groupings of population and natural geographical units. A historian in a recent volume devotes a chapter to "Natural Areas and Boundaries." The geographer talks of production in terms of natural "regions." Gras, in his *Introduction to Economic History,* reminds us that a stable banking system must be based not on units of administrative convenience, but upon the basis of natural "metropolitan" areas of financial service. We are just beginning, however, to take account of the natural areas of the city.

Students of municipal affairs are coming to appreciate the relationship of the cultural individuality of the natural areas of the city to the problem of city government. For one thing, the theory and practice of American municipal government, evolved to meet the needs of village communities, make no allowance for the existence of distinct areas within the city, each with an individuality, and unequally adapted to function politically under our present system. On the Lower North Side of Chicago, for example, is a rooming-house area affording dormitories to 25,000 people. This population is exceedingly mobile. It turns over every four months. There are no permanent contacts in such an area. No one knows anyone else. There are no permanent interests in the area, and no public opinion. The population are not "citizens" of the locality. There are few votes, and many of these are sold. Local self-government is a myth. The area is administered by the social agencies and the police, though this fact is but imperfectly recognized by these agencies. The situation should be frankly faced. Such an area should be disfranchised and administered from the city hall. Natural areas are unequally adapted to function politically under our present system of municipal government.

Again, administrative units cut across natural areas. Ward lines divide a "Little Sicily," or ward lines encompass a number of natural areas and natural groups. As a result, the ward vote frequently represents a stalemate among conflicting natural areas; and large parts of the city are politically impotent. The real issues of the areas that make up the city rarely get into politics; municipal government becomes a concession, a state of affairs that is rapidly assuming the proportion of a national scandal. One remedy would seem to be the political recognition of the natural areas of the city, and at least a geographical pluralism in city government.

There have been numerous extra-political attempts to solve the problems of local self-government in the city. Among these is the community organization movement. Looking to the village as a "golden age" of social life, and believing that if the neighborliness of the village could be restored in the city the city's problems would take care of themselves, the community organizers have set out to make "villages" of areas within the city. But in selecting the areas for the experiments they have usually but substituted one administrative area for another, totally oblivious of the existence and significance of natural areas and natural groups. The Lower North Community Council of Chicago set out to make a "community" of a section of the city including a colony of 15,000 Sicilians, a colony of 6,000 Persians, a belt of some 4,000 Negroes, a colony of 1,000 Greeks, a rooming-house population of 25,000, "Towertown"—Chicago's Greenwich village—and Chicago's much-vaunted "Gold Coast."

A further complicating factor is introduced by the fact that the natural areas of a city are only relatively stable, either in respect to values or in respect to the cultural segregation upon them. Particularly is this true in a new or growing city. In older cities residence is more permanent; a historical sentiment enters in to stabilize residence, inclining people to cling to the old community. And in a city that is not growing competition for position tends to cease and values and groupings of the population to reach an equilibrium. But in the growing city, expanding as it grows, natural areas are only relatively stable. They seem to change in a predictable manner, a succession like that observable in plant communities. The laws of this succession are imperfectly known, however. One of the purposes of the studies of the Community Research Fund of the University of Chicago has been to analyze this succession. Chicago's "Gold Coast," again, offers an interesting example of succession in process. As more and more of Chicago's industrial kings achieve incomes worthy of evasion of the government tax, they crowd in upon the "Gold Coast." Chicago's first families find themselves increasingly aliens in their own land. And we view the spectacle, not without its pathos, of the perambulators of the leaders of future assemblies disappearing from the Esplanade to reappear along Sheridan Road.

These ecological facts—natural areas within the city, competition for position, segregation over natural areas, succession—are facts that must be taken into account by those who would control the city's growth as well as by those who would administer the city's government. We are interested here not in cities planned from their origin—though there seems to be limits to what can be done in such instances. Berlin, for example, like Amsterdam and many other European cities, has grown since the time when it was a small city according to a carefully directed plan. The scheme is not called zoning in Berlin, but there is a city architect and everything is planned in advance. The city is solidly built; there are no vacant spaces that may serve as speculative holdings. There is absolute standardization of buildings—squares, fountains, apothecaries' shops are located in advance. Houses have shops on the first floor, with the rooms of the tradesmen in the rear. The well-to-do have the apartments above, facing the street. The lower middle class have the back apartments. All classes are represented in a block. It is known how many people will be in each block, and what shops will be needed. Yet with all this careful planning Berlin has gotten out of bounds. The wealthy want to live on the parks and boulevards. They get located on certain streets. These streets acquire reputation and prestige, become distinctive regions not called for in the city plan. Values rise. Speculation goes on. The city gets out of control. Especially is this true since the war, with its sudden turnover of fortunes and breaking down of class distinctions.

The experience of the Chicago Zoning Commission affords an interesting example of an attempt to control the growth of a new, rapidly growing, unplanned city. The Chicago zoning ordinance has been approximately two years in operation. Mr. H. J. Frost, formerly of the engineering staff which gathered the data on which the ordinance is based, and now of the board of appeals, has kindly given me data on the Chicago situation. His data would seem to indicate that it is futile to impose a plan upon a city which involves the attempt to control land values and the natural groupings of the population. Where use districts cut across natural areas of the city there is a constant pressure upon the board of appeals, which invariably necessitates revision. That is, use districts are merely another form of administrative area where they ignore natural areas. In attempting to control a city's growth we are not merely rearranging our "block," refashioning an artifact, but are working with a natural organization and natural groupings within that organization. The ordinance can neither control this organization of the city nor the inevitable succession of the city. It can, however, taking this organization and succession into account, stabilize the processes of city growth and prevent the waste involved in scattering and uncontrolled speculation.

Whatever we may think such evidence indicates, certainly it is apparent that city planning and zoning, which attempt to control the growth of the city, can only be economical and successful where they recognize the natural organization of the city, the natural groupings of the city's population, the natural processes of the city's growth. An ideal city is not likely to be the mold of a real city.

Natural areas and significant statistics

One of our crying needs in planning for and administering the city is significant statistics of city life. But statistics, to be significant, must be based not only upon accurately defined and comparable units but upon units that are actual factors in the process under examination. Our statistics of city life are based, at the present time, upon administrative areas, which have no real correspondence with the natural areas of the city. Consequently, our statistics are of little significance for the problems of city life. Mowrer, in his recent study of family disorganization in Chicago, found that statistics of family disorganization meant nothing until they were prepared for natural areas. Similarly, Shaw, studying the problem of juvenile delinquency, found that statistics, revealing when compiled for the natural areas of the city, meant nothing when compiled for wards.

The natural areas of the city are real units. They can be accurately defined. Facts that have a position and can be plotted serve to characterize them. Within the areas we can study the subtler phases of city life—politics, opinion, cultural conflicts, and all social attitudes. As this data accumulates it becomes possible to compare, check, and fund out knowledge. With natural areas defined, with the processes going on within them analyzed, statistics based upon natural areas should prove diagnostic of real situations and processes, indicative of real trends. It is not improbable that statistical ratios might be worked out which would afford a basis for prediction beyond the mere agglomeration of population, making it possible to apply numerical measurement to that collective human behavior in the urban environment which is the growth of the city.

The Distribution of Commercialized Vice in the City: A Sociological Analysis*

Walter C. Reckless

Segregation and personal disorganization

The commercialized vice areas of the city represent a natural segregation of individuals on the basis of their interests and attitudes. They attract, on the one hand, persons who seek sexual excitement, and on the other, those who exploit sex as a business or profession. Indeed, the very development of vice areas is dependent upon the conditions making for personal disorganization, since under these circumstances the impulses and desires get released from the socially approved channels and consequently find an outlet in the pattern of vice.

Concerning the more or less temporary population of the vice areas it may be said that to a large extent the patrons of commercialized vice, and to a lesser extent amateur and clandestine prostitutes, fit into the category of dual persons who circulate between two conflicting worlds, namely, a world of respectability in the residential neighborhoods and a world of disrespectability in the downtown districts. The former offers them a life of shelter and security according to the sanctioned definitions of society; the latter, a life of adventure and romance in the realm of the disapproved. Again, a large quota of the more or less permanent habitués of the commercialized vice areas consists of persons whose demoralization has made them outcasts from respectable society, and also of those individuals who, growing up amid great neglect, have developed a disorderly, wild, unregulated scheme of life which makes them unfit to enter organized society without passing through a rather complete re-education.

The moral and geographical isolation of vice

But vice is usually censored by the mores of the community. It is not merely defined as immoral; it is also conceived as pestilential. And its open patrons and entrepreneurs are relegated to a social pariah existence. Vice has, therefore, been forced to hide from the moral order of society in order to flourish.

Because of this moral isolation vice gets spatially separated from wholesome family and neighborhood life in the community. The moral attitudes operate as barriers to isolate geographically this peculiar form of human activity.

*Reprinted from *Publications of the American Sociological Society*, 20 (1926), 164–76, by permission of the author and The American Sociological Association.

Accordingly, commercialized vice has assumed two characteristic locations in the community: one at the center, the other at the circumference. It is well known that the central parts of the city, because of the decaying neighborhoods, have very little resistance to the invasion of vice resorts. Furthermore, commercialized vice on the fringe of the city, lodged at inns, taverns, and roadhouses, meets with practically no opposition, since the *hinterland* of the urban community, due to its sparsely settled condition and its decadent rural culture, is really unorganized.

But the vice resorts are usually prevented from assuming this most central location. In the first place legitimate business, such as large retail stores, financial establishments, skyscraper office buildings, is able to pay the high rents necessary in the competition for space. In the second place the public generally exerts pressure to drive vice out of the community market, although, as will be pointed out later, a large part of it is able to evade suppression and surveillance through subterfuge and camouflage. But commercialized vice can assume a decentralized location without threatening its existence. The very urgency of its demand, namely, this desire for sexual thrill, means that patrons will seek the supply even to the most remote places of the city. In fact, the delay entailed to this pursuit adds to the intensity of the urge as well as to the excitement of the chase.

The central position of commercialized vice may be said to represent the natural, unimpeded play of economic forces. The decentralized or outlying location signifies, in the main, a reaction to political factors, namely, those of legal control and public suppression. However, rapid transit and the automobile have made these ordinarily remote sections readily accessible, and consequently commercialized vice has gone with the tide of an outgoing pleasure traffic.

Vice areas related to the natural zones of the city

A study of the particular regions of the city in which commercialized vice flourishes will reveal more definitely the factors that determine the distribution and location of this activity throughout the community. In order to get an accurate picture of the exact regions in which commercialized vice exists, a spot map was made from the cases dealt with by the Committee of Fifteen of Chicago during 1922.[1] The vice resorts handled by this law-enforcing agency extended radially from the center into the surrounding residential areas, principally along the important traffic arteries. Transferred to E. W. Burgess' chart describing the natural organization of the city,[2] the commercialized vice areas as revealed by this spot map are found to be implanted upon the central business zone (Zone I), the zone of transition (Zone II) with its slums, immigrant and racial colonies, lodging- and rooming-house area, and the restricted residential zone (Zone IV), which includes apartment houses as well as single homes.[3] It may be said, therefore, that commercialized vice areas represent a parasitic formation, since they thrive upon the natural organization of the city.

The adaptation of commercialized vice to natural areas

A closer examination of the Committee of Fifteen data in reference to the economic and cultural order of the city shows that this agency was dealing with assignation hotels in the central business district, brothels in the slum, and "immoral flats" in the high-class residential area. It is clear, therefore, that commercialized vice makes special adaptation to the type of neighborhood invaded. The peculiar conditions characterizing these regions in which commercialized vice is located constitute very definite factors in the distribution and segregation of this parasitic activity.

Prostitution, supposedly excluded from the center of the city, actually, however, is able to evade surveillance by certain camouflages. While the brothel type of prostitution in most instances cannot exist in the central business district, not merely because of its open, public character, but

---

[1] The year 1922 was selected to show the more recent tendencies in the distribution of vice in the modern American city. Ten years earlier, before public repression had produced its noticeable effects, the vice resorts, if plotted, would probably show a greater concentration in the near central regions and less dispersion into the more decentralized neighborhoods.

[2] [See Chart 2 in E. W. Burgess, "The Growth of the City," p. 38 of this volume.]

[3] In Chicago the rooming-house district of Zone II and the apartment-house area of Zone IV merge into one another on the direct south, west, and north sides, a fact which is due primarily to the high value of land resulting from favorable locations and good transportation facilities. The zone of workingmen's homes (III) in Chicago is found largely on the northwest and southwest sides of the city, outside the lines of greatest mobility, and consequently outside the regions in which commercialized vice flourishes best. However, it is doubtful whether the vice resorts in any city can successfully invade Zone III because of the strong family and neighborhood organization found there.

also because of its inability to command a site in face of competition from financial, retail, and wholesale establishments, the freer and more clandestine form of commercialized vice surmounts these obstacles. Streetwalkers have never been eliminated from the downtown districts. Moreover, the activities of the streetwalker in very recent times are not so easily distinguished from the rather wide-spread practice of making casual acquaintances. A large number of these clandestine prostitutes have access to the cheaper hotels, many of which are used for assignation purposes.

Prostitution is frequently an insidious adjunct to the downtown "high life," the social whirl centering about the restaurants, the cafés, the theaters. The existence of commercialized vice in the central business district is an inevitable part of the flux and flow of the region. Besides being a market place for thrill, the downtown district is a region of anonymity, where conduct either remains uncensored or is subject merely to the most secondary observation and regulation. Under such conditions personal taboos disintegrate and appetites become released from their sanctioned moorings.

But streetwalking and assignation hotels by no means exhaust the adaptations which commercialized vice makes to the central business district. It frequently insinuates itself under the protective coloration of massage parlors and bathhouses. In these instances the "vice interests" are exploiting a very natural relationship of bathing and massage to sexual excitement.

The slum as the habitat of the brothel

The area of deterioration encircling the central business district furnishes the native habitat for the brothel type of prostitution. All the conditions favorable to the existence of this flagrant, highly organized form of commercialized vice are to be found there. In the slums the vice emporia not only find very accessible locations, but also experience practically no organized resistance from the decaying neighborhoods adjacent. And, furthermore, they are located in a region where the pattern of vice is an inevitable expression or product of great mobility and vast social disorganization.

Unorganized prostitution in rooming-houses

The rooming-house sections and, to some extent, the tenement districts harbor an unorganized form of prostitution. The free-lance clandestine prostitutes, unattached to brothels, resort frequently to furnished rooms as a place to live and "bring tricks." The landlords or landladies either demand high rents from them or require a special room tax on each service. Because of the great anonymity in these rooming-house areas the activities of these prostitutes go on relatively unnoticed and consequently undisturbed. Here again the location is one of proximity to the demand, for it is a matter of common observation that the rooming-house and lodging-house areas quarter the hordes of homeless men in the community.

Immoral flats in apartment-house areas

Commercialized vice has recently invaded the livelier apartment-house districts of the city and has appeared at this location in the form of "immoral flats," "buffet flats," and "call flats." The presence of vice in this decentralized part of the city, such as in the rooming-house sections and even on the fringe of the community, is due partly to a reaction to public repression. But the prostitution which has fled the slum for the apartment-house area has materially changed its external dress. Commercialized vice in the apartment house, as a rule, seems to be much less organized and much more refined than it is in the brothel.

The immoral flats are really only accessible by taxicab or automobile, since they hug the boulevards rather than the street-car lines. They attract, therefore, a high-class patronage, a sporting element that does not subscribe to the cheaper entertainment provided by the brothel. The apartment areas in which this externally changed form of prostitution is found present a very inviting field to commercialized vice, not merely because of the lively and mobile character of these regions, but also because of the anonymity and individuation produced by the highly mechanized living conditions.

Indexes of commercialized vice areas

Certain of the factors and forces that determine the distribution of vice throughout the community are reducible to indexes, which help to delimit, as well as explain, the distribution of vice in the city. It may be said that commercialized vice is found in those regions characterized by burlesque shows, rescue missions, crime and other major social problems, immigrant and racial colonies, disproportion of sexes, declining population, and high land values and low rents.

The burlesque shows

The burlesque shows of large American cities, if plotted on a map giving the distribution of vice resorts, would fall within the areas in which flourish the most open, public forms of prostitution. This part of the larger commercialized vice areas of the city is really the homeless man's playground, for, besides these cheap theaters, the brothels, saloons, gambling-dens, fortune-tellers, "dime museums," and lady barbers compete with one another in catering to the play and sex interests of the non-family men of the slum. The burlesque show, or "border drama," is symbolic of the fact that a veritable man's community, with all its characteristic patterns of disorder, exists at the core of the city.

The rescue missions

It is well known that the rescue mission has pioneered among the brothels and vice resorts of the urban community. From a spot map showing the characteristic institutions of hobohemia in Chicago it is quite evident that these rescue missions are located on, or adjacent to, the notorious rialtos of the underworld. In fact, the "church on the stem" has grown up to reclaim the "lost souls" of the city's slums, and consequently points to social forces at work in the community to counteract those making for demoralization.

Crime and other social problems

The underworlds of vice and crime have usually been inseparable. The distribution of crime throughout the urban community portrays, in the main, the location of commercialized vice. A spot map of felony cases, giving the place of the crime and the address of the criminal, which were reviewed by the Chicago Crime Commission during 1921, describes about the same territorial distribution for crime as the spot map of the cases dealt with by the Committee of Fifteen of Chicago in 1922 does for vice.[4] On analysis it appears that both crime and vice depend upon mobility and collections of people; both forms of activity are legally and morally isolated and consequently must hide in the disorganized neighborhoods in order to thrive. It is also interesting to note that commercialized vice exists in the same general regions of the city characterized by the distribution of the cases of poverty, divorce, desertion, suicide, abandoned infants. Indeed, these problems, considered ecologically, indicate the areas of greatest social disorganization within the city.

[4]There are certain discrepancies between the two maps. As would be expected, crime shows a somewhat wider distribution than vice. Furthermore, a large proportion of burglaries occur in the wealthier residential districts, which are usually free from commercialized vice.

Immigrant and racial colonies

Since commercialized vice thrives amid the vast social disorganization of the urban community, the major part of which is localized in the slum, it is to be expected that the underworld intrudes itself in the immigrant and racial colonies. The relationship of Chinatown to the commercialized vice areas of American cities is too well known to need elaboration. It is only fair to say, however, that the assumption of the usual parasitic activities by the Chinese in the Western World is probably to be explained by their natural segregation at the center of cities, as well as by their uncertain economic and social status.

The "black belts" of American cities have usually been located in or adjacent to the vice areas, while the Negroes themselves, in face of limited occupational opportunity, have of necessity found work as maids and porters in the vice resorts.[5]

Vice resorts are also found in the settlements of the most recent foreign immigration, which must generally take over the most undesirable sections of the slum in order to gain a foothold in the community. But commercialized vice does not invade all immigrant settlements. Those like Little Italy and the Ghetto, with a strong family and neighborhood organization, are relatively free from prostitution.

Vice is more characteristic of the cosmopolitan areas of the city, which represent a sediment of caught families and individuals from the various classes and nationalities. Since group controls in such regions have practically disintegrated, social life tends to be unregulated and often disorderly.

While burlesque shows, rescue missions, crime and other major social problems, [and] immigrant and racial colonies are valuable as rough indicators of the location and ecological setting of commercialized vice, the disproportion of sexes, declining population, and the correlation of high land values and low rents more nearly approximate indexes as used in the scientific sense; for in the first place, they are capable of mathematical formulation, and in the second place, they reveal factors and forces fundamentally related to commercialized vice in the chain of causation.

The disproportion of sexes

The drift and gravitation of innumerable casual workers, tramps, hobos, bums, into the twilight zone between the central business district and the

[5]See the report of The Commission of Race Relations, *The Negro in Chicago,* pp. 342–43.

area of deterioration surrounding it has stimulated the development of so-called "womanless slums," and consequently has created a very marked disproportion of sexes.

The disproportion of sexes, on analysis, discloses certain conditions which underlie the very existence of commercialized vice. Men's communities and "hobohemias" have ever been characterized by the presence of prostitution. Westermarck has shown that a primitive sort of prostitution existed in Easter Island, where the men greatly outnumbered the women.[6] Bloch, in his study of *Die Prostitution*, specifically states that the men's communities of classical antiquity, namely, the university towns and the military camps, provided a fertile soil for the activities of prostitutes.[7] According to Bancroft, vice ran amuck in the mining camps of California's Gold Rush when, in 1850, the female population constituted less than 2 percent of the total in the mining counties.[8] To take a more recent example, attention has been called to the fact that commercialized vice is rampant in Pekin of the present day, where the male population amounts to 63.5 percent of the total number of inhabitants for that city.[9]

The disproportion of sexes acquires greater significance as an index of commercialized vice when taken in connection with marital status. The homeless man is not merely footloose; he is usually unmarried. In his study of *The Hobo*, Nels Anderson makes the following pertinent statement:

> Of the one thousand men studied by Mrs. Solenberger (1911), 74 percent gave their marital status as single. Of the four hundred interviewed by the writer, 86 percent stated they were unmarried. Only 8 percent of the former, and 5 percent of the latter, survey claimed they were married. The others claimed to be widowed, divorced, or separated from their wives.[10]

As a result of the personal disorganization incident to this detachment from family life, the sex impulses seek outlets in the unapproved channels, not merely in prostitution, but also in perversion.

Furthermore, the homeless man of the city's slums usually suffers from sex isolation, due to his great mobility, his low economic status, and his unpresentable appearance. About the only accessible women are the lower order of prostitutes. The vagrant men of all time, because of their social-pariah existence and their resulting sex isolation, have of necessity subscribed to commercialized vice.

Declining population

The density of population is frequently used as a criterion to explain the major problems of city life. And, offhand, it would seem that this principle would apply to commercialized vice. For prostitution flourishes in the areas of highest density within the city, namely, in the slum, where there is great concentration, while it is conspicuously absent from decentralized neighborhoods with a comparatively low density. This general relationship can be shown by a transposition of the Committee of Fifteen data on a density base map of the city.

But there are sections of the downtown environs which are outside the radial distribution of commercialized vice and yet are within the circle of the most thickly populated areas in the city. Certain immigrant colonies are cases in point. Foreign settlements are frequently protected against a wholesale invasion of commercialized vice not merely by virtue of their semi-remote location, but also by a strong family and neighborhood organization. Furthermore, on the outskirts of the city commercialized vice is very often lodged at roadhouses, which flourish in the most sparsely settled regions of the urban community.

It is the type of community organization, rather than the density of population, that has the direct bearing on the presence and distribution of vice. This is the reason why declining population, rather than sheer density of population, is the more satisfactory index, since it points to a lack or a disintegration of community organization, and consequently to a condition in which commercialized vice can exist best. According to maps showing the comparative density of the census districts in Chicago, it was found that certain sections contiguous to the central business section revealed a marked decline in the number of inhabitants in 1920 as over against 1910. These areas of declining population are precisely the ones which harbor the brothels, according to the Committee of Fifteen cases for 1922. Indeed, commercialized vice, as already noted, is merely one of the many symptoms of the intense social disorganization in these twilight neighborhoods at the core of

---

[6] See *History of Human Marriage*, 3d ed., I, p. 157.
[7] See *Die Prostitution*, I, p. 252.
[8] See *History of California*, IV, pp. 221–39, for account of rampant vice conditions; pp. 221–22 for statement of disproportion of sexes in 1850.
[9] Sydney David Gamble, *Pekin: A Social Survey* (New York, 1921), pp. 243–44.
[10] *The Hobo*, p. 137n.

the city, neighborhoods which are decaying in the inevitable transition from residence to business.

The correlation of high land values and low rents

Indicative also of this transition and disorganization is the correlation of high land values and low rents which describes a condition of neighborhood deterioration in the slum area about the center of the city. It is known that high land values appear at the traffic centers. In fact, they are a product of mobility of population, which in turn creates a situation of social instability and flux—a setting in which the pattern of vice thrives. Furthermore, commercialized vice almost inevitably develops in these areas of great mobility which, after all, become the natural marketplace for thrill and excitement.

The slum, which has ever sheltered the most blatant forms of commercialized vice, has generally been noted for its fluidity and kaleidoscopic life, and the high land values in this zone of deterioration certainly indicate this condition of great mobility and disorganization. The land here not only has a relatively high value because of its centralized, and thereby accessible, location, but also has a speculative value, due to the approach of business itself.

The improved property in these mobile, decaying neighborhoods that are in direct line of business expansion is allowed to run down, to deteriorate, for upkeep generally results in a total loss to the owner, since business only ordinarily demands the site. These deteriorated dwellings of the slum, because of their undesirability, can command but very low rents. It is unavoidable that the poor and vicious classes share the same locality in the city's junk heap.

The relationship of the distribution of commercialized vice to neighborhood deterioration and the value of the correlation of high land values and low rents as an index of the vice areas may be indicated by the following statement of findings:

> By actual count in the city of Seattle over 80 percent of the disorderly houses recorded in police records are obsolete buildings located near the downtown business section, where land values are high and new uses are in process of development.[11]

It is clear that the distribution of commercialized vice in the city comes about through the working of factors determined by the economic, political, and cultural organization of the community as well as through the operation of forces lodged in human nature. The segregation of vice into characteristic urban areas is, therefore, the result of a natural process of distribution rather than—as is so often thought—a sheer artifice of legal control.

The propositions expounded in the foregoing analysis are not presented in terms of absolutes, especially in view of the fact that the factual material for this paper was drawn from an intensive study of the growth and development of vice areas in Chicago.[12] They are merely working hypotheses which invite the challenge of future investigation.

[11]R. D. McKenzie, "The Ecological Approach to the Study of the Human Community," *American Journal of Sociology*, 30 (Nov. 1924), p. 229 n.

[12]See Walter C. Reckless, *The Natural History of Vice Areas in Chicago*. Unpublished doctoral dissertation, University of Chicago, 1925.

Vicksburg: A Study in Urban Geography*

Preston E. James

The personality of Vicksburg is well expressed in the contrast between two of the town's buildings which stand not three blocks apart on the slopes of the valley bluff overlooking the broad flood plain of the Mississippi. One of these is the old courthouse, which has watched over the varying fortunes of the town since the days before the Civil War and which is representative of a host of traditions dating back to the period when river towns could look with disdain on their inland neighbors. The other building is the new twelve-story hotel, Vicksburg's skyscraper, looking oddly out of place among the architectural types of the last century and representing a spirit of rejuvenation in the community life and the beginning of a renovation of the urban scene.

A river town

The attachment of a nucleus of population to some site in the vicinity of Vicksburg seems to have come, almost inevitably, from the advantages of the situation. As the Mississippi River winds southward from Memphis, it leaves its eastern valley bluff to follow a course through the middle of its fertile flood plain. After more than 150 miles, the great river again impinges upon the eastern

*Reprinted from the *Geographical Review,* 21 (April 1931), 234–43, by permission of the author and The American Geographical Society.

bluff. South of this point the meanders approach but do not touch the bluff for another 100 miles to the vicinity of Natchez, where two swings of the river reach the flood plain margin. The immediate site of Vicksburg is one of considerable relief, with a number of steep slopes. The valley bluff at this point rises about 140 feet above the flood plain, but, unlike the bluff at Natchez, it is not too steep for building. It has been cut in the deep accumulation of loess which borders most of the lower Mississippi flood plain and which makes up the material of most of the site of Vicksburg. Back of the valley bluff small tributary streams in dendritic pattern have dissected the loess into early mature ravines and ridges, with a relief of from 60 to 100 feet. As a result of the peculiar quality of loess which permits it to stand in steep slopes, the valley sides are characteristically steep, and the ridges relatively narrow. The valley bottoms are flat and in places swampy. Two streams, or bayous, Glass Bayou and Stouts Bayou, trending from northeast to southwest drain the site of Vicksburg. Where the ridge which separates Glass Bayou from Stouts Bayou reaches the top of the valley bluff it broadens out into a fairly extensive area of relatively level land (Fig. 1). This favorable surface, located at the end of a ridge which gives easy access to the east, was occupied by a part of the original nucleus of the town.

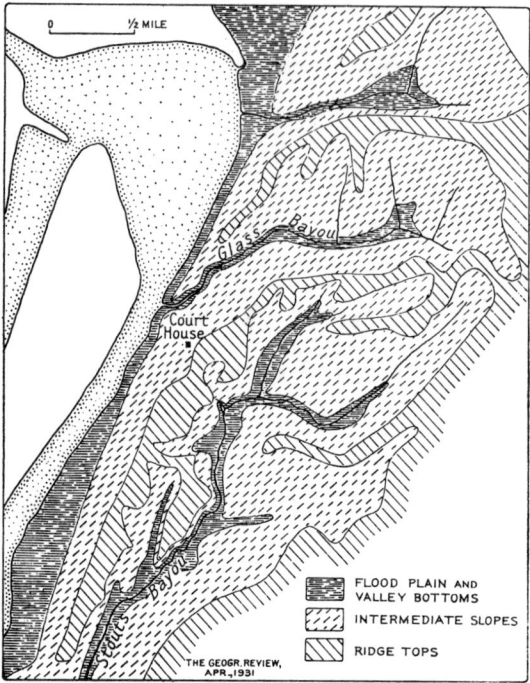

Figure 1—The land surface of the site of Vicksburg.

While Vicksburg was originally located on the immediate banks of the Mississippi, such is not the case today. During a flood in 1876 (April 26) the river suddenly broke through the narrow neck of land inside its meander and left Vicksburg high and dry. For a number of years the town could not be reached by boat, but in 1903 government engineers again brought navigable water to its landing place by diverting the channel of the Yazoo River (Fig. 2).

The foundation of Vicksburg

Urban patterns come into being in various ways; but once established it is only rarely that they can be fundamentally changed. Each part of a city reflects to a certain degree the period of its development, not only in the architecture of its buildings but also in its pattern of arrangement. In Vicksburg, for example, the rectangular pattern of the northern part of the city had its origin in a very different period from the confused array of patterns in the southern part.

Early use of sites in the vicinity of Vicksburg for the establishment of forts by the French and Spanish was not accompanied by any extensive settlement. During the closing years of the eighteenth century, however, while the ownership of this territory was still in dispute between Spain, England, and the United States, many settlement schemes were laid out, at least on paper. Among the paper claims on the present site of Vicksburg was one by Andrew Glass, which was confirmed in 1803 when the area came definitely under the control of the United States. This was a rectangular tract oriented approximately parallel to the course of the river, which at that time ran about 25° west of south. The land grant was laid out 18° east of true north. This orientation placed it obliquely on the more permanent valley bluff, the presence of which was apparently disregarded. This tract of land later came into the possession of Mr. Newet Vick, whose plantation lay several miles to the northeast and who made use of the Glass tract as a location for his slave quarters.

The idea of establishing a town apparently originated with Mr. Vick. Before his death in 1819 he had made plans for the platting of his river property, and in his will he directed his heirs to carry out the project. In 1824 the Glass tract was marked off, and Vicksburg was established. A rectangular pattern, conforming to the outlines of the original land grant, was adopted, which made possible the sale of lots of uniform size and shape—a highly practical consideration. The blocks were two acres in area, separated by streets of 66 feet, or one chain, in width. However, two streets that intersect at the site of the courthouse were made 100 feet in width; and the road that approaches Vicksburg from the northeast, from Vick's old plantation, was allowed to interrupt the rectangular pattern to follow the trend of the ridge. The area between the westernmost street and the river was set aside as a common for use as a community landing place.

The application of this pattern to the site has had some interesting results. It misses the trend of the ridges, and as a result it is difficult to travel through the town without either making sharp ascents and descents or else a series of right-angle turns. The two wider streets that meet at the courthouse present an interesting anomaly, for they have never been used for commercial purposes and today stand in strange contrast to the narrow congested thoroughfare that runs through the commercial core.

Vicksburg at the close of the Civil War

The period of growth along the lines of this pattern came to an end with the Civil War. General

Grant's engineers have left valuable evidence regarding the extent of Vicksburg in 1863 in the detailed map drawn by them after the siege. This map shows that while the eastern margin of the original town had never developed, there were a few new streets, also on a rectangular pattern, which had been added to the northeast and the southwest, and that the beginnings of a new subdivision had been made in the valley bottom some distance to the east. There was a small commercial section along the streets north and east of the courthouse, but the chief commercial core was located, as now, along Washington Street, the only connected route of travel north and south through the town. The commercial core was surrounded by a ring of residences, with superior residences localized in two places.

A second period of development began at the close of the Civil War. In 1860 the town contained some 4500 people, with whites predominating nearly three to one. In 1870 there were about 12,500 people, with 55 percent of them Negroes. These figures tell a story of radical change. The well-ordered life of an *ante bellum* southern town was suddenly disrupted by an influx of Negroes, newly freed, and of several thousand northerners who came seeking a fortune in the turmoil of reconstruction. This was a period of confusion, of readjustment with the uncouth elements predominating. And the results are plainly visible on the face of the town: a chaos of small subdivisions, chiefly south of the original nucleus, each with its own street pattern and poorly integrated to the larger pattern, and streets of irregular width, even including Washington Street, the main north-and-south highway.

Vicksburg throughout this time was a unifunctional commercial town; although even before 1850 it contained railroad shops and cotton gins, which gave it somewhat of an industrial flavor. Commerce, however, was its dominant function; and this was at first largely retail, serving the plantations within a radius of not more than fifteen or twenty miles on the eastern side of the river. But the position of the town on the chief line of communication later gave it a strategic hold on the wholesale activities throughout a large portion of central Mississippi. Wholesaling consisted in part of supplying the retail stores of smaller urban communities; but in large part it consisted of buying, selling, collecting, and shipping the crop of cotton.

A number of events, however, combined to diminish the value of Vicksburg's position and to cut short its decade of rapid expansion. River traffic probably reached its height, if we may judge by the commerce of New Orleans, just before the Civil War. Even in the period after the war people

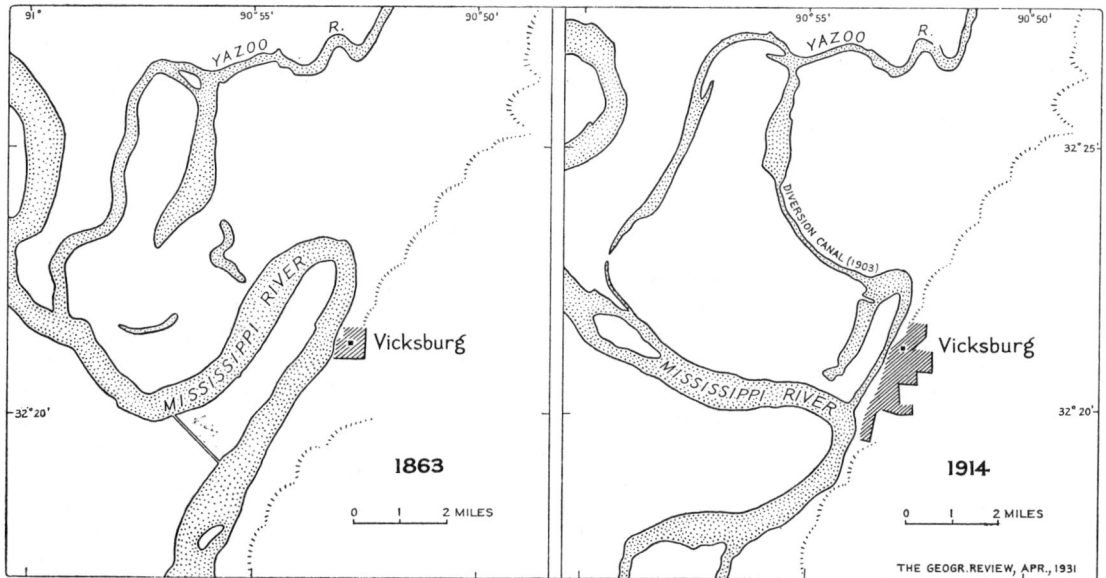

Figure 2—The Mississippi and Yazoo rivers in 1863 and in 1914.

did not quite realize the effect that rail transportation would have. The river towns, Natchez and Vicksburg, were at a decided disadvantage as railroad centers. Along with the decline in river transportation came the shift of the river in 1876, not as serious in its consequences, however, as if it had taken place two decades earlier. Add to this a severe epidemic of yellow fever in 1878, which had its chief effect among the Negroes, and it is easy to understand the decline in population from 12,443 in 1870 to 11,814 in 1880.

The last quarter of the nineteenth century was a period of stagnation for Vicksburg. Jackson, strategically located at a focus of railroad lines, was slowly growing, with nothing but the prestige of Vicksburg to keep it from usurping the wholesale functions. Furthermore, the increase of transportation facilities was leading to a greater diversification of the cotton markets; and Vicksburg, after playing a leading rôle, was becoming just one market among many others. Instead of reflecting the productivity of a large area of the cotton belt, especially of the fertile Yazoo-Mississippi delta to the north, Vicksburg came to depend more and more on the local cotton produced in Warren County.

The present town

Between 1900 and 1910 Vicksburg increased in population from about 14,800 to about 20,800. By 1920 a slight decrease had taken place, but the 1930 census shows a population of 22,943. Most of the gains and losses of this period are in the Negro population, while the whites show a gradual and steady increase. In the meanwhile Jackson has forged definitely into the lead, especially in the 1930 census (48,282), and Natchez has dropped far behind (13,422). This increase in both Vicksburg and Jackson has resulted from a rapid increase in heavy industries.

Vicksburg at the present time (1929) contains some 39 industrial plants. Local industries account for 22 of these: of the remainder, nine manufacture lumber and timber products, such as furniture, flooring, boxes and crates, cooperage, and oars. To these Vicksburg presents something of an advantage in that the larger part of the timber is brought in by river. The other eight industries include cotton gins and compresses, a cottonseed oil plant, a mattress factory, and railroad and engineering shops for two railroads and for the government levee building and repairing equipment. The addition of these heavy industries in significant number to the unifunctional commercial town of the nineteenth century has produced the plurifunctional commercial and manufactural city of the present day (see Fig. 3).

Commercial and industrial areas

The commercial core, which makes up three percent of the city, is and has been, since the first period of development, located along Washington Street, ascending the valley bluff at an angle. The commercial core includes the retail store district, which has for its function the service of the larger urban community and even of the outlying rural areas. It is characterized by buildings two, three, and four stories in height, although this general level is surmounted by an eight-story bank and office building and by the new twelve-story hotel. Brick is the prevailing material, and the architectural types are those characteristic at least of the eastern United States in the seventies, eighties, and nineties. With these types of buildings facing each other across a narrow street, now crowded with automobiles and street cars and overhung with a network of wires, the aspect of the commercial core is one which can be duplicated in its essential features in many of our older cities that have not added new construction in recent years. A larger percentage of the core is covered with buildings than any other district, and their height is greater.

The wholesale district, comprising two percent of the total area of the city, lies along the lower slopes of the valley bluff and the narrow flood plain along the water front. For the most part it shows unmistakable signs of decadence. Most of its structures are one-story brick or corrugated iron sheds built during the period of thriving wholesale activity. At present a large number, perhaps half, of these older buildings are unoccupied.

The maintenance of the wholesale function at Vicksburg has been accompanied by a renewal of activity in river transportation under government subsidy. The Mississippi-Warrior Barge Line maintains a floating dock on the Mississippi, south of the city, and rail connections between the dock and the wholesale and industrial districts. Rates are fixed by the Interstate Commerce Commission so that water shipments, including the cost of rail transshipment in the city, are lower than competing all-rail rates. Shipments by water increased from 80,000 tons in 1919 to 294,000 tons in 1928.

Cheap land and accessibility to lines of communication are the chief desiderata in the location of

Figure 3—Vicksburg in 1929. A quantitative analysis of the area occupied by various utilities give the following result in percentages: industries, 7; wholesale and storage, 2; commercial core, 3; lodging and boarding-house zone, 3; superior residence, 7; ordinary residence, 27; Negro residence, 41; vacant or agricultural land within urban limits, 9; playgrounds and parks, 1. The total area in reference to which these percentages are measured is the area of the geographical city and does not include the agricultural and vacant land beyond the limits of urban development even though within the political city limits.

heavy industries within a city. These conditions are found, in general, on the urban periphery and on low, swampy lands not suitable for other utilities. In the case of Vicksburg, one group of industries is located on the river flood plain just south of the wholesale district, within the political city limits. Another group, consisting of five of the nine lumber and timber products plants, is located two or three miles north, beyond the bend of the Yazoo; and several plants are located south of the city, one as much as five miles distant. These detached pieces, or ectochores, are distinctly a part of the city and owe their physical separation to the unfavorable terrain that intervenes. The numerous local light industries are scattered in characteristic positions throughout the city, but chiefly on the periphery of the commercial core.

The residence districts and vacant land

Of the total area of Vicksburg 78 percent is taken up with residences. These are divided in this study into four districts—lodging-house, superior, ordinary, and Negro districts. The arrangement of these four classes may be described in general terms as a series of concentric circles around the commercial core, this simple circular pattern being considerably modified by adjustment to the ridges and valleys. In general the better classes of residence take the ridge tops and the poorer classes the valley bottoms. Of the Negro district, for instance, 20.4 percent is located in the ravines, while only 12.5 percent is on the ridge tops.

Residence subdivisions, chiefly occupied by Negroes with some ordinary white residences, are scattered eastward along the ridges and are connected to the main portion of the city by ridge-top roads. The political city limits are formed by the inner boundary of the National Military Park. Beyond this park, attracted by the cheaper land outside and by the lower taxes beyond the city limits, there are several urban ectochores of fairly dense settlement, of a character similar to the ridge-top extensions within the city limits. These are all included within the geographical city. Excepting for the inclusions of vacant land in the south, Vicksburg is relatively compact, the few vacant areas within the main body of the city being made up of near-vertical walls of loess on which occupation is almost impossible, or of land close to the railroads or on the river flood plain valuable chiefly for industrial utilization. Perhaps it is a safe generalization that rapidly growing cities tend to include more such undeveloped land within the urban area, while slowly growing or static cities tend to fill in such spaces without expanding the limits.

In conclusion it may be remarked that the future of Vicksburg is bound up with the development of river transportation and with the more productive utilization of the loess uplands to the east and the Yazoo flood plain to the north.

The Gold Coast and the Slum*

Harvey W. Zorbaugh

The Near North Side

"North Town" is divided into east and west by State Street. East of State Street lies the Gold Coast, Chicago's most exclusive residential district, turning its face to the lake and its back upon what may lie west toward the river. West of State Street lies a nondescript area of furnished rooms: Clark Street, the Rialto of the half-world; "Little Sicily," the slum.

The Lake Shore Drive is the Mayfair of the Gold Coast. It runs north and south along Lake Michigan, with a wide parkway, bridle path, and promenade. On its western side rise the imposing stone mansions, with their green lawns and wrought-iron-grilled doorways, of Chicago's wealthy aristocracy and her industrial and financial kings. South of these is Streeterville, a "restricted" district of tall apartments and hotels. Here are the Drake Hotel and the Lake Shore Drive Hotel, Chicago's most exclusive. And here apartments rent for from three hundred fifty to a thousand dollars a month. Indeed, the Lake Shore Drive is a street more of wealth than of aristocracy; for in this Midwest metropolis money counts for more than does family, and the aristocracy is largely that of the financially successful.

South of Oak Street the Lake Shore Drive, as it turns, becomes North Michigan Avenue, an avenue of fashionable hotels and restaurants, of smart clubs and shops. North Michigan Avenue is the Fifth Avenue of the Middle West; and already it looks forward to the day when Fifth Avenue will be the North Michigan Avenue of the East.

On a warm spring Sunday "Vanity Fair" glides along "the Drive" in motor cars of expensive mark, makes colorful the bridle-paths, or saunters up the promenade between "the Drake" and Lincoln Park. The tops of the tan motor busses are crowded with those who live farther out, going home from church—those of a different world who look at "Vanity Fair" with curious or envious eyes. Even here the element of contrast is not lacking, for a mother from back west, with a shawl over her head, waits for a pause in the stream of motors to lead her eager child across to the beach, while beside her stand a collarless man in a brown derby and his girl in Sunday gingham, from some rooming-house back on La Salle Street.

For a few blocks back of "the Drive"—on Belleview Place, East Division Street, Stone, Astor, Banks, and North State Parkway, streets less pretentious but equally aristocratic—live more than a third of the people in Chicago's social register, "of good family and not employed." Here

*Reprinted from Harvey W. Zorbaugh, *The Gold Coast and the Slum* (Chicago: University of Chicago Press, 1929), pp. 7–12, 69–72, and maps opposite pp. 50, 83, and 132, by permission of The University of Chicago Press.

are the families that lived on the once fashionable Prairie Avenue, and later Ashland Boulevard, on the South and West sides. These streets, with the Lake Shore Drive, constitute Chicago's much vaunted Gold Coast, a little world to itself, which the city, failing to dislodge, has grown around and passed by.

At the back door of the Gold Coast, on Dearborn, Clark, and La Salle streets, and on the side streets extending south to the business and industrial area, is a strange world, painfully plain by contrast, a world that lives in houses with neatly lettered cards in the window: "Furnished Rooms." In these houses, from midnight to dawn, sleep some twenty-five thousand people. But by day houses and streets are practically deserted. For early in the morning this population hurries from its houses and down its streets, boarding cars and busses, to work in the Loop. It is a childless area, an area of young men and young women, most of whom are single, though some are married, and others are living together unmarried. It is a world of constant comings and goings, of dull routine and little romance, a world of unsatisfied longings.

The Near North Side shades from light to shadow, and from shadow to dark. The Gold Coast gives way to the world of furnished rooms; and the rooming-house area, to the west again, imperceptibly becomes the slum. The common denominator of the slum is its submerged aspect and its detachment from the city as a whole. The slum is a bleak area of segregation of the sediment

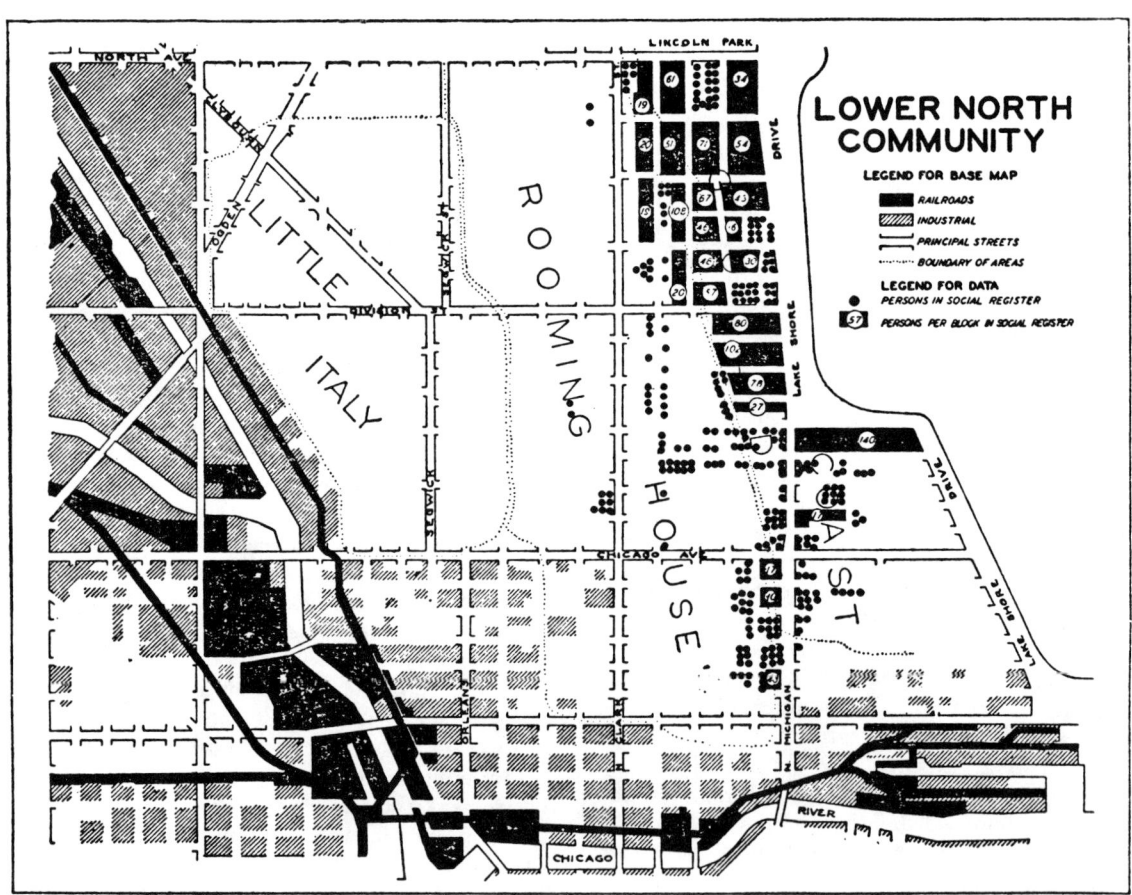

The Gold Coast—Chicago's society is concentrated along the strip of lake shore north of "Streeterville," with a scattering on La Salle, lower Rush, Huron, Superior, Ohio, and Cass streets, fashionable a generation ago. This map brings out strikingly the segregation characteristic of the life of the great city. In this and succeeding maps, the solid black areas indicate that the dots—here representing the residences of persons whose names appear in the *Social Register* (1923)—cluster too thickly to be individually represented.

of society; an area of extreme poverty, tenements, ramshackle buildings, of evictions and evaded rents; an area of working mothers and children, of high rates of birth, infant mortality, illegitimacy, and death; an area of pawnshops and second-hand stores, of gangs, of "flops" where every bed is a vote. As distinguished from the vice area, the disintegrating neighborhood, the slum is an area which has reached the limit of decay and is on the verge of reorganization as missions, settlements, playparks, and business come in.

The Near North Side, west of Clark Street from North Avenue to the river, and east of Clark Street from Chicago Avenue to the river, we may describe as a slum, without fear of contradiction. For this area, cut off by the barrier of river and industry, and for years without adequate transportation, has long been a backwater in the life of the city. This slum district is drab and mean. In ten months the United Charities here had 460 relief cases. Poverty is extreme. Many families are living in one or two basement rooms for which they pay less than ten dollars a month. These rooms are stove heated, and wood is sold on the streets in bundles, and coal in small sacks. The majority of houses, back toward the river, are of wood, and not a few have windows broken out. Smoke, the odor from the gas works, and the smell of dirty alleys is in the air. Both rooms and lots are overcrowded. Back tenements, especially north of Division Street, are common.[1]

Life in the slum is strenuous and precarious. One reads in the paper of a mother on North Avenue giving away her baby that the rest of her children may live. Frequently babies are found in alleyways. A nurse at the Passavant Hospital on North La Salle tells of a dirty little gamin, brought in from Wells Street, whose toe had been bitten off by a rat while he slept. Many women from this neighborhood are in the maternity ward four times in three years. A girl, a waitress, living at the Albany Hotel on lower Rush Street, recently committed suicide leaving the brief note, "I am tired of everything. I have seen too much. That is all."[2]

Clark Street is the Rialto of the slum. Deteriorated store buildings, cheap dance halls and movies, cabarets and doubtful hotels, missions, "flops," pawnshops and second-hand stores, innumerable restaurants, soft-drink parlors and "fellowship" saloons, where men sit about and talk, and which are hangouts for criminal gangs that live back in the slum, fence at the pawnshops, and consort with the transient prostitutes so characteristic of the North Side—such is "the Street." It is an all-night street, a street upon which one meets all the varied types that go to make up the slum.

The slum harbors many sorts of people: the criminal, the radical, the bohemian, the migratory worker, the immigrant, the unsuccessful, the queer and unadjusted. The migratory worker is attracted by the cheap hotels on State, Clark, Wells, and the streets along the river. The criminal and underworld find anonymity in the transient life of the cheaper rooming-houses such as exist on North La Salle Street. The bohemian and the unsuccessful are attracted by cheap attic or basement rooms. The radical is sure of a sympathetic audience in Washington Square. The foreign colony, on the other hand, is found in the slum, not because the immigrant seeks the slum, nor because he makes a slum of the area in which he settles, but merely because he finds there cheap quarters in which to live, and relatively little opposition to his coming. From Sedgwick Street west to the river is a colony of some fifteen thousand Italians, familiarly known as "Little Hell." Here the immigrant has settled blocks by villages, bringing with him his language, his customs, and his traditions, many of which persist.

Other foreign groups have come into this area. North of "Little Sicily," between Wells and Milton Streets, there is a large admixture of Poles with Americans, Irish, and Slavs. The Negro, too, is moving into this area and pushing on into "Little Hell." There is a small colony of Greeks grouped about West Chicago Avenue, with its picturesque

---

[1] A five-room house on Hill Street, the rooms in which are 9 × 12 × 10 feet high, has thirty occupants. A nurse told the writer of being called on a case on Sedgewick Street and finding two couples living in one room. One couple worked days, the other nights; one couple went to bed when the other couple got up. Mrs. Louise De Kowen Bowen (*Growing Up with a City*), reminiscing of her United Charities experiences, tells of a woman who for three years existed on the food she procured from garbage cans and from the samples of department store demonstration counters. She adds: "Sometimes fate seems to be relentless to the point of absurdity, as in one case I remember of an Italian family. . . . The man was riding on a street car and was suddenly assaulted by an irate passenger. . . . His nose was broken and he was badly disfigured. . . . A few days later, on his way home from a dispensary where he had gone to have his wound dressed, he fell off a sidewalk and broke his leg. The mother gave birth to a child the same day. Another child died the following day, and the eldest girl, only fourteen years old, who had been sent out to look for work, was foully assaulted on the street." Such is the life of the slum!

[2] *Chicago Evening American,* December 21, 1923.

coffee houses on Clark Street. Finally, there has come in within the past few years a considerable colony of Persians, which has also settled in the vicinity of Chicago Avenue. The slum on the Near North Side is truly cosmopolitan.

In the slum, but not of it, is "Towertown," or "the village." South of Chicago Avenue, along East Erie, Ohio, Huron, and Superior Streets, is a considerable colony of artists and of would-be artists. The artists have located here because old buildings can be cheaply converted into studios. The would-be artists have followed the artists. And the hangers-on of bohemia have come for atmosphere, and because the old residences in the district have stables. "The village" is full of picturesque people and resorts—tearooms with such names as the Wind Blew Inn, the Blue Mouse, and the Green Mask. And many interesting art stores, antique shops, and stalls with rare books are tucked away among the old buildings. All in all, the picturesque and unconventional life of "the village" is again in striking contrast to the formal and conventional life of the Gold Coast, a few short blocks to the north. . . .

The world of furnished rooms

Back of the ostentatious apartments, hotels, and homes of the Lake Shore Drive, and the quiet, shady streets of the Gold Coast, lies an area of streets that have a painful sameness, with their old, soot-begrimed stone houses, their none-too-clean alleys, their shabby air of respectability. In the window of house after house along these streets one sees a black and white card with the words "Rooms To Rent." For this is the world of furnished rooms, a world of strangely unconventional customs and people, one of the most characteristic of the worlds that go to make up the life of the great city.[3]

This nondescript world, like every rooming-house district, has a long and checkered history.

> The typical rooming-house is never built for the purpose; it is always an adaptation of a former private residence, a residence which has seen better days. At first, in its history as a rooming-house, it may be a very high-class rooming-house. Then, as the fashionable residence district moves farther and farther uptown, and as business comes closer and closer, the grade of the institution declines until it may become eventually nothing but a "bums' hotel" or a disorderly house.[4]

We have seen, in reading the history of the Near North Side, that after the fire this was a wealthy and fashionable residence district. But as business crossed the river and came north it became less and less desirable as a place to live. Gradually the fashionable families moved out of their old homes. Less well-to-do, transient, and alien groups came in. As the city has marched northward, however, land values and rentals have been slowly rising, until now the families who would be willing to live in this district cannot pay the rentals asked. As a result the large old residences have been turned into rooming-houses—another chapter in the natural history of the city.

This lodging- and rooming-house district of the Near North Side lies between the Gold Coast on the east and Wells Street on the west, and extends northward from Grand Avenue and the business district to North Avenue. South of Chicago Avenue the district merges with the slum; its rooming- and lodging-houses sheltering the laborer, the hobo, the rooming-house family, the studios of the bohemian, the criminal, and all sorts of shipwrecked humanity, while some of its small hotels have a large number of theatrical people—and others, the transient prostitute. The whole of the district is criss-crossed with business streets. The area north of Chicago Avenue, however, save for Clark Street, is not a slum area. And it is in this area, with its better-class rooming-houses in which live, for the most part, young and unmarried men and women. . . .

An analysis of the *Illinois Lodging House Register* reveals the fact that there are 1,139 rooming- and lodging-houses on the Near North Side, and that in these houses 23,007 people are living in furnished rooms of one kind and another. Ninety blocks in the better rooming area north of Chicago Avenue were studied intensively, by means of a house-to-house census. This study revealed the additional facts that 71 percent of all the houses in this district keep roomers; and that of the people who live in these rooms, 52 percent are single men, 10 percent are single women, and 38 percent are couples, "married," supposedly with "benefit of clergy."[5] The rooming-house area is a

---

[3]This rooming-house district of the Near North Side is one of three such districts in Chicago. Similarly, on the South and West sides there are areas of furnished rooms, wedging their way along the focal lines of transportation, from the apartment areas into the slum and the business district. Rooming-house districts will be found similarly situated in every large city.

[4]Trotter, *The Housing of Non-Family Women in Chicago*, p. 5.

[5]The schedules of this rooming census are filed with the Committee on Social Research of the University of Chicago.

childless area.[6] Yet most of its population is in the productive ages of life, between twenty and thirty-five.[7]

The rooming-house is typically a large, old-fashioned residence, though many apartments are converted into rooming-houses as well.[8] And the population living in these rooming-houses is typically what the labor leader refers to as the "white collar" group—men and women filling various clerical positions—accountants, stenographers, and the like, office workers of various sorts. There are also students from the many music schools of the Near North Side. Most of them are living on a narrow margin, and here they can live cheaply, near enough to the Loop to walk to and from their work if they wish.[9]

The constant comings and goings of its inhabitants is the most striking and significant characteristic of this world of furnished rooms. This whole population turns over every four months.[10] There are always cards in the windows, advertising the fact that rooms are vacant,[11] but these cards rarely have to stay up over a day, as people are constantly walking the streets looking for rooms. The keepers of the rooming-houses change almost as rapidly as the roomers themselves. At least half of the keepers of these houses have been at their present addresses six months or less.[12]

> Most people on La Salle street are forever moving. The landlords move because they think they will do better in another house or on another street. I think many landlords might have been gamblers or inventors—they see visions, or are of the temperament which is always looking for a new stroke of luck. The tenants also keep moving, because they hope for something better in another house or on another street. They are always looking for a place more like home, or more comfortable, or cheaper.[13]

So the scenes shift and change in the drama of the rooming-house world—change with a cinema-like rapidity.

---

[6]School census, 1920. The small number of children in this area is in striking contrast to the number in the slum area to the south and west; even to the number on the Gold Coast.

[7]This age grouping is based on the opinion of social workers, and on the findings of A. B. Wolfe, *The Lodging-House Problem in Boston*.

[8]Part of the story of the rooming-house is told in the fact that these great old residences can never be converted into tenements. See Breckinridge and Abbott, "Chicago's Housing Problem," *American Journal of Sociology*, 16, 295–96.

[9]An accompaniment of the rooming-house is the cheap cafeteria and restaurant, scores of which are found along Clark and State streets, and Chicago Avenue, Division Street, and North Avenue.

[10]Census of Rooming Houses (see above). Of course there are people who have lived in the same room for years. But there are hundreds of others who live in a given house but a month, a week, or even a day. . . .

[11]As of November 1, 1923, 30 percent of the houses in the district had cards in the windows. Census of Rooming Houses (see above).

[12]*Illinois Lodging House Register*. With respect to length of residence, 117 consecutive registrations for the Lower North Side were distributed as follows:

0–1 Mo.	1–3 Mo.	3 Mo.–6 Mo.	6 Mo.–1 Yr.	1–5 Yr.	5 Yr. and More
9	9	26	25	35	13

[13] . . . Interview with a resident of a rooming-house on La Salle Street.

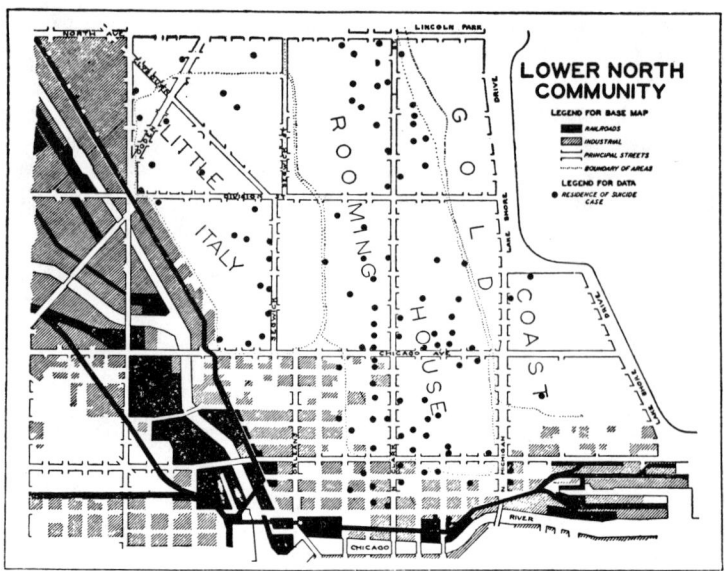

Suicides—Suicide often seems the only escape from the isolation of rooming-house life. Suicide tends to concentrate in rooming-house areas, as this map of the addresses of Near North Side residents who committed suicide within a four-year period indicates (data, after Earle and Cavan, from the coroner's reports for 1919–22).

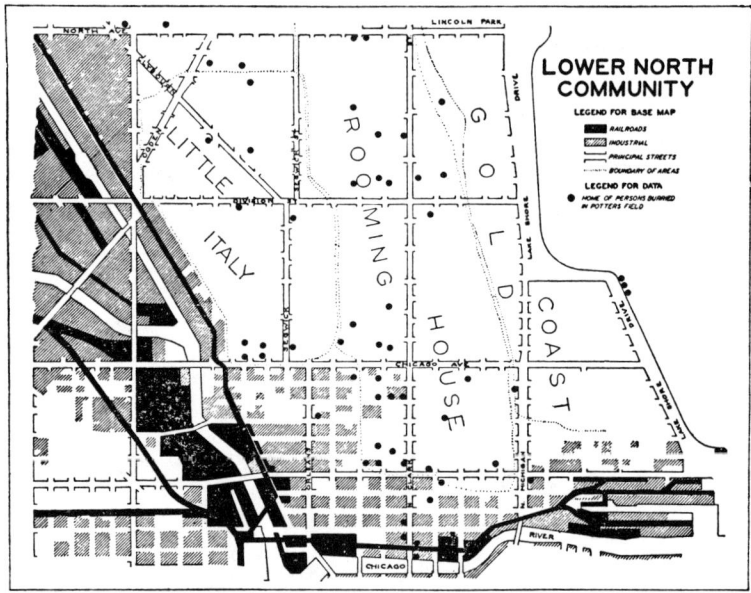

The Potter's Field—Interestingly enough, burial in the "potter's field" seems indicative not so much of the poverty of the slum as of the isolation and lack of group contacts of the rooming-house areas. The above map shows the addresses of persons buried from the morgue when their bodies were left unidentified and unclaimed.

Section B *The Decline of Classical Ecology*

"Community" and Ecological Studies*

Milla A. Alihan

The fundamental assumption of ecologists is that every action or phenomenon or move of living beings is territorially based. The essential attributes of the concept "community" are the territorial basis and organic, spontaneous reactions on the animal level. However, in the distinction between the concepts "community" and "society" there is an inference that some human actions have a more specific relation to territory than others. In fact, in the correlation of ecological distribution and animal or organic behavior it is more or less generally assumed that the more rational or conscious activities are less dependent on territorial factors. This assumption is apparent in most of the theoretical formulations of the school and is explicitly given in Wirth's statement that the willed and contractual relationships between men in "society" are "less directly affected than their organic relationships by their distribution in space."[1] It would seem to follow, therefore, that a study made on ecological premises would be directed more to the asocial or purely organic activities of man. It is this logical conclusion that the ecologists themselves reach when they occasionally point out the "unsocial" character of the concept "community." However, the actual "ecological" studies cannot follow out this distinction, because if we take a territorially demarcated unit as a basis for study we do not discriminate between certain activities carried on within the area as those of "society" and others which are those of "community." When ecologists themselves subdivide "community" into various communities, economic, political, cultural, and ecological, we seem to be tracking the territorially determinant factor down to narrower bounds. But this conclusion becomes untenable, since these other "communities" are indifferently included in the scope of ecological studies. Nor do the researches conform to the initial distinction, upon which they are supposed to depend, any more than does the theoretical application of the ecological principle. To make the matter even more obscure, "community" is taken in its broadest scope and mingled with what was first defined as "society." The only help we have in this regard is that we are sometimes warned that certain studies are not "purely" ecological.

To cite only a few examples: Nels Anderson's study, *The Hobo,* while admittedly approaching the type from the point of view of locomotion,

*Reprinted from Milla A. Alihan, *Social Ecology* (New York: Columbia University Press, 1938), pp. 81–91, by permission of the author and the publisher.
[1] Louis Wirth, "The Scope and Problems of the Community," *Publications of the American Sociological Society,* 27 (1933), 62.

deals to a great extent with the cultural background and the social activities of the hobo. In fact, no distinction is attempted on the lines suggested above. The same may be said of *The Gold Coast and the Slum,* by Harvey W. Zorbaugh. The areas are taken as the frame within which the social phenomena occur, but the phenomena themselves are neither those of community as ecologists conceive it, nor those of society in their specific sense. Rather are these monographs general sociological studies in which territorial distribution is taken account of in reference to sociological data.

It might be said that Clifford Shaw's delinquency study[2] more nearly bears upon ecological distribution, but he neither investigates symbiotic and competitive relations, nor does he probe into the organic, natural reactions of the delinquents. In fact, there is little in Shaw's study to suggest Park's and other ecologists' interpretation of the biotic substructure. Instead, Shaw confines himself to the finding of correlations between the frequency of a social phenomenon, delinquency, in various areas and the relative distance of these areas from the center of the city, following the ideal zonal pattern into which every city supposedly tends to fall. This done, he interprets delinquency primarily in terms of social, cultural, and economic factors. He does not submit any evidence to show that the different frequencies of delinquency in different areas are ecological adaptations to the particular areas.

Such studies as *The Gang,* by Frederick Thrasher, and *The Taxi Dance-Hall,* by Paul G. Cressey, are subject to the same observation. The gang, for example, is assumed to be a natural group and is then studied in its sociological setting. We have no glimpse of what the biotic aspect of this group represents, unless we assume that it is natural in the sense that it constitutes a "moral region." But we do not find this group described in terms of the natural behavior independent of communication and expressing, in particular, tendencies common to all living organisms rather than strictly human characteristics. In fact, the gang as treated by Thrasher is not specifically and essentially an ecological group, but rather a social grouping exhibiting a certain type of social behavior. The only "ecological" fact considered by the author is the delimitation of the territorial area in which the gang concentrates. Excellent though this study may be from the sociological point of view, it does not confine itself to community as formulated by the ecologists.

[2]Clifford R. Shaw, *Delinquency Areas* (1929).

When we consider studies such as R. D. McKenzie's *Metropolitan Community* and "Ecological Succession in the Puget Sound Region,"[3] they are seen to fall more within the ecological scope, save that McKenzie includes in the ecological community the economic organization in its broadest and "most evolved" sense. He makes no attempt to abstract the natural economy from the highly complex economic relationships any more than he tries to isolate biotic competition from conscious competition. In fact, he strongly emphasizes communication as a determining factor as well as a process inherent in the organization with which he deals. It is true that he studies the movements and distribution of populations and goods and on the whole approaches the subject of economic relations from an external point of view, but in essence his works cover the field of distribution economics rather than the less evolved aspect of human processes and structure. Nor is McKenzie concerned with the product of free, as opposed to controlled, competition or with the natural division of labor as distinguished from any other type.

Conclusions

So intertwined are the two aspects which ecologists would sever into two distinct entities that the treatment of them invariably results in their fusion, while the theoretical statements come to the contradictions just dealt with. Taking their clue from the processes in plant and animal colonies, the ecologists discover common elements between these and the human community and try to delimit this simpler aspect of behavior. At the same time, however, they find that the presence of "group economy," which defines a plant or an animal colony, is in human community so highly evolved and so intrinsically a part of other social phenomena that the analogy becomes worthless and the ecologists are forced to define the very concepts intended to describe this organic aspect of life in terms of the assumed social psychological concept "society." Consequently, what is said to be unsocial in one instance is asserted to be social in another.

Assuming that symbiotic relationships correspond to an existing phase of group life, it is still difficult to perceive how in the study of the "biologic economic" organization of the "community" one can abstract the "organic interdepen-

[3]*Publications of the American Sociological Society,* Vol. 23, 1928.

dence" and the "common life based upon the mutual correspondence of interests" characteristic of "community" from the "willed and contractual relationships between men" which define "society." Nor is it clear where the line is drawn between free and controlled or limited competition nor, for that matter, between biotic and social interdependence. Although the economic organization is claimed to be the product of competition, it is also said to be the effect of accommodation: "The equilibrium based on accommodation . . . is not biological; it is economic and social and is transmitted, if at all, by tradition."[4] Accommodation is in turn the outcome of conflict. Thus the three processes are closely related and serve as determinants of the economic organization while, at the same time, conflict and accommodation are determined by competition. However, conflict and accommodation are non-ecological processes and, in fact, they restrict the ecological process of competition. We are not told how these processes of "society" are separated from the ecological process of competition, nor is there any definite means of severing them except by the arbitrary assumption that a particular aspect of the economic organization is natural and another is not natural.

When we come to the factual ecological studies, there seems to be no distinction between the "natural" aspects of the economic organization, which result from competition, and the cultural, or those which are the product of accommodation and conflict—nor, for that matter, between any unconsciously effected phenomena and those brought about consciously.[5] Even when ecologists do distinguish between the natural and the planned, it seems that the planned phenomena eventually take their natural course of development. For example, Park states:

> The newspaper has a history; but it has, likewise, a natural history. The press, as it exists, is not, as our moralists sometimes seem to assume, the wilful product of any little group of living men. On the contrary, it is the outcome of a historic process in which many individuals participated without foreseeing what the ultimate product of their labors was to be. . . . *In spite of all the efforts of individual men and generations of men to control it and to make it something after their own heart, it has continued to grow and change in its own incalculable ways.*[6]

The two levels of human behavior, "that which is common to all organic life" and "that which is strictly human," have never yet been successfully separated by science, perhaps for the very reason that one is found to be a continuation of and a development from the other; not only is the more evolved aspect rooted in the less evolved, but one imperceptibly shades into the other, the two interact continuously and in fact are one. C. O. Whitman's remark illustrates the realization of this merging of the two aspects of life:

> We are apt to contrast the extreme of instinct and intelligence to emphasize the blindness and inflexibility of the one and the consciousness and freedom of the other. It is like contrasting the extremes of light and dark and forgetting all the transitional degrees of twilight. . . . Instinct is blind; so is the highest human wisdom blind. The distinction is one of degree. There is no absolute blindness on the one side, and no absolute wisdom on the other.[7]

Division such as the ecologists make of human behavior is born of the absolutist assumption of a constant, unchanging nature, where evolution is approached as a simple additive process. But does not the cumulative process of natural and social evolution mean transformation as well as addition? Can reality, social reality in particular, be reasonably interpreted in terms of "more or less," especially if we seek to interpret the "more" by the "less" and not vice versa? The very distinction which ecologists make between the struggle for existence on the animal level and the struggle for livelihood on the human level would necessarily imply change in the quality of the so-called struggle. There may be a difference of degree, but there is also a difference in kind, not only in the struggle itself, but also in the total situation of which the struggle is an expression. Human beings differ from plants and animals, not only in that they control their environment, but also in that they *desire* to control it and that with this desire they consciously seek and find the means to create a new environment. The processes which are

---

[4]Park and Burgess, *Introduction to the Science of Society* (Chicago: Univeristy of Chicago Press, revised edition, 1924), p. 664.

[5]See McKenzie, *The Metropolitan Community;* Wirth, *The Ghetto;* Ruth S. Cavan, *Suicide.*

[6]Park, "The Natural History of the Newspaper," in Park, Burgess, et al., *The City,* p. 80; italics mine. This point is further elaborated in Chapter VII, section on "Historical versus Logical Sequence."

[7]Quoted in C. C. Adams, "The Relation of General Ecology to Human Ecology," *Ecology,* 16 (July 1935), 322.

unconscious on the plant or animal level become infused with consciousness of various degrees on the human level. The competition of plants for soil and water has no corresponding factual process in the competition of human beings within any physical area. Since any crowding is translated into conscious struggle in human groupings, where the very expression and the methods of competition are not only conditioned and complicated by this consciousness but also are actually determined by it, is it possible to speak of biotic competition among men?

Although the division of community and society carries the implication of the abstraction of the biotic from the civilized form of competition, and although the ecologists themselves admit this implication, they have nowhere succeeded in effecting it. What they often do, is to concentrate their attention upon the external factors of human behavior, such as distribution and movements of populations, utilities, and physical structures—these usually being interpreted with a particular stress upon the economic and technological factors as the determining ones. There is not a vestige of the "natural" order in the movements and allocation of these elements, in the highly developed forms and methods of communication and transportation, and in the elaborated design and framework of organization within which these mobile forces function. Likewise, nothing specifically natural is revealed either in the scheme of motivations or in the technical economic and other external conditions determining the distribution of the elements in human organization.

There is no objection to the abstraction of these external phenomena to serve the ends of scientific investigation, but the relevant point is that these are not the intrinsic factors of the community as we have found it defined by ecologists.

If ecologists intend to abstract certain external manifestations in human groupings, even if based upon an analogy with organic life, they might develop a discipline which would treat of the civilizational aspect of society as against its cultural facets; or, with a difference in emphasis, they might found a discipline dealing with spatial phenomena; in any case, human ecology would mark either a specific approach or a selected subject matter.

But the ecologists' theory, particularly their distinction between "community" and "society," is based upon a priori assumptions, and it is upon the relevance, validity, and application of these assumptions that the scientific claims of human ecology rest. It is also these same a priori assumptions that, although not always consistently, direct the development and the scope of human ecology. Some of them have already been suggested in the course of the preceding discussions. Others will appear later. Here we may point to the assumption of an intrinsic association between the less evolved organic structure and the external physical phenomena. There is also the assumption of an inherent nexus between spatial and economic phenomena, between locomotion and competition, and again between locomotion and freedom. If we take the last of these, the supposed intrinsic relation of locomotion and freedom, thinking of these two concepts in their naked and simplest sense, as probably the ecologists have done, then the one factor can properly be regarded as the obverse of the other. Yet freedom, as ecologists deal with it in their theory, ceases to be freedom on an animal level but becomes freedom to live in a society as we know it today—the freedom which has been the subject of innumerable queries by philosophers and political theorists and of endless strife between soldiers and between priests.

For such a widening and enriching of the concept "freedom," we have another a priori assumption by Park that "mind is an incident of locomotion" and that "the first and most convincing indication of mind is not motion merely but . . . locomotion."[8] It seems superfluous to inquire into this relation of locomotion and mind in the case of animals or, even more to the point, in the case of birds. If the intrinsic association of freedom and locomotion is assumed to remain constant in spite of the change of the meaning of the concepts and of the reality which each represents, then surely by the same principle locomotion would be expected to have an identical relation to mind whether among plants, animals, or human beings. Any concept which may qualify locomotion on the human level, such as purpose, may, for example, be equally applied to animals and birds. There is no wider difference between the freedom among plants or animals and the freedom among human beings than there is between purpose among the former and purpose among the latter.

Another a priori assumption inherent in the ecological theory is that in a civilized society the division of labor has a more organic basis than customs or mores which have grown from the

[8]"Community Organization and Juvenile Delinquency," in Park, Burgess, et al., *The City,* p. 156.

intimate relationships of human beings. If as Park and Burgess state "custom is group habit,"[9] there is as much basis for speaking of custom among animals, such as the songs belonging to any species of birds, as there is for regarding the division of labor in the animal world as being the equivalent of the division of labor in the human organization. Moreover, to take only one aspect, it is still a matter of speculation as to whether it is the survival of the individual or the survival of the species which is the "concern of nature." We have no single theory which has finally proved and pigeonholed the derivation of customs as against the derivation of division of labor, of social as against economic phenomena. If the intricate pattern of the modern division of labor is a product of organic needs of human beings, so are the manifold modern customs and mores, whatever the worth of either, whatever the consistencies or inconsistencies of both. If consciousness characterizes the social aspect, it also characterizes the economic; if division of labor has its roots in nature, so have customs and traditions. Yet the differences and similarities existing between these categories are differences and similarities which must perforce be discerned and interpreted on the level on which they are found—the human organization. Finally, the designation of competition as the primary, the universal, and the fundamental process, as against assimilation, accommodation, co-operation, or, for that matter, any other process, is also a matter of a particularistic ideology. The approach to human organization from the angle of competition is one thing, but a claim that competition is the process basic to all other processes is another. Cooley's discussion of Tarde's treatment of the process of imitation brings out this point:

> I think, that other phases (than imitation) of social activity, such for instance, as communication, competition, differentiation, adaptation, idealization, have as good claims as imitation to be regarded as the social process, and that a book similar in character to M. Tarde's might, perhaps, be written upon any one of them. The truth is that the real process is a multiform thing of which these are glimpses. They are good so long as we recognize that they are glimpses and use them to help out our perception of that many-sided whole which life is; but if they become doctrines they are objectionable.
>
> The Struggle for Existence is another of these glimpses of life which just now seems to many the dominating fact of the universe, chiefly because attention has been fixed upon it by copious and interesting exposition. As it has had many predecessors in this place of importance, so doubtless it will have many successors.[10]

[9] Park and Burgess, op. cit., p. 799.
[10] Charles Horton Cooley, *Human Nature and the Social Order* (New York, 1922), p. 272.

The Concept of Natural Area*

Paul Hatt

There are two general emphases in the definition of the concept, *natural area*. One of these views the natural area as a spatial unit limited by natural boundaries enclosing a homogeneous population with a characteristic moral order. The other emphasizes its biotic and community aspects and describes the natural area as a spatial unit inhabited by a population united on the basis of symbiotic relationships.

It is the purpose of this paper to examine a series of data with a view to studying whatever natural areas may be present within a larger urban residential district. The area studied has been described in an earlier paper as the central residential area of Seattle, Washington.[1] It seems particularly suited to the problem because it possesses the variation in population necessary to the existence of mutually exclusive natural areas. Rental values range in their block averages from five dollars to one hundred and fifty dollars per month. Many racial, national and religious categories are located here, and the residences have been built at various periods of time. Thus in the considerable spread in rental value, age of structure, and ethnic background of the population, the basic elements for the production of either functional or homogeneous local units are present.

Figure 1 represents the areal pattern of rental values by a series of sub-areas. These sub-areas were constructed in accordance with the technique used in the W.P.A. Real Property Surveys of the late nineteen-thirties.[2] They are thus areas relatively homogeneous in rental value and race. Figure 1 shows a pattern of areas that would indicate the presence of five, or possibly six, natural areas, rental value alone considered. Further, in accord with accepted ecological generalization, these natural areas (belts of sub-areas homogeneous in rental value) are arranged in a gradient series of the clearest sort. Rental values rise, step by step, along a line leading from the southwest corner (near the zone in transition) to the northeast (a relatively new construction area overlooking Lake Washington).

---

*Reprinted from the *American Sociological Review*, 11 (August 1946), 423–27, by permission of The American Sociological Association.

[1] Paul Hatt, "The Relation of Ecological Location to Status Position and Housing of Ethnic Minorities," *American Sociological Review*, 10, 4 (August 1945). Briefly, this area is the central residential district of Seattle, Wash., lying between the central business district and the city boundary of Lake Washington.

[2] These data are from the W.P.A. Survey conducted in 1939. The technique described in the handbook involves the use of four or more contiguous blocks and surrounding blocks of the same rental value class-interval ($15) and either containing or not containing (depending on the area being constructed) more than 10 percent race other than white.

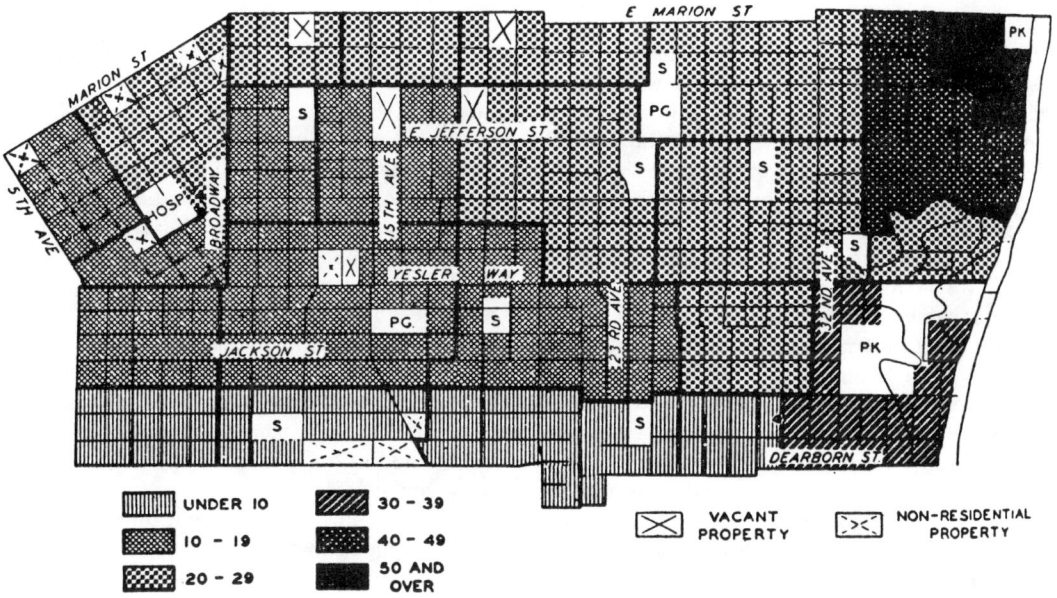

Figure 1—Mean rental value in dollars for tenant-occupied dwelling units, by sub-areas.

It would be possible to quarrel with the statement that these are natural areas especially since the belts of sub-areas may include racially divergent populations. Certainly the problem of bounding natural areas is a complex one. Hagood[3] and Leiffer,[4] among others, have shown the necessity of using several rather than one criterion. However, the most useful single criterion seems to be rental value, with race perhaps the next most important.[5] These constructions in Figure 1 represent one possible approach to the delineation of natural areas, conceived as homogeneous units.

Before examining the relation of the distribution of the ethnic pattern to the rental pattern apparent in Figure 1, the block data from which the sub-areas in Figure 1 were constructed must be considered. These are shown in Figure 2. Here the pattern of natural areas is by no means so clear as in Figure 1. Even the gradient pattern is less obvious. The outstanding characteristics of this distribution are the presence of a concentration of blocks low in rental value at the extreme southwest corner, and of high value blocks at the northeast corner. Thus, the general pattern of the gradient is preserved. However, this map allows the immediate identification of only two very small areas homogeneous in rental value. The remainder must be assigned to the category of *interstitial area*, i.e., an area showing mixed characteristics.

Other research in this area shows that this same mixed pattern is characteristic of any of several housing indexes when these are portrayed as block data.[6] The northeastern corner and the southwestern corner show concentrations of "good" and "bad" housing, respectively. In the case of every ecological index, there exists a very large middle ground of confused values.

The comparison of block data maps with those made by using sub-areas reveals the superiority of the former for most purposes of research. The latter smooth out considerable variations within the sub-areas. This is, of course, the inevitable result of using any measure of central tendency as a figure representative of a total distribution. Obviously, this procedure is not only sometimes

---

[3]Margaret Jarman Hagood, "Statistical Methods for Delineation of Regions Applied to Data on Agriculture and Population," *Social Forces*, 21, 3 (March 1943).

[4]Murray H. Leiffer, "A Method for Determining Local Urban Community Boundaries," *Publications of the American Sociological Society*, 26, 3 (August 1932).

[5]Cf. Homer Hoyt, *The Structure and Growth of Residential Neighborhoods in American Cities* (F.H.A., 1939); and Calvin F. Schmid, "Land Values as an Ecological Index," *Proceedings of the Pacific Sociological Society* (1940).

[6]Paul Hatt, *Natural Areas in the Central Residential District of Seattle.* Unpublished Ph.D. dissertation, University of Washington, 1945.

justified, but frequently necessary. In this case, however, the result seems to be to force data, with considerable violence, into a pattern which conforms to ecological theory. In short, even from the point of view of rental values alone, the natural areas deduced from Figure 1 are fictitiously homogeneous and intensify the gradient and natural area pattern; and this to the point of almost creating a reality where none exists. This is a point to which we shall return.

It is not adequate, however, in the light of the definition of natural area, merely to examine these distributions for homogeneity. We must examine the utility of the concept as descriptive of a functional symbiotic unity. In conformance with the literature, it seems safe to assume that such a symbiotic unity would be associated with clusterings of ethnic types. Thus the Chinese, the Japanese, the Negro and the Jewish populations, all have different needs and utilities. The distributions of these populations within natural areas should produce units organized around these utilities. In this area, however, such a procedure does not produce areas which are homogeneous either in ethnic type or in rental value. In an earlier paper, the distribution of ethnic types was seen to be markedly overlapping and so dispersed as to cover several rental value belts for each ethnic category.[7] However, again in the pattern of ethnic populations, the two phenomena noted in connection with Figure 2 were apparent. That is, the southwest corner appears as the most completely mixed area, and the northeast corner is occupied by an almost completely white and gentile population. Thus again, two natural areas are apparent, with the remainder classed as interstitial. With the gradient concept in mind, it might be more accurate to say that two points of highest frequency are observable. One, the highest frequency of "good" housing; the other, the highest frequency of "bad" housing. Upon inspection of these data, the concept of natural area, as it is understood in this paper, does not seem to fit the data with satisfactory accuracy.

However, this analysis is not interpreted by the writer to mean that the concept of natural area is totally inapplicable in such an area. Rather, it is interpreted to mean that the concept should be used cautiously and critically. The analysis of ethnic patterns referred to above revealed the existence of certain consistent and significant patterns in the housing indexes for each of the

[7]Paul Hatt, "Spatial Patterns in a Polyethnic Area," *American Sociological Review,* 10, 3 (June 1945).

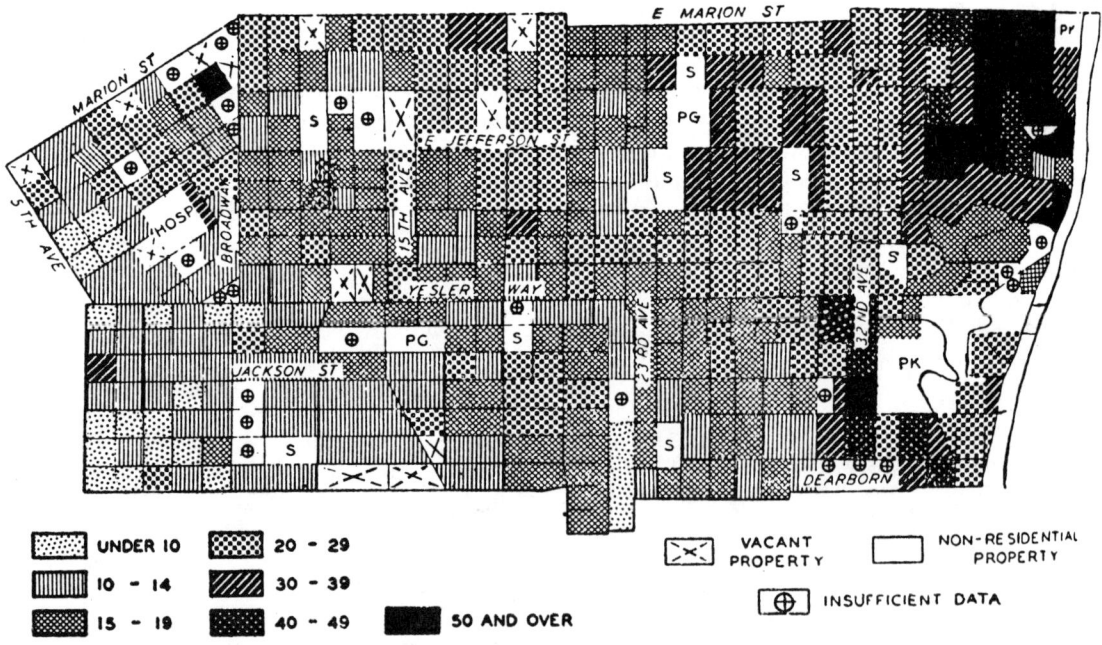

Figure 2—Mean rental value in dollars for tenant-occupied dwelling units, by blocks.

ethnic types. It further showed these to be different, in every case, from the character of the area in which the ethnic type lived. Significant spatial patterns for each ethnic category are present. This might indicate that in considering each ethnic type separately, and only incidentally considering the other populations in the area, that a clearer pattern of natural areas and a gradient could be established. The "moral order" of a given area would apply only to a specific population within that area and would show a greater validity than if applied to the total population of the area. The dangers in this latter view are clearly seen in the case of Seattle's Japanese population. This ethnic type inhabits an area characterized by the highest indexes of disorganization in the city. The indexes of disorganization for the Japanese, however, are very low. Clearly then, the Japanese within this area constitute a different community, and a different moral order, from that formed by other residents of the same area.

Since these divisions are never actually homogeneous within the categories of rent and race, and since these two categories disturb the homogeneity of areas when both variables are considered, it appears that "natural areas" can be found in this section of Seattle but it is also evident that *the* natural areas of "classical" ecology do not exist in this district. The distinction is made here between natural areas as logical, statistical constructs integrated with a plan for research (or administration) and the concept of natural areas as a series of spatial and social factors which act as coercive influences upon all who inhabit the geographically and culturally defined area.

This distinction is an important one inasmuch as it impinges upon what the writer regards as the dangerous practice of the reification of concepts so abundant in Human Ecology. If ecological phenomena are viewed as a frame of reference, and the concept of natural area as a construct within that frame, then the data presented in this paper could be assigned to natural areas. The location of the areas would depend entirely on the problem defined by the student. The fact that these areas might overlap would be a matter of no great consequence since the utility of these areas would be determined by the point of view of the investigator and the problem at hand. However, if the concept of natural area is reified, and these areas sought as actual entities, then their very existence is in doubt, if the data presented in this paper are at all representative of urban patterns.

This insistence upon ecological areas as being primarily ecologically organized units, and only incidentally as areas, implies a very critical attitude toward a theory of ecology as expressed by such writers as Quinn[8] and Hawley,[9] and the similar implications of Park[10] and McKenzie.[11] These views of ecology seem to imply the existence of a *real kind* of data, or a *real* series of *forces,* which then produce *real* areas and other ecological phenomena. The conclusion suggested by analysis of the data in this paper, however, is that ecology consists in a way of looking at data without assuming any inherent qualities of those data. The "orthodox" ecologists, by reification, set for themselves the task of delimiting human interaction into levels that are mutually exclusive; traditionally, *community* and *society,* the *symbiotic* and the *social.* This series of distinctions was convincingly demolished by Alihan's work.[12] Her work was a valuable contribution toward clarifying the concept of natural area. Orthodox ecology defined such an area as "natural" chiefly because it was unplanned. This again emphasized the familiar dichotomy of ecology, expressed here as the *unplanned* versus the *planned.* It should be unnecessary here after Alihan's careful work to point out the difficulties inherent in this dichotomy.

Obviously then the point of view in this paper is that the techniques of ecology can be fruitfully applied within any logically related framework defined by an investigator, without requiring that different levels of interaction be considered. No obeisance need be made to *the* natural areas of a city, but only those natural areas logically determined by the data and the problem need be constructed, used and defended. Conceived in this way, natural areas seem to offer an excellent framework within which any researcher may study a problem. In every case, however, it would seem wise to recognize that areal units of any kind are essentially a short cut substitute for case by case study, and caution in their use is indicated.

[8]James A. Quinn, "Culture and Ecological Phenomena," *Sociology and Social Research,* 25, 4 (March 1941), and "Ecological versus Social Interaction," *Sociology and Social Research,* 18, 6 (July 1934).
[9]Amos H. Hawley, "Ecology and Human Ecology," *Social Forces,* 22, 4 (May 1944).
[10]Especially in R. E. Park, "Human Ecology," *American Journal of Sociology,* 42, 1 (July 1936).
[11]R. D. McKenzie, "The Scope of Human Ecology," *American Journal of Sociology,* 32, 1, Part 2 (July 1926).
[12]M. A. Alihan, *Social Ecology* (Columbia University Press, 1938).

A Re-examination of Ecological Theory*

A. B. Hollingshead

I

Two recent articles focus attention upon the breach which lies between ecological theory and investigation.[1] Both Firey and Hatt took as their point of departure the examination of an ecological concept in a specific communal area; like many other ecological investigators they found their research data were only partially congruous with theory. To explain them they had to resort to nonecological concepts. This disconsonance between theory and investigation has been the subject for considerable debate, but to date little has been accomplished toward the re-examination of ecological theory in the light of the changed emphasis in general sociological theory that has become an accepted part of the funded knowledge of the social sciences since the early ecologists formulated their theories a generation ago. In the last decade, Davie, Alihan, Gettys, and Quinn—to mention only a few—have raised fundamental questions about the concepts, definitions, and terms ecologists have used. It is not the purpose of this paper to explore, interpret, criticize, or defend their works. Each can do this for himself if he chooses. On the contrary, three things are attempted: First, attention is called to the problem the human ecologist faces as a student of man; second, his method of attacking the problem is discussed very briefly; third, the necessity is stressed to take into consideration the factor of culture in the analysis of data ecologically minded sociologists have marked as their own.

The ecologist's problem appears to be posed by the unique position man occupies in the realm of living things: to wit, his place in nature as an animal and his role in societies as a possessor of culture. Biologists have demonstrated beyond question that man, the animal, is inextricably tied into the web of life along with all other creatures. Unlike them, though, he is connected with the nonorganic system of his sociocultural heritage. The fact that man is in this singular position has placed ecologists in a difficult position among the students of man. Ecologists by the nature of their interests are compelled to view man both as an animal in the system of nature and as a part of the sociocultural system he has developed through learning. Anthropologists, being either wiser or less versatile, have been concerned primarily with man as an entity in a culture. Sociologists have tended to study man in society. More recently some anthropologists and sociologists have been

*Reprinted from *Sociology and Social Research*, 31 (January–February 1947), 194–204, by permission of *Sociology and Social Research*.

[1] Walter Firey, "Ecological Considerations in Planning for Rurban Fringes," *American Sociological Review*, 11 (1946), 411–21; also "Discussion," by Amos H. Hawley, ibid., pp. 421–23; and Paul Hatt, "The Concept of Natural Area," ibid., pp. 423–27.

interested in studying man in his social and cultural contexts. Biologists, by way of contrast, have viewed man almost exclusively as an animal member of the vast, complex sphere of animate existence. Ecologists, as pseudo biologists, economists, and sociologists, have tended to view him as a competitive existence struggling for a place in the web of life along with other forms. Man, from the ecological viewpoint then, may be represented diagrammatically as holding membership in two spheres: the animate and the sociocultural. He is, in turn, connected with physical nature. The fact that he is connected with and influenced by these three types of factors has set him apart from all other creatures. Because the ecologists are interested in the nature of the interactions between man and man, man and other species, and man and the physical environment, their problem is highly complex; and to date has hardly been appreciated in all its implications. Moreover, the human ecologists have not developed methodological tools or a conceptual framework which squares theory with investigation. Only a few ecologists, however, appear to be aware of this situation, as Hatt has recently indicated.[2]

II

Park, who, it is generally agreed, founded the discipline, insofar as American sociology is concerned, saw man's unique position in relation to the total structure of reality dealt with by social scientists and biologists. He tried to solve the dilemma created by man's position in the scheme of things by assuming there are two basic processes that underlie and organize human life: competition and communication. He viewed competition as the organizing process which connects man to man, to animate and inanimate nature in the struggle for existence. Furthermore, he believed that men's ability as competitors distributed them spatially and functionally over the landscape. As such, competition was an abstract, impersonal, nonsocial process common to all living things. Communication, on the other hand, was looked upon as the human process that ties men into society.[3] Man as animal is organized competitively in the scheme of nature, but man as social being is organized cooperatively into groups through communication. By this reasoning Park envisaged two basic types of data in human affairs: the ecological and the social. Park as philosopher attempted to connect the one with the other through his version of the timeworn concept of a hierarchy of orders—the ecological, the economic, the political, and the moral. This type of reasoning assumed that man in society is basically, ever and always, man the primordial animal. Society and its concomitant culture are only excrescences, not integral parts of the animal man. This, it is believed, is why Park, with equanimity, stated the moral order (social order) is a superstructure built upon the competitive order.

The way in which Park stated his analysis led the "classical" ecologists, as Hatt has called them, to take as their point of departure the assumption that human beings are organized ecologically in the same impersonal, competitive way which appears to be the condition entailed in plant and nonhuman animal life. This observation may be verified in the writer's early thoughts on the subject,[4] in Alihan's thoroughgoing criticism,[5] and in Gettys'[6] and Quinn's writings.[7] Quinn's restatement of this point will be used here for illustrative purposes, because he has labored diligently to define the nature of the data the ecologically bent sociologist should focus upon, as well as to define the nature of the interactions which take place between the data to produce ecological structures. In two theoretical papers he specifically stated:

> Human ecology may now be defined tentatively as a specialized field of sociological analysis which investigates (1) those impersonal subsocial aspects of communal structure—both spatial and functional—which arise and change as the results of interaction between men through the medium of limited supplies of the environment, and (2) the nature and forms of the processes by which this subsocial structure arises and changes.[8]

[2]Hatt, op. cit., pp. 426–27.
[3]Robert E. Park, "Reflections on Communication and Culture," *The American Journal of Sociology,* 44 (1938), 187–205; also his "Symbiosis and Socialization: A Frame of Reference for the Study of Society," *The American Journal of Sociology,* 45 (1939), 1–25.
[4]A. B. Hollingshead, "Human Ecology," in R.E. Park (Ed.), *Principles of Sociology* (New York: Barnes and Noble, 1939), pp. 63–168; also "Human Ecology and Human Society," *Ecological Monographs,* 10 (1940), 354–66.
[5]Milla Aissa Alihan, *Social Ecology: A Critical Analysis* (New York: Columbia University Press, 1938).
[6]Warner E. Gettys, "Human Ecology and Social Theory," *Social Forces,* 18 (1939), 469–76.
[7]James A. Quinn, "The Nature of Human Ecology," *Social Forces,* 18 (1939), 161–68; "Human and Interactional Ecology," *American Sociological Review,* 5 (1940), 713–22.
[8]Quinn, "The Nature of Human Ecology," op. cit., p. 167; "Human and Interactional Ecology," op. cit., p. 721.

No attempt is made to infer what Quinn means here, but merely to point out that it is in the "classical" ecological tradition and appears to ignore or minimize the cultural factor in the organization and structure of human communities.

By casting their thought within this frame of reference, the "classical" ecologists have largely ignored the peculiarly human factor, culture, which differentiates human from plant and animal societies. The result has been their failure to realize that human relations—ecological, as defined above, as well as social, if you choose to speak in these terms—take place within a sociocultural matrix. Therefore, the first proposition to which the writer wishes to call attention is that *human activities are organized within a sociocultural framework, and ecological analysis needs to face this fundamental fact.* The assumption here is that competition in human society is regulated by the prevailing institutions, beliefs, values, and usages of the society rather than vice versa, which appears to be the position ecologists who have followed the Parkian tradition have taken. Nowhere has the anthropologist found man living as a beast, without culture. Furthermore, no proof exists that the sociocultural complex has evolved out of competition as it has been defined above. The assumption that competition is primordial and that consensus based on communication is derived from it may have speculative value, but empirically it appears to lack verification. That *a priori* assumption rested implicitly on the old belief that man was originally in a state of nature which was different in kind and degree from the state of society. In nature his relations were believed to be purely competitive in the sense described ideally by the plant ecologists and to a less extent by the animal ecologists. Out of this condition consensus was supposed to have evolved. In this scheme the principal function of consensus, or the moral order, was to limit competition.

This order of thought inevitably created the dilemma of the dichotomy of man in nature and man in society—a belief which arose from the dual assumptions that man in nature was presocial and struggled competitively for existence, and that society evolved out of this competitive condition when men developed consensus. In the process he did not lose his essential animality. Thus, it was easy to posit the hypothesis that there are two levels of phenomena in human affairs: the subsocial, or ecological, which is essentially competitive-impersonal, and the social, which is communicative-cooperative and personal. This is believed to be a case where oversimplified abstraction has elevated an assumption into a universal, all-pervading arbiter which distributes people and things in time and space. In the process the verbal symbol, competition, has become reified, and the "classical" human ecologist has worshipped before its shrine.

The thesis presented here is that upon close analysis impersonal competitive relations as defined by classical ecologists are so intertwined with personal cooperative ones it is only by abstraction that we are able to separate the one from the other. Everywhere, behind the apparently impersonal activities the ecologist has assumed were regulated by this semimystical mechanism, competition, lies a people's sociocultural system which organizes and regulates the form of the division of labor, man's relation to man, to things, and to nature. From the evidence accumulated by anthropologists, sociologists, economists, and historians, it is believed safe to say that men do not compete as abstractions; they compete and communicate as persons in a sociocultural system; as such they are foci of a complex of values and usages which limit and direct the form of their interrelations with one another, whether these relations are personal or impersonal. Therefore, our second proposition is this: *Cultural values and usages are tools which regulate the competitive process.* Once this principle is grasped, the ecologists, instead of studying spatial distributions and structures *per se,* can go beyond the mere manifestation of spatial and structural distributions and analyze out of his data the factors which have regulated the processes that produced the transsubjective structures and movements observable in space and time. To illustrate this point, two well-known studies are cited: Stouffer's[9] analysis of spatial movement in Cleveland and Hawley's[10] paper on the distribution of urban service institutions. Stouffer found not only that people move short distances, and that their movements can be predicted within calculable limits, but also that the factors of race and ethnic affinity had to be taken into account. In effect, he came to the conclusion, "a redefinition of opportunities to take direct account of the ethnic," race,

---

[9] Samuel A. Stouffer, "Intervening Opportunities: A Theory Relating Mobility and Distance," *American Sociological Review,* 5 (1940), 845–67.

[10] Amos H. Hawley, "An Ecological Study of Urban Service Institutions," *American Sociological Review,* 6 (1941), 629–39.

and occupational factors entailed in spatial movement was necessary for each type of group one might examine. He states: "Members of an ethnic group might fall into two types in their movement, namely those who follow the trend in movement within their nationality group and those who deliberately seek to dissociate themselves from the group."

This statement would apply equally well to families where the head changes occupational connections which involve moves from one city to another, or where people move from the farm to the city. To handle the problems involved in such cases, it is necessary to define opportunity in each class of case in terms of the cultural influences inherent in the situation. When spatial mobility is attacked in this way, the problem for investigation is not just to find out how movements occur, but what factors condition the movement. To get at and evaluate such data, the migrant must be studied in a sociocultural context if we are to discover "the pushes and pulls" in migration as they impinge upon the persons involved in the process. It would appear that these are social, cultural, and personal, rather than "impersonal," "subsocial," and "competitive," as ecologists have been prone to assume.

Hawley, on the other hand, studied the distribution of urban service institutions in relation to population. In his preliminary analysis of "1,780 cities ranging in size from 2,500 to 1,000,000 population," he discovered there was no consistent and direct relation between population and institutional volume. In this phase of the study he saw that there are differences among cities with respect to their institutional structures which are not accounted for by variations in size of population. This observation compelled him to assume that different types of populations live in different ways, and that the way in which a population lives is reflected in the institutions which serve it. He then explored this possibility and found the relation between institutional volume, location, and size of city insofar as they are related to income, industrial occupation, age, sex, nativity, and race. It would appear to be worth noting that Stouffer and Hawley came to the point where they recognized, first, that cultural factors were the dynamic elements in the processes they were studying; and, second, that the analyst had to understand the cultural factors operating in the world of reality which lay beyond his statistics if he would understand the meaning of his figures. Had they started with a clear understanding of the role of culture as a competitive tool, as did Hughes in his study of the distribution of English and French-Canadians, we might have been given a different insight into the entailed processes.[11] The "classical" ecologists by definition ruled the cultural factor from their theoretical purview. However, they recognized its existence when they turned from speculation to investigation. Therefore, as a consequence, there has developed in the literature a cleavage between speculative assertion and investigative results. Alihan thoroughly discussed this problem without reaching a satisfactory conclusion, possibly because, as Gettys has stated, "to examine is not to explain." The explanation of this impasse, it is believed, inheres in the history of the development of scientific discipline. Scientific thought, in its early phases, is characterized generally by philosophical speculation on the relations transsubjective data bear to one another that is not justified by the facts. As research continues, new facts are discovered and new concepts are formulated which challenge the old. Thus, scientific theory develops from vague speculation on the cause-effect relation between variables, to precise statement, either in the form of verbal or mathematical propositions. In this process each scientist needs to check the theoretical conceptions he has learned from his predecessors and colleagues against the facts as he observes them.

A third of a century ago, when the loosely connected ideas that were formulated into the now "classical" body of ecological theory were impinging upon Park's thoughts, the concept of the integral relation between culture, society, and human behavior was still germinal. At that time most sociologists were interested in social origins, social evolution, social welfare, social reform, and like philosophical questions. What must not be forgotten is that the body of literature we now speak of as "ecological" in the "classical" tradition and which issued mainly from the pen of Dr. Park and his students was a product of their thoughts only in a secondary sense. Behind them lay the dominant ideas of the eighteenth and nineteenth centuries, just as their experience and the experiences of sociologists since about 1915 lie behind the present empirical investigators.

III

Before a body of ecological theory can be formulated where conceptualization is congruous with

[11]Everett C. Hughes, *French Canada in Transition* (Chicago: University of Chicago Press, 1943), pp. iii +227.

investigative results, ecologists must recognize explicitly that culture and socially organized behavior forms, transmitted through learning subsequent to birth, are the elementary factors which differentiate human from animal society. Culture, therefore, is a surrogate among men for the structural and instinctive mechanisms, which have been set forth as an explanation for the type of collective activities observed among animal societies. If this fundamental assumption is taken as a point of departure in ecological analysis, the dilemma of trying to bridge the gap between man in nature and man in society is dissolved. Furthermore, if we assume that competition is the basic organizing process in the formation of ecological structure and distributions, as Park and other orthodox ecologists have taken for granted, it operates perforce within rather than outside a people's sociocultural system. If these two points are taken on faith, as all assumptions must be, then instead of talking about "ecological substructures" and "social superstructures," the ecologist can go into the field and analyze his data without worrying whether his research is "ecological" or "social." In this sense, his research will be focused upon the study of human distributions and structures which have been produced by the competition of man with man, with other species, and with physical nature; but ever and always he must keep in mind the cardinal principle that competition occurs within the purview of a sociocultural matrix.

If this is accepted as a point of departure in ecological research, the next proposition would appear to be that cultural usages and values are competitive tools which operate in the organization and distribution of institutions and people in the community. This fact was recognized by Bossard and Dillon in their study of the distribution of divorced women in Philadelphia.[12] Almost a decade ago the present writer tried to make it explicit in his study of succession in land ownership in Nebraska.[13] Hughes[14] has perceived clearly that ecological relations take place within the matrix of the sociocultural systems of the English, the more or less indigenous Canadians, and the French. He has emphasized that the French family system has enabled the French to survive as a society and a culture when the French are in competition with the English for the land. The reason for this is that, put categorically, the French desire land, the English social status. The Frenchman's lack of social ambition, coupled with his desire for security, enables him to win out. In the rural succession cycle, land ownership is the first step in the introduction of the French sociocultural system into an area. However, when the English industrial system invades an area dominated by the French-Canadian culture, a different set of competitive conditions obtain. In the latter case, the French-Canadian folk have management, capital, and machinery shipped into their villages. There it is put into operation by the use of their labor under the direction of English managers and American technicians. In this industrial structure there is no direct competition between French and English workers, simply because the English are the managers and the French the laborers. Likewise, there is little or no competition between the English and the Americans, as each group pursues a specialized function. The French do not compete with the Americans or the English, since Frenchmen are not accorded any ability as engineers, technicians, or mechanics. The French compete with Canadians for minor office jobs; but here stenographers, clerks, and department heads represent the best French families in the community, while the white-collared Canadians come from a lower social status.

Hughes' data show succinctly how a person or a group's position in the sociocultural structure controls its competitive opportunities. Within the worker group there are other controls of a cultural nature which limit the person's economic behavior. Hughes has reported to the writer in conversation that there are some families in this town who are considered a bad lot; therefore, not a single member of those in this category will be hired by the industry. There are other families whose members are looked upon as reliable, and any member who asks for a job is put to work as soon as possible. Among this group the members work on one another to see that they do their work and behave themselves according to the expectancies set by the company and the family. In both cases the French family system acts as a competitive instrument to help the person survive either in or out of the industry.

A few additional illustrations, from a Middle Western community of some 6,000 population, will have to suffice to give substance to the point on the role of cultural values as competitive tools. In this

---

[12]James H. S. Bossard and Thelma Dillon, "The Spatial Distribution of Divorced Women—a Philadelphia Study," *American Journal of Sociology,* 40 (1935), 503-07.

[13]A. B. Hollingshead, "Changes in Land Ownership as an Index of Succession in Rural Communities," *American Journal of Sociology,* 43 (1938), 764-77.

[14]Everett C. Hughes, op. cit.

community there are 80-odd Hungarian families at the present time. The first ones were brought in by a mining company in 1910 as strikebreakers. As such, they began their residence under the dual handicap of being "scabs" and "foreigners." As a consequence, "Hunkey" is synonymous with everything that is undesirable. Currently, the older generation has died off almost completely, and the third generation is reaching adulthood, but many of the old attitudes toward the "Hunkeys" persist in the non-Hungarian population. What is more important from the ecological viewpoint is that these values have acted as controls to channelize the Hungarian's occupational choices; and, in turn, they have limited his competitive ability to acquire pecuniary and favorable status values. A few prosperous families have sent their children away to college, but these young people have not returned to the community. Last summer the writer was talking with the postmaster in this town; he asked him about "X," who had made Phi Beta Kappa at the state university and had been a star member of one of the recent national intercollegiate championship basketball teams. The postmaster remarked, "You know, it's funny, a few years ago X came in here after his university had voted him its most valuable athlete. I asked him why he didn't come back home to coach. X said, 'Around here, people never forget I am a "Hunkey." There's only one thing I'm afraid of.' He stopped and I asked him in my dumb way, 'What's that?' and d'you know what he said?" "Why no," the writer answered. " 'To be a Hunkey.' For a while I couldn't figure out what he meant, but then I caught on."

Prejudice against the Hungarians was so strong until the war labor shortage that Hungarian girls experienced difficulty in obtaining positions in offices. In fact, until the end of 1942, only two Hungarian girls had been able to obtain positions as secretaries. In one case, the priest talked a prominent layman into hiring a devout girl whose two older sisters had become prostitutes, after finishing high school, because they could not obtain work during the depression. This appointment created considerable gossip directed toward both the priest and the girl's employer, relative to their moral relations with the girl. No evidence could be found by careful investigation to substantiate the gossip. It is believed the gossip was the community's technique of draining its aggression toward the Hungarians. In the other case, an elected official appointed the daughter of the leader of a prominent veteran's organization as a subdeputy in his office, in return for the Hungarian vote. This, too, was accompanied by a wave of stories against the politician. To date no Hungarian has entered the professions, or found work in the banks, trust companies, and leading businesses of the community. To summarize the situation, the attitudes of the old American stock toward the Hungarians have limited their job opportunities largely to unskilled and semiskilled occupations. To be sure, a patent element in this situation is the real and attributed cultural usages and values of the Hungarians.

By way of conclusion, two ideas should be stressed: First, ecological theorists appear to have overemphasized the role of competition as a subsocial phenomenon in the organization of economic functions and the physical distribution of population and services; second, they have more or less ignored the influence of culture and social evaluations on the phenomena the ecologist studies. The basic assumption our theorists have followed is that competition is all-powerful and operates according to laws of its own which lie outside the realm of the sociocultural complex. It is believed that we must substitute investigation into the role culture plays in the ecological process for this canon. Once this position is taken, we soon see that men live within a sociocultural construct which organizes and controls their activities. This is a departure from the position posited by the early ecological theorists who envisaged an abstract ecological man motivated by physiological appetites and governed in his pursuit of life's goals by competition with others who sought the same things he sought because physiologically they were like him. To accept the position advocated, we must assume that the sociocultural complex of a people is a surrogate for the drives and physiological specializations found in the lower forms of life. The time has come when we should study the influence of the cultural factor on the phenomena sociologists have defined as ecological. This calls for a withdrawal of the defense of the old position and an exploration of the newer one.

Ecological Correlation Reexamined: A Refutation of the Ecological Fallacy*

Donald J. Bogue
and Elizabeth J. Bogue

Introduction

A quarter of a century ago (1950), W. S. Robinson called into question the validity of a vast amount of social research that had been performed using correlations in which the unit of statistical measurement was groups of persons. Robinson used the term "ecological correlations" to refer to measures of association derived in this way. He contrasted these measures with "individual correlations" for which persons are the unit of measurement and declared:

> The relation between ecological and individual correlations which is discussed in this paper provides a definite answer as to whether ecological correlations can validly be used as substitutes for individual correlations. They cannot. While it is theoretically possible for the two to be equal, the conditions under which this can happen are far removed from those ordinarily encountered in data. From a practical standpoint, therefore, the only reasonable assumption is that an ecological correlation is almost certainly not equal to its corresponding individual correlation.
>
> I am aware that this conclusion has serious consequences, and that its effect appears wholly negative because it throws serious doubt upon the validity of a number of important studies made in recent years. The purpose of this paper will have been accomplished, however, if it prevents the future computation of meaningless correlations between the properties of individuals.[1]

Menzel (1950), Duncan and Davis (1953), Goodman (1953, 1956), and Shively (1969) have done much to clarify the problem and to point to conditions under which valid inferences concerning the behavior of individuals can be made from grouped data. The two articles by Goodman are especially innovative. Nevertheless, Robinson's curse upon ecological correlation remains, and the taboo on generalizing about the behavior of individuals from grouped data is obeyed almost religiously by all but unsophisticated researchers.

---

*Reprinted from Donald J. Bogue and Elizabeth J. Bogue, *Essays in Human Ecology I* (Chicago: Community and Family Study Center, University of Chicago, 1976), pp. 1–42, by permission. [Shortened]

[1] William S. Robinson, "Ecological Correlations and the Behavior of Individuals," *American Sociological Review,* 15 (June 1950), 357.

Table 1. The Individual Correlation Between Nativity and Illiteracy for the United States, 1930 (population 10 years old and over)*

Literacy Status	Total	Nativity	
		Foreign-born white	Native white and Negro
Illiterate	3918	1304	2614
Literate	93354	11913	81441
Total	97272	13217	84055
		Percent Distribution	
Illiterate	4.0	9.9	3.1
Literate	96.0	90.1	96.9
Total	100.0	100.0	100.0

Source: W. S. Robinson, "Ecological Correlation and the Behavior of Individuals," ASR, 15 (June 1950), 351-57.
*Numbers are in thousands
Measure of association: $\phi = .118$

The procedures described here are intended to help reestablish the validity of ecological correlation as a general tool for making inferences and testing hypotheses about the behavior of individuals. Instead of Robinson's outright prohibition, this study will propose a range of conditions within which valid inferences can be expected.

Robinson's examples of the "ecological fallacy"

Robinson arrived at his set of conclusions by submitting two particularly devastating examples employing the (then) forty-eight states as units of observation. The first was the correlation between percent of population illiterate and percent of population foreign born; the second between percent illiterate and percent Negro. The same data employed by Robinson (also used by the Duncans and Goodman) will be re-analyzed here to illustrate the invalidity of Robinson's downright prohibition against making inferences about individuals from grouped data.

The Pearsonian correlation for the data in the first example, i.e., the ecological correlation, is $-.526$. This implies that there is an inverse relationship between foreign birth and illiteracy. This, of course, is contrary to the individual correlation that existed at the time. (Educational attainment of immigrants to the U.S. between 1900 and 1930 was lower than for the native white population, and therefore a positive individual correlation should be expected between foreign birth and illiteracy.) Table 1, which reports the data for individuals, confirms this expectation. The individual correlation (ϕ)—explained below—computed from this table is .118. Robinson submitted this example as a climactic demonstration of the futility of trying to predict individual behavior from grouped data.

In his second example, which is less dramatic, Robinson found highly discrepant results between the ecological and the individual correlations between race (Negro, non-Negro) and illiteracy. The ecological correlation was .773 when states were used as units of analysis, whereas the value of the individual correlation was only .203. Table 2 provides the data for individuals.

By procedures described below, both of Robinson's examples of ecological atrocity are reexamined to yield results amazingly close to the corresponding individual correlations.

Methodology for comparing individual and ecological correlations

Because the variables of literacy, nativity, and race are qualitative (categorized) variables, measures of their association ("correlation") call for one of the statistical measures of association for cross-classified data. Robinson used the fourfold-point correlation (ϕ coefficient) to measure the individual correlation from cross-classified data. The observations of ecological data, by contrast, are quantitative (percent illiterate, percent foreign-born, percent Negro) for each of the states. For ecological correlation Robinson specified conventional bivariate product-moment (Pearsonian) correlation in which observations for areas (forty-eight states) are used to form the sums of squares and cross-products required to calculate r.

Goodman demonstrated (1953, 1956) that the correct test of how well an ecological correlation approximates the individual correlation is *not* to compare the ø coefficient (qualitative) with a Pearsonian (quantitative) correlation, as Robinson did. He pointed out that there is no mathematical necessity for the values of Pearsonian correlation coefficients computed from grouped data to be exactly equal to the ø coefficient, even if the ecological and the individual relationships were identical. He argued that the correct test is a measurement of how precisely it is possible to use data (observations for groups) to predict the (unknown) frequencies of the interior cells of the statistical table (Table 3) from which the corresponding individual correlation would be computed, given only data for the marginal totals. Stated in terms of Robinson's examples, the question becomes one of discovering how precisely it is possible to estimate the interior cells of Tables 1 and 2 using only areal data to measure the association between the traits of literate—non-literate, Negro—non-Negro, native-born—foreign-born, while knowing only the correct values for the row and column totals for these tables.

In the presentation which follows, this criterion is also accepted as valid. (It is possible to use ecological data to estimate the interior cells of polytomous cross-classifications [R × C tables]. Once these cells have been estimated, individual correlation may be computed by using any measure of association the researcher may choose. The only restrictions are: (a) that the categories of the R × C table match exactly those of the table for which the observed data for individuals are tabulated, and (b) that after the interior cells of the R × C table have been estimated by ecological procedures, the same statistical tool be used to measure the association between the traits for the ecologically estimated data as for the observed individually tabulated data.)

Use of bivariate ecological regression for estimating individual correlations

In 1956 Goodman demonstrated that data for aggregates could be used to write the equation for a valid relationship between individual and grouped data. Where this equation is valid, data for ecological aggregates may be used to estimate its parameters and thereby make inferences about the behavior of individuals. Using Robinson's example of the correlation between illiteracy and race, he reasoned (see Figure 1):

> The proportion, y, of individuals in the Negro-white population who are illiterate may be written as $y = xp + (1 - x) r$, where x is the proportion in the population who are Negro, p is the proportion of Negroes who are illiterate, $(1 - x)$ is the proportion in the population who are white, and r is the proportion of whites who are illiterate. Thus $y = r + (p - r) x = a + bx$, where $a = r$ and $b = p - r$. Hence, if different populations or areas are considered where the proportion p is the same for each of these populations, and also the proportion r is the

Table 2. The Individual Correlation Between Color and Illiteracy for the United States, 1930 (population 10 years old and over)*

Literacy status	Total	Color	
		Negro	White
Illiterate	3918	1512	2406
Literate	93354	7780	85574
Total	97272	9292	87980
		Percent distribution	
Illiterate	4.0	16.3	2.7
Literate	96.0	83.7	97.3
Total	100.0	100.0	100.0

Source: W. S. Robinson, "Ecological Correlation and the Behavior of Individuals," ASR, 15 (June 1950), 351-57.
*Numbers are in thousands.
Measure of association: ø = .203.

Table 3. Format for Tabulating Data for Computation of Individual Correlation ∅

Trait A (illiteracy)	Trait B (foreign birth) Absent	Present	Total
Present	a	b	a+b
Absent	c	d	c+d
Total	a+c	b+d	a+b+c+d

Note: The row and column totals are known from aggregate tabulations. Values of the interior cells a, b, c, d are unknown. If one of these cells can be estimated, the remaining cells can be determined by subtraction from the known totals.

$$\emptyset = \frac{bc - ad}{\sqrt{(a+b)(c+d)(a+c)(b-d)}}$$

same for these populations, then there will be an exact linear relationship, $y = a + bx$, between the values of y and x for the different populations (assuming that not all the values of x are equal), where the slope will be $b = p - r$, and the y-intercept will be $a = r$. This straight line could be used to determine $r = a$ and $p = b + a$.

In practice, the actual values of p and r will not be constant, but it may be the case that the average $E(p \mid x)$ of the values of p, for populations with the same proportion x of Negroes, is constant [i.e., $E(p \mid x)$ is the same for different values of x] and the average $E(r \mid x)$ of the values of r, for populations with the same x value, is also constant. In this situation, the main assumption of linear regression analysis, $E(y \mid x) = A + Bx$, holds true, where $A = E(r \mid x)$ and $B = E(p \mid x) - E(r \mid x)$. Thus, standard methods of linear regression can be used to estimate A and B.[2]

Figure 1 uses Goodman's notation to provide a numerical illustration using Robinson's data on the association between race and illiteracy.

Stated more simply, Goodman suggested the use of a conventional bivariate regression equation to estimate the value of one variable (y = percent illiterate) given pre-selected values of the other variable (x = percent Negro). This regression equation is computed from grouped data where areas (states) are the units of observation. This procedure is valid if the slope of the line and the y-intercept that are estimated from aggregate data, by the procedure specified in the next section, are both equal to those that would be established from individual data. Goodman then solves the problem

[2] Leo Goodman, "Some Alternatives to Ecological Correlation," *American Journal of Sociology,* 64 (May 1956), 612.

by in effect substituting into this equation values of $x = 0$ (absence of Negroes) and $x = 1.0$ (all Negro). In short, the solution is arrived at by substituting extreme values into a regression that has been computed from more "normal" or "actually occurring" values. It asks, "What would be the value of the dependent variable in two extraordinary populations, not observed, where (a) the independent variable is completely absent, and (b) the independent variable is at its maximum possible value?" This is a form of extrapolation, and hence is subject to the assumptions and dangers of any extrapolation beyond the range of actual observation. Being cognizant of this, Goodman warns repeatedly that this procedure is valid only if:

(a) The relationship is genuinely linear (extrapolation of curvilinear relationships is far more hazardous than extrapolation of linear ones), and

(b) The regression equation validly expresses the unknown relationship that is being estimated. (The estimated values of p and r derived by substituting into an equation derived from ecological data are reasonably close to those that may be computed from the data for individuals as illustrated in Figure 1. This requires that the unknown values of p and r are the same within each aggregate or the regression provides valid estimates of these parameters.)

By applying the regression/extrapolation procedure, based on the ecological data for the nine geographical divisions, Goodman obtained a value of .38 for the fourfold-point correlation between race and illiteracy. Inasmuch as the ecological (Pearsonian) correlation was .95 and the estimated ("true") correlation based on actual census data for individuals was .20, Goodman concluded that the procedure, though not exact, was a marked improvement over previous methods. He concluded, "These methods are simple, but they cannot be expected to lead to an accurate estimate as those obtained from relevant data on individual behavior."[3]

A graphic illustration of the ecological regression argument

Reference to Figure 1 will assist in clarifying the argument supporting the Goodman principle of ecological regression. The x-axis of this figure represents the percentage of the total population that is Negro (or foreign-born white). The y-axis represents the percentage of the population that would be expected to be illiterate under varying percentages of Negro or foreign-born composition.

[3]Ibid., p. 615.

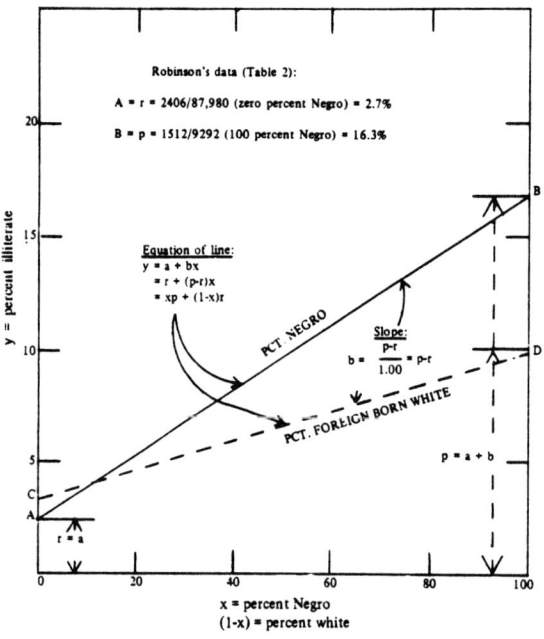

Figure 1—Regression of percent illiteracy on percent Negro (foreign-born) derived from data for individuals (Goodman's notation and Robinson's data).

A separate regression line is plotted for each of Robinson's two examples, using Goodman's notation (the variable x has been assigned to the percent Negro). The solid line AB of Figure 1 is based upon the data provided for individuals in Table 2. Dotted line CD is based upon the data of Table 1. From these tables we are able to establish only the *endpoints* of these lines, and assume a linear distribution between the endpoints. When zero percent of the population is Negro, we expect only the illiteracy rate that is characteristic of the white population. When zero percent of the population is foreign-born, we expect only the illiteracy rate that characterizes the native-born population. When 100 percent of the population is Negro, we expect the much higher rate of illiteracy that is characteristic of this group. A similar statement holds for the foreign-born.

This regression solution assumes that the proportion of illiteracy is everywhere the same for foreign-born and Negro subgroups—a condition which Goodman indicates should be true if the regression approach is to be valid. If native-born population is progressively substituted for foreign-born, a linear decline in illiteracy would be expected, and would culminate in 2.73 percent of illiteracy when zero percent of the white population is foreign-born. Thus, a cross-tabulation for individuals provides all the information needed to construct an ecological regression equation of the type described by Goodman. Therefore, *if the relationship between foreign birth and illiteracy were exactly the same in every state, and if foreign birth were the only factor that determined illiteracy, then the points of a scattergram of percent illiteracy for each state plotted against the percent of the total population that is foreign-born would all fall exactly on line CD.* If the information in Table 1 were not available, under the above conditions we could reproduce it exactly by regressing the percent illiteracy in each state against the percent foreign-born and then substituting into the regression the values of 100 percent and zero percent foreign-born. Thus, when the conditions specified in Goodman are met, one may pass easily from individual regression to ecological regression and from ecological regression to individual regression.

The hazards of ecological correlation arise because the slopes of the regressions computed from grouped data often do not conform to the slopes of the lines derived from individual data. Goodman's bivariate procedure provides no protection against these hazards. Shively comments, "In the decade since his [Goodman's] method was presented, it has remained virtually unused in published research. This has probably been due to its propensity to produce badly biased estimates when its basic assumption is violated.[4]

Assumptions which can invalidate bivariate ecological regression

There are three important assumptions of the simple bivariate regression described above, each of which is capable of distorting or invalidating the procedure outlined by Goodman. Fortunately, however, there is a methodological corrective for each of them.

1. *Multiple independent variables.* Simple bivariate regression assumes either (1) that no other independent variables are present to affect the regression of dependent variables upon the independent variable, or (2) that the two or more independent variables which are correlated with the dependent variables are themselves uncorrelated. Both of these assumptions are highly unrealistic in most "real life" ecological regression problems. For example, it would be absurd to presume that illiteracy in 1930 was a function solely of foreign birth. Robinson's own data show that it was also highly correlated with race. Moreover, there is ample reason to suspect a strong inverse relationship between these two variables: where highly illiterate Negroes were concentrated (the South), the foreign-born population tended to be highly literate. Where illiterate foreign-born populations were concentrated (the Northeast and the North Central states), the level of literacy among Negroes tended to be higher.

When confronted with a prediction problem wherein two or more variables are independently capable of explaining variation in the outcome to be predicted, it is conventional to use multiple-regression instead of bivariate regression. This permits a measurement of the regression of the dependent variable upon each of the independent variables individually, while "holding constant" the regression effects of the other independent variables. For example, a multiple regression equation of the proportion of illiteracy that simultaneously involves both the proportion of the population that is Negro and the proportion that is foreign-born should produce estimates superior to those that would result from a single bivariate relationship of either, leaving the other uncontrolled.

This study takes the position that the researcher who wishes to make use of ecological regression in empirical research should *assume* that several variables are simultaneously influencing the behavior of the dependent variable, and that the multiple linear regression method should *always* be utilized unless there is special evidence that it is not needed. In theory, if the researcher is able to introduce indicators of other independent variables (significantly correlated with y) into a multiple regression equation, the resulting estimates of the individual correlation will never be worse and usually will be superior to the results of the simple bivariate regression. Thus, multiple regression should be the standard approach to ecological regression analysis.

2. *Unequal size of ecological units.* Geographical units used in arriving at the regression equation are treated by simple regression procedures, as if each unit is of the same importance as every other unit. This, of course, can be highly unrealistic. In studying illiteracy and foreign birth do we give equal weight to North Dakota and New York? to Wyoming and Pennsylvania? to Vermont and Illinois? The size and direction of the individual correlations (Tables 1 and 2) are contributed primarily by the data for the more populous states, whereas the simple ecological regression procedure gives equal weight to each state, irrespective of size. In other words, Tables 1 and 2 are "weighted" totals of ecological units. The larger ecological units contain more information for arriving at a correct estimate of $E(p \mid x)$ than the smaller.

A simple corrective, of course, is to use *weighted regressions,* rather than unweighted regressions, in making ecological correlation estimates. This permits the various ecological units of observation to receive a weight that is proportional to their influence in arriving at the individual correlation with which the regression estimates are to be compared.[5]

The position taken in this paper is that weighted regressions are the correct procedure, as a matter of fundamental theory. Unweighted regressions are allowable only under the very special circumstances where they would yield the same results as

---

[4] W. Phillips Shively, "Ecological Inference: The Use of Aggregate Data to Study Individuals," *American Political Science Review,* 63 (December 1969), 1190.

[5] See Arthur H. Robinson, "The Necessity of Weighting Values in Correlation Analysis of Areal Data," *Annals of the Association of American Geographers,* 46 (1956), 233–36; and Edwin N. Thomas and David L. Anderson, "Additional Comments on Weighting Values in Correlation Analysis of Areal Data," *Annals of the Association of American Geographers,* 55 (September 1965), 492–505.

a weighted regression. The "standard" regression methods for ecological regressions are the use of weighted regressions; they should be used because they conform more precisely to the specifications of the regression model. Modern computers make it easy to comply with this mandate.

3. *Clustering of ecological units into types.* Instead of possessing a single homogeneous regression (single set of mean values, variances, and regression coefficients), the dependent and the independent variables in a given set of ecological units may tend to be related to each other in *clusters* of units, each of which has a fairly homogeneous regression relationship different from that of the other clusters, with unique combinations of mean values, variances, and regression coefficients. For example, for historical reasons, one might expect a lower percentage of Negroes in the Northeast region of the U.S. to be illiterate than would be the case in the South. Conversely, one might expect much more illiteracy among the foreign-born population in the Northeast and North Central regions than in the South and the West, because the heavy migrations from eastern and southern Europe flowed primarily to big industrial cities.

The solution to this problem is a *stratification* procedure. Because all of the ecological units in the universe of discourse contribute to some extent, and in an unknown way, toward the individual correlation, all of them should be employed in arriving at the ecological regression estimate. If there is reason to suspect heterogeneity among the ecological units, *an effort should be made to stratify them according to some criterion* which is believed would make the regression relationships within each stratum more homogeneous. The grouping of observations for states into regions would be one example of stratification. Alternatively, they could be classified according to type of economy, urban and rural composition, or any other criterion that would maximize internal homogeneity. (It would, of course, be improper to stratify according to any principle that would arbitrarily determine the elements of the regression equation.) Once the strata are established, a separate ecological regression equation is prepared for each stratum. Using these equations, an estimate is made of the individual relationship (estimates of the interior cells of Table 3) for each stratum. Each class of interior cells is then accumulated across strata to obtain a final estimate of the individual correlation in the total set of ecological units. If the stratification significantly reduces the variances of p and r, and/or their covariance, the resulting estimates will tend to be closer to the unknown individual effects than one made for all ecological areas without stratification. If stratification by a particular set of criteria has no effect, the resulting estimates will, in general, be no worse than those arrived at by using all ecological areas in a single regression. Although stratification uses up some degrees of freedom, it can be a useful way of imposing control over congeries of variables for which more explicit data are unavailable.

SUMMARY. The argument of this section may be summarized as follows:

(a) Simple unweighted linear regression does not realistically conform to the specifications of the research problems where ecological regression might usefully be employed. It is deficient on three counts: multiplicity of independent variables, unequal size of ecological units, and the tendency for ecological units to be clustered into types or strata, within each of which regression tends to be more homogeneous than among all units as a pool.

(b) An effective antidote or corrective for each of these is available.

(c) Robinson gave no attention whatever to any of these three deficiencies.

(d) The researcher who wishes to make use of ecological regression methods to arrive at predictions of individual behavior should assume that all three of these factors are operating simultaneously.

INFERENCES. Instead of becoming discouraged and defeatist at the apparently nonsensical results of bivariate ecological correlations which sometimes emerge, even with Goodman's reinterpretation of the problem in terms of regression, a much more optimistic view is adopted here. *For many situations where the researcher wishes to employ ecological correlation as an alternative to individual correlation, valid results can be obtained by simultaneously applying the standard corrective for the three major weaknesses noted above. This calls for the substitution of stratified weighted multiple regression for simple bivariate regression. Moreover, when this recommended procedure fails to yield valid results, there is abundant warning that this has occurred.*

If this proposition is correct, then stratified weighted multiple regression should be acknowledged to be the "normal" or "standard" research design for practical efforts to predict individual behavior from ecological or aggregated data. The next section of this paper is dedicated to an empirical test of the above proposition.

A stratified weighted multiple regression re-analysis of the Robinson problems

Robinson's coup de grâce example of illiteracy among the foreign-born is a highly instructive problem on which to test the assertion arrived at above, for it involves in rather extreme form almost all of the pitfalls that can be encountered in this endeavor. By reanalyzing Robinson's example in detail and reporting the results of several alternative solutions, it is possible to appreciate both the power and the limitations of the procedures proposed.

It is not difficult to state several hypotheses concerning the factors which may be expected to have a high correlation with illiteracy. A study of a dozen variables was made from the data for the 1930 census, from which Robinson drew his example. Six variables were identified which appeared to have the greatest independent explanatory effect in explaining what proportion of the population is illiterate at the simple bivariate level. They were:

X_1 percent of population ten years old and over that is foreign-born white
X_2 percent of population ten years old and over that is Negro
X_3 percent of population ten years old and over that is of Mexican ancestry
X_4 percent of population ten years old and over that is American Indian
X_5 percent of population living in urban areas
X_6 percent of population that is under five years of age

Using multiple regression equations, all six of these variables will be used simultaneously to predict the dependent variable,

Y = percent of population ten years old and over that is illiterate.

REGIONAL STRATIFICATION. It is suspected beforehand that there will be substantial regional variations in the correlations among these six variables and in their correlations with the dependent variable. Therefore, instead of making one grand ecological regression for the entire nation, a separate multiple regression equation involving all of the variables was computed for each of four regions:

Northeast–New England and Middle Atlantic states
North Center–East North Central and West North Central states
South–South Atlantic, East South Central, and West South Central states
West–Mountain and Pacific states

The estimating procedure was performed separately for each region and the results summed to get national totals.

UNITS OF OBSERVATION. Because states are very large and heterogeneous ecological entities, and this heterogeneity is a potential source of error, counties were selected as the unit of ecological observation. (All the data used by Robinson to compute his ecological correlation by states were also available for counties.) This yields 3,101 units of observation instead of 48, and permits the stratification-by-region approach to have a sufficient number of observations for each region to support the multiple regression procedure.

WEIGHTED REGRESSIONS. To adjust for variations in the size of the counties, the regressions used here are weighted regressions, in which the 1930 total population age ten years or over of each county is used as its weight.

It is proposed to use ecological multiple regressions (stratified and weighted) to estimate the values of a, b, c, and d—the interior cells of schematic Table 1—to determine what proportion of the foreign-born and native-born populations are illiterate, and the consequent individual correlation between foreign birth and illiteracy. In order to do this, the multiple variables in the equation are subdivided into three classes:

Dependent variable—Y, the phenomenon under study—percent illiterate.

Principal independent variable—the variable whose ecological relationship (correlation with Y) it is desired to establish—percent foreign-born.

Subordinate independent variables—variables that are correlated both with the dependent and with the principal independent variables, whose effects must be controlled in establishing the correlation between Y and the principal independent variable. For example, variables X_2, X_3, X_4, X_5, and X_6 are subordinate variables in the task of predicting the individual correlation between illiteracy and foreign birth.

Illiteracy and foreign birth

This re-analysis will undertake to arrive at a stratified weighted multiple regression estimate of the individual correlation between foreign birth and illiteracy, the relationship reported in Table 1. This is a particularly difficult problem for the following reasons:

(a) The sign of the bivariate unweighted ecological correlation (for states) is the opposite of that of the individual correlation.

(b) The individual correlation is not very large ($\phi = .18$). Thus, even at the individual level the correlation is a weak one.

(c) The foreign-born population is a heterogeneous population, representing several waves of migration from different countries to different regions of the U.S. Some of these waves were more literate than others. Consequently it has a very peculiar age and country-of-origin composition which varies substantially from state to state.

(d) The foreign-born illiterate population is concentrated in a few states with large industrial cities. The many less populous and more rural states have smaller fractions of foreign-born population, and their foreign-born populations tend to be more literate.

(e) At least one other variable—the percent of population Negro—is an even more powerful factor in determining the prevalence of illiteracy.

(f) In several states there are other ethnic groups which have very high rates of illiteracy—Mexicans and American Indians. Though not important nationally, these populations are important in determining the proportion of illiteracy in these states.

THE MULTIPLE REGRESSION EQUATION. An equation in which all six of the independent variables are simultaneously regressed on the dependent variable Y is needed for each of the four regions of the United States, using counties as units of observation. These equations must be weighted regressions, in which the total populations age ten and over in 1930 are the weights. Table 4 reports the results of such computations. The partial regression coefficients are reported in metric units. In mode of calculation, this equation is identical with any conventional multiple regression equation; its only distinguishing characteristic is that the units of observation are groupings of population (counties) rather than individuals. The means and standard deviations and standardized regression coefficients (beta values) are all meaningful, as are the values of the multiple correlation R and the coefficient of determination R^2. This table provides all of the data needed to write the prediction equations for making estimates of illiteracy among both the foreign-born and the Negro populations. It should be noted that the multiple correlations are quite high, ranging from .61 to .83.

The next step is to make appropriate substitutions into the multiple regression equation for each region in order to estimate a value of Y that will provide an estimate of the interior cells of a statistical table that refers to individuals (Table 3). In other words, a weighted multiple regression equation is used as a predictor for each region instead of simple bivariate regression for the entire nation equation. In order to proceed it is necessary to:

(a) Specify what value of the principal explanatory variable (in this case, X_1) is to be substituted into the regression equations.

(b) Specify what values of the subordinate independent variables (X_2, X_3, X_4, X_5, X_6) are consistent with the value selected for the principal independent variable (X_1).

A number of options, to be explained later, are available at this point. The simplest and most direct of these is to assume that the principal explanatory variable will take on its maximum possible value. For the present problem this calls for substituting $X_1 = 100$ into the regression equation, in order to estimate what proportion of the population would be illiterate if it were comprised entirely of foreign-born whites.

The values of X_2, X_3, X_4 (proportions Negro, Mexican, and Indian) must obviously be zero under these conditions, and hence all would be set equal to 0.0 in the estimation computations.

Because the regression coefficients of X on the principal independent variable were measured while taking account of variables X_5 and X_6, it is appropriate to substitute the mean values of these variables into the regression equation. An example of the substitution for the Northeast is as follows:

$$Y = (100.0)(0.1176) + (8.437)(0.3923) - (77.576)(0.0154) - (1.7666)$$

This substitution yields an estimate of the percent of foreign-born population in the region that is illiterate, which in this case is 12.1092 precent. By multiplying the proportion Y, estimated in the above manner, by the total number of foreign-born residents age ten and over in the region, we obtain an estimate of the number of illiterate foreign-born whites in the region.

EXAMPLE. The Northeast region contained 7,018,000 foreign-born whites, of whom 12.1092 percent were estimated to be illiterate. The esti-

Table 4. Weighted Regression Coefficients and Constants for Each of Four Regions and the United States. Measuring Relationship Between Proportion of Illiteracy and Variables X_1 to X_6

Independent variable	Symbol	U.S. Total	Region			
			North-east	North central	South	West
A. Partial Regression coefficients						
Percent foreign-born	X_1	.0841	.1176	.0512	− .0288	.0577
Percent Negro	X_2	.2054	.0668	.1779	.1868	.1831
Percent Mexican	X_3	.2228	3.8155	.1287	.2447	.1560
Percent American Indian	X_4	.2484	− .0467*	.1308	− .0307	.4838
Percent urban	X_5	− .0199	− .0154	− .0087	− .0422	− .0058
Percent of population under 5	X_6	.5781	.3923	.0949	.5506	.3431
B. Regression constant	A	−3.211	−1.766	.252	−1.130	−2.007
C. Mean values of variables						
Percent foreign-born	X_1	13.558	24.689	13.468	1.686	18.982
Percent Negro	X_2	9.776	3.292	3.191	25.157	1.173
Percent Mexican	X_3	1.111	.020	.258	1.867	4.683
Percent American Indian	X_4	.273	.030	.178	.312	1.163
Percent urban	X_5	56.238	77.576	58.662	33.194	59.280
Percent of population under 5	X_6	9.351	8.437	8.956	10.944	8.245
Percent illiterate	Y	4.538	3.546	1.905	8.591	3.079
D. Multiple correlation	R	.848	.690	.606	.817	.829
E. Explained variation in Y	R^2	.719	.476	.368	.667	.688
F. Number of counties		3100	217	1057	1414	413

All regressions are significant at the .05 level or greater if not asterisked.

mated number of illiterate foreign-born whites in the Northeast is therefore:

$$(7{,}018{,}000)(12.1092) = 849{,}824$$

This same procedure is carried out individually for each of the four regions.

The following tabulation reports the value of Y estimated for each region and the implied number of illiterate foreign-born whites in each region. The sum of the four regions equals 1,184,663, an estimate of the number of illiterate foreign-born whites in the entire nation in 1930.

Using this item of information, the values of all other cells in Table 3 can be obtained by subtraction from the row and column totals of Table 1, which are already known.

Table 5 reports the results of carrying out the procedure of estimating the interior cells of the table. By comparing this table with Table 1, it is possible to learn how precisely the regional multiple regression equations were able to estimate the individual relationship between literacy and nationality. The following comparisons are useful:

	Actual	Estimated
Percent of foreign-born illiterate	9.87	8.96
Value of ø, individual correlation	+0.118	+0.10

The estimated values are amazingly close to the actual census values. The multiple regression procedures not only were able to "turn around" the ecological correlation from a negative to a positive value, but succeeded in estimating the proportion of foreign-born illiterates within 0.9 percent of the census tabulation. The value of the individual correlation differed from the tabulated value by an almost negligible amount. The comparatively small error of the estimates in handling this apparently impossible problem suggests that the proposed methodology is valid. However, further demonstration and exploration is needed. [Table 6 reports the weighted zero-order correlations be-

	Region			
Item	Northeast	North Central	South	West
Estimated percent illiterate. . . .	12.1092	5.7117	.6150	6.2481
Total foreign-born population . . .	7,018,000	4,238,622	527,946	1,432,360
Estimated illiterate foreign-born .	849,824	242,097	3,247	89,495

tween Y and each of the six variables for each region.]

Guidelines for valid use of ecological multiple regression to measure individual correlations

The regression principle introduced by Goodman, here expanded to the routine use of weighted multiple regression (with or without stratification), provides both mathematical and empirical evidence that it is possible to generalize about the behavior of individuals from grouped data under a much wider range of conditions than many researchers may now realize. This does not mean, however, that researchers may return to their past uncritical use of grouped data, provided only that they throw in a few additional independent variables, use weighted regressions, and possibly stratify the ecological units along some hastily contrived basis. Instead, an effort should be made to formulate some guidelines which would curtail improper use while encouraging greater use of these techniques under conditions favorable to obtaining valid results. Although these guidelines have been arrived at pragmatically, based primarily upon the explorations with Robinson's examples, reported above, it is nevertheless believed that if adhered to they will succeed in discriminating between valid and invalid results.

All of these guidelines are based on the assumption that the estimates will be derived from weighted multiple regressions. Whether or not to add stratification will be an option controlled by the following guidelines:

1. The researcher should develop hypotheses about all factors that might have a substantial effect in explaining variation in the dependent variable Y. He must devise a multiple regression or stratification system (or both) which refers to the behavior of individuals that, for the ecological units, will bring these variables under maximum control. Those which have a small independent effect in all of the strata can be discarded without loss of precision.

2. The ecological units used in the regression should be the smallest aggregates for which it is possible to obtain data, keeping in mind the need for compromising internal homogeneity with sampling error of the cell frequencies. For example, it is desirable to shift from states to counties, from wards to census tracts, wherever it is possible. The larger number of ecological units makes it possible to obtain more information about the interrelationships among the independent variables as well as of each independent variable with the dependent variable.

3. The greater the area-to-area variation in the dependent and the independent variables, the more valid the results of the ecological analysis will tend to be.

4. Maximum relative precision in predicting individual behavior occurs where the categories representing the dependent variable divide the population evenly between them, 50 percent in each (assuming the wide variation cited in 3 above). Conversely, the smaller the percentage of the total population that any category represents (or the more nearly universal it is) the more difficult it is to find a valid correlation between individual behavior and that variable. Extreme caution is advised when the dependent variable comprises less than 2 percent or more than 98 percent of the total population and is rather evenly distributed among the ecological units.

5. The multiple regression equation (weighted) must account for a highly significant and meaningful proportion of the total variance in the dependent variable. As a rule of thumb, it is suggested that ecological regression estimates should not be undertaken unless the value of R is at .35 (at least 10 percent of all variance is accounted for). The higher the value of R, the greater the likelihood that the ecological correlations are valid for predicting individual behavior. This, however, is not a sufficient condition to guarantee validity.

Table 5. Estimated Individual Correlation Between Nativity and Illiteracy for the United States, Using Stratified Weighted Multiple Regression

(population 10 years old and over, in thousands)

Literacy status	Total	Nativity	
		Foreign-born white	Native white and Negro
Illiterate	3,921	1,185	2,736
Literate	93,352	12,032	81,320
Total	97,273	13,217	84,056
		Percent distribution	
Illiterate	4.0	9.0	3.3
Literate	96.0	91.0	96.7
Total	100.0	100.0	100.0

Measure of association: $\phi = .10$
Note: The totals shown in this table differ slightly from those reported by Robinson, but are correct.

6. The smaller the variance in the regression coefficient for the principal independent variable (variable whose individual correlation with Y is being estimated), the greater the probability that the ecological estimates will be valid. The greater the significance level of the partial regression coefficients for the principal independent variable, the more likely the regression equation will be a valid predictor of individual behavior.

7. The regression constant (intercept) should receive careful study. As the principal independent variable is set to zero (or maximum value for negative regressions), the resulting estimate of Y should not be negative nor exceed 100 percent. If no valid prediction of individual correlation from the grouped data can be derived from a particular regression equation, this test will help to discover that fact.

8. Use as few independent variables as possible while controlling all of the hypothesized factors that appear to have a strong explanatory effect.

9. If the ecological units are stratified, construct

Table 6. Zero-Order Correlation Between Proportion of Illiteracy (Y) and Variables X_1 to X_6 and Mean Values of These Variables, Four Regions and U.S.

Variable Name	Symbol	U.S. Total	Region			
			Northeast	North Central	South	West
Zero-order correlation with Y						
% foreign-born	X_1	-.186	.594	.389	-.241	.237
% Negro	X_2	.727	.114	.423	.651	.228
% Mexican	X_3	.226	.192	.182	.193	.384
% American Indian	X_4	.139	-.070	.187	-.056	.665
% Urban	X_5	-.326	.318	.219	-.377	-.197
% population under 5	X_6	.455	-.008	.065	.313	.376

the strata in accordance with carefully developed and explicitly stated hypotheses. The following are some guidelines for constructing strata:
  (a) Units where there is a distinctive economy, culture, religion, language, racial composition, ethnic composition, or particular historical, geographic, or legal traits which would be expected to be related to the dependent variable in a unique or atypical way in comparison with the rest of the nation should be treated as a separate stratum.
  (b) If there are a large number of ecological units for which the prevalence of the dependent variable (or the principal independent variable) is very low (zero or nearly zero), consider isolating these units from the remainder and treating them as a separate stratum. (Examples: counties where illiteracy is almost zero, if there are a large number of them.) Shively (1969) points this out.[6]
  (c) Keep the number of strata to a minimum, consistent with the hypotheses for delineating them.
  (d) Keep in mind that the multiple regression procedure tends to reduce inter-strata differences. If the basis for stratification is already expressed by one of a combination of the dependent variables, fewer strata (or none) may be needed.

10. As in other forms of multiple regression analysis, much can be learned by computing an estimated value for each unit of observation and studying the deviations from regression.

11. Compute the regressions in several ways—both weighted and unweighted, with and without stratification, with alternative criteria of stratification, with alternative combinations of independent variables. Study the differences that make a significant change in sign (from being moderately positive to moderately negative, or vice versa) as a result of different and equally plausible alternatives.

12. All regression estimates that yield an estimate of more than 100 percent or less than zero percent must be used at their maximum or minimum value (100 or zero). As point 7 above emphasizes, estimates which exceed these limits by a wide margin at the national level should be regarded as unsatisfactory. If this happens within one stratum or more, it should be taken as a sign of weakness in that particular set of estimates. (Where strata have almost zero percent of the dependent variable, small negative estimates are not uncommon.)

13. Try to minimize the number of variables for which the mean value is substituted while substituting values of zero and 100 percent for the principal independent variable. Insofar as possible, introduce independent variables which may also be set to zero or 100 percent when making these estimates.

14. Compare the signs and magnitudes of regression coefficients among strata, and especially for the principal independent variable. The researcher should be able to formulate a plausible hypothesis to account for each difference in sign and magnitude between them.

15. As in all multiple regression estimates, transforming the variables to normalize them or increase homoscedacity can greatly improve the precision of estimates. Every effort should be made to make the ecological data conform to the specifications for valid multiple regression estimation.

IMPLICATIONS. If all fifteen of the guidelines specified above are adhered to faithfully, the instances in which an individual correlation derived from an ecological regression is declared to exist when in fact one does not exist (or vice versa) should be no greater than the one-in-twenty risk which statisticians routinely accept. The confidence which the researcher develops in his generalizations should be based upon how well his data conform to these guidelines. The reader will have already appreciated that these guidelines are highly interdependent. In many instances, strong compliance with some of them can compensate for low compliance with others.

Implications for future social research

It is believed that the number of worthwhile research projects now being neglected because of excessive conservatism about the use of ecological correlation but which are admissible under the guidelines given above is quite large. Some possible avenues of research that may be accelerated because of this re-analysis are listed below. In particular cases, the data available may not conform adequately to the guidelines, of course.

1. *Rates and ratios which combine locally collected data as numerators and census tract or county data as denominators.* Among these are:
  (a) School attendance rates—elementary, high school and college

[6]Shively, "Ecological Inference."

(b) Vital rates—birth, death, marriage, divorce
(c) Employment and unemployment rates
(d) Voter registration, voting results
(e) Rates of being on welfare, receiving food stamps, etc.
(f) Church membership, church attendance
(g) Admissions to hospitals, by cause
(h) First admissions to mental hospitals
(i) Arrest rates, conviction rates by type of offense—adults and juveniles
(j) Registration of automobiles, traffic violations, age and type of car
(k) Sales of appliances (televisions, refrigerators, etc.)
(l) Subscriptions to telephones, magazines, newspapers
(m) Memberships in formal, informal community, recreational, work-oriented organizations
(n) Home ownership, home loans for renovation, building permits
(o) Place of work, place of shopping, residence of friends

The data for one year often will provide too few cases to meet the requirements of the guidelines; in such cases the data may be aggregated for several consecutive years.

2. *Intensive analysis of percents, rates, and ratios computed from small area data published by the census.* Individual behavior within particular cities can be studied by multiple regression estimation procedures. Among such items are:

School attendance
Mobility and migration
Household composition
Divorce, separation
Working wives with small children
Income
Occupation

3. *Studies in which organizations are treated as individuals,* for which multiple items of data are available—the dependent variables may represent policy decision by administrators. Such organizations include:

Hospitals
Churches
School districts
Voter precincts, wards
Universities
High schools
Grammar schools
Police districts, township governments
Departments of a city or state government

In all of the above examples it is assumed that the ecological (aggregate) data contain independent variables which the researcher is willing to accept as indicators of the hypotheses he wishes to test.

Concluding comments

The analysis and empirical research presented here warrant a recommendation that the term "ecological fallacy" or similar derogations of the use of grouped data to predict the behavior of individuals be dropped from the lexicon of social research methodology.[7] There is nothing inherently fallacious about the methodology discussed above.

The branch of research known as "contextual analysis" or "structural effects" can benefit a great deal from the approach recommended here.[8] Most studies of structural effects have been based on comparatively simple univariate classifications of the environments within which the studies of individual behavior are examined. With multiple-variable procedures it should be possible to estimate more precisely what impact a particular multi-faceted environment appears to have upon the behavior of individuals in comparison with other multi-faceted environments.

Sociologists such as Boudon[9] have become

[7] See Hayward R. Alker, Jr., "A Typology of Ecological Fallacies," in *Quantitative Ecological Analysis in the Social Sciences*, ed. M. Doggan and S. Rokkan (Cambridge: M.I.T. Press, 1969), pp. 69–86; and Edward L. Thorndike, "On the Fallacy of Imputing the Correlations Found for Groups onto the Individuals or Smaller Groups Composing Them," *American Journal of Psychology*, 52 (January 1939), 122–24.

[8] See James Davis, J. Spaeth, and C. Hudson, "Analyzing the Effects of Group Composition," *American Sociological Review*, 26 (April 1959), 215–25; Goodman, "Some Alternatives"; Peter M. Blau, "Structural Effects," *American Sociological Review*, 25 (April 1960), 178–92; E. Campbell and C. Norman Alexander, "Structural Effects and Interpersonal Relationships," *American Journal of Sociology*, 71 (November 1965), 284–89; James S. Coleman, "Relational Analysis: The Study of Social Organization with Survey Methods," *Human Organization*, 17 (Winter 1959), 28–36; James S. Coleman, *Introduction to Mathematical Sociology* (Glencoe, Ill.: The Free Press, 1964); Paul Lazarsfeld and Herbert Menzel, "On the Relations between Individual and Collective Properties," in *Complex Organizations*, ed. Amitai Etzioni (New York: Holt, Rinehart and Winston, 1961), pp. 422–40; Juan J. Linz, "Ecological Analysis and Survey Research," in *Quantitative Ecological Analysis in the Social Sciences*, ed. M. Doggan and S. Rokkan (Cambridge: M.I.T. Press, 1969), pp. 91–132; and Tapani Valkonen, "Individual and Structural Effects in Ecological Research," in Doggan and Rokkan, pp. 53–68.

[9] Raymond Boudon, "Proprietes Individuelles et Proprietes Collectives: Une Probleme d'Analyse Ecologique," *Revue Francaise de Sociologie*, 4 (July–September 1963), 275–99.

sensitized to the problem of "disaggregation" or "decomposition"—the division of some given aggregate total into a number of component or constituent parts, such as predicting the behavior of individual households, firms, other decision-making units, specific commodities, prices of specific commodities, or changes over time in any of the above. "Ecological regression" offers new avenues for exploring the decomposition problem. One possible avenue for improving upon predicting the behavior of individuals from aggregate data would lie in substituting multiple characteristics of individuals into equations which represent conditional probabilities of a particular behavior, derived by ecological multiple regression and stratification which involve these same characteristics.

Several writers have insisted that sociological research typically involves the use of aggregates as units of analysis, and that such studies often have no interest or desire to measure individual correlations or relationships.[10] This certainly is a valid research perspective and nothing in the arguments developed here challenges it.

Bibliography

Alker, Hayward R., Jr. "A Typology of Ecological Fallacies." In M. Doggan and S. Rokkan (eds.), *Quantitative Ecological Analysis in the Social Sciences*, pp. 69–86. Cambridge: M.I.T. Press, 1969.

Allardt, Erik. "Aggregate Analysis: The Problem of Its Informative Value." In M. Doggan and S. Rokkan (eds.), *Quantitative Ecological Analysis in the Social Sciences*, pp. 41–52. Cambridge: M.I.T. Press, 1969.

Blau, Peter M. "Structural Effects." *American Sociological Review*, 25 (April 1960): 178–92.

Boudon, Raymond. "Proprietes Individuelles et Proprietes Collectives: Une Probleme d'Analyse Ecologique." *Revue Francaise de Sociologie*, 4 (July–September 1963): 275–99.

Campbell, E.; and Alexander, C. Norman. "Structural Effects and Interpersonal Relationships." *American Journal of Sociology*, 71 (November 1965): 284–89.

Coleman, James S. "Relational Analysis: The Study of Social Organization with Survey Methods." *Human Organization*, 17 (Winter 1959): 28–36.

———. *Introduction to Mathematical Sociology*. Glencoe, Ill.: The Free Press, 1964.

Curry, Leslie. "A Note on Spatial Association." *The Professional Geographer*, 18 (March 1966): 97–99.

———. "Quantitative Geography, 1967." *The Canadian Geographer*, 11, No. 4 (1967): 265–79.

Davis, James; Spaeth, J.; and Hudson, C. "Analyzing the Effects of Group Composition." *American Sociological Review*, 26 (April 1959): 215–25.

Duncan, Otis Dudley; and Davis, Beverly. "An Alternative to Ecological Correlation." *American Sociological Review*, 18 (December 1953): 665–66.

Duncan, Otis Dudley; Cuzzort, Ray P.; and Duncan, Beverly D. *Statistical Geography*. Glencoe, Ill.: The Free Press, 1961.

Fisher, Walter D. *Clustering and Aggregation in Economics*. Baltimore: Johns Hopkins Press, 1969.

Goodman, Leo. "Ecological Regression and the Behavior of Individuals." *American Sociological Review*, 18 (1953): 663–64.

———. "Some Alternatives to Ecological Correlation." *American Journal of Sociology*, 64 (May 1956): 610–25.

Green, H. A. *Aggregation in Economic Analysis*. Princeton: Princeton University Press, 1964.

Hammond, John L. "Two Sources of Error in Ecological Correlations." *American Sociological Review*, 38 (1973): 764–77.

Hannan, Michael T. *Aggregation and Disaggregation in Sociology*. Lexington, Mass.: D.C. Heath and Company, 1971.

Klein, Lawrence R. "Remarks on the Theory of Aggregation." *Econometrica*, 14 (October 1946): 303–12.

Lazarsfeld, Paul; and Menzel, Herbert. "On the Relations Between Individual and Collective Properties." In Amitai Etzioni (ed.), *Complex Organizations*, pp. 422–40. New York: Holt, Rinehart and Winston, 1961.

Linz, Juan J. "Ecological Analysis and Survey Research." In M. Doggan and S. Rokkan (eds.), *Quantitative Ecological Analysis in the Social Sciences*, pp. 91–132. Cambridge: M.I.T. Press, 1969.

Menzel, Herbert. "Comments on Robinson's 'Ecological Correlation and Behavior of Individuals.'" *American Sociological Review*, 15 (1950): 674–77.

Nataf, Andre. "Aggregation." In D. Sills (ed.), *International Encyclopedia of the Social Sciences*, Vol. I, pp. 162–68. New York: Macmillan, 1968.

Robinson, Arthur H. "The Necessity of Weighting Values in Correlation Analysis of Areal Data." *Annals of the Association of American Geographers*, 46 (1956): 233–36.

---

[10]See Herbert Menzel, "Comments on Robinson's 'Ecological Correlation and Behavior of Individuals,'" *American Sociological Review*, 15 (1950), 674–77; Otis Dudley Duncan, Ray P. Cuzzort, and Beverly D. Duncan, *Statistical Geography* (Glencoe, Ill.: The Free Press, 1961); Erik Allardt, "Aggregate Analysis: The Problem of Its Informative Value," in *Quantitative Ecological Analysis in the Social Sciences*, ed. M. Doggan and S. Rokkan (Cambridge: M.I.T. Press, 1969), pp. 41–52; and Leslie Curry, "A Note on Spatial Association," *The Professional Geographer*, 18 (March 1966), 97–99.

Robinson, William S. "Ecological Correlations and the Behavior of Individuals." *American Sociological Review,* 15 (June 1950): 351–57.

Scheuch, Erwin K. "Social Context and Individual Behavior." In M. Doggan and S. Rokkan (eds.), *Quantitative Ecological Analysis in the Social Sciences,* pp. 133–56. Cambridge: M.I.T. Press, 1969.

Shively, W. Phillips. "Ecological Inference: The Use of Aggregate Data to Study Individuals." *American Political Science Review,* 63 (December 1969): 1183–96.

Slatin, Gerald T. "Ecological Analysis of Delinquency." *American Sociological Review,* 34 (December 1969): 894–906.

Theil, Henri. *Linear Aggregation in Economic Relations.* Amsterdam: North Holland Publishing Company, 1954.

———. *Statistical Decomposition Analysis.* Amsterdam: North Holland Publishing Company, 1972.

Thomas, Edwin N.; and Anderson, David L. "Additional Comments on Weighting Values in Correlation Analysis of Areal Data." *Annals of the Association of American Geographers,* 55 (September 1965): 492–505.

Thorndike, Edward L. "On the Fallacy of Imputing the Correlations Found for Groups onto the Individuals or Smaller Groups Composing Them." *American Journal of Psychology,* 52 (January 1939): 122–24.

Valkonen, Tapani. "Individual and Structural Effects in Ecological Research." In M. Doggan and S. Rokkan (eds.), *Quantitative Ecological Analysis in the Social Sciences,* pp. 53–68. Cambridge: M.I.T. Press, 1969.

Section C Reconciling Conflicting Conceptions of Social Ecology: Neo-Orthodox Approach vs. Sociocultural Approach

Ecology and Human Ecology*

Amos H. Hawley

Human ecology, from its inception to a comparatively recent date, is reminiscent of Alice's curious experience in the rabbit hole when she, after consuming the pretty little cake, opened out "like the largest telescope that ever was." Emerging abruptly in the early 1920's, human ecology quickly became, as an otherwise unkind commentator puts it, "one of the most definite and influential schools in American sociology. . . ."[1]

It is now beginning to appear, however, that the period of burgeoning growth has given way to a second phase in which sober criticism rather than feverish application is the prevailing note. Re-examination and reappraisal are the order of the day. This cannot be anything but welcome, for it is a necessary preface to the sorely needed reconstruction of human ecological thought.

Hence the addition of still another voice to the developing symposium may not be amiss.[2]

Perhaps it is to be expected that the sudden ascent to popularity of an innovation in scientific thought should be accompanied by a certain amount of confusion as to its specific connotation. If so, human ecology has satisfied expectations, for after twenty years it remains a somewhat crude and ambiguous conception. A perusal of the literature that has accumulated under the name can hardly fail to produce bewilderment. One finds it variously argued that the study deals essentially with "sub-social" phenomena, with the effects of competition, with spatial distributions, with the influence of geographic factors, and with still other more or less intelligibly delineated aspects of human behavior. There are some writers who would have human ecology encompass the whole field of social science, and there are others who prefer to relegate it to the status of a mere

---

*Reprinted from *Social Forces*, 22 (May 1944), 398–405, by permission of the author and The University of North Carolina Press.

[1] M. A. Alihan, *Social Ecology, A Critical Analysis* (New York, 1938), p. xi.

[2] I am indebted to the late Professor R. D. McKenzie for most of the ideas set forth in this paper, but responsibility for their statement here is entirely mine.

sociological research technique. Between these wide extremes the subject can be found identified in turn with biology, economics, human geography, sociology, and, as if not to overlook a possibility, it is sometimes described as marginal to all other life sciences. Indeed, the sole point of agreement among the many diverse conceptions of human ecology seems to be that it pertains to some phase of man's relation to his physical universe. This, unfortunately, is no distinction, since most of the sciences of man may be characterized in the same manner.

Whatever may be said regarding the confusion as to the nature of the study, it cannot be charged to a lack at the outset of careful attempts at systematic theoretical formulation. The success of human ecology in attracting and holding the large share of attention it has enjoyed is largely a result of the ingeniousness, simplicity, and utility of the early definitive statements. But these seem to have been accepted as dogmas rather than, as intended, as suggestions of the possibilities of an ecological approach to the study of human social life. Subsequent work in the field, with very few exceptions, was not aimed at exploring the full implication of ecology as applied to man. Instead there was a wholesale application of a little understood point of view and in consequence the theoretical development of the discipline received scarcely any attention. In fact, most so-called ecological studies have been occupied with incidentals and by-products of the approach, and not a few are totally irrelevant to the caption under which they appear in print.

But to be more specific, responsibility for the existing chaos in human ecology, it seems to me, rests upon certain aberrant intellectual tendencies which have dominated most of the work that has been done. The more significant of these may be described as: (1) the failure to maintain a close working relationship between human ecology and general or bioecology; (2) an undue preoccupaton with the concept competition; and (3) the persistence in definitions of the subject of a misplaced emphasis on "spatial relations." Whether such habits of thought originated from one source or another is unimportant. What is important is that they have consistently confused the issue, thereby hampering the progress of the discipline. The purpose of the present paper is to indicate the deficiencies of these elements of human ecological thought and thus to aid in clearing the way for a reorganization of the subject.

Probably most of the difficulties which beset human ecology may be traced to the isolation of the subject from the mainstream of ecological thought. Although it seems almost too elementary to mention, the only conceivable justification for a human ecology must derive from the intrinsic utility of ecological theory as such. Obvious as this may seem, it is not a fact that is generally taken seriously. Exponents of human ecology, despite their steadfast adherence to the name, tend to view with indifference or regret the fact that their subject has any connection with the parent discipline. This is indeed a paradox. If a person chooses to call this work ecology, it would appear reasonable to assume that his studies are intended to parallel, at least in some particulars, those of others working under the same general title. However, very few persons who regard themselves as human ecologists indicate an awareness that they are logically committed to follow out in the study of man the implications of ecology.

In general, students are divided into two camps with respect to the relation of human to general ecology. One group, taking the position that ecology offers an essentially biological approach to the study of the human community, has recognized a close association between the two.[3] But while this admission has been accompanied by a relatively free borrowing of terminology, it has yielded very little in the way of theoretical unity. The second group expresses a somewhat reactionary viewpoint. Its representatives strongly oppose even a suggestion of similarity between the two phases of the discipline on the ground that any assumption of analogy as between social and biological phenomena is invalid and impractical.[4] Human ecology, according to this view, should be developed independent of other branches of ecology.

Without entering into a detailed consideration of either of these positions, it will be sufficient to point out that the conception of ecology contained therein is actually a misconception. The widespread belief that ecology is a biologism, as it were, has no logical support, not even in the conventional academic distinction between sociology and biology. That ecology is basically a social science has long been clear to most serious students of the subject. It is apparent, moreover, in almost every aspect of the discipline: in the root

---

[3] See R. E. Park, "Human Ecology," *American Journal of Sociology*, 42 (July 1936), 1–15; and A. B. Hollingshead, "Human Ecology," in *An Outline of the Principles of Sociology*, ed. R. E. Park (New York, 1939), pp. 65–74.

[4] See W. E. Gettys, "Human Ecology and Social Theory," *Social Forces*, 18 (1940), 470–71.

of the term ecology; in the historical details of the subject's development; in the large place given to sociological concepts such as community, society, niche, commensalism, symbiosis, dominance, succession, etc.; and in the manner in which problems for investigation are stated. But all of this appears to have escaped the majority of so-called human ecologists; they have proceeded without benefit from the theoretical position they believe themselves to have adopted. Evidently it is for such reasons that the concept competition and the interest in spatial analysis have absorbed so much of the energies of students of the subjects.

The assignment of the concept competition to a key role in human ecology is, in fact, premised largely on the biological interpretation of the subject. The steps which lead to this inference may be simply stated. Struggle, of which competition is but a refined expression, is the law of biological nature and the circumstances out of which all order arises. Competition is therefore a biological phenomenon. Moreover, since competition is definable as a process in which individuals or other units affect one another through affecting a common limited supply of sustenance materials, it does not presuppose consciousness or social concensus in the units concerned,[5] and what is not social must therefore be biological. Hence, it is concluded, to base human ecology on the concept competition is to carry through to the study of man the distinctive ecological approach. Thus has competition come to be regarded as the necessary hypothesis of the study—as the efficient cause, so to speak, in the development of ecological phenomena. "Human ecology," writes one author, in what may be considered a representative statement, "deals with society in its biological and symbiotic aspects, that is, those aspects brought about by competition and by struggle of individuals, in any social order, to survive and to perpetuate themselves."[6]

The defects in this line of reasoning are manifold. The desire on the part of human ecologists to achieve a thorough-going natural science treatment of human behavior undoubtedly lies at the roots of their theorizing relative to competition. But the question as to whether the struggle for existence is categorically a natural, in the sense of biological, phenomenon is seldom considered. To insist that it is, for no other reason than that the conception was first extensively used in connection with a biological problem and later became recognized as a part of the language of biology, would appear to indicate a stronger addiction to words than to thoughts. As a matter of fact, a cogent argument can be made in favor of the inherent sociological quality of the idea of struggle. Unless I am mistaken, "struggle for existence" pertains primarily to the behavior of organisms relative to one another. If this be the province of biology, then *ipso facto* all social science resolves itself into biology.

Further difficulty in this respect arises from the belief, not limited to human ecologists, that a natural science must seek causation outside the sphere of consciousness. Competition, because of its essentially unconscious or asocial character, is assumed to provide a definitely natural science, i.e., objective and impersonal, avenue of approach.[7] Why the natural and the conscious should be regarded as mutually exclusive categories it is impossible to say. Surely it is as natural for a man to think and act accordingly as for a squirrel to store nuts or for a rock, when loosened, to roll down the mountain slope.

However, and more to the point, the distinction between conscious and unconscious activity is difficult if not impossible to maintain in practice. It presents problems of observation for which there is no yardstick. Whether competition does or does not include conscious elements is a matter of definition and therefore subject to individual opinion. What is important, if true, is that individuals do affect one another through affecting the available supply of required materials. This is all that need concern the ecologist. In any event, as economists, anthropologists and others have amply shown, an objective or so-called natural science approach does not stand or fall on an exclusive use of unconscious behavior as data.

The application of competition as a hypothesis also involves a number of serious problems. For example, it presupposes a knowledge—not always at hand—of the intrinsic qualities of the individuals or other units concerned, i.e., in regard to homogeneity or similarity of life requirements. Frequently individuals who at first glance might be considered competitors turn out to be so differentiated, through the operation of genetic processes and early conditioning, as not to be competitors at all. Braun–Blanquet states: "It has further been said that certain species [of plants] are in general confined to certain soils, but when they come into competition one wins on calcareous soil, the other

---

[5]R. E. Park and E. W. Burgess, *An Introduction to the Science of Sociology* (Chicago, 1929), p. 506.

[6]A. B. Hollingshead, loc. cit., p. 70.

[7]Cf. C. A. Dawson and W. E. Gettys, *An Introduction to Sociology* (New York, 1935), p. 122.

on siliceous soil." But, he continues, "the life requirements of these pairs of species are so different that the question of competition cannot arise."[8] This illustrates the ecologist's need for an adequate taxonomy, a need which has been sadly neglected in the social sciences. The utility of competition as an explanatory tool will remain in doubt until a fuller knowledge of functional or social types is developed.

A related problem exists with regard to the observability of the operation of competition. The specific sequence of changes by which a homogeneous aggregate is converted into a differentiated and interdependent population has not been described in detail. Consequently it is almost impossible to indicate what to look for in order to see competition in action. The situation is not improved by pointing out that the process is a type of interaction, that is, a process of mutual internal modification. Ecologists, unfortunately, lack the technique for the observation of internal phenomena. Defined in terms of competitive interaction, ecology amounts to little more than the contemplation of a concept.[9] This, parenthetically, seems to be the net result of interactional theory in general so far as its use by sociologists is concerned. It would appear that psychologists are better equipped to deal with such a matter.

There would be no cause to mention this problem had human ecologists actually treated competition as a hypothesis to be tested and demonstrated. However, in no instance, so far as I am aware, has a student of the subject applied himself to such a task. The truth of the matter is that the concept serves in practice as a *post hoc* interpretation. This being the case, the question whether the concept describes what it is supposed to describe remains unanswered. Doubt will linger on this point until the prerequisites for observation have been fulfilled.

It has been fairly well established, however, that the competitive hypothesis is a gross oversimplification of what is involved in the development of pattern, structure, or other manifestation of organization. As a matter of fact, the customary interpretation of the Darwinian "struggle for existence" to mean that the primary and dominant relationship in animate nature is opposition, whether clamorous combat or the more subtle competition, forms one of the neatest illustrations of the "fallacy of misplaced concreteness" that may anywhere be found. Darwin used the phrase in "a large and metaphorical sense,"[10] subsuming under it all expenditures of effort to maintain and expand life. Combination and cooperation as well as competition and conflict are embraced in the concept. That mutual aid is just as fundamental and universal as opposition has been abundantly shown in numerous field and laboratory studies by students of plants and animals.[11] There seems to be no reason to assume that human collective life is any more amenable to monistic explanation.

These remarks should not be taken to imply that competition has no place in ecological thought. The criticism is directed solely at the loose and extravagant use of the concept which enabled it to become accepted as the basic theoretical element in human ecology. The significance of competition may better serve as a topic for a separate discussion and hence will not be taken up here. Certainly competition is not the pivotal conception of ecology; in fact, it is possible to describe the subject without even an allusion to competition.

Another persistent inconsistency in human ecology, which also reflects the failure of the discipline to develop in close relation to general ecology, exists in the emphasis put upon spatial relations or spatial aspects of human interdependencies. The origin of this peculiarity may be found in early definitions of the subject, such as, human ecology is "a study of the spatial and temporal relations of human beings as affected by the selective, distributive, and accommodative forces of the environment."[12] While such a statement has the advantage of concreteness and was highly useful in the "absence of any precedent," it seemed to indicate a subordination of interest in symbiotic relations to a concern for the spatial pattern in which such relations are expressed. Thus it permitted human ecology to be construed as merely the description of distributions of social phenomena.

Accordingly, much of the research identified as

---

[8] J. Braun-Blanquet, *Plant Sociology*, trans. G. D. Fuller and H. S. Conrad (New York, 1932), pp. 15–16 (brackets mine).

[9] Cf. James A. Quinn, "Human Ecology and Interactional Ecology," *American Sociological Review*, 5 (October 1940), 721–22.

[10] *Origin of Species* (New York, 1925), p. 78.

[11] For a brief but excellent summary of this literature, see W. C. Allee, *The Social Life of Animals* (New York, 1937), chap. III. See also M. L. McAtee, "The Malthusian Principle in Nature," *The Scientific Monthly*, 42 (May 1936), 453 ff.

[12] R. D. McKenzie, "The Ecological Approach to the Study of the Human Community," in *The City*, ed. R. E. Park, E. W. Burgess, and R. D. McKenzie (Chicago, 1925), pp. 63–64.

human ecology has consisted in compiling inventories of the observable characteristics of community life and in plotting their distributions on maps. It is sometimes difficult to understand why this kind of work should be called anything other than geography, except possibly—out of deference to the geographers—because of the inferior cartographic skill which is often exhibited. The mapping of phenomena, however, is usually a first step in the establishing of correlations between crime, delinquency, domestic discord, mental disorders, etc., on the one hand, and housing conditions, recreational facilities, proximity to city center, and other physical features, on the other hand. But so far as the determination of the degree of correlation is the sole aim of the study, which seems to be the rule rather than the exception, it is not ecological; it is rather more in the nature of a statistical study in psychological behaviorism. The prevalence of the use of the word ecology in connection with such work as this has been so great that it has come to be regarded, in some quarters, as a "method" to be compared and contrasted with so-called statistical, case-study, and historical methods.[13] In other words, one of the techniques employed in ecological research—mapping—has been mistaken for the discipline itself.

That space and time are merely convenient abstractions by which to measure activities and relationships has been rather consistently overlooked. To contend that human behavior is bound by such dimensions is but to insist that it occurs in an experiential universe and is therefore subject to observation and measurement. This is what is meant, fundamentally, when it is asserted that human ecology is a natural science. But it is important to note that every enterprise which may be called science is a natural science in at least this sense of the term. Every science, that is, must deal with the spatial and temporal aspects of its own subject-matter. The differences between scientific disciplines arise not in respect to method but rather in respect to problems. And in the case of human ecology as elsewhere the problem is the distinguishing feature. Spatial and temporal considerations are incidental to the investigation of the ecological problem.

Now it may be asked: What remains of human ecology, if its usual mainstays—the concept competition and spatial analysis—are removed to positions of minor importance? Before entering into a discussion of this question, it may be well to give some thought to the matter of preference of one definition or another for a given study. By what prerogative may one say that human ecology is this or that? The answer, of course, depends on how the criteria of appropriateness of a discipline happen to be regarded. Probably few will deny, however, that the problem with which a study is to be concerned must not only be significant but must also be a problem that is not already preempted by other disciplines. It is no easier to defend a needless duplication of effort than it is a preoccupation with irrelevant issues. Unless human ecology has a problem of its own, then, it is nothing and may as well be forgotten. But just as urgent is the necessity that a discipline be coherent within itself and consistent with the point of view it pretends to represent. There is no basis, in other words, for calling a study human ecology, if it is not ecological. Both of these considerations should be kept in the foreground in any definition or redefinition of the nature and scope of a subject for study. It is desirable, then, in returning to the original question, to begin with a review of the rudiments of general ecology.

Briefly stated, ecology is concerned with the elemental problem of how growing, multiplying beings maintain themselves in a constantly changing but ever restricted environment. It is based on the fundamental assumption that life is a continuous struggle for adjustment of organism to environment. However, the manifest interrelatedness of living forms, which leads students to speak of the "web of life," suggests that adjustment, far from being the action of independent organisms, is a mutual or collective phenomenon. Drawing together the relevant facts, it seems that the inevitable crowding of living forms upon limited resources produces a complex action and reaction of organism with environment and organism with organism in the course of which individuals become related to one another in ways conducive to a more effective utilization of the habitat. As the division of labor which thus develops approaches equilibrium, such that the number of organisms engaged in each of the several activities is sufficient to provide all the needs that are represented, the aggregate of associated individuals assumes the aspect of a compact viable entity, a superorganism, in fact. The (biotic) community, as such a functionally or symbiotically integrated population may properly be called, is in effect a collective response to the habitat; it constitutes the adjustment, in the fullest sense of the term, of organism to environment.

---

[13]Calvin F. Schmid, "The Ecological Method in Social Research," in *Scientific Social Surveys and Research,* ed. P. V. Young (New York, 1939), chap. XII.

The subject of ecological inquiry then is the community, the form and development of which are studied with particular reference to the limiting and supporting factors of the environment.[14] Ecology, in other words, is a study of the morphology of collective life in both its static and its dynamic aspects. It attempts to determine the nature of community structure in general, the types of communities that appear in different habitats, and the specific sequence of change in community development.

Two elements, one implicit and the other explicit, in the conception as outlined here merit special emphasis. Not immediately evident perhaps, though nevertheless of basic importance, is the fact that the units of observation, i.e., the data, are neither physiological processes nor anatomical structures but are rather the activities of organisms. Taxonomic characteristics are relevant only so far as they serve as indexes of behavior traits.[15] "When an ecologist says 'there goes a badger'" writes Elton, "he should include in his thoughts some definite idea of the animal's place in the community to which it belongs, just as if he had said 'there goes the vicar.'"[16] Thus if the term species and species designations recur frequently in ecological discussion, it is simply because that is the most convenient way of referring to the expected or observed occupations of the organisms denoted.

Secondly, as already indicated, life viewed ecologically is an aggregate rather than an individual phenomenon. The individual enters into ecological theory as a postulate and into ecological investigation as a unit of measurement; but as an object of special study he belongs to other disciplines, e.g., physiology, genetics, psychology, etc. The focus of attention in ecology is upon the population which is either organized or in process of becoming organized. This cannot be too strongly emphasized, for it places ecology squarely in the category of social science.

Human ecology, like plant and animal ecology, represents a special application of the general viewpoint to a particular class of living things. It involves both a recognition of the fundamental unity of animate nature and an awareness that there is differentiation within that unity. Man is an organism and as such he is dependent on the same resources, confronted with the same elementary problems, and displays in essential outline the same mode of response to life conditions as is observed in other forms of life. Thus the extension of patterns of thought and techniques of investigation developed in the study of the collective life of lower organisms to the study of man is a logical consummation of the ecological point of view. One important qualification is necessary, however; the extraordinary degree of flexibility of human behavior makes for a complexity and a dynamics in the human community without counterpart elsewhere in the organic world. It is this that sets man apart as an object of special inquiry and gives rise to a human as distinct from a general ecology.

While to reason from "pismires to parliaments" would do violence to the facts, it is nevertheless necessary to keep the phenomenon of culture in proper perspective. When man by virtue of his culture-producing capacity is regarded as an entirely unique type of organism, the distortion is no less acute than if this quality were completely ignored. Human behavior, in all its complexity and variability, is but further evidence of the tremendous potential for adjustment inherent in life. Culture is nothing more than a way of referring to the prevailing techniques by which a population maintains itself in its habitat. The component parts of human culture are therefore identical in principle with the appetency of the bee for honey, the nest-building activities of birds, and the hunting habits of carnivora. To argue that the latter are instinctive while the former are not is to beg the question. Ecology is concerned less now with how habits are acquired, than with the functions they serve and the relationships they involve.

Thus despite the great difference between the

---

[14] This definition differs but slightly from others. For example: (1) Ecology is the science of "the correlations between all organisms living together in one and the same locality and their adaptations to their surroundings." (Ernest Haeckel, *The History of Creation, II* [New York, 1896], p. 354); (2) "Ecology is the science of the relation of organisms to their surroundings, living as well as non-living; it is the science of the 'domestic economy' of plants and animals." (R. Hesse, W. C. Allee, and K. P. Schmidt, *Ecological Animal Geography* [New York, 1937], p. 6); (3) "... the essence of ecology lies in its giving the fullest possible value to the habitat as cause and the community as effect, the two constituting the basic phases of a unit process." (F. E. Clements and V. E. Shelford, *Bio-Ecology* [New York, 1939], p. 30); and (4) "The descriptive study of the interrelations between co-existing species, and, more generally, their environment, is the province of ecology." (A. J. Lotka, "Contact Points of Population Study with Related Branches of Science," *Proceedings of the American Philosophical Society*, 80 [Feb. 1939], p. 611).

[15] Cf. H. C. Cowles, "An Ecological Aspect of the Conception of Species," *The American Naturalist*, 42 (1905), 265–71.

[16] Charles Elton, *Animal Ecology* (New York, 1927), p.64.

behavior of men and that of lower forms of life—a difference which appears to be of degree rather than of kind—the approach described as general ecology may be applied to the study of man without radical alteration. In simplest terms, human ecology is the descriptive study of the adjustment of human populations to the conditions of their respective physical environments. The necessity that life be lived in a specific place and time, operating upon man as it does upon other organisms, produces an inescapable compulsion to adjustment which increases as population increases or as the opportunities for life decrease. And out of the adaptive strivings of aggregated individuals there develops, consciously or unconsciously, an organization of interdependencies which constitutes the population a coherent functional entity. The human community, in other words, is basically an adaptive mechanism; it is the means whereby a population utilizes and maintains itself in its habitat. Human ecology, then, may be defined more fully as the study of the development and the form of communal structure as it occurs in varying environmental contexts.

The human community, of course, is more than just an organization of symbiotic relationships and to that extent there are limitations to the scope of human ecology. Man's collective life involves, in greater or less degree, a psychological and a moral as well as a symbiotic integration. But these, so far as they are distinguishable, should be regarded as complementing aspects of the same thing rather than as separate phases or segments of the community. Sustenance activities and interrelations are inextricably interwoven with sentiments, value systems, and other ideational constructs. Human ecology is restricted in scope then not by any real or assumed qualitative differences in behavior but simply by the manner in which its problem is stated. The question of how men relate themselves to one another in order to live in their habitat yields a description of communal structure in terms of its overt and visible features. It does not, however, provide explanations of all the many ramifications of human interrelationships. The external and descriptive approach of ecology is ill-suited to the direct study of the psychological counterpart of symbiosis, although it may serve as a fruitful source of hypotheses concerning that aspect of the community.

It may be helpful to call attention to the fact that the problems of human ecology, and ecology in general, are basically population problems. The broad question, as previously indicated, concerns the adjustment of population to the resources and other physical conditions of the habitat. This resolves itself into a number of related problems, such as: (1) the succession of changes by which an aggregate passes from a mere polyp-like formation into a community of interdependencies; (2) the ways in which the developing community is affected by the size, composition, and rate of growth or decline of the population; (3) the significance of migration for both the development of the community and the maintenance of community stability; and (4) the relative numbers in the various functions composing the communal structure, together with the factors which make for change in the existing equilibrium and the ways in which such change occurs.

Clearly, human ecology has much in common with every other social science. The problem with which it deals underlies that of each of the several specialized studies of human social life. Its data are drawn from the same sources and it employs many of the same techniques of investigation. The points of convergence are, in fact, too numerous to detail in this paper. There is no basis therefore to conclude from what has been said that human ecology is an autonomous social science: it is quite unlikely that there is any autonomy in science. The distinctive feature of the study lies in the conception of the adjustment of man to habitat as a process of community development. Whereas this may be an implicit assumption in most social science disciplines, it is for human ecology the principal working hypothesis. Thus human ecology might well be regarded as the basic social science.

Human Ecology, Space, Time, and Urbanization*

Amos H. Hawley

Human ecology

The study of territorially based systems, of which the urban community is a prime example, is known as *human ecology*. More abstractly, this designation refers to a concern with the processes and the form of man's adjustment to environment.[1] The community is a generalized form of that adjustment. In the reasoning of human ecology, adjustment is accomplished, not by each individual's acting independently, but by means of a division of labor developed among a number of individuals. Adjustment is a collective achievement. The unit of observation, therefore, is a population; the adaptive organization, i.e., the division of labor with its various ramifications in social relationships, is the property of a population.

*Reprinted from Amos H. Hawley, *Urban Society: An Ecological Approach* (New York: Ronald Press, 1971), pp. 11–16, by permission of the author and John Wiley and Sons, Inc.

[1] See Amos H. Hawley, *Human Ecology: A Theory of Community Structure* (New York: Ronald Press, 1950); Leo F. Schnore, "Social Morphology and Human Ecology," *American Journal of Sociology,* 63 (1958), 620–29; O. D. Duncan and Leo F. Schnore, "Cultural, Behavioral and Ecological Perspectives in the Study of Social Organization," *American Journal of Sociology,* 65 (1959), 132–53; June Helm, "The Ecological Approach in Anthropology," *American Journal of Sociology,* 67 (1962), 630–39; Julian Steward, "Cultural Ecology," *International Encyclopedia of the Social Sciences* (New York: Macmillan and The Free Press, 1968), Vol. 4, pp. 328–36.

Environment, on the other hand, embraces all of the externalities that impinge on the life chances of a definable population. It includes not only the physical and biotic elements of an occupied area, but also the influences that emanate from other organized populations in the same and in other areas. In certain circumstances, the latter acquires a more critical importance than the former.

The basic hypothesis of human ecology, then, is that, as a population develops an organization, it increases the chances of survival in its environment. The emphasis is upon organization. In its ecological application, organization is an inclusive concept. It refers to the entire system of interdependences among the members of a population which enables the latter to sustain itself as a unit. The parts of such a system—families, clubs, shops, industries, for example—cannot be self-sustaining; they can only survive in a network of supporting relationships. As easy as this is to say, it does not fully dispose of the question of how to define the boundaries of a system. The identification of the effective limits of organizational units is one of the persistent problems in human ecology. It is sufficient at the moment, however, to state that ecology is committed to dealing with the most inclusive unit manageable.

Thus the problems selected for study are of the order commonly characterized as macroscopic. This choice is in contrast to a microscopic type of problem, such as that concerned with a part of a system or, more particularly, an individual taken separately. There is no reason to infer *a priori* that either level of investigation is more or less preferable; the issue must be decided entirely on the basis of the problem at hand. But it is assumed that the variables applicable to an analysis of phenomena at one level have very little explanatory utility for phenomena that lie at the other level. Human ecology seeks its explanations among variables that are structural properties, demographic attributes, and features of environment, including interactions with other systems. The import of this theoretical position will become more understandable as one follows the exposition in succeeding [pages].

Space, time, and organization

It should be apparent that the system of interdependences we have described as urban is a territorially based phenomenon. Indeed, that is its most obvious aspect. Spatial considerations in one respect or another will be, therefore, a continuing concern in this volume. For that reason, it is desirable that we put the concept *space* in sociological perspective.

Man and his behavior are facts of nature. As such they share certain characteristics with other facts of nature. They occupy space, for example. Moreover, man relies on a great many things that are distributed over space: food, water, raw materials of all kinds. Were these necessities everywhere present in abundant supply, it is conceivable that the idea of space might never have occurred to man. They are not, of course, if for no other reason than that two or more finite things cannot exist in the same place at the same time. The very irregular distribution of the requisites for life is a fact that provokes in man's behavior a constant sensitivity to distances and locations.

Of more immediate importance is a second fact, namely, that space is an ever-present element in the fundamental interdependence among men. There should be no need to document the inescapability of interdependence; it begins at biological conception and is omnipresent throughout life. That is why we describe man as a "social" animal. Stripped to its essential content, the word *social* means interdependence. It follows with simple logic that interdependence presupposes accessibility between the parties concerned. There can be no mutual support or exchanges unless individuals can come together with whatever frequency the business at hand may demand. Conversely, the distances that separate people affect the kinds of joint activities they can pursue. A family operates as a family only when its members are close together. When they are widely separated, the family becomes something else, a kin group, a clan, or nothing at all. The universal tendency for human settlement to be nucleated, to live in clusters, which reaches its highest form in urban centers, maximizes interpersonal accessibility and therefore the opportunity for devising elaborate cooperative arrangements.

But space, as it is experienced in human collective life, is not a fixed quality. It is rather a function of the time consumed in movement. Man is a time-bound creature. He must eat and rest at regular intervals. These elementary requirements set limits on the time and energy available to him for other activities, including movement. Space may be thought of as a friction measurable by the time and the effort expended in getting from point to point. Where the friction is great, the number of different things that can be done in the span of a day are few; conversely, the number and variety of activities can be increased as the friction is reduced.

Many of our notions of distance and space seem to have begun as measurements on a time dimension. In times past, long distances were calculated in number of days of travel: Herodotus described distances by the number of days required to move between points by "a man traveling light." The Roman *league* may have been the distance legionnaires could march in an hour's time. On the other hand, an acre originated as the area a man with an ox could plow in a single day. It is of interest to note that, despite revolutionary advances in overcoming distance, the 60-minute radius still defines the ambit of local life. Seldom has man been willing to devote more than an hour's time to trips that must be made every day. The distance that can be traveled in that interval has been extended from about 3 miles for pedestrian to approximately 30 miles for the vehicular-borne rider. Beyond the 60-minute radius, interactions with a center are reduced to less than a daily periodicity. Through most of the nineteenth century and the early part of the twentieth, the team-haul distance—the distance a horse-drawn wagon could travel and return in the course of a single day, or about 12 to 15 miles—measured the scope of weekly trips to a center. More recent improvements in transportation, of course, have enormously extended the radius of occasional travel. Moreover, efficiency in the overcoming of distance has been heightened by the development of means of sustaining indirect or mediated relationships.

The only means at man's disposal for pressing back the limits that time presents is through enlarging upon a division of labor with his fellows so that an increased number of activities can be carried on simultaneously. Such specialization produces a countereffect: the greater the number of complementary activities carried on by different people, the greater is the amount of time that must be devoted to exchanging and communicating. A characteristic solution to that problem has been a close concentration of the people engaged in high-frequency exchanges, forming villages, towns, or cities. As organization develops, it becomes possible to economize further on time by delegating responsibility for movement to specialists: messengers, teamsters, dispatchers, and the like. The effect of organizational development in this respect rests in large part on the capability it affords for the acquisition and utilization of technological advances. Thus the man mounted on horseback can travel farther in an hour than can a pedestrian, but the maintenance and equipment of the horse imposes larger technical demands and consumes the time and effort of more people. The animal-drawn cart, though somewhat slower than the mounted rider, can carry much larger loads. Again, however, the cart presupposes skills in manufacturing and a knowledge of harnessing the animal that do not occur in very simply organized groups. Similarly, the four-wheeled wagon, the railway, the motor vehicle, and numerous other innovations that have contributed to the reduction of the friction of space have followed from the growth and elaboration of organized life.

For the time being, perhaps enough has been said to indicate that the meaning of space for man, the way in which it enters into his activity, grows out of his experience with time. As he has learned to economize on his use of time, he has been able to spread his activities over larger amounts of space. The conservation of time depends in turn on the extent to which man is able to elaborate his organization of complementing activities. Thus space, in a very large and important sense, is a derivative of organization. Urbanization is but one of the more conspicuous ways in which man has, through the development of organization, enlarged upon the territory from which he obtains the materials of daily life.

Urbanization as a growth process

In its most visible aspect, urbanization is a process of increasing territorial scope of organization. The process is much more involved than that, of course. There are other factors that are also vital to a system; change must move concurrently in those respects as well. In short, growth in a phenomenon as complex as a community or a social system entails reciprocal effects among culture, population, territory, and organization. Unless change in one is accompanied by changes in the others, the results are not cumulative, that is, growth does not occur. Instead, the tendency to change is short-lived and the unit in question reverts to its original state.

The interrelations among the four factors or dimensions of a social system become apparent in analysis. Cultural accumulation, with which change begins, cannot proceed far without an increase in population. Additional people are needed to retain and to put to use a diversifying repertory of ways of acting. At the same time, more territory is required to supply the increased amounts of food needed to nourish an enlarging population and to provide the increasing variety of materials used in fabricating the multiplying items of material culture. Furthermore, the growing number, variety, and spread of activites presuppose an elaboration of organization to assure coherence among complementing parts. The more finely drawn the division of labor, the more imperative is the requirement for centralized mediating and coordinating institutions.

This paradigm outlines a rather generalized conception of an expansion process. At first glance urbanization might seem to be a special case of the general principle. Yet, although the process may be mounted on various scales, it seems that in all instances it unfolds from a center strategically situated with reference to access to its tributary area, on the one hand, and to external regions, on the other hand. The center is the locus of the organization that knits together dispersed activities and links them to what is being done in other places. Center and territory advance together, each supporting the other.[2]

[2]See R. D. McKenzie, "Industrial Expansion and the Interrelations of Peoples," in E. B. Reuter (ed.), *Race and Culture Contacts* (New York: McGraw-Hill, 1934), pp. 19–33; Amos H. Hawley, *Human Ecology,* pp. 348–70; and Otis D. Duncan, "Social Organization and the Ecosystem," in R. E. L. Faris (ed.), *Handbook of Modern Sociology* (Chicago: Rand McNally, 1964), pp. 36–82.

It is not meant to imply that the growth of a territorial system follows a smooth or uninterrupted path. Historically, the process has ebbed and flowed. On some occasions it has subsided after a brief career. Elsewhere the process has advanced to a grand scale, only to end in dissolution. Invariably, however, expansion has sprung up in another quarter and has surged to still larger dimensions. The vagaries of the process are of interest. . . .

Nor is expansion in any sense a simple matter. To speak of the growth of organization is to refer in a very cryptic way to a highly involuted sequence of events. Urbanization is a transformation of society, the effects of which penetrate every sphere of personal and collective life. It affects the status of the individual and his opportunities for advancement, it alters the types of social units in which people group themselves, and it sorts people into new and shifting patterns of stratification. The distribution of power is altered, normal social processes are reconstituted, and the rules and norms by which behavior is guided are redesigned.

Social Morphology and Human Ecology*

Leo F. Schnore

Introduction

Émile Durkheim, of course, was not himself a human ecologist. The ecological viewpoint did not develop within sociology until near the end of Durkheim's life, and then in America.[1] There is no evidence that this new approach to social phenomena exerted any profound influence upon his thought, despite the fact that he regarded "social morphology" as one of the major branches of sociology. In Durkheim's scheme, this field was to be devoted to two major inquiries: (1) the study of the environmental basis of social organization, and (2) the study of population phenomena, especially size, density, and spatial distribution.[2] These areas of interest obviously converge with those of human ecology as it was originally formulated.

This paper consists of an exegesis . . . of one of his major theoretical contributions and a consideration of the broad implications of his "morphological" analysis for contemporary human ecology. It is concerned, for the most part, with Durkheim's doctoral dissertation, *De la division du travail social: étude sur l'organisation des sociétés supérieures,* first published in 1893.[3] More particularly, the discussion is largely limited to Book II, where he dealt with the "causes" of division of labor and where the morphological approach was most explicitly used. The brief exegesis is based on a selective restructuring of his main argument, which is unfortunately scattered through many pages. We trust that taking up the crucial elements in his thought in somewhat different order does no violence to the essential logic of his position. This

---

*Reprinted from *The American Journal of Sociology,* 63 (May 1958), 620–24, 629–34, by permission of the author and The University of Chicago Press.

[1] Durkheim died in 1917, and the first use among sociologists of the term "human ecology" did not appear until 1921, in Robert E. Park and Ernest W. Burgess (eds.), *Introduction to the Science of Society* (Chicago: University of Chicago Press, 1921), pp. 161–216. However, Durkheim was familiar with the work of Ernst Haeckel, who coined the word "ecology" in 1868 and who is often described as the father of plant ecology.

[2] Taken from Durkheim's essay, "Sociologie et sciences sociales" (1909); cited in Harry Alpert, *Émile Durkheim and His Sociology* (New York: Columbia University Press, 1939), p. 51. Durkheim's own discussion of social morphology appears in scattered essays and reviews in *L'Année sociologique* (old series), e.g., "Note sur la morphologie sociale," II (1897–98), 520–21.

[3] Paris: Alcan, 1893; translated by George Simpson as *The Division of Labor in Society* (New York: Macmillan, 1933; Glencoe, Ill.: Free Press, 1947). All citations to *Division* hereafter refer to the 1947 edition. Occasional reference will also be made to *Les Règles de la méthode sociologique* (Paris: Alcan, 1895), a collection of essays that had appeared in *Revue philosophique* in 1894. *Les Règles* was translated by Sarah A. Solovay and John H. Mueller and edited by George E.G. Catlin as *The Rules of Sociological Method* (Chicago: University of Chicago Press, 1938; Glencoe, Ill.: Free Press, 1950). All subsequent citations to *The Rules* refer to the 1950 edition.

procedure has been adopted in order to point up the contrasts between his morphological theory of differentiation and the alternative explanations that were available at the time he wrote.

Exegesis

First, it must be emphasized that Durkheim's intention in Book II of *Division* was to account for differentiation and its obvious increase in Western societies. The very subtitle is the key: "A Study of the Organization of Advanced Societies." Second, it is necessary to preserve the historical context of his work. The division of labor had long interested social philosophers, especially in the West. As early as 1776, Adam Smith had pointed to division as the main source of "the wealth of nations," and the concept itself can be traced at least to the Greeks. Unfortunately, these earlier writers gave scant attention to the determinants of differentiation, contenting themselves with analyses of its nature and its implications for economic efficiency and productivity.

In the latter half of the nineteenth century, however, increasing effort was given to explaining the process, with special reference to the "advanced" societies of the time. Comte dealt with the matter at some length, discussing the nature of differentiation as a generic social phenomenon.[4] Tönnies and Simmel also examined the problem in publications that preceded Durkheim's by only a few years.[5] By the time that Durkheim began his work, however, the dominant views in intellectual circles were still a peculiar admixture of utilitarian and evolutionary "explanations," both best represented in the works of Herbert Spencer. In large part, Durkheim's analysis must be seen as a reaction against the Spencerian view.

Durkheim's own analysis actually began in Book I, with a distinction between two forms of organization somewhat similar to the types sketched by Maine and Tönnies. The first type ("mechanical") was used by Durkheim to describe the relatively undifferentiated or "segmented" mode of organization characteristic of small and isolated aggregates, in which little control has been achieved over the local environment. The basis of social unity is likeness or similarity. There is minimal differentiation, chiefly along age and sex lines, and most members are engaged most of the time in the same activity—collecting, hunting, fishing, herding, or subsistence agriculture. The "social segments" of the community (families and kinship units) are held together by what they have in common, and they derive mutual support from their very likeness. Unity is that of simple "mechanical" cohesion, as in rock forms, and homogeneity prevails.[6]

Durkheim was fully aware that structural differentiation is a variable characteristic of aggregates, for he recognized another and fundamentally different mode of organization. He saw that modern Western society was based increasingly upon differentiation, and his concept of the "organic" type of organization was designed to describe the complex and highly differentiated structural arrangements of his own time. According to Durkheim, a complex and heterogeneous society, like all but the most rudimentary organisms, is based on an intricate interdependence of specialized parts. Labor is divided; all men do not engage in the same activities, but they produce and exchange different goods and services. Moreover, not only are individuals and groups differentiated with respect to functions, but whole communities and nations also engage in specialized activities. In short, there has been a breakdown of internal "segmentation" *within* communities and societies and a reduction of isolation *between* them, although mechanical solidarity never completely disappears.[7]

With this distinction between major types of or-

---

[4] August Comte, *The Positive Philosophy*, translated and edited by Harriet Martineau (New York: D. Appleton, 1853).

[5] Ferdinand Tönnies, *Gemeinschaft und Gesellschaft* (1887), translated and edited by Charles P. Loomis as *Fundamental Concepts of Sociology* (New York: American Book Co., 1940). Although it is not cited in *Division*, Durkheim had previously reviewed *Gemeinschaft und Gesellschaft* in highly favorable terms (see *Revue philosophique*, 27 [1889], 416–22). Georg Simmel's *Über soziale Differenzierung* appeared in 1890, but Durkheim indicated that he did not see it until after 1893, when *Division* first appeared. For a general critique of Simmel, see Durkheim's "La Sociologia ed il suo dominio scientifico," *Rivista italiana di sociologia*, 4 (1900), 127–48.

[6] "We say of these societies that they are segmental in order to indicate their formation by the repetition of like aggregates in them" (*Division*, p. 175). This type is not to be understood as somehow lacking any differentiation whatsoever (see pp. 129, 173, 177, 180). As Redfield has suggested, homogeneity in simpler societies is more than merely "occupational," extending to biological characteristics and even to outlook. Small size and extreme isolation appear to be crucial factors in the development of both genetic and cultural homogeneity (see Robert Redfield, "The Folk Society," *American Journal of Sociology*, 52 [1947], 292–308).

[7] See *Division*, p. 229, and Durkheim's assertion that "mechanical solidarity persists even in the most elevated societies" (p. 186). Some critics erroneously accuse him of failing to see that both forms of integration can be found in every society.

ganization in mind, Durkheim's task in Book II was to explain the conditions under which "mechanical" organization is superseded by the "organic" form. According to the mode of analysis that prevailed at the time that he wrote, Durkheim viewed this change in social organization as comprising a kind of "evolutionary" sequence, and much of his theory was cast in these terms. However, it would be extremely misleading to portray his work as that of an uncritical evolutionist, for Durkheim possessed a sensitive, critical mind and he considered and rejected a number of alternative hypotheses that had been widely accepted as explanations of increasing differentiation.[8]

With respect to the popular utilitarian version, Durkheim vigorously attacked the idea that differentiation was somehow the product of man's rational desire to increase his own happiness. In fact, he rejected all individualistic interpretations. The notion that social structure is merely the product of the motivated actions of individuals was apparently almost repugnant to him. It ran directly counter to his conception of society as an entity *sui generis,* and it obviously violated his most famous principle: that "the determining cause of a social fact should be sought among the social facts preceding it and not among the states of the individual consciousness."[9]

Durkheim then turned his attention to the evolutionary portion of the Spencerian argument. The organismic analogy, of course, was in vogue at the time, and Spencer had used it brilliantly. As to the division of labor in society, Spencer had held that "along with increase of size in societies goes increase of structure. . . . It is also a characteristic of social bodies, as of living bodies, that while they increase in size they increase in structure. . . . The social aggregate, homogeneous when minute, habitually gains in heterogeneity along with each increment of growth; and to reach great size must acquire great complexity."[10] In other words, Spencer's theory of differentiation—despite its cosmic overtones and utilitarian underpinnings—reduced to an explanation based on sheer population size. At the very least he pointed to a universal association between size and differentiation.

Durkheim recognized the potential role of population increase in bringing about further differentiation. Along with Adam Smith, he was aware of the permissive effect of sheer size.[11] Large aggregates allow greater differentiation to emerge, but Durkheim concluded that the population-size factor was a necessary, but not a sufficient, cause. His reasons for this conclusion are particularly instructive. In contrast to Spencer, who exemplified the deductive method of proceeding from first principles, Durkheim was very much the inductive analyst. In fact, he showed the underlying weakness of Spencer's theory by pointing to "deviant cases." Concretely, he called attention to large, densely settled areas in China and Russia clearly characterized, not by extreme differentiation (organic solidarity), but by homogeneity (mechanical solidarity).[12]

Having thus rejected the Spencerian argument on empirical grounds, Durkheim tried to explain the absence of any marked differentiation in these places in the face of great size and density. It is at this point that Durkheim introduced a series of essentially sociological concepts, the first of which must be seen as an "intervening variable." First, he noted that social segmentation had not broken down (i.e., that there was minimal contact between the constituent parts of Chinese and Russian society). In the face of limited contact, these parts remained homogeneous, very much like each other with respect to structure and functions, representing a proliferation of essentially similar village units. Durkheim asserted that this "segmentation" disappears and that division increases only with an increase in "moral" or "dynamic density." In contrast to physical density—the number of people per unit of space—"dynamic density" refers to the density of social intercourse or contact or, more simply, to the rate of interaction—the number of interactions per unit of time. Until this rate of interaction reaches a high (although unspecified) level, the constituent social segments or parts remain essentially alike. Accord-

[8] Durkheim usually tried to dispose of competing hypotheses before setting out his own views. *Division* contains a perfect example of his didactic style, which Alpert calls the method of "argumentum per eliminationem" (op. cit., pp. 84–87).

[9] *The Rules,* p. 110. In the course of his argument, Durkheim cited comparative suicide rates as "proof" to the contrary. Whatever the merits of this argument, it is interesting to note that Durkheim here anticipated his later work in this area. At another point, he dealt with religious phenomena (Book I). A number of writers have observed that *Division* contained the seeds of all his later work.

[10] Herbert Spencer, *Principles of Sociology* (London, 1876; New York: D. Appleton & Co., 1884 and 1892), I, 459 (1892 ed.).

[11] See Smith's famous aphorism to the effect that "the division of labor is limited by the extent of the market" (*The Wealth of Nations* [New York: Modern Library], p. 17).

[12] *Division,* p. 261. This thought is further developed in *The Rules,* p. 115.

ing to Durkheim: "The division of labor develops . . . as there are more individuals sufficiently in contact to be able to act and react upon one another. If we agree to call this relation and the active commerce resulting from it dynamic or moral density, we can say that the progress of the division of labor is in direct ratio to the moral or dynamic density of society."[13] In other words, differentiation tends to increase as the rate of social interaction increases.

Durkheim then asked the next logical question: Under what conditions does this rate of interaction increase? In answer, he first observed that dynamic density "can only produce its effect if the real distance between individuals has itself diminished in some way."[14] He then pointed to two general ways in which this might come about: (1) by the concentration of population, especially in cities, i.e., via increases in *physical density;* (2) by the development of more rapid and numerous means of transportation and communication. These innovations, "by suppressing or diminishing the gaps separating social segments . . . increase the [dynamic] density of society."[15]

Thus, to demographic factors (essentially the Spencerian explanation), Durkheim added a technological emphasis. An increase in population size and density *plus* more rapid transportation and communication bring about a higher rate of interaction. However, the crucial questions still remain: What brings about differentiation? Why should a simple increase in the rate of interaction produce greater division of labor? If social units (whether individuals or collectivities) are brought into more frequent contact, why should they be obliged to specialize and divide their labor? A simple identification of "factors" obviously was not enough; Durkheim was also compelled to indicate the mechanism that would produce further differentiation under the prescribed circumstances. As it turns out, he had in mind a particular type of interaction, viz., competition.

It is in his identification of competition as the vital mechanism that Durkheim borrowed most heavily upon Darwinian thought, and it is this part of his theory that has been most widely distorted. Durkheim's argument was based on Darwin's observation that, in a situation of scarcity, increased contact between like units sharing a common territory leads to increased competition. Being alike, they make similar demands on the environment. Inspired by the Malthusian account of population pressure on limited resources, Darwin had been led to stress the resultant "struggle for existence" as the essential condition underlying the differentiation of species. In the human realm, Durkheim reasoned in turn, individuals or aggregates offering the same array of goods or services are potential, if not active, competitors. Thus, according to Durkheim,

> If work becomes divided more as societies become more voluminous [i.e., larger in size] and denser, it is not because external circumstances are more varied, but because struggle for existence is more acute. Darwin justly observed that the struggle between two organisms is as active as they are analogous. . . . Men submit to the same law. In the same city, different occupations can co-exist without being mutually obliged to destroy each other, for they pursue different objects. . . . The division of labor is, then, a result of the struggle for existence, but it is a mellowed dénouement. Thanks to it, opponents are not obliged to fight to a finish, but can exist one beside the other. Also, in proportion to its development, it furnishes the means of maintenance and survival to a greater number of individuals who, in more homogeneous societies, would be condemned to extinction.[16]

The division of labor is thus seen by Durkheim as essentially a mode of resolving competition and as an alternative both to Darwinian "natural selection" and to Malthusian "checks.". . .

Implications for human ecology

The very first point to be made is that ecologists concern themselves with precisely the same problem as that attacked by Durkheim in Book II of *Division.* Just as he tried to explain one aspect of structure, contemporary ecologists attempt to identify the factors determining variations in structure. Hawley, for example, defines human ecology as the study of the form and development of the

---

[13]*Division,* p. 257. Later, Durkheim graciously credited Comte with this basic idea (see ibid., pp. 262–63). In the quoted passage and elsewhere, Durkheim spoke as if the individual were the referent. However, the treatment of change that he applied to interindividual relations appears to be even more appropriate in the analysis of the changing relations between areal units or whole aggregates in the process of differentiation.

[14]Ibid., p. 257.

[15]Ibid., pp. 259–60. Durkheim went on to say that one can usually substitute physical density ("this visible and measurable symbol" or index) for dynamic density, but that they are not inevitably correlated (see n. 11, p. 260). The point apparently troubled Durkheim, for in *The Rules* he repeats this idea in the form of an apology for having confused the two types of density (see *The Rules,* p. 115).

[16]*Division,* pp. 266–70.

community. At one point, he adopts Durkheim's exact phraseology and describes the ecologist's objective as the elucidation of "the morphology of collective life in both its static and its dynamic aspects."[17] Although he represents a more traditional ecological viewpoint, Quinn also declares that the logic of ecological inquiry points to the study of "the occupational pyramid" as essential subject matter, despite the unfortunate preoccupation of some ecologists with spatial distributions.[18] Thus modern human ecology deals with the Durkheimian problem of "morphology" and takes the same dependent variable (structure) as its *explanandum*. This is despite the fact that ecologists of Hawley's persuasion frequently limit themselves to discussing community structure, avoiding Durkheim's broader concern with society.

Second, once the environment is brought into the picture, modern ecology can be regarded as working with essentially the same array of *independent* variables—most broadly, population, technology, and the environment. Building on Hawley's theory, Duncan has labeled the resulting scheme "the ecological complex."[19] Although it tends to be implicit rather than explicit, Hawley's own effort seems to consist of treating community structure as the product of the interaction of these broad factors. The structure of a given community is viewed as a collective adaptation on the part of a population to its total environment (including other organized populations, as well as physical features), an adaptation that is strongly modified by the technological equipment in use and by certain "purely" demographic attributes of the population itself, notably its size, rate of growth, and biological (age-sex) composition.[20]

Thus the general relevance of Durkheim's thought to modern ecology is clear. He worked with essentially the same broad factors, taking one of them (structure) as his dependent variable. Moreover, his general mode of analysis is highly similar to that employed in current ecological theory. This becomes particularly apparent when one considers Hawley's treatment of differentiation, which clearly follows Durkheim in its major outlines.[21] Moreover, there are obvious formal parallels between Durkheim's *mechanical-organic* typology and the concepts of *commensalism* and *symbiosis, categoric* and *corporate groups,* and *independent* and *dependent communities* in Hawley's work.[22] Both writers point to *(a)* two modes of relationship, or forms of interaction, between like and unlike unit parts and to *(b)* two major forms of organization, depending upon which type of relationship is most prominent. Also deserving stress here is their common search for the factors that explain the progressive breakdown of isolation, the welding-together of larger and more inclusive functional units, and the emergence of a more complex structure.

An even more recent variety of ecological thought—Julian Steward's "cultural ecology"—is amenable to interpretation along the lines suggested here. In other words, the "ecological complex" appears to be in use throughout much of Steward's work, despite the fact that he does not consciously focus upon organization as the *explanandum,* preferring to work with "culture," a much broader dependent variable, and despite the fact that he gives a much larger role to the physical environment than either Durkheim or Hawley.[23] Durkheim's influence on Steward is apparently more indirect, via Durkheim's contribution to the development of "functional anthropology."

But we need not confine ourselves to the most recent statements of the ecological position to see the relevance of Durkheim's thought. A Durkheimian approach has informed human ecology since its

---

[17] Amos H. Hawley, *Human Ecology: A Theory of Community Structure* (New York: Ronald Press, 1950), p. 67.

[18] James A. Quinn, *Human Ecology* (New York: Prentice-Hall, 1950), p. 14.

[19] Otis Dudley Duncan, "Human Ecology and Population Studies," in Philip M. Hauser and Otis Dudley Duncan (eds.), *The Study of Population* (Chicago: University of Chicago Press, forthcoming).

[20] In contrast to earlier ecological emphases, spatial distributions of population and human activities enter into Hawley's thinking only as convenient indexes of organizational form; in this view, space is of interest only to the extent that it reflects structure. The same thing can be said for temporal patterns, which also have value as indexes of organization (see Amos H. Hawley, "The Approach of Human Ecology to Urban Areal Research," *Scientific Monthly,* 73 [1951], 48–49).

[21] Hawley, *Human Ecology,* chap. xi.

[22] Ibid., chap. xii.

[23] Steward criticizes Hawley as "uncertain in his position regarding the effect of environmental adaptations on culture" and indicates that he prefers to give this factor a larger causal role (Julian H. Steward, *Theory of Culture Change* [Urbana: University of Illinois Press, 1955], p. 34). This greater stress on physical-environmental factors is undoubtedly related to the fact that ethnologists are more frequently concerned with simpler societies, where the physical environment is literally a more fundamental determinant, pressing upon small and stable local populations that survive by means of relatively simple and unchanging technology and organization. To use Duncan's "ecological complex" once again, where technology and population are relatively constant, the environment assumes the position of the dynamic causal variable with respect to organization.

inception. In one of his most influential essays—"The Urban Community as a Spatial Pattern and a Moral Order"—Robert E. Park identified the subject matter of human ecology as "what Durkheim and his school call the morphological aspect of society."[24] It has probably also occurred to the reader that the use of the concept of competition in Durkheim's work is highly similar to Park's. To quote Park: "Competition determines the distribution of population territorially and vocationally. The division of labor and all the vast organized economic interdependence of individuals and groups of individuals characteristic of modern life are a product of competition."[25] Thus both Durkheim and Park saw structure as ultimately emerging out of competition in a context of scarcity, although Park was no more helpful than Durkheim in providing a detailed account of the process as a whole.[26]

In addition, it should be pointed out that Durkheim anticipated much of McKenzie's theoretical work, especially the latter's treatment of the rise of "metropolitan" communities. In Durkheim's analysis, we have seen that great stress is given to advances in transportation and communication technology, which lessen isolation and break down "social segmentation." McKenzie showed that this theory can be readily given an areal referent, since formerly isolated and territorially distinct populations are frequently brought into more intimate contact by virtue of improvements in transportation and communication. McKenzie saw the key feature of metropolitan development as the emergence of an intricate territorial division of labor between communities that were formerly almost self-sufficient, and he viewed the whole process as mainly due to technological improvements. In fact, McKenzie went so far as to characterize the metropolitan community as "the child of modern facilities for transportation and communication."[27]

Although Durkheim's analysis was largely at the societal level and dealt mainly with occupational differentiation, McKenzie used an essentially similar model in treating communities and regions, analyzing the problem of territorial differentiation. The process of differentiation is presumably the same in each case. Units that are brought into contact via technological improvements become competitors; such units necessarily compete to the extent that they offer the same goods and services to the same population. In the communal or regional context, the resolution of this competitive situation is frequently effected by territorial differentiation. Certain areal units, including whole communities, then give up certain functions and turn to new specialties. A case in point is the historical "flight" of certain specialties, particularly infrequently purchased goods and services, from nearby smaller cities to the metropolis, following the development of the automobile. In the process, formerly semi-independent centers, which once offered a rather full range of services, came to take up more narrowly specialized roles in a larger and more complex division of labor—the metropolitan community as a whole.[28]

At any rate, whether we examine earlier or more recent versions of human ecology, Durkheim's stamp is clearly imprinted.[29] In order to provide maximum utility in ecological analysis, Durkheim's theory needs certain modifications, particularly along the lines of bringing the environment into the schema as a factor worthy of recognition. As a result of its conceptual heritage from biology, human ecology has a rather full appreciation of the role of the physical environment as it affects social structure. This is not to say, however, that the ecologist is an environmental determinist; rather, he points to the relevance of the environment as it is modified and redefined by the organized use of technology. To paraphrase

---

[24]Originally published in 1925 as "The Concept of Position in Sociology" (see Robert E. Park, *Human Communities: The City and Human Ecology* [Glencoe, Ill.: Free Press, 1952], p. 166).

[25]Park and Burgess, op. cit., p. 506.

[26]On the use of competition as an all-explanatory concept in the earlier ecological literature, see Amos H. Hawley, "Ecology and Human Ecology," *Social Forces*, 23 (1944), 398–405.

[27]See R. D. McKenzie, *The Metropolitan Community* (New York: McGraw–Hill, 1933); see also Leslie Kish, "Differentiation in Metropolitan Areas," *American Sociological Review*, 19 (1954), 388–98.

[28]That differentiation is not the only mode of resolution of competition is again dramatically demonstrated by the experience of many rural service centers and hamlets after the coming of the automobile. With their markets usurped by larger centers now within easy access, a great number of these smaller places literally disappeared.

[29]This review has been confined to American developments. In France, Durkheim's morphological interests were carried on by his students, especially Maurice Halbwachs. In addition to extending Durkheim's analysis of suicide and "collective representations," Halbwachs' *Morphologie sociale* (Paris: Armand Colin, 1938) drew heavily upon his mentor's views and—at the same time—incorporated an ecological perspective that is often strikingly similar to Park's. Halbwachs visited the United States and taught at the University of Chicago in 1930 (see his "Chicago, expérience ethnique," *Annales d'histoire économique et sociale*, 4 [1932], 11–49).

a recent compendium of valuable ecological data, man has a key role in changing the face of the earth.[30] Although the human ecologist's initial concern may be with the interaction between "man and his total environment," as a sociologist he inevitably turns to a study of the organized relations between man and man in the environmental setting, i.e., to morphological considerations. As Park said for ecology, it is "not man's relation to the earth which he inhabits, but his relations to other men, that concerns us most."[31] And in following out the interaction of a given aggregate with other organized populations, the ecologist necessarily concerns himself with what Durkheim called "the social environment."

Conclusions

The only American sociologists to make any intensive use of Durkheim's earliest and most ambitious work are those who have adopted the ecological perspective. Very little attention has been given to Durkheim's "social morphology," and his theory of differentiation has been widely misunderstood. Most American writers who have discussed *Division* have drawn upon Book I, where Durkheim treated the effects of division with his customary insight. His later works, especially those dealing with suicide and religion, have been much more influential in this country. In these later studies Durkheim was more frequently dealing with individual behavior, especially as it is "normatively defined" and modified by group ties.

This selective emphasis by American writers is probably related to the main drift of American sociology in this century (i.e., toward increasing concern with social-psychological considerations). Instead of taking social structure as the phenomenon to be explained—the dependent variable—most American sociologists habitually deal with social structure as an independent variable with respect to individual behavior. More particularly, structure is usually treated as it is perceived by the individual.

Now it must be made very clear that this procedure is an entirely legitimate enterprise; the variables with which one works and their analytical status depend upon the problem to be investigated. Moreover, this approach has vastly illuminated the human situation. Since the individual is somehow regarded as a less abstract unit than the organized aggregate and as a more interesting subject for study, social psychology has grown rapidly and has made giant strides toward acceptance in the scientific community. Witness the present status of "behavioral science." For all its past progress and future promise, however, the social-psychological sector of sociology still deals with some of the consequences of structural arrangements, leaving the determinants of structure to someone else.

In the light of these considerations, Durkheim's conception of *collective representations*—"shared norms and values" in the contemporary lexicon—provides an interesting sidelight on the position of social psychology within sociology. Durkheim regarded these social phenomena as mere "emanations" of underlying social morphology or structure.[32] If one accepts this position, then he holds that the social psychologist be concerned with little more than the derivative manifestations or passive reflections of underlying structural arrangements. Such a view clearly poses the analysis of structure itself as a logically prior problem. However, if current sociological output is any measure, few of us are inclined to grant any kind of priority to a morphological approach.

It is true that Durkheim himself turned more and more to the analysis of individual behavior in his later years, but he rarely departed from his original position regarding the undesirability of attempting to explain "social facts" by reference to individual characteristics.[33] This is in dramatic contrast to the direction taken in American sociology: toward the view that has been labeled "voluntaristic nominalism." As the most significant characteristic of American sociology, our fundamental postulates have recently been identified as follows: "The feeling, knowing, and willing of individuals—though limited by cultural prescriptions and social controls—are taken to be the ultimate source of human interaction, social structure, and social change. . . . Social behavior is interpreted voluntaristically. Social structures are real only as

---

[30]William L. Thomas, Jr. (ed.), *Man's Role in Changing the Face of the Earth* (Chicago: University of Chicago Press, 1956).

[31]Park, *Human Communities*, p. 165.

[32]See Durkheim's "Représentations individuelles et représentations collectives," *Revue de métaphysique et de morale*, 6 (1898), 273–302.

[33]Durkheim's shifting interests are mirrored not only in the subjects of his later books but also in his writings in the old series of *L'Année sociologique*. Although references to "social morphology" are less frequent after about 1905, it should be noted that it was maintained as a major caption as long as Durkheim himself held the editorship.

they are products of individuals in interaction."[34] One must be impressed by the fact that so many American theorists now acknowledge a heavy indebtedness to Durkheim. If this voluntaristic position is actually dominant, however, we have only succeeded in turning Durkheim upside down.

Be that as it may, Durkheim's conception of "social morphology" suggests that one of the most promising areas of structural analysis lies in the development of a general taxonomy of aggregates and collectivities. Few sociologists seem to have addressed themselves to this task in recent years. To the extent that "types of society" are used today, they represent minor variants of the dichotomies presented long ago by Tönnies, Durkheim, and other writers of the nineteenth century. More important, most of the refinements and reformulations of these typologies in recent years have been left to writers like Redfield and Steward. In other words, a genuinely sociological tradition is being kept alive by the efforts of anthropologists.

With respect to "types of community," the initiative has been taken by economists and geographers, despite the fact that many areas of current sociological interest absolutely require close attention to the community context. To choose only the most obvious example, community studies of stratification would probably be enormously improved if the over-all structure and functions of the selected research sites were indicated with some precision according to their taxonomic types. For one thing, the overgeneralizations that seem to emerge from many such studies might be far less frequent.[35] It is probably unfortunate that the few sociologists currently attempting to develop a systematic taxonomy of communities appear to be those who employ an ecological framework.[36]

As for types of groups within communities and societies, we have not advanced very far beyond the rather rudimentary notions of "in-" and "out-groups" and "primary" versus "secondary" groups. Both of these dichotomies, of course, tend to be employed within a social-psychological context. The only notable recent addition to this limited array of group types is the notion of "membership" versus "reference" groups. However, the latter turn out not to be groups at all, for the distinction rests not upon structural or functional attributes of aggregates but upon the identifications and aspirations of individuals. It would be difficult to find a better index of just how far we have gone in bartering our sociological heritage for a mess of psychological pottage.

Morphological problems, including the development of fundamental structural taxonomies, deserve far greater attention than they have received in recent years. These are the tasks that have been largely ignored since Durkheim's day. Moreover, Durkheim's earliest work offers a challenge to those interested in the most neglected area of sociology—the analysis of the determinants of structure. As we have tried to suggest, Durkheim also provided a fascinating view of the problematics of social psychology. We leave it to the other *Journal* contributors to point out the influence of his later work and its implications for contemporary methodology and social psychology. Given the current division of labor within American sociology, Durkheim's morphological theory of structural differentiation is probably of greatest value to ecologists, although not without relevance to other students of social organization. In this age of specialization, that he saw developing so rapidly, the sheer breadth and scope of Durkheim's achievement becomes all the more impressive with the years.

---

[34] Roscoe C. Hinkle, Jr., and Gisela J. Hinkle, *The Development of Modern Sociology* (New York: Doubleday, 1954), p. 73.

[35] See Seymour M. Lipset and Reinhard Bendix, "Social Status and Social Structure: A Re-examination of Data and Interpretations," *British Journal of Sociology*, 2 (1951), 150–68 and 230–54; Harold W. Pfautz and Otis Dudley Duncan, "A Critical Evaluation of Warner's Work in Community Stratification," *American Sociological Review*, 15 (1950), 205–15; and Ruth Kornhauser, "Warner's Approach to Stratification," in Reinhard Bendix and Seymour M. Lipset (eds.), *Class Status and Power* (Glencoe, Ill.: Free Press, 1953), pp. 224–55.

[36] See Otis Dudley Duncan and Albert J. Reiss, Jr., *Social Characteristics of Urban and Rural Communities, 1950* (New York: John Wiley & Sons, 1956), pp. 215–370; for a more limited set of sub-community types, see Leo F. Schnore, "Satellites and Suburbs," *Social Forces*, 36 (1957), 121–27.

From Social System to Ecosystem*

Otis Dudley Duncan

Levels and systems

All science proceeds by a selective ordering of data by means of conceptual schemes. Although the formulation and application of conceptual schemes are recognized to entail, at some stage of inquiry, more or less arbitrary choices on the part of the theorist or investigator, we all acknowledge, or at least feel, that the nature of the "real world" exercises strong constraints on the development of schemes in science. Some schemes, used fruitfully over long periods of time, come to seem so natural that we find it difficult to imagine their being superseded. One type of scheme is deeply ingrained by our training as social scientists, to wit, the organization of data by *levels*. Kroeber is only voicing the consensus of a majority of scientists when he writes:

> The subjects or materials of science . . . fall into four main classes or levels: the inorganic, organic, psychic, and sociocultural. . . . There is no intention to assert that the levels are absolutely separate, or separable by unassailable definitions. They are substantially distinct in the experience of the totality of science, and that is enough.[1]

*Reprinted from *Sociological Inquiry,* 31 (1961), 140–49, by permission of the author and the publisher.

[1] A. L. Kroeber, "So-Called Social Science," ch. vii in *The Nature of Culture* (Chicago: University of Chicago Press, 1952), pp. 66–67.

MacIver gives substantially the same classification, but instead of using the relatively colorless term "levels," he chooses to segregate the several "nexus of causation" into "great dynamic realms."[2]

It is significant that scientists, insofar as they do accept the doctrine of levels, tend to work *within* a level, not *with* it. The scheme of levels does not itself produce hypotheses; it can scarcely even be said to be heuristic. Its major contribution to the history of ideas has been to confer legitimacy upon the newer scientific approaches to the empirical world that, when they were emerging, had good use for any kind of ideological support.

Quite another type of conceptual scheme, the notion of *system,* is employed by the scientist in his day-to-day work. Conceptions of interdependent variation, of cause and effect, or even of mere patterning of sequence, derive from the idea that nature (using the term broadly for whatever can be studied naturalistically) manifests itself in collections of elements with more than nominal properties of unity.

No doubt there are many kinds of systems, reflecting the kinds of elements comprising them and the modes of relationship conceived to hold among these elements. The point about this

[2] R. M. MacIver, *Social Causation* (Boston: Ginn & Co., 1942), pp. 271–72.

diversity that is critical to my argument is this. When we elect, wittingly or unwittingly, to work *within* a level (as this term was illustrated above), we tend to discern or construct—whichever emphasis you prefer—only those kinds of systems whose elements are confined to that level. From this standpoint, the doctrine of levels may not only fail to be heuristic, it may actually become anti-heuristic, if it blinds us to fruitful results obtainable by recognizing *systems that cut across levels*.

One such system, probably because it is virtually a datum of immediate experience, is rather readily accepted by social scientists: personality. Manifestly and phenomenologically an integration of non-randomly selected genetic, physiologic, social, and psycho-cultural elements, personality has a kind of hard reality that coerces recognition, even when it can be related to other systems only with difficulty or embarrassment. If I am not mistaken, however, the concept of personality system enjoys a sort of privileged status. We do not so readily accede to the introduction into scientific discourse of other sorts of system concepts, entailing integration of elements from diverse levels. The resistance to such concepts is likely to be disguised in charges of "environmental determinism" or "reductionism." An example: The working assumption of some human ecologists that the human community is, among other things, an organization of activities in physical space is criticized (though hardly refuted!) by the contention that such a conceptual scheme is contrary to "essentially and profoundly social" facts, i.e., "conscious choice of actors who vary in their ends and values."[3] We must resist the temptation to comment here on the curious assumption that the "essentially and profoundly social" has to do with such personal and subjective states as "ends and values," rather than with objective relations among interdependent living units. (Surely the latter is the prior significance of the "social," in an evolutionary if not an etymological sense.) The point to emphasize at present is, rather, that such a reaction to ecological formulations is tantamount to a denial of the crucial possibility that one can at least conceive of systems encompassing both human and physical elements. The "dynamic realm" of the psychosocial has indeed become a "realm," one ruled by an intellectual tyrant, when this possibility is willfully neglected or denied.

[3]Arnold S. Feldman and Charles Tilly, "The Interaction of Social and Physical Space," *American Sociological Review,* 25 (December 1960), p. 878.

The ecosystem

Acknowledged dangers of premature synthesis and superficial generalization notwithstanding, ecologists have been forced by the complexity of relationships manifested in their data to devise quite embracing conceptual schemes. The concept of ecosystem, a case in point, has become increasingly prominent in ecological study since the introduction of the term a quarter-century ago by the botanist, A. G. Tansley. "The *ecosystem,*" according to Allee and collaborators, "may be defined as the interacting environmental and biotic system."[4] Odum characterizes the ecosystem as a "natural unit . . . in which the exchange of materials between the living and nonliving parts follows circular paths."[5] The first quotation comes from an enlightening synthesis of information now available on the evolution of ecosystems; the second prefaces an exposition of principles concerning the operation of "biogeochemical cycles" in ecosystems. Social scientists whose acquaintance with general ecology is limited to gleanings from the essays of Park[6] or the polemic by Alihan[7] might do well to inform themselves concerning current developments in ecological theory by consulting such sources as these. Even more readily accessible is the statement of Dice:

> Ecologists use the term ecosystem to refer to a community together with its habitat. An ecosystem, then, is an aggregation of associated species of plants and animals, together with the physical features of their habitat. Ecosystems . . . can be of any size or ecologic rank. . . . At the extreme, the whole earth and all its plant and animal inhabitants together constitute a world ecosystem.[8]

Later in his text (ch. xv) the same author undertakes a classification of "human ecosystems." This classification presents in elementary fashion much material familiar to social scientists; but it also conveys an unaccustomed emphasis on the "di-

[4]W. C. Allee, Alfred E. Emerson, Orlando Park, Thomas Park, and Karl P. Schmidt, *Principles of Animal Ecology* (Philadelphia: W. B. Saunders Co., 1949), p. 695.

[5]Eugene P. Odum, *Fundamentals of Ecology* (Philadelphia: W. B. Saunders Co., 1953), p. 9.

[6]Robert E. Park, *Human Communities: The City and Human Ecology* (Glencoe, Ill.: The Free Press, 1952).

[7]Milla Aïssa Alihan, *Social Ecology: A Critical Analysis* (New York: Columbia University Press, 1938).

[8]Lee R. Dice, *Man's Nature and Nature's Man: The Ecology of Human Communities* (Ann Arbor: University of Michigan Press, 1955), pp. 2–3.

verse relationships" of human societies "to their associated species of plants and animals, their physical habitats, and other human societies."[9]

Popularization of the ecosystem concept is threatened by the felicitous exposition by the economist K. E. Boulding of "society as an ecosystem."[10] The word "threatened" is well advised, for Boulding uses "ecosystem" only as an analogy, illustrating how human society is "something like" an ecosystem. His ecosystem analogy is, to be sure, quite an improvement over the old organismic analogy. But ecosystem is much too valuable a conceptual scheme to be sacrificed on the altar of metaphor. Human ecology has already inspired a generation of critics too easily irritated by figures of speech.

If the foregoing remarks suggest that general ecologists have come up with cogent principles concerning the role of human society in the ecosystem, then the discussion has been misleading. Actually, the writing of Dice is an exception as a responsible attempt to extend general ecology into the human field. Most biological scientists would probably still hold with the caution of Clements and Shelford, that "ecology will come to be applied to the fields that touch man immediately only as the feeling for synthesis grows."[11] There is abundant evidence in their own writing of the inadvisability of leaving to biological scientists the whole task of investigating the ecosystem and its human phases in particular. As a discipline, they clearly have not heeded the plea of the pioneer ecologist, S. A. Forbes, for a "humanized ecology":

> I would humanize ecology . . . first by taking the actions and relations of civilized man as fully into account in its definitions, divisions, and coordinations as those of any other kind of organism. The ecological system of the existing twentieth-century world must include the twentieth-century man as its dominant species—dominant, that is, in the sense of dynamic ecology as the most influential, the controlling member of his associate group.[12]

[9] Ibid., pp. 252–53.
[10] Kenneth E. Boulding, *Principles of Economic Policy* (Englewood Cliffs, N.J.: Prentice–Hall, 1958), pp. 14–16.
[11] Frederic E. Clements and Victor E. Shelford, *Bioecology* (New York: John Wiley & Sons, 1939), p. 1. Cf. F. Fraser Darling, "Pastoralism in Relation to Populations of Men and Animals," in *The Numbers of Man and Animals*, edited by J. B. Cragg and N. W. Pirie (Edinburgh: Oliver & Boyd, 1955).
[12] Stephen A. Forbes, "The Humanizing of Ecology," *Ecology*, 3 (April 1922), p. 90.

Symptomatically, even when discussing the "ecology of man," the biologist's tendency is to deplore and to exhort, not to analyze and explain. The shibboleths include such phrasings as "disruption," "tampering," "interference," "damage," and "blunder," applied to the transformation of ecosystems wrought by human activities. Such authorities as Elton, Darling, and Sears state very well some of the dilemmas and problems of human life in the ecosystem.[13] They evidently need the help of social scientists in order to make intelligible those human behaviors that seem from an Olympian vantage point to be merely irrational and shortsighted. Insofar as they recommend reforms—and surely some of their suggestions should be heeded—they need to be instructed, if indeed social science now or ultimately can instruct them, in "The Unanticipated Consequences of Purposive Social Action."[14] If social science falls down on its job, a statement like the following will remain empty rhetoric: "Humanity now has, as never before, the means of knowing the consequences of its actions and the dreadful responsibility for those consequences."[15]

Illustration

Now, it is all very well to assert the possibility of conceptual schemes, like ecosystem, ascribing system properties to associations of physical, biological, and social elements. But can such a scheme lead to anything more than a disorderly collection of arbitrarily concatenated data? I think the proof of the ecosystem concept could be exemplified by a number of studies, ranging from particularistic to global scope, in which some such scheme, if implicit, is nevertheless essential to the analysis.[16] Instead of reviewing a sample of these

[13] Charles S. Elton, *The Ecology of Invasions by Animals and Plants* (London: Methuen & Co., Ltd., 1958); F. Fraser Darling, *West Highland Survey: An Essay in Human Ecology* (Oxford: Oxford University Press, 1955); Paul B. Sears, *The Ecology of Man*, "Condon Lectures" (Eugene: Oregon State System of Higher Education, 1957). See also, F. Fraser Darling, "The Ecology of Man," *The American Scholar*, 25 (Winter 1955–56), pp. 38–46; Donald F. Chapp, "Ecology—A Science Going to Waste," *Chicago Review*, 9 (Summer 1955), pp. 15–26.
[14] Title of an early essay by Robert K. Merton, *American Sociological Review*, 1 (December 1936), pp. 894–904; a recent statement, pertinent to ecology, is Walter Firey's *Man, Mind and Land: A Theory of Resource Use* (Glencoe, Ill.: The Free Press, 1960).
[15] Sears, op.cit., p. 50.
[16] The following are merely illustrative: A. Irving Hallowell, "The Size of Algonkian Hunting Territories:

studies, however, I would like to sketch a problematic situation that has yet to be analyzed adequately in ecosystem terms. This example, since it is deliberately "open-ended," will, I hope, convey the challenge of the concept.

The framework for the discussion is the set of categories suggested elsewhere[17] under the heading, "the ecological complex." These categories—population, organization, environment, and technology (P, O, E, T)—provide a somewhat arbitrarily simplified way of identifying clusters of relationships in a preliminary description of ecosystem processes. The description is, by design, so biased as to indicate how the human elements in the ecosystem appear as foci of these processes. Such an anthropo-centric description, though perfectly appropriate for a *human* ecology, has no intrinsic scientific priority over any other useful strategy for initiating study of an ecosystem.

The example is the problem of air pollution, more particularly that of "smog," as experienced during the last two decades in the community of Los Angeles. Southern California has no monopoly on this problem, as other communities are learning to their chagrin. But the somewhat special situation there seems to present a configuration in which the role of each of the four aspects of the ecological complex, including its relation to the others, is salient. I have made no technical investigation of the Los Angeles situation and have at hand only a haphazard collection of materials dealing with it, most of them designed for popular rather than scientific consumption. (The personal experience of living through a summer of Los Angeles smog is of value here only in that it permits sincere testimony to the effect that the problem is real.) The merit of the illustration, however, is that ramifying influences like those postulated by the ecosystem concept are superficially evident even when their nature is poorly understood and inadequately described. I am quite prepared to be corrected on the facts of the case, many of which have yet to come to light. I shall be greatly surprised, however, if anyone is able to produce an account of the smog problem in terms of a conceptual scheme materially *less* elaborate than the ecological complex.

During World War II residents of Los Angeles began to experience episodes of a bluish-gray haze in the atmosphere that reduced visibility and produced irritation of the eyes and respiratory tract (E→P); it was also found to damage growing plants (E→E), including some of considerable economic importance, and to crack rubber, accelerating the rate of deterioration of automobile tires, for example (E→T). In response to the episodes of smog, various civic movements were launched, abatement officers were designated in the city and county health departments, and a model control ordinance was promulgated (E→O). All these measures were without noticeable effect on the smog. At the time, little was known about the sources of pollution, although various industrial operations were suspected. By 1947, a comprehensive authority, the Los Angeles County Air Pollution Control District, was established by action of the California State Assembly and authorized to conduct research and to exercise broad powers of regulation. Various known and newly developed abatement devices were installed in industrial plants at the instance of the APCD, at a cost of millions of dollars (O→T).

Meanwhile, research by chemists and engineers was developing and confirming the "factory in the sky" theory of smog formation. Combustion and certain other processes release unburned hydrocarbons and oxides of nitrogen into the atmosphere (T→E). As these reach a sufficiently high concentration and are subjected to strong sunlight, chemical reactions occur that liberate large amounts of ozone and form smog. In particular, it was discovered that automobile exhaust contains the essential ingredients in nearly ideal proportions and that this exhaust is one of the major sources of the contaminants implicated in smog formation. It became all the more important as a source when industrial control measures and the prohibition of household open incinerators (O→T) reduced these sources (T→E). Also implicated in the problem was the meteorological situation of the Los Angeles Basin. Ringed by mountains and enjoying only a very low average wind velocity, the basin frequently is blanketed by a layer of warm air moving in from the Pacific. This temperature inversion prevents the polluted air from rising very far above ground level; the still air hovering over

---

A Function of Ecological Adjustment," *American Anthropologist,* 51 (January–March 1949), pp. 34–45; Laura Thompson, "The Relations of Men, Animals, and Plants in an Island Community (Fiji)," *American Anthropologist,* 51 (April–June 1949), pp. 253–76; Edgar Anderson, *Plants, Man and Life* (Boston: Little, Brown & Co., 1952); Fred Cottrell, *Energy and Society* (New York: McGraw-Hill, 1955); Harrison Brown, *The Challenge of Man's Future* (New York: Viking Press, 1954).

[17]Otis Dudley Duncan, "Human Ecology and Population Studies," ch. 28 in *The Study of Population,* edited by Philip M. Hauser and Otis Dudley Duncan (Chicago: University of Chicago Press, 1959).

the area is then subject to the aforementioned smog-inducing action of Southern California's famous sunshine (E→E).

The problem, severe enough at onset, was hardly alleviated by the rapid growth of population in the Los Angeles area, spreading out as it did over a wide territory (P→E), and thereby heightening its dependence on the already ubiquitous automobile as the primary means of local movement (T↔O). Where could one find a more poignant instance of the principle of circular causation, so central to ecological theory, than that of the Los Angelenos speeding down their freeways in a rush to escape the smog produced by emissions from the very vehicles conveying them?

A number of diverse organizational responses (E→O) to the smog problem have occurred. In 1953 a "nonprofit, privately supported, scientific research organization, dedicated to the solution of the smog problem," the Air Pollution Foundation, was set up under the sponsorship of some 200 business enterprises, many of them in industries subject to actual or prospective regulatory measures. The complex interplay of interests and pressures among such private organizations and the several levels and branches of government that were involved (O→O) has not, to my knowledge, been the subject of an adequate investigation by a student of the political process. Two noteworthy outcomes of this process merit attention in particular. The first is the development of large-scale programs of public health research and action (O→P, E) concerned with air pollution effects (E→P). Comparatively little is known in this field of epidemiology (or as some research workers would say nowadays, medical ecology), but major programs have been set up within the last five years in the U.S. Public Health Service (whose interest, of course, is not confined to Los Angeles), as well as in such agencies as the California State Department of Public Health. Here is a striking instance of interrelations between medical ecology and the ecology of medicine illustrating not merely "organizational growth," as studied in conventional sociology, but also an organizational response to environmental-demographic changes. Second, there has been a channeling of both public and private research efforts into the search for a "workable device," such as an automatic fuel cutoff, a catalytic muffler, or an afterburner, which will eliminate or reduce the noxious properties of automobile exhaust. California now has on its statute books a law requiring manufacturers to equip automobiles with such a device if and when its workability is demonstrated (O→T).

Some engineers are confident that workable devices will soon be forthcoming. The Air Pollution Foundation has gone so far as to declare that the day is "near when Los Angeles' smog will be only a memory." Should the problem be thus happily resolved, with reduction of pollution to tolerable levels, the resolution will surely have to be interpreted as the net result of an intricate interaction of factors in the ecological complex (P, O, T→E). But if the condition is only partially alleviated, how much more growth of population and increase in automobile use will have to occur before even more drastic technological and organizational changes will be required: redevelopment of mass transit, introduction of private electric automobiles, rationing of travel, limitation of population expansion, or whatever they may be? What will be the outcome of experience with increasing air pollution in other communities, whose problems differ in various ways from that of Los Angeles? And the question of questions—Is the convulsion of the ecosystem occasioned by smog merely a small-scale prototype of what we must expect in a world seemingly destined to become ever more dependent upon nuclear energy and subject to its hazards of ionizing radiation?

Conclusion

I must assume that the reader will be kind enough to pass lightly over the defects of the foregoing exposition. In particular, he must credit the author with being aware of the many complications concealed by the use of arrows linking the broad and heterogeneous categories of the ecological complex. The arrows are meant only to suggest the existence of problems for research concerning the mechanisms of cause, influence, or response at work in the situation so sketchily portrayed. Even the barest account of that situation, however, can leave no doubt that social change and environmental modification occurred in the closest interdependence—so close, in fact, that the two "levels" of change were *systematically* interrelated. Change on either level can be comprehended only by application of a conceptual scheme at least as encompassing as that of ecosystem.

The reader's imagination, again, must substitute for documentation of the point that smog, though a spectacular case and full of human interest, is no isolated example of how problems of human collective existence require an ecosystem framework for adequate conceptualization. I do not intend to argue, of course, that sociologists must somehow shoulder the entire burden of research suggested by such a conceptualization. Science, after all, is one of our finest examples of the advantages of a division of labor. But labor can be effectively divided only if there is articulation of the several sub-tasks; in scientific work, such articulation is achieved by employment of a common conceptual framework.

Sociologists may or may not—I am not especially optimistic on this score—take up the challenge to investigate the social life of man as a phase of the ecosystem, with all the revisions in their thought patterns that this kind of formulation will demand. If they shirk this responsibility, however, other disciplines are not unprepared to take the leadership. Anthropology of late has demonstrated its hospitality to ecological concepts.[18] Geography, for its part, cannot forget that it laid claim to human ecology as early as did sociology.[19]

Of even greater ultimate significance may be the impending reorientation of much of what we now call social science to such concepts as welfare, level of living, and public health. Programs to achieve such "national goals" (to use the former President's language), like the studies on which such programs are based, are finding and will find two things: first, each of these concepts is capable of almost indefinite expansion to comprehend virtually any problem of human collective life; and, second, measures or indicators of status or progress in respect to them must be multi-faceted and relational. Public health, to take that example, is surely some sort of function of all elements in the ecological complex; it is observable in any sufficiently comprehensive sense only in terms of interrelations of variables located at all levels of the ecosystem. Extrapolation of current trends over even a short projection period is sufficient to suggest the future preoccupation of the sciences touching on man with much more macroscopic problems than they now dare to set for themselves. It is perhaps symptomatic that spokesmen for the nation's health programs now declare that the "science of health is a branch of the wider science of human ecology,"[20] and that expositions of the problem of economic development have come to emphasize the necessary shift "From Political Economy to Political Ecology."[21] Even the literati proclaim that the "fundamental human problem is ecological."[22] (Cf. the similar remark of Kenneth Burke: "Among the sciences, there is one little fellow named Ecology, and in time we shall pay him more attention."[23]) If one holds with Durkheim that the basic categories of science, as well as the interpretive schemes of everyday life, arise from the nature and exigencies of human collective existence, it cannot be long before we are forced to conjure with some version of the ecosystem concept. The question is whether sociology will lead or lag behind in this intellectual movement.

[18]Marston Bates, "Human Ecology," in *Anthropology Today,* edited by A. L. Kroeber (Chicago: University of Chicago Press, 1953); J. G. D. Clark, *Prehistoric Europe: The Economic Basis* (New York: Philosophical Library, 1952); Julian H. Steward, *Theory of Culture Change* (Urbana: University of Illinois Press, 1955).

[19]H. H. Barrows, "Geography as Human Ecology," *Annals of the Association of American Geographers,* 13 (March 1923), pp. 1–14; William L. Thomas, Jr., editor, *Man's Role in Changing the Face of the Earth* (Chicago: University of Chicago Press, 1956).

[20]President's Commission on the Health Needs of the Nation, *America's Health Status, Needs and Resources,* Vol. 2: *Building America's Health* (Washington: Government Printing Office, 1953), p. 13.

[21]Title of an essay by Bertrand de Jouvenel, *Bulletin of the Atomic Scientists,* 8 (October 1957), pp. 287–91.

[22]Aldous Huxley, *The Devils of Loudon,* "Torchbook edition" (New York: Harper & Bros., 1959), p. 302.

[23]Kenneth Burke, *Attitudes Toward History,* Vol. I (New York: The New Republic, 1937), p. 192.

Sentiment and Symbolism as Ecological Variables*

Walter Firey

Systematization of ecological theory has thus far proceeded on two main premises regarding the character of space and the nature of locational activities. The first premise postulates that the sole relation of space to locational activities is an impeditive and cost-imposing one. The second premise assumes that locational activities are primarily economizing, "fiscal" agents.[1] On the basis of these two premises the only possible relationship that locational activities may bear to space is an economic one. In such a relationship each activity will seek to so locate as to minimize the obstruction put upon its functions by spatial distance. Since the supply of the desired locations is limited it follows that not all activities can be favored with choice sites. Consequently a competitive process ensues in which the scarce desirable locations are pre-empted by those locational activities which can so exploit advantageous location as to produce the greatest surplus of income over expenditure. Less desirable locations devolve to correspondingly less economizing land uses. The result is a pattern of land use that is presumed to be most efficient for both the individual locational activity and for the community.

Given the contractualistic milieu within which the modern city has arisen and acquires its functions, such an "economic ecology" has had a certain explanatory adequacy in describing urban spatial structure and dynamics. However, as any theory matures and approaches a logical closure of its generalizations it inevitably encounters facts which remain unassimilable to the theoretical scheme. In this paper it will be our purpose to describe certain ecological processes which apparently cannot be embraced in a strictly economic analysis. Our hypothesis is that the data to be presented, while in no way startling or unfamiliar to the research ecologist, do suggest an alteration of the basic premises of ecology. This alteration would consist, first, of ascribing to space not only an impeditive quality but also an additional property, *viz.,* that of being at times a symbol for certain cultural values that have become associated with a certain spatial area. Second, it would involve a recognition that locational activities are not only economizing agents but may also bear

---

*Reprinted from the *American Sociological Review*, 10 (April 1945), 140–48, by permission of the author and The American Sociological Association.

[1] See Everett C. Hughes, "The Ecological Aspect of Institutions," *American Sociological Review*, 1 (April 1936), 180–9.

sentiments which can significantly influence the locational process.²

A test case for this twofold hypothesis is afforded by certain features of land use in central Boston. In common with many of the older American cities Boston has inherited from the past certain spatial patterns and landmarks which have had a remarkable persistence and even recuperative power despite challenges from other more economic land uses. The persistence of these spatial patterns can only be understood in terms of the group values that they have come to symbolize. We shall describe three types of such patterns: first, an in-town upper class residential neighborhood known as Beacon Hill; second, certain "sacred sites," notably the Boston Common and the colonial burying grounds; and third, a lower class Italian neighborhood known as the North End. In each of these land uses we shall find certain locational processes which seem to defy a strictly economic analysis.

The first of the areas, Beacon Hill, is located some five minutes' walking distance from the retail center of Boston. This neighborhood has for fully a century and a half maintained its character as a preferred upper class residential district, despite its contiguity to a low rent tenement area, the West End. During its long history Beacon Hill has become the symbol for a number of sentimental associations which constitute a genuine attractive force to certain old families of Boston. Some idea of the nature of these sentiments may be had from statements in the innumerable pamphlets and articles written by residents of the Hill. References to "this sacred eminence," "stately old-time appearance," and "age-old quaintness and charm," give an insight into the attitudes attaching to the area. One resident reveals rather clearly the spatial referability of these sentiments when she writes of the Hill:

> It has a tradition all its own, that begins in the hospitality of a book-lover, and has never lost that flavor. Yes, our streets are inconvenient, steep, and slippery. The corners are abrupt, the contours perverse.... It may well be that the gibes of our envious neighbors have a foundation and that these dear crooked lanes of ours were indeed traced in ancestral mud by absent-minded kine.³

²Georg Simmel, "Der Raum und die raumlichen Ordnungen der Gesellschaft," *Soziologie* (Munich: 1923), pp. 518–22; cf. Hughes, ibid.
³Abbie Farwell Brown, *The Lights of Beacon Hill* (Boston, 1922), p.4.

Behind such expressions of sentiment are a number of historical associations connected with the area. Literary traditions are among the strongest of these; indeed, the whole literary legend of Boston has its focus at Beacon Hill. Many of America's most distinguished literati have occupied homes on the Hill. Present day occupants of these houses derive a genuine satisfaction from the individual histories of their dwellings. One lady whose home had had a distinguished pedigree remarked:

> I like living here for I like to think that a great deal of historic interest has happened here in this room.

Not a few families are able to trace a continuity of residence on the Hill for several generations, some as far back as 1800 when the Hill was first developed as an upper class neighborhood. It is a point of pride to a Beacon Hill resident if he can say that he was born on the Hill or was at least raised there; a second best boast is to point out that his forebears once lived on the Hill.

Thus a wide range of sentiments—aesthetic, historical, and familial—have acquired a spatial articulation in Beacon Hill. The bearing of these sentiments upon locational processes is a tangible one and assumes three forms: retentive, attractive, and resistive. Let us consider each of these in order. To measure the retentive influence that spatially referred sentiments may exert upon locational activities we have tabulated by place of residence all the families listed in the Boston Social Register for the years 1894, 1905, 1914, 1929, and 1943. This should afford a reasonably accurate picture of the distribution of upper class families by neighborhoods within Boston and in suburban towns. In Table 1 we have presented the tabulations for the three in-town concentrations of upper class families (Beacon Hill, Back Bay, and Jamaica Plain) and for the five main suburban concentrations (Brookline, Newton, Cambridge, Milton, and Dedham).... The most apparent feature of these data is, of course, the consistent increase of upper class families in the suburban towns and the marked decrease (since 1905) in two of the in-town upper class areas, Back Bay and Jamaica Plain. Although both of these neighborhoods remain fashionable residential districts their prestige is waning rapidly. Back Bay in particular, though still surpassing in numbers any other single neighborhood, has undergone a steady invasion of apartment buildings, rooming houses, and business

Table 1. Number of Upper Class Families in Boston, by Districts of Concentration, and in Main Suburban Towns, for Certain Years

	1894	1905	1914	1929	1943
Within Boston					
Beacon Hill	280	242	279	362	335
Back Bay	867	1166	1102	880	556
Jamaica Plain	56	66	64	36	30
Other districts	316	161	114	86	41
Suburban Towns					
Brookline	137	300	348	355	372
Newton	38	89	90	164	247
Cambridge	77	142	147	223	257
Milton	37	71	106	131	202
Dedham	8	29	48	69	99
Other towns	106	176	310	403	816
Total in Boston	1519	1635	1559	1364	962
Total in Suburbs	403	807	1049	1345	1993
Totals	1922	2442	2608	2709	2955

Tabulated from: Social Register, Boston

establishments which are destroying its prestige value. The trend of Beacon Hill has been different. Today it has a larger number of upper class families than it had in 1894. Where it ranked second among fashionable neighborhoods in 1894 it ranks third today, being but slightly outranked in numbers by the suburban city of Brookline and by the Back Bay. Beacon Hill is the only in-town district that has consistently retained its preferred character and has held to itself a considerable proportion of Boston's old families.

There is, however, another aspect to the spatial dynamics of Beacon Hill, one that pertains to the "attractive" locational role of spatially referred sentiments. From 1894 to 1905 the district underwent a slight drop, subsequently experiencing a steady rise for 24 years, and most recently undergoing another slight decline. These variations are significant, and they bring out rather clearly the dynamic ecological role of spatial symbolism. The initial drop is attributable to the development of the then new Back Bay. Hundreds of acres there had been reclaimed from marshland and had been built up with palatial dwellings. Fashion now pointed to this as the select area of the city and in response to its dictates a number of families abandoned Beacon Hill to take up more pretentious Back Bay quarters. Property values on the Hill began to depreciate, old dwellings became rooming houses, and businesses began to invade some of the streets. But many of the old families remained on the Hill and a few of them made efforts to halt the gradual deterioration of the district. Under the aegis of a realtor, an architect, and a few close friends there was launched a program of purchasing old houses, modernizing the interiors and leaving the colonial exteriors intact, and then selling the dwellings to individual families for occupancy. Frequently adjoining neighbors would collaborate in planning their improvements so as to achieve an architectural consonance. The results of this program may be seen in the drift of upper class families back to the Hill. From 1905 to 1929 the number of *Social Register* families in the district increased by 120. Assessed valuations showed a corresponding increase: from 1919 to 1924 there was a rise of 24 percent; from 1924 to 1929 the rise was 25 percent.[4] The nature of the Hill's appeal, and the kind of persons attracted, may be gathered from the following popular write-up:

> To salvage the quaint charm of Colonial Architecture on Beacon Hill, Boston, is the object of a well-defined movement among writers and professional folk that promises the most delightful opportunities for the home seeker of moderate means and conservative tastes. Because men of discernment were able to visualize the possibilities presented by these architectural landmarks, and have undertaken the gracious task of restoring them to their former

[4] *The Boston Transcript,* April 12, 1930.

glory, this historic quarter of Old Boston, once the centre of literary culture, is coming into its own.[5]

The independent variable in this "attractive" locational process seems to have been the symbolic quality of the Hill, by which it constituted a referent for certain strong sentiments of upper class Bostonians.

While this revival was progressing there remained a constant menace to the character of Beacon Hill, in the form of business encroachments and apartment-hotel developments. Recurrent threats from this source finally prompted residents of the Hill to organize themselves into the Beacon Hill Association. Formed in 1922, the declared object of this organization was "to keep undesirable business and living conditions from affecting the hill district."[6] At the time the city was engaged in preparing a comprehensive zoning program and the occasion was propitious to secure for Beacon Hill suitable protective measures. A systematic set of recommendations was drawn up by the Association regarding a uniform 65-foot height limit for the entire Hill, the exclusion of business from all but two streets, and the restriction of apartment house bulk.[7] It succeeded in gaining only a partial recognition of this program in the 1924 zoning ordinance. But the Association continued its fight against inimical land uses year after year. In 1927 it successfully fought a petition brought before the Board of Zoning Adjustment to alter the height limits in one area so as to permit the construction of a four million dollar apartment-hotel 155 feet high. Residents of the Hill went to the hearing en masse. In spite of the prospect of an additional twenty million dollars worth of exclusive apartment-hotels that were promised if the zoning restrictions were withheld the petition was rejected, having been opposed by 214 of the 220 persons present at the hearing.[8] In 1930 the Association gained an actual reduction in height limits on most of Beacon Street and certain adjoining streets, though its leader was denounced by opponents as "a rank sentimentalist who desired to keep Boston a village."[9] One year later the Association defeated a petition to rezone Beacon Street for business purposes.[10] In other campaigns the Association successfully pressed for the rezoning of a business street back to purely residential purposes, for the lowering of height limits on the remainder of Beacon Street, and for several lesser matters of local interest. Since 1929, owing partly to excess assessed valuations of Boston real estate and partly to the effects of the depression upon families living on securities, Beacon Hill has lost some of its older families, though its decline is nowhere near so precipitous as that of the Back Bay.

Thus for a span of one and a half centuries there have existed on Beacon Hill certain locational processes that largely escape economic analysis. It is the symbolic quality of the Hill, not its impeditive or cost-imposing character, that most tangibly correlates with the retentive, attractive, and resistive trends that we have observed. And it is the dynamic force of spatially referred sentiments, rather than considerations of rent, which explains why certain families have chosen to live on Beacon Hill in preference to other in-town districts having equally accessible location and even superior housing conditions. There is thus a non-economic aspect to land use on Beacon Hill, one which is in some respects actually dis-economic in its consequences. Certainly the large apartment-hotels and specialty shops that have sought in vain to locate on the Hill would have represented a fuller capitalization on potential property values than do residences. In all likelihood the attending increase in real estate prices would not only have benefited individual property holders but would have so enhanced the value of adjoining properties as to compensate for whatever depreciation other portions of the Hill might have experienced.

If we turn to another type of land use pattern in Boston, that comprised by the Boston Common and the old burying grounds, we encounter another instance of spatial symbolism which has exerted a marked influence upon the ecological organization of the rest of the city. The Boston Common is a survival from colonial days when every New England town allotted a portion of its land to common use as a cow pasture and militia field. Over the course of three centuries Boston has grown entirely around the Common so that today we find a 48-acre tract of land wedged directly into the heart of the business district. On three of its five sides are women's apparel shops, department stores, theaters, and other high-rent locational activities. On the fourth side is Beacon Street, extending alongside Beacon Hill. Only the activities of the Hill residents have prevented business from invading this side. The fifth side is

[5]Harriet Sisson Gillespie, "Reclaiming Colonial Landmarks," *The House Beautiful,* 58 (September 1925), 239–41.
[6]*The Boston Transcript,* December 6, 1922.
[7]*The Boston Transcript,* March 18, 1933.
[8]*The Boston Transcript,* January 29, 1927.
[9]*The Boston Transcript,* April 12, 1930.
[10]*The Boston Transcript,* January 10 and January 29, 1931.

occupied by the Public Garden. A land value map portrays a strip of highest values pressing upon two sides of the Common, on Tremont and Boylston streets, taking the form of a long, narrow band.

Before considering the ecological consequences of this configuration let us see what attitudes have come to be associated with the Common. There is an extensive local literature about the Common and in it we find interesting sentiments expressed. One citizen speaks of:

> ... the great principle exemplified in the preservation of the Common. Thank Heaven, the tide of money making must break and go around that.[11]

Elsewhere we read:

> Here, in short, are all our accumulated memories, intimate, public, private.[12]

> Boston Common was, is, and ever will be a source of tradition and inspiration from which the New Englanders may renew their faith, recover their moral force, and strengthen their ability to grow and achieve.[13]

The Common has thus become a "sacred" object, articulating and symbolizing genuine historical sentiments of a certain portion of the community. Like all such objects its sacredness derives, not from any intrinsic spatial attributes, but rather from its representation in people's minds as a symbol for collective sentiments.

Such has been the force of these sentiments that the Common has become buttressed up by a number of legal guarantees. The city charter forbids Boston in perpetuity to dispose of the Common or any portion of it. The city is further prohibited by state legislation from building upon the Common, except within rigid limits, or from laying out roads or tracks across it. By accepting the bequest of one George F. Parkman, in 1908, amounting to over five million dollars, the city is further bound to maintain the Common, and certain other parks, "for the benefit and enjoyment of its citizens."

What all this has meant for the spatial development of Boston's retail center is clear from the present character of that district. Few cities of comparable size have so small a retail district in point of area. Unlike the spacious department stores of most cities, those in Boston are frequently compressed within narrow confines and have had to extend in devious patterns through rear and adjoining buildings. Traffic in downtown Boston has literally reached the saturation point, owing partly to the narrow one-way streets but mainly to the lack of adequate arterials leading into and out of the Hub. The American Road Builders Association has estimated that there is a loss of $81,000 per day in Boston as a result of traffic delay. Trucking in Boston is extremely expensive. These losses ramify out to merchants, manufacturers, commuters, and many other interests. Many proposals have been made to extend a through arterial across the Common, thus relieving the extreme congestion on Tremont and Beacon streets, the two arterials bordering the park. Earlier suggestions, prior to the construction of the subway, called for street car tracks across the Common. But "the controlling sentiment of the citizens of Boston, and of large numbers throughout the State, is distinctly opposed to allowing any such use of the Common."[14] Boston has long suffered from land shortage and unusually high real estate values as a result both of the narrow confines of the peninsula comprising the city center and of the exclusion from income-yielding uses of so large a tract as the Common. A further difficulty has arisen from the rapid southwesterly extension of the business district in the past two decades. With the Common lying directly in the path of this extension the business district has had to stretch around it in an elongated fashion, with obvious inconvenience to shoppers and consequent loss to business.

The Common is not the only obstacle to the city's business expansion. No less than three colonial burying grounds, two of them adjoined by ancient church buildings, occupy downtown Boston. The contrast that is presented by 9-story office buildings reared up beside quiet cemeteries affords visible evidence of the conflict between "sacred" and "profane" that operates in Boston's ecological patterns. The dis-economic consequences of commercially valuable land being thus devoted to non-utilitarian purposes goes even further than the removal from business uses of a given amount of space. For it is a standard principle of real estate that business property derives added value if adjoining properties are occupied by other busi-

---

[11]Speech of William Everett, quoted in *The Boston Transcript*, March 7, 1903.

[12]T. R. Sullivan, *Boston New and Old* (Boston, 1912), pp. 45–46.

[13]Joshua H. Jones, Jr., "Happenings on Boston Common," *Our Boston*, 2 (January 1927), 9–15.

[14]*First Annual Report of the Boston Transit Commission* (Boston, 1895), p. 9.

nesses. Just as a single vacancy will depreciate the value of a whole block of business frontage, so a break in the continuity of stores by a cemetery damages the commercial value of surrounding properties. But, even more than the Common, the colonial burying grounds of Boston have become invested with a moral significance which renders them almost inviolable. Not only is there the usual sanctity which attaches to all cemeteries, but in those of Boston there is an added sacredness growing out of the age of the grounds and the fact that the forebears of many of New England's most distinguished families as well as a number of colonial and Revolutionary leaders lie buried in these cemeteries. There is thus a manifold symbolism to these old burying grounds, pertaining to family lineage, early nationhood, civic origins, and the like, all of which have strong sentimental associations. What has been said of the old burying grounds applies with equal force to a number of other venerable landmarks in central Boston. Such buildings as the Old South Meeting-House, the Park Street Church, King's Chapel, and the Old State House—all foci of historical associations—occupy commercially valuable land and interrupt the continuity of business frontage on their streets. Nearly all of these landmarks have been challenged at various times by real estate and commercial interests which sought to have them replaced by more profitable uses. In every case community sentiments have resisted such threats.

In all these examples we find a symbol-sentiment relationship which has exerted a significant influence upon land use. Nor should it be thought that such phenomena are mere ecological "sports." Many other older American cities present similar locational characteristics. Delancey Street in Philadelphia represents a striking parallel to Beacon Hill, and certain in-town districts of Chicago, New York, and Detroit, recently revived as fashionable apartment areas, bear resemblances to the Beacon Hill revival. The role of traditionalism in rigidifying the ecological patterns of New Orleans has been demonstrated in a recent study.[15] Further studies of this sort should clarify even further the true scope of sentiment and symbolism in urban spatial structure and dynamics.

As a third line of evidence for our hypothesis we have chosen a rather different type of area from those so far considered. It is a well known fact that

[15]H. W. Gilmore, "The Old New Orleans and the New: A Case for Ecology," *American Sociological Review*, 9 (August 1944), 385–94.

immigrant ghettoes, along with other slum districts, have become areas of declining population in most American cities. A point not so well established is that this decline tends to be selective in its incidence upon residents and that this selectivity may manifest varying degrees of identification with immigrant values. For residence within a ghetto is more than a matter of spatial placement; it generally signifies acceptance of immigrant values and participation in immigrant institutions. Some light on this process is afforded by data from the North End of Boston. This neighborhood, almost wholly Italian in population, has long been known as "Boston's classic land of poverty." Eighteen percent of the dwellings are eighty or more years old and sixty percent are forty or more years old.[16] Indicative of the dilapidated character of many buildings is the recent sale of a 20-room apartment building for only $500. It is not surprising then to learn that the area has declined in population from 21,111 in 1930 to 17,598 in 1940.[17] To look for spatially referable sentiments here would seem futile. And yet, examination of certain emigration differentials in the North End reveals a congruence between Italian social structure and locational processes. To get at these differentials recourse was had to the estimation of emigration, by age groups and by nativity, through the use of life tables. The procedure consists of comparing the actual 1940 population with the residue of the 1930 population which probably survived to 1940 according to survival rates for Massachusetts. Whatever deficit the actual 1940 population may show from the estimated 1940 population is a measure of "effective emigration." It is not a measure of the actual volume of emigration, since no calculation is made of immigration *into* the district between 1930 and 1940.[18] Effective emigration simply indicates the

[16]Finance Commission of the City of Boston, *A Study of Certain of the Effects of Decentralization on Boston and Some Neighboring Cities and Towns* (Boston, 1941), p. 11.

[17]Aggregate population of census tracts F1, F2, F4, F5: *Census Tract Data, 1930 Census,* unpublished material from 15th Census of the United States, 1930, compiled by Boston Health Department, table 1; *Population and Housing—Statistics for Census Tracts, Boston,* 16th Census of the United States, 1940, table 2.

[18]By use of *Police Lists* for two different years a count was made of immigration into a sample precinct of the North End. The figure (61) reveals so small a volume of immigration that any use of it to compute actual emigration by age groups would have introduced statistical unreliability into the estimates. Survival rates for Massachusetts were computed from state life tables in: National Resources Committee, *Population Statistics,* 2.

extent of population decline which is attributable to emigration rather than to death. Computations thus made for emigration differentials by nativity are shown in Table 2. The second generation, comprising but 59.46 percent of the 1930 population, contributed 76.42 percent of the effective emigration from the North End, whereas the first generation accounted for much less than its "due" share of the emigration. Another calculation shows that where the effective emigration of second generation Italians represents 27.08 percent of their number in 1930, that of the first generation represents only 12.26 percent of their number in 1930.

Equally clear differentials appear in effective emigration by age groups. If we compare the difference between the percentage which each age group as of 1930 contributes to the effective emigration, and the percentage which each age group comprised of the 1930 population, we find that the age groups 15–24 account for much more than their share of effective emigration; the age groups 35–64 account for much less than their share. In Table 3 the figures preceded by a plus sign indicate "excess" emigration, those preceded by a minus sign indicate "deficit" emigration.

In brief, the North End is losing its young people to a much greater extent than its older people.

These differentials are in no way startling; what is interesting, however, is their congruence with basic Italian values, which find their fullest institutionalized expression in the North End. Emigra-

State Data (Washington, 1937), Part C, p. 38. The technique is outlined in C. Warren Thornthwaite, *Internal Migration in the United States* (Philadelphia, 1934), pp. 19–21.

tion from the district may be viewed as both a cause and a symbol of alienation from these values. At the core of the Italian value system are those sentiments which pertain to the family and the *paesani*. Both of these put a high premium upon maintenance of residence in the North End.

Paesani, or people from the same village of origin, show considerable tendency to live near one another, sometimes occupying much of a single street or court.[19] Such proximity, or at least common residence in the North End, greatly facilitates participation in the *paesani* functions which are so important to the first generation Italian. Moreover, it is in the North End that the *festas,* anniversaries, and other old world occasions are held, and such is their frequency that residence in the district is almost indispensable to regular participation. The social relationships comprised by these groupings, as well as the benefit orders, secret societies, and religious organizations, are thus strongly localistic in character. One second generation Italian, when asked if his immigrant parents ever contemplated leaving their North End tenement, replied: "No, because all their friends are there, their relatives. They know everyone around there." It is for this reason that the first generation Italian is so much less inclined to leave the North End than the American-born Italian.

Equally significant is the localistic character of the Italian family. So great is its solidarity that it is not uncommon to find a tenement entirely occupied by a single extended family: grandparents, mature children with their mates, and grandchildren. There are instances where such a family has overflowed one tenement and has expanded into

[19]William Foote Whyte, *Street Corner Society* (Chicago, 1943), p. xix.

Table 2. Effective Emigration From the North End, Boston, 1930 to 1940, by Nativity

Nativity	1930 Population	Percent of 1930 Pop. in each Nativity Group	Effective Emigration 1930–1940	Percent of Emigration accounted for by each Nativity Group
American-born (second generation)	12553	59.46	3399	76.42
Italian-born (first generation)	8557	40.54	1049	23.58
Totals	21110	100.00	4448	100.00

Calculated from: Census tract data and survival rates

Table 3. Difference Between Percentage Contributed by Each Age Group to Effective Emigration and Percentage It Comprised of 1930 Population

Age Groups as of 1930	Differences between Percentages	
	Male	Female
Under 5	−1.70	−0.33
5–9	+0.38	+0.04
10–14	+0.21	+2.66
15–19	+4.18	+3.01
20–24	+2.04	+2.35
25–34	−0.97	−0.07
35–44	−2.31	−1.09
45–54	−1.43	−1.17
55–64	−2.29	−1.19
65–74	−1.13	−0.59
75 and over	uncalculable	

Calculated from: Census tract data and survival rates

an adjoining one, breaking out the partitions for doorways. These are ecological expressions, in part, of the expected concern which an Italian mother has for the welfare of her newly married daughter. The ideal pattern is for the daughter to continue living in her mother's house, with she and her husband being assigned certain rooms which they are supposed to furnish themselves. Over the course of time the young couple is expected to accumulate savings and buy their own home, preferably not far away. Preferential renting, by which an Italian who owns a tenement will let apartments to his relatives at a lower rental, is another manifestation of the localizing effects of Italian kinship values.

Departure from the North End generally signifies some degree of repudiation of the community's values. One Italian writes of an emigrant from the North End:

> I still remember with regret the vain smile of superiority that appeared on his face when I told him that I lived at the North End of Boston. *"Io non vada fra quella plebaglia."* (I do not go among those plebeians.)[20]

As a rule the older Italian is unwilling to make this break, if indeed he could. It is the younger adults, American born and educated, who are capable of making the transition to another value system with radically different values and goals.

Residence in the North End seems therefore to be a spatial corollary to integration with Italian values. Likewise emigration from the district signifies assimilation into American values, and is so construed by the people themselves. Thus, while the area is not the conscious object of sentimental attachment, as are Beacon Hill and the Common, it has nonetheless become a symbol for Italian ethnic solidarity. By virtue of this symbolic quality the area has a certain retentive power over those residents who most fully share the values which prevail there.

It is reasonable to suggest, then, that the slum is much more than "an area of minimum choice."[21] Beneath the surface phenomenon of declining population there may be differential rates of decline which require positive formulation in a systematic ecological theory. Such processes are apparently refractory to analysis in terms of competition for least impeditive location. A different order of concepts, corresponding to the valuative, meaningful aspect of spatial adaptation, must supplement the prevailing economic concepts of ecology.

[20] Enrico C. Sartorio, *Social and Religious Life of Italians in America* (Boston, 1918), pp. 43–44.

[21] R. D. McKenzie, "The Scope of Human Ecology," in Ernest W. Burgess, ed., *The Urban Community* (Chicago, 1926), p. 180.

Communities in the Salt Lake Basin*

Albert L. Seeman

A pioneer company of Mormons headed by Brigham Young entered the Salt Lake Valley on July 24, 1847, through the pass now known as Emigration Canyon. Among the first undertakings after reaching the present site of Salt Lake City was the diversion of water from City Creek and the planting of a few acres of potatoes, buckwheat, and turnips. Because of extreme dryness of the soil, plowing was at first difficult, but a dam was put across the Creek and the soil was well flooded, making plowing comparatively easy. In spite of late planting they harvested a fairly successful crop.

During his life Brigham Young insisted that the Mormon people should make agriculture their chief occupation, rather than prospect for mineral resources, which in their richness and extent offered shortcuts to material wealth. He argued that quick wealth was unwholesome and tended to break down the ideal of sturdy character which he had set up as a mold for his people. His advice that they remain on their farms was religiously followed.

As rapidly as new arrivals followed the first colonizers, exploring parties were sent out to find new sites for homes. Selection was exercised in the personnel of both the exploring parties and the settlers for the sites which had been chosen. It was Brigham Young's plan to settle the country as rap-

*Reprinted from *Economic Geography,* 14 (July 1938), 300–308, by permission of *Economic Geography.*

idly as possible and thus to form a new Empire in the west. In a short time along the entire east side of the Jordan River, wherever the wall of the Wasatch Mountains is cut by a canyon from which flows a mountain stream, was to be found a settlement. The same plan is being used today. Three miles from Malta, Idaho, one may see modern Mormon pioneering. The Mormon Church purchased a 4,000-acre ranch as a colony site and a group of Mormons, as pioneers, have moved in on the land. When the writer visited the settlement in September 1937, it was apparent that Mormon pioneering of 1937 was virtually identical in philosophy and motive with Mormon pioneering of 1847.

The colonizers went as agriculturists and yet they did not follow the practice of other agricultural communities in the settlement of the west. Instead of selecting and settling on a farm, they selected a townsite and settled in the town, arranging their farms around the townsite. Murray King (*The Nation,* June 28, 1922, p. 769) made the following observation: "The Church, the desert, and the canyon stream have conspired to produce a village concentration. The Church has created so many religious activities, and so monopolized social activities that it cannot carry out its program except in organized communities. There is little isolated rural living in Utah. The man who cultivates the soil is the main pillar of a highly structured town life. . . ." While the church may have wanted or desired such communities yet it is the environment which made this demand and desire possible. Their complete isolation from other settled parts of the United States made it necessary for them to furnish their own protection from the Indians. Necessity of artificially watering the soil made it imperative, at that time, that they coöperate in the construction of all irrigation projects. Individual efforts would have been futile in irrigation farming with conditions such as are found along the west side of the Wasatch Mountains. Consequently, the Mormon community of the Salt Lake Basin is a peculiar social structure and a unique American institution. It is the outgrowth of a religious and social ideal, influenced and patterned by the environment.

Mormon communities depart from general pattern

The economic town in its simplest form is a market place where goods may be exchanged. Exchange of goods has characterized all stages of known human history. It is an obvious means of getting what one does not have in return for what one already possesses in quantity beyond one's immediate needs. In nomadic stages the exchanges are made at *various* favorable places such as at boundaries of grazing lands of each tribe. In early phases of the agricultural stage, the market was *definitely* located at one place accessible to many people. At the market, farmer exchanged products with farmer and each gave of his surplus to the other in return for part of the other's surplus, whatever it might be. Trade was direct. Original producer exchanged with original producer.

When there came into being a class of traders who resided in a market village, traders who had stores which not only supplemented the market but which became its rivals, then the economic town began to develop into a modern complex institution. The town economy was an organization of consumers and producers where townsmen worked out their dependence on one another and the outside world through the agency of a town. The early towns of necessity laid stress on fortification and therefore tended to be compact, but otherwise they developed without any plan; their growth was merely accretion. As towns grew in size and number, more manufacturing and thereby more commerce developed; consequently towns became more and more important.

Towns and cities generally originated and developed according to this trend. But, as is the case with all generalities, there are exceptions. Towns of the Salt Lake Basin offer a good example of this exception. Towns founded by the Mormons in the Salt Lake Basin depart from the general pattern. Towns were definitely planned and their primary purpose was not to exchange goods. The original plan and purpose of the Mormons can be seen in all three types of communities in the Salt Lake Basin.

Types of communities

The Mormon communities of the Salt Lake Basin may be classified into three main groups: (1) Rural Village, (2) Rural Town, and (3) Urban.

Rural village

Rural Village refers to that community made up largely of farm people, residing in a relatively compact geographical area, with farms outlying and adjacent to the community; the farmers must travel to their farms for their daily duties. These *Rural Villages* are usually incorporated but the fields are not included in the corporate limits. The following *Rural Villages* were included in the writer's field study: Snowville, Bear River City, Corinne, Willard, Pleasant Grove, Lindon, Orem, Salem, Santaquin, Mona, Fountain Green, Moroni, and Ephraim.

Such a community is composed of a main street with side streets running perpendicular and parallel to the main street. The houses are usually of brick surrounded by well-kept lawns. Behind the house will be found a general farmyard with barns and other outbuildings, haystacks, and livestock pens. The church, social hall, and school are usually juxtaposed and in the center of the community. Thus the community becomes a series of farmsteads grouped around a church, school, and social hall. The stores and other small service establishments may be tended by the women while the men attend to the fields.

The holdings of the individual farmers around the community are all small. This has been brought about by two factors. The first of these is that irrigation farming results in relatively smaller holdings than are found in humid areas of the United States. The second factor is the influence of the church.

An outstanding characteristic of the Mormon groups with their strong religious motivation was their passion for equality. In many cases it became so strong that it resulted in a communistic organization. Under conditions of private ownership the land was divided into lots of equal size and quality as far as possible, and possession determined by drawing lots. In some settlements, however, no attempt was made to have a common ownership of the land, private property having been instituted from the beginning. At all times great care was taken to insure an equal distribution of the land. Generally the initial allotments were twenty acres. As Mormon settlers joined the community later the land would be redivided, reducing the original holdings of each farmer. Most of the settlers were Europeans, accustomed to small farms, who had little or no conception of the usual size of land holdings in the new world. Also, in the settlement of land by Mormons, emphasis was placed upon community building. This was based on the tenet of the church which required its members to live simply and not to seek individual wealth. There was, accordingly, a strong tendency away from large individual holdings, and land was divided into small parcels to encourage many to settle in the respective communities.

EPHRAIM: AN EXAMPLE OF A RURAL VILLAGE. Ephraim is located in Sanpete County about 120 miles south of Salt Lake City. It is situated near the mouth of Ephraim Canyon on the east side of the valley. The source of irrigation water is Ephraim Creek and Sanpitch River.

Two years after the Mormon pioneers arrived in Salt Lake Valley, Brigham Young sent a colony to settle Sanpete Valley. They founded the settlement of Manti in 1849. In 1850 an effort was made by a settler from Manti to establish a colony in what is now Ephraim, but the Indians forced him to return to Manti. It was not until the early spring of 1854 that a colony of fifteen families succeeded in founding a settlement there.

The first necessity confronting the pioneers was a means of defense against the Indians. Accordingly a fort was constructed as rapidly as possible. For this purpose men were organized as a military unit under the general direction of the Bishop. Some were assigned to guard duty while others worked on the fort. In the center of the enclosure was the "meeting house" for church services.

In addition to the necessity of preparing a defense, there were the crops to be planted and cared for, canals to be dug, and roads and bridges to be constructed. New settlers were continually coming who had to be fed and housed; although they increased the supply of labor they reduced the supply of food which was not too plentiful.

The community was settled originally by American families, but almost immediately the first immigrants arrived from Denmark and the Scandinavian Peninsula. Common religion and the absolute necessity of united group action brought about an amalgamation which otherwise might have been most difficult.

First title to the lands around Ephraim, as with most early Utah holdings, was obtained under the Preëmption Act. One member of the group was selected to file on the necessary 160-acre claim and he in turn divided the tract and gave a deed to each member of the group for the subdivisions. In return the others would pay him a certain amount to compensate him for his trouble and expense, and for the fact that he was spending his "homestead right." Every precaution was taken to insure a fair division of the land from the standpoint of quantity and of quality. The best land was divided among all of the farmers as was the poorer land. As a result more than 50 percent of the farmers have their holdings in more than one piece of land and some farmers may have as many as twelve pieces of ground; yet their holdings are not large. The average size of an irrigated farm around Ephraim is about fifty acres.

Rural town

The *Rural Town* differs from the *Rural Village* in the generally accepted distinction made by geographers between a town and a village, i.e., a village is a community engaged in concentrating agricultural production and distributing goods to that same agricultural group. A town is a community that carries on similar activities but in addition has some facilities for processing those commodities. Thus the determining factor in making the distinction between a town and a village is not size but rather the economic activity. These *Rural Towns* differ from towns in other parts of the United States in that the bulk of their population consists of farmers who are actually engaged in tilling the soil but who live in juxtaposition as was the case in the *Rural Village*.

The following *Rural Towns* were included in the field study: Tremonton, American Fork, Springfield, Spanish Fork, Payson, and Nephi.

All of the *Rural Towns* investigated were found to be larger than the *Rural Village,* and had a well-defined business district which consisted of general stores, restaurants, hotels, banks, and other service establishments together with some processing plants of one type or another. These processing plants were canneries of fruits and vegetables, meat packing plants, sugar beet plants, and creameries.

The urban community

All of the *Urban* communities, the third type, are patterned after the first city, Salt Lake City. The third type of community differs fundamentally from the other two communities in that the population is not composed of farmers actually engaged in tilling the soil; there are some who manage farms, but they are primarily engaged in some other occupation. These communities resemble most other cities in the rest of the United States in that they have definite police and fire departments, educational centers, large centers of religious activities, and are focal points of industry and commerce, but they differ in that they are not a mere accretion of population. Instead the pattern of the city was planned before the community was started.

The following *Urban* communities were included in the field study: Salt Lake City, Ogden, Provo, Brigham, and Logan.

From their social concepts the Mormons drew certain conclusions which were considered to be essential in bringing into existence the Plat of the City of Zion. Among these concepts were those of nationalism, communism, and millennialism. The *nationalistic* ideology of the period, as is given in the Book of Mormon, was that the Almighty held the American continent in favor and that this continent should be the gathering place in the last days. The Salt Lake Basin was revealed to the leaders of the church as the site for the "gathering" and so the city was planned as a monument and a center for the "last days." The *communistic* doctrines of the Mormons are embodied in their system called the "United Order." This doctrine can be summarized as follows: the earth is the Lord's and His people can be but stewards of that portion of goods which the agents of the Lord assigns to each. Each person is entitled to as much as he needs to support himself and his family but no more. If there accrues a surplus from the operations of any church member over and above these needs, such surplus is to be turned back to the Bishop's storehouse—the Bishop being the agent of the Lord. But above all it was the dominant passion of the Mormons to build a perfect city in preparation for the *millennium* which was imminent. The City of Zion was planned as a Utopia. It was a community designed for the dwelling place of the Saviour and those human beings freed from selfishness, greed, and vanity and thus perfected after the order of their tradition.

The essential features of this city pattern were that the streets should be wide, intersecting each other at right angles, and running in north-south and east-west directions; and that this community should become the residence establishment of the farmers who cultivate the farms adjacent to the community.

The plan provided that all the people should live in the city. The city should be a mile square made up of blocks containing ten acres each, cut into half-acre lots, allowing twenty houses to the block. The streets should be eight rods wide and the middle tier of blocks 50 percent wider than the others because they were to be used for schools, churches, and public buildings. Stables and barns should be on the edge of the city. Not more than one dwelling house should be put on a lot. Such were the general specifications of the first city planned by the Mormon leaders.

From the first, Salt Lake City was the center—political, religious, cultural, and commercial—of the Mormon activities and of the Salt Lake Basin. It was the pattern for all communities established by the Mormons in Utah. Today, it is only in other *urban* communities that a replica of the pattern is apparent since much of the original pattern and ideals of Salt Lake City have become vague because of its economic activities and size. Each *urban* community, however, has become a religious, cultural, and commercial center.

Conclusion

The Mormon communities in the Salt Lake Basin today are the result of the convergence of the following facts and influences: (1) the plan of the City of Zion; (2) the development of extraordinary group solidarity; (3) the favorable environment of the Salt Lake Basin for such a community.

The physical and social environment of the Salt Lake Basin was particularly favorable to this unique village plan. This type of village facilitated settlement because it met the following needs: (1) it provided security; (2) it facilitated coöperative efficiency, by placing the members of the community in ready touch with the directing officers of the group; (3) it made for contentment, in that social intercourse was enhanced—even in the pioneer stages, these villages were of sufficient size to make possible the maintenance of religious, educational, and other social institutions; (4) by the separation of residence area from arable lands, a more advantageous utilization of lands was made possible. Common pasturing of the fields after harvest, common fencing at first, were made possible by the fact that crops were stored and stacked in the village.

While the Mormon influence has spread to other western states through immigration from the Salt Lake Basin and through conversion to Mormonism, in no other region have they been able to establish these Mormon types of communities. In other states the Mormons were additions to a social and an economic pattern already established, rather than pioneers planning and making their own type of community. While they have been able to carry the doctrine of the Mormon church to other states they have not been able to transplant the types of communities established by the pioneers in the Salt Lake Basin.

It is a question of considerable interest whether or not this type of community in Utah is likely to persist as a dwelling place of active farmers. The limitation of irrigation systems in the early days made a certain amount of congestion necessary; but extension of canals, increased activity on the part of the Federal Government to promote the use of surface and sub-surface waters, the installation of cisterns and the use of wells for culinary water, all have largely negated the influence of this factor. Finally, the handicaps of separation have been greatly reduced by the radio, automobile, and improved roads.

At present some of the farmers have established their residence on the farm for the summer but they move back to the village or town in the winter; the town or village home is always regarded as their official residence.

Most of the towns and villages of Utah are situated in valleys at or near the mouths of canyons, and the very nature of this topography will tend to preclude, to some extent, satisfactory open country settlement.

It seems safe to say, however, that farmers who have settled in villages or towns will be loath to change. The mere inertia of the initial type of community will tend to make it persist with slight changes.

The Origin and Spread of the Grid-Pattern Town*

Dan Stanislawski

Many geographers have concerned themselves with the study of towns, their distribution, position, site, function, and anatomy, and yet, of the innumerable articles and books written on this subject, none, to my knowledge, has been devoted to the origin and spread of the design that is now standard throughout much of the world—the grid pattern with straight streets (parallel or normal to one another) and rectangular blocks. It is true that some writers have casually considered this pattern, concluding that it spontaneously recommended itself to the town builder whoever or wherever he might be. I likewise made this assumption at first. But the obviousness of the grid is more apparent than real. In the record of its use it seems to have been no more obvious than, for example, the wheel.

My interest started in the Spanish towns of the New World, where I soon found that not only did native towns fail to exhibit such a pattern but during the earliest period of Spanish settlement it was lacking also, and subsequent Spanish cities, except when constructed under direct orders, were likely to vary greatly from the simple rectangular design. It was this that indicated the need for further inquiry into the background of grid towns. My investigation led me into the Middle East and into the third millennium before Christ. That the grid may have an even longer history awaits further archeologic investigation. It may have been a one-time invention which has spread from its source region until at present it encompasses the globe.

Arguments for and against the grid

The casual assumption that the grid almost automatically becomes the pattern of a new settlement cannot hold up in the light of the history of its distribution. Only those regions directly associated with, or accessible to, areas of earlier use have shown evidence of its existence. I know of no region in the world that will clearly contradict this thesis. But when once known and recognized and fitted into the culture pattern, the grid has both obvious advantages and some disadvantages. Let us consider the disadvantages first. From the point of view of the individual there are many reasons for a man to place his building, whether it be dwelling or workshop or temple, at an angle with buildings near by and at some distance from them rather than directly in line and adjoining. Such placement offers advantages in terms of circulation

*Reprinted from the *Geographical Review,* 36 (January 1946), 105–20, by permission of the author and The American Geographical Society.

of air and exposure to sunlight, as well as accessibility of the various parts, whereas in the grid efficiency is largely lost without the alignment and juxtaposition of buildings. Secondly, again as regards the individual, there are other plans that would have greater utility. For example, the radial plan with streets leading out from a center like spokes from the hub of a wheel offers certain advantages over the grid in communication from the periphery to the center. Thirdly, the topography very frequently indicates easier street planning than the insistence upon straight lines mounting hills and falling steeply into valleys.

To consider the advantages of the grid plan is to consider a longer, and from many points of view, a superior list. Perhaps its greatest single virtue is the fact that as a generic plan for disparate sites it is eminently serviceable, and if an equitable distribution of property is desirable, there is hardly any other plan conceivable. It can be extended indefinitely without altering the fundamental pattern or the organic unity of the city. Property can be apportioned in rectangular plots fitting neatly into a predetermined scheme of streets and plazas. It can be sketched on the drawing board and, within certain obvious limitations, made serviceable. It is also far the easiest plan to lay out with crude instruments of measurement. For a *compact* settlement of rectangular buildings this scheme is the only one that lends itself to the efficient use of space. Moreover, a distinct advantage for the grid-plan town under certain political conditions is that of military control. This would apply in the case of subject towns to be held under control; for it has been clearly recognized, not only by the Spaniards in the New World[1] but by the Romans and early Greeks before them,[2] that a tortuous street facilitates defense by individuals and a straight street lends itself to control from without.

Theories of origin

One theory as to the origin of the grid is based on its obvious efficiency in the use of space where rectangular buildings are involved. The reasoning is seductive but not borne out by facts. Examples of strict rectangularity of buildings with a highly irregular street pattern are far too common. They long predate the first use of the grid and continue to the present in large areas of the world.

Another point of interest with regard to theories of the origin of the grid-pattern town concerns the straight processional street. Another far too casual assumption was likewise made here that such a street would suggest the advisability of others parallel or at right angles to it. This also fails to be borne out, both in Egypt and through the long history of early Mesopotamia.[3]

The theory that the grid stemmed from an orientation toward the points of the compass, probably based on religion, has proved equally inadequate. In Mesopotamia, Egypt, and early Greece the orientation of a building and even a street was common, but it did not lead to the laying out of other streets in accordance.[4] On the other hand, in Mohenjo–Daro, in northwestern India, there was obvious orientation of all the streets and rectangularity of blocks, yet excavation has shown no temple, and there may have been none.[5] It seems, then, that religious significance as basic to the grid can likewise be written off as inapplicable.

In weighing these advantages and disadvantages of the grid pattern certain things seem clear:

1. It is possible only in either a totally new urban unit or a newly added subdivision. This pattern is not conceivable except as an organic whole. If the planner thinks in terms of single buildings, separate functions, or casual growth, the grid will not come into being; for with each structure considered separately the advantage lies with irregularity. History is replete with examples of the patternless, ill-formed town that has been the product of growth in response to the desires of individual builders. Nor is it simple to rectify an older city. The difficulty, and probably the impossibility, of this has been demonstrated by Von Gerkan.[6]

[1]"Fundación de pueblos en el siglo XVI," *Bol. Archivo Generál de la Nación,* 6 (1935), p. 350, Sec. 116.
[2]Rex Martienssen, "Greek Cities," *South African Architectural Record* (Johannesburg, Jan. 1941), p. 25 (quoting Aristotle); Vitruvius, *The Ten Books on Architecture,* translated by M. H. Morgan (Cambridge and Oxford, 1914), p. 22.
[3]Armin Von Gerkan, *Griechische Städteanlagen* (Berlin and Leipzig, 1942), p. 31; T. H. Hughes and E. A. G. Lamborn, *Towns and Town-Planning, Ancient & Modern* (Oxford, 1923), p. 2.
[4]G. Maspero, *Life in Ancient Egypt and Assyria* (New York, 1892), p. 196; S. H. Langdon, "Early Babylonia and Its Cities," in *The Cambridge Ancient History,* Vol. 1 (New York, 1923), p. 374; E. A. Speiser, *Excavations at Tepe Gawra,* Vol. 1 (Philadelphia, 1935), p. 24; Von Gerkan, op. cit., pp. 31 and 78.
[5]Sir John Marshall (ed.), *Mohenjo-Daro and the Indus Civilization,* Vol. 1 (London, 1931), pp. 22 and 283.
[6]Von Gerkan, op. cit., pp. 114 and 115. This fact was recognized by the Spanish king in his instructions to Cortes (see "Colleccíon de documentos in éditos relativos aldescubrimiento, conquista y organización de las antiguas posesiones españolas de ultramar," Ser. 2, 17 vols. (Madrid, 1885–1925), Vol. 9, p. 177.

2. Some form of centralized control, political, religious, or military, is certainly indicated for all known grid-plan towns. When centralized power disintegrates, even if the grid pattern has been established it disappears. This is indicated clearly by medieval Europe as compared with Europe under Roman rule.

3. It may indicate colonial status, not necessarily a situation in which the younger settlement is bled by the older, but more frequently an amiable association for mutual benefit between mother and daughter settlements.

4. Desire for measured apportionment of land.

But none of the foregoing can be said to indicate that a strongly organized political group desirous of founding a colony will, because of its obvious virtues, set up a grid town. The virtues are obvious only when demonstrated. This is confirmed by history. According to the evidence, only those exposed to the idea will utilize this pattern. Hence another requirement must be added:

5. Knowledge of the grid.

The city of Mohenjo-Daro

The earliest record we have of this street pattern is that of Mohenjo-Daro, a city which flourished in the first half of the third millennium before Christ. This city was not casually built. The precision of its plan could not have been accidental. It was a well-rounded concept designed to fit the needs of a highly organized, highly urbanized people. The streets were straight and either parallel or at right angles to one another, as far as the inaccurate instruments of the time permitted. This was not a placing of buildings merely with the idea of the individual in mind. The concept was that of an organic city in which all parts were designed to function within the whole.

Trade was of enormous importance to the people of the city. The very high quality of the manufactures makes evident that it was indubitably the home of men of skill with a long background of training and organization. That Mohenjo-Daro does not represent the earliest settlement of this people may be indicated by the fact suggested above, that the grid city is completely planned and established as a new unit. We can, therefore, postulate that the ancestors of the people inhabiting Mohenjo-Daro had a long history of social organization in this region or elsewhere.

For the next known example we must seek much later times, although there may be Oriental material that will, when known, alter opinion with regard to this intervening period. There is at present no reason to suppose that any Oriental settlement with anything suggesting a grid pattern could rival Mohenjo-Daro in antiquity. However, Creel[7] has some interesting though inconclusive statements on early Chinese-planned buildings and streets. The next record of the grid is found at the eastern Mediterranean in the eighth century before Christ. Sargon of Assyria, tiring of his old capital, decided to perpetuate his glory by the establishment of a new one, Dur-Sarginu. For its site he chose the unimportant and formless little village of Magganuba, where he laid out his new capital precisely in terms of the grid. This was not destined to last, but the gap in time was not to be long until Hippodamus, and undoubtedly his predecessors, would take up the idea in Greece and Greek lands and establish it in such fashion that it was not again to be lost to the record.

A continuing tradition in India?

The question may be raised why one should attribute to a single invention a plan that has appeared in places far distant from one another with a gap of long centuries between. The question is not an unreasonable one. A further inquiry into Indian sources might yield the answer. The data that we have at hand, although inaccurate as to dating, seems to indicate a strong possibility that the tradition of Mohenjo-Daro has been continued in India, perhaps unbrokenly. If one were to accept the claims of recent Indian writers with regard to town planning in their country, one would need to seek no further; for it is their contention that town planning existed in India long centuries before the Christian Era. The brilliance and completeness of Indian town planning indicated in the Śilpa Śastra is not an overnight creation. It is the outcome of the thought of many men and must have evolved through many centuries. The casual assumption that Indian town planning derives from Alexander's generals, the Greek Bactrian kings, or Vitruvius and the Romans may not be a fair one, in view of the great elaboration of Indian thought and in view of the indicated contribution of architec-

[7] H. G. Creel, *The Birth of China* (London, 1936). Creel says that buildings in a settlement of the fourteenth century before Christ (p. 57) were carefully oriented but that their arrangement otherwise has not yet been determined (p. 68). He quotes a poet of a later period who, in describing the city, said that land was distributed in predetermined plots and that, under central supervision, houses were planned along streets that presumably were straight (p. 64).

tural types from the Iranian plateau, as well as of the possible development of town planning even among the Dravidians.[8] India may have carried the tradition of this town pattern for all later ages to accept at their leisure.

One regrets that Sargon, in the eighth century before Christ, did not record why he chose the grid or where he found his sources for such a plan, but again, eyes may have been turned to the East. The trade from Mesopotamia through Persia and even into India cannot be questioned. That the East was contributing ideas, specifically in architecture, which might suggest a contribution to broader planning, is shown by Herzfeld's demonstration that even the Ionic pillar was a product of lands to the east of Greece.[9]

To those who question the assumption of Indian derivation it can be asked: "Where has the pattern of the grid town appeared without possible connections with India?" No part of Europe or Asia except those regions that had contact with this area of oldest appearance has given evidence of the grid pattern. Nor did any part of Africa exhibit this pattern until Alexander introduced it as derived from eastern Mediterranean lands.

No proven New World examples

Nowhere did this plan appear in the New World, statements to the contrary notwithstanding. The Chimu city of Chan Chan on the Peruvian coast certainly was, it is true, one of straight lines and right angles.[10] Some of these lines were maintained for a notable length, but they did not carry on through; the organic quality of the grid plan was broken by irregularities. It was rather a series of blocks, many rectangular, but not communicating with other blocks in the functional way necessary to the grid.

Many contentions have been made concerning the use of the grid in Mexican towns, but here again the evidence does not support it. The famous "Plano en Papel de Maguey," despite some theories to the contrary,[11] is obviously a post-Conquest design drawn to the order of Europeans.[12] The theory that Tenochtitlán had rectangular blocks because of the rectangularity of its temples and temple squares does not stand up, in view of the fact that so many places in the Old World had square temples and corresponding courtyards with the remainder of the settlement clearly at variance. Certainly Cortes and Bernal Díaz remarked about the straightness of the passageways leading into Mexico, but nowhere did they suggest more than the straightness of single streets.[13] It might also be indicated that in their apparent surprise at first sight of this straight passageway these Spaniards, who were used to the tortuous streets of sixteenth-century Spain, surely should have been even more struck with the rectangularity of blocks. Failure to mention such a condition may well be taken to indicate that it did not exist.

According to present evidence, the rectangular grid was nowhere a casual, spontaneous thing. In spite of its apparent obviousness, it would seem that it was not put into practice by any except those who had known it previously or who had access to regions of its occurrence.

The Greek record

The continuous record starts in the sixth century before Christ, in Greek lands. Before this time the regular pattern was clearly not a typical feature of Greek settlement.[14] There are many examples of earlier Greek cities showing anything but the regularity of the grid plan, and a definite record of cities settled at least as late as the middle of the seventh century before Christ shows that irregularity was typical. In fact, according to Von Gerkan, as late as the early part of the fifth century some cities were settled without a standard pattern. Hippodamus, a Milesian, is credited by Aristotle with being the planner of the grid-pattern harbor

---

[8]C. P. Venkatarama Ayyar, *Town Planning in Ancient Dekkan* (Madras, 1916).

[9]E. E. Herzfeld, *Archaeological History of Iran* (London, 1935), p. 15.

[10]J. L. Rich, "The Face of South America: An Aerial Traverse," *American Geographical Society Special Publication No. 26* (1942), photograph 277; Otto Holstein, "Chan-Chan: Capital of the Great Chimu," *Geographical Review,* 17 (1927), pp. 36–61, Fig. 26.

[11]George Kubler, "Mexican Urbanism in the Sixteenth Century," *Art Bulletin,* 24 (1942), pp. 160–71, footnotes 3 and 59.

[12]M. Toussaint, F. Gomez de Orozco, and J. Fernandez, *Planos de la ciudad de México* (Mexico, 1938), p. 36.

[13]Hernán Cortés, *Cartas de relación,* Vol. 1 (Madrid, 1932), p. 98 (Map 2): "Son las calles della [referring to Temixtitan-site of present Mexico City], digo las principales, muy anchas y muy derechas." By his limiting phrase he specifically excludes all but the main streets as being wide or straight. Also, Bernal Díaz del Castillo, *Historia verdadera de la conquista de la Nueva España,* 3 vols. (Mexico, 1939), Vol. 1, pp. 309 ff.

[14]Martienssen, op. cit., pp. 19 and 33; Joseph Gantner, *Grundformen der europäischen Stadt* (Vienna, 1928), p. 37; H. V. Lanchester, *The Art of Town Planning* (London, 1925), p. 9; Hetty Goldman, *Excavations at Eutresis in Boeotia* (Cambridge, 1931), p. 50.

of Athens, the Piraeus,[15] but even earlier than the Piraeus—probably also of the fifth century—was the grid design of Hippodamus' own birthplace, Miletus. There can be no doubt that the plans of Hippodamus were not born in his brain but derived from earlier times—there is at least one clear example, Olbia—and perhaps distant places. It is interesting to note that the earliest plans are associated with Ionic Asia Minor and settlements by Ionians on the Black Sea, and not nuclear Greece. So the first appearance among Greeks was in the western extension of Asia, where it could have been based on earlier knowledge and use.

Long before the time of Hippodamus, Greeks had been expanding their knowledge of the world through their growing trade connections. For several centuries these connections were those of "tramp" traders, who either settled among "barbarian" peoples, taking more and more control of the region by reason of their superior training, or merely came temporarily to these regions to exchange merchandise. This did not involve planning. It was simple contact for the purposes of exchange and profit.

The Greeks pursued their course westward through the Mediterranean, making contacts with the present Italian mainland and islands. Many different groups were involved in this trade until the latter part of the eighth century, when Corinth was infected with the virus of what might be termed precocious imperialism. Whereas before the founding of Syracuse all Greeks, so far as can be determined, traded with the west, after 734 B.C. Corinthian goods became dominant in the market and eventually were far more important than the materials from all other Greek traders combined.[16]

Corinth was operating according to a plan. Whereas theretofore settlements had, presumably, been made rather casually, Syracuse was founded under authority from the mother city and by settlers who were dispatched to the place with orders based on careful planning. These orders included instructions for the division of land for use by the settlers. Here was a clear indication of the growth of centralized control. It was likewise an indication of increasing importance of trade as well as of a possible pressure of population at home.

To the east the Greeks were making other contacts. Miletus sent out secondary colonies, particularly after the middle of the eighth century, to take over the trade of the Black Sea. The "great Asiatic mother of colonies," like Corinth, was not averse to the use of force to maintain trade supremacy. She was, however, the greatest center of Oriental influence, and the attainments of these Asiatic Greeks are thought by some to have been far superior to those of the European homeland. Their contacts with the interior of Asia Minor and the countries of high civilization to the east of the Mediterranean were a liberal education.

The drive of colonization both in the Mediterranean and in the Black Sea was temporarily reduced during the period of the Tyrants. During their regime, however, there was an even greater centralization of control, and part of this remained to contribute to the colonization that increased again after their decline. After the epoch of the Tyrants, the trade of the Asiatic Greeks spread greatly through the lands of the friendly Lydian king Alyattes. This monarch controlled a considerable part of the interior of Asia Minor and had alliances with Mesopotamia, Egypt, and others. His son, Croesus, likewise a Hellenophile, offered continuing opportunities for Greek traders, which were only partly broken by his defeat at the hands of the Persians just after the first half of the sixth century.

The planning that is called Hippodamic was a product of the period following that of the concentration of power in the hands of the Tyrants, and also following the period of greatly expanded trade through Lydian country into Mesopotamia and other eastern lands where examples of the grid were to be seen. Olbia was laid out in grid form at the end of the sixth century, Miletus not long afterward, in the fifth century, after the destruction of the old city by Cyrus of Persia.

By this time all of the factors favoring the grid had come into being: (1) there was centralized control, and a background of town planning; (2) totally new units were being founded, with dependent—"colonial"—status; (3) knowledge of the grid was available from the East; (4) desirability of the grid as a general plan would have been apparent, especially with regard to the distribution of land, which was important to the land-hungry Greeks.

---

[15]"A Treatise on Government," translated from the Greek of Aristotle by William Ellis (New York, 1912); Percy Gardner, "The Planning of Hellenistic Cities," *Transcript, Town Planning Conference, London, October 10–15, 1910,* Royal Institute of British Architects (London, 1911), p. 113.

[16]Alan Blakeway, "Prolegomena to the Study of Greek Commerce with Italy, Sicily and France in the Eighth and Seventh Centuries, B.C.," *British School at Athens Annual,* No. 33, Session 1932–1933 (London, 1935), pp. 170–208; reference on p. 202.

It is likewise interesting to note, and perhaps it is the explanation of the Greek acceptance of this plan, that its methodical regularity and orderly quality well suited the Greek philosophic view of worldly order created out of variety. The idea of a corporate whole is typical of Greek thought of the period. During this period the settlement of towns was widespread, and the grid was used by Greeks not only in their homeland but in western places as well. For example, there is Thurii in southern Italy, commonly attributed to Hippodamus; there is Selinus in Sicily, and Naples on the peninsula. These undoubtedly made their contribution to, and saw their continuance in, Roman planning of a somewhat later date.

Effects of Alexander's conquests

After the period of Hippodamus the next striking development of the use of the grid plan is in the Alexandrian age, when it was spread so widely by the conqueror and his heirs. Again it is of interest to speculate whether the strengthening of interest in the plan at this time was not a product both of the background in Greek lands and of further knowledge acquired in eastern lands. Alexander brought in his entourage not only fighting men but men of intellectual attainment who might easily have been struck by urban developments in the lands they visited.

The cities that remain from the time of Alexander or his successors present us with excellent examples of the planning of the period. Many were founded in Anatolia. Priene, the best known, through the work of Wiegand and his associates,[17] is a perfect example of the grid pattern, its buildings precisely oriented and carefully aligned with the streets. In the more distant lands are the cities of the Greek Bactrian kings and those of India proper, which, although commonly accepted to be of Alexandrian age or later, may indeed show an earlier history.

The transfer of knowledge along the Mediterranean by Greeks, however, was a matter of early centuries—long preceding the Alexandrian age—and the technique of town planning was carried into lands that were to become Roman. Here it became basic to later Italian settlement form. Greek traders were in Italy centuries before the rise of Rome. During this early period the Etruscans arrived from the east and settled in the peninsula.

The early Etruscan settlements were certainly not neatly plotted grids, though within the exigencies of the hill locations which they chose for their settlements they may have striven for greater regularity than appears at first glance.

Greek influence was felt throughout Etruria from the outset. After the middle of the seventh century, however, the influence became more strongly Ionic.[18] This was the period of the first definite Etruscan grid town, Marzabotto, built at the end of the sixth century, and perhaps the first real grid town in Italy. Here the *cardo* and *decumanus* of later Roman cities clearly appear.[19]

It is to be recalled that Ionic influence in Italy coincides not only with Marzabotto but with Olbia, and roughly with Miletus, all of which used the grid plan that had had earlier exemplification in parts of western Asia. It is to be recalled likewise that Ionian Greeks had wide experience and knowledge of these regions of western Asia.

In the early period of Roman development there is little, if any, evidence of awareness of the grid—or of town planning at all. It was the late Republic and the early Empire that saw the rapid development of the form. Then it spread through Roman colonies to near and distant points in the Empire.

The Roman grid

The grid plan as used by the Romans was not precisely that of the Greeks. It was an adjustment of the plan used by Greek traders to the demands of a Roman order, perhaps with influences derived from Etruscan practices, with an interesting association of the Roman block and the *jugerium,* the rural unit of surveying.[20] The town block was clearly rooted in history, and linked with the distribution of agricultural plots.

This rigorous, clear pattern lent itself smoothly to the necessities and point of view of the Roman state. Here was an intense centralization of power in the hands of men faced with the pressure of population and the necessity of protecting exposed frontiers of the Empire. For both these problems daughter colonies were an obvious solution. Particularly after the civil wars of Sulla, Caesar, and Octavian, who had amassed great armies to

---

[17]Theodor Wiegand and Hans Schrader, *Priene* (Berlin, 1904).

[18]Hans Mühlestein, *Die Kunst der Etrusker* (Berlin, 1929). One of the major divisions of this book is "Epoche des Überganges vom orientalisierenden zum ionisierenden Stil: *ca.* 650–550."

[19]Pericle Ducati, *Storia dell' Arte Etrusca* (Florence, 1927), Vol. 1, pp. 372–74.

[20]R. C. Bosanquet, "Greek and Roman Towns," *Town Planning Review,* 5 (1915), pp. 286–93 and 321.

support their causes, there was a pressing necessity for the absorption of these soldiers into a peacetime economy. This was largely achieved through the establishment of newly planned urban units in various parts of the Empire. Given the necessities of the Roman state, the psychology of its rulers, the background of its history, what was more logical than to establish the grid wherever new urban units were planned?

Following the downfall of the Western Empire the era of city planning came to a close, and, more important, even those cities that were completely planned and built before the dissolution of the Empire fell into other ways and forms, so that the end of the medieval period saw hardly an example of Roman planning in the cities that she had established.[21]

The medieval collapse

Following upon the organized control and planning of the Romans, the early medieval period saw a degree of collapse in which the factors militating against the serviceability of the grid-pattern town became dominant. Centralized power, basic to its establishment, no longer existed. Division of power and localization of authority came into being. No longer was the broad power present which tends to maintain a single pattern. Secondly, as has been indicated, defense of the local unit was facilitated by tortuous lanes; straight thoroughfares lent themselves to control by centralized power. Thirdly, with local control each unit used its topography as individuals saw fit. There was no necessity for following the rigorous grid plan. Indeed, for many topographic situations it would have been costly and excessively difficult, and it served no real purpose in this feudal period. Fourthly, this was a period in which trade was greatly restricted, and the grid plan, which had functioned well for a trading center, was no longer needed for that purpose. Perhaps more important than all others is the fact that there was no longer the idea of equitably distributed plots of ground. This was not a period of small holders asserting their rights over definite recognized portions of territory. The feudal order operated on an entirely different basis.

However, in spite of all these tendencies toward breakdown, the pattern was never completely lost in the former Roman lands. Several examples remain in northern Italy—Turin, for example. Traces remain in such places as Braga in Portugal, Chester in England, Tarragona and Mérida in Spain, and Cologne and Trier in Germany. Some would place Oxford in this category, though this now seems dubious.[22] It has been fortunate for the planners of later centuries that these examples remain.

The Renaissance

If the early part of the Middle Ages saw the decline and almost the obliteration of this pattern, the later Middle Ages saw its adumbration again, and the Renaissance its establishment. Again political conditions had changed so that central power, planning, and trade re-emerged and local units existing in the feudal structure began to lose their dominance—in short, the trend again contributed to the utility of the grid.[23]

Particularly was there a striking advance in the use of the pattern in the thirteenth century. In this century at least one urban unit using the grid was made by Italians in Sicily. The Germans, in establishing cities on the Slavic frontiers and beyond, such as some of those in Prussia, Breslau, and Cracow, used this plan as their basis.

The grid in France and England

But most important during this period was the establishment in France of the bastides, the *villesneuves*. The record is clear. The site was plotted into rectangular blocks, divided by streets parallel to one another or at right angles, in which the main roads running from the gates led to a large square or market place at the center. Around this square were the homes of the more important residents, with arcades giving shade to the walk.[24]

The most important founders of the French bastides were St. Louis and his brother, Alfonse of Poitiers. Kings of England who possessed French territory at this time also built towns of a similar order in France.

Again the function and desirability of the pattern are apparent with the change in the

---

[21]Oskar Jürgens, *Spanische Städte: Ihre bauliche Entwicklung und Ausgestaltung* [Hamburgische Universität Abhandl. aus dem Gebiet der Auslandskunde, Vol. 23, Ser. B, Vol. 13] (Hamburg, 1926), p. 1; Ramón Menéndez Pidal, *Historia de España, Vol. 2: España Romana* (Madrid, 1935), p. 607.

[22]Hughes and Lamborn, op. cit., p. 73.
[23]A. E. Brinckmann, "The Evolution of the Ideal in Town Planning since the Renaissance" (translated from the German), *Transcript, Town Planning Conference* (see footnote 15), p. 171.
[24]Félix de Verneilh, "Architecture civile au Moyen Age," *Annales Archeologiques* (Paris), 6 (1847), pp. 71–88; reference on pp. 74–75.

political and social order. Again power was centralized, and it was those individuals that exerted power over a large area of land who were responsible for the establishment of the towns. Again it is to be noted that it was not the replotting of existing towns. This is virtually impossible. These were completely new units founded under the direction of centralized power, and all at one time. They were under military control and functioned as military centers. Also, the plots in town were distributed on the basis of standardized units, and again it is to be noted that the agricultural plots beyond the city were likewise distributed in terms of standard units.

The situation in England is probably not as clear as that in France, but perhaps it is even more interesting. Although English settlement of this period was clearly influenced more strongly by France than by any other source, there are still the yet undetermined possibilities of an earlier development within England itself. As was mentioned above, Oxford is thought by some to be a Roman foundation. This seem a dubious postulate. It appears now that Oxford was clearly later than Roman times, but almost as clearly it seems indicated that it may have been earlier than the period of the French bastides, and may perhaps have reached back into Saxon times. Ludlow is another example of a town that was clearly earlier than the period of the bastide. It is a foundation of early Norman time, settled during the twelfth century, and probably using the grid plan. This, of course, suggests knowledge brought in from the continent. It may well have served as a partial inspiration for later models.

The real development of towns in England, mostly in the pattern of the grid, began with Edward I. It should be noted that Edward possessed territories in France, his training was French, his language was French. He knew well the town planning of thirteenth-century France, and it is clear that this was the model he had in mind in setting up the so-called Welsh bastides and other towns in England. Again, the factors contributed to the utility of this plan; for now England with a centralized authority felt itself in need of totally new units and had the experience of France before it.

With one exception, nothing more need be said regarding the grid in Western Europe. Its serviceability in the period of expanding settlement both within Europe and in European colonies was obvious. Never has it been lost since the time of its redevelopment toward the end of the medieval period.

Spain and the New World

The exception to be noted is Spain. Isolated from the rest of Europe during the long period when she was involved in internecine warfare, she failed for the most part to take part in developments of neighboring countries. It is unfortunate that she lacked their experience with Renaissance planning; for it was she that conquered the New World and established thousands of completely new settlements there. As she was uninitiated in the methods of town planning, her settlements were amorphous for about three decades after the beginning of her control. Finally she realized the necessity for a plan, and for this she turned to her neighbors, and beyond them to the Roman and Greek sources from which they had profited. But this is a subject in itself and must be treated in a separate paper.

Issues in Sociocultural Ecology*

Walter Firey
and Gideon Sjoberg

Consensus and quiescence are not among the characteristics of an expanding field of knowledge. In this respect there is every reason for optimism concerning the prospects for human ecology. Even the most casual browse through a sample of publications bearing the word "ecology" in their titles will convey to the reader the sense of a field that is marked by intellectual diversity and vigor.

Yet the study of human ecology is not without its repertoire of generally accepted concepts. Such expressions as "community," "differentiation," "environment," and "adaptation," for example, have much the same range of denotations for the ecologist whose research interests lie in geography as they have for the ecologist whose concerns are sociological, or anthropological, or psychological. And few of these investigators would take exception to a characterization of their field as " . . . the study of the form and the development of the community in human population" (Hawley, p. 68). A fair measure of assent, too, might be gained for the proposition that the term "community" entails a reference to certain features of a population and its environment, including the feature of localization which attaches to so many of the relationships that hold among the members of a population

(Janowitz and Street, p. 92). In general, then, the subject matter of human ecology will be rather widely acknowledged to reside in the phenomenon of *environmental organization,* understood as a people's codification of the various properties of its physical environment in their relation to processes of community formation. Those properties will generally be recognized as differing considerably with respect to the degree to which they have been socially constructed, ranging from such features as accessibility and reputation, on the one hand, to those of climate and terrain, on the other hand.

Paradigms in human ecology

It is when human ecologists go beyond these preliminary definitions and endeavor to link their observations to a theoretical frame of reference that they come to a parting of the ways. For, as is the case with most fields of inquiry, the data of ecological research, and the general form of the relations which those data exhibit, are partially dependent on the conceptual schema that has been employed for their description. Different conceptual schemata, for their part, reflect different paradigms, i.e., different " . . . fundamental image(s) of the subject matter within a science" (Ritzer, p. 7). Two such paradigms, along with the

---

*This article was written for this volume and has not been published previously.

overlap which may be defined between them, can be identified within the field of human ecology.

The two paradigms

Adherents of the first of these paradigms find themselves committed to a natural science image of their field. For them the paramount task is that of establishing theories whose generalizations would link together various properties of a localized population and its environment. The terms of those generalizations—i.e., the basic data of the discipline—would consist of *indicators,* or "external" measures, of the properties in question. Their substantive reference is conceived to be an *objective* order of ecological phenomena—objective in the sense of freedom from any reference to the meanings which the members of a population might attach to those properties.

Adherents of the second ecological paradigm find themselves committed to a rather different view of their field, one that is fundamentally interpretive in character.[1] For them the fundamental task of human ecology lies in a reinterpretation of the interpretations which human beings have already made of their environments and of their own organized actions relative thereto. The terms of those reinterpretations—i.e., the basic data of the discipline—would consist of *symbols* whose meanings depend on their use in varying contexts of everyday communication. The substantive reference of an observer's reinterpretation of the interpretations of some human beings is conceived to be a *subjective* order of ecological phenomena—subjective in the sense of freedom from any contingency upon the actual fulfillment of the meanings in question.

A third contingent of ecologists is composed of those investigators who, with varying degrees of sensitivity to the philosophical problems that are involved, seek to avoid the theoretical anomalies which are generated by the first two paradigms. For them the principal task of the ecologist lies in a description of ecological systems which will be faithful to the objective relations that hold among various properties of those systems, and faithful, too, to the interpretations which the participants in those systems have themselves made of those systems. Fulfillment of this task should yield neither a pure objective nor a pure subjective order but, rather, an integrally experienced system

[1] Certain scholars have limited the term "ecology" to the neo-orthodox framework (e.g., Goodwin). In our judgment, this is not useful for an understanding of land-use patterns.

(Sorokin, pp. 16–25, 91–96)—one whose description will answer to both subjective and objective criteria of validity. The image of such a project would correspond to the overlap that can be defined between the two paramount paradigms.

Human ecology, then, can be described as a multiple-paradigm science (Masterman, cited in Ritzer, p. 12), a characterization which it shares with most of the other social sciences. Each of its paradigms has its own philosophical rationale, each of them counts its own adherents, and each lends itself, unknowingly as a rule, to rather distinct kinds of ideological exploitation on behalf of planning and policy-making interests.

Two research traditions in human ecology

It is appropriate, then, to identify two principal traditions of research within the field of human ecology. Each of them is based on one of the two paramount paradigms just described; yet each is capable of extension into the common overlap of those paradigms. The two traditions have been commonly characterized as neo-orthodox ecology and sociocultural ecology. They find their parallel in two fundamentally different research traditions that pervade all the social sciences, one of them emphasizing "hard data," the other emphasizing what Geertz has called "thick description" (Social Science Research Council, pp. xvii–xviii).

Among neo-orthodox ecologists there is a basic concern with the " . . . description of community structure in terms of its overt and measurable features" (Hawley, p. 74). Findings are to be expressed "in quantitative terms" so as to "make verification and objective comparisons possible" (Gibbs, p. 3). Their description "in terms familiar to the layman" is regarded as inconsistent with the requirements of scientific objectivity (ibid., p. 4). In this research tradition, then, the predicates that have figured in the descriptions and generalizations of human ecologists have been *indicators*— i.e., operationally defined terms that denote various properties of populations and their organized environments. In actual practice there has been a marked tendency to limit the range of acceptable indicators to terms which have physical, demographic, or behavioral referents. This practice, however, is in no way dictated by the logical requirements of the paradigm in question. In its own terms there is no reason for excluding indicators that would refer to such interpretive data as the attitudinal and motivational properties of a population, so long, of course, as these have been defined and measured by appropriate scales.

Such an extension of scope would be consistent with the requirements of a natural science paradigm and yet would permit the incorporation of terms that have been central to the sociocultural tradition of research. In this way it would contribute to the sort of synthesis which can only be achieved within the overlap of the two paramount paradigms of human ecology.

It is in the nature of an indicator, of course, that the relation which it bears to its referent is strictly one of naming or designating. The connotations of the terms to which an indicator corresponds will extend well beyond the semantic reference of the indicator in question. That is to say, the meanings which those terms have for the members of a society, the evaluative significance which they possess, and the expressive functions which they serve, are all too variable from one occasion of use to another to be captured by the indicator in question. It is precisely these kinds of data which hold the greatest interest for the sociocultural ecologist.

The adjective "sociocultural" points up the dual orientation which marks this approach to human ecology. In the first place, there is a concern with the meanings which the members of a particular society have relative to various aspects of their physical environment—meanings which, of course, form part of that people's culture. Such meanings, ranging over cognitive, evaluative, and expressive modes of consciousness, find their expression in symbols, which thereupon become the basic data for the sociocultural ecologist. Among such symbols are the cosmological constructions of a cultural tradition (Wheatley; Wright). In the second place, there is a concern on the part of the sociocultural ecologist with the ways in which environmentally relevant meanings get built into the structures of people's social relationships, especially in the form of norms and institutions. Among such norms and institutions are those which pertain to property, social class, bureaucratic organization, and politics—structures that are associated with some of the more important processes of community formation.

In accepting as his basic data the environmentally relevant symbols of a people, the human ecologist finds himself confronted with some problems of interpretation. His initial task must be that of determining what it is that particular symbols mean to the people who use them in the course of their communicative, expressive, and introspective activities. But his task is more than this. It is also one of augmenting those first-order interpretations of laypersons by a second-order interpretation of his own—a reinterpretation, that is to say, which must be made in the context of a more inclusive frame of reference than that appropriate to the affairs of everyday life (cf. Schutz, pp. 6, 38–44). Such a reinterpretation, however, should not violate the understanding of the members of the society in question, and yet it has to be consistent with knowledge which is not normally accessible to most of those individuals. The difficulties which reside in these seemingly contrary requirements are not to be underestimated, but they are not insurmountable.

Further difficulties can be anticipated with the gradual extension of interpretive methods into the overlap which exists between the sociocultural and the neo-orthodox paradigms—a development, however, which is as imperative as it is possible. Such an extension can become one of the more propitious of the developments that may lead toward a dialectical synthesis of the two paradigms. Indicators, after all, are symbols. The fact that they are the tools of scientific investigators, and as such are once removed from the everyday symbols employed by the laity, in no way gainsays the fact that they sustain systematic relationships with those symbols. The investigation of such relationships may call for more innovative research procedures than those that have been generally employed in sociocultural ecology—including, certainly, the techniques of the ethnomethodologist.

Both of the two paramount paradigms of human ecology, and the research traditions which correspond to them, have generated substantial contributions to our knowledge of community-forming processes. Each has been a well-spring of descriptive accounts of a variety of such processes, and each has generated a fund of theoretical generalizations that help to explain a number of the relations that hold among those processes. In the pages that follow, it will be our purpose to review some of the descriptive accounts, hypotheses, and theoretical generalizations that have emerged from the research of sociocultural ecologists. The objective of the review will be to point up a few of the issues which have figured in that research and to note some of the formulations that might follow from a resolution of those issues. The relative merits of alternative modes of explanation that can be employed in ecological analysis—i.e., causal, functional, or historicist (Stinchcombe, chaps. 2–3)—will not be at issue in the present review, nor indeed will the relative merits of explanation as distinct from description. All of these will be regarded as admissible forms for the organization of the findings of ecological research.

In the discussion which follows it will be convenient to consider, first, those issues in which the role of power, treated as a fundamental, *generalized* sociocultural variable, is particularly salient for ecological research, and, second, those issues in which more specific sociocultural variables—those whose behavior is channeled through that of power—appear to be more immediately salient for ecological research.[2] Both classes of variables can, in principle, be fitted into a comprehensive framework for the study of variations in people's orientation to environment.

A schema of orientations to environment

Such a framework has recently been proposed by Cohen. In his schema, four principal orientations are identified, viz., the instrumental (environment as means), territorial (strategic or political control over environment), sentimental (attachment to environment), and symbolic (aesthetic or sacred significance of environment). Each of these four orientations, in turn, is divided into suborientations, to which corresponding types of regulative processes, organizations, and institutions are associated. The overall schema possesses a certain generalized significance by virtue of the correspondence which it posits between the four orientations and the four basic functions which Parsons finds in social action systems generally.

In terms of this schema, Cohen identifies a number of issues which deserve the attention of human ecologists. The relative importance of the four orientations, he notes, both as between different societies and in a given society over time, can vary considerably. The grounds for such variations constitute a critical area for investigation. Likewise, as Cohen notes, the mechanisms by which priorities get assigned to the various orientations, and trade-offs established, call for systematic investigation. The consequences of alternative priorities and trade-offs for processes of succession in a given type of environment are still poorly understood features of ecological dynamics whose analysis can be enhanced by recourse to an interpretation of people's environmentally relevant meanings.

Power as a generalized ecological variable

In terms of Cohen's schema, it is possible to see in the concept of "power" a complex property, or

[2] In no sense does the following discussion purport to be a comprehensive "review of the literature." Rather, it presents a purposive sample of studies selected in terms of their representativeness of particular issues.

"master variable," whose variations will subsume many of the variations in each of the four orientations which comprise that framework. As conceived by Parsons, power is the *generalized* capacity of a social system, or any of its units, to mobilize the resources of a society with respect to the attainment of some socially defined goals (Parsons, pp. 121–27). Among those resources will be land, tools, knowledge, prestige, aesthetics, psychological identifications, and sacredness—all comprising elements that fall within the categories of Cohen's schema of orientations. Accordingly, the concept of power takes on a particular salience for sociocultural ecology, despite the relative neglect which it has thus far received within the discipline (Cohen, p. 50).

The following paragraphs attempt to fill in this gap by examining, successively, power relations and ecology within a historical perspective, the relationship between social planning and contemporary ecological patterns, the impact of bureaucracy on urban ecology, and the role of class conflict in land use.

Power and cities in historical perspective

In examining the urban community historically, as well as in a comparative perspective, it becomes clear that cities, particularly the ecological patterns which they display, have been shaped to a considerable degree by shifting power alignments. More specifically, the rise and fall of empires has been associated both with the expansion of city life into previously nonurbanized or urbanized regions and with the disappearance, decline, and at times eventual resurgence of such communities (Sjoberg, b).

Consider, at the outset, the following working hypothesis: At any given stage of technological development, the more substantial the empire, the greater the size and number of its cities. There are several reasons for giving due importance to the broader political structure which supports city life. Preindustrial cities, for instance, were often protected from marauders through the construction of walls or other fortifications, as well as by means of some type of military force. In addition, their political control over a broad region stabilized trade routes and facilitated commercial exchanges between different cities and between urban areas and the countryside.

Social scientists have been engaged in an ongoing debate concerning the relative priority of economic vs. political forces in the construction of social orders in general, and of cities in particular. There will be no attempt to resolve this intellectual debate in the present paper. Rather, the point will lie in making clear to human ecologists that they cannot isolate themselves from this controversy, nor can they ignore the role of political power in the development and flowering of cities. The urban area has always been highly dependent upon the countryside (or region) for food and other supplies without which city dwellers cannot exist. The political structure, with its particular symbol or value system, has been oriented toward undergirding urban life. Even the taxation of ruralites has been a means of furthering this end, for farmers have often been forced to sell their produce to the cities in order to pay their taxes.

As noted above, the rise and fall of empires has been associated with the diffusion of urban arrangements into new areas. For instance, the widest expansion of the Roman empire led to the construction of small urban communities in areas as far north as present-day England and Southern Germany. In recent centuries, European empires have transplanted their urban civilization—including its ecological patterns—into their colonies in the Americas, Africa, and Asia. The Spanish, for instance, destroyed the existing urban centers in Meso–America and established their own urban ecological patterns with certain "Spanish" features such as the grid arrangement of streets about a central plaza. So, too, the British constructed cities in many parts of their farflung empire, and the ecology of these cities differed to some degree from those already in place. In India, for example, the British established new cities such as Calcutta, Bombay, and Madras, which contrasted in some distinct ways from existing Muslim centers such as Delhi or Lucknow, or an older Hindu city such as Benares.

Although European empires carried new ecological forms into such areas as Meso–America and India, it should also be realized that there were similarities between the indigenous preindustrial cities and the new urban centers which were constructed. An awareness of both similarities and differences is thus essential.

Still another pattern associated with the expansion and decline of empires has been a radical shift in the role and function of cities—political centers, in particular. Certainly, with the fall of the Roman empire the capital shrank to the size of a small community and remained so for several centuries. Often the decline of empire has meant the complete demise of the capital city. The ruins of Ur, Babylon, Knossos, Thebes, Angkor, Anuradhapura, Karakorum, and many others provide mute testimony of their once vital role as imperial capitals. In South India, for instance, Vijayanagar was the center of a powerful empire and, in the eyes of fourteenth- and fifteenth-century European travelers, surpassed in splendor any city to be found in the West. Yet by the middle of the following century the empire had been conquered and its capital destroyed, never to be rebuilt. In cases such as these, the conquerors were unable (or unwilling) to establish a political system that would support a thriving urban existence in these localities.

In some cases, however, empire builders have sustained urban life in a conquered area. Thus, under Roman dominance Athens for a time sustained a degree of vitality as an intellectual and cultural center. And cities such as Jerusalem and Benares have been destroyed and rebuilt on the same general site, for their symbolic significance as religious centers has (especially in the case of Benares) overridden any counterposing political and economic considerations.

But it is not necessary to merely look to the past to recognize the importance of shifts in large-scale political systems for urban ecology. Wars in the twentieth century have surely influenced the nature of cities. The nuclear bombing of Nagasaki and Hiroshima testify to the capacity of modern warfare to destroy particular urban communities. On a somewhat lesser scale, warfare and the rise of new power arrangements have affected the urban ecology of Pnom-Penh, the capital of Cambodia. Recent accounts indicate that it has become virtually a ghost town, in part because of the strong anti-urban bias of one of the factions which controlled the city for a period of time. And in Afghanistan we might expect changes in urban patterns as a result of the Soviet invasion and the likely diffusion of new urban patterns into the area.

Perhaps the most dramatic illustration in recent decades of the impact of nation-state power relations upon urban ecology has been the Berlin Wall. The construction of this barrier between East and West Berlin drastically changed, almost overnight, the ecological structure of the city.

Contemporary patterns within nations

In viewing the consequences for urban ecology of power arrangements within nation-states, one is confronted by the considerable diversity among nations which holds in the realm of political structure. Moreover, different political systems tend to exhibit different ecological patterns—a fact that is too often overlooked by ecologists of various persuasions.

One major reason that political systems lead to divergent ecological patterns is the fact that different political structures and ideologies lead to variations in the kinds of social planning or policies with respect to urban land use. As urban ecologists direct attention to this problem area, they must of necessity become more familiar with the nature of the planning process itself.

Social planning involves the purposive effort to stabilize, revise, or reconstitute a social arrangement—in this instance, urban ecology—along preconceived lines. National and local governments, as well as large-scale corporations, are continually engaged in a variety of planning efforts.

Although prior to the twentieth century there were some purposive efforts by governments to intervene in the patterns of urban ecology (note, for instance, the role of the U. S. Government in the development of railroads in the nineteenth century and its effects upon urban ecology), the twentieth century marks a watershed in human history. The Soviet Union, and later Nazi Germany, initiated centralized planning on a massive scale. Even democratic nations in the West began planning programs, particularly as a result of the great Depression, though their efforts have not been as centralized or as coercive as those in authoritarian orders.

Indeed, it is possible to construct two ideal types of social planning: One of them involves centralized and highly focused planning efforts; the other can be termed "piecemeal planning." The latter is characterized by a variety of programs where the goals are loosely defined and where the efforts are loosely coordinated. Even within each of these types, however, there may be plans that rest upon quite different ideological foundations. For example, the highly centralized planning efforts of Hitler can certainly be contrasted with those under Stalin. Or in the case of India during the past few decades, the country's loosely centralized planning efforts have vacillated between an orientation reflecting Gandhian views (with emphasis on the rural sector) and that patterned on the Soviet model (with emphasis upon the industrial sector) (Noble and Dutt).

It seems clear that, whereas many twentieth-century societies are engaged in planning, disagreements do arise as to how plans are to be carried out and what their overall impact is upon the urban scene. Urban ecologists should realize that frequently there is a considerable difference between an initial plan and its eventual outcome. In recent decades, students of planning and policy-making have concerned themselves with the implementation process. There have been a number of studies that have examined the changes that occur between the initial directives and the final result. This pattern emerges for both centralized and piecemeal forms of planning.

To clarify the impact of planning decisions and practices upon urban ecology, it is instructive to draw upon selected case materials from the Soviet Union, South Africa, and Brazil. A convenient point of departure is a geographer's discussion of planning in Leningrad. Shaw finds little evidence of decay or decline in the central area of the city, or for the existence of ghettos or massive slums. And he observes that Soviet planners have sought to reduce urban sprawl and the need for widespread commuting. It may be that Shaw's image of Soviet planning is too optimistic. Yet even he recognizes the existence of tensions between the planning authorities of the central government and the planners in Leningrad, and he concedes that Soviet planners have encountered difficulties in coping with the housing crisis and the demands for social and cultural services. Still, overall, some important contrasts do exist between Soviet cities and those in the industrial West, and these in part can be attributed to basic differences in the planning process and its content (cf. French and Hamilton).

A different set of issues comes to the fore in South Africa (Lemon; Mayer). Here the kinds of planning programs with which this essay is concerned are related not to economic development but to governmental segregation policies (apartheid). With the rise to power of the Afrikaners in the late 1940s, programs were enacted to exclude the tribal peoples from permanent establishment in South African urban centers. Although shifts in plans and policies have occurred, there have been some marked effects upon urban ecology. For example, during the 1950s, the government forcibly transported many blacks (or natives) from central Johannesburg to the community of Soweto, which had been set aside for them. More generally, the natives cannot own their own homes in the urban centers in which they live. In theory, though not necessarily in practice, the natives are bound to their native "homeland." Those born in urban centers are viewed as residing there, but many of the natives are only temporary residents. And in recent years many of the latter have been sent back to their homelands.

The strength of these controls over residential patterns in South Africa indicates that urban segregation is not a natural process but one predetermined, at least in this instance, by social planning. In addition, the South African situation presents urban ecologists with an extreme case of conflict between ideological and economic imperatives. The question can be raised as to whether it will be possible to expand upon a complex industrial-urban order without revamping the existing segregation program, for much of the future labor force, even of skilled workers, will be drawn from the native population.

Still another pattern emerges from the data on Brazil. Here it is useful to turn to the informative study made by Perlman of the favelas of Rio de Janeiro. She documents in some detail the manner in which planners have pushed these slum dwellers from the central city into certain suburban areas. As far back as 1947, a Commission had recommended the return of the favela residents to the rural areas from which they came. With the military takeover in 1964, the central government commanded the power and resources to implement their resettlement efforts. Working through various agencies, especially the National Housing Bank, the government summarily removed slum dwellers from their homes and transplanted them to special housing projects in the suburbs. Many persons had their few possessions destroyed in the process.

The stated objective of this planning effort was to improve the economic, social, and moral life of slum dwellers, as well as to upgrade the aesthetic appeal of life in Rio de Janeiro. But, as Perlman observes, "one of the strongest motivations for favela removal is freeing the valuable inner-city land for more 'profitable' uses." Also the new housing provided a boost for the private construction industry.

The consequences for many of the people have been far-reaching. Their position on the city's outskirts (as in the case of the movement of natives to Soweto, South Africa) has meant much longer journeys to work. Also, their isolation from the job market has seriously damaged the job possibilities for women. The separation of these people from recreational attractions and other activities that urban living offers has led to an even more barren existence.

Overall, data from various parts of the world strongly suggest that urban ecologists, and those in America are no exception, need to take a keen interest in the planning process.

Bureaucracy and the city

Implicit in the foregoing discussion of social planning and its effects upon urban ecology is the need to understand the bureaucratic context within which planning takes place. It seems clear that bureaucratic organizations have implemented the planning process in the Soviet Union, in South Africa, and in Brazil, as well as elsewhere. There is also evidence to suggest that struggles among bureaucratic organizations (or sectors) make a difference in which plans come to be carried out. This seems to hold true for even the Soviet Union—with its highly centralized control.

Although it is essential to understand bureaucratic politics in order to interpret the planning process, bureaucracy seems to have additional effects upon urban patterns that are somewhat independent of the planning process. In order to underscore the impact of bureaucracies upon the spatial arrangements or cities, it will be useful to briefly survey certain ecological patterns in American society during the past several decades.

Not more than a few decades ago, social scientists, as well as governmental officials, began to speak seriously of the demise of the central city as increasingly more middle-class families—as well as many kinds of service organizations, especially those associated with large shopping malls—gravitated to the suburbs. But a re-examination of the implications of suburbanization is in order.

A case can tentatively be made for the proposition that the central city is not disappearing but is only changing its function.[3] Admittedly, many central city areas continue to lose population, and major retail establishments have been abandoning the central city for suburbia. Yet, certain highly specialized economic functions have become especially prominent in the central city. The home offices of banks as well as other financial institutions (e.g., life insurance companies) loom large in the central city—San Francisco, New York, etc. Likewise, government office buildings and the

[3]Data in newspapers and magazines lend credence to the revitalization of the central city in Detroit, Philadelphia, Washington, D.C., and elsewhere. Still, the data produced by social scientists at times point in other directions or are inconclusive. (See, for example, Burns & Pang; Rees; Rosenthal; and Schexnider; for England, see Daniels). Actually, the role of the central area of the city in the preindustrial era (Sjoberg, a) and the industrial period requires critical re-evaluation (e.g., Cutler).

headquarters of large multinational corporations (in addition to financial ones) are in many cases located in the downtown area of major American cities. Finally, there are the hotels that cater to the visitors to these central offices, as well as to casual tourists and conventioneers.

Then, too, there is some evidence (though the data are contradictory) that there has been a revitalization of certain inner city neighborhoods—often by privileged members of society. In New York, Chicago, Washington, D.C., Philadelphia, and elsewhere, some affluent persons have been returning to the central city, apparently to minimize the inconvenience of commuting and to take advantage of the special cultural advantages which the central area offers. It is accordingly highly questionable to look upon cities such as Los Angeles as representing the models of the future.

A reasonable interpretation of these ecological patterns runs as follows: Recent marked changes in the function of the central city are a reflection of important modifications in bureaucratic structures. Those structures that have recently become dominant, or have maintained their earlier dominance in terms of social power, have often retained (or even expanded) their head offices in the central city. The retail outlets of businesses, and the branch offices of other organizations that have emerged in suburban shopping centers, are much less the authentic reflection of the power bureaucracies of modern America. Although it is true that some head offices have become suburbanized, many major central cities still sustain a high degree of their vitality.

Why have many of the leading bureaucratic power centers remained in the central city? Is it because the central city still enjoys a major symbolic significance? Does the prestige of the central city reflect a continuity with earlier preindustrial forms? Or does this centralization pattern reflect the fact that the corporate and governmental elite must sustain a degree of concentration in order to preserve personal face-to-face interaction which is still so essential for decision-making? Individuals may still have to sustain personal ties if they are to negotiate or broker issues that involve high stakes.

If either or both of these interpretations (or perhaps others) are warranted, one can, with the continued expansion of multinational corporations—not only in the United States, but in many other industrialized nations as well—expect a still further centralization of the managerial sector of bureaucracies. The central areas of cities such as New York, San Francisco, Atlanta, etc., will become the hubs of a worldwide economic apparatus, in ways that ecologists have yet to anticipate. Then, as these corporations gain a degree of autonomy within nation-states—for instance, a freedom to manipulate "stateless money"—it may be that the centers of many leading cities in the United States, Europe, and elsewhere will find themselves inhabited by a managerial group that owes little loyalty to the city or to the nation but looks mainly to the world community. What kind of impact will such a pattern have on urban ecological processes? It is not too early to raise this issue.

To speculate on a less grandiose scale—it seems clear that with the growing energy crisis, it becomes necessary to revise many older notions concerning the impact of technology upon the ecology of cities. It should always be kept in mind that, although technology indirectly influences values and organizations, these latter also shape the course of technology. Will the hassles that are associated with commuting lead more of the affluent to re-establish residence in the central city? What then will happen to the poor, many of whom have in recent decades been concentrated in the city center? Will many of them be pushed to the suburbs? Urban ecologists will need to pay heed to these kinds of questions, and readjust their theories and methodological orientation in order to be in a position to monitor the major changes that are likely to occur in urban ecological patterns in the United States and in other parts of the world. Urban ecologists must be prepared to assess alternative futures.

Class conflict and land use

A related set of issues arises out of a renewed appreciation of the role which class structure plays in the dynamics of urban land use. In their study of Birmingham, England, Rex and Moore have observed that conflict over access to housing can be viewed as one aspect of a more basic conflict between social classes. McAdams and Feagin have investigated " . . . the ways in which the inequality of power or the class structure shapes the decisions which originate and arrange the physical face of cities" (cf. Snow). They have developed a systematic typology of the "land-interested actors" whose activities exert a significant influence upon urban land use. Feagin has also examined the specific role of land speculators in the determination of land-use decisions, finding it to be highly relevant to processes of urban site selection, internal community differentiation, the development of blight, and suburbanization. The *modus operandi* of speculators in converting socially generated increases in land value to their own profit is shown to be correlative to "a capitalistic political-economic system." Likewise Molotch has described the phenomenon of urban growth in contemporary America as an "areal expression of the interests of some land-based elite." The city, in his words, has become "a growth machine" in which new industries, a greater volume of business, additional labor, and a larger population provide the principal basis for political consensus for a substantial part of local business elites. All of these investigators see urban land-use processes as an expression of certain features of the class structure of American cities.

Variations in environmental orientations

Variations in the complex property known as power are likely to be accompanied by variations in other more specific variables. The nature of those variables, and their place in the description and explanation of ecological processes, may now be considered.

Descriptive approaches

Perhaps the least refractory of the issues that confront a sociocultural approach to ecology are those which are associated with the straightforward description of "natural areas," "social space," and other such mixed cultural-environmental phenomena. It is part of the very nature of such phenomena that they are both subjective and objective in character. Hence their description requires some recourse to the methods of interpretation as well as those of natural science. The resulting descriptions postulate an interdependence "between the internal subjective order . . . and the external spatial order" (Buttimer, b, p. 134), without, however, seeking to formally spell out the causal relations that might be involved in that interdependence. Their presentation takes the form of holistic, "ideographic" (Buttimer, a, p. 178) accounts of regions, communities, and neighborhoods, including in those accounts a description of soils and terrain, landscapes, life styles, and history.

Yet straightforward description raises issues of its own. To posit an interdependence between society, culture, and environment is to call for a "study [of] the forces which maintain this interdependence" (Buttimer, a: on Friedrich Ratzel, p. 31). It is to suggest the need for identifying the components that define a natural area: physical, economic, political, religious, etc.—a task for which Cohen's typology of environmental orientations should be singularly helpful. It is to point to the theoretical problem that is raised by instances of disharmony between culture and environment, particularly those of which the members of an affected population are aware. Finally, it is to remind the investigator of the truth that all description is necessarily selective of the reality with which it is concerned. Awareness of this truth is an injunction to examine the assumptions which reside in the concepts that one has employed in his description.

A special case of the descriptive approach to cultural-environmental phenomena may be found in the correspondence that has been frequently noted between spatial structure and social structure. First explored in the classic paper of Durkheim and Mauss, the phenomenon has been abundantly documented in the work of other ethnographers, most notably that of Lévi–Strauss (chaps. 7–8, 15–16). Its essential nature, however, is far from clear. The patterns which emerge in a people's account of its own village and farm layouts, for instance, are often a projection, onto space, of the patterns which that people sees in its kinship, class, and religious institutions. But the relation of each of these terms of the correspondence—the spatial and the sociocultural—to the ethnographer's own description of them will vary from one society to another. That is to say, the description of village and farm layouts offered by natives may or may not be the same as that which the ethnographer reports, and the natives' description of their institutions may or may not agree with the ethnographer's. Yet it would be a mistake to set this general problem aside as a matter of fortuitous variations in human perception. The identification of systematic interrelations between spatial structure and social structure, whether in the natives' version or the ethnographer's version, is a sufficiently plausible prospect to warrant further descriptive research.

Previous studies have yielded only the most tentative results. Among the Bororo, Lévi–Strauss has noted, the actual configuration of a settlement pattern corresponds, not to the ethnographer's account of the social structure, but to a fictional image which those people have of their society (Lévi-Strauss, p. 292). In other cases the correspondence is between actual spatial patterns and unconscious structures that reside in the minds of a people (ibid., p. 332). In either case, as Lévi–

Strauss observes, the study of the visible arrangement of houses and roads can afford a clue to people's images, conscious or unconscious, of their society (ibid., pp. 331–32).

Further complications arise when natives give differing accounts of their spatial structures. Even here, however, an unexpected order can emerge out of seeming contradiction, as Lévi–Strauss demonstrates in his analysis of Radin's report on the Winnebago. The descriptions of village layout which were offered by members of that society showed a tendency to vary with the kinship status of those individuals. To the members of one lineage group, the spatial design was that of a circle, with the huts of the chiefs being distributed around the periphery. To members of the other lineage group, the design was that of two concentric circles, with the chiefs' huts being situated in the center. Figuratively speaking, one group saw the doughnut, the other saw the hole in the doughnut. In Lévi–Strauss' words, the two images " . . . correspond to two different ways of describing one organization too complex to be formalized by means of a single model . . . " (pp. 134–35). Clearly, the conceptual resources of psychology will be indispensable to a further clarification of cultural-spatial relationships of this kind.

One of the more intriguing variants on the theme of cultural orientation to environment is afforded by the Dogon. In that society there is, on the one hand, a correspondence between actual village and farm layouts and the images which people have thereof, and, on the other hand, a correspondence between those layouts and the people's philosophical system. According to Dogon cosmology the universe has developed in the form of a spiral. This has become the visual model which provides the format for a subjective map of houses, shrines, and fields. According to Griaule and Dieterlen, " . . . the nature of the country, the contours, and the siting of watercourses necessitate compliance with the rule" (p. 94n). In this observation, description has turned into explanation. Whatever the validity of that explanation may be, the observation illustrates the potential fruitfulness, for causal analysis, of essentially descriptive studies of the meanings which peoples attach to their physical environments.

The usefulness of an ethnographic approach to human ecology can be seen in a recent study by Perin of real estate development in two American cities, Philadelphia and Houston. In her words, "My fieldwork was intended to produce a 'natives' account' of the American land-use system from the points of view of those largely in control of it" (p. 16). Those native accounts involved an "everyday vocabulary" by which developers, mortgage bankers, and civic leaders discussed such matters as home ownership, suburbs, privacy, race relations, and life styles, all in relation to their conceptions of such basic American values as success, citizenship, status, etc. Perin concludes that " . . . land-use ideas and practices translate American principles of social order into settlement patterns" (p. 210). The single-family dwelling, for example, is symbolic of the successful American who is at once an owner, a suburbanite, an insider, an old-timer, and a stable citizen (Perin, pp. 211–15).

The meanings thus revealed, of course, are strictly those of "native" developers and civic leaders, as inferred by the ethnographer from her interviews with those individuals. Their possible relation to the meanings of native laypersons (*non*-developers and *non*-leaders) remains problematic. Its systematic investigation, based on appropriate techniques of informant interviewing, might illuminate a wide range of important issues stemming from the existence of diverse sets of environmentally relevant meanings across the different subgroups of a population. In the meantime it is well to appreciate the role of the straightforward descriptive study in confirming expected findings but, more than that, in pointing up unexpected and variant findings concerning cultural-environmental relations. Such research can at times take the form of the explicitly negative case, as such providing a corrective to oversimplified explanations of particular features of environmental organization.

The quest for universal, or at least widely distributed, patterns of land use has been a longstanding and legitimate concern of human ecology (including, in this context, land economics and social geography). Sectors and rings figure prominently in that quest. Various observers have noted, for instance, the widespread prevalence of "autoconstruction" on the outskirts of industrial cities, the resulting areas forming segments of a concentric ring often described as a slum. Cities as different as Budapest, a socialist city, and Rio de Janeiro, a dependent capitalist city, are both characterized by such outlying fringes of shanty settlements. Yet the underlying causes of this seemingly widespread land-use pattern may be quite different from one case to another. In Budapest the phenomenon has been attributed to under-urbanization (Konrad and

Szelenyi, p. 157), whereas in Rio de Janeiro it has sometimes been linked with over-urbanization. There remains a further possibility that both are the spatial correlates of malsynchronized investment as between productive facilities and consumption goods in rapidly industrializing countries (cf. ibid., pp. 157–58).

It is possible that the universals which in all likelihood do reside in ecological processes are too complex to be visualized in cartographic forms. Perhaps such symbolic dualisms as "inner-outer," "central-peripheral," "sacred-profane," and "near-far," more nearly capture whatever universals there are in the settlement forms of human communities (cf. Lévi–Strauss, p. 142). But it is more likely that the level of conceptualization will have to be still more abstract and analytic than that. In any event there is everything to be gained, and nothing lost, by encouraging the descriptive study of variant community forms and processes. The discovery of *intra*-cultural "universals" would in itself be no mean achievement. For any given society, that is to say, there may be "a national type of city, a basic theme on which the individual cities are variations" (Wissink, quoted in Pahl, b, p. 137). Some of the features of American cities, for example, may stem "from a particular historical pattern of urbanization . . . that . . . has been infrequently replicated elsewhere in the world" (Roberts, p. 619).

Causal approaches

The salience of the concept of "culture" for an understanding of processes of environmental organization is by no means limited to its usefulness in descriptive research. Its place in a causal analysis of such processes is sufficiently well-founded to deserve systematic investigation. Its general nature is apparent in a variety of ecological phenomena. The most obvious of these, perhaps, is the phenomenon of historical "survivals" (Stinchcombe, p. 104). Anachronistic land-use patterns, reflecting the requirements of a previous period of history, are familiar features of many of the older cities of the world. As "fixed costs" their economic value, residing mainly in buildings and streets, may bear little relation to their current "productivity," and yet, under conditions that have yet to be theoretically specified, they persist. One component of such a theoretical specification appears to lie in the meanings which people attach to these survivals—meanings that involve class and ethnic identity, nationhood, religion, etc.

The formal structure of this type of causal relation is that of a "historicist explanation" (Stinchcombe, chap. 3). In such an explanation, according to Stinchcombe, " . . . an *effect* created by causes at some previous period *becomes a cause of that same effect* in succeeding periods" (p. 103). Thus, buildings and streets that were built in one period of a city's history—as such, representing an effect of circumstances prevailing at that time—sometimes outlive those circumstances, nevertheless remaining as assets that require repair and maintenance.

The mechanisms through which this self-renewal takes place are still imperfectly understood and call for systematic investigation. Among them, certainly, must be counted the transmission of property rights from one generation to another. The nineteenth-century Jewish ghetto of Rome, for example, while by then inhabited by just a fraction of the adherents of that faith in Rome, remained where it had been almost three centuries earlier, still occupied by preponderantly low-income, orthodox Jews, long after the religious discrimination which first gave rise to it had disappeared. An adequate explanation of this survival would have to include, in its formulation, some statement concerning the *de facto* tenure rights which its residents, most of them renters, had historically acquired. Those rights traced back to the seventeenth century when the government imposed a rent freeze on the Christian landlords who had been exploiting their Jewish tenants. Thus deprived of an incentive to repair and maintain their properties, the landlords, in effect, turned that responsibility over to the tenants who, in return for their trouble, retained tacit rights of lease to the buildings which they occupied— "property" rights which were subsequently transmitted from one generation to the next (Dunn and Dunn). A satisfactory conceptualization of mechanisms like this remains one of the central problems in sociocultural ecology.

The truly ancient cities of the world—Athens, Rome, Jerusalem, and Cairo, to name but a few of them—present a special opportunity for ecological investigation along these lines. In her account of the thousand-year history of Cairo, for instance, Abu–Lughod presents detailed descriptive materials concerning successive phases of the history of that ancient city and its various districts. She notes, too, that " . . . the ecological organization of the contemporary city appears confused and utterly capricious to an observer who fails to understand its previous natural and cultural history" (p. 9). The one-time boundaries of medieval Cairo, for instance, match almost perfectly those of one of the "natural areas" that emerge in a factor analysis of modern Cairo—an area that is still distinguishable from other parts of the city by its physical structure and population characteristics (Abu-Lughod, pp. 188–93). Historical studies of this kind, in which descriptive and explanatory functions have been joined in an integrated historical inquiry, can go far toward broadening the fund of empirical knowledge on the basis of which historicist causal explanations may be further developed (see, for example, Weinstein, pp. 23–29).

It is not only as anachronisms, however, that antecedent land-use patterns can acquire their causal significance. Under certain conditions the meanings which attach to particular land-use patterns may undergo a geographic extension or generalization. An expanding city, for example, may find itself moving into rural areas that have already acquired some kind of reputation. In his study of Belfast, Jones observed that outlying land once occupied by country houses and parks, having become invested with a certain prestige, tended to attract urban residents of a corresponding character, thus giving rise to a sectoral pattern (a, pp. 278–80). Community-forming processes like these exemplify what Sorokin has called the "margin of autonomy" of a sociocultural system. Once established as a going concern, a culture tends to acquire a certain "margin of autonomy vis-à-vis the external forces amidst which it functions" (p. 26). It is this feature which accounts for the "self-replicating causal loops" which Stinchcombe sees as the distinctive mechanism that gives warrant to the historicist mode of explanation (pp. 103–6).

The processes by which environmentally relevant meanings emerge, and then enter as causal factors into the dynamics of community formation, are an invitation to collaborative research by human ecologists and social psychologists. The ubiquitous sense of territorial identity is a case in point. That various types of sentiments can in fact attach to a locality, along with its artifacts and inhabitants, is an ecological commonplace (Firey, a). That such sentiments bear some relationship to processes of succession is also clear, though the interpretation of that relationship may be open to dispute. But the suggestion, advanced by Hunter, that localistic sentiments can be something more than survivals of older community formations, is an insight that deserves to be further explored. Under certain conditions, Hunter has suggested, there may be forces at work within a community which tend to continually regenerate sentiments of a localistic character. Indeed, it appears that " . . . mass society has permitted propinquity itself to become an important basis for defining relationships" (Hunter, b, p. 155). Stone, too, has suggested that a sense of community affiliation on the part of some urban residents may represent a psychological substitution for nonexisting primary relationships (cited in Hunter, b).

This longing for communal roots, the "grieving for a lost home" (Fried), and other such manifestations of "false consciousness," are psychological realities whose consequences for community-forming processes deserve more study than ecologists have generally been willing to give to them (but see Hunter, a; Suttles). The manner in which such expressions are employed in everyday conversation, the contexts in which they occur, and the community imagery which they articulate, represent a fruitful topic for joint investigation by ecologists and ethnomethodologists. So, too, the subjective experience of "territorial injustice" (Pahl, c, pp. 51–52; cf. Harvey, a, pp. 101–18) might well become a topic for discourse between the disciplines of ecology and ethics. Personality defense mechanisms, such as fantasy, isolation, identification, and displacement, when projected onto the physical environment and its artifacts, offer further opportunities for interdisciplinary research for human ecologists. The pertinence of this kind of conceptualization is particularly evident in some of the more extreme cases of ethnic segregation, such as that which presently holds between Protestant and Catholic working-class residents of urban areas in Ulster.

Sentiments of territorial identity are, of course, only one of the many kinds of environmentally relevant meanings which may, under certain conditions, assume causal significance with respect to processes of community formation. The specification of those conditions should be one of the basic concerns of human ecology. For, as Pickvance has observed, value orientations " . . . refer to tendencies for action . . . which in actual situations are subject to constraints which fully or partially prevent their realization" (Pickvance, p. 177). Thus, the decision by a number of residents to leave a particular locality at a particular time may be an expression of meanings which come into play only at a certain income level, at a certain point in the life cycle, at a certain stage of housing obsolescence, at a certain level of awareness of alternatives, etc. Such variations in the causal influence of environmentally relevant meanings, over different categories of other variables, point to the importance of comparative research in which these interaction effects would form the principal topic of the investigation.

Functional analysis

Another perspective in terms of which to examine the relevance of sociocultural factors for community-forming processes is that of functional analysis. In this type of explanation, at least part of the causation of a process is regarded as lying in the consequences which that process has for its own self-maintenance (Stinchcombe, p. 80). Such consequences may be positive or they may be negative—functional or dysfunctional. The spatial segregation of the residents of a community in terms of their relative wealth will illustrate some of the issues that are involved in an application of this perspective to human ecology. Segregation of subpopulations according to wealth can have, among other effects, that of reinforcing the very system of differential access to privilege which it reflects. This it accomplishes through the operation of at least two mechanisms. First, by limiting the occasions for contact between the more affluent and the less affluent, it forestalls the cognitive dissonance that might otherwise develop out of frequent encounters between the extremes of wealth within the context of an ideologically egalitarian society, in this way contributing to the "legitimacy" of that system of privilege (cf. Gans, p. 187). Second, by symbolically representing the identity of the wealthy, it reassures the members of that subpopulation in their claims to that identity—claims that might otherwise be eroded by the indifference of the nonwealthy members of an open-class society which affords its wealthy members few of the other kinds of symbolic representation that are normally available in a closed stratification system (cf. Sopher, p. 113).

An explication of such mechanisms as these, and a determination of the conditions under which they operate, will be essential to an assessment of the full potential of functionalism for human ecology. Such an explication will have to address itself to the problems which Merton has codified in his "paradigm for functional analysis" (b, pp. 50ff). This will entail an identification of the specific processes to which functions are being imputed, an indication of the consequences which those processes have for a specified system, a description of the mechanisms that lead to those consequences, and an account of "the sociological procedures of analysis" (ibid., p. 54) employed in a given study (Firey, b).

There are, of course, other kinds of community-forming processes for whose study the functional perspective may also be appropriate. The spatial correlates of an economic system are a case in point. Many underdeveloped countries, for example, have what is commonly known as a dual economy—an economy that is characterized by a modern capitalist sector on the one hand, and a traditional subsistence sector on the other. In the larger cities of such countries, these two sectors will often exhibit competing "space demands" (McGee, p. 278). "Blocks of space" (ibid., p. 279) occupied by small shops and bazaars will find themselves from time to time invaded by larger capitalist firms, following which there may be substantial changes in the social and economic character of the affected area (ibid., pp. 273–74). Such processes of succession invite an application of the functional perspective to a study of comparable data for countries distributed on a scale of economic development.

So, too, there can be "unanticipated consequences of purposive social action" (Merton, a) among the processes of environmental organization—a reminder of the salience for ecological theory of the notion of "functional requirements." In his study of the development and consequences of federal urban renewal in fifteen American cities, Greer has observed that that program " . . . succeeded in materially reducing the supply of low-cost housing in American cities" (Greer, p. 3). Yet one of the objectives of the program had been "to remedy the serious housing shortage" (legislation, cited in ibid., p. 4), as well as to effect slum clearance and overall community development (ibid., p. 165). Part of the explanation for this ironic outcome lay in the heterogeneousness of the program's goals, but part of it lay in the program's commitment to a local initiation of all projects, and a reliance upon the market system (ibid., p. 6), both of which were part of the very problem which urban renewal was supposed to remedy. In time the program came to place more of its emphasis upon the revitalization of the central business districts of its participating cities. Thus, in meeting the functional requirements of a system of localized decision-making and a market economy, urban renewal found itself in default of part of its original mission (cf. Greer, pp. 118–23 and chap. 7).[4]

These, then, are among the issues which are likely to engage the attention of sociocultural

---

[4]The functionalist terminology is ours, not Greer's.

ecologists for some time to come. They are, to be sure, the immediate concerns of only one of the two paradigms that presently characterize the field of human ecology. Yet progress toward their resolution should enhance the practical usefulness of the discipline as a whole, and it might well eventuate in theoretical initiatives at "paradigm bridging" (cf. Ritzer, p. 212).

Bibliography

Abu-Lughod, Janet L. *Cairo: 1001 Years of 'The City Victorious'.* Princeton: Princeton University Press, 1971.

Burns, Leland S.; and Pang, Wing Ning. "Big Business in the Big City: Corporate Headquarters in the CBD." *Urban Affairs Quarterly,* 12 (June 1977):533–44.

Buttimer, Anne. *Society and Milieu in the French Geographic Tradition.* Chicago: Rand McNally, 1971. (a)

Buttimer, Anne. "Social Space in Interdisciplinary Perspective." In *Readings in Social Geography,* edited by Emrys Jones pp. 128–37. Oxford: Oxford University Press, 1975. (b)

Castells, Manuel. "Toward a Political Urban Sociology." In *Captive Cities,* edited by Michael Harloe, pp. 61–78. New York: John Wiley & Sons, 1977.

Cohen, Erik. "Environmental Orientations: A Multidimensional Approach to Social Ecology." *Current Anthropology,* 17 (March 1976):49–70.

Cutler, William W., III. "The Persistent Dualism: Centralization and Decentralization in Philadelphia, 1854–1975." In *The Divided Metropolis: Social and Spatial Dimensions of Philadelphia, 1800–1975,* by William W. Cutler, III, and Howard Gillette, Jr. Westport, Conn.: Greenwood Press, 1980.

Daniels, P. W. "Office Location in the British Conurbations: Trends and Strategies." *Urban Studies,* 14 (October 1977):261–74.

Dunn, Leslie C.; and Dunn, Stephen P. "The Jewish Community of Rome." *Scientific American,* 196 (March 1957):118–28.

Durkheim, Emile; and Mauss, Marcel. "De quelques formes primitives de classification: Contribution à l'étude des représentations collectives." *L'année sociologique,* 6 (1901–1902):1–72.

Feagin, Joe R. "Urban Real Estate Speculation: Implications for Social Science and Urban Planning." Forthcoming.

Firey, Walter. *Land Use in Central Boston.* Cambridge: Harvard University Press, 1947. (a)

Firey, Walter. *Man, Mind and Land.* New York: Free Press, 1960. (b)

French, R. A.; and Hamilton, F. E. Ian, eds. *The Socialist City: Spatial Structure and Urban Policy.* New York: John Wiley & Sons, 1979.

Fried, Marc. "Grieving for a Lost Home." In *The Urban Condition,* edited by Leonard J. Duhl. New York: Basic Books, 1963.

Gans, H. J. "The Balanced Community: Homogeneity or Heterogeneity in Residential Areas?" In *Readings in Social Geography,* edited by Emrys Jones, pp. 179–90. Oxford: Oxford University Press, 1975.

Gibbs, Jack P., ed. *Urban Research Methods.* Princeton: Van Nostrand, 1961.

Goodwin, Carole. *The Oak Park Strategy.* Chicago: University of Chicago Press, 1979.

Greer, Scott. *Urban Renewal and American Cities.* Indianapolis: Bobbs–Merrill, 1965.

Griaule, Marcel; and Dieterlen, Germaine. "The Dogon of the French Sudan." In *African Worlds,* edited by Daryll Forde. London: Oxford University Press, 1954.

Harloe, Michael, ed. *Captive Cities.* New York: John Wiley & Sons, 1977.

Harvey, David. *Social Justice and the City.* London: Arnold, 1973. (a)

Harvey, David. "Government Policies, Financial Institutions, and Neighbourhood Change in United States Cities." In *Captive Cities,* edited by Michael Harloe, pp. 123–40. New York: John Wiley & Sons, 1977. (b)

Hawley, Amos H. *Human Ecology.* New York: Ronald Press, 1950.

Hunter, Albert. *Symbolic Communities.* Chicago: University of Chicago Press, 1974. (a)

Hunter, Albert. "Persistence of Local Sentiments in Mass Society." In *Handbook of Contemporary Urban Life,* by David Street et al., pp. 133–62. San Francisco: Jossey–Bass, 1978. (b)

Janowitz, Morris; and Street, David. "Changing Social Order of the Metropolitan Area." In *Handbook of Contemporary Urban Life,* by David Street et al., pp. 90–128. San Francisco: Jossey–Bass, 1978.

Jones, Emrys. *A Social Geography of Belfast.* New York: Oxford University Press, 1960. (a)

Jones, Emrys, ed. *Readings in Social Geography.* Oxford: Oxford University Press, 1975. (b)

Konrad, György; and Szelenyi, Ivan. "Social Conflicts of Underurbanization." In *Captive Cities,* edited by Michael Harloe, pp. 157–74. New York: John Wiley & Sons, 1977.

Lemon, Anthony. *Apartheid.* Westmead, Farmborough, Harts., England: Saxon House, 1976.

Lévi-Strauss, Claude. *Structural Anthropology.* New York: Basic Books, 1963.

Masterman, Margaret. "The Nature of a Paradigm." In *Criticism and the Growth of Knowledge,* edited by Imre Lakatos and Alan Musgrave, pp. 59–89. Cambridge: Cambridge University Press, 1970.

Mayer, Philip. "Class, Status, and Ethnicity as Perceived

by Johannesburg Africans." In *Change in Contemporary South Africa,* edited by Leonard Thompson and Jeffrey Butler, pp. 138–67. Berkeley: University of California Press, 1975.

McAdams, D. Claire; and Feagin, Joe R. "A Power-Conflict Approach to Urban Land Use: Toward a New Urban Ecology." Forthcoming.

McGee, T. G. "Planning the Asian City: The Relevance of 'Conservative Surgery' and the Concept of Dualism." In *The Outlook Tower,* edited by J. V. Ferreira and S. S. Jha, pp. 266–81. Bombay: Popular Prakashan, 1976.

Merton, Robert K. "The Unanticipated Consequences of Purposive Social Action." *American Sociological Review,* 1 (December 1936):894–904. (a)

Merton, Robert K. *Sociological Ambivalence and Other Essays.* New York: Free Press, 1976. (b)

Mingione, Enzo. "Theoretical Elements for a Marxist Analysis of Urban Development." In *Captive Cities,* edited by Michael Harloe, pp. 89–110. New York: John Wiley & Sons, 1977.

Molotch, Harvey. "The City as a Growth Machine: Toward a Political Economy of Place." *American Journal of Sociology,* 82 (September 1976):309–32.

Musil, J. "The Development of Prague's Ecological Structure." In *Readings in Urban Sociology,* edited by R. E. Pahl (a), pp. 232–59. New York: Pergamon Press, 1968.

Noble, Allen G.; and Dutt, Ashok K., eds. *Indian Urbanization and Planning: Vehicles of Modernization.* New Delhi: Tata McGraw–Hill, 1977.

Pahl, R. E., ed. *Readings in Urban Sociology.* New York: Pergamon Press, 1968. (a)

Pahl, R. E. *Whose City?* London: Longman Group Ltd., 1970. (b)

Pahl, R. E. "Managers, Technical Experts and the State: Forms of Mediation, Manipulation, and Dominance in Urban and Regional Development." In *Captive Cities,* edited by Michael Harloe, pp. 49–60. New York: John Wiley & Sons, 1977. (c)

Parsons, Talcott. *The Social System.* New York: Free Press, 1951.

Perin, Constance. *Everything in Its Place.* Princeton: Princeton University Press, 1977.

Perlman, Janice E. *The Myth of Marginality.* Berkeley: University of California Press, 1976.

Pickvance, C. G. "From 'Social Base' to 'Social Force': Some Analytical Issues in the Study of Urban Protest." In *Captive Cities,* edited by Michael Harloe, pp. 175–86. New York: John Wiley & Sons, 1977.

Rees, John. "Manufacturing Headquarters in a Post-Industrial Urban Context." *Economic Geography,* 54 (October 1978):337–54.

Rex, John A.; and Moore, R. *Race, Community and Conflict.* London: Oxford University Press, 1967.

Ritzer, George. *Sociology: A Multiple Paradigm Science.* Boston: Allyn & Bacon, 1975.

Roberts, Bryan R. "Comparative Perspectives on Urbanization." In *Handbook of Contemporary Urban Life,* by David Street et al., pp. 592–627. San Francisco: Jossey–Bass, 1978.

Rosenthal, Donald B., ed. *Urban Revitalization.* Beverly Hills, Calif.: Sage Publications, 1980.

Schexnider, Alvin J. "Blacks, Cities, and the Energy Crisis." *Urban Affairs Quarterly,* 10 (September 1974):5–16.

Schutz, Alfred. *Collected Papers, I.* The Hague: Martinus Nijhoff, 1962.

Shaw, Denis J. B. "Planning Leningrad." *The Geographical Review,* 68 (April 1978):183–200.

Sjoberg, Gideon. *The Preindustrial City: Past and Present.* New York: Free Press, 1960. (a)

Sjoberg, Gideon. "The Rise and Fall of Cities: A Theoretical Perspective." *International Journal of Comparative Sociology,* 4 (September 1963):107–20. (b)

Snow, David A.; and Leahy, Peter J. "The Making of a Black Slum-Ghetto: A Case Study of Neighborhood Transition." *Journal of Applied Behavioral Science,* forthcoming.

Social Science Research Council. *Annual Report, 1978–1979.*

Sopher, David E. "Place and Location: Notes on the Spatial Patterning of Culture." In *The Idea of Culture in the Social Sciences,* edited by Louis Schneider and Charles Bonjean. Cambridge: Cambridge University Press, 1973.

Sorokin, Pitirim A. *Sociocultural Causality, Space, Time.* Durham, N.C.: Duke University Press, 1943.

Stinchcombe, Arthur L. *Constructing Social Theories.* New York: Harcourt, Brace & World, 1968.

Stone, Gregory P. "Urban Identification and the Sociology of Sport." Cited in Albert Hunter (b).

Street, David, et al. *Handbook of Contemporary Urban Life.* San Francisco: Jossey–Bass, 1978.

Suttles, Gerald D. *The Social Order of the Slum.* Chicago: University of Chicago Press, 1968.

Weinstein, Jay. "Subjective Components of Urban Ecological Structure: A Theoretical Statement and an Application to Two Indian Cities." Presented at the 42nd Annual Meeting of the Southern Sociological Society, Atlanta, April 6, 1979.

Wheatley, Paul. *The Pivot of the Four Quarters.* Edinburgh: The University Press, 1971.

Willhelm, Sidney M. *Urban Zoning and Land Use Theory.* New York: Free Press, 1962.

Wissink, G. A. *American Cities in Perspective.* Assen, The Netherlands: 1962.

Wright, Arthur F. "The Cosmology of the Chinese City." In *The City in Late Imperial China,* edited by D. William Skinner, pp. 33–73. Stanford, Calif.: Stanford University Press, 1977.

Sociocultural Versus Neoclassical Ecology: A Contribution to the Problem of Scope in Sociology*

Kenneth D. Bailey
and Patrick Mulcahy

Criticisms of human ecology are common; internecine bickering is the rule (cf. Alihan, 1938; Gettys, 1940; Firey, 1947; Willhelm, 1962; Willhelm, 1964; Sjoberg, 1965). In addition to this internal criticism, a recent paper has criticized human ecology as a whole: Michelson (1968) expressed his dissatisfaction with human ecology's progress in an article entitled "A Parsonian Scheme for the Study of Man and Environment, or What Human Ecology Left Behind in the Dust." It is our feeling that what the critics of human ecology have "left behind in the dust" is a number of distinct contributions and potential contributions to sociology.

Many sociological researchers seek to explain individual behavior using the individual as their basic unit of analysis. Their interests include individual decision-making, motivations, and attitudes. Other researchers study social structure, using as their basic unit of analysis a human aggregate, e.g., a group, city, population, or society. Rather than study properties of individuals *per se,* they study properties of aggregates. These two types of studies obviously differ greatly in scope. For heuristic purposes we will call *micro* those studies in which the basic unit of analysis is an individual, and *macro* those which utilize an aggregate of individuals as the basic unit. There is often much competition between those who favor micro analyses and those who prefer macro analyses. Each researcher seems to feel that the scope of his study is "best" for sociology. We feel that these two approaches are generally complementary, and that both are necessary for sociology.

This micro-macro controversy also exists in human ecology. The purpose of this paper is to reconcile the sociocultural (micro) and neoclassical (macro) schools of human ecology, two schools of thought with several points of disagreement. Chief among these is the role of human volition in ecology. We contend that much of this disagreement dissolves when the difference in the scope of these two schools is made clear. By showing that the two schools of thought complement rather than com-

*Reprinted from *The Sociological Quarterly,* 13 (Winter 1972): 37–48, by permission of the authors and the publisher.

pete with one another, this paper contributes to an understanding of the more general micro-macro debate in sociology. We will show that in ecology it is not necessary to argue reductionism, to be only a socioculturalist, or only a neoclassicist. Rather, one chooses the frame of reference (either sociocultural or neoclassical) which is best suited for the study of a particular research problem.

The methodological contributions of both the neoclassicists and the socioculturalists stem largely from the fact that each has developed a frame of reference. The sociocultural frame of reference emphasizes individual decision-making in accordance with social norms and values, while the neoclassical frame of reference emphasizes the population as a basic unit of analysis.

A frame of reference

A frame of reference is a set of definitionally interrelated concepts, defining the theoretically relevant properties of an object. It provides the context within which the construction and testing of hypotheses can take place. As such, it is pretheoretical. As an example consider the concepts of "status," "role," and "norm." These concepts are interrelated by definition; an individual's position in society is his *status;* the *norms* of society prescribe the proper behavior for a person in his position; and this proper behavior constitutes his *role*. These three interrelated concepts provide a context for the analysis of the individual in society. Such a frame of reference is very important, for it provides a framework in which research findings can be accumulated. Without a common context, the work of various researchers cannot be compared. More studies would simply result in more confusion, rather than in an accumulation of research findings.

It is impossible to theorize without the definitional framework that a frame of reference provides. This does not mean, however, that the frame of reference *is* the theory. A theory provides an explanation; it can be true or false, tautologous or not tautologous. A frame of reference cannot be tautologous; it is only a set of definitions which *precedes* the construction of theory. For example, a theory could be written which utilizes the frame of reference (status, norm, and role) described above. A theory is a set of interrelated propositions. The concepts of "status," "norm," and "role" were shown to be interrelated by definition. However, no propositions in the form of testable statements have been provided. Until this is done, no theory has been provided. Such propositions could be written. For example, one could hypothesize that "the higher an individual's status, the lower is the number of norms prescribing his conduct in that position (i.e., his role)." This would be a testable proposition and could be considered to be in the realm of theory. As such, it can be judged by the criteria used to judge all theory. However, a frame of reference *per se* cannot be judged by the criteria used to judge theory.

This point is important because some socioculturalists contend that the ecological complex represents tautological reasoning (Willhelm, 1962:22–23; Willhelm, 1964:243), that it *explains itself*. But as Homans (1964:811) says, "The . . . office of theory is to explain." Theories explain; frames of reference do not. The ecological complex as generally presented by the neoclassical ecologists (cf. Duncan and Schnore, 1959) is a frame of reference only, a set of definitions, and definitions cannot be true or false. Therefore, charges of tautology are relevant for theory, but not for frames of reference such as the ecological complex.

The sociocultural frame of reference

For the present we may dichotomize scope. Many ecologists take a *population* as their basic unit of analysis; others take one or more *individual members of a population* as their basic units. Socioculturalists generally choose the latter course. They often study land use not only within a population but within a single city.

One recent study of land use within a city is Willhelm's (1962) study of the zoning process in Austin, Texas. According to Willhelm (1962:30), socioculturalists feel that " . . . volition must be taken into consideration in developing ecological theories." To socioculturalists, culture and social values are important explanatory variables in ecology. In his presentation of a sociocultural frame of reference, Willhelm (1962:36; italics added) states:

> In locating over *space,* individuals seek to attain a certain *end*. To accomplish the goal, a specific *norm* must be selected from the *means* in accordance with *social conditions* that prescribe the course social action is to take. Furthermore, such performance requires cognition on the part of actors concerning the properties—actual or potential—of the circumstances in a situation. But because the *goal, means, normative conditions,* and *cognitive data* must be given specificity to arrive at concrete decisions and

conclusions or to voice opinions and viewpoints, actors must develop and employ evaluative standards, that is, *value systems*.

The italicized words represent concepts in Willhelm's frame of reference. This sociocultural frame of reference is similar to the "status-norm-role" example given above. An individual actor tries to achieve some specific end with regard to land usage. The action he takes to achieve this end is similar to the concept of "role" discussed above. He is guided in his action by factual information (cognitive data) and by a value system. However, his action is also guided by norms. (The reader is referred to Willhelm [1962:28–49] for more definitions and discussion.)

In Willhelm's work, interrelated concepts are defined which provide a framework for the accumulation of research findings on land usage. Firey's (1947) findings on Boston can be compared with Willhelm's findings on Austin. As far as scope is concerned, the sociocultural frame of reference is clearly concerned with one or more individuals working within the context of a larger population. The frame of reference emphasizes the effects of individual decision making on land use and other ecological problems. Further, it emphasizes the fact that individuals have alternatives, and make their decisions in a definitely cultural context. Such a frame of reference is designed to avoid the determinism (economic or biological) that socioculturalists feel is inherent in the classical and neoclassical approaches. Within the sociocultural frame of reference, land use is not "adaptation," with that concept's connotation of an individual changing so as to meet the requirements of the land. The individual controls the land rather than "adapting" to it.

The sociocultural frame of reference has proven effective for the study of urban land use, but its effectiveness has not been proven for other very interesting and important ecological problems. For example, imagine two hypothetical countries, A and B. Imagine that the distribution of cities by size in Country A fits the rank-size rule. This means that the country contains a few large cities, more intermediate-sized cities, and many small cities. Imagine that Country B exhibits a primate distribution. This means that the majority of the urban population resides in only one or a few large cities. These two patterns are common in the world today. They can also be viewed as different patterns of adaptation to space. One research question is, "What variable or variables account for the differences in the urban size distributions of these two countries?"

The sociocultural approach has not proven its effectiveness in solving such problems. This does not mean that the approach is incapable of solving such problems, but only that socioculturalists have not attempted to solve such problems. It seems that decisions made by individuals in the society surely could affect whether the country exhibits the rank-size distribution or the primate distribution. But, neither the rank-size distribution nor the primate distribution represents an "end" in the sociocultural sense.

The neoclassical frame of reference

Such relevant ecological variables as the type of city-size distribution of a country, or the proportion of a country's population that is urban, are properties of *populations*, not of individuals. Here the scope is not the same as for the socioculturalists' study of land use within one city. Here the relevance of individual decision-making is more difficult to assess. It seems that when ecological research involves properties of whole populations, a different frame of reference is needed from one such as the sociocultural which is designed for the analysis of individual action *within* populations. The neoclassical ecologists have formulated such a frame of reference, known as the "ecological complex." This frame of reference is different in scope from the sociocultural frame of reference. Its basic unit is a spatially delimited population. As such, the sociocultural approach and the ecological complex are not incompatible. They may be used to complement each other. In certain situations where one is studying adaptation of individuals *within* a population, he may utilize the sociocultural approach. When he is studying differences in adaptation of one population at two or more points in time, or differences in adaptation between two or more populations at one point in time, he may utilize the neoclassical frame of reference.

The ecological complex has been presented by Duncan and Schnore (Duncan and Schnore, 1959; Duncan, 1959; Duncan, 1961; Schnore, 1958; Schnore, 1960; Schnore, 1961; Schnore, 1962; Schnore, 1964). A similar formulation was provided earlier by Ogburn (1951) and Park (1936).[1]

---

[1]The ecological complex as presented by its principal proponents Duncan and Schnore (1959) has been severely criticized by Willhelm (1962; 1964). The ecological complex is presented here in a manner designed to answer critics. The ecological complex as presented in this paper may thus differ in fact or emphasis from the

Use of this approach is not limited to contemporary human ecology. Similar formulations can also be seen in Mott (1965) and in Durkheim's (1933) *Division of Labor in Society*. Gibbs and Martin (1959) present a neoclassical approach that shares the scope of the ecological complex.

The ecological complex is a frame of reference which consists of population, organization, environment, and technology. These components each represent groups of variables rather than single variables (just as the components' "cognitive data" or "values" in the sociocultural frame of reference represent groups of variables and not single variables). Each of these components represents a set of *properties* of a population. Duncan (1959:681) says, "That at least some spatially delimited population aggregates have unit character is one of the key assumptions of human ecology, as is the premise that there are significant properties of such an aggregate which differ from the properties of its component elements." Ecologists generally delimit a "population" by reference to national boundaries, although other types of boundaries could be specified.

It should also be noted that the term "environment" refers to all properties of the land area and water area (and the air over it) contained within national boundaries. One example of an environmental property is land area in square miles. In this paper there is by definition a separate environment for each population. Other populations are not considered part of the environment.[2]

---

portrayal of other writers. For example, Stephan (1970:224) suggests a radically different definition of a population.

[2] Duncan and Schnore (1959:136) include other human populations in the environment. This is apparently to recognize the fact that a population can be affected by other populations. However, this practice is confusing if one defines population as spatially delimited (as does Duncan, 1959:681). If a population is defined as the group of people who inhabit a certain space, and if all persons within that space are by definition in the same population, the only way one can include other populations within the environment is to either consider them subpopulations (e.g., the black population of the United States), or to extend the limits of the environment past that space used to define the population. If this is done, the definition of environment becomes arbitrary since the environment is not spatially delimited. For example, one could say that the environment of the population of the United States is the entire world. However, this represents a change of scope. It should be noted that the task of defining all components of the ecological complex is far beyond the scope of the present paper. All of the reasoning in support of these definitions cannot be included in this paper, but the definitions are designed to provide a frame of reference satisfactory to critics of the ecological complex.

The four components of the ecological complex comprise a frame of reference helpful in the study of adaptation. Adaptation can be characterized in terms of the "level of living" or the "standard of living" of a population. As Ogburn (1951:314) says, "In this paper four factors that affect differences in the standard of living of peoples will be considered: population, natural resources, organization, and technology. Actually, these factors are interrelated and not independent." Duncan (1959:707) symbolizes the level of living by "L," and writes it as a function of the ecological complex: $L=f(P,E,T,O)$. Gibbs and Martin (1959:30) feel that a basic ecological question is "How does this species survive?" The "survival level" or "subsistence level" can be symbolized by L_s. This is clearly a measure of level of living. But it is the lower limit of a population's level of living.

We may characterize the neoclassical frame of reference by paraphrasing Willhelm's (1962:36) statement of the sociocultural frame of reference. In locating within a certain *environment*, populations seek to attain a certain *level of living*. To accomplish this goal they must regulate their *population size*. They must also select a certain form of *organization* from the alternative forms available. The form of *organization* they choose may also depend upon the *technology* they possess, and on the nature of this *environment*.

Level of living does not have to be treated as the dependent variable. Propositions can be written which treat any of the components as dependent variables. Many conservationists are interested in treating the environment as the dependent variable. They are interested in how societies have changed their environment by maintaining a population of a certain size at a certain level of living with a particular organization and technology. Also, organization is often used as a dependent variable (Duncan and Schnore, 1959:136). However, much of the interest in organization probably stems from its effect on level of living. This is seen in Gibbs and Martin's (1959:30) discussion of organization for sustenance.

In defense of the ecological complex

Sociocultural ecologists have criticized several features of the ecological complex. We feel that these criticisms are not valid for the ecological complex as presented in this paper, although they may be valid for some earlier formulations.

1. *Tautology*. It has been pointed out above that a frame of reference cannot be tautological.

Neoclassicists would like to move beyond the pretheoretical stage of using the ecological complex as a frame of reference and develop theoretical statements. Some socioculturalists feel that neoclassicists would use the ecological complex to explain the ecological complex (Willhelm, 1962:23; Willhelm, 1964:243). This would of course be tautological. We do not propose using the ecological complex to "explain" all of its components simultaneously. Usually we will use the four components to explain level of living. This is certainly not tautological. Further, we might, for example, attempt to explain the differences in some organizational variable among 20 different countries by examining the differences in populational, technological, and environmental variables among these countries. We would "explain" only the variation in the organizational variable.[3] The dependent variable does not have to be chosen from the organizational component. In other analyses we could choose the dependent variable from one of the other components of the ecological complex. Such usage would not be tautological.

2. *"Mixed order of data."* Some socioculturalists say that "organization" is only a mental construct. As such it should not be analyzed on the same level as the material components of population, environment, and technology (Willhelm, 1962:24). It would seem that the classical ecologists' treatment of the "biotic" and "social" as separate levels would satisfy the sociocultural critics. However, critics attacked this separation into levels as artificial and unworkable (Alihan, 1938; Gettys, 1940).

Michelson (1968) thinks that ecologists need to use an *intersystem congruence model*. In an intersystem congruence model there is no determinism or dominance of any one component (Michelson, 1968:207). In addition, the environment should be specifically included (Michelson, 1968). We think that the ecological complex fits the requirements of such a model. No hierarchy of components is assumed. The environment is defined as being on the same "level" as organization, technology, and population in affecting the level of living.

When we say the components of the ecological complex are all on the same "level," we simply mean that they are interrelated and do not have to be studied separately. They can all be studied simultaneously. Some socioculturalists feel that the environment is not on the same level of analysis as the other components. Sjoberg (1965:166) says that changes in technology and organization result in a redefinition of the environment, but the opposite does not occur. That is, changes in the environment will not influence organization and technology. At the most, Sjoberg is saying that the relationships between variables in the environmental component and variables in the other components are not symmetrical. This surely does not mean that the environment cannot be studied at the same time that the other components are being studied. Further, Sjoberg seems to be saying, for example, that if people in two different countries have two different physical environments, they will not adopt two different technologies to adapt to these environments. This is a claim which he does not empirically demonstrate.

3. *Determinism.* Socioculturalists have also charged that neoclassicists " . . . feel that social values are psychological and therefore must be excluded from an ecological inquiry" (Willhelm, 1962:25–26). We feel that if some neoclassicists have rejected concepts they have considered to be individualistic or psychological, it is chiefly because their particular research problem was macro in scope, and their basic unit of analysis was a population rather than an individual. We certainly would not preclude the use of values in our formulation. In our framework men control the ecological complex, not vice-versa.

When we say that the environment should be included as a relevant component in an ecological frame of reference, this does not mean that we are espousing "environmental determinism." We do not mean that the environment *causes* man to act a certain way without giving him any choice in the matter. We only mean that man's decisions can be based upon knowledge of environmental factors. As Barker (1960:5) says:

> Furthermore, it is common knowledge that behavior is responsive to climatic, economic, and technological changes. When electricity and electrical appliances come to a farm family, behavior is revolutionized in a matter of hours. The social history of recent times can leave little doubt that ecological variables

---
[3] Actually we do not "explain" the variation in a variable. What we really do is find the correlates of the dependent variable. For example, we might find that other variables chosen from the ecological complex are correlates of a particular organizational variable. The variation in the dependent organizational variable is explained in the statistical sense, as when we say that R^2 can be interpreted as the proportion of variation in the dependent variable that is "explained" by the independent variables. From a theoretical (as opposed to a statistical) point of view, we would call the relationship an empirical generalization. The role of an empirical generalization is treated in detail below.

are dominant in the energetics of much behavior change.

Individual action in the ecological complex

We have contended that the ecological complex and the sociocultural frame of reference differ in scope but complement each other. This means that ecologists are able to bridge the micro-macro gap by moving from the analysis of *properties of individuals within a population* (sociocultural) to the analysis of *properties of populations* (neoclassical). Although the ability to move between levels represents a potential contribution to the micro-macro dilemma in sociology, some ecologists may not see the value of such inter-level analysis.

It is true that an ecological researcher will usually find that either the neoclassical or the sociocultural framework best suits his research problem, and will not find it necessary to use both simultaneously, or to move between levels. *The ability to move between levels is important primarily because it demonstrates the complementarity of the two schools and shows that (1) any sociocultural study must make assumptions about macro phenomena, and (2) any neoclassical study must make assumptions about individual behavior.* Such assumptions often remain implicit.

Since the sociocultural frame of reference studies individual action within a population, and the ecological complex studies that population in comparison with other populations, it follows that the study of values, decision-making, etc., is not alien to the ecological complex. However, users of the ecological complex cannot study everything at once. Thus, they make *assumptions* about individual actions while comparing populations. How this is done is demonstrated in the remainder of this section.

Gibbs and Martin (1962:669) hypothesize that "The degree of urbanization in a society varies directly with technological development." They measure technological development as the per capita consumption of energy expressed in per capita tons of coal. Gibbs and Martin (1962:674) find a *rho* value of +.84 for the relationship in a study of 46 countries. Socioculturalists and neoclassicists alike can agree that the relationship between urbanization and technology found by Gibbs and Martin is an *empirical generalization*. Gibbs and Martin have accomplished something that is scientifically relevant. They have stated a proposition specifying a relationship between urbanization and technology and have tested this proposition. But it is true that such an empirical generalization is not a complete *explanation* of differences in the degree of urbanization for a sample of countries.

There are other possible correlates of the degree of urbanization. Since decisions about what technology to use or where to live are made in accordance with norms and value systems, it is entirely possible that correlations between certain values or norms and degree of urbanization could be found for a sample of countries. However, socioculturalists have not done this. We have seen no empirical generalizations presented that link values to the degree of urbanization in a country that are comparable to the empirical generalizations presented to link urbanization and division of labor or urbanization and technology.

The most promising avenue of investigation for anyone wishing to link values and urbanization would probably be to study values that prescribe the proper level of living or quality of life for each country. Such empirical generalizations are obviously difficult to establish by comparing countries on a world-wide basis because the main sources of comparative data such as federal censuses generally do not include data on values. Thus, ecologists should not be criticized for using data such as those on urbanization and division of labor that are available from the censuses.

But the basic point here is that *the neoclassicists can establish empirical generalizations such as the one about technology and urbanization without studying individual decision-making*. We do not say that they can fully explain the process through which such relationships develop, but we do say that they can establish the correlation. They can establish such generalizations without denying that men make decisions about where to live and about what technology to use. The mere establishment of empirical generalizations certainly does not represent economic or physical determinism. Socioculturalists need to demonstrate how the fact that men make decisions is relevant to the neoclassicists' empirical generalizations. We think that this relevance can be shown.

For example, the degree of urbanization is at least partially determined by the number of persons who migrate from rural areas to the city. Studies have shown that for most persons the decision to migrate is based upon economic factors (Lansing and Mueller, 1967). Thus, one could generalize about the decisions of individuals by assuming a rational "economic man" who lives wherever he can secure the highest level of living.

Of course, when one makes such a generalization, he realizes that it does not hold in every case, and that negative instances could be found. He simply makes this assumption as a means of simplifying his research task, and not as a means of securing perfection.

Such a generalization about individual action would help to provide an explanation of the relationship between technology and urbanization. We can see that changes in technology enable fewer farms to provide more food. Changes in technology also create new jobs (e.g., the computer industry). Urban areas grow because individuals decide to move to the city to work in the computer industry or some other nonfarm job. Thus, our explanation of the relationship between technology and urbanization does not exclude individual decision-making at all, but rather depends on it. The empirical generalization about urbanization and technology exists whether we study human action or not. But the explanation of this empirical generalization depends upon the study of human action. Notice that this is not "economic determinism" either. As the socioculturalists point out, economic materialism is itself expressive of a value system (Willhelm, 1964:247).

Perhaps another researcher does not think that economic factors predominate in decisions about where to live or what technology to use. This researcher does not feel that "economic man" represents a practical simplifying assumption for studying changes in one country over time, or for studying a number of countries at one point in time. The researcher knows that, since decisions about such things as where to live can be made for countless reasons, one has to generalize to simplify the study of entire populations. But he does not want to assume that economic factors predominate. This researcher feels that most people are hedonists. If the degree of urbanization in a country increases, it is because people put fun before money. They move to the city even if it means a loss of income because there is more entertainment available there. But notice that the existence of this hedonistic value system would not explain the relationship between technology and urbanization. Rather, *the existence of the hedonistic value system would be dependent upon the relationship between technology and urbanization.* A country full of urban pleasure seekers could not exist without a technological level capable of supplying food for them.

But if assumptions about individuals are helpful in studying the properties of populations, the converse is no less true. Sociocultural studies of land use within a city must also make assumptions about properties of populations. For example, suppose that a sociocultural ecologist studies land use in City A in Country Z in Time T. He finds that the value system an individual utilizes along with the cognitive data available determines the particular decision he makes about land use. What goes unsaid is that this is true for a particular *population,* with a particular *organization* and *technology* in a particular *environment.* It may well be that changes in such things as the country's population size and technology could drastically alter the cognitive data available to the individual by Time $T + 1$, so that a decision made at Time T would be different from one made at Time $T + 1$. For example, the new emphasis on preservation of the environment throughout America may very likely affect the values of land use decision-makers within individual cities.

As stated above, our conception of the ecological complex does not exclude such things as values and decision-making. In fact, the level of living was said to be determined by a country's population, organization, environment, and technology. This is done in part by governmental policy decisions which regulate such things as population size when overpopulation threatens to lower the level of living. The recent interest in birth control is an illustration of this. Also, the variable "level of living" is itself reflective of values. If societies did not value life, they could all become extinct. In this case, the level of living would sink to its lowest possible level as the population disappeared. Decisions are definitely relevant in the ecological complex, even though the exact relevance of various decisions is often difficult to evaluate at the present time.

Discussion and conclusion

In the last analysis, the sociocultural and neoclassical frames of reference require each other. Although the ecological complex can be used as a context to establish correlations between such properties of populations as degree of urbanization, division of labor, etc., these correlations cannot be fully explained without making some generalizations about the motives of individuals. A neoclassicist may say that a certain population size is a necessary condition for a certain degree of division of labor. This does not mean that he is denying the fact that individuals make decisions about whether to have babies, and that such decisions contribute to population size. Similarly, any sociocultural analysis of land use assumes a certain level of living, population size, organization, technology, and environment. Changes in these variables may necessitate changes in the socioculturalists' conclusions.

The ecological complex is not an arena constructed for a debate over free will versus determinism. For example, we do not presume that humans have no control over their organizations. However, it is claimed that the result of such control depends upon decisions made by humans in which they take factors such as technology, environment, and population into account. But the primary concern of neoclassical ecologists is not the study of decision-making by individuals. This is undoubtedly a fruitful area of study. However, it is also an indisputable fact that the macroecological analyses have been sorely neglected. Such study is fruitful for sociology as the science of society. In this nonindividualistic approach the emphasis is on the structural or contextual aspect of human existence.

Even though ecology is the generalizing science, one researcher cannot study all aspects of adaptation simultaneously. Thus, socioculturalists should not criticize neoclassicists for failing to study decisions made by individuals at the same time that they are studying other important problems concerning properties of populations. Similarly, neoclassicists working with entire populations should not criticize socioculturalists for studying individual adaptation within cities. The sociocultural frame of reference and the neoclassical frame of reference are essentially complementary.

References

Alihan, Milla Aissa. 1938 *Social Ecology.* New York: Columbia University Press.
Barker, Roger G. 1960 "Ecology and Motivation." In *Nebraska Symposium on Motivation, 1960,* edited by Marshall R. Jones, pp. 1–49. Lincoln: University of Nebraska Press.
Duncan, Otis D. 1961 "From Social System to Ecosystem." *Sociological Inquiry,* 31 (Spring):140–49.
———. 1959 "Human Ecology and Population Studies." In *The Study of Population,* edited by Philip M. Hauser and Otis D. Duncan, pp. 678–716. Chicago: The University of Chicago Press.
Duncan, O. D., and L. Schnore. 1959 "Cultural, Behavioral, and Ecological Perspectives in the Study of Social Organization." *American Journal of Sociology,* 65 (September):132–46.
Durkheim, Emile. 1933 *The Division of Labor in Society.* New York: Macmillan.
Firey, Walter I. 1947 *Land Use in Central Boston.* Cambridge: Harvard University Press.
Gettys, W. E. 1940 "Human Ecology and Social Theory." *Social Forces,* 18 (May):469–76.
Gibbs, J. P., and W. T. Martin. 1962 "Urbanization, Technology, and the Division of Labor: International Patterns." *American Sociological Review,* 27 (October):667–77.
———. 1959 "Toward a Theoretical System of Human Ecology." *Pacific Sociological Review,* 2 (Spring): 29–36.
Homans, G. C. 1964 "Bringing Men Back In." *American Sociological Review,* 29 (December):809–18.
Lansing, John B., and Eva Mueller. 1967 *The Geographical Mobility of Labor.* Ann Arbor: University of Michigan, Institute for Social Research.
Michelson, W. 1968 "A Parsonian Scheme for the Study of Man and Environment, or What Human Ecology Left Behind in the Dust." *Sociological Inquiry,* 38 (Spring): 197–208.
Mott, Paul E. 1965 *The Organization of Society.* Englewood Cliffs, N.J.: Prentice-Hall, Inc.
Ogburn, W. F. 1951 "Population, Private Ownership, Technology, and the Standard of Living." *American Journal of Sociology,* 56 (January):314–19.
Park, R. E. 1936 "Human Ecology." *American Journal of Sociology,* 42 (July):1–15.
Schnore, L. F. 1964 "Urbanization and Economic Development: The Demographic Contribution." *American Journal of Economics and Sociology,* 23 (January):37–48.
———. 1962 "Social Problems in an Urban Industrial Context." *Social Problems,* 9 (Winter):228–40.
———. 1961 "The Myth of Human Ecology." *Sociological Inquiry,* 31 (Spring):128–39.
———. 1960 "Social Problems in the Underdeveloped Areas: An Ecological View." *Social Problems,* 8 (Winter):182–201.
———. 1958 "Social Morphology and Human Ecology." *American Journal of Sociology,* 63 (May):620–34.
Sjoberg, Gideon. 1965 "Theory and Research in Urban Sociology." In *The Study of Urbanization,* edited by Philip M. Hauser and Leo F. Schnore, pp. 157–89. New York: Wiley.
Stephan, G. E. 1970 "The Concept of Community in Human Ecology." *Pacific Sociological Review,* 13 (Fall):218–28.
Willhelm, Sidney M. 1964 "The Concept of the 'Ecological Complex': A Critique." *The American Journal of Economics and Sociology,* 23 (July):241–48.
——— 1962 *Urban Zoning and Land Use Theory.* Glencoe: The Free Press.

Part II **Human Ecology as a Framework for the Study of the City**

Introduction

While Part I dealt with the history and theoretical issues in human ecology, Part II emphasizes the application of the ecological perspective to the analysis of urban structure. In seeking to understand the structure of the city today, the ecology of racial and ethnic groups plays a crucial role. We have in Section A five studies dealing with three major issues in racial and ethnic ecology.

Frazier's study deals with the ecology of a racial subcommunity, segregated within the larger community. He shows that the black community of Harlem has an ecological structure similar in many respects to that which is characteristic of most American cities as a whole. After dividing Harlem into five zones, he finds a fairly characteristic distribution of building structures, age, sex, marital status, fertility rate, and desertion. Crime and delinquency appear to follow a less clear-cut pattern.

Allen and Farrell examine the ecological development of Harlem in the forty years since Frazier's study was published to see if the patterns found in the mid-1930s have persisted despite the tremendous social, political, and economic changes that have occurred since then. They find that the basic tendency for economic well-being to improve from the inner to the outer zones has persisted and, in fact, is also found in other black communities in New York City. Beyond this they examine Frazier's conception of black urbanization and find it surprisingly accurate in predicting patterns of development, despite some paradoxes.

While the study of black Harlem focuses on a community reacting to external constraints, the study by Jonassen of Norwegian–Americans deals with the ecological impact of influences arising within the ethnic community itself. Jonassen attempts to show that the movement of the Norwegian–American community in New York City can only be understood by taking into account their modified rural values.

Hoover, in studying Norwegian–American ecological patterns in Metropolitan New York in the years since Jonassen's study, finds that ethnic values have been modified by increasing assimilation into general American culture by the younger generation. Other factors that have modified earlier patterns include decreased immigration from Norway (resulting in an older ethnic community), renewed contacts with Norway by the older generation as a result of World War II and postwar improvements in travel, and changed economic circumstances.

The question of ethnic community persistence versus the forces of dispersion and assimilation, implicit in the previous study, is the specific focus of Driedger's comparison of two neighborhoods in Winnipeg. One of these has followed the typical pattern of invasion and succession, with former residents moving to the suburbs, while the other has retained the same ethnic identity for 160 years. Driedger explores possible reasons for these contrasting patterns.

In Section B we turn to the ecology of neighborhoods, beginning with O'Brien's article on Beale Street, a temporal study of a black neighborhood. O'Brien illustrates not only how a neighborhood may change over time by discussing the long-term changes that have occurred on Beale Street in the city of Memphis in the last hundred years, but he also shows how cyclical patterns affect the neighborhood's character by contrasting the daytime and nighttime populations and the differences in population on certain nights of the week. Following the original study is a chapter containing two papers written by O'Brien and O'Brien. The first looks at Beale Street today to see if the earlier patterns persist, while the second deals with the significance of temporal ecology.

In the next article, Suttles focuses our attention on a dimension of greatly increased importance in the ecology of many urban neighborhoods today—redevelopment through the combined efforts of large-scale governmental and private organizations. Because such redevelopment may attempt to control not only the type of housing but other community characteristics as well, such as the race and class of residents, Suttles uses the term "contrived community" to contrast with the notion of the "natural community." In this paper he reexamines a Chicago neighborhood he had studied earlier,[1] analyzing the problems and interplay of forces in the last decade to see to what extent earlier expectations of the contrived community have been met.

It is interesting to compare Suttles's contrived community with the community in New York City studied by Rosen. Both are located in a large urban setting and both contain apartment complexes built by large developers. The community studied by Rosen, however, evinces many more characteristics of a true neighborhood. This may be explained by the area's greater homogeneity in ethnic identity, economic level, and type of housing, highlighted by its distinctiveness from the surrounding area. These neighborhood characteristics, however, are not immediately apparent to the casual observer, or even fully conceptualized as such by the residents, but emerge from this study. Thus, Rosen shows that what seems at first glance to be an anonymous apartment-house area of the city is, in fact, an urban neighborhood.

Hudson's study deals with a very different type of urban neighborhood. Here we see a pattern of invasion and succession in which residential land use has invaded an industrial area. Moreover, this is not an instance of an old industrial area becoming a slum, nor does it follow the pattern of urban redevelopment in which old buildings have been torn down and replaced by new housing complexes. In an interesting deviation from these patterns, an unplanned invasion of an outmoded industrial area by middle-class artists resulted in the adaptation and modification of existing structures. Hudson analyzes the conditions that led to this invasion, the series of changes in the area that resulted, and the conflicts that emerged. We see an interesting interplay of economic factors, technology, cultural values, and political influence in shaping the ecology of an urban neighborhood.

To conclude this Section, Downs turns our attention to a very different aspect of urban neighborhoods—the mental or cognitive aspect. Downs begins his discussion by analyzing the basic notion of the "sense of place," which he divides into a sense of identification, a sense of boundary, a sense of attachment, and a sense of responsibility and control. Next, he considers the phenomenon of cognitive mapping or mental maps. Finally, he deals specifically with the question of the cognition of urban neighborhoods, relating the type of neighborhood to the nature of the residents' sense of place.

In Section C we turn to the ecology of mental illness. The selection by Levy and Rowitz, taken from their book, provides an excellent review of the literature in this field, starting with the earliest studies and including the classic study of Chicago by Faris and Dunham and the research since that time. The authors then examine the issues raised by these studies, including the basic question of whether conditions in certain areas cause a higher incidence of mental illness (social selection) or if the mentally ill drift to these areas. Finally, the major findings of their own study are presented and the implications for a drift or social-selection hypothesis examined.

Commuting, the theme of Section D, has become a phenomenon of major importance in the ecology of large metropolitan areas. The separation of place of residence from place of work results in marked differentiation of daytime and nighttime population in many sections of a metropolis. Attention initially was focused on this facet of urban ecology by Foley and Breese, who studied the daytime movement of population into the central business districts of American cities.[2] We lead off this Section with an article by Foley, which, while written a good many years ago, still poses the basic issues in the study of the daytime

---

[1] Gerald D. Suttles, "The Contrived Community," in *The Social Construction of Communities,* by Gerald Suttles, pp. 82–107 (Chicago: University of Chicago Press, 1972).

[2] In addition to the article by Foley included in this volume, see Gerald W. Breese, *The Daytime Population of the Central Business District of Chicago* (Chicago: University of Chicago Press, 1949); and Donald L. Foley, "The Daily Movement of Population into Central Business Districts," *American Sociological Review,* 17 (October 1952), pp. 538–43.

movement of urban population as an ecological subfield. Foley also discusses the extent and importance of this daily movement of population, the factors that bring it about, and the findings of some of the studies that had been conducted up to that time.

Guest reviews some of the major research in this field in the twenty years following the publication of Foley's article. Guest's article, however, shifts the focus from the central business district to the suburbs. Typically, the suburbs have been regarded as a place of residence from which workers commute to the central city. In a study of 129 American suburban communities, Guest finds this conception is an oversimplification. He studies the relationship between the distribution of residential (nighttime) and work force (daytime) populations in these suburbs, analyzing the importance both of demographic characteristics (such as race, sex, age, and social status) and of community characteristics (such as size, housing, retail sales, and municipal expenditures) in these patterns of distribution.

The selection by Manning turns our attention to Australia, particularly Sydney. Many of the issues involved in commuting are similar to those found in American cities, such as the desire for individual houses which tend to be cheaper and more available in the outlying suburbs versus the cost of long-distance commuting and the pressures of an energy crisis. In comparing Manning's and Guest's studies, it would appear that employment is more centralized in Australian cities. Manning suggests that decentralization of employment might be a means to reduce commuting, but it should be noted that, even with decentralization, workers may not necessarily live near their places of employment but may travel from one part of the metropolitan area to another. Decentralization of employment, if widely scattered, could make mass transit less effective, and so Manning proposes that employment be concentrated in suburban centers.

Section E, the last Section of Part II, deals with social area analysis and factorial ecology, closely related methods for studying the areas of a city in terms of a few crucial variables. Social area analysis emerged from the work of Eshref Shevky and other West Coast ecologists, and first received national attention with the publication in 1949 of *The Social Areas of Los Angeles* by Eshref Shevky and Marilyn Williams.[3] A major methodological work presenting the theoretical reasoning behind the development of social area analysis, as well as a description of the techniques used, was published by Shevky and Bell in 1955.[4] Basically, the method rested upon the classification of the census tracts of a city according to the three variables of social rank, urbanization, and segregation.[5] These variables were selected because they were felt to measure crucial factors in the development of modern, industrialized society; and they, in turn, are specifically measured by indexes obtained from census tract data. In the 1960s factorial ecology increasingly replaced social area analysis. Also based on the study of census tract data, the main difference between the two is that, rather than using a three-dimensional model derived from theoretical assumptions, factorial ecology seeks to determine empirically the crucial dimensions differentiating urban areas by subjecting a wide range of variables to factor analysis.

Johnston, in the first selection in this Section, presents a methodological examination of social area analysis and factorial ecology, including a careful consideration of the techniques, criticisms, and uses of each. Interestingly, Johnston reports that, although factorial ecology replaced social area analysis largely because it was felt the latter's tri-dimensional schema was too limited and arbitrary, the majority of studies in factorial ecology in fact have found the three Shevky–Bell dimensions to be the major determinants of residential area differentiation. After examining in some detail the advantages and disadvantages of factorial ecology, Johnston concludes it is a powerful tool in the study of urban ecology.

Taking a different perspective, Hamm examines social area analysis and factorial ecology theoretically and substantively. He relates these two approaches to certain of the basic principles of classical ecology, particularly those associated with Burgess's concentric-zone hypothesis. Hamm rejects the two prevalent notions that classical ecology is outdated and that factorial ecology adequately replaces social area analysis. In a highly significant analysis, he shows the interrelationships among these three approaches and suggests that all three taken together are needed to understand fully the structure and processes of urban, socio-spatial differentiation. Hamm also

[3] Berkeley: University of California Press, 1949.

[4] Eshref Shevky and Wendell Bell, *Social Area Analysis: Theory, Illustrative Application and Computational Procedure* (Stanford: Stanford University Press, 1955).

[5] These were the terms originally used by Shevky. They were renamed economic status, family status, and ethnic status by Bell. The variables measuring "urbanization" were family-related.

provides tables summarizing the major research that has been conducted in social area analysis and factorial ecology and a comprehensive bibliography of relevant literature.

Finally, in Abu-Lughod's article we see an example of the application of social area analysis and factorial ecology to a non-Western city. Abu-Lughod uses the techniques of factorial ecology to evaluate the extent to which the theoretical framework of social area analysis applies to Cairo, Egypt, a "partially modernized" city. She relates differences between Cairo and typical Western cities to patterns found in Rome, some African, and a few southern U.S. cities and suggests theoretical propositions to account for these differing patterns.

Section A Ethnic and Racial Groups

Negro Harlem: An Ecological Study*

E. Franklin Frazier

In a study published a few years ago, the writer was able to show, by means of an ecological analysis, that the organization and disorganization of Negro family life in the northern city were closely tied up with the economic and social structure of the Negro community.[1] Specifically, in the case of Chicago, it was found that, as a result of the selection and segregation incident to the expansion of population, the Negro community had assumed a definite spatial pattern. This spatial pattern bore the impress of the ecological organization of the larger community and could be represented by seven zones indicating the outward expansion of the community from the slum area about the central business district. On the basis of these seven zones, it was possible to measure the selection and segregation, as revealed in the distribution of occupational classes, in the proportion of males, mulattoes, and illiterates in the population, and in other indexes to the economic and social structure of the community. Family disorganization—measured in terms of family dependency and desertion—nonsupport, illegitimacy, and juvenile delinquency were found to diminish in the successive zones marking the progressive stabilization of community life.

With the results of the Chicago study in mind, the writer undertook, on the basis of materials collected while making a survey of Harlem for The Mayor's Commission on Conditions in Harlem, to determine to what extent the Negro community in Harlem had assumed a natural or ecological order during its expansion.[2] The results of this study are embodied in the present article.

I. Origin, growth, and expansion of the Negro community

Reports differ concerning the historical events leading up to the settlement of the Negro in Harlem; but it seems fairly well established that Harlem had already deteriorated as a residential area when Negroes began finding homes there at the opening of the present century. As is usually the case when Negroes first enter neighborhoods occupied by whites, the movement of Negroes into Harlem provoked a storm of protest. The *New York Herald* of July 10, 1906, reported indignation meetings "throughout the neighborhood of West 135th Street; where thirty-five white families" were to be ejected to make room for Negro tenants. The article ended with the following

---

*Reprinted from *The American Journal of Sociology*, 43 (July 1937), 72–88, by permission of The University of Chicago Press.

[1] See *The Negro Family in Chicago* (Chicago: University of Chicago Press, 1932).

[2] The Commission was appointed by Mayor La Guardia following the outbreak in Harlem on March 19, 1935.

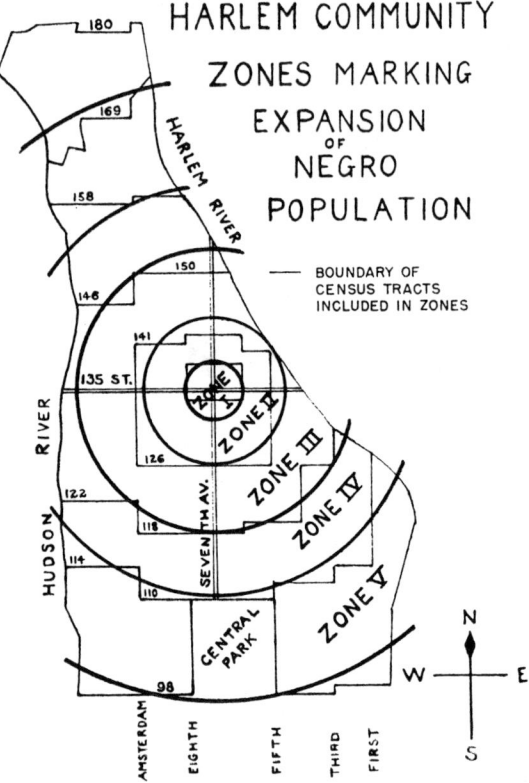

Map 1—Harlem Community zones marking expansion of Negro population.

comment: "It is generally believed by the residents, however, that the establishment of the Negroes in 135th Street is only the nucleus of a Negro settlement that will extend over a very wide area of Harlem within the next few years."[3]

The prophecy contained in the concluding com-

[3]Quoted, Clyde Vernon Kiser, *Sea Island to City* (New York, 1932), p. 21.

ment has been fulfilled by the subsequent growth of the Harlem Negro community. From the small settlement in the block referred to above, the Negro community has gradually spread out in all directions. While the expansion of the Negro community in Harlem has been governed largely by social and economic forces similar to those that have determined the growth of the Negro community in Chicago, an important difference in the growth of these two communities is observable. Whereas the growth of the Negro community in Chicago was dominated, as we have indicated, almost entirely by the ecological organization of the city of Chicago, the Harlem Negro community has shown a large measure of autonomy in its growth and, as we shall see, has assumed the same pattern of zones as a self-contained city.

The radial expansion of the Negro population from the area about One Hundred and Thirty-fifth Street and Seventh Avenue may be represented ideally by drawing concentric circles about the census tract in which the intersection of these two main thoroughfares is located (see Map 1). In 1910 there were 15,028 Negroes or 54 percent of the Negroes in the Harlem area concentrated in the first two zones (see Table 1). At that time Negroes comprised less than a fifth of the entire population of these two zones; while in the three remaining zones marking the outward expansion of the Negro community, they became less and less significant in the population (see Diagram 1). By 1920 Negroes constituted over three-fourths of the population of the first zone, over half of that of the second zone, and about a seventh of the population of the third. Up to 1920, whites in the two outlying zones still resisted the expanding Negro population. However, by 1930, the Negro had not only taken over almost the entire first zone and increased to seven-eighths and two-fifths of the populations of the second and third zones, respectively, but had

Table 1. Negro Population in the Five Zones of the Harlem Community, New York City, 1910, 1920, 1930, and 1934

Zone	1910	1920	1930	1934*
I	1,856	9,053	12,585	7,661
II	13,172	43,734	72,214	59,783
III	6,145	22,661	64,368	67,304
IV	1,879	2,058	40,312	55,337
V	5,775	6,742	14,415	13,397
Total	27,827	83,248	203,894	203,482

*Census by the New York Housing Authority.

become a significant element—22.7 percent—in the population of the fourth zone. Even in the fifth zone Negroes had increased from 2.5 to 6.2 percent. This was due chiefly to the movement of Negroes into the area between Fifth Avenue and the Harlem River (see Map 1). The bulk of the Negro population in the fifth zone had hitherto been concentrated in the neighborhood of Amsterdam Avenue and Ninety-eighth Street, this settlement being an extension of the West Side Negro community rather than an expansion of the Harlem community.

Although the five zones indicate the general tendency of the population to expand radially from the center of the community, the Negro population has not expanded to the same extent in all directions. Because of economic and social factors, the expansion of the Negro population has followed many tortuous paths. It has been held in check until residential areas have deteriorated and therefore have become accessible not only to Negroes but to Italians and Puerto Ricans who live in areas adjacent to those inhabited by Negroes. In some instances white residential areas, when almost surrounded by the expanding Negro population, have put up a long and stubborn resistance to the invasion of the Negro. This was the case with the area about Mount Morris Park; but when this area lost its purely residential character, and brownstone fronts became rooming-houses, the eventual entrance of the Negro was foreshadowed. Then, too, the advance of the Negro had been heralded by the location of light industries, as in the western section of Harlem where, after the establishment of a brewery doomed the area as a residential neighborhood for whites of foreign extraction, signs inviting Negro tenants began to appear on houses. But it seems that the westward expansion of the Negro population has been definitely halted at Amsterdam Avenue and will not be able to invade the exclusive residential area on Riverside Drive.[4]

We can get some idea of the relation between the expansion of the Negro population and the character of the areas into which it has spread by considering the predominant types of structures located in the five zones. First, we note in Table 2 that the Negro population predominates in those zones where the majority of the structures are nonresidential in character. Then, if we consider more closely the character of these nonresidential structures, we find that the Negro population is

[4]Since 1920 there has been a decrease in the number of Negroes west of Amsterdam Avenue.

concentrated in those zones where rooming- and lodging-houses comprise a relatively large proportion of the nonresidential structures. If further analysis were made of the various zones, it would probably reveal an even closer relationship between the expansion of the Negro population and the location of nonresidential structures.

Further light on the relation between the expansion of the Negro community and the physical character of the areas into which Negroes have moved is afforded by data on the type, age, and condition of the residential structures in the five zones. In respect to type of residential structures, the third zone showed a comparatively large

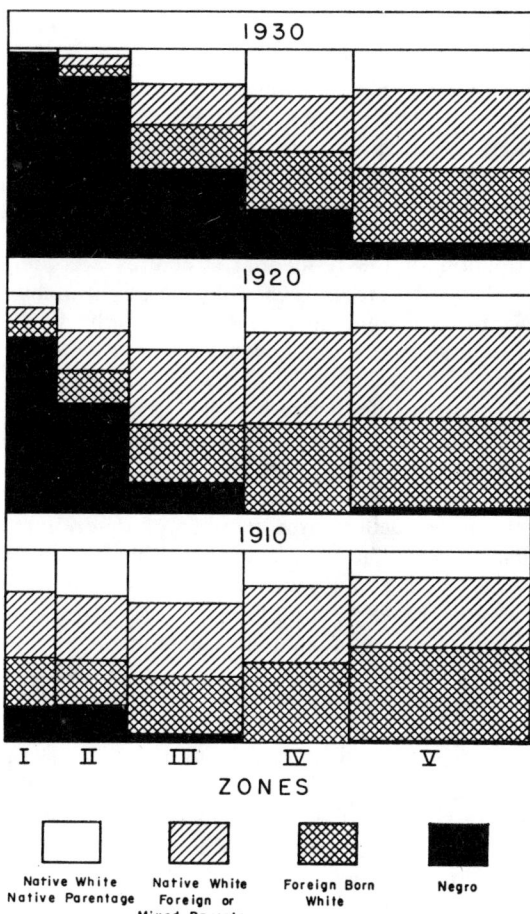

Redrawn from the original.

Diagram 1—Percentage distribution of the four nativity groups in the population of each of the five zones of the Harlem Community, New York City, 1930, 1920, and 1910.

Table 2. Percentage of Population Negro and Types of Structures in Five Zones of the Harlem Community, New York City

	Zones				
	I	II	III	IV	V
Percentage of population Negro in 1930	99.0	87.8	41.4	22.7	6.2
Percentage of structures that were nonresidential in 1934	83.8	78.2	59.8	42.5	28.0
Percentage of nonresidential structures that were rooming- and lodging-houses in 1934*	34.2	32.0	31.5	23.0	18.5

*Rooming- and lodging-houses are classified as nonresidential structures.

proportion of one-family dwellings. This was due to the fact that the western section of the third zone (see Map 1) includes a large part of the Riverside Drive area. The most noteworthy difference between the zones appeared in the proportion of hotels, boarding-houses, and institutions which were simply classified as "other." The proportion of this type of residential structure declined sharply from 51.7 percent in the first zone to 14.9 percent in the fifth. In the distribution of the residential structures according to their age, the differences in the physical character of the zones stand out more clearly. In the first and second zones, where 99 and 87.8 percent of the residents, respectively, were Negroes, 90 percent of the residential structures were thirty-five years of age and over. For the remaining three zones the proportion of older structures declined significantly except in the fourth zone which included a large number of deteriorated tenements in the eastern section. This is the very section of the fourth zone in which Negroes have settled. However, the relation between the condition of the residential structures in the various zones and the expansion of the Negro population is obscured by the fact that the zones are far from homogeneous in physical character. While the third and fifth zones showed the greatest proportion of first-class residential structures, the fourth zone had the highest proportion of fourth-class dwellings. Nevertheless, there was a smaller proportion of first- and second-class dwellings in the first zone than in any of the other four zones. This was true despite the fact that there was a higher proportion of first-class dwellings in the first zone than in either the second or the fourth zone. The comparatively large proportion of first-class structures in the first zone was due to the rehabilitation of this area.

II. Age and sex distribution of the population

The selection and segregation which have taken place as the Negro population has expanded are seen, first, in the variations in the proportion of grown people in the five zones. Practically four out of five persons in the first zone were adults in 1930 (see Table 3). In the second zone the proportion of adults in the population declined to three out of four, and in the next three zones, from about seven to six out of ten persons in the population. The tendency on the part of older persons to become segregated toward the center of the community is reflected also in the relative number of children in the population of the various zones. In the first or central zone only 3.8 percent of the entire population in 1930 was under five years of age. The proportion of children in this age group increased in each of the successive zones until it reached 12.3 percent in the fifth zone. There was also a slight increase in the proportion of females in the successive zones marking the outward expansion of the population. Here, too, the influence of selective factors was apparent. For, although there was an excess of females in the total population of the community, the excess of females in the first zone was counterbalanced by the tendency on the part of males to concentrate there. . . .

III. Marital status of the population

The tendency on the part of family groups to move toward the periphery of the community was shown in the increasing proportion of married men and women in the successive zones. In the first zone or center of the community only half of the men and women were married. From this zone outward the percentage of both men and women married increased until it amounted, in the fifth or outermost zone, to 64.2 percent for the men and 60.1

percent for the women. Correlated with the increase in the proportion of men and women married was the gradual decline not only in the proportion of men and women single in the successive zones but also in the proportion of widowed persons in these five zones. Although these figures do not give an absolutely correct picture of the marital condition of the men and women in the community, it is interesting to note that the proportion of men and women widowed was highest in the center of the community where one would expect to find considerable family disorganization. The decline in the percentage of widowed among the males was even greater than among the females. At the same time there was an increase in the proportion of divorced persons in the successive zones as one left the center of the community. A possible explanation of the comparatively larger number of divorced persons in the outer zones is that it may indicate a greater regard for legal requirements in the breaking of marital ties.

IV. Ratio of children, births, and deaths

The low fertility of Negroes in northern cities has been revealed in a number of important studies. For example, Thompson and Whelpton have shown that there has been a marked tendency for the ratio of children to Negro women of child-bearing age to vary inversely with size of city. According to these same authors, Negroes in large cities, including Chicago and New York, "were not maintaining their numbers on a permanent basis in either 1920 or 1928."[5] The extremely low fertility of Negroes in Chicago has been clearly demonstrated by Philip M. Hauser, of the University of Chicago, in an unpublished study. In the case of Chicago the present writer has shown in a study of the Negro family how selective factors within the Negro community affected the relative fertility of different sections of the Negro population.[6] Lately, Kiser found in a study of Negro birth-rates in a health area of Harlem that the fertility of Negro women was lower than that of white women of a similar or even a higher occupational level in Syracuse and two other urban communities. Kiser indicated in his study that the low fertility of Negroes was "due partly to selective processes with reference to residence in Harlem as indicated by higher birth rates among the colored population in other parts of the city."[7] As a matter of fact, even within Harlem itself important differences are revealed if the fertility of Negro women is studied in relation to the selective processes within the community. These differences became apparent when the children under five to women of child-bearing age were calculated for the five zones by which we have indicated the expansion of the

[5]Warren S. Thompson and P. K. Whelpton, *Population Trends in the United States* (New York, 1933), pp. 280–81.
[6]See *The Negro Family in Chicago,* pp. 136–45.
[7]Clyde V. Kiser, "Fertility of Harlem Negroes," *Milbank Memorial Fund Quarterly,* 13 (July 1935), p. 284.

Table 3. Percentage Distribution of Males and Females in the Negro Population of Each of the Five Zones of the Harlem Community, New York City, 1930

Age Period	Zone I M	Zone I F	Zone II M	Zone II F	Zone III M	Zone III F	Zone IV M	Zone IV F	Zone V M	Zone V F
75+	0.1	0.2	0.0	0.2	0.0	0.2	0.0	0.2	0.0	0.2
65–74	0.5	0.6	0.3	0.6	0.3	0.6	0.2	0.5	0.4	0.6
55–64	1.6	1.7	1.4	1.6	1.2	1.4	0.9	1.2	1.3	1.5
45–54	6.3	5.8	5.3	5.0	4.4	4.3	3.6	3.8	4.0	4.3
35–44	11.4	11.2	11.0	10.7	10.0	10.0	8.8	8.5	8.5	8.7
30–34	6.5	6.0	6.5	6.8	6.8	6.7	6.7	6.8	5.0	5.4
25–29	8.0	8.6	7.5	8.4	7.4	9.0	7.6	8.5	5.5	6.9
20–24	5.8	7.1	5.1	6.9	4.9	6.9	5.5	7.1	4.4	5.9
15–19	2.7	3.6	2.6	3.4	2.6	3.5	2.9	3.8	3.2	3.7
10–14	1.9	2.2	2.3	2.5	2.4	2.3	2.7	2.9	3.7	3.7
5–9	1.9	2.0	2.8	2.9	3.3	3.6	3.8	3.9	4.8	5.4
–5	1.9	1.9	2.9	2.9	3.7	3.7	4.9	4.8	6.3	6.0
Total	48.6	50.9	47.7	51.9	47.0	52.2	47.6	52.0	47.1	52.3

Table 4. Percentage of Negro Males and Females 15 Years of Age and Over Single, Married, Widowed, and Divorced in the Five Zones of the Harlem Negro Community, New York City, 1930

Marital Status	Sex	Zone I	Zone II	Zone III	Zone IV	Zone V
Single	M	42.6	38.5	35.3	34.0	31.1
	F	30.9	27.6	26.3	25.6	23.5
Married	M	49.8	56.0	60.3	62.3	64.2
	F	50.5	54.8	57.6	59.8	60.1
Widowed	M	7.3	4.7	3.6	2.9	3.8
	F	17.6	16.4	15.0	13.0	14.4
Divorced	M	0.2	0.5	0.4	0.6	0.5
	F	0.6	0.8	0.7	1.1	1.6

Harlem Negro community. We find that both in 1920 and in 1930 there was, with one exception, a regular increase in the ratio of children from the first to the fifth zone. In 1930 the ratio of children in the fifth zone was 462 or four times that in the first zone. The exception to the general trend, observable in the fourth zone in 1920, was probably due to the fact that at that time only a small number of economically better-situated families had moved into this zone. On the other hand, the changes between 1920 and 1930 in the ratio of children in the three outer zones seem to indicate a movement toward or settlement in the peripheral zones by the more fertile groups.

We can get further light on the relation between the fertility of Negro women and residence in the various areas of the community by studying the ratio of children to women fifteen years of age and over who were married, widowed, and divorced and the number of births to married women fifteen to forty-four years of age. Here, again, we find the ratio of children increasing regularly in the successive zones marking the expansion of the Negro community. The same trend was apparent in regard to birthrates in 1930. In the first zone there were only 66.1 births per one thousand Negro married women fifteen to forty-four years of age. But, as in the case of the ratio of children, the fertility of the women in the successive zones increased according to their distance from the center of the community. The fertility of the women in the fifth zone was slightly over two and one-half times as great as it was in the first.

Because of the differences in the age and sex composition of the five zones, the crude death-rates were not significant. However, when the ratio of births to deaths was calculated, important differences appeared. In the first zone deaths were in excess of births, while in the second zone they almost balanced the births. In the next three zones the number of births per one hundred deaths increased from 149 to 225 and declined to 167 in the outermost zone. In respect to infant mortality there was little difference between the zones, the highest infant death rate—10.8—being in the

Table 5. Number of Children Under 5 to 1000 Negro Women 20-44 Years of Age in Five Zones of the Harlem Community, New York City, 1920 and 1930

Zone	1930			1920		
	Women, Age 20-44	Children Under 5	Ratio of Children to Women	Women, Age 20-44	Children Under 5	Ratio of Children to Women
I	4,141	476	115	3,083	336	109
II	23,612	4,160	176	15,021	2,793	186
III	21,107	4,749	225	7,217	1,858	257
IV	12,498	3,940	315	805	173	214
V	3,872	1,790	462	2,262	621	274

Table 6. Number of Children Born to 1000 Negro Married Women, 15-44 Years Old, and Ratio of Children Under 5 to Negro Women 15 and Over, Married, Widowed, and Divorced in Five Zones of the Harlem Community, New York City, 1930

Zone	Married Women, Age 15-44 (Estimated)	Number of Births	Births Per 1000 Married Women, Age 15-44	Women Age 15 and Over, Married, Widowed, and Divorced	Children Under 5	Ratio of Children to Women, Age 15 and Over, Married, Widowed, and Divorced
I	2,495	165	66.1	3,883	476	123
II	15,087	1,230	81.5	22,670	4,160	184
III	13,883	1,276	91.9	20,246	4,749	234
IV	8,552	1,211	141.6	12,120	3,940	325
V	2,833	477	168.4	4,104	1,790	436
Total	42,850	4,359	101.7	63,023	15,115	240

second zone, and the lowest—7.8—being in the fourth zone.

V. Crime, delinquency, and dependency

When we study such phenomena as crime and delinquency in their relation to the ecological organization of the Harlem Negro community, it appears that economic and cultural factors affect their distribution to a far greater extent than the distribution of the population with respect to age, sex, marital condition, and fertility. First, we note (Table 7) that, during the first six months of 1930, the highest number of arrests in proportion to men in the population occurred in the second zone just outside of the center of Negro Harlem's economic and cultural life. The rate of adult delinquency measured in terms of arrests declined gradually in the next two zones. But we find that the rate in the outermost zone equaled that in the center of the community. As we have already indicated, the southern portion of this outermost zone included a slum section and therefore manifested many of the characteristics of a slum area. The juvenile delinquency rates for the five zones were even more difficult to explain on the basis of the general community pattern without a knowledge of the variations in the character of these zones. In 1930—and the same held true for the five-year period from 1930 to 1934—the juvenile delin-

Table 7. Crime and Delinquency Rates in the Five Zones of the Harlem Negro Community, New York City

Zone	Number of Males 17 Years and Over 1930*	Number of Arrests of Males First Six Months, 1930	Arrests Per 100 Males 17 Years and Over	Number of Boys 10-16 Years of Age, 1930*	Number of Boys Arrested, 1930	Number of Boys Arrested Per 100 Boys, Age 10-16
I	5,329	333	6.2	329	18	5.5
II	28,256	2,264	8.0	2,326	105	4.5
III	23,065	1,350	5.8	2,088	120	5.7
IV	14,274	597	4.2	1,462	70	4.8
V	4,508	284	6.2	716	31	4.3

*Number of men and boys seventeen and sixteen years of age, respectively, estimated.

Table 8. Dependency and Desertion in the Five Zones of the Harlem Negro Community, New York City

Zone	Total Negro Families 1930	C.O.S. Under Care Families				Total Negro Families 1934	Families on Home Relief Sept., 1935	
		All Cases		Desertion Cases				
		Number	Rate Per 1000 Families	Number	Rate Per 1000 Families		Number	Rate Per 1000 Families
I	2,221	80	36	20	9.0	2,110	1,497	709
II	15,793	448	28	83	5.2	16,321	9,560	585
III	16,145	533	33	78	4.8	18,875	7,473	395
IV	9,558	343	35	34	3.5	14,945	4,658	311
V	3,717	140	37	15	4.0	3,886	1,104	284

quency rate, measured in terms of boys arrested in proportion to boys ten to sixteen years of age, was practically as low in the second zone as in the outermost zone.[8]

Although dependency as represented in the comparatively few cases handled by the Charity Organization Society in 1930–31 did not indicate the influence of selective factors in the ecological organization of the Negro community, selection was apparent in the desertion rates and more especially in the proportion of families on Home Relief in the five zones (see Table 8). Desertion rates, based upon desertion cases handled by the Charity Organization Society, declined from 9.0 per one thousand families in the first zone to 4.0 in the fifth. On the basis of the census made by the New York City Housing Authority in 1934, the number of families on Home Relief declined from 709 per one thousand in the first zone to 284 in the fifth zone. However, it should be noted that the highest percentage—91.2—of families on relief was found in a census tract in the second zone, south of the census tract which constitutes the first zone in our scheme. But, in spite of this variation from the general pattern, the percentage of families on Home Relief in the poorer sections of the fifth zone varied slightly from the average for the entire zone. In view of what our statistics indicate concerning the nature of group life in these various zones, it seems reasonable to conclude that the large number of families on relief in the zones close to the center of the community was associated with the breakdown of group life as represented by normal family life in these areas.

VI. Distribution of institutions

The distribution of institutions in the Harlem Negro community reflects in a visible form the general community pattern. The concentration of institutions in the first zone was vividly portrayed by Rudolph Fisher in a story of Negro life there. "In a fraction of a mile of 135th Street," he wrote, "there occurs every institution necessary to civilization from a Carnegie Library opposite a public school at one point to a police station beside an undertaker's parlor at another."[9] A recent survey of this area revealed the extent to which the economic life of the Negro community, especially with respect to Negro business enterprises, is centered about One Hundred and Thirty-fifth Street and Seventh Avenue. There were in this area, in 1935, 321 business establishments, two-thirds of which were conducted by Negroes, in addition to 53 offices of Negro professional men and women. Because of the economic dependence of the community, whites owned the bank and more than 80 percent of the retail food stores, while Negroes controlled practically all the businesses providing personal services and other types of enterprises not requiring large outlays of capital. In this area were also located the two principal Negro newspapers in Harlem and the offices of four Negro insurance companies.

As the center of Negro Harlem has come to play a specialized role in the organization of the

[8] It might be mentioned in this connection that adult and juvenile delinquency in Chicago fitted into the much-simpler ecological pattern of the Negro community; see *The Negro Family in Chicago*, chap. X.

[9] "Blades of Steel," *Anthology of American Negro Literature*, ed. V. F. Calverton (New York, 1929), p. 53.

community, the area affected by the process has extended beyond the limits of the single census tract which constitutes the first zone. For example, as an indication of this process, since 1930 the population of the second zone has declined as well as that of the first zone. Hence, in our consideration of the distribution of institutions with reference to zones, we shall regard as a single area Zones I and II, which have a total population about equal to that in each of the two next zones, III and IV (see Table 1). In 1935 there were in the central area seventy-five churches, not including spiritualists, psychologists, and Father Divine's "Kingdoms." Forty-two of the churches were of the "storefront" type, three so-called "spiritualist" churches, and the remaining thirty were denominational churches housed in regular edifices. The number of all types of religious institutions declined in the three zones outside of this central area. For example, in Zone V, there were only one regular church edifice and nine "storefront" churches. As the focus of the political life of the community, the central area contained ten of the eighteen political clubs in the community, while Zone III had six such clubs. Although about 40 percent of the recreational institutions serving primarily Negro Harlem were located in the central area, they were more widely distributed in the five zones than other types of institutions. This fact is of special interest because it indicates how, in regard to the cultural superstructure, the main arteries of travel—Lenox, Seventh, and Eighth avenues—running the entire length of the community, and the "satellite loops" at One Hundred and Sixteenth, One Hundred and Twenty-fifth, and One Hundred and Forty-fifth streets tend to mar the symmetry of the community pattern.

However, this does not affect in any important manner the conclusion to which our study of the Harlem Negro community leads. Although our analysis provides additional substantiation of the general ecological hypothesis that the distribution of human activities resulting from competition assumes an orderly form, it introduces at the same time an important extension of the theory. It appears that, where a racial or cultural group is stringently segregated and carries on a more or less independent community life, such local communities may develop the same pattern of zones as the larger urban community.

Black Harlem Revisited: Patterns of Ecological and Social Organizational Change, 1940-1970*

Walter R. Allen
and Walter C. Farrell, Jr.

As the title implies, black Harlem has undergone significant changes during the forty years since Frazier's classic study. We have seen major shifts in the social, political, and economic circumstances of Afro-Americans generally, and black Harlemites in particular. Many of these transitions were anticipated by Frazier. Equally important are those that were not. This brief paper attempts to accomplish several goals, foremost of which is to update Frazier's study of ecological patterns in black Harlem. In this regard, questions of how well the study's premises, methods, findings, and conclusions have stood the test of time seem most appropriate.

The paper begins with an overview and analysis of Frazier's "Negro Harlem: An Ecological Study." We then seek to place the study in its proper social and historical contexts. More specifically, what combination of historical, socio-economic, and political processes culminated in black Harlem of the mid-1930s? Our attention next focuses on significant, observable patterns in Harlem's demographic data from the 1940, 1950, 1960, and 1970 censal periods. The paper concludes with a discussion of findings and implications for future patterns of ecological change in black Harlem.

Overview and analysis of "Negro Harlem"

The distinctive contributions of E. Franklin Frazier's work to the field of human ecology is evidenced not only by its inclusion in this volume, but also by frequent citations in the literature. As a sociologist, Frazier was vitally interested in processes of social organization and social change. In particular, he was concerned with urbanization and its effects on Afro-Americans. For him, the rapid urbanization of blacks during the early twentieth century represented the most momentous

*This article was written for this volume and has not been published previously. The authors thank the following people for their helpful comments: Joe Darden, Donald Deskins, Amos Hawley, Harold Rose, Ernest Spaights, and George Theodorson. Mark Lundgren provided valuable assistance in the collection and analysis of data.

change in their experience since emancipation.[1] Thus, he sought to understand the interplay between processes of disorganization-reorganization, cultural conflict-cultural accommodation, and social well-being-social pathology among Afro–Americans as they moved from a largely rural to a largely urban existence.[2] Frazier's preferred framework for the study of these complex processes was a combination of the case study and human ecology methods.[3]

"Negro Harlem" provided an empirical test of Frazier's "natural evolution" assumptions regarding black urbanization processes (as did his earlier study of blacks in Chicago).[4] Simply stated, this perspective equated the distinctive characteristics of successive concentric zones in black communities with contrasting levels of social adjustment or acculturation to the urban environment. Thus, as one moved outward from the center zone, overall socio-economic status, family stability, community vitality, and other quality-of-life variables in the black populace could be expected to improve. Two important challenges to conventional views of blacks and black communities were explicit in this perspective. First, Frazier argued for a view of black urban social organization as being internally differentiated, rather than monolithic. He amassed data that provided clear-cut evidence of variation of levels of social disorganization, social pathology, and value orientations among blacks. Second, Frazier sought to explain the varying levels of community organization (and disorganization) in terms of empirical relationships between culture, historical circumstances, socio-economic status, and political power, *rather than* in terms of biological predispositions. In short, "Negro Harlem" represented an extension of Frazier's research interests in ". . . the influence of the larger societal and community forces upon the institutions of Negro life, and reciprocally, the impact of the Negro presence upon the larger society."[5]

A major contribution of "Negro Harlem" to our understanding of urban ecology lies in its substantiation of Burgess's theory of urban expansion using a racially-specific urban community. Since Frazier's studies, the applicability of the Burgess "Concentric Zone Theory" to black urban community development has been reconfirmed under varying conditions.[6] An equally important contribution of "Negro Harlem" was its introduction of cultural variables into Concentric Zone Theory. Frazier notes that the restrictive effects of forced segregation combined to produce in the case of black Harlem what was essentially a separate (although parallel) community life and structure. Subsequent research supports his conclusions on this point; most notably, the Kerner Commission's observation that "our nation is moving toward two societies, one black, one white—separate and unequal."[7] Having addressed the importance of the contribution of "Negro Harlem" to urban ecological theory, our understanding of the study might be further enhanced by a slight digression. In this regard, we think a brief examination of the socio-historical antecedents of black Harlem, as found by Frazier in 1935, would help to better gauge the full significance of his study.

The historical roots of black Harlem to 1935

Harold Cruse, in acknowledging Harlem as the cultural capitol of the black world, concluded that it "is in several ways the most unique and, in terms of 'The Theory of Black Cities,' it is the most crucial community of all."[8] Why was Harlem destined to achieve such importance for Afro–America? Of all the segregated black communities in the years before and after the First World War, Harlem was unique. Whereas, other black urban areas within large cities were referred to as "Bronzevilles, Niggertowns, Black Bottoms, Smoketowns, and Chinchrows . . . ," Harlem was a name of class.[9] The name and proud history, unlike their equivalents throughout the nation, were symbols of elegance and distinction, not derogation.[10] The Harlem streets were wide, well-paved, well-landscaped, and free of litter; the

---

[1] E. Franklin Frazier, "The Impact of Urban Civilization Upon Negro Life," *American Sociological Review*, 2 (August 1937), p. 609.

[2] G. Franklin Edwards, "E. Franklin Frazier," in *Black Sociologists: Historical and Contemporary Perspectives*, ed. James E. Blackwell and Morris Janowitz (Chicago: University of Chicago Press, 1974).

[3] Ibid., p. 111.

[4] E. Franklin Frazier, *The Negro Family in Chicago* (Chicago: University of Chicago Press, 1932).

[5] G. Franklin Edwards, op.cit., p. 86.

[6] Donald R. Deskins, *Residential Mobility of Negroes in Detroit, 1837–1965* (Ann Arbor: Michigan Geographical Publication No. 5, Department of Geography, University of Michigan, 1972), p. 168.

[7] U. S. Riot Commission, *Report of the National Advisory Commission on Civil Disorders* (New York: Bantam Books, 1968).

[8] Harold Cruse, "Harlem's Special Place in the Theory of Black Cities," *Black World*, 20 (May 1971), p. 10.

[9] John Henrik Clarke, ed., *Harlem, U.S.A.* (New York: Collier Books, 1971), p. 17.

[10] Gilbert Osofsky, *Harlem: The Making of a Ghetto* (New York: Harper & Row, 1963), pp. 71–77.

homes were roomy, "replete with the best of modern facilities, finished in high style."[11] Harlem was not a slum, but an ideal place in which to reside. For the first time in the history of New York City, and most of the nation as well, blacks were able to move into decent homes in a respectable neighborhood.

As late as the 1880s, the major part of New York City's black population lived in lower Manhattan.[12] But by the 1890s, in an attempt to secure better housing, blacks moved as far north as 53rd Street. This location ushered in a new phase of life among black New Yorkers.[13] But, the 53rd Street area reached its utmost development during early 1900, and blacks began searching for housing space in upper Manhattan.

At the same time, as a result of speculative building, Harlem had an excess of relatively new apartment houses. The white landlords were unable to fill their empty houses because transportation was inadequate and subways had not been extended to the area.[14] Phillip A. Payton, a black entrepreneur, persuaded the tenantless landlords that he could fill their vacant houses with black occupants.[15] The economic desperation of the landlords overcame their prejudices and Payton, negotiating as a representative of the Afro-American Realty Company, was authorized to bring blacks to Harlem.[16] He encountered little difficulty in recruiting tenants, as the opportunity to move to Harlem was met with considerable enthusiasm by the ebony segment of the New York population.

By 1910, the black population of New York had grown to 92,000, with the overwhelming majority located in Harlem.[17] At this time, the average black, New York City working man earned approximately $7.00 a week, which afforded a comfortable standard of living.[18] It was also during this era that many Harlem blacks amassed fortunes—the most celebrated being the cosmetologist, Madame C. J. Walker.[19] When news of these earnings reached blacks of the rural South, the migration to Harlem—a journey to the "promised land"—was increased. By 1920, the black population in New York had risen to more than 152,000, and during the years 1920–24, an additional 50,000 black Southerners came to the city[20]—most of whom settled in Harlem. This large black influx produced a heavily concentrated and spatially compact population pattern in Harlem.

The relative prosperity and the movement of the 1920s known as the Harlem Renaissance[21] brought this black community to the attention of the world. Nevertheless, the subsequent depression of the 1930s stimulated its ghettoization, and this process was in full bloom when Frazier undertook his study in 1935. It should be noted that this research was in conjunction with Frazier's official status as head of the commission studying conditions in Harlem, appointed shortly after the Harlem riot of 1935.

Ecological trends in black Harlem, 1940–1970

This paper, like the study it seeks to update, restricts its analysis of black Harlem's ecology to descriptive statistics. While the original study utilized far richer, more detailed data (from case studies), this study addresses a broader span of time. The units of analysis are zones, and these are tracked over four censal periods.[22] Nine measures of ecological and social organizational patterns are

---

[11]Ibid., p. 18.
[12]James Weldon Johnson, " 'Harlem': The Cultural Capitol," in *The New Negro*, ed. Alain Locke (New York: Arno Press, 1978), p. 302.
[13]James Weldon Johnson, *Black Manhattan* (New York: Knopf, 1946), p. 147.
[14]J. Johnson " 'Harlem': The Cultural Capitol," pp. 302–4.
[15]Roi Ottley and William Weatherby, *The Negro in New York* (Dobbs Ferry, N.Y.: Oceana Publishers, 1967), p. 182.
[16]J. Johnson, " 'Harlem': The Cultural Capitol," pp. 303–4.
[17]J. Johnson, *Black Manhattan*, p. 144.
[18]Ottley and Weatherby, *The Negro in New York*, p. 182.

[19]Paul Morand, *New York Negroes* (New York: Henry Holt, 1930), pp. 237–41.
[20]Charles Johnson, "Black Workers and the City," *Survey Graphic: Harlem*, 6 (March 1925), p. 641.
[21]Nathan Huggins, *Harlem Renaissance* (New York: Oxford University Press, 1971).
[22]Definition of zones by year of census is as follows—*1940 (33 health areas)*: Zone 1 = no. 10; Zone 2 = nos. 7.2, 8, 9, 11, 12, 13; Zone 3 = nos. 6.1, 6.2, 7.1, 14, 15, 16; Zone 4 = nos. 3, 4, 5, 17–22, 23, 24, 25–31. *1950 (83 tracts)*: Zone 1 = no. 228; Zone 2 = nos. 208, 210, 212, 213, 214, 217, 221, 224, 226, 227, 230, 232; Zone 3 = nos. 196, 198, 200, 204, 206, 207, 209, 211, 215, 219, 220, 222, 223, 225, 229, 231, 234, 235, 236; Zone 4 = nos. 182, 184, 190, 194, 202, 218, 201, 203, 205, 233, 237, 239; Zone 5 = nos. 162–180, 185–189, 191–193, 195, 197, 199, 216, 241–255. *1960 (83 tracts)*: Zone 1 = no. 228; Zone 2 = nos. 208, 210, 212, 213, 214, 217, 221, 224, 226, 227, 230, 232; Zone 3 = nos. 196, 198, 200, 204, 206, 207, 209, 211, 219, 220, 222, 223, 237, 239; Zone 5 = nos. 162–180, 185–189, 191–193, 195, 197, 199, 216, 241–255. *1970 (85 tracts)*: Zone 3 = nos. 196, 198, 200, 204, 206, 207, 209, 211, 219, 220, 222, 223, 225, 229, 231, 234–236; Zone 4 = nos. 182, 184, 190, 194, 201–203, 205, 218, 233, 237, 239; Zone 5 = nos. 162–180, 185–189, 191–193, 195, 197, 199, 216, 241–255.

used (see Table 1 for a list).[23] Table 1[24] reveals several significant patterns in our data on black Harlem. Spatially, one observes a continued outward expansion over the thirty-year period, as the percent black population steadily increases in each of the five zones. Were data for the other New York City boroughs included, this outward expansion of the black population from Harlem would appear all the more dramatic. Interestingly, Zone I retained its racial identity as the original center of New York City's black populace. Over the years, this zone has consistently been just short of all black (99.5 percent black in 1940 and 1970). The negative relationship between distance from Harlem's center and percent black (in Manhattan) has also held over time, although the predictive strength of this association has decreased. Changes in black Harlem's spatial density are made apparent by our indices of household crowding and size. In both instances, large declines are evident across all zones from 1940 to 1970. While New York City's black population increased eight-fold, black Harlem actually decreased in size. The 195,000 residents in 1970 represented roughly only 11 percent of New York City's total black population, a drastic change from 1935 when the overwhelming majority of blacks in New York City lived in Harlem.

When our attention turns to economic variables, we see that from 1940 to 1970 major improvements in economic status accompanied the spatial shifts of blacks in Harlem. During this time period, unemployment rates were reduced from the high levels of the post-Depression years (1940, average 19 percent vs. 5 percent in 1970, Zones 1–4). The proportions of blacks in white-collar occupations (1940 zone average of 22 percent vs. 40 percent in 1970) and annual family incomes also increased (by an average 360 percent, 1950–1970). From 1940–1970, as in 1935, the positive association between distance from Zone I and overall economic well-being remained constant. Thus, more blacks from Zone 5 were in white-collar occupations and, of these, fewer fell into the lower-echelon clerical category than was true for the other zones. Likewise, family income was, on an average, much higher in the outer zones. Finally, the steadily increasing median monthly rent as one moves outward from the center provides a reliable proxy measure of improvements in the overall quality of housing.

Marital disruption rates, contrary to Frazier's expectations, did not decline from 1935 to the present. Instead, the proportions of marriages ending in divorce, separation, or death from 1940 to 1970 increased across the zones. Thus, this indicator of community disorganization (according to Frazier) was negatively related to black Harlemites' spatial redistribution and economic status enhancements. Our general impressions from other research suggest that the additional indices used by Frazier to measure community disorganization (e.g., crime rate, proportion of populace on welfare, and health status) also reflect poorly on contemporary Harlem.[25]

Black Harlem in contemporary perspective

From a historical view, the conceptions of black urbanization processes set forth in "Negro Harlem" seem surprisingly accurate. As Frazier expected, reorganization followed the initial disorganization experienced by black migrants to Harlem. Moreover, substantial differentiation in rates of disruption and recovery has been observed among black Harlemites. Apparently, differential individual and family resources, combined with differential opportunities at the community level, determined the rates of adaptation by different black Harlemites to the urban environment and its demands. In most cases, with successful adaptation came upward social mobility and the move outward from the center of Harlem.

---

[23]Data sources for census statistics are as follows: *Sixteenth Census of the United States: 1940, Population and Housing Statistics for Health Areas: New York City; U.S. Census of Population: 1950, Census Tract Statistics: New York City*, Vol. III, Ch. 37; *U.S. Census of Population and Housing: 1960, Census Tracts: New York, N.Y. SMSA*, Part I, PHC (1)–104; *1970 Census of Population and Housing Report, Census Tracts: New York, N.Y. SMSA*, Parts 1–3. All volumes are published by U. S. Department of Commerce, Bureau of the Census.

[24]Unless self-explanatory, calculation information for variables follows:

white collar workers

$$= \left(\frac{\text{professional} + \text{proprietor/manager} + \text{clerical workers}}{\text{total labor force}} \right) \times 100;$$

crowded households

$$= \left(\frac{\text{households with} \geq 1.01 \text{ persons per room}}{\text{total occupied dwelling units}} \right) \times 100;$$

marital disruption

$$= \left(\frac{\text{widowed, separated, divorced persons}}{\text{total married persons}} \right) \times 100.$$

[25]Thomas Vietorisz and Bennett Harrison, *The Economic Development of Harlem* (New York: Praeger, 1970); Frank Hercules, "To Live in Harlem . . . ," *National Geographic Magazine*, 151 (February 1977), p. 178.

Table 1. Ecological and Social Organizational Change, 1940-1970, in Frazier's Five Zones of the Harlem Community

	Zone I				Zone II				Zone III				Zone IV				Zone V			
	1940	1950	1960	1970	1940	1950	1960	1970	1940	1950	1960	1970	1940	1950	1960	1970	1940	1950	1960	1970
% Black	99.5%	99.5%	99.3%	99.5%	71%	97.3%	96.7%	96.4%	42.1%	62.7%	67.8%	70%	20.2%	56.6%	61.9%	63.3%	----	20.3%	28.6%	36.8%
Unemployment Rate	20.1%	9.4%	4.9%	4.4%	18.3%	9.8%	7.0%	5.3%	18.6%	9.5%	7.3%	5.3%	18.5%	10.4%	7.1%	6.2%	----	8.5%	7.3%	5.7%
White Collar Workers	10.1%	17.9%	21.2%	30.9%	17.4%	18.2%	23.3%	36.7%	25.9%	26.3%	26.5%	40.3%	33.5%	29.4%	28.9%	40.3%	----	37%	36.9%	48.2%
Clerical Workers	50.1%	61.2%	63.2%	76.9%	52.2%	62.7%	68.2%	68%	55.1%	55.9%	61.1%	64.4%	51.2%	51.6%	55.1%	63.2%	----	53.6%	54.2%	53.8%
Crowded Households	22.6%	17.8%	16.7%	11.7%	19.3%	20.4%	20.8%	11.9%	19.8%	20%	18.4%	12.8%	21.5%	19.3%	18.9%	14.4%	----	20.8%	18.3%	14.4%
Persons per Household	3.5	2.6	1.6	1.8	3.2	2.8	2.2	1.9	3.2	2.8	2.2	2.1	3.3	3.1	2.4	2.3	----	3.1	2.6	2.5
Monthly Rent	$40	$38	$53	$67	$37	$38	$58	$81	$39	$41	$56	$76	$39	$46	$58	$83	----	$43	$62	$86
Family Income	----	$1276	$4126	$6181	----	$1929	$4357	$6270	----	$1939	$4276	$6564	----	$1888	$4757	$7370	----	$2244	$4729	$7151
Marital Disruption	----	25%	46.8%	53.1%	----	21%	41.5%	46.6%	----	21%	35.2%	39%	----	19.5%	33.3%	37.3%	----	21%	27.4%	32.8%

It would appear, therefore, that Frazier correctly predicted urbanization patterns in black Harlem; nevertheless, several paradoxes remain. Perhaps the most striking is the apparent reality of multiple black communities in contemporary New York City. The gradients of what Frazier labeled social disorganization, and what we choose to label "quality of life indicators," are duplicated for black communities in other New York City boroughs. For example, within Queens, the Bronx, and Brooklyn, there are noticeable gradations in the quality of life in black communities based on the distance from some center. Although such patterns are most evident among blacks living in Manhattan, specifically in Harlem, the similarity in structure of black communities elsewhere in New York City is significant. Contrary to Frazier's notion, then, one detects separate concentric zonal patterns for black communities in each of the different New York City boroughs. This finding presents somewhat of a contradiction for Frazier's theory of an improved quality in black life with increased distance from Harlem's center. Additional contradictions are posed by observed increases in rates of marital disruption, births outside marriage, crime, and welfare dependency, despite overall increases in socio-economic status (i.e., urban acculturation) among black Harlemites. In these instances, time fails to bear out hypotheses in which Frazier placed great faith.

This brief revisitation of "Negro Harlem" convinces us of the need for a revival of earlier research on processes of black urbanization and spatial relocation. Among the many important issues to be addressed by such studies are the numerous questions raised by "Negro Harlem." In addition, we wonder whether the idea of multiple black communities, each with its own center and surrounding concentric zonal patterns, can be empirically validated for other major U.S. cities.[26] To what extent are the observed patterns of ecological and social organizational change in such black subcommunities differentiated on the basis of origins, historical time period, or economic circumstances of the initial and subsequent black migrants? What are we to make of the increasing return of upper- and middle-class citizens to the central city and the resultant displacement of low-income blacks? How far has internal social differentiation progressed among black communities?

In sum, a variety of hypotheses have been generated by our cursory look backward to "Negro Harlem." Of these, perhaps the most fruitful group are those revolving around the cultural component introduced by Frazier to Burgess's theory of urban expansion. There is reason to believe that racial discrimination fundamentally alters the model's applicability to the analysis of black community patterns of urban expansion. Purely and simply, black communities are not allowed to grow, expand, and evolve in a natural fashion. Instead, their growth and development is channeled in predictable directions by largely external factors. What is required now is research which specifies the sources and goals of such influences. We must establish how differences in individual and/or family resources merge with differences in access to societal opportunity structures to create the clear-cut social differentiation found in contemporary black communities. In a related sense, we must also seek to specify the direct and indirect impacts of community of residence on quality-of-life variables and, indeed, life chances. To the extent Frazier's study serves as a springboard to new discoveries about black urbanization processes, then its purposes would have been well served. As a student of social change, he fully expected his perspectives to become dated, since the reality with which he was concerned was certain to evolve to another level of development. However, Harlem's prominent position in the Afro–American ethos has not changed; now, as before, ". . . Harlem remains a key fortress for black people in the United States."[27]

[26]Our examination of black population trends and patterns in several North Central region cities (Chicago, Detroit, and Milwaukee) revealed, for example, a closer adherence to the concentric-zones urban-expansion model. Black community patterns in the Kansas City and Los Angeles SMSA's, however, seem—like New York City—to depart from this model in significant ways, being characterized, instead, by multiple, black subcommunities with distinct concentric-zones patterns.

[27]Hercules, "To Live in Harlem . . . ," p. 180.

Cultural Variables in the Ecology of an Ethnic Group*

Christen T. Jonassen

The attempt to discover, describe, and explain regularities in man's adaptation to space has long been a matter of concern to social scientists and sociologists. In the United States the ecological school of sociology, depending primarily on the observation of man in an urban environment, has concerned itself with this problem. Since Alihan's shattering critique[1] of the Parkian ecological theory a decade ago, two schools of thought seem to have emerged. Their discussions have sought to determine whether or not a science of ecology is possible without a socio-cultural framework of reference. The crux of the problem seems to center around the relative influence of "biotic," strictly economic, "natural," and "sub-social" factors on the one hand, and socio-cultural elements on the other hand. Those stressing the former as causative factors have been referred to as the "classical" or "orthodox" ecologists, while those emphasizing the latter factors might be called the "socio-cultural" ecologists.

Perhaps the best if not the only way to determine where the correct emphasis should lie is by empirical research. It is hoped that the results of a research project[2] reported in this paper may contribute toward that end.

One writer suggests "that the time has come when we should study the influence of the cultural factor in the phenomena sociologists have defined as ecological."[3] The study of an ethnic group in an American urban environment seems particularly suitable for such a project. Such a group has a distinct culture which can be described and characterized, and the reaction of such a group to the American environment is more readily observed since it is set apart from the general population in the Census and other governmental reports.

The Norwegians in New York have a continuous history as a group since about 1830 when they formed their first settlement and community in Lower Manhattan. Since that day the community has moved until it is now located in the Bay Ridge section of Brooklyn.[4] The first location was .25 mile from City Hall, the center of the city; the

---

*Reprinted from the *American Sociological Review*, 14 (February 1949), 32–41, by permission of the author and The American Sociological Association.

[1] Milla A. Alihan, *Social Ecology* (New York, 1938).

[2] See C. T. Jonassen, *The Norwegians in Bay Ridge: A Sociological Study of an Ethnic Group* (Ann Arbor, Michigan: University Microfilms, 1947).

[3] A. B. Hollingshead, "A Re-examination of Ecological Theory," *Sociology and Social Research*, 31 (January 1947), pp. 194–204.

[4] Smaller settlements have also been formed in suburban sections of Staten Island, Queens, and the Bronx, but the main group is located in Bay Ridge. See Figure 1.

present location is about ten miles from that point. From 1830 to the present time six fairly distinct areas of settlement may be observed.

I. The problem

We shall be primarily interested in the mobility of the Norwegian community. Why did the group first settle where it did, and why did it move from this area to another? We shall want to know why it moved in one direction and not in another, and we shall be interested in the rate and type of movement. And if we are able to suggest some answers to these questions, we shall be able to ascertain if the distribution of the Norwegian group in New York and the movement of its community can be explained in terms of factors that are "non-cultural," "sub-social," "impersonal," and "biotic," as the classical ecologists and their followers would contend; or if causality must be referred to cultural and social factors to explain the movement of this community in New York, as the "socio-cultural" ecologists would maintain.

II. Cultural background of the settlers

If we are to ascertain the comparative influence of culture in determining spatial distribution, it becomes necessary to sketch briefly the cultural background of this group so that their values and cultural heritage may be indicated. The Norwegians who created this settlement, unlike those who pioneered in the Western states, came for the most part from the coastal districts of Norway. Norway was in those days underdeveloped industrially and its main means of livelihood were agriculture, lumbering, fishing, and seafaring. Many individuals would combine all of these occupations and especially fishing and agriculture which were carried on in the innumerable fjords and inlets of the long indented shoreline of Norway. In these districts a culture based primarily on the sea as a means of transportation and a source of food combined with a little farming has flourished for centuries since the Viking days. The people are trained from their earliest youth in skills necessary to make a living in such an environment. The men and women who founded and continued the Norwegian settlement in New York originated in such environments, and many men joined the colony by the simple expedient of walking off the ships on which they worked as sailors.

Norway, of all the civilized countries in the world, has one of the most scattered populations, the density being only 23.2 persons per square mile as compared to 750.4 for England and 41.5 for the United States.[5] Norway does not have very large cities and its people never live far from the mountains, the fjords, and the open sea. They are for the most part nature lovers and like green things and plenty of space about them.

III. Settlement and movement of Norwegians in New York

The first Norwegian community which has an unbroken connection with the present one was located about 1830 in the area now bounded by the Brooklyn Bridge, the Manhattan Bridge, and the East River.[6] At that time, along this section of Manhattan were located docks where ships from all parts of the world loaded and unloaded, and here were also located the only large drydocks in New York, capable of repairing large ocean-going vessels. Here also were found the offices of shipping masters, vessel owners, and other seafaring occupations. In this atmosphere of salt water and ships, men familiar with the sea could feel at home. And within walking distance of their homes they found plenty of work as carpenters, shipbuilders, sailmakers, riggers, and dock and harbor workers.

Across the East River lay Brooklyn, a town of some 3,298 inhabitants in 1800. It grew rapidly and became an incorporated city in 1834, and by 1850 it had grown to 96,850 inhabitants. In 1940, the Borough—according to Census figures—had a population of 2,698,284. Brooklyn gradually superseded New York as a shipbuilding, ship repairing, and docking center. There was the New York Navy Yard in Wallabout Bay. But the center of shipping activity became Red Hook, that section of Brooklyn jutting into the New York harbor, across from the Battery. The Atlantic Docks were completed here in 1848. It also became the terminus of the great canal traffic that tapped the vast resources of the American continent. Here large grain elevators were built to hold grain for ships that came from all parts of the world to load and discharge. In 1853, the famous Burtis Shipyard already employed 500 men, and in 1866 a great celebration was held when the John N. Robbins Company opened two huge graving docks and three floating docks in Erie Basin. These docks could build and float the largest vessels, and they were the only such docks in New York outside

[5]As of 1938.
[6]See Figure 1.

those in the Navy Yard. Red Hook became a humming yachting, shipping, and shipbuilding center.

The Norwegians living in New York found the journey by horsecar and ferry tedious and time-consuming. They soon began to settle in Red Hook and the next Norwegian settlement developed in the area immediately adjacent to and north of Red Hook, where a small group of Norwegians settled in 1850. By 1870 the invasion of Brooklyn was gathering speed.

A horsecar, travelling along South Street in Manhattan, took Norwegian ship workers to Whitehall. Here they boarded the Hamilton Ferry to Hamilton Avenue, Brooklyn. Between 1870 and 1910, Hamilton Avenue became the most Norwegian street in Brooklyn and New York.

The colony developed to the north of Hamilton Avenue. The churches moved over from New York and new churches were established. In the nineties, this section was one of large beautiful homes and tree-shaded streets. The section became better as one went north and became very exclusive at Brooklyn Heights where the grand old families lived. This section occupied in those days a functional relation to the downtown section of Manhattan that Westchester, Connecticut, and Long Island do today. A contemporary wrote, "... the greater part of the male population of Brooklyn daily travels to Manhattan to work in its offices.... The very fact that Brooklyn is a dwelling place for New York ... a professional funny-man long ago called it a 'bed chamber.' "[7] It was actually as the saying went "a city of homes and churches."

Norwegian immigrant girls coming to New York found jobs as domestics in these beautiful homes and Norwegian men, skilled in the building, repair, and handling of ships of all kinds, found plenty of work for their hands in Erie and Atlantic Basins a short distance to the south. The section therefore became a logical location for the development of a Norwegian immigrant community. It offered them everything they needed. The Irish and Germans also moved into this neighborhood, and as it grew more and more crowded the old families moved out. Just as the New Englanders had forced out the Dutch, so now Norwegians, Irish, and Germans were forcing out the New Englanders. The stately old homes were converted to two- and three-family houses, and some to boarding houses. In this neighborhood the Norwegian colony flourished for some decades up to the beginning of the twentieth century.

At the time that certain members of the New York community moved away to settle in this area of Old South Brooklyn, others migrated across the river to Greenpoint in another part of the Brooklyn waterfront.[8] This section was also connected to the old Manhattan community by ferry. There was some shipping activity along this side of the waterfront but it offered only limited opportunity for the particular skills of the Norwegian immigrants. The area was soon invaded by new immigrants from the south of Europe and by factories of various kinds. It is rather significant that unlike the settlement in Old South Brooklyn this community did not move to adjacent territory, and after some years it ceased to exist, its inhabitants scattering in all directions.

The inexorable growth of the city continued. In Old South Brooklyn, open places became fewer and fewer and green grass and trees disappeared. Old large one-family houses were torn down to give place to tightly packed tenements. Then it came the turn of the Norwegians, Irish, and Germans to be invaded and succeeded by the southern Europeans, mostly Italians from Sicily.

By 1890, many old downtown families purchased fashionable homes a little further out near Prospect Park, in the Park Slope section, as "a means of getting away from the thickly populated section of Brooklyn," the incentive being the scarcity of houses, plenty of wide open spaces and an abundance of trees and garden spots in the Park Slope area.[9] The residents of the area used to be known as the brownstone people who lived in beautiful mansions, paid their bills monthly, and ordered from the store by telephone. In the beginning of the century, the Norwegians also started to move out of the downtown area and into this section. This became the next center of the Norwegian colony in Brooklyn.

But the city continued to push its rings of growth further and further out and the same process repeated itself all over again. By 1910, the Norwegians were on the move again, this time to the adjoining area of Sunset Park. The docks and shipyards were extended all the way out to Fifty-ninth Street. And in 1915, the Fourth Avenue subway was completed. Electric cars running on Ninth and Fifteenth Streets and Third Avenue and Hamilton Avenue provided transportation to the shipping center at Red Hook.

[7]Edward Hungerford, "Across the East River," *Brooklyn Life, 1890–1915*, p. 81.

[8]See Figure 2.
[9]*Brooklyn Daily Eagle,* December 5, 1926.

The center of the Norwegian colony remained in Sunset Park district up to about 1940. The exodus of Norwegians from this section and into Bay Ridge and other outlying sections is now in progress. It is the sections of Sunset Park and Bay Ridge which now constitute the area of the settlement. The present Norwegian settlement is located on the high ground overlooking New York Harbor. For the most part it is a section of one- and two-family houses with small lawns, backyards, and tree-planted streets. The nature of this area was determined by indices which have proved reliable in characterizing urban neighborhoods. Indices of economic status, rents, condition of housing, density of population, mobility rates, morbidity and mortality rates, demographic characteristics, standardized rates of crime and juvenile delinquency, dependency, poverty, and desertion rates were also employed. From the cumulative evidence of such data it is apparent that the area in which the Norwegians live is, when compared with other areas of New York and Brooklyn, one of the best, and no part of this area according to this study displays the characteristics of a slum district. However, a detailed study of the various parts of the area shows that it can be divided into areas that, on the basis of the indicated indices, may be designated as "poor," "medium," and "best." The distribution of Norwegians living within these areas is as follows: ten percent live in "poor" sections, fifty-four percent in "medium," and thirty-six percent in "best" areas. The "best" areas include parts of Bay Ridge and Fort Hamilton which contain some of the best residential areas in New York, while the "poor" areas in the northwestern part of the Sunset Park district have some sections that border perilously on slum conditions.

An analysis of population movements within the area of the Norwegian settlement indicates that the Norwegians are moving out of the northern and western census tracts of the Sunset Park district and into the southwestern census tracts of Bay Ridge and Fort Hamilton. Italians and Poles are moving in from the northeast and Russian Jews are taking over the sections of the northern and eastern periphery of the area vacated by the Norwegians and other Scandinavians.

From the ecological and historical study of the characteristics of the Norwegian community over a period of more than a hundred years, it appears that it has maintained rather consistent characteristics and a functional position in New York since the community was established. Like all other groups, native and foreign, the Norwegians were unable to prevent change of the character of their neighborhood, nor were they able to prevent invasion by other land use and lower status groups; they could maintain the things they valued only by retreating before the inexorable development of the city to new territory where conditions were more in harmony with their conceptions of a proper place to live.

It is apparent from the data of this study that numerous causative factors have operated in determining the location and movement of the Norwegian community in New York: the economic and social conditions of Norway, the economic and social conditions of the United States, the rate and direction of New York City growth, the condition of the neighborhood, available lines of communication between the cultural area and the location of the economic base, and attitudes and values of the Norwegian heritage. Where they were to settle and the rate and direction of movement were thus largely determined by elements of the immigrants' heritage and the character and needs of the host society of the United States.

Neither one of these factors was *the* determining one. The Norwegians' reaction to this urban environment resulted rather from a judicious balance of all these factors. It is clear from old maps that transportation to Bay Ridge was available as early as 1895, if they had wanted to live there. But this was slow transportation by horsecar in the early days, and the downtown area evidently presented agreeable enough conditions. As the city grew, however, these conditions became less desirable to people who valued plenty of space around them and nearness to nature.

It is apparent that the Norwegian immigrants broke away from the original economic base to a certain extent later. This development depended on the advance of lines of transportation and new technological and economic development and on the fact that the Norwegian culture was becoming even more industrialized, which gave later emigrants new skills and knowledge that they could apply here. The erection of skyscrapers and use of steel construction in New York gave Norwegian sailors jobs as structural steel and iron workers. They were used to working aloft and their experience as riggers made them particularly valuable for this work. In the twenties, the great building boom provided skilled carpenters with plenty of work.

Figure 1 shows the sections the Norwegians have inhabited at various times. The dotted lines represent lines of transportation. Figure 2 is the same

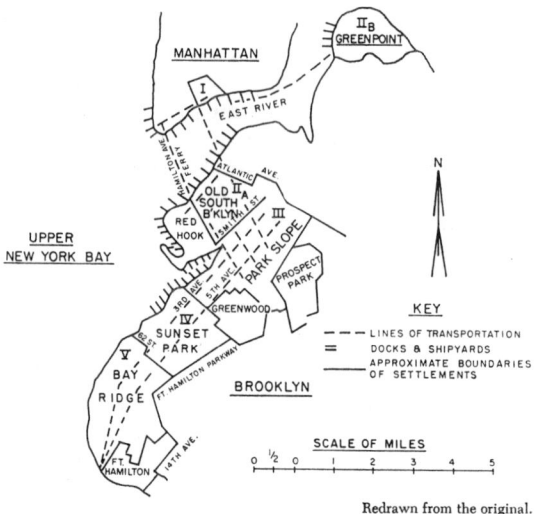

Figure 1—Norwegian settlements, 1850–1947.

map with is salient factors consolidated and simplified. The progression of the Norwegian cultural area, as can be seen, may be represented as a series of interlocking circles, the centers of which are the centers of the cultural areas at the times specified. The path of this progression is the locus of the centers of the interlocking circles, and it represents in reality the lines of communication.

At each stage, the cultural area presents definite ecological characteristics. It has a center, a clustering of ethnics. The center attracts and repels;[10] it repels some who move out and establish the basis for a new center farther out along the path of advance, and it attracts others to it who are lagging behind. The lagging areas, shaded on the map (Fig. 2), are created at the stage when the colony is breaking up to advance again; they represent transitional areas in process of invasion by other land use and lower status groups. They are therefore the least desirable sections of the settlements to which those who are economic failures gravitate. The advance guard of the new cultural area settled new territory or mingled with native Americans, and these Norwegians in turn formed a center for a new Norwegian cultural area. The process is a continuous one, and change from one area to another must be measured in decades rather than in years. It is a seepage-like movement rather than a sudden mass change.

[10]Repulsion and attraction here are considered as functions of the choice of individuals in relation to the realities of the environment.

IV. Some implications of this study for ecological theory

The change of location of the Norwegian community was produced by persons breaking away from the old area and individually choosing a new habitat. Because of its concerted progression in a certain direction to a certain place, the illusion of a directed mass movement is created. But this ecological behavior arises out of the interaction of the realities of the New York environment with the immigrants' attitudes and values. The resulting actions of many individuals are very much alike since they are motivated by very similar attitudes created in conformity with the cultural pattern of Norway. *It is therefore indicated that the movement of these people must be referred to factors that are volitional, purposeful, and personal, and that these factors may not be considered as mere accidental and incidental features of biotic processes and impersonal competition.*

It has been stated that immigrant colonies are to be found in the slums or that immigrants make their entry into the city in the area immediately adjacent to the central business district. From the data of this study we are fairly certain that the Norwegian colony has not existed in an area with the characteristics of a slum, and we can be certain that it does not occupy such an area today even though it is the habitat of recently arrived immigrants. It would therefore appear that the statements referred to above cannot be taken as generalizations, but apply to certain ethnic or racial groups only.

The cause of the Norwegians' settling where they did and in no other place around New York is not at all clear if we refer the explanation to biological, sub-social, and non-cultural factors. It is logical to assume that as biological creatures interested primarily in sustenance and survival, the Norwegians could have survived in any number of other places. But if we refer the explanation of the location of their community to cultural factors it becomes so obvious as to be banal. It is clear that their cultural heritage had given them the tools whereby they were able to elicit meaning and values from this particular environment. Other sections of New York, for example the financial section, the clothing manufacturing sections, etc., had little meaning for them in terms of survival or satisfaction. To the Jew from a crowded ghetto in the center of Poland the realities of the harbor district would probably have no values and meanings, or they might have different values and meanings, perhaps negative values. But to the Norwegian, socialized in the coast culture of Norway, this environment had meaning and value in terms of sustenance and psychological satisfaction. The very method by which he could compete and sustain himself was inherent in the cultural heritage which he brought to this country, and whether or not this cultural heritage should ever find expression and be useful to him depended on the cultural pattern of the United States and the cultural artifacts of that country.

The objective realities of New York thus presented the Norwegians with a multitude of environments to which they might have reacted. It is significant that they reacted primarily to those aspects of the New York milieu that had meaning in their value system. Thus the environmental facts were of little significance *per se* and only as they were incorporated into the value-attitude systems of the Norwegian immigrants.

The movement of the group, when compared with the movements of other ethnic groups in New York and other American cities, assumes some significance. Studies of Italians[11] and Jews[12] reveal different developments. The usual situation in these groups is one in which an area of first settlement is established which stays in one place, and continues to receive new arrivals. As the old immigrants become assimilated and the second generation grows up, they move out to an area of "second settlement," usually far removed from the first in space and time. Thus Italian and Jewish communities in New York are still found in many of the areas, such as the Lower East Side of Manhattan and downtown Brooklyn where they were first established. But there is hardly a trace of any Norwegians in the areas of New York and Brooklyn which they originally inhabited. Furthermore, the development and progression of Norwegian cultural areas in New York show a continuum of space and time and result from the unique character of their heritage in interaction with their new environment. It does not therefore seem possible to generalize as to the type of movement that all immigrant groups in urban areas will exhibit; rather the type of movement, its rate and direction will depend on the interaction of the particular heritage of each immigrant group with the urban environment in which the immigrants live. The different rates of movement of different ethnic groups[13] from the center of cities might find a more satisfactory explanation on this basis.

The area of the Norwegian community was described in terms of indices of various kinds. These might be regarded as objective measures of the values which Norwegians have in regard to the environment in which they want to live. Thus the amount of crowding within the home and congestion without, and other conditions indicated by crime, delinquency, health, and population statistics, have for Norwegians apparently reached an intolerable point in certain census tracts. Other tracts present them with conditions that they find more favorable, and it is to these areas that they move as soon as they are able to do so.

It is probable that the Norwegian community has been able to maintain its solidarity for over a hundred years, and in spite of constant moving, because the variable factors that determined its existence were favorable. The dissolution of the Greenpoint settlement indicates what happened when the factors that sustained it were unfavorable. But for the community that did survive and move, there was, when conditions reached an intolerable state, always an appropriate area immediately adjacent to the old area; so the commu-

[11]Cf. Leonard Covello, *The Social Background of the Italo–American School Child: A Study of the Southern Italian Family Mores and Their Effect on the School Situation in Italy and America* (New York: New York University School of Education, Ph.D. thesis, 1944).

[12]Cf. Louis Wirth, *The Ghetto* (Chicago, Illinois, 1929).

[13]Cf. Paul F. Cressey, "Population Succession in Chicago: 1898–1930," *American Sociological Review* (August 1938), pp. 59–69.

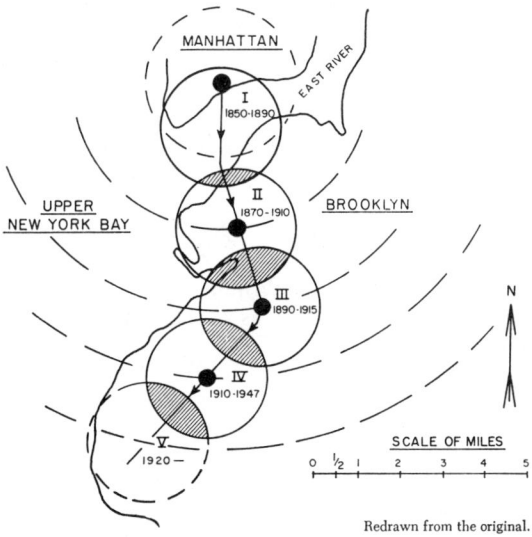

Figure 2—Movement of Norwegians, 1850–1947.

nity was able to move from Manhattan to Old South Brooklyn, to Park Slope, to Sunset Park, and finally to Bay Ridge. Norwegians have not been segregated from native whites, nor is there any evidence that they have been discriminated against in any way as far as choosing a home is concerned. The clustering within the area is therefore voluntary.

However, there is no place having the characteristics which Norwegians require adjacent to the present settlement in Bay Ridge. The city is moving in on them from north and west, and there is only water to the east and south. The area is also being invaded by other ethnic groups. Nor is the type of buildings within this area entirely to their liking. It is still predominantly a neighborhood of single- and two-family houses, but a great number of large, high-class apartment houses have been built, and the land value has increased so tremendously that, wherever zoning permits, this is the type of housing that is erected. It would seem that the Norwegian community in Brooklyn is making its last stand in Bay Ridge with its back to the sea. Its final dissolution is a matter of years and will be brought about because the balance of variables that determined its development cannot be maintained much longer. As long as the values of their heritage could be integrated and harmonized with conditions of the developing city, the community grew and flourished; when this integration is no longer possible it will disintegrate and its members disperse.

This development has already commenced. Census figures and the changes of addresses for subscribers to *Nordisk Tidende*[14] indicate that many Norwegians are moving to Queens, Staten Island, New Jersey, and Connecticut, where new settlements are forming in environments which are more in harmony with the values of their heritage. Some of these settlements have started as colonies of summer huts, and finally developed into all-year-round communities.

The peculiar interplay of a plurality of motives that goes into the determination of ecological distribution of Norwegians is well illustrated by these informants:

> I like it here (Staten Island) because it reminds me of Norway. Of course, not Bergen, because we have neither Fløyen nor Ulrik, nor mountains on Staten Island, but it is so nice and green all over in the summer. I have many friends in Bay Ridge in Brooklyn, and I like to take trips there, but to tell the truth when I get on the ferry on the way home and get the smell of Staten Island, I think it's glorious. However, I'm taking a trip to Norway this summer, and Norway is, of course, Norway—and Staten Island is Staten Island.[15]

A man states:

> I arrived in America in 1923, eighteen years old. I went right to Staten Island because my father lived there and he was a ship-builder at Elco Boats in Bayonne, New Jersey, right over the bridge. I started to work with my father and I am now foreman at the shipyard where we are building small yachts—the best in America. I seldom go to New York because I don't like large cities with stone and concrete. Here are trees and open places. . . .[16]

Another tells what he likes about his place in Connecticut:

> I like the private peace up here in the woods. There is suitable space between the cabins so that we do not have to step on each other's toes unless we want to get together with someone once in a while. Since I started to build this house, it is as if I have deeper roots here than in the city. This is my *own* work for myself. . . .[17]

And a woman says:

> . . . It is a real joy to get out of the city with all its wretchedness. I go down to the brook where I have a big Norwegian tub. There I sing lilting songs and wash and rinse clothes. Everything goes like play, and before you know it, the summer is over, and all this glorious time is gone and I could almost cry.[18]

One who has moved to Staten Island weighs the advantages and disadvantages:

> It is countrylike and quiet here with plenty of play room for the children. But I must admit I am homesick for Brooklyn once in a while, perhaps often. Then I take the ferry and visit friends and acquaintances there.[19]

The assumption that "in general, living organisms tend to follow the line of least resistance in obtaining environmental resources and escaping environmental dangers" has been used as the basis for hypotheses of human distribution in space.[20]

---

[14]The newspaper of the Norwegian community.

[15]*Nordisk Tidende,* March 3, 1947.
[16]Loc. cit.
[17]Ibid., September 5, 1946.
[18]Loc. cit.
[19]Ibid., March 3, 1947.
[20]James A. Quinn, "Hypothesis of Median Location," *American Sociological Review* (April 1943).

Such a statement in the light of this study seems too mechanistic, too simple, and therefore inadequate as an explanation of the distribution of this group in New York. Men need not merely to survive, require not only shelter or just any type of sustenance; they want to live in a particular place, in a particular way. A better description of man's distributive behavior might be: *men tend to distribute themselves within an area so as to achieve the greatest efficiency in realizing the values they hold most dear.*[21] Thus man's ecological behavior in a large American city becomes the function of several variables, both socio-cultural and "non-cultural."

One writer has pointed out that the early ecologists "envisaged an abstract ecological man motivated by physiological appetites and governed in his pursuits of life's goals by competition with others who sought the same things he sought because physiologically they were like him."[22] It is quite evident now that this ecological creature was the product of the same intellectual miscegenation which begot the now somewhat extinct "economic man." The men and women observed in this study are not abstract entities; they are very real persons with physical needs. But they are also governed and motivated in the pursuit of culturally determined goals by culturally determined habits and ways of living. They compete for things high in the hierarchy of their value system which may or may not be the same things for which other individuals and groups strive. It hardly seems possible to achieve a systematic theory of ecology that squares with empirical observation and meets the needs of logical consistency without the cultural component as an integral part of such formulations.

---

[21]This conclusion is essentially in agreement with the "theory of proportionality" as proposed by Walter Firey, *Land Use in Central Boston,* p. 328.

[22]A. B. Hollingshead, op. cit., p. 204.

The Ecology of Norwegian-Americans in Metropolitan New York from 1940 to 1980*

Knight E. Hoover

The controversy between the classical and neo-orthodox theoretical positions on the one hand and the socio-cultural on the other has ebbed. Few analysts would argue against the view that "urban spatial organization is the product of human relations and social values rather than of subsocial and impersonal forces."[1] Ecological distribution is multidimensional and involves the interplay of demographic, organizational, environmental, technological, and social psychological factors through time.

The Norwegian ethnic group and its migration in Brooklyn is viewed here in the broader context of the New York metropolitan region. This approach is taken because Norwegians lived not only in Manhattan and Brooklyn, but also along the waterfront of the upper New York Bay in northeastern New Jersey and on northern Staten Island at least since 1900.[2] Changes within the group are due to the changing environment, as well as to the unique and changing characteristics of the people themselves. One major consideration, however, is that, in spite of three-quarters of a century of change, the area of greatest Norwegian concentration in fifteen counties of the New York region remained in the Brooklyn Borough sections of Sunset Park and Bay Ridge (i.e., Viking Village) by 1970.[3]

One purpose of this paper is to describe the post-World War II Norwegian migration. A second aim is to describe both the modifications which have occurred in Viking Village, as well as in the Norwegian population of the broader New York region, during the past four decades. The

---

*This article was written for this volume and has not been published previously.

[1] Dennis E. Poplin, *Communities,* 2nd ed. (New York: Macmillan, 1979), p. 117.

[2] Knight E. Hoover, "Organizational Networks and Ethnic Persistence" (Ph.D. dissertation, City University of New York, 1979), p. 32. Information is based upon analysis of *Nordisk Tidende* (Norwegian-language press in Brooklyn, N.Y., 1891–present) organizational pamphlets and brochures.

[3] U. S. Department of Commerce, Bureau of the Census, *Detailed Characteristics, New York State,* PC (1)-D 34, 1970 (Washington, D.C., 1970). Brooklyn contained 14,300 first- and second-generation Norwegians in 1970, approximately 70 percent (about 10,000) of whom lived in Viking Village. Thus, about 20 percent of the total Norwegian population in fifteen New York region counties resided in Viking Village in 1970.

primary goal is to suggest explanations for the present-day settlement patterns of Norwegians[4] in Viking Village and in the New York metropolitan region.[5]

Post-World II era: The watershed

Clearly the post-World War II period (ca. 1945 to 1960) did provide a watershed for Norwegian ecological distribution between relatively old, inner city core and relatively new, suburban non-core areas in the region. Ethnic dispersion from core areas, however, was not limited to Norwegians, nor was it entirely a post-World War II phenomenon. Norwegian suburban migration from urban centers in New York (the Bronx, Brooklyn, and Manhattan) and New Jersey (Bayonne, Hoboken, Jersey City, Elizabeth, and Newark)[6] had begun during the first decades of the twentieth century.[7] The Depression of the 1930s and World War II halted the suburban flow. The cessation of war led to the exodus of thousands of white, middle-class persons from New York City and other urban areas in the region in the 1940s and 1950s. This migration was unique because the participants included those persons forty-five years of age and over, an age group which did not usually migrate. Improved and expanded highway systems and the increased personal wealth which led to the acquisition of automobiles and homes were a few of the factors which led to suburbanization. Among the migrants from the region's central cities were Norwegians from Viking Village.

Norwegian merchant marine sailors returning to Norway represented a second migration from Viking Village in the late 1940s.[8] The Norwegian government in exile in London during the five-year Nazi occupation of Norway used the New York–New Jersey harbor as a "home port" for its large merchant marine fleet. Many of the sailors resided with friends and relatives in Viking Village. Their presence added to the density of Norwegian population and social activities. Migration from Viking Village, the departure of the sailors coupled with the suburban migration, must have appeared unending in the late 1940s. The suburban migration, however, reached a zenith in the late 1950s. Follow-up studies in Viking Village between 1966 and 1979 revealed a smaller but persisting Norwegian population.

Viking Village in 1980: The hub of the New York metropolitan region Norwegian community

The largest single area of Norwegian-born concentration in New York City was in Brooklyn in both 1930 and 1940.[9] Despite three decades of change, the single largest concentration of first- and second-generation Norwegians in the New York region still remained in Viking Village in 1970. It was not until the 1970 census that a majority of the Norwegian population had shifted to Bay Ridge.[10] Nonetheless, the greatest concentration still remained where it had been in 1940, in the southeastern sector of Sunset Park, despite some major changes in that community.[11]

The physical and social environments of both Sunset Park and Bay Ridge declined absolutely with the increases in housing deterioration, crime, assistance, and poverty rates during the 1950s and 1960s. Relative to other sections of Brooklyn, however, the two areas contain some of the borough's best residential housing with relatively little housing dilapidation, deterioration, and

---

[4]Norwegian is defined here as first- and second-generation, that is, born in Norway or born in the U. S. of Norwegian parents, unless otherwise noted. This definition was used in the 1960 and 1970 U. S. Census reports.

[5]The term "New York metropolitan region," as used here, refers to a fifteen-county area in New York and New Jersey. It includes the five boroughs of New York City (Bronx, Brooklyn, Manhattan, Queens, and Richmond), the New York State counties of Nassau and Suffolk (on Long Island), and the suburbs north of New York City (Rockland and Westchester). New Jersey counties included are Hudson, Bergen, Passaic, Essex, Morris, and Union.

[6]Elizabeth and Newark are urban centers in non-core Essex County in proximity to the New Jersey Bay, which is a part of the New York–New Jersey harbor.

[7]*Nordisk Tidende,* March 31, 1927.

[8]The history of Viking Village is closely related to the development of the Port of New York and to Norwegian shipping.

[9]The Norwegian-born in five New York City boroughs totaled 38,130, of which 26,142 (68.5 percent) were in Brooklyn in 1930 (*Nordisk Tidende,* October 29, 1931). By 1940, 65.7 percent (N = 20,214 of 30,750) of the Norwegian-born lived in Brooklyn. Of these 20, 214, 72 percent (N=14,561) lived in Viking Village—Christen T. Jonassen, "The Norwegians in Bay Ridge" (Ph.D. dissertation, New York University, 1947), pp. 271–72. Presumably, this represented the greatest concentration of Norwegian-born in the New York region.

[10]In 1960, 7,273 (45 percent) first- and second-generation Norwegians lived in Bay Ridge, whereas 8,871 (55 percent) lived in Sunset Park. By 1970, 5,435 (55.1 percent) lived in Bay Ridge and 4,435 (44.9 percent) lived in Sunset Park (1960 and 1970 U. S. Census reports).

[11]As in 1940, the densest areas of Norwegian concentration in Viking Village were in three census tracts (numbers 104, 106, and 108). They were, however, considerably thinned out in terms of their Norwegian populations since 1940.

crowding.¹² The northwestern sector of Sunset Park underwent further physical deterioration and ethnic and class composition change. In 1968 it was designated a "poverty area" by New York City. The greatest physical deterioration was restricted primarily to five non-contiguous census tracts in one health area,¹³ and although the poverty area boundary extended to, it did not include the three census tracts of greatest Norwegian concentration. Sunset Park also had greater social change than Bay Ridge in the 1950s and 1960s.

Between 1950 and 1969, Brooklyn lost 17 percent (416,000) of its 1950 white (non–Hispanic) population. At the same time, Brooklyn gained 139,800 Puerto Ricans, a 347 percent increase over the 1950 Puerto Rican population.¹⁴ The newcomers to Sunset Park settled largely in the poverty area. Changes in both the physical and social composition were apparently part of the reason fewer Norwegians resided in northwestern Sunset Park by 1970.

Bay Ridge did not substantially change its social or population composition between 1940 and 1970, although it did experience a small population growth primarily due to an increase in apartment complexes. Nonetheless, Bay Ridge remained as it had since at least the 1950s, that is, an area dominated by white, Roman Catholic ethnics. The Protestants, including the Norwegians, remained a numerical minority, although southwest Brooklyn remained one of New York City's most Protestant areas. Although Norwegians continued to move away from some dissimilar ethnic groups, it had always been the case that Viking Village was located in a multi-ethnic area. Norwegians never constituted more than, at most, a third of the area's total population.¹⁵ As time passed, the Norwegian concern with ethnic diversity was moderated by the structural and cultural assimilation of previously dissimilar groups such as Italians. Neighborhood deterioration, fear of crime, and sharp social-class contrasts, as well as a relative lack of assimilation, still result in Norwegian migration. Movement based on such factors is, however, not restricted to the Norwegian group only.

Suburban migration and settlements

Ethnic diversity within a broader framework of cultural and structural assimilation is also a characteristic of suburban, non-core areas where Norwegians have settled in the New York region. Once migration from original ethnic settlements commences, Norwegians, like other ethnic groups, are residentially dispersed.

With regard to other indices said to be important to the Norwegian heritage—such as open spaces relatively free of poverty, crime, and congestion—the non-core areas represented more tolerable living environments. Suburbs were especially praised as conducive to child-rearing. While there were immigrant persons in the child-bearing stage of the life cycle following World War II, many of the migrants were American-born. As one minister noted, the young did not always want to leave Brooklyn. The suburbs, however, had housing more readily available and at less expensive prices than one would find in Brooklyn. Given these conditions in conjunction with the middle-class status of many Norwegians, one would expect to find considerable migration. Three-fifths of the Norwegians within the New York region do live in non-core areas (30,200 or 60.8 percent), whereas 19,500 (39.2 percent) of the 49,700 in the fifteen-county area in 1970 reside in core zones.¹⁶ Certainly, Norwegians did leave the core areas; however, there are no figures assembled on the core/non-core distribution of Norwegians in 1940 or earlier.

There are a number of factors which have to be considered before definite conclusions can be made about mobility stemming from socio-cultural values based on ethnic heritage. First, the Norwegian suburban migration is not unique. One cannot definitely conclude that being Norwegian is any more important than participation in the American Dream; both are socio-cultural factors. Further, many of the migrants were American-born, with-

---

¹²Knight E. Hoover, "The Norwegian 'Colony' in Brooklyn, New York" (M.A. thesis, Brooklyn College, City University of New York, 1970), pp. 16–65.

¹³*United States Census of Housing, 1960*, Series HC (3)–274, pp. 3–19.

¹⁴The 1957 Special Census indicated a "sharp gain in residents of Puerto Rican origin . . ." in Sunset Park (The Community Council of Greater New York, *Brooklyn Communities*, p. 50).

¹⁵Based on Norwegian-born figures, Norwegian percentages of the total population in ten health areas which constituted Viking Village ranged from 3.2 percent to 10.6 percent (Jonassen, "Norwegians in Bay Ridge," p. 274). Scandinavian percentages in the same areas, however, ranged from 14.2 percent to 38.1 percent. In 1940, Scandinavians were the dominant group in eight of ten health areas. By 1970, Scandinavians (Danish, Finnish, Swedish, and Norwegian) were not dominant in any of the health areas, ranging from 3.7 percent to 10.8 percent of the area's total population (1970 Census).

¹⁶U. S. Department of Commerce, *Census of Population, New York City, 1970*, computer print-out, PS-2.

out memories of Norway. Given the Norwegians' middle-class status, they participated in suburban migration not unlike other white, middle-class ethnics during the 1940s and 1950s.

Norwegians did choose to by-pass Staten Island, the only suburban-like borough of New York City.[17] One might attribute this to socio-cultural factors. Staten Island's population did increase between 1940 and 1960, but its greatest growth did not occur until the 1960s when the first bridge link to New York City was opened in 1964. Perhaps Norwegians did not move to Staten Island because it was in New York City or because it was not convenient for commuting. The group's economic base was less and less connected to the harbor areas. Why the Norwegians chose the suburbs rather than Staten Island remains an unanswered question. One point is certain—some Norwegians chose to remain in Brooklyn. Some of the reasons can be gleaned from residents of Viking Village.

Ethnic persistence in Viking Village

Sentimental attachment to the southwestern area of Brooklyn appears to play a role in some Norwegians' decision to remain. The Viking Village "colony," with its memories and ethnic organizations, was and remains, as one minister stated, a "state of mind" in a sea of Italian and Irish Catholics. The state of mind to some extent attaches to Norwegian organizations. As with earlier moves in Brooklyn, organizations slowly moved into Bay Ridge in the 1970s, although some shops and churches remain in Sunset Park in 1980. Certainly Viking Village is in part a social construct. Viking Village also represents the hub of an infrastructure of business, religious, and social organizations which is now maintained through links to the broader metropolitan region. The organizational network has to some extent replaced the former spatial settlement of Viking Village as Norwegian ethnicity is expanding to include national and international ties.[18]

Ties to Norway represent another important change, not as visible in the late 1940s, which have contributed to ethnic persistence in Brooklyn. Somewhat ironically, the war that enhanced the Norwegian ethnic group's economic status—which in part led to its further ecological dispersion—also resulted in a significant strengthening of ties between the United States and Norway. According to the late editor emeritus of the Brooklyn Norwegian-language paper, World War II enhanced the life of the colony in Brooklyn by decades. This resulted from Norwegian–Americans and Norwegian nationals coalescing to solicit support for a "free Norway" in New York City.[19] Viking Village organizational activity flourished, revitalizing ethnic interest in the homeland.[20] This contact was maintained after the war as air travel brought the two nations even closer.

Thus in 1980, Viking Village still represents a link to the past for many Norwegians. Suburban churches and social clubs, for example, often had their origins in Brooklyn and these ties are maintained. It is not uncommon for suburbanites and those from other city boroughs to commute to Brooklyn to participate in ethnic events. Norwegian Constitution Day, for example, is the main event on the ethnic calendar and is celebrated annually with a parade. The parade was held primarily in Sunset Park until the early 1970s, when the marching route was moved entirely to Bay Ridge, reflecting the population shift. The parade's aim, as it has been since its World War II origination, is to show solidarity with Norway. It is supported by ethnic organizations from the entire New York region. In 1977, two-thirds (N = 147) of the 218 participating organizations were from Brooklyn, but more than a quarter came from New Jersey, New York, and Connecticut suburbs.[21] The hub of the New York metropolitan region Norwegian community, as an organizational and demographic base, is still centered in Viking Village. This is one sustaining factor of the Norwegian presence in southwest Brooklyn.

Viable organizations in Viking Village are important to their memberships. Research conducted in Viking Village in the late 1960s suggests that Norwegians remained in Brooklyn due to organizational affiliations, as well as to the group's changing age structure, commuting convenience, and other factors having little to do with aesthetic features of the physical area. Although tree-lined streets and park areas exist, no one in a 1969 questionnaire noted the beauty of the area as a retentive factor. Fifty-seven Norwegian residents (of 94 or 60 percent) checked "no" when asked if

---

[17]Norwegian stock (first and second generation) on Staten Island between 1940 and 1970 was as follows—1940: 5,883; 1950: 5,076; 1960: 5,300; and 1970: 4,387.

[18]Hoover, "Organizational Networks and Ethnic Persistence"—the entire study was concerned with this theme.

[19]Interviews with Carl Soyland, Editor Emeritus, *Nordisk Tidende*.

[20]Hoover, "Organizational Networks and Ethnic Persistence," p. 52.

[21]Ibid., p. 319.

they want to move to the suburbs. Sixteen stated "yes" and twenty-one were "not sure". The major reasons for remaining in Brooklyn were: work and commuting convenience (N = 10); "like" living in the area and having the cultural advantages of the city (N = 10); and age prohibits move (N = 5)—church, clubs, friends, and relatives were noted by the remainder.[22] These reasons, with the exception of age, are positive, that is, the people chose to remain.

There was a relationship between age and the desire to remain in Brooklyn. Forty-four percent of the 25–44 age category (11 of 25); 52 percent of the 45–64 age group (24 of 46); and 59 percent of the 65 plus age group (23 of 39) did not wish to move to the suburbs.[23]

A further factor possibly leading to the persistence of the Norwegian population in Brooklyn is home ownership. A Norwegian-language newspaper survey conducted in 1941 stated that seventeen percent (N = 100) owned their own homes.[24] This figure was lower than the owner-occupied units for the total population in most (8 of 10) health areas in southeast Brooklyn. Figures for 1950 indicate that home ownership for the total population was 21 percent in Sunset Park and 28 percent in Bay Ridge.[25] Fifty-seven percent (61 of 107) of the Norwegians answering a 1969 questionnaire indicated home ownership.[26] One hypothesis for the high figure relates to the fact that the questionnaire was distributed primarily through organizations. This may suggest a relationship between organizational participation, home ownership, and continued residence in Viking Village.

Advanced age and fixed incomes, coupled with current high interest rates and home ownership, are possible factors adding to the continued persistence of Viking Village in 1980. Nonetheless, the Norwegian population in Brooklyn is steadily declining. Death and movement to retirement areas as well as to the suburbs, without new immigration,[27] are presently accounting for the population decline among first- and second-generation Norwegians in Viking Village.

By the late 1960s, discussion with organizational leaders noted a move of some Norwegians into Bay Ridge, mostly from other sections of Brooklyn including Sunset Park. Although no systematic survey has been made, some Norwegians have moved into apartment complexes in Bay Ridge. As parents age and children leave home, the convenience of apartment living is preferred by some. An apartment complex for the retired, for example, was built along the shore in the 1970s; some of the residents are Norwegian. Although the Norwegians in Viking Village may prefer a suburban-like environment, factors such as those just discussed may limit choice. On the other hand, it is clear that some have the option to leave, but do not.

The persistence of Viking Village suggests that social researchers must reevaluate theories which categorize people primarily on the basis of ethnic category and values which were relevant to a past age. This is not to say that ethnic hertiage and long-held values are not important, but to recognize that socio-cultural values do not operate in a vacuum unaffected by changes in the broader social environment. From an ecological perspective, continued residence among first- and second-generation Norwegians in Viking Village is based on a complex of interrelated factors which cannot be abstracted from time or place. These factors include the modification of an immigrant group's values through time and the importance of the group's changing age structure and its effect on life styles. The death of immigration and the changing age structure of the Norwegian population have presented a variety of options and constraints which differed from those of the 1940s. The 1980 census should once again indicate a reduction of the Norwegian population in Viking Village, but for reasons other than those related to their socio-cultural heritage. The numbers in Viking Village could be as few as 5,000, assuming that "Norwegian" will continue to be defined as first and second generation. Despite the decline in Viking Village, however, a Norwegian community, based on a network of institutional arrangements with links to broader national and international contexts, will persist in the New York metropolitan region well into the twenty-first century.

---

[22]Hoover, "The Norwegian 'Colony' in Brooklyn, New York," pp. 70–72.
[23]Ibid., p. 71.
[24]Jonassen, "The Norwegians in Bay Ridge," p. 286.
[25]Hoover, "The Norwegian 'Colony' in Brooklyn, New York," p. 29.
[26]Ibid., p. 147.

[27]Vincent N. Parrillo, *Strangers to These Shores* (Boston: Houghton Mifflin, 1980), pp. 470–72. The last substantial immigration to the United States from Norway occurred in the 1921–1930 period, when 68,531 arrived. Figures for the decades following—1931–1940: 4,740; 1941–1950: 10,100; 1951–1960: 22,935; 1961–1970: 15,484. Between 1971 and 1977, the maximum annual number was 413 in 1974.

Ethnic Boundaries: A Comparison of Two Urban Neighborhoods*†

Leo Driedger

Although Park and Burgess's (1967) early work on "natural areas" drew undue criticism, we wish to compare two "natural" neighborhoods in Winnipeg with a focus on ethnic boundaries. Suttles (1968) has shown that ethnicity and territory provide the setting for an interesting study of social networks. We have found similar ethnic patterns in metropolitan Winnipeg. The community of St. Boniface has remained essentially a French urban neighborhood for 160 years, while the East European (Jewish, Ukrainian, Polish) Winnipeg North End in its ninety years has experienced very different forms of segmentation and mobility. These two segmented "natural areas" of distinct ethnic qualities are known to all Winnipegers and are referred to as such frequently. Northenders and the French of St. Boniface identify intensely with their respective neighborhoods.

*Reprinted from *Sociology and Social Research*, 62 (January 1978), 193–211, by permission of the author and *Sociology and Social Research*.
†Paper presented to the 1977 Annual Meeting of the Midwest Sociological Society held in Minneapolis, April 13–16. This research was made possible by Grants S69-1445 and S72-0331 received from the Canada Council, whose assistance is gratefully acknowledged. I wish to thank Gerald Suttles for his helpful comments.

In the first part of this paper we shall compare the two communities by focusing on rural-urban continuity, ethnic concentration, isolation, segregation, and mobility. In the second part religious, educational, and voluntary organizations will be examined to show the differential ethnic institutional completeness in these two neighborhoods. Variations of ethnic cultural identity (language, endogamy, choice of friends, parochial education, religion, and ethnic organizations) will be discussed in part three. Territory, institutions, and culture form important ethnic neighborhood boundaries.

I. Ethnic ecological segmentation

Ethnic neighborhood networks of the French in north St. Boniface and the East European (Jewish, Ukrainian, Polish) in North End Winnipeg are very different. In the first part we shall describe the rural-urban setting in which these two communities are located; in the second section we shall look at the ethnic concentration of four groups within these two communities; in the third section minority isolation of these four ethnic groups from the majority (British) will be examined; in the fourth section we shall see the extent to which these ethnic groups are segregated from other ethnic groups; and finally, we shall focus on intra- and inter-area mobility of ethnic groups within these two neighborhoods.

Rural-urban continuity

The two urban ethnic communities of metropolitan Winnipeg can better be understood if we place them in a larger rural-urban ecological context.

THE FRENCH IN ST. BONIFACE. The little settlement at the forks of the Assiniboine and Red Rivers (now Greater Winnipeg) was located in a key and influential ecological position in the Canadian West. These rivers and their tributaries provided easy communication routes by which the Indians and traders brought furs to the forks, and the water system also provided a ready means for the Catholic missionaries to fan out from the forks. As long as the chief means of travel was by boat, St. Boniface remained the natural focal point for Catholic organization in Western Canada (Page, 1958). When the railways came, they, to a lesser extent, had the same effect. In the summer of 1818, Lord Selkirk received an answer to repeated requests made to the Bishop of Quebec for a Catholic priest. Provencher began the first Roman Catholic church on a plot of land on the east side of the Red River opposite Point Douglas. This place now forms the heart of St. Boniface.

At first, in the early 1800s, the St. Boniface parish was dominantly a French parish, though it served both English and French Catholics in the surrounding area which was largely rural. The French are still heavily represented in the southwestern section of the province, with a series of parishes beginning with the urban St. Boniface and St. Norbert parishes, and spreading southeastward into the rural hinterland parishes of Ste. Anne, Ste. Agathe, Ste. Elizabeth, St. Adolphe, St. Pierre, St. Jean Baptiste (Figure 1). Thus, St. Boniface has historically and ecologically been supported by a strong French traditional rural hinterland which feeds into the urban focal point. The rural-urban continuity has been maintained to this day, and we propose that it is a factor in French urban boundary maintenance as contrasted to that of the North End.

Since their beginning the French in St. Boniface have been fairly isolated territorially as the Red River forms a natural boundary on the west. At the turn of the century, another boundary, the Canadian Pacific Railroad on the north, separated St. Boniface from the Kildonans, while the coming of the Canadian National Railroad which followed the Seine river reinforced the eastern boundary. To the southeast were the many rural French parishes which are a natural extension of French influence and support. The French form a vigorous group with a provincial population in 1971 of 86,500 (8.8%), a metropolitan Winnipeg population of 46,200 (8.6%), and a St. Boniface population of 16,700 (36%). In north St. Boniface, which was the first French settlement in the west (census tracts 49 and 50), the French represent 64.6 percent of the population.

THE EAST EUROPEANS IN THE NORTH END. From its early beginnings in the 1900s, the North End of Winnipeg was occupied by East Europeans (Jews, Ukrainians, and Poles). As we will show later in this paper, this is changing drastically. While the Poles are still fairly concentrated in the North End, the Ukrainians are beginning to move out, and the Jews have almost all left for suburban areas. The invasion-succession process is taking place as more recent immigrants from Yugoslavia, southern Europe, and native Indians come into the older southern parts of the North End.

The North End, although a part of the city of

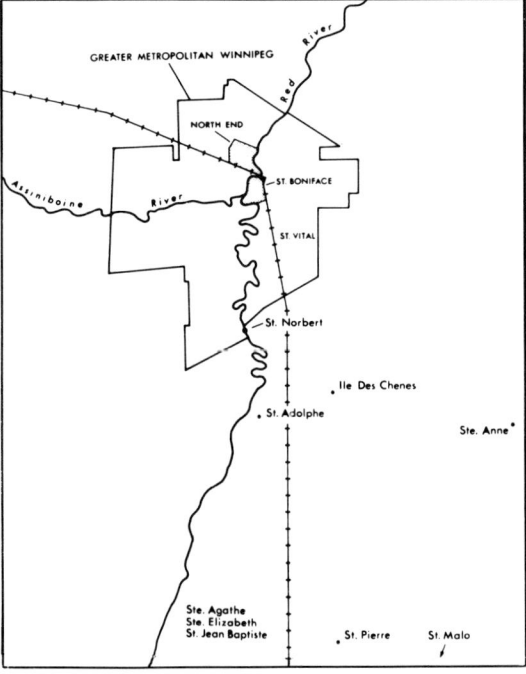

Figure 1—North End Winnipeg and St. Boniface urban communities and the surrounding area.

Winnipeg proper, is isolated from the rest of Winnipeg by the Red River on the east, which separates it from the Kildonans, and on the south by the Canadian Pacific Railroad, which acts as an iron curtain since there are few outlets to southern Winnipeg (Figure 1). To the north and west are the rural Ukrainian and Icelandic settlements of the interlake region.

The North End territory was dominated by East Europeans—the number of Ukrainians has been substantial as well as that of the Poles and Jews. The Poles came most recently and in smaller numbers.

The Ukrainians, the third largest (the British and Germans are the largest) ethnic group in Manitoba, have formed about 12 percent of the population of the province for the past thirty-five years. Until recently a majority of the Ukrainians were rural, but in 1971 over half (64,300) lived in Winnipeg, of which 17,300 (27.5%) lived in the North End. This is a slight drop from earlier years when Ukrainians made up about one-third of the population. Whereas the Ukrainians formed a majority in some of the southerly North End census tracts (4, 11, 12) until 1961, by 1971 they had dropped to about one-third of the population.

The Jews were always mostly urban in Manitoba, so that they have not depended on rural/urban continuity for their identity. While in 1941 69 percent of the Jews in greater Winnipeg resided in the Winnipeg North End, by 1971 this had declined to 19 percent (3,530 out of 18,700). Until 1941 the Jews were the entrepreneurs of the North End; many of them were merchants who served the East Europeans. Since then they have changed from relatively poor immigrants to the highest socio-economic status group in Winnipeg. Most of them have moved out of the North End into the suburbs.

In 1971 the Poles made up 10 to 15 percent of the population in most of the census tracts of the North End. They were not dominant in any area, but tended to be fairly evenly distributed throughout the territory. They came to Winnipeg more recently and many seem to have settled in the North End, attracted by the East European cultural groups already present there, and by the older cheaper housing which was available to new immigrants. In 1971 there were 26,000 Poles residing in metropolitan Winnipeg, 14,000 of whom lived in the city of Winnipeg, and 6,440 (25%) in the North End.

ETHNIC CONCENTRATION. To establish a baseline from which to measure ethnic segregation and mobility, historical documents were searched to plot the original Winnipeg settlements of ethnic groups. Two major areas of concentration similar to Joy's (1972) Quebec area were found in St. Boniface (French) and the North End (Jews, Ukrainians, and Poles).

We found that in 1941 four ethnic groups were still largely concentrated in their original areas of what is now greater Winnipeg. The concentrations of 1941, 1951, 1961, and 1971 were cartographed, with the 1941 and 1961 results shown in Figures 2 and 3, respectively (Driedger and Church, 1974).

In 1941, 39 percent of the French in greater Winnipeg were concentrated in north St. Boniface, an area corresponding roughly to census tracts 49 and 50. This group was clustered about French institutions.

While St. Boniface was dominated by one ethnic group, the North End included concentrations of Jews, Ukrainians, and Poles. The probability of these different ethnic groups reinforcing the maintenance of minority identity was considerable. These groups also had much in common: the three came from eastern Europe, two were largely Catholic, and all of them were of lower-class education and occupation. The North End to this day is politically radical and often elects a Communist member to local office (Chiel, 1955; Woycenko, 1967).

In 1941, 69 percent of the Jews in greater Winnipeg lived in eight social areas roughly located in tracts 5, 8–10, and 13. They were still highly concentrated in the North End, where they originally settled in the late 1800s (Chiel, 1955).

In 1941, 42 percent of the Ukrainians in greater Winnipeg were also concentrated in the North End—the location of their original settlement—though by this time they had moved somewhat farther west in six social areas roughly designated by tracts 2–5 and 12. Ukrainians began settling in Winnipeg just after the turn of the century (Woycenko, 1967:11).

The Poles, more recent immigrants to Winnipeg, also settled in the North End in the same areas as the Ukrainians. Their East European heritage attracted them to the North End. The Poles, however, began arriving later and in smaller numbers; only 18 percent of those who lived in greater Winnipeg concentrated in the North End in 1941.

As indicated in Figure 3, by 1961 all four groups were less concentrated. Over half of the Jewish group (58%) was still concentrated in five census tracts, but the population in each of the other

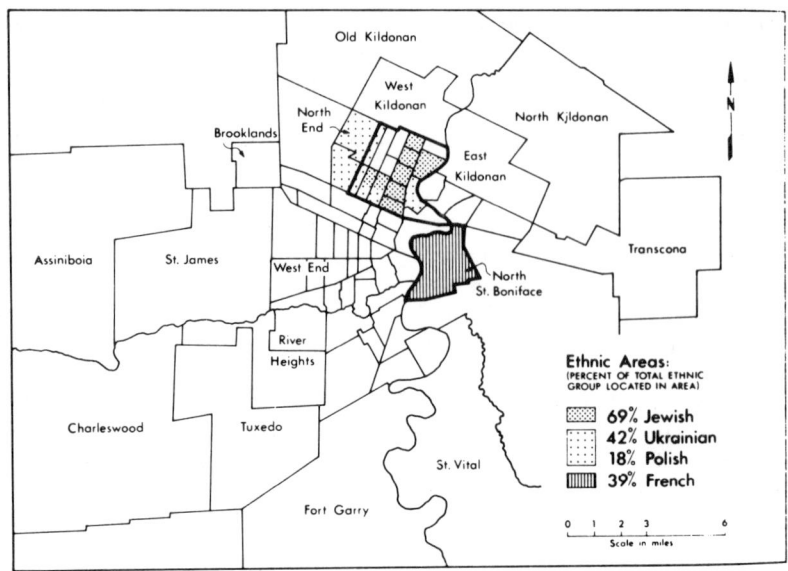

Figure 2—Ethnic group concentrations in metropolitan Winnipeg, 1941.

groups had declined, with fewer than 5 percent of the Polish group remaining in any one census tract. Because of this small percentage, the Polish population does not appear in our analysis.

In 1961, the Ukrainians and French were still concentrated in their 1941 areas. The Jews, however, had changed greatly. By 1961 the Jews had left large areas of the North End for West Kildonan as is clear in tracts 72 and 73, and for River Heights, as emerges in tract 48 (33% and 17% of all Jews, respectively).

The small synagogues that were originally established in the North End were largely abandoned because of age and size. They gave way to larger, modern synagogues which were built in West Kildonan and River Heights. The 1951 map, not presented here, shows concentrations similar to those appearing in 1941, with the additional invasion of tracts 72 and 73, where concentrations later became evident in 1961 and also 1971.

Figure 3 shows that the original settlement areas of the four ethnic groups were largely unchanged by 1941, except for small shifts.

By 1961 and 1971, however, the Jewish population concentration of greater Winnipeg had changed drastically and there were indications that the other groups were beginning to shift slightly as well. A more detailed presentation of ethnic isolation and segregation is needed to develop some indices in order to measure these phenomena.

Ethnic isolation

Assuming that an ethnic group has a better chance of developing its own social institutions when it is in proximity to another minority, we used the Shevky and Bell isolation index to measure residential concentration of ethnic groups in Winnipeg (Shevky and Bell, 1955). The results are shown in Table 1, which compares the four Winnipeg ethnic groups over four decades, 1941, 1951, 1961, and 1971.[1]

It is apparent that the French, Jews, and Ukrainians of greater Winnipeg had the best opportunities of meeting their own group in 1941, and that this opportunity prevailed to a considerable degree until 1971. While the French values remained fairly steady, the Jewish and Ukrainian values declined. The Polish group, which experienced only slight changes, scored much lower in 1941 and 1971 than the other groups.

The Jews and Urkrainians were the most isolated in 1941, but by 1971 this had changed considerably. In the meantine, French isolation remained almost the same; the Jews, French, and Ukrainians were almost equally isolated, with values ranging from 0.14 to 0.19. The Polish group, which never was very isolated in 1941, declined to insignificant values in 1971.

[1] See Shevky and Bell (1955:44) and Driedger and Church (1974) for more details on the formulae used to calculate the isolation index.

Polish isolation was minimal largely because of its small numbers, although the Poles were concentrated in the North End. The French maintained their cluster in St. Boniface by increasing their relative population size. Although the Jews resettled in their two new cluster areas, they lost some isolation strength because of a relative decline in their population from 6 to 4 percent.

Ethnic segregation

It is assumed that, when ethnic groups are concentrated in areas where the numerically dominant British are less heavily represented, the non-British minorities will have greater influence on social institutions, thus, allowing these minorities a greater opportunity to maintain their culture (Breton, 1964). The Shevky and Bell segregation index was adopted to measure the probable interaction between members of a subordinate group (defined as non-British) with members of all subordinate groups in Winnipeg.[2] All groups in the two neighborhoods had a better chance of meeting a minority person than a British person. The findings are shown in Table 1 also.

The segregation index shows the French, Jewish, Ukrainian, and Polish groups in relatively

[2]See Shevky and Bell (1955:47) and Driedger and Church (1974) for more details on the formulae used to calculate the segregation index.

highly segregated areas. The Jews in their transfer from the North End to West Kildonan and River Heights maintained high segregation through their high concentrations and through their alliance with many Ukrainians in West Kildonan, the suburban extension of the North End.

Ecological mobility

While ethnic groups in the two areas in Winnipeg were differentiated by residential segregation, it seemed necessary to examine ecological mobility patterns to assess the direction of relocation as well.

Four 1951 samples of each of the ethnic groups were taken in the two principal areas where ethnic groups were concentrated in 1951. These samples were grouped according to those who had remained in the area ("non-movers"), those who had moved ("movers"), and those who did not show ("no-shows") (Driedger and Church, 1974).[3]

About 7 of 10 French movers (69.1%) moved within the north St. Boniface area, and most of the remaining 30 percent moved to the southern part of St. Boniface, confirming Joy's (1972) bilingual-

[3]For more details on how the samples were selected and how mobility patterns were established, see Driedger and Church (1974).

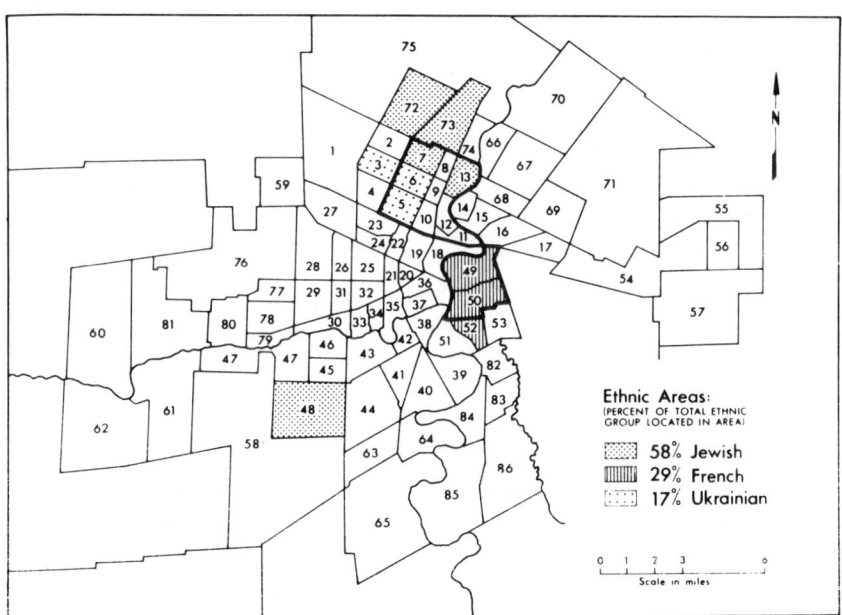

Figure 3—Ethnic group concentrations in metropolitan Winnipeg, 1961.

belt extension hypothesis.[4] Further evaluation of non-French movers showed that large numbers moved out of the St. Boniface area, making the area more exclusively French. This pattern seems to follow French–English mobility patterns in Quebec (Joy, 1972; Lieberson, 1963:64).

In the North End the Jewish inter-area mobility was unique. Figure 4 shows two massive Jewish movements, one to West Kildonan in the north, and the other to River Heights in the south. The extensive inter-area movement to the suburbs was probably due to upward social-class mobility, as was suggested by Latif and Hunter's (1972) factor analysis of Winnipeg Jews moving in large numbers into university education and managerial occupations in 1961, and Nicholson and Yeates' (1969) principal component analysis which placed Winnipeg Jews in the above $10,000 income bracket, in managerial occupations, and in university education. Of interest here is the West Kildonan Jewish extension, which follows Joy's Soo–Moncton bilingual-belt theory, while the southern River Heights extension into Anglo-Saxon territory separates them from their original territory.

The Ukrainians and Poles seemed to remain generally in their North End territory, though there was some extension of the Ukrainians into West and East Kildonan. Less suburban mobility among these groups may be due to the fact that East Europeans remain largely in the lower class and, being less educated, in the craft and laboring occupations (Latif and Hunter, 1972).

[4]Richard Joy (1972:27) suggests that "the perpetuation of the French language seems assured within the Soo–Moncton limits." The Soo–Moncton line includes the French who live north of a line between Sault Ste. Marie, Ontario (between Lakes Superior and Huron), running through Ottawa and on to Moncton, New Brunswick. They will be able to retain their language and culture because they hug the solid French core located in the province of Quebec. The core will act as a supportive feeder for those who live within the Soo–Moncton line, adjacent to Quebec.

Only two of the mobility patterns are shown in Figure 4. The French maintained high segregation and high intra-area mobility, while the Jews represent high inter-area mobility re-establishing high residential segregation.

Comparing the St. Boniface and North End neighborhoods, we found them to be similar in that they are both highly non–British ethnically, and they represent evidence of high ethnic concentration, isolation, segregation, and mobility. There are also differences. St. Boniface is dominated by the one French ethnic group with very few other ethnic groups near them; they have been dominant in the area since its beginning with little evidence of invasion and succession; French mobility is largely within St. Boniface; and continuity with their rural traditional hinterland is preserved. In contrast the North End ethnic neighborhood is more heterogeneous (Ukrainian, Jewish, Polish); although they originally settled in the area, considerable invasion and succession is taking place so that it is now less East European than it was; although mobility of the Poles is within the area, Jewish inter-area mobility is high; and Jewish and Polish continuity with a rural hinterland is not now, and never was, established. We conclude that although the two ethnic neighborhoods show similarities, their differences in ecological segmentation seem even greater.

II. Ethnic institutional completeness

In addition to territorial segmentation as a form of boundary maintenance in the two ethnic neighborhoods in Winnipeg, we wish to evaluate the ethnic institutional completeness of the four ethnic groups in St. Boniface and the North End (Lieberson, 1970; Driedger and Church, 1974). Breton (1964) argues that "the direction of the immigrants' integration will to large extent result from the forces of attraction (positive and negative) stemming from three communities; the community of his ethnicity, the native (receiving) community,

Table 1. A Comparison of Ethnic Group Isolation and Segregation in Winnipeg, by 1941, 1951, 1961, and 1971 Decades

Ethnic Group	Isolation				Segregation			
	1941	1951	1961	1971	1941	1951	1961	1971 (Estimated)
French	.19	.20	.18	.15	.43	.52	.57	.60
Jewish	.31	.21	.18	.19	.71	.64	.62	.60
Ukrainian	.23	.16	.12	.14	.64	.58	.56	.53
Polish	.05	.04	.02	.02	.61	.60	.59	.58

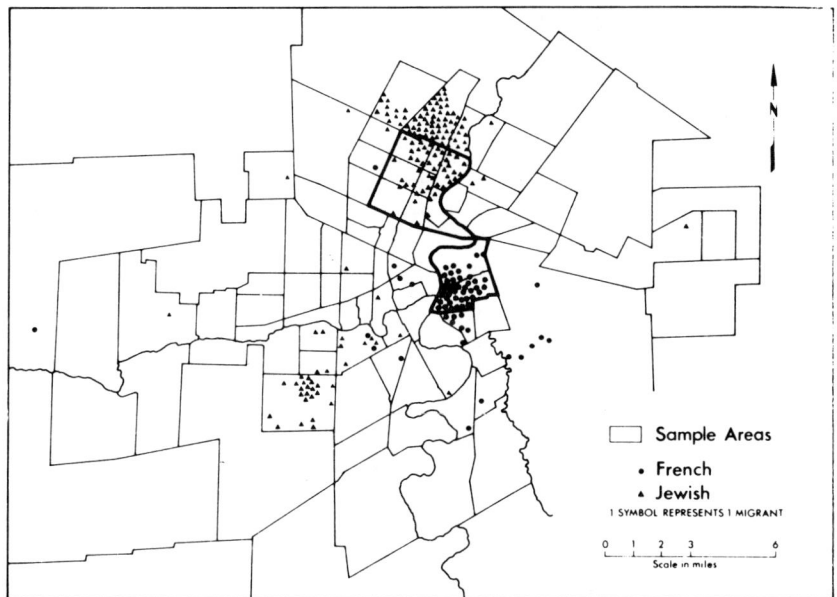

Figure 4—French and Jewish intra- and inter-area mobility.

and the other ethnic communities" (p. 193). These forces are generated by the social organization of ethnic communities and their capacity to attract and hold members within their social boundaries. French, Jewish, Ukrainian, and Polish integration into their own ethnic communities supported by the institutional completeness of their group would reinforce solidarity.

The rationale for institutional completeness is that when a minority can develop a social system of its own with control over its institutions, then the social interaction patterns of the groups will take place largely within the system. Breton (1964) suggests that religious, educational, and welfare institutions are crucial, while Joy (1972) adds the importance of political and economic institutions. Vallee (1969) confirmed Breton's claims by summarizing the need for organization of group structures and institutions which influence socialization and ethnic community decision-making.

The French in St. Boniface established a Catholic parish early in the nineteenth century. Since then a series of institutions have been built within a few blocks of each other, including St. Boniface College, the French Roman Catholic Archdiocese, and le Grand Seminaire, the Mother House of the Missionnaries Oblates, Tache and St. Boniface Hospitals originally started by the Grey Nuns, a museum, the new French cultural center, the CBC French radio and TV station, and the offices of La Liberte, the first French weekly in the West founded in 1913 (Bonin, 1974). St. Joseph's Academy for girls and the Provencher School for boys have been closed. Within this cluster of historical French institutions lies the grave of Louis Riel, French Metis leader of the Manitoba provisional government (1869–70), as well as the graves and monuments of Jean–Baptiste Lagimodiere, a famous coureur des bois, and his wife Marie–Anne Gaboury, the first white woman on the plains. There is also a monument erected to Bishops Provencher, Tache, Langevin, and Beliveau. Most of the streets in this area are named after important French religious or political families. There is a most impressive cluster of historical institutions, which are symbolic of French history and life.

As illustrated in Table 2, some 4,300 students attend ten schools in which French is the major language of instruction (Driedger and Church, 1974).[5] Nine churches are located in the area with about 11,000 members largely of French origin, and there are numerous voluntary associations in the area, which was discovered when a thorough survey was made by contacting leaders of all of these institutions. French institutions are highly

[5]For more details on the method by which the ethnic institutions were calculated, see Driedger and Church (1974).

developed, and many of them have existed for a long time. Much activity takes place within this territory and these institutions relate to French boundary maintenance. The new Canadian emphasis on bilingualism in Canada, and the rise of separation in Quebec tends to create additional hope and support.

The most visible institutions of the Winnipeg North End are the Ukrainian Greek Orthodox onion-shaped church domes. Although not quite as concentrated as the social institutions of the French in St. Boniface, the Ukrainian churches are visible everywhere in the North End. As illustrated in Table 2, the Ukrainians are distinguished by their numerous voluntary associations (31) and their numerous churches (27). The Greek Orthodox and Ukrainian Catholic parish system seems to help tie members to their territory. Although Saturday language schools are still held in many churches, their parochial elementary and high school system is negligible with 200 students enrolled in two parochial schools.

Jewish institutional strength appears to be most consistently complete. Although the eight synagogues are somewhat fewer in number than their population would suggest when compared to the Ukrainians, their membership is the highest proportionately. The fact that Jews comprise the only non–Christian group adds to their distinctiveness. The four Jewish parochial schools serve somewhat fewer than half as many students as French schools do, but proportionately they are a strong second. Jewish voluntary associations are the most numerous (except for the Ukrainians), even when compared to the much larger ethnic groups. However, whereas most of these institutions were located within the North End in 1941, now almost all of them have been moved to North Kildonan and River Heights. Only a few synagogues and schools remain in the northern part of the North End where some of the Jews still reside. As the Jews move out, they take their institutions with them.

There are very few Polish institutions. Their institutional strength is confined to a few Roman Catholic churches and a few associations. Since they arrived late, since they are few in number, and since they are not highly concentrated in any part of the North End, they seem to find it difficult to maintain ethnic institutions.

A comparison of the ethnic institutional patterns in the two neighborhoods shows that French institutional completeness is high in St. Boniface; these institutions are clustered together in a symbolic traditional area; and they are maintaining these institutions within old St. Boniface to a large extent. In contrast, in the North End, Jewish institutional completeness is also high, but Polish institutions are minimal; the Jews are taking their institutions with them as they move out and these ethnic institutions are not replaced by other groups who succeed them; and, while Ukrainian churches are highly visible, they are not as intensely clustered. All in all the ethnic institutional strength in the North End seems more diffuse, less ordered, not as dominant, and more mobile. The institutional boundaries in French St. Boniface remain intact, while invasion and succession is shifting and weakening institutional boundaries in the North End.

III. Ethnic cultural identity

Kurt Lewin (1948) proposed that the individual needs to achieve a firm clear sense of identification with the heritage and culture of the ingroup in order to find a secure "ground" for a sense of well-being. We plan to show that minority cultures are well groomed and watered within the two territorial enclaves we have discussed where these groups have built a concentration of ethnic institutions. French territorial segregation and institutional completeness in St. Boniface would suggest that they have laid important ground for their French culture. Although the strength of cultural identity is less influential in the North End because of ethnic fragmentation and out-mobility, it is nevertheless an important factor.

Backeland (1971) compared an adult French sample and a French high school student sample taken from St. Boniface and found that adult identification with their culture was higher than that of their adolescents. Parents scored high on regular church attendance (90%), French language usage in the home (84%), high religiosity (61%), use of French radio (48%), use of French television (37%), and reading French papers (22%). The number of adolescents who scored high on the same variables was considerably lower: language use (42%), religiosity (26%), French radio use (13%), French television use (11%), and French newspapers' use (28%, the only one higher than adults). Backeland's (1971) study would suggest that French cultural identification declines with younger generations although it may also be partially due to the fact that adolescents tend not to identify with the adult French culture until later.

A multidimensional measure of ethnic identity is needed to distinguish the components that different individuals and groups stress (Lazerwitz,

Table 2. A Comparison of Institutional Completeness (Religion, Parochial Education, Voluntary Associations) by Ethnic Groups

Ethnic Groups (Population, 1961 Census)	Religion				Social Institutions						Associations	
	Members Per Capita	(No.)	Churches/Synagogues Per 1000 Population	(No.)	Education						Per 1000 Population	(No.)
					Students Per 10 Population	(No.)	Schools Per 1000 Population	(No.)				
French (39,777)	.27	(10,700)	.22	(3)	1.08	(4,300)	.25	(10)			.56	(22)
Jewish (19,376)	.36	(7,000)	.42	(8)	.70	(1,365)	.20	(4)			1.18	(23)
Ukrainian (53,918)	.23	(12,500)	.50	(27)	.04	(200)	.04	(2)			.58	(31)
Polish (24,904)	.16	(4,000)	.12	(3)	.00	(0)	.00	(0)			.24	(6)

Table 3. Comparison of Ethnic Group Behavioral Rankings by Composite Mean Scores, and Six Identity Factors (Religion, Endogamy, Language, Organizations, Parochial Education, Friends).

Composite Mean Behavioral Identity Rank	Behavioral Identity Factors (Percentages)					
	Religion	Endogamy	Language	Organizations	Parochial Education	Friends
French	55.0	53.5	65.4	60.5	22.6	79.1	48.8
Jewish	44.2	7.2	91.3	1.8	28.6	74.0	62.5
Ukrainian	36.8	43.6	75.6	21.8	22.9	41.1	15.9
Polish	31.5	46.4	53.2	14.3	12.7	57.1	5.4

1953). Our studies (Driedger, 1975) of cultural identity by comparing ethnic groups in Winnipeg indicated that French university students scored higher than other ethnic groups (with the exception of the Jews) on many of six cultural identity factors (language use, endogamy, choice of friends, and participation in religion, parochial schools, and voluntary organizations).[6] As indicated in Table 3, French university students ranked highest on attendance in parochial schools (79%), and use of the French language while speaking to their parents (61%). Large numbers of French students also reported no marriage in their family outside of their ingroup (65.4%), participated in religious services twice a month or more (53.5%), and chose a majority of French friends (48.8%). The overall identification of French students with their culture appeared to be fairly high (highest of all the groups studied).

We did not take a sample of adults in the North End in order to inquire about the extent to which they identified with their ethnic culture as Backeland did in St. Boniface. However, the data presented in Table 3 is taken from a survey of Jewish, Ukrainian, and Polish university students (Driedger, 1975), many of whom reside in the North End of Winnipeg. This survey shows that, overall, Jewish (44.2%) students identified somewhat less with their culture than French (55.0%) students, but that they identified most strongly in the areas of endogamy (91.3%) and choice of a majority of Jewish friends (62.5%). Jewish students also scored well in attendance of Jewish schools (74%) and participation in Jewish organizations (28.6%); Jewish student attendance of synagogue services (7.2%) and use of their language at home (1.8%) was negligible (Driedger, 1975).

Ukrainian university students scored moderately high in comparison to other ethnic groups. As indicated in Table 3, Ukrainian students did not rank highest on any of the six factors but three-fourths (75.6%) indicated only endogamy in their family, and almost half (43.6%) attended Ukrainian religious services twice a month or more. The overall identification of Ukrainian students (36.8%) with their culture is somewhat greater than that of the Poles but less than that of the French and Jewish students (Driedger, 1975). Only about one-fifth (21.8%) spoke Ukrainian to their parents at home; three times as many (60.5%) French students did so.

[6]For more details on the Driedger Ethnic Cultural Behavioral Identity (ECBI) Index, see Driedger (1975) for more methodological information.

Polish students identified less with their culture, although almost half (46.4%) attended religious services, over half (57.1%) had attended parochial school, and over half (53.2%) reported no intermarriage in their family. As expected, lack of territorial concentration and institutional maintenance seems to be taking its toll in diminished Polish cultural identification.

Although our ethnic cultural data for the two urban neighborhoods are not adequate, the data we do have seem to confirm a distinct pattern. The French neighborhood in St. Boniface with its distinct territorial boundaries, French institutions, and French cultural identification represents a fairly homogenous ethnic community which has maintained itself for 160 years, and will most likely continue to do so for many more. In contrast, the Winnipeg North End, although also a distinct territorial neighborhood, still has an East European presence with the Poles and Ukrainians, but the invasion-succession process has resulted in out-mobility of the Jews and the influx of other groups. The East European North End is in cultural transition.

IV. Conclusions

Our study of St. Boniface and the Winnipeg North End shows that they are two "natural ethnic areas" (Park and Burgess, 1967), with distinct urban boundaries (Suttles, 1968). Territory, institutions, and culture were important ethnic boundary maintenance factors.

The community of North St. Boniface has remained essentially a French urban neighborhood for 160 years. The urban French community, by means of residential segregation, with limited mobility, has maintained a French culture within a fairly complete ethnic institutional framework.

The North End Winnipeg community, originally dominated by East European Jews, Ukrainians, and Poles, is in the process of change. The Jews have moved to the suburbs, taking their culture and institutions with them; the Ukrainians, while still heavily concentrated in the North End with their visible ethnic institutions, are beginning to move away; and the Poles never were heavily concentrated, nor very strong culturally and institutionally. The community is in the process of invasion and succession with new groups moving into the area. The multi-ethnic North End is definitely changing—the East European territorial, institutional, and cultural boundaries are giving way to a new complex of ethnic networks.

The invasion and succession process in the Winnipeg North End is very similar to findings in many other cities. The community changes as various ethnic groups come and go. Why is the French community in St. Boniface different? How were they able to continuously inhabit their original community for 160 years? We have already found that the territorial, institutional, and cultural ethnic boundaries were strongly maintained, but their unique stability seems to call for additional reasons. The fact that the French settled in Canada first, and also settled in Manitoba first, certainly gives them a pioneer advantage. The very place where they have been located for 160 years is sacred ground. Secondly, the French are one of the two founding charter groups who have been given legal language and cultural rights, an advantage that other ethnic groups do not have. In the third place, the Federal Commission on Bilingualism and Biculturalism made numerous recommendations to promote the French language and culture which has resulted in special grants and federal support. Finally, the separatist movement in Quebec seems to provide a model for greater loyalty to the French cause, which has given the French in St. Boniface more courage. French local boundary maintenance factors are strong, but national external factors seem to have given the St. Boniface ethnic neighborhood a special boost. They are a unique community which does not follow the numerous other invasion-succession patterns.

References

Backeland, Lucille L. *The Franco-Manitobans: A Study on Cultural Loss.* Unpublished M.A. Thesis, University of Manitoba, 1971.

Bonin, Adrienne. "St. Boniface: A Study into the Institutional Completeness of a French Community." Unpublished manuscript, 1974.

Breton, Raymond. "Institutional Completeness of Ethnic Communities and Personal Relations to Immigrants." *American Journal of Sociology* 70 (1964):193–205.

Chiel, Arthur. *Jewish Experiences in Early Manitoba.* Winnipeg: The Cornet Press, 1955.

Driedger, Leo. "In Search of Cultural Identity Factors: A Comparison of Ethnic Students." *Canadian Review of Sociology and Anthropology,* 12 (1975):150–62.

Driedger, Leo. "Ethnic Self Identity: A Comparison of Ingroup Evaluations." *Sociometry,* 39 (1976): 131–41.

Driedger, Leo. "Toward a Perspective on Canadian Pluralism: Ethnic Identity in Winnipeg." *Canadian Journal of Sociology,* 2 (1977):77–95.

Driedger, Leo; and Church, Glenn. "Ethnic Segregation and Institutional Completeness: A Comparison of Ethnic Minorities." *Canadian Review of Sociology and Anthropology,* 11 (1974):30–52.

Joy, Richard J. *Languages in Conflict.* Toronto: McClelland and Stewart, 1972.

Latif, A. H.; and Hunter, A. A. "Some Problems of Statistical Artifact in Factorial Ecology." Unpublished manuscript, 1972 (forthcoming).

Lazerwitz, Bernard. "Some Factors in Jewish Identification." *Jewish Social Studies,* 15 (1953):24.

Lewin, Kurt. *Resolving Social Conflicts.* New York: Harper and Brothers, 1948.

Lieberson, Stanley. *Ethnic Patterns in American Cities.* New York: The Free Press of Glencoe, 1963.

Lieberson, Stanley. *Languages and Ethnic Relations in Canada.* New York: John Wiley and Sons, 1970.

Nicholson, T. G.; and Yeates, Maurice H. "The Ecological and Spatial Structure of the Socio-Economic Characteristics of Winnipeg, 1961." *Canadian Review of Sociology and Anthropology,* 6 (1969):162–78.

Page, John Edward. *Catholic Parish Ecology and Urban Development in the Greater Winnipeg Region.* Unpublished M.A. Thesis, University of Manitoba, 1958.

Park, Robert E.; Burgess, Ernest; and McKenzie, Roderick D. *The City.* Chicago: University of Chicago Press, 1967.

Shevky, Eshref; and Bell, Wendell. *Social Area Analysis.* Stanford: Stanford University Press, 1955.

Suttles, Gerald D. *The Social Order of the Slum.* Chicago: University of Chicago Press, 1968.

Vallee, Frank G. "Regionalism and Ethnicity: The French Case." *Perspectives on Regions and Regionalism,* edited by B. Y. Card, pp. 19–25. Edmonton: Western Association of Sociology and Anthropology, 1969.

Woycenko, Ol'ha. *The Ukrainians in Canada.* Winnipeg: Trident Press, 1967.

Section B Neighborhoods

Beale Street, Memphis: A Study in Ecological Succession*

Robert W. O'Brien

The process of succession which emphasizes temporal phases of ecology has, in studies heretofore carried out, emphasized change in use of land, institutional services, and occupation type over a long period of time. This emphasis has caused us to lose sight of the role of rhythm and timing in ecological changes. The modification in the use of a street on different days of a week or by day and by night is also of great significance in the local study of a community.

An analysis of racial succession on Beale Street in Memphis, Tennessee, yields definite stages of succession with one land utilization, one institutional service, or one population type giving way to another. This long-time cyclical activity, however, cannot explain daily and weekly variations in population and activity. These latter are temporal succession in the limited sense. The street serves a different set of institutions and a different group of persons at night from those in the daytime. This is not a situation peculiar to Beale Street, but characteristic of busy streets in most American cities.

Memphis is a particularly good city for the study of both long-time stages of succession and short-time rhythmic movements, for it has had a vivid history of racial and class structure. From the beginning the local racial patterning took the ecological form called "marble-caking." Instead of a division of the city into halves or quarters according to color, as found in many Southern communities, or of complete dispersion, which would characterize a casteless society, the races lived in numerous white and Negro communities adjacent to each other.

Beale Street, settled by upper-class white people one hundred years ago, is one of these communities which best illustrates the stages of succession. Here lived members of the white aristocracy of the mid-South. Some of their homes are still show places of today. The Hunt–Phelan home, erected in 1835, was occupied by the confederate General Polk in 1861, used by General Grant as headquarters in 1862, and converted into a federal hospital in 1863. To it and other homes on Beale Street came Andrew Johnson, Jefferson Davis, and other notables of their period.

The first known invasion of Negroes into this area came in 1865, when Beale Street Baptist Church was erected. Built by Negro labor, it was the first Negro Missionary Baptist church in America. Five years later the American Missionary Association founded a school at Beale and

---

*Reprinted from *Sociology and Social Research*, 26 (May 1942), 430–36, by permission of the author and *Sociology and Social Research*.

Orleans. No large-scale displacement of whites occurred, however, until the 1870s, when a series of yellow-fever epidemics threatened to destroy the city.

The year 1860 may have been a critical date for most Southern cities, but for Memphis 1880 was the beginning of a new epoch. The ecological succession which followed can best be interpreted in terms of the changes of class dominance within both the caste groups. Gerald Capers, social historian of Memphis, believes that the

> social and economic consequences of the yellow-fever epidemics [of the 1870's] were so far-reaching as to warrant the conclusion that there have been two cities on the lower Chickasaw [Memphis] bluff: one which existed prior to the pestilence, and a second which sprang up on the ruins of the first.[1]

The Memphis before 1880 was heterogeneous, cosmopolitan, and unique. Although it was located in the South, it was never entirely a Southern town in the strictness of its caste relationships. When its first mayor, Marcus Winchester (1829), married "a beautiful French Quadroon" in Louisiana, the small town disapproved of the marriage, but the couple returned to Memphis to live. When he ran for the state legislature, his opponents made no mention of the biracial union. This would scarcely have occurred in a less cosmopolitan community such as the Memphis of today.

The complete change of population type after the fever epidemic is revealed in a census taken in 1918 by the National Bureau of Education. Of the 11,781 white parents residing in Memphis, only 183, less than 2 percent, had been born there. The plague almost annihilated the Irish. Most of the Germans moved to St. Louis. The white upper class either migrated out of the city or moved from Beale Street to the highlands east of the city. Of 3,801 Negro parents only 171, or approximately 5 percent, were born in Memphis.

Consequently, the new Memphis, although modern in physical aspect, was rural in background, rural in mores, and rural in prejudice. Being without either tradition or an aristocracy, white Memphis defended its own lower middle-class status by insisting upon the tightening of the caste system. There was thus an increasingly rigid caste situation, caused in part by the Reconstruction period, but accentuated by the emigration of the upper and middle classes of the white group.

As Beale Street came increasingly under Negro domination, the area passed into the second stage of succession. The small missionary school became an accredited Negro college. Small shops and saloons gave way to banking establishments and large stores. Negro doctors, dentists, financiers, lawyers, politicians, writers, real-estate promoters, and insurance men flocked to the area. It became the one central location open to Negro business and activities. Soon Beale Street was the meeting place for all classes within the Negro caste.

Beale Street began to attract national attention through the activities of its band and orchestra leaders.

> Up from the docks of the Mississippi River, up from the saloons, the bawdy houses of Beale, up from the honky tonks of the sawmill towns, up from the white cotton fields of Dixie, accompanied by banjo strumming and hand clapping, rose the sorrow songs of the Negro toiler.[2]

These songs, the blues, were first played and recorded on Beale Street by men like W. C. Handy, William Bailey, Johnny Dunn, Wes Dukes, and Jim Turner. White as well as colored Americans came to know Beale Street by its songs—"Memphis Blues," "St. Louis Blues," "Beale Street Blues," "A Good Man is Hard to Find."

With the enforced segregation of races on streetcars in 1906, many upper-class Negroes, unwilling to play a lower-caste role, migrated to the North. Others remaining in Memphis still refused to use streetcar service. This loss was partially compensated for by the migration to Beale Street in the early 1920s of the entire upper-class Negro population of Dollard's "Southern Town."[3] Nevertheless, the contacts between traditionally established upper-class families of both caste groups were limited.

This interesting process of racial succession with its change in population type and its change in land use and institutional services is paralleled by rhythmic changes.

Anderson and Lindeman pointed out this rhythmic trend in New York in their volume on *Urban Sociology* fourteen years ago.[4] To them the population breaks up into "motile aggregates moving

[1] Gerald M. Capers, Jr., *The Biography of a River Town* (Chapel Hill: University of North Carolina Press, 1939), p. 204.

[2] George W. Lee, *Beale Street* (New York: Robert Ballou, 1934), p. 119.

[3] John Dollard, *Caste and Class in a Southern Town* (New Haven: Yale University Press, 1937). Same community as studied by Hortense Powdermaker in *After Freedom* (New York: Viking Press, 1939).

[4] Nels Anderson and Edward Lindeman, *Urban Sociology* (New York: Alfred A. Knopf, 1930), pp. 146–47.

rhythmically past one another in their daily and hourly activities." The segregation in point of time of the different classes using the transportation system of any large city illustrates this characteristic. From six to eight o'clock the factory workers and lower-paid group of clerks go to work. Then follow in succession: the white-collar men and women, the managers, the shoppers, the matinee crowd, the home-bound workers, the watchmen and scrubwomen, the theater crowd, and the "hot-spot" patrons. This last group together with the scrubwomen and watchmen will meet, on their way home, the early shift of factory workers and the milkmen and morning newsboys.

In Memphis a similar pattern is followed. It is one, however, which is modified by the ecological history of the area.

On six nights a week Beale Street is a Negro street, but on the seventh (Thursday) it belongs to the white people. White people of all classes, some in overalls, others in evening dress, attend what are locally called "rambles." Here they see "scantily clad brown beauties dancing across the stage," and they hear and see the weaknesses of men, both white and colored, subjected to burlesque.

Saturday night the street belongs to the cooks and maids, the factory hands and houseboys. Although quiet and peaceful in daytime, at night "we stomps the daylights into the flo'." The sidewalks are crowded; the working folks are on parade. Lower-class and some middle-class Negro society is on the avenue. Some will see the western films at the Palace or Daisy. Some will crowd Atkinson's Stag Poolroom. Others will go to Simms' Beer Garden. Middle-class members may attend a fraternity dance at the Hotel Men's Improvement Club or the auditorium. Streetwalkers and blind guitar players mix with the crowd. From honky tonks come both "sweet" and "hot" music.

In the daytime, Beale Street is a world of lunchrooms, beauty parlors, pawnshops, grocery stores, fish and meat markets, fruit stands, print shops, and service stations. It is the home of one of the largest Negro enterprises in the country, the Universal Life Insurance Company. Its office buildings house Negro lawyers, doctors, dentists, pharmacists, photographers, newspaper men, undertakers, real-estate operators, Scout executives, social workers, political leaders, and financiers. Traders and merchants display goods on sidewalks. Barkers entreat passers-by to stop and inspect bargains. "Conjure" doctors sell good luck bags, love powders, and graveyard dust charms.

The daytime Beale Street is particularly active on Saturdays, when the country farm hands from Arkansas, Northern Mississippi, and Western Tennessee throng to the avenue to bargain for clothing, groceries, and entertainment. This group usually leaves before the Saturday night activities begin.

While these rhythmic changes on different days of the week as well as by day and by night have been developing, the area has been moving toward another stage in long-time succession. During the last twenty years there has been a migration away from Beale Street of certain types of persons and institutions. The largest Negro-owned drugstores are near a new elite residential area. A two-thousand-pupil Negro high school has been built in the new area. In 1923 Le Moyne College moved from Beale Street three miles south. Since that time the leading upper- and middle-class churches have located or rebuilt within a radius of four blocks. Four years ago a Catholic school was established in the neighborhood. In 1940 a federal housing project eliminated the only lower-class residences in the area. Upper- and middle-class parents discouraged attendance at dances on Beale Street, preferring to have their children associate only with those in the elite residential district.

Depression years have brought many changes to Beale Street. Negro businesses have failed, and, for the most part, they are now owned or operated by white men. Chain drugstores, chain groceries, and chain ice-cream parlors have forced most of their competitors out of business. Beginnings of a new stage of succession are evidenced with the establishment of an aspirin factory and a municipal T.V.A. power station in the area.

Beale Street as an area in transition continues to meet more and more the needs of the lower class and lower middle class, rather than of the entire Negro group. Upper-class Negroes find embarrassment in the poorer shops, the conjure doctors, and the vice of Beale Street. They look with shame upon the street's famous intoxicating ten-cent drink known as "fight-your-mammy." This disapproval on the part of the upper-class Negroes brings the street into greater disrepute, and thereby accelerates the ecological changes that gradually force the middle class to seek residences elsewhere.

Nevertheless, Beale Street of 1941 not only illustrates historical succession of land utilization, institutions, and population types, but also, like many other communities, is illustrative of rhythmical fluctuation of great significance on various days of the week and various hours of the day.

Notes on Beale Street Forty Years Later, and "Beale Street" and Temporal Ecology*

Robert W. O'Brien
and Robert M. O'Brien

Notes on Beale Street Forty Years Later

In the forty years since the field work on Beale Street was completed, many changes have occurred. The small missionary school which had moved three miles south from its Beale Street location to become LeMoyne College (an accredited institution of more than 300 Negro students) has grown into a predominantly black, mixed student body of 1,200. It now is the anchor of a middle-class residential and business section, which is served by a nearby black-owned and operated shopping center. The upper-class Negroes who moved to Memphis from Dollard's "Southern Town" (Indianola, Mississippi) in the 1920s have become, a generation later, a strong economic force in the community. When their Universal Life Insurance Company in 1935 hired a few white agents to serve their white policyholders, the site of the experiment was in Texas. Today, in Memphis and elsewhere in the South, they openly follow their program of crossing racial lines in both patronage and employment. Their assets are nearly a hundred million dollars, and it was their affiliate, the Tri-State Bank, which financed the new black shopping center near LeMoyne College.

The first impression on walking up Beale Street from the waterfront is the rows of abandoned cotton warehouses. Cotton may or may not be King in Memphis, but the Cotton Exchange is functioning and one-third of the nation's cotton crop is sold there every year. What has happened is that the dominance of cotton and the river trade has been challenged by the growth in other sectors of the economy. During the past twenty years, more than 300,000 new jobs have been created in industrial plants. More recently the popularity of the "Memphis Sound," a blend of folk, rock, and church music, has resulted in the building of many new recording studios in the city. Although "Memphis Sound" is essentially "soul" music, and elsewhere in America "soul" is synonymous with black, in the Beale Street area this ability to go back to the "roots" is not considered the exclusive property of one race. The current Memphis recording artists are bands with both black and white musicians. Their musical touchstone is the

---
*This article was written for this volume and has not been published previously.

ability to communicate to the listener a deep and sincere feeling. It is a return to the W. C. Handy tradition. This sound has made Memphis fourth nationally in the recording industry, behind only New York, Los Angeles, and Nashville.

As recently as 1970, the rhythm of Beale Street was surprisingly like that of the 1930s. During the weekday, it was a world of traders, markets, stores, and skyscraper professional offices, plus a block or two of decaying buildings devoted to loan offices, pawn shops, cafes, and pool halls. Today tourists, both black and white, come to the Street to visit memorials to two black leaders. One is the W. C. Handy Park, now a national historical monument, named for the father of the "blues" and dominated by his statue with his coronet in hand. The other is the Martin Luther King memorial at the Lorraine Hotel where he was shot on April 4, 1968. A plaque carries this quotation from Genesis: "They said one to another 'Behold here comes the dreamer. Let us slay him and we shall see what becomes of his dreams.' "

In 1970 they were still bargaining for clothing and groceries, patronizing conjure doctors, seeing films at the Daisy Theater, visiting beer gardens, or playing games of chance. When the sun went down, the sidewalks were still crowded with night people enjoying night-time activities: gambling, drinking, and the many traditional forms of "low-life." This longstanding pattern was modified in 1972, when the two remaining blocks of decaying buildings gave way to an urban renewal project. The street that gave the world the "blues" has become a mere plastic imitation of itself. Urban renewal has designated the area as a so-called "blue-light" district with restaurants, night clubs, and high-rise buildings. On Beale Street, the old patterns of temporal ecology are gone, and new ones have taken their place.

"Beale Street" and Temporal Ecology

The field work for "Beale Street" was completed over forty years ago. Although it is substantively a study of a black neighborhood in Memphis, it is also a study of ecological succession on Beale Street in Memphis in the mid 1930s. The focus is on the temporal, as opposed to the spatial, aspect of human ecology. The importance of the temporal components of human ecology was noted quite early by, for example, Anderson (1928), Lynd (1929), Engel–Frisch (1943), and Hawley (1950); and, more recently, by Melbin (1978) and Cohen and Felson (1979). However, these aspects have been neglected relative to the spatial aspects of human ecology.

Hawley (1950) in his now classic *Human Ecology: A Theory of Community Structure*[1] emphasized that human activities are performed over both space and time. In fact, in addition to two chapters concerning the spatial aspects of human ecology, he devotes an entire chapter to the "temporal aspects of ecological organization." In that chapter, he identifies three important temporal components of community structure: rhythm, tempo, and timing. Rhythm refers to the regular temporal patterns in which events occur, tempo to the number of events per unit of time, and timing to the temporal coordination among different activities. For example, *rhythm* is apparent in the diurnal cycle of activities such as travel to and from work; *tempo* may be seen in such diverse social phenomena as the number of criminal violations per day in a given area or the number of cars passing a specific location per hour; and *timing* is evident in the scheduling of extra workers in restaurants to coincide with traditional lunch hours or the opening and closing of extra lanes on a highway to meet the demands of a sporting or cultural event.

However, these concepts, and the interest of sociologists in the temporal aspects of ecology, predate Hawley's book. Gladys Engel–Frisch, in a paper entitled "Some Neglected Temporal Aspects of Human Ecology," details some of this history:

> The concepts of *rhythm, timing,* and *tempo,* as conceived by the late R. D. McKenzie, were first met by the present writer in a lecture given by Dr. Amos H. Hawley in a course in human ecology at the University of Michigan, in the first semester of the school year 1939–40. A year later, at the University of Washington, Dr. Jesse F. Steiner presented a paper to a seminar in human ecology of which the writer was a member. This paper discussed more minutely the ramifications of the temporal aspects of human ecology. Being intrigued by the concepts, the writer began some sporadic writing on the subject during the spring of 1941, and attempted a statistical study of tempo in Atlantic City, New Jersey, with some advice from both Dr. Steiner and Dr. Hawley.[2]

The senior author of this paper was also trained in the McKenzie–Steiner tradition at the Univer-

---

[1] Amos H. Hawley, *Human Ecology: A Theory of Community Structure* (New York: Ronald Press, 1950), p. 289.

[2] *Social Forces,* 22, 1 (October 1943), p. 44.

sity of Washington, and wrote up his 1935 observations of Beale Street at the prompting of his major professor, Jesse F. Steiner. This author's first instructor to suggest charting events by time periods was W. Lloyd Warner in a social anthropology seminar at Harvard in 1932–1933. The writer was working Saturdays from 6 a.m. to 10 p.m. in the produce section of a Boston market. Warner proposed recording the activities of customers by quarter-hour units, listing three common areas of change: change in either age, sex, or social class of the shoppers at different time periods; variation in the prices asked for identical units of food related to the time of day and the age, sex, or social class of the customers; and changes in the types and volume of products sold at different times of the day. Although Warner's major interest appeared to be in finding a social-class hierarchy of the market patrons, he recognized the importance of temporal changes during the sixteen-hour work shift.

The writer's first awareness of the temporal aspects of human ecology came even earlier, in 1929, with the reading of Robert and Helen Lynd's *Middletown*. Of particular importance were their comments on the periodic dislocations of the rhythms of living when three or four thousand heads of families were moved from day-time work to the night shift. The shift to night work was followed by a change in "the normal relations between husband and wife, the children's customary noisy play around the home, family leisure-time activities, lodge life, jury duty, civic interest and other concerns."[3]

That these concepts can, and do, have a role in modern social theory, and in the understanding of social behavior is evident in two recent articles. The first, "Night as Frontier" by Murray Melbin, is based upon the hypothesis that "night has become a new frontier . . . [and] that time, like space, can be occupied and is so treated by humans."[4] Melbin goes on to described rhythm, tempo, and timing in a new frontier (Boston at night) and to compare this frontier with the spatial frontier of the "Old West." The second essay, "Social Change and Crime Rate Trends: A Routine Activity Approach," by Cohen and Felson also emphasizes rhythm, tempo, and timing. As a starting point, they ". . . take criminal inclination as given and examine the manner in which the spatio-temporal organization of social activities help to translate . . . criminal inclination into action."[5] This framework is used to explain the circumstances and location of criminal offenses, as well as crime rate trends in the United States from 1947 to 1974. The concepts used in temporal ecology remain a rich source of insights for social scientists.

[3] Robert and Helen Lynd, *Middletown* (New York: Harcourt, Brace & Co., 1929), pp. 145–46.
[4] *American Sociological Review,* 43, 1 (February 1978), p. 3.
[5] *American Sociological review,* 44, 4 (August 1979), p. 590.

The Contrived Community: 1970–1980*

Gerald D. Suttles

To a large extent, our understanding of American cities has been guided by the image of a multiplicity of home builders, residents, businessmen, and public employees going about their lives independently and only crescively, unintentionally creating a local community. This is the way Park, Burgess, and Wirth thought of the growth of Chicago in the 1920s and 1930s; and it was a reasonable perspective, for Chicago, like all American cities, was built on virgin land, free from the obstruction of prior installation. By 1970, however, it was clear that the redevelopment of our older communities would not exactly retrace their previous development.

This was especially apparent in the Douglas Park Community area on Chicago's Near South Side. Of all the city's community areas, this one had experienced the most clearance and redevelopment. Originally initiated by Stephen Douglas, who had sold off the land to smaller developers, it had included the city's first Gold Coast before Potter Palmer led the way to the Near North Side by relocating his home there. Downwind from the old stockyards, the area underwent a regular process of succession until the early 1920s, when it became the city's first black ghetto, widely known from the studies of St. Clair Drake and Horace Caton.[1] Until recently, the area was never known as the Douglas Park community, but it did have an illustrious place in Chicago's history, especially among blacks who sometimes referred to it as the "Mother Ward" and knew of its separate neighborhoods as the residence of several famous black writers and politicians.[2]

This memory of the area and the area's physical structure itself were practically erased shortly after the Second World War. Two local and powerful institutions, a hospital and a university, urged the clearance of the area and its redevelopment along startling new lines. At first the cleared land was used for low-income, high-density, public housing, but hopes for better public safety were quickly disappointed after three large developments were located at its margins (see Map 1). From that point on, the attempt was to encourage middle-class housing, mixed according to race and ethnicity. By 1970, seven such developments were completed or underway. Only a small part of the old slum was

---

*This article was written for this volume and has not been published previously. For an earlier study of this community, see Gerald D. Suttles, "The Contrived Community," in *The Social Construction of Communities*, pp. 82–107 (Chicago: University of Chicago Press, 1972).

[1] *Black Metropolis* (London: J. Cape, 1946). See also Allen H. Spear, *Black Chicago* (Chicago: University of Chicago Press, 1967).

[2] Harold F. Gosnell, *Negro Politicians* (Chicago: University of Chicago Press, 1935); James Q. Wilson, *Negro Politics* (Glencoe, Ill.: Free Press, 1960).

left and appropriately called the "Gap": a lapse in the encroaching wall of high-rises. Students who visited me in the area were so struck by the mixture of glass, concrete, and remaining vacant land that one of their appellations, "moonscape," drew general agreement. After living in the area from 1968 to 1970, I came to call it the "contrived community" to sharpen the contrast with what Park had called the "natural community."[3] This article retraces many of the reasons given for that characterization, but it is informed as well by the opportunity to revisit the area and to reexamine it ten years later. In this way, I hope not only to uncover some of the limitations of my original conceptualization, but to point up its relevance for the more widespread redevelopment which has become apparent in the last decade.

Changes in scale

The most noticeable alternation in the redevelopment of the Douglas Park Community was a

[3]Suttles, op cit. The brevity of this article may make it useful for the reader to consult the earlier publication.

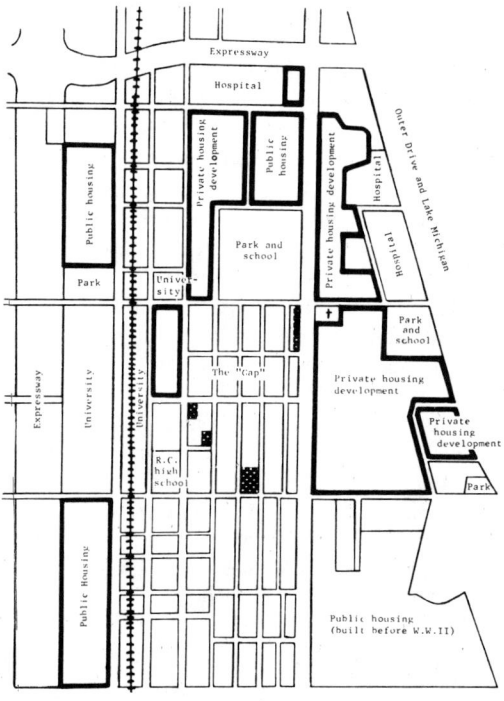

■ Areas developed since 1970

Map 1—Douglas Park Community Area.

change in scale. Developers were not simply faced with the most economical usage of cleared land, but especially concerned with what they could expect to happen to adjacent land. Their tendency, then, was to hang back until they could control a block of land large enough that the context of their development was unproblematic. Necessarily, such land purchases, from a city block to forty acres, eliminated almost all small developers, financiers, realtors, construction firms, or local politicans. Redevelopment had to be carried out by large and highly visible developers: the city government, the local hospitals and university, and some of the largest firms in the city or nation. Government was also involved in establishing standards for private builders, since the proposed developments had to be judged on the basis of quality and contribution to public welfare, rather than the size of bids alone. Thus, many of the private builders regarded their investments as a sort of public service on which they would make money but forego higher profits elsewhere.

One obvious consequence of this was that decisions made among so small a group of developers were no longer independent of one another. This is not to say that they were in collusion, for the scale of development and the stakes were so great that competition was intense. But developers, public or private, needed to keep a close eye on one another; and, as one of them made a decision, he might start a trend quite apart from the most advantageous economic usage of the land. Thus, the earlier development of public housing along portions of the lake front had tended to dissuade the location of high-cost housing on prime land nearby. Similarly, the construction of one restricted-income development on a single block in the area's interior has warned off some prospective builders interested in adjacent land.

Examples of this sort were among the primary reasons for a prevalent insistence on planning and coordination among the various parties affected by development. A comprehensive plan for the area was widely desired by developers, because it might protect them from further "down usage" in the community. Moreover, the continuation of government controls and favored rates of property exchange was seen as crucial to future development because it was the main assurance for the prospective value of properties. But developers were also ambivalent about the further expansion of governmental controls because they were not confident that such controls would be adhered to.

Indeed in 1973, a comprehensive plan for the area and the city ("Chicago 21" plan) was published, but subsequent instabilities in the Cook County Democratic Party have undermined the plan's credibility.

Among the residents themselves there was also a loud outcry for planning and coordination. School officials, parents, retailers, and builders all complained about the sudden influx of population without prior warning or adequate preparation for service. Such planning and coordination, however, presented not only practical, but moral, dilemmas. Coordinating the plans of different developers violated the rules of free enterprise and laid each developer open to frequently made charges of conspiracy and collusion. Even more suspect was the coordination between politicians, developers, realtors, federal government, and financial institutions. Collaboration among these organizations was seemingly without legitimate precedent in America, and, despite its growth over the past years (frequently at the insistence of the federal government), it became a cause for untold suspicion. This suspicion has not abated and by 1972 had grown to such proportions that the prior pattern of development in the area had almost come to a halt. In 1980 the physical plant of the community looks almost, though not quite, like it did in the early 1970s.

Cast of characters

This merging of large-scale governmental and private organizations in urban redevelopment changed the cast of characters, in both degree and kind, from that observed by Park and Burgess in the 1920s and 1930s. Political bodies, often at the insistence of private ones, had come forth with plans and had to bear the initial brunt of criticism for the dislocations that ensued. Civil rights groups residential aggregates, community organizations, local politicians, churchmen, and professional planners rose up and entered the controversy. The private developer entered later to find a hornet's nest already in swarm. With one hand, he had to assuage indignant citizens, while with the other he had to juggle governmental and financial contracts, realtors, chain stores, and an occasional researcher. No doubt some of these characters were present at earlier times and, to a limited extent, in contemporary suburban development on vacant land. Yet, there was a difference in the visibility of the public and private agencies called into redevelopment on the scale proposed for the Douglas Park Community Area. The merging of large governmental agencies and private firms gave credence to the idea that together they should be able to do something about the general quality of life throughout the community. In the Douglas Park Community Area, this often took the form of local community groups attempting to develop a grievance structure to redress wrongs now thought correctable because control seemed so centralized and monolithic. Tenents' groups, PTA's, and concerned parents—altogether 22 groups in 1970—took up a flock of issues and pressed them for consideration by developers or public officials: police harrassment, the treatment of children at school, eviction notices, welfare payment, parking, recreational services, etc.

Builders were often bewildered by the demands for public services made upon them (e.g., due process in police arrests), but just as frequently public officials were baffled by efforts to have them enforce what were felt to be the promises of local builders. Many of these efforts at lateral coordination were clandestine in 1970, and pursued as a kind of "favor," although a reasonable favor, because of the obscure boundaries between organizations. These kinds of demands have relented somewhat in recent years—in part, because the lines of authority have become routinized, but also because of the sharp decline of construction in the area and a lowering of goals of developers. Earlier, some developers had embraced important social goals. Several had taken "affirmative action" to attract white residents and assure integration. One had constructed a community center, including an integrated elementary school run by the public schools. Several developers included a local community center or exceptional amounts of shared space. Although three of the developments were initially integrated, only one remains so. The aspiration for an integrated school has passed out of existence, in large part because the public school system failed to enrich the program to meet the diversity of demands placed on it. The community centers and shared space are still there but have moved to a commercial form where they are available for a fee. The area is regarded as a successful, largely middle-class black community, but the high aspirations of the late 1960s have been subdued and, with that, the high expectations of the new residents. One developer expressed what seems to be a widespread view: "I'm glad we did it, but never again."

The population in the area has peaked at about 30,000, far short of the 80,000 once expected. This is not because redevelopment in the city has slackened, but because of two other trends. One is the inclination of builders of large-scale projects to move toward areas completely vacant, especially the old railyards to the north of the Douglas Park Community. The other tendency seems to be to scale down some from the ambitious projects of the 1960s and to move in the direction of a smaller number of units somewhat more dispersed across the city, although focused on the Lake. Three small developments, including about 400 units, have been recently erected in the Douglas Park Community itself. Much vacant land remains, and rumors of new development still abound, but with less urgency and plausibility than in 1970. Ironically, one result of this inactivity is that some residents of the "Gap" have begun to make personal investments in the old period-piece housing left among the cleared lots. The blacks are beginning to gentrify this part of the community and, although it occurs at a snail's pace, there is scarcely a block without fresh signs of reinvestment.

What is most notable, however, is not the slackening of citizen protest and demands in the Douglas Park Community itself, but the diffusion of similar demands across the entire city. Increasing federal, state, and municipal participation in redevelopment has elicited widespread, not just local, reaction to such issues as displacement, provision for low-income housing, the availability of shared space, and the need for new housing to be a general purpose solution to old problems: redlining, tenants' rights, police-community relations, school integration, and so on. Even those projects being undertaken on completely vacant land are not immune from such demands, although they often originate from an unexpected distance.

Within the Douglas Park Community, the unavailability of a high-quality integrated school is still an obstacle to a more family-oriented community. Yet, developers here and elsewhere have been surprised at the ease with which new buildings can be filled with people who make few demands upon the public school system. One estimate is that over 70 percent of the dwelling units in Chicago's lakeside communities currently have no school-age children. Even those areas not undergoing obvious redevelopment seem to be moving toward low family status.

Heritage of distrust and pessimism

In 1970, a heritage of pessimism and distrust had left many residents of the Douglas Park Community with a widespread credibility gap between themselves and the social claims of developers and city officials. Many had a past history of displacement, others a dependency on public welfare, and most of them a thorough grounding in "river ward politics." This led, in some instances, to vague threats or rumors of violence and, in others, to a vociferous and exclamatory style of negotiation. Security precautions were, and still are, formidable, with security guards, wire fences, peepholes, TV monitors, buzzers, and watchdogs in wide usage. Police reports, however, now indicate a slight decline in crime, especially those forms of crime that might be interpreted as conflict between the low-income residents of public housing and the more affluent ones in the private developments. Now that most developments are overwhelmingly black in composition, informants report that there is more of a tendency to see street crime as individual predation rather than class or race conflict. In this respect, the perception of violence has moved more nearly toward the American norm.

This movement toward racial homogeneity also seems to have diminished the tendency of middle-class residents, especially some whites, to sponsor the outrage of nearby, low-income blacks. Apparently, the middle-class blacks in the area feel less necessity to dramatize their militancy and greater confidence in speaking up for their separate interests and security. Still, there is a broad feeling of moral indignation about the plight of local low-income blacks and a deep cynicism about the motives of developers, city politicians, and "well-meaning" private organizations. The lack of construction in the area, however, has not given much exercise to such a heritage of pessimism and distrust.

Again, what is noteworthy is not so much the decline of militancy in the Douglas Park Community itself, as a general scaling up of the adversary relationship between community groups and those involved in redevelopment. Recent announcements and hearings on new residential developments have attracted "commentary" and sometimes blistering criticism from across the entire city, rather than from the affected neighborhoods only. This follows a general pattern of aggregation among community groups that was anticipated by Albert Hunter in his 1970 study.[4] As some of the main features of the contrived community, especially its organizational scale and the merging of

[4] Albert D. Hunter, *Symbolic Communities* (Chicago: University of Chicago Press, 1974).

private and public sponsorship, have become city-wide, the reaction to redevelopment seems to have moved toward a similar scale.

New metes and bounds

Aside from the "Gap," the new metes and bounds of the Douglas Park Community are the development-island, outlined by some of the major arteries of the old grid pattern but enclosing a community that aims for a complete complement of facilities for weekday life. This centripetal pattern of drawing back into one's own development continues. Shopping areas are nested in the interior of most of the private developments and the lack of shopping in the public housing creates problems in inter-community relations. The internal streets which dead end and twist about have acquired more lawn space and the trees have grown to surprising heights to relieve the ground level monotony of concrete, glass, and brick. But it is still surprising how little "green space" seems to attract people unless some regular activity can be imposed upon it.

Residents remain rather closely confined to their respective developments, and it is the development that gives them their residential identity. This has made it difficult for specialty shops to survive in any of the shopping centers, and the lack of additional development after 1972 has worked a hardship on some merchants. One shopping center has moved sharply downward in quality and become a greater source of complaint than before. The fragmentation of the area persists and the developments are sharply delineated by income and reputation. One integrated development and three small areas of continuing white residency stand out sharply in the cognitive map of informants. This ecological legibility remains so available that, even where consensus might exist, its discovery is precluded by the presumption that differences outweigh similarities.

Community organization

Community organization in the Douglas Park Community remains the self-conscious, deliberate association with explicit aims, rather than the natural group with diffuse ties. The tenants' association, which aims primarily to represent its demands and desires to "management," was the most common associational form in 1970 and is more exclusively so today. There seem to be even fewer special interest groups, something surprising at least to me, given the longer duration of residency. No local group of residents seems to be organized around broad social goals for the entire community. The lack of much development in the area or new, special programs aimed at community development seems to have enervated further community organization. Of course, there are still lots of complaints: about the schools, the police department, snow removal, political insensitivity, etc. But the community seems to have lost organizational momentum in doing anything about them. One suspects that the contrived community needs some form of external, perhaps professional, stimulation to arouse it, organizationally.

One potential source for this external stimulation is a new area-wide planning board which includes representatives from some of the area's major institutions and businesses. So far, this organization has worked primarily to promote further economic development of the area, rather than community betterment among the residents. If successful, this planning board may at least inadvertently arouse the residents to think in terms of broader issues.

Another potential source of stimulation to local organization is the volatility of local politics. After one upset, the machine candidates have continued to win local offices, but narrowly, and largely because of low turnout. Currently, one incumbent committeeman is being strongly challenged, and this may excite at least temporary political activism. Nonetheless, traditional precinct politics, or the more recent forms of community activism, are poorly developed. Where realtors and formal city bureaucracies play such a large role in one's residential life, elected political leaders simply are not perceived as the people that count.

Conclusion

Much of the distinctiveness of the Douglas Park Community Area has been retained over the decade. It is still the contrived community, but there are marginal changes, some of which might have been anticipated, some more surprising. The scale and magnitude of development in the area has dropped off dramatically, largely because no adequate formula has been found to reduce the moral dilemmas associated with the coordination of large private and public organizations and the suspicions this provokes. With the decline in new construction and the lowering of goals by developers and realtors, there is less activism and a closer focus on tenant-management relations. Surprisingly, after a decade, small, more nearly primary, groups seem less evident than earlier. Apparently, external stimulation is needed for these groups, as well. It is more nearly a contrived community than I expected.

While the local controversies that focused themselves in the area have abated, similar controversies have escalated to become city-wide, scaling up the adversary relationship between community groups and developers. Increasingly, this aspect of the contrived community has become part of what Park and Burgess once called the natural community.

Children of the Ghetto: The Evolution of an Urban Neighborhood*

Laurence S. Rosen

Introduction

Sociological interest in the nation's urban neighborhoods has taken several forms and has been couched in divergent theoretical perspectives over the past sixty years (see Burgess, 1925; Zorbaugh, 1926, 1929; Green, 1932; Davie, 1938; Janowitz, 1952; Greer, 1962; Hillery, 1968; Suttles, 1968). From these divergent perspectives a common view of the contemporary urban neighborhood as a fundamental ecological building block of metropolitan society has been drawn.

In recent years this consensus has come under attack as it has become increasingly difficult to clearly identify viable, thriving neighborhoods (Greer and Greer, 1974:4). Many of the immigrant ethnic neighborhoods of years past have disappeared, and many of the children of these ghettos have discretely established less visible enclaves as they became dispersed throughout the suburbs. Moreover, the neighborhoods which still survive in the public eye are often portrayed through the news media as deteriorated slums or as reactionary ethnic backwaters, all of which are largely outside of the mainstream of contemporary metropolitan life.

In fact, some of the old neighborhoods have survived, and new neighborhoods have been established. Many of these neighborhoods have thrived over the years. These communities may seem to be less visible as their residents have become more acculturated and have established social forms and structures which are more consonant with the dominant society. This paper focuses on the evolution of one such neighborhood in New York City from the time of its establishment in 1952 until the present.

Early years: The pioneers

The Mitchell–Linden neighborhood was established in 1952 in the northern part of Queens, one of the five Boroughs which comprise the City of New York. It was built in response to pent-up demands for housing created through the depression, World War II, and the post-war baby boom. By the end of 1952, the neighborhood's first fourteen elevator-equipped buildings, with eighty-four apartments each, were fully occupied.

These first apartments (Mitchell Gardens) represented a relatively new type of housing in New York City. Unlike most multi-unit housing built

*This paper is based on a master's thesis entitled "Mitchell–Linden: A Communal Neighborhood" (The Pennsylvania State University, 1971), and has not been published previously. The committee was chaired by George Theodorson.

at that time, these buildings were located in a planned environment which included play areas for children, sitting areas for adults, parking areas adjacent to the buildings, and landscaped green spaces between the buildings. More importantly, these apartments were cooperatively owned. Built under the rules of Section 213 of the Housing Act of 1949, these apartments are owned by corporations composed of the families who occupy them. Later additions—Linden Hill (1953–55, 672 apartments), Embassy Arms (1957–58, 144 apartments), and Linden Towers (1958–60, 884 apartments)—were built along much the same lines and they, too, are cooperatively owned by their residents.

This neighborhood, like many others of its time, was built on a swampy site in what was then a relatively sparsely populated outer area of the city. Most of the more than 1,100 families who moved there in the early 1950s had young children, and many of these families had been living with parents or in cramped apartments in the older sections of Brooklyn and the Bronx. Quite a few were veterans, since veterans were given preference in obtaining apartments; some were idealists drawn by the quasi-socialist prospect that cooperative ownership offered them; most, however, regardless of ideology or previous military service, were second generation Americans attempting to enter the middle-class mainstream of American life in the most suburban-like setting they could afford or which they were willing to try.

Approximately 60 to 75 percent of these first families were Jewish, primarily of eastern European origin. Although quite a few Jews were attracted to the traditional and newly-built suburban communities of Long Island, Westchester, and New Jersey during the early 1950s, it is important to note that Jews were not welcome in all of these communities, and many Jews could not afford to live there either. Many young Jewish families of that period were just emerging from the ghettos and first achieving admittance to a wide variety of white-collar and professional occupations. Many could initially afford little more than the nominal cost of buying into the cooperative corporations (which entitled them to apartments), and many were not willing to isolate themselves any further from relatives and friends in the "old neighborhood" by moving to the more distant and largely unknown suburbs. In effect, many of these families were neither ready nor willing to take the great leap towards assimilation that dispersal throughout the suburbs would have fostered. Mitchell Gardens, instead, offered them an opportunity for acculturation without assimilation in a relatively close-by and not totally unfamiliar environment.

Isolation

Isolation was the neighborhood's premiere feature during its formative years, and isolation—along with cooperative ownership and the neighborhood's ethnic plurality—stamped the area's evolving character. Initially, physical and social isolation prompted an inward orientation and interdependency somewhat reminiscent of the pioneer communities of nineteenth century America (see Smith, 1968).

The lack of easy access to shopping, work, school, and entertainment was stressful for many of the neighborhood's earliest residents. Unlike the contemporary suburban household, two-car families were not ubiquitous, and public transportation in this less built-up part of the city was severely limited. Moreover, television was still a novel form of entertainment in 1952, and even telephones were in short supply due to the unexpected demand created by an expanding populace. One resident remembers "when there was a phone booth out in the street and in the afternoons housewives would line up to call their husbands in order to tell them to bring a loaf of bread home with them."

At first, an informal cooperative spirit blossomed. Car pools were formed for shopping and entertainment, and there was extensive neighboring. Some of the men formed on-going Friday night poker games, while many of the women regularly met in the afternoons to play mah-jong. Those who had television sets found they did an inordinate amount of entertaining. Indeed, the proximity of the apartments to one another encouraged and sustained a great deal of informal visiting among adults and children. This was so common that many residents never locked their doors except when they were ready for bed.

The physical structure of the buildings and their facilities also served to ameliorate some of the problems of isolation. The playgrounds and sitting areas adjacent to the buildings were designed to encourage children's play and socializing among housewives. During warm-weather evenings, whole families congregated in the "backyards" with their neighbors until 10 or 11 p.m. The laundry rooms located in each building's basement were also important centers for informal contacts. They served as important arenas for gossip, and bulletin boards were installed there to hold notices of upcoming social events, items for sale, and memoranda from each building's management.

Of even greater importance, however, were the recreation rooms located in the basement of each building. In response to their isolation and their need for indoor as well as outdoor meeting places and social centers, each building's residents joined together to paint, lay flooring, and brighten up these jointly-owned rooms. The largest recreation rooms became the sites for monthly "building meetings," as well as a wide variety of other activities. In several of the buildings, they were used for after-school activity centers, for scout meetings, and, occasionally, for showing children's films on Saturday afternoons. Evening activities for adults, including dancing, card parties, and even an occasional cultural program, were held in these rooms. These rooms could also be rented by residents for various family celebrations that could not be accommodated in an apartment.

In addition to its physical isolation, the neighborhood was socially isolated at this time. This isolation took several forms: it was found in terms of the class, ethnic, and cultural identity of the neighborhood's residents; their religious identification and practices; and their residence type and its mode of ownership.

Although physically distinct, Mitchell Gardens was built within an established and stable area of single-family dwellings occupied by solidly middle-class families. Few of the surrounding area's residents were Jews. The fact that Mitchell Gardens was an apartment complex further isolated the neighborhood from its home-owning neighbors. Common interests which may have been derived from the daily responsibilities of home ownership simply did not exist in Mitchell Gardens. Even the basic interest in protecting and enhancing property values, which has often drawn diverse neighborhood residents together, was missing due to the unique cooperative ownership mechanism operating in Mitchell Gardens. Similarly, nearby home owners could little appreciate the collective responsibilities of cooperative apartment ownership and management.

By definition, cooperative ownership requires the establishment of formal mechanisms for self-government. Formal corporate structures were immediately established to accomplish this, while residents of each building also regularly conducted "house meetings" at which they gathered to discuss problems and air grievances. During the early years, these meetings were always well attended. As one resident put it, "there was a genuine fraternal atmosphere. Everyone wanted to do everything."

Everything, of course, could not be done within the corporate structure. Furthermore, because of its newness, the neighborhood had neither political identity nor recognition from the city administration or local politicians. In response, residents formed the Mitchell Gardens Civic Association in 1953 to deal with neighborhood-wide issues and to represent the neighborhood's interests in dealings with local government. Membership was limited to residents of the neighborhood and most families joined during the early years, thus providing the Civic Association with a clear mandate to work on the neighborhood's behalf. That all of the Association's initial goals, particularly the establishment of an elementary school within the neighborhood, were achieved by the end of the 1950s provides testimony to the Association's success. In addition, the Civic Association strove to establish neighborhood-wide integration and indentity by establishing its own monthly newspaper, the *Mitchell Gardens News,* and by encouraging the creation of other neighborhood organizations. In the first edition of the *Mitchell Gardens News* (March 1953: 2), for example, the editors stated that

> . . . we hope that other groups will take the plunge. There is need for more social and other groups to set up shop in our community. We are thinking particularly of such groups as the Legion, the VFW, the JWV and other veteran outfits, as well as some of the religious and civic organizations that are usually found in communities like ours.

The middle years: Growth and change

During the following fifteen years substantial changes occurred as the neighborhood passed through a transitional period of growth, physical expansion, and institutional diversification. What emerged was a socially distinct and culturally identifiable community which continues to reflect, as well as shape, the daily lives of its residents.

Although this transition would have undoubtedly occurred anyway, its initial impetus was the neighborhood's expansion. Construction of Linden Hill, Embassy Arms, and Linden Towers quickly added 1,700 new apartments to the neighborhood. Each of the new complexes was built quite similar to Mitchell Gardens, and the occupants of these buildings were similar to most of the neighborhood residents who had preceded them. The new families were mainly young and of modest size (approximately two children per household); they, too, were escaping the old and crowded neighborhoods of Brooklyn, the Bronx, and Manhattan's Lower East Side; most of them were newly middle class; and they were overwhelmingly Jewish.

Institutional diversity and neighborhood boundaries

As the neighborhood's population grew and became more ethnically homogeneous, its institutional and organizational structure became more diverse. In time the neighborhood gained many of the commercial, educational, religious, and governing structures that are typical of residential communities. In 1954, as the result of intensive lobbying by the Civic Association, a primary school was built on land adjacent to Mitchell Gardens. The student body consisted almost exclusively of children living in the neighborhood, and the school was viewed as the neighborhood's own. At about the same time, a row of stores was built alongside the neighborhood to serve the residents' daily shopping needs. In addition to a supermarket, barber, pharmacy, and a few other standard services, several businesses catered specifically to the neighborhood's Jewish majority. A kosher butcher and a delicatessen provided meats and sausages, another merchant offered a variety of traditional Jewish delicacies and dairy products, and a Jewish-style bakery provided the traditional breads and familiar pastries—all made without animal fats and other prohibited products. The neighborhood even had an old-fashioned New York City "candy store" and soda fountain which offered the egg-creams and Yiddish-language newspapers which had been daily staples in the "old neighborhood." Later additions to the neighborhood's commercial facilities included several banks, a post office, a library, and a bagel bakery.

Two Jewish congregations were established during this transitional period. The Garden Jewish Center was founded in 1953. After several years of conducting religious services and classes in rented basement quarters, the congregation built a synagogue in 1958 to serve its 850 member families. A second, more orthodox, congregation was founded in 1960 by several residents of Linden Towers. This congregation also built a synagogue which replaced rented facilities in basement recreation rooms.

By 1965, construction of residential facilities and all adjacent commercial, religious, and recreational facilities which form the neighborhood had been completed, and the neighborhood was no longer physically isolated. Over the years the neighborhood had expanded outward until it had reached natural boundaries or other land uses; other adjacent parcels had been filled in by the construction of row houses and duplexes. But while physical isolation had been eliminated, social demarcations remained; and the neighborhood's boundaries were as clearly delineated as ever. By 1965, the neighborhood was a classic "natural area" characterized ". . . both by a physical individuality and by the cultural characteristics of the people who live in it" (Zorbaugh, 1926:192). Informants within the neighborhood and in surrounding areas have confirmed that the neighborhood's borders delineate the limits of informal interaction and relationships, as well as the generally accepted service area, for most of the neighborhood's organizations, and even for many of its merchants.

While differing land uses and housing types visibly mark the neighborhood's boundaries, they also mark the settlement areas of groups which differ by ethnicity, race, and social class from those who reside in the neighborhood. To the west, the neighborhood is bounded by an eight-lane expressway, a pocket of old industrial buildings, and a small area of deteriorated housing occupied mainly by impoverished blacks and Puerto Ricans. To the south, the neighborhood is bounded by a public athletic field. Although located no more than fifty feet from some of the neighborhood's residences, this field is viewed as alien territory which is avoided except for occasional, late-night beer guzzling by neighborhood adolescents.

The neighborhood's southern boundary also includes a swim club built by neighborhood residents, the public school, and a paved recreation area adjacent to the school building. From here the boundary moves north along the neighborhood's major thoroughfare to include all of the commercial facilities on the west side of the street and exclude several single-family dwellings (mainly physician's and dentists' offices) on the eastern side. The boundary meanders further northward to surround the Linden Hill complex. The northern boundary separates Mitchell Gardens from Whitestone Terrace, an amalgam of single-family and two-family homes built during the 1950s and 1960s. Also, the neighborhood's two synagogues—one located a block north of the neighborhood and the other two blocks south—are functionally part of the neighborhood, although physically separate from it.

Cultural mobilization

In 1952, informants estimate that up to three-fourths of the neighborhood's residents were Jewish. By 1965, the best estimates of knowledgeable informants indicate that Jews constituted approximately 95 percent of the population. With this growth, a pervasive Jewish character became well recognized both as the central cultural focus within the neighborhood and its main identifying feature to those outside; by the mid-1960s, the neighborhood was frequently referred to as "the ghetto."

Although definitive understanding of why this particular location initially attracted a large Jewish population may never be achieved, it is clear that a series of connected events and processes which occurred after the neighborhood's "pioneer" phase directly contributed to establishing its particular cultural identity and its growing homogeneity. These may be described as *cultural mobilization*. That is, acculturation promoted the establishment of new, or at least modified, cultural forms which were highly adapted to the neighborhood's sociocultural environment but which were based strongly on the familiar traditions of the previous generation's ghetto experiences in Europe and America. Mobilization of selected traditional cultural patterns within this environment produced a unique Jewish-American neighborhood in which religious identity and cultural affiliation are of paramount importance.

Barely six months after the neighborhood's establishment, for example, religious services for the Jewish high holidays of *Rosh Hashonah* and *Yom Kippur* were held in a makeshift synagogue in one of the buildings; almost all of the neighborhood's Jewish residents attended. Some months later a Jewish community center was established. By the fall of 1953, a rabbi had been hired, high holiday services were held again, fund raising for a synagogue building commenced, and the congregation's monthly newsletter began publication. Six hundred Mitchell Gardens families (more than half of the neighborhood's total) joined the congregation.

Several factors—including the neighborhood's isolation and the changing religious orientation of its residents—led to this crucial series of events. Physical isolation was a particular problem for the neighborhood's religious Jews, since travel on the sabbath, even travel to the synagogue, is prohibited. The two nearest synagogues, located several miles away, not only were too far away to walk to, by they offered the unfamiliar and even alien liturgy of the Jewish Reform and Conservative movements.

The Reform synagogue had been founded more than thirty years earlier and was well established. Its emphasis on English rather than Hebrew, the absence of many of the familiar religious traditions, and its "high class" and assimilated ways did not suit their tastes. The other nearby congregation was part of the Jewish Conservative movement (see Sklare, 1972), which has striven to modify many traditional practices without wholly abandoning them. This acculturated, middle-ground, congregation would have been suitable for many of the neighborhood's less religious but tradition-oriented families were it not so distant. Since synagogues tend to be neighborhood institutions, the solution was to establish a new Mitchell Gardens congregation.

This was not easily done, as competing factions strove to dominate the nascent organization. The smaller faction favored the establishment of an orthodox congregation focused mainly on strict religious observance and daily adherence to the Jewish code of law (see Poll, 1969). The majority favored establishment of a conservative synagogue. Unlike the male-dominated orthodox tradition, the conservative movement is more family-oriented (Landes and Zborowski, 1950: 449–50) and has attempted to establish identifiably Jewish institutions which cater to many of the recreational and cultural needs of Jewish-American families.[1] Conservative synagogues emphasize Jewish identity through youth groups, clubs, theatricals, and other activities which are built around Jewish cultural themes, rather than through seclusion from or conflict with the secular society. These activities, along with more familiar religious and educational practices, have been the prime attractions of conservative congregations for Jewish families in the United States.

Since neither faction was strong enough to establish its own congregation, and since there was considerable sentiment within the neighborhood to prevent any divisive action, a compromise was reached. The Jewish Center was officially established as a conservative congregation which required that its religious leader be a formally recognized orthodox rabbi.

---

[1]The Young Israel movement among orthodox Jews has, in recent years, been relatively successful in maintaining faithful adherence to traditional orthodox practices, while incorporating some of the additional social and cultural interests and activities that many Jews today expect from synagogue membership.

Once established, the Garden Jewish Center became the central focus of neighborhood life. This was particularly true during the "middle years," as the mass of the neighborhood's male baby-boom generation attended Hebrew School in preparation for Bar Mitzvah at age thirteen. The Jewish Center's founding firmly established the Jewish identity of the neighborhood and served as a beacon attracting other Jewish families to the area. Yiddish conversations on the streets and familiar Jewish-style commercial facilities helped to further solidify the neighborhood's identity. Moreover, when normal turnover made apartments available and construction created new apartments in Linden Hill, Linden Towers, and Embassy Arms, neighborhood residents often encouraged their friends and relatives to join them there.

These same factors also served to discourage the concentration of non-Jews in the neighborhood. Several non-Jewish residents moved out after the first few years, as New York's ethnic and religious minorities continued to sort themselves out during the great exodus from the inner city in the decade following World War II (Handlin, 1959:67–69). The neighborhood's non-Jews were of diverse ethnic and religious affiliation, and any potentially unifying mechanism was undermined as families were siphoned off to participate in various denominational churches and other organizations located nearby. For many, too, Mitchell Gardens was merely a way-station to suburban homes and gardens, according to several informants. Not surprisingly, even the rules established by the cooperative corporations to allocate newly available apartments were employed to discourage any dilution of the neighborhood's Jewish identity. Occupancy rules established in Mitchell Gardens and the other cooperatives, for example, limited one-bedroom apartments to families with no more than one child; two- and three-bedroom apartments to families with no more than two and three children, respectively. Large families, which were not common among New York's Jews (Horowitz and Kaplan, 1959), were thus excluded as potential residents. Furthermore, although families which grew after they had occupied apartments were not required to leave, the limited size of the apartments did not favor their continued residence. Catholic families, as well as others who had large numbers of children, frequently left the neighborhood as soon as they could find more spacious quarters.

Formal mechanisms controlled entry into the neighborhood, and these served to limit the volume of non-Jewish migration, even as the desire to move into the neighborhood was sustained by a continuing housing shortage in New York and by the attractive financial aspects of cooperative ownership. Those wishing entry into the neighborhood were placed on waiting lists on a first-come, first-served basis. This rule was modified, however, by the establishment of two subsidiary lists: an "insiders" list and an "outsiders" list.

The insiders list included those already living within the complex who wished to obtain a different apartment. Residents were also allowed to place the names of close relatives on this list. The outsiders list contained the names of anyone else who wanted an apartment there. Both lists were established for each apartment type (i.e., one, two, or three bedrooms) available within each cooperative. When an apartment became available, it was first offered to whomever headed the insiders list for that size apartment. If the apartment was refused, it was subsequently offered twice more, usually at one-month intervals. If it was refused all three times, then the apartment would be offered to whomever was at the head of the outsiders list. As this rarely happened, the next time an apartment became available it, too, would be offered to the first person on the insiders list. In effect, this system was highly biased towards those already living in the neighborhood and towards their relatives, most of whom were Jewish.

Two further aspects of this procedure also helped to maintain the neighborhood's cultural identity. Residents of the neighborhood who had placed their relatives' names on one insiders list (say, for a three-bedroom apartment) quickly learned to place those names on each insiders list with the intention that they would accept any available apartment just to get into the neighborhood. Once they had moved in, it was assumed that there would only be a relatively short wait—this time as true "insiders"—until something more suitable became available. Neighborhood residents also systematically abused the rules, according to several informants. Friends or more distant relatives who wished to live in the neighborhood were fraudulently placed on the insiders list of several cooperatives simultaneously under the guise that they were immediate relatives of residents in each of those cooperatives. Name differences were explained as the result of marriages or relations on the female side of the resident's family.

Manipulation of this system and other manifestations of the process of cultural mobilization resulted in a neighborhood which, externally, is most obviously characterized by its "Jewishness." Numerous Jewish fraternal, civic, and charitable organizations—including more than twenty Jewish women's groups—operate within the neighborhood. It is important to recognize, however, that the "Jewishness" which permeates the neighborhood is neither a traditional nor a particularly religious way of life. Rather, it is a cultural pattern which has adapted traditional and religious norms to the secular patterns of the social and physical environment in which the neighborhood is located.

Much of the neighborhood's Jewish life revolves around the synagogues. While only 30 or 40 percent of the neighborhood's families maintain memberships in these organizations at any one time, most neighborhood families have been members at one time or another, their four or more years of membership usually coinciding with the years their male children are training for Bar Mitzvah. Similarly, although religious services are held each day, few residents attend these services or the Sabbath services held each Saturday. On the other hand, attendence at Jewish high holiday services is quite high, and even those who do not attend these most important religious services are still expected by local custom to make the appearance of observing the holidays. Schools are closed and adults do not go to work. In observance of the religious proscription against riding on holidays—a proscription which is ignored at all other times—cars remain parked and city busses pass by the empty bus stops. Even if they are not Jewish, most of the local merchants close their shops in observance of the holidays.

As these holidays fall during the early autumn, much of the day is spent outdoors where behavior may be publicly observed. Among the more easily observed norms are those concerning appropriate dress. Those who attend services, of course, dress attractively for the occasion. Those who do not are nonetheless expected to dress suitably. Teenagers and children, for example, usually wear suits and dresses which are reserved for just such occasions. Activities are restricted as well. Teenagers still congregate on street corners but the ubiquitous basketball games are absent, and cruising the neighborhood in cars is strongly discouraged. Children are not expected to refrain totally from their sidewalk games, but they are expected to be somewhat restrained. Indeed, older residents have been observed disciplining children they would otherwise ignore who were playing more enthusiastically than was appropriate to the occasion.

Religious education also touches almost every neighborhood family. For most families, synagogue membership is directly related to the requirement that males between the ages of nine and thirteen attend twice-weekly Hebrew classes in preparation for Bar Mitzvah. Participation has been so great over the years that it is futile to attempt to schedule any other activities for boys of this age during the times when Hebrew classes are in session.

Traditionally, the Bar Mitzvah ritual has marked attainment of adult status in the religious community and, thus, the beginning of adult responsibilities for and participation in synagogue activities. Today, in Mitchell-Linden and most other neighborhoods like it, Bar Mitzvah represents the end of religious commitment and a cessation of religious participation; it is the end goal, rather than the beginning of more intensive religious training. Nonetheless, Bar Mitzvah remains one of the most crucial cultural events for neighborhood families as it affirms their identification with the neighborhood's cultural orientation in an expected and acceptable manner. Neighborhood norms require that the ritual be held during Saturday morning Sabbath services (although religious law does not require this) and that it be celebrated later that day with a catered dinner to which large numbers of relatives, friends, and neighbors are invited. In return, the Bar Mitzvah boy is showered with gifts. Moreover, the quality and size of these events are carefully scrutinized by those attending, as these factors are often used to judge and impart status within the neighborhood to the family.

Neighborhood residents regularly participate in the synagogue's recreational and entertainment activities. Youth activities, for example, have been provided within the Garden Jewish Center over the years. The most successful of these were "lounges," sponsored each week by the YM-YWHA during the 1960s. These "lounges" consisted of games, dancing, folk-singing, and cultural programs for those between the ages of twelve and sixteen, and they regularly attracted between 100 and 200 youngsters to the Jewish Center. The "Y" also held occasional dances and field trips for teenagers and summer day camps for children age six through twelve. Together, these activities attracted almost all of the neighborhood's children at some time or other. Parents approved their children's participation not only because of the activities' Jewish sponsorship and occasional content, but also because they served to focus on interaction of neighborhood children with other Jews.

The synagogues sponsor men's and women's clubs which are extensively involved in neighborhood-wide entertainment and fund-raising activities. The Sisterhood of the Garden Jewish Center, for example, has approximately 250 to 300 members who meet each month and are entertained with a "program." Programs usually consist of a culturally oriented presentation such as an Israeli singer, a Chanukah party, or a talk by a Jewish author. The meetings always conclude with refreshments and minor fund-raising through a raffle or the sale of various small items of clothing and kitchen utensils.

The Sisterhood's main fund-raising activities reach out beyond the confines of its membership to embrace the entire neighborhood. Raffle tickets are sold throughout the neighborhood, as are tickets to the theatre parties they sponsor. Funds collected are used to supplement the congregation's budget and to make contributions to various pet charities. Men's club activities parallel those of the Sisterhood, although the congregation's Men's Club is smaller and less socially influential. Men's Club activities revolve almost exclusively around fund-raising through the sponsorship of several dinner-dances and other similar events which are open to the entire community.

Of all the functions neighborhood congregations attempt to serve, maintenance of their fiscal integrity appears to be the most important. Informants report that a congregation can no longer merely offer a sanctuary for prayer and study, as in the "old country," in order to survive. Today, they must offer a fully staffed Hebrew School, catering facilities, entertainment for adults, after-school activities for children, dances for teenagers, clubs for older people, and even athletic facilities. Thus, in order to pay for these services, fund-raising has become the central focus of many congregations.

Highly visible fund-raising techniques which reward the donor with prestige and which signify the donor's commitment to this key neighborhood institution are employed, for example, by the Garden Jewish Center. For particularly large amounts, the congregation will erect a plaque in honor of the donor who paid for the facility. Windows, prayer books, and other equipment are also frequently marked with the donor's name for all to see. Where that is not feasible, plaques listing the names of those contributing smaller amounts are erected. Thus, the entrance to an addition to the Jewish Center contains three plaques: (1) the "Founders" plaque lists those who contributed at least $1,000 to the construction of the facility; (2) "Master Builders" are those who contributed $500 to $1,000; and (3) "Builders" are those who contributed at least $300.

A more dynamic technique is the "status auction" which occurs when many of the neighborhood's residents are attending high holiday services. Calls for funds made during breaks in the day-long services generate vocal responses identifying the individual and the amount pledged. The public nature of these appeals invariably results in competition over who can and will contribute the most, thus gaining the greatest prestige among one's peers. These public appeals also serve to inform the community of the individual's ability and willingness to support local facilities and services.

The effectiveness of these techniques and other mechanisms of identification and integration which operate through the congregations ultimately depends on communications. This is assured through the Garden Jewish Center's monthly bulletin, the *Menorah*. This bulletin regularly contains the lists of donors to various funds and pledge drives, as well as announcements of upcoming events, both religious and secular, which will be held at the Jewish Center. It also lists Bar Mitzvahs, weddings, births, and deaths among members of the congregation, and it periodically reports on the progress of students in the Hebrew School. Unlike most organizational newsletters, however, the *Menorah* is not mailed or handed to the congregation's members. Instead, it is placed on the doorstep of each apartment within the neighborhood, regardless of whether the residents belong to the congregation or not and regardless of whether they are even Jewish or not. Thus, even those who have no other contact with this institution are regularly informed of their neighbors' and friends' activities within the community.

Recent years: Diversity and stability

The Mitchell-Linden neighborhood entered its most recent phase of development in the mid–1960s. In contrast to the preceding ten years, this period has been marked by institutional and structural maturity, along with greater diversity among the neighborhood's residents, their interests, organizations, and activities. Perhaps the most important consquence of the adaptive mechanism of cultural mobilization within the neighborhood has been the diversification of its population.

Jews drawn to the neighborhood during its "middle years" were an increasingly diverse lot, and their various interests and needs diminished the neighborhood's singular life-cycle orientation by diluting its preoccupation with young children. Both older and younger families, as well as childless couples, moved to the neighborhood. Several of the elderly people who moved there did so at their children's insistence that they leave older, changing areas for Mitchell–Linden's relative comfort and safety.

Those who moved in later were mainly middle class, as were the predecessors, but some of the vacancies were filled by those who earned their livings as skilled workers in the printing, clothing, and various other industries. They, too, were Jewish, but their interests and values did not always coincide with the white-collar majority that had preceded them. This growing diversity in the neighborhood prompted the establishment of even more specialized interest groups and voluntary associations and, in time, distinct factions concerned with a variety of local civic and political issues emerged from among these organizations. In addition, more highly structured patterns of social interaction, along with more rigid patterns of age-grading, developed among the residents as the neighborhood grew in size and complexity.

Dozens of voluntary associations have been established, including chapters of several American, Jewish, and Israeli charitable organizations, veterans organizations, service and civic groups, local self-help organizations (such as the community ambulance service), school associations, scouts, and political associations. Most residents are affiliated with one or a few of these associations, but neighborhood-wide levels of participation have tended to be low. Leadership and strength for most local associations are provided by small groups of interested and outspoken members who participate more regularly and intensively than their neighbors. Moreover, those who are intensive participants in any one association often generalize this behavior and become key participants in several similar or related organizations. This overlapping membership has resulted in the emergence of two distinct concentrations of power and moral authority within the neighborhood, and the distinction between them falls clearly within Merton's (1957:393) localite-cosmopolite formulation.

The neighborhood's liberal cosmopolites are concentrated about the Mitchell–Linden Civic Association. They include several of the neighborhood's original residents and others who have always viewed the neighborhood's establishment and its cooperative form of ownership as politically advanced forms of social action. They run the *Mitchell–Linden News,* and they tend to be aligned with the North Shore Democratic Club and the Free Synagogue of Flushing. The North Shore Club is the local arm of the liberal movement in New York's Democratic Party, and the Free Synagogue is the nearby area's leading Reform congregation. Over the years this faction has promoted the more universalistic perspective on a variety of local issues, including support for the civil-rights movement and concern over the discriminatory nature of the neighborhood's system for allocating apartments.

In contrast, localites represent the neighborhood's cultural "establishment." They are more politically and religiously conservative than the cosmopolites, and they are oriented primarily towards local ethnic and religious concerns. Localites are often associated with the "regular" Democratic Party organization, belong to the Jewish Center, and participate in various Jewish civic and charitable associations, such as the B'nai B'rith and Hadassah, which cosmopolites rarely join. Localites are considerably less ecumenical in their outlook and they tend to ignore issues of social or political importance which do not have strong and immediate implications for the neighborhood or their special concerns with Zionism, anti-Semitism, or other Jewish affairs.

Social relations

In contrast to the hyper-neighborliness and informality of the neighborhood's earliest years, more highly structured patterns of social interaction govern the informal relations of many neighborhood residents today. Among adults, socializing is less a function of propinquity, as in the past, than it is the result of couples sorting through and finding others in the neighborhood who are reasonably similar in age, lifestyle, social status, and interests. Similarly, there are dozens of tight-knit groups of neighborhood women who regularly meet to play cards or mah-jong. They include well-defined sets of participants who adhere to rigid schedules regarding where games will be played, who will host them, the refreshments to be served, and other similar concerns. Several of these groups have existed for years with the same members.

Age-grading is particularly important among neighborhood children and teenagers. For young children, play groups are usually composed exclusively of those residing within the same building who are of the same age and sex. Although these propinquitous patterns break down by the third or fourth grade as children gain more opportunities to socialize throughout the neighborhood, the age-graded structure of the schools and religious education continue to exert considerable influence on their social relations through high school.

On any night during the warmer months (and on weekends during the winter), groups ranging in size from 20 to almost 200 teenagers may be found "hanging around" the neighborhood. These groups are sexually mixed and exhibit a range in age from about twelve to eighteen years old.[2] Age is an important factor here, as these groups are composed of numerous small cliques which are sexually segregated and which usually vary in age by no more than two years. Neighborhood and ethnic identity is also maintained throughout these groups and cliques in that non-residents and non-Jews rarely participate.

Social patterns among the neighborhood's older residents are not as structured as those among younger people. Weather permitting, the neighborhood's older residents converge daily on many of the small sitting areas which dot the neighborhood, and they socialize (both in Yiddish and English) with those sitting nearby, as well as with passers-by. In addition, the neighborhood has two senior-citizens clubs which are quite popular and have waiting lists to join.

For the most part, the elderly are well integrated with the daily life of the neighborhood. However, some differences in religious observance among the elderly are evident, and they are particularly valuable in that daily participation in traditional religious services by a handful of older residents serve important symbolic functions for the entire neighborhood. Their participation provides an important symbolic link to the cultural and religious traditions which have been modified and might otherwise be forgotten by their children who established this new neighborhood. Moreover, this activity helps to maintain the religious legitimacy and importance of the neighborhood's core institution. Symbolically, the neighborhood's Jewish character is continuously reaffirmed through performance of these daily rituals on behalf of all those others in the neighborhood who cannot or will not perform them but who value them nonetheless.

---

[2] Eighteen is the legal driving age in New York City.

Conclusion

Identified by inwardly focused communalism, hyper-neighborliness, and a strong pioneering spirit during its earliest years, which was then followed by the mobilization of its ethnic and cultural resources, the Mitchell–Linden neighborhood entered a new and distinct phase of development during its third decade of existence. This phase of the neighborhood's ecological history provides the casual observer with the appearance of stability and continuity that mark it as a mature, "established" community. Beneath this surface of stability, however, the neighborhood may best be characterized by its social complexity and structural diversity. Indeed, during a span of twenty years, the Mitchell–Linden neighborhood evolved from an area solidified by its isolation and the life-cycle homogeneity of its residents—like Park Forest Village (Gans, 1951; Whyte, 1956) and hundreds of other post-war bedroom communities—to a more complex and diverse community organically integrated and ecologically identifiable by its interlocking network of families, institutions, social groups, and voluntary associations.

The neighborhood's original settlers were mostly young families, and many of them were Jewish. Although not completely pervasive, these similarities nonetheless served to focus attention and communal action towards widely shared needs within the neighborhood. Adaptation to the physical and cultural isolation they perceived was required by the neighborhood's young, middle-class, child-rearing families, and this entailed developing local institutions and cultural responses to a challenging environment. They responded by mobilizing the cultural resources they had inherited as children and brought with them to Mitchell–Linden as adults. In so doing, however, they made compromises with the larger, secular environment that resulted in the establishment of hybrid Jewish–American institutions and social patterns which remain identified with this neighborhood today.

References

Burgess, Ernest W. "The Growth of the City: An Introduction to a Research Project." In *The City,* edited by Robert E. Park, Ernest H. Burgess, and R.D. McKenzie. Chicago: University of Chicago Press, 1925.

Davie, Maurice R. "The Pattern of Urban Growth." In *Studies in the Science of Society,* edited by George

P. Murdock. New Haven: Yale University Press, 1938.

Gans, Herbert J. "Park Forest, Birth of a Jewish Community." *Commentary,* 11 (April 1951): 33–39.

Green, H. W. "Cultural Areas in the City of Cleveland," *American Journal of Sociology,* 38 (1932):356–67.

Greer, Scott. *The Emerging City.* New York: The Free Press, 1962.

Greer, Scott, and Greer, Anne Lennarson. "Introduction." In *Neighborhood and Ghetto: The Local Area in Large Scale Society,* edited by S. Greer and A. L. Greer. New York: Basic Books, 1974.

Handlin, Oscar. *The Newcomers.* Garden City, N.Y.: Anchor Books, 1959.

Hillery, George A., Jr. *Communal Organizations: A Study of Local Societies.* Chicago: University of Chicago Press, 1968.

Horowitz, C. Morris, and Kaplan, Lawrence. *The Jewish Population of the New York City Area, 1900–1955.* New York: B'nai B'rith, 1959.

Janowitz, Morris. *The Community Press in an Urban Setting.* New York: The Free Press of Glencoe, 1952.

Landes, Ruth, and Zborowski, Mark. "Hypotheses Concerning the Eastern European Jewish Family." *Psychiatry,* 12 (1950): 447–60.

Merton, Robert K. *Social Theory and Social Structure.* Glencoe, Ill.: The Free Press, 1957.

Mitchell Gardens News. Volumes I–III. Flushing, N.Y.: Mitchell Gardens Civic Association, 1953–1955.

Poll, Solomon. *The Chasidic Community of Williamsburg.* New York: Schocken Books, 1969.

Sklare, Marshall. *Conservative Judaism,* augmented edition. New York: Schocken Books, 1972.

Smith, Page. *As a City Upon a Hill: The Town in American History.* New York: Alfred A. Knopf, 1968.

Suttles, Gerald. *The Social Order of the Slum.* Chicago: University of Chicago Press, 1968.

Whyte, William H., Jr. *The Organization Man.* Garden City, N.Y.: Anchor Books, 1956.

Zorbaugh, Harvey W. "The Natural Areas of the City." *Publications of the American Sociological Society,* 20 (1926):188–97.

Zorbaugh, Harvey W. *The Gold Coast and the Slum.* Chicago: University of Chicago Press, 1929.

Changing Land-Use Patterns in SoHo: Residential Invasion of an Industrial Area*

James R. Hudson

Despite some methodological and theoretical reservations, urban sociologists in general have accepted Burgess's seminal formulation of the outward spread of commercial and industrial activities to replace residential uses in their path. Commercial and industrial activities make more intensive use of land and, therefore, can absorb the higher costs of these more desirable locations than can low-density single- or multiple-residential dwelling units. This general formulation still appears to be accurate for the rims of our urban areas. (Hawley, 1956). The outward migration of commercial and industrial activities continues to replace other uses even in suburban areas. Shopping malls become surrounded by service activities, while garden apartments, as well as high-rise apartments, replace single residences and densities of all kinds increase.

As the process of deconcentration of urban areas has taken place in the past several decades, it has left in its wake deteriorating inner-city neighborhoods, empty plants and factories, boarded-up stores, and other instances of inner-city decay. While the data are scattered and fragmentary, there is evidence that several older inner-city residential areas are being revitalized and converted to less intensive usage (Bradley, 1978). There are even some cases in which old commercial and industrial areas are undergoing alterations that have made them attractive, both as places for new commercial uses, as well as desirable residential locations for upper- and middle-class populations (Bunnell, 1977; *Economic Benefits* . . . , 1976).

The analysis presented here will focus on one such area, SoHo. SoHo is an acronym that stands for a twenty-nine-block area south of Houston Street in the Borough of Manhattan in the City of New York. Its boundaries are roughly the Avenue of the Americas on the west, Houston Street on the north, Lafayette Street on the east, and Canal Street on the south. Its precise boundaries were

---

*This article was written for this volume and has not been published previously. Funds for the study were provided, in part, from the Capitol Campus Fund for Research, The Pennsylvania State University. The author is indebted to Professor Melvin L. Reichler for a number of suggestions that facilitated this research and his generous hospitality during part of the field work. The cooperation of the Office of Community Development and the Office of Economic Development of the City of New York was indispensable in this study. The critical comments of Professor Gerald D. Suttles were most useful in preparation of this paper.

Table 1. Number of Business Establishments and Employees in a Twelve-Block Area in SoHo in 1962 and 1973[a]

1962[b]		1973[c]	
Business Establishments	Employees	Business Establishments	Employees
651	12,671	459	8,364

[a] Neither Rapkin nor the City of New York included Broadway, the most industrial and commercial street passing through SoHo.
[b] Rapkin (1963:11-12).
[c] Estimate by the City Planning Commission, Department of City Planning, City of New York.

established by a zoning change in 1971 which gives the area a legal status not usually associated with inner-city neighborhoods.

This analysis will employ the ecological concept of invasion-succession as it has been applied to research on changing land-use patterns in cities. Simply stated, invasion-succession refers to the sequential changes in land-use patterns. The city is viewed as composed of a number of zones with distinctive functions that are reflected in the type of population, activities, and structures located in these zones. Alterations in technologies, among other variables, produce changes in population, activities, and structures situated in these zones. The precise changes in these variables depend on their original characteristics and the nature of those that replace them. One of the most consistent findings has been that, when one residential population succeeds another, densities tend to increase—sometimes because the incoming population has larger nuclear families than the population being displaced or the existing structures are subdivided because the incoming population cannot economically maintain the existing housing patterns (Aldrich, 1975).

The process of invasion-succession that is taking place in SoHo is unique in the literature on the subject since it represents a succession from industrial and certain types of commercial activities to a residential population and other types of commercial activities. And while the succession in SoHo of this residential population represents a shift in density, we must recognize the caution about measuring density—namely, that density may be a function of daily and weekly rhythms. Commercial areas, for example, may have a high daytime density but low nighttime and weekend densities. If we measure the daytime density of SoHo, it can be demonstrated (as shown in Table 1) that the daytime population has declined. The conversion of buildings to residential use has reduced density.

On the other hand, if we measure the population in terms of residents and nighttime and weekend populations, the density has increased.

Nevertheless, a closer examination of the changes in the area suggest that they are due to the same considerations that Burgess presented in his formulation of urban growth. In the first place, industry has seldom, if ever, been concentrated in the very central areas of cities. Because industrial activities make such extensive use of land, their competitive position is less than that of other activities. The location of industrial SoHo represents an earlier pattern when the city was smaller and the existing technologies demanded a more central location to markets and workers. The swift dispersion of commerce and industry, both on Manhattan Island and to other New York City boroughs, simply by-passed SoHo, leaving it an anachronistic industrial zone. Nor was it feasible to remove the existing structures and replace them with industrial plants that incorporate newer technologies, since the outlying areas are in a much better competitive position to accomodate the one-story horizontal building that has become the model structure for manufacturing and distributing functions.

Finally, in this analysis of invasion-succession, the thesis will be that the shift in land-use patterns followed a definite sequence and form. The initial conversion from industrial to residential use of existing structures began as a surreptitious and furtive movement. The zoning laws prohibited the use of these manufacturing spaces for residential use—and some of the consequences of this will be dealt with later in this paper. As the rates of conversion of industrial spaces to residential use increased and the transition that was taking place in SoHo became more open, there was an increasing institutionalization of the process. Available spaces were openly advertised for residential use. At the same time, there was also a shift from less intensive residential usage to more intensive residential usage. At the beginning, conversions were scattered among floors and among buildings. The pattern soon emerged where entire buildings (usually accomplished by buying the building through a cooperative) were converted to residential use, with the exception of the ground floors which were occupied by commercial activities that reflected the changing character of the area.

Some historical considerations

The area of SoHo has undergone several transformations as New York City has developed. By the middle of the nineteenth century, SoHo was one of the most fashionable neighborhoods in the city, with the best department stores, restaurants, and hotels (Koch, 1976). This period lasted no more than twenty years, as the middle-class populations began to move toward mid-town Manhattan. In the next twenty years, the area was transformed from a middle-class enclave, with appropriate services and conveniences, to a deteriorating slum, which, in turn, was razed and replaced "with loft buildings which provided space for the developing industries" (Rapkin, 1963:10).[1] These new loft buildings were built in a manner that reflected the growing dominance of the American entrepreneurial spirit. Many of the buildings were built by using prefabricated parts. The facades were made of cast iron and elaborately decorated. The design permitted large areas of glass windows. The effect was very impressive. As one observer noted, "One was supposed to walk down West Broadway to see the beauty of the renaissance, in modern iron" (Koch, 1976:117).

From the middle of the 1870s until the beginning of World War I, SoHo and other areas of lower Manhattan enjoyed economic prosperity. Two major industries in the area were the production of women's and children's garments and the wholesale fur trades. There was, in addition, a scattering of other industrial and commercial activities. As the garment district in mid-town Manhattan began its ascent just before World War I, the space these activities vacated in SoHo were replaced "by a variety of establishments with a concentration of firms dealing with low value paper and textile wastes" (Rapkin, 1963:105).

SoHo's reputation as a viable commercial and industrial area came under serious attack. To many it was considered a wasteland, a sore upon the healthy economic body of Manhattan. A study made in 1963 demonstrated that such an evaluation was overstated. According to this study, SoHo could still be considered a strong commercial and industrial area. It occupied a desirable location, since it was well placed in New York City in terms of access to transportation (its western boundary opens on the entrance to the Holland Tunnel and on the east there is the Manhattan Bridge and Brooklyn) and it is well served by New York's subways and bus lines. The report was able to claim that:

> In some sections, of which this study is typical, the loft buildings form an unrelieved facade block after block, which gave the narrow streets a canyon-like and dismal appearance even on a bright day. These dingy exteriors, however, conceal the fact that the establishments operating within them are, for the most part, flourishing enterprises of considerable economic value to the City of New York (Rapkin, 1963:9).

Although the data for this study was gathered in 1963, we can see in Table 1 that this same area, as late as 1973, still contained a large number of firms employing a substantial number of workers. Rapkin's recommendations that this area be supported as a commercial and industrial area (1963:278–98) found continued support among some of the city planners and economists interviewed during this study.

To understand the changes taking place, it is necessary to consider several developments and some of SoHo's unique features. A key phrase in the quotation above is "for the most part." Rapkin was well aware that a number of businesses in SoHo were not flourishing but were surviving because the rents were minimal and some of the businesses were being operated by an aging managerial population whose heirs were seeking their fortunes elsewhere. As some of these more marginal businesses began to expire in the sixties, the owners of the buildings in SoHo found it increasingly difficult to replace their former commercial and industrial tenants. This was particularly true of the narrower buildings that could not accommodate the new technologies that demanded large horizontal spaces.

The shift in manufacturing technology since World War II has favored large horizontal spaces. Horizontal spaces are much more readily converted from one activity to another. It is possible

---

[1] A loft building is defined as one in which there are no internal load-bearing partitions. If the size of the building requires additional internal support, load-bearing columns are employed, leaving the floor spaces more or less open.

to change the flow of production lines, their length and distribution of material, much more readily in horizontal space than in vertical space (Rapkin, 1963:64–88). Thus, those lofts in the SoHo area that were narrow were much more likely to be those most readily abandoned for commercial and industrial use. At the same time, these narrow lofts with 1,500 to 2,000 square feet of space were very attractive to a different population.

The artist pioneers

The residential population that first began to enter these lofts was very distinctive. It was almost exclusively artists—painters, sculptors, dancers, video- and film-makers, print-makers. Artists, no matter what their medium of expression, have never been able to achieve financial security in large numbers. Thus, they have always lived and worked at the margins of the society. The appeal of SoHo was the availability of large spaces at very low rents—spaces that could be used for both living and working. Also, some of the artistic movements of the sixties demanded large spaces. The products of the Minimalists, Fluxus, Op, and Pop movements were frequently large scale. The large, open spaces of the SoHo lofts permitted experiments with works that were thirty or more feet in execution. Such spaces also provided opportunities to employ mixed media presentations. Indeed, that the floors of the SoHo lofts had such great bearing capacities meant that large sculptures could be undertaken on the upper floors, while the freight elevators permitted easy movement of these works. The very dynamics of the art world require that artists who seek to remain current with the latest developments need to be close to the important galleries and accessible to others working in their field. And Manhattan is the art center of the world.

SoHo also has several other distinguishing characteristics that made it appealing to artists. The area is located just south of Greenwich Village, which has long been a haven for those interested in the arts and an unconventional lifestyle. It is also adjacent to both Chinatown and Little Italy, which provided a cosmopolitan character to lower Manhattan.

In addition, the proximity of some industrial and commercial establishments supported artistic work. Canal Street, the southern boundary of SoHo, has been described as a "ten-block-long hardware store and junk shop for the surrounding industry. For the artists of SoHo, it has become a vast, luxurious, tangled garden of the real.... Here on Canal Street, lofts are furnished, works and careers are conceived, souls are soothed" (Koch, 1976:121). While this image may be exaggerated, those interviewed during this study claim that much of the materials that they use can be obtained cheaply in these stores. Furthermore, some of the waste by-products of the industries located in SoHo have provided materials for the artists. Others have provided services to the artists in the way of technical assistance, e.g., techniques in welding, print making, and so forth.

The exact size of the artistic population of SoHo is hard to determine, although there is a requirement in the zoning laws that those who live in the lofts of SoHo be certified as artists. There are a few general estimates. In 1963, Millstein claimed that there were between 5,000 and 7,000 artists in residence in downtown Manhattan businesses as early as 1962 (p. 24). Another estimate placed 2,000 to 3,000 artists in SoHo itself (*Time Magazine*, 1970:82). After the zoning change of 1971 occurred, making it legal for artists to live in the lofts in which they worked, approximately 3,000 artists applied for such status within the first year (Koch, 1976:107). No matter which figure is taken, the official figures probably underestimate the number of artists living there, since—to gain official status—the lofts must meet certain housing standards, and it is clear that many do not.

Subsequent changes

There are a number of features that distinguish what happened (and is happening) in SoHo from the usual process of invasion-succession. In the scheme outlined by Burgess, changing land-use patterns are usually accompanied by changes in architecture. Dwelling units are replaced by commercial buildings—buildings increase in horizontal and vertical dimensions, for example. In residential areas, dwelling units are frequently subdivided for intensive use and zoning variances are granted so that certain commercial establishments can be located there, e.g., small grocery stores, barber shops, and other small shops of various kinds. In the shift in SoHo from industrial to residential use, the original buildings remained. In fact, it was the very character of these buildings that attracted the artist pioneers, and there is a great deal of energy being expended to maintain their architectural integrity. Where it has been financially and architecturally feasible, there have been some restorations to return the buildings to their original status.

Unlike the usual patterns of residential invasion-succession, the original population that established residences within the existing structures did so illegally. The area had been zoned exclusively for manufacturing and commercial use. Although residential occupancy was illegal, there was a certain degree of official tolerance for the incoming population (Millstein, 1963; Koch, 1976:331). Efforts continue to be made to restrict residential conversions by limiting the size of the lofts eligible for conversions. In that part of the Broadway Corridor that passes through SoHo, there have been vigorous efforts on the part of the city to maintain the commercial and industrial uses of the buildings—and these efforts appear to have been relatively successful. There are also prohibitions against subdividing floors into apartments in order to maintain the large loft buildings as viable industrial and commercial spaces. Such efforts on the part of the city are actively supported by most residents, who fear that such subdivisions would so increase property values that continued residency in lofts would be jeopardized if developers were able to convert buildings to more intensive usage.

The movement of artists into SoHo generated a new set of commercial establishments. At the same time, there was a shift in the product mix of the original businesses that indicates the changes in the area. This shift was generated by two conditions. The first reflected the change produced by the new residents themselves. A local liquor store, for example, reported that it had increased its line of wines to meet the tastes of the new residents, and a paint store located in the area altered its stock in order to cater to the requests made by local artists. The second condition was the increase in the tourist traffic that took place as the area became known as a haven for artists, thus gaining a certain degree of glamour for those seeking to enjoy a "bohemian" milieu.

Among the new establishments that were opening in SoHo were a number of galleries displaying the works of local artists. The first gallery in SoHo was established in 1968. It was primarily a local operation, even though the owner had run a gallery in a more fashionable uptown location (Gardner, 1974). This gallery opening was quickly followed by a series of galleries, some with ties to those in the more traditional locations, as well as those that were strictly artist co-ops—artists living and working in SoHo. As the number of galleries grew, encouraging an increased number of tourists to enter the area, there were other changes in the retail establishments of SoHo. Several restaurants opened that cater to these tourists, a number of bars with artistic or bohemian motifs meet the needs of the tourist trade, and a number of shops feature chic clothing, custom jewelry, and other services geared to the tourist rather than the local population. Residents have been successful in excluding certain kinds of commercial activities, such as discos and "juice bars," that have given parts of Greenwich Village its tawdry appearance. There is a strong commitment on the part of many to maintain the area as a place for serious artistic work, although these efforts are threatened by its very reputation as an artistic community.

These changes in street-level uses have not occurred equally in all areas of SoHo. As with many inner-city neighborhoods, there are clear distinctions between areas of commercial and residential use. Hunter (1974) has made use of these differences in his analysis of symbolic communities in Chicago. As he shows, certain focal points come to convey not only a dominating physical presence, as with cathedrals in the walled cities of the Middle Ages, but also "a convergence of significant functions in the life space of local residents" (Hunter, 1974:89). In SoHo the focal points are employed not only by the residents, but also serve as centers of attraction for tourists and tourist activities. The area of highest concentration of these tourist facilities represents another pattern of the succession process occurring in SoHo.

An important distinction was introduced in SoHo by the zoning changes of 1971. The area that has received the official status by the zoning changes does not include all the area that is circumscribed by the natural boundaries formed by the two major arteries on its northeast corner. Both of these arteries are six lanes wide and clearly delineate the area from adjacent territories. In fact, on the northern border of SoHo there are only two business enterprises that open onto it and these are gasoline stations. All other buildings have no entrances on this street. The streets, however, that are in the most extreme northwest corner are populated by an old Italian community with all the institutions associated with such a population, including store-front social-athletic clubs, parochial schools, and churches. This area has been largely residential for several generations and contains within its boundaries a number of small stores and shops that service the local population. In addition, this population sees itself as belonging to Greenwich Village—banners hang over the streets proudly announcing "Welcome to Greenwich: The Cradle of Democracy."

The small stores and shops of this area have been particularly vulnerable to the pressure generated by the changes in SoHo. Most of these shops and stores are those adjacent to the cluster of art galleries that attract tourists to SoHo. Their size makes them more amenable to conversions to boutiques, craft shops, and small restaurants than the larger floor spaces that typify the industrial lofts of SoHo. There have been approximately twenty-five conversions on the five streets in the Italian neighborhoods from businesses that had served the local population to those aimed at the tourists and the artists.

Unlike the residential conversions of the lofts, the conversion of these stores from local use to tourist use has not been without a certain degree of conflict. A number of informants within the Italian neighborhood claim that the invasion of these activities has "ruined" their neighborhood, a sentiment shared by a number of artist residents who preferred the less commercial character of the older, Italian-run businesses. In one instance, a neighborhood restaurant was displaced by a very fashionable "bistro" when its lease ran out. While the old restaraunt simply moved across the street, the reaction to the new "bistro" was openly antagonistic.

On the one hand, while there is some antagonism between the artist population and the developing tourist trade, there is, on the other, a symbiotic relationship. The bars and the restaurants that have opened have provided a source of employment for some of the residents of SoHo or have provided employment for artists. For example, one restaurant owner describes her service of food as a "performance," and employs only actors as waiters. In another restaurant, the waiters are all dancers. There are a number of employment niches that cater to the odd working hours and habits of artists and this complementary relationship gives SoHo part of its unique character.

Conclusions and implications

The major portion of this discussion has been on the differences in the invasion-succession experience of SoHo as contrasted to those studies that have examined the process in terms of the replacement of one residential population by another. But there are also certain similarities to be commented on. Aldrich, in discussing Cressey's analysis of the process, comments that the invasion usually begins with a "few 'pioneers' who are upwardly mobile and have the resources to translate their desire for a more prestigious address into an actual move" (1975:332). This analogy of the invaders as pioneers finds support in what occurred in SoHo. Although the artists who were moving in cannot be considered to be upwardly mobile in the usual sense of a change in socioeconomic status, their move represented for them an improvement in the living and working conditions they had enjoyed in their previous dwellings.

Millstein, in assessing the move into SoHo on the part of the early arrivals, uses the pioneer analogy: "In an oddly paradoxical way living in a loft, apart from the presumptive spiritual toughening, is a kind of pioneering in which is unquestionably the most built-up city on earth" (Millstein, 1963:26). He continues this analogy by suggesting that the pioneers were arriving only after "the lofts had been sacked and abandoned by the white settlers—anyone from paper-box manufactures to fabricators of brassiere hooks . . ." (Millstein, 1963:28). What is even more interesting in Millstein's analysis is his assessment that the invasion by the artists began to change the property values and the character of the area.

The very presence of the artists launched another chain of events. Without stretching the pioneer analogy too far, it is apparent that, once the artists began to move in and began to convert the industrial lofts to residential use, they had the effect of civilizing the area. In fact, one of the problems that the initial pioneers confronted was that in renting these spaces for both living and working, they made substantial investments in making them habitable. Since their living arrangements were illegal, they found it difficult to obtain suitable leases from landlords. As a consequence, there developed a familiar pattern of renting a loft, investing in improvements, and then having their landlords ask them to vacate. The landlord would then turn around and rent it at higher rates to the next artist.

To meet this particular problem, the artists developed another form of adaptation, namely, purchasing the entire building as a cooperative effort. This strategy was feasible, given the prices of the buildings. It is now estimated that there are approximately eighty such co-op buildings in the twenty-nine-block area that comprises SoHo. Although data are very incomplete on the prices of these co-ops, information from the late sixties gives some indication of the prices. Of the seven co-ops for which complete data are available, the prices ranged from $60,000 to $210,000; and these were able to accomodate between nine and sixteen artist-tenants, respectively.

In their drive to gain a change in the legal status of their occupancy in the lofts, the artists called the attractiveness of these spaces to the attention of other groups in the city—a city that has notoriously high rents. Even though there is the requirement that only an artist can live and work in the lofts of SoHo, the definition of what an artist is offers such a wide interpretation that almost anyone with some pretense toward artistic activity can slip into the classification.

In conclusion, this analysis has applied the concept of invasion-succession to an area that was distinctive on a number of dimensions from most studies of this ecological concept. Employing Burgess's general scheme of urban growth, the point of departure in this investigation was the changing technology that rendered some of the nineteenth-century loft buildings of SoHo obsolete for modern manufacturing technologies. While a number of manufacturing and commercial establishments continue to operate, others have either moved or closed their SoHo operations. These were not replaced with similar functions, but rather became available for new and different uses.

The spaces in these loft buildings had a particular attraction to a special population—artists. While the use of loft space by artists for living and working is not intrinsically unique to SoHo, it did represent a large concentration of such spaces that were adjacent to other areas of Manhattan and provided opportunities for a cosmopolitan, bohemian lifestyle. The invasion by artists of these loft spaces produced several alterations in the area. As the number of artists increased, there emerged a distinct community that developed social networks that contributed to the ambiance of the area. Their special needs and lifestyles were met by the adaptation of some indigenous stores and shops, and at the same time provided the impetus for the development of more specialized facilities catering to their needs, e.g., art galleries, art-supply stores, bookstores, and so forth.

The complement of functions and activities enumerated are not based completely on the internal consumption of the SoHo community. Their continued existence and prosperity depend upon attracting a tourist clientele. These tourists patronize the expensive restaurants, bars, and cafes that have located in SoHo. Some of these tourist facilities have been developed by the indigeneous population; others are the creations of outside entrepreneurs. In either case, they provide some employment to community residents. This complementary usage of space, combining a special residential population and an industry based upon its unique characteristics, is one of the distinguishing features of SoHo.

Finally, the transformation of SoHo from an industrial area with unique architectural features to a residential enclave with special housing characteristics has attracted a new population. More prosperous than the artist pioneers and less involved in the older social networks that were based upon shared work and the struggle to establish the legitimacy of loft living that combines living and working space, their presence has also created new markets. For example, a specialized food store reminiscent of those found in other prosperous middle-class bohemias of New York City has recently opened. Other consumer services that were absent in the early stages of the invasion process have also appeared, indicating the changing population of SoHo.

The casual observer who enters SoHo at some of its boundaries would be hard pressed to discover the changes that have taken place in some of its areas. Much of SoHo remains a heavily industrial and commercial area, with trash-littered streets whose building facades are unchanged since Rapkin's 1963 study. The process of a complete conversion to a residential area, with the tourist facilities and specialized stores serving a prosperous residential population, may never be reached. SoHo still contains cheap loft space. It still provides niches for those marginal to the economic system, and the social networks of the original artist pioneers continue to flourish. But what has occurred to date reflects the process of invasion-succession that has typified urban growth, even if the process and emerging structure have unique features.

References

Aldrich, Howard. "Ecological Succession in Racially Changing Neighborhoods: A Review of the Literature." *Urban Affairs Quarterly*, 10 (1975): 327–48.

Bradley, Donald S. "Neighborhood Transition: Middle-Class Home Buying in an Inner-City, Deteriorating Community." *American Communities Tomorrow*, 1 (1978): 80–106.

Bunnell, Gene. *Built to Last: A Handbook on Recycling Old Buildings*. Washington, D.C.: The Preservation Press, 1977.

Economic Benefits of Preserving Old Buildings. Proceedings of a Conference. Washington, D.C.: The Preservation Press, 1976.

Gardner, Paul. *Art News,* 73 (April 1974): 56.

Hawley, Amos H. *The Changing Shape of Metropolitan America.* Glencoe, Ill.: The Free Press, 1956.

Hunter, Albert. *Symbolic Communities: The Persistence and Change of Chicago's Neighborhoods.* Chicago: University of Chicago Press, 1974.

Koch, Stephen. "Reflections on SoHo." In *SoHo.* Berlin: Academie der Kunste–Berliner Festwochen, 1976.

Millstein, Gilbert. "Portrait of a Loft Generation." *New York Times Magazine,* January 7, 1963.

Rapkin, Chester. *The South Houston Industrial Area.* New York: City Planning Commission, Department of City Planning, City of New York, 1963.

Time Magazine. "Bohemia's Last Frontier." May 25, 1970, p. 82.

Cognitive Dimensions of Space and Boundary in Urban Areas*

Roger M. Downs

Introduction

Underpinning an understanding of the processes of urban governance and administration are assumptions about human behavior. To the extent that political scientists, for example, can account for and manipulate governing processes, the assumptions remain implicit. To the extent that governing processes seem unable to resolve an increasing number of urban problems, assumptions are explicitly questioned.

Many social science disciplines share assumptions about the individual's knowledge of the spatial environment. Pre-eminent among the concepts derived from these psychological assumptions is the sense of place (or community, or neighborhood). This attachment to a bounded space is assumed by urban planners, community organizers, and politicians; knowledge of the names and boundaries of the parts of a city is assumed by the communications media and by realtors. Despite such pre-eminence, understanding of spatial cognition is recent, fragmentary, and often speculative.

This paper will consider knowledge of spatial cognition as a source of psychological underpinnings for commonly accepted ideas in planning, sociology, political science, and geography. Working from a cognitive perspective, it explores: (1) the sense of place; (2) the nature of cognitive space; and (3) the spatial cognition of urban neighborhoods.

The sense of place

Boundaries express spatial limits to the exercise of control and responsibility. At the most basic level, boundaries include a "within" and exclude a "without," partitioning space and society. Urban boundaries are an uncomfortable mixture of partitions generated by two distinct processes:

(1) *A community of interaction,* evolving as a function of shared residence, generates and defines a cohesive, lived-in, or experienced space. It has no legal standing, but depends on the forces of custom and consensus for its survival. Neighborhood, territory, and turf are expressions of this idea.

(2) *A community of interest* is a unit of space

---

*This article has not been published previously. An extended version of this article was presented at the Annual Meeting of the American Association for the Advancement of Science, Washington, D.C., 1978, in the symposium entitled "Neighborhoods, Cities, and Regions: Governing the Future of Urban Spaces." The author would like to acknowledge the assistance of James T. Meyer in preparing the original manuscript.

legally established and delimited in order to exercise specified responsibilities. In contrast to the community of interaction, it is an imposed, often arbitrary, unit. Most, if not all, municipal governing units, such as school or sewer districts, are of this form.

With increasing urban size and social complexity, the likelihood of congruence between the boundaries of these two types of space decreases. At the same time, the control exercised by communities of interest increases at the expense of communities of interaction, and the potential for misunderstanding and conflict is magnified.

Both communities necessarily incorporate statements about the members' views of the world. Most frequently, these statements are encapsulated in the idea of the sense of place. For analytic purposes, we can divide the sense of place into: (1) a sense of *identification,* via the process of naming; (2) a sense of *boundary;* (3) a sense of *attachment* to, or identification *with,* the community; and (4) a sense of *responsibility* and *control.*

Ideally, the sense of place should emerge "naturally," almost in some mysterious organic form, although in practice we assume that it can be induced and manipulated. Moreover, the sense of place is shared: groups of people residing in an area necessarily develop a communal sense of identity, boundary, attachment, responsibility, and control. Generalized agreement is presumed to underlie communities of interaction and interest. Both types of community are identified and bounded, although the nature of the attachment and the expressions of responsibility and control differ between the two. The final step in the argument posits a hierarchical structure. Individuals belong to a variety of communities simultaneously. Thus, a person is part of a neighborhood, city, county, state, etc. Each spatial partition in this scalar structure generates, and is validated by, a sense of place.

A belief in the existence of the sense of place forms part of the assumptive base of political science, sociology, planning, geography, and even law. At issue here is the empirical support for this base. Does the sense of place have psychological validity? More particularly, what can the field of spatial cognition contribute to an understanding of the links between urban neighborhoods, boundaries, and the governing process? Such questions are appropriate for a series of reasons. First, several disciplines have begun to explore the role of cognitive maps in the governing process; for example, in political science, see Warren and Wechsler (1976); in sociology, Suttles (1972); in planning, Lynch (1976); and in geography, Saarinen (1976). The existing work in spatial cognition can guide this exploration. It can guide what we look for, how we look for it, what we can expect to find, and how we interpret research. Second, many problems of urban government are ascribed to such phenomena as voter apathy, lack of involvement in urban issues, and lack of the knowledge necessary to reach educated decisions. Can we generate a sense of place for those communities of interest that affect urban residents? Third, there are suggestions that we should aim for congruence between communities of interest and interaction. Is this desirable, or possible?

The nature of cognitive space

Central to the understanding of the nature of cognitive space are twin concepts: cognitive map and cognitive mapping. A cognitive map is an organized representation of the spatial environment. Cognitive mapping is the capacity to collect, organize, store, recall, and manipulate information about the spatial environment (Downs and Stea, 1977). Mapping is an activity that we engage in rather than an object that we have. It is the way in which we come to grips with and comprehend the world around us. A map is a cross-section representing part of the world at one instant in time. The map can be an internal cognitive construct, or we can externalize it. There is no assumption of formal identity between internal and external representations.

These simple process-product terms are often misunderstood. Put simply, the constructivist view of cognition argues that knowledge of the world is not given. The ability to cope with the spatial environment necessitates the acquisition and processing of information, the formation of representations. Thus, representations of the environment function *as if* they are maps, the cognitive process *as if* it is mapping. Cognitive mapping and cognitive map are metaphors of function, *not* of form. Failure to recognize this distinction can lead to serious problems. The mistaken beliefs that the form of the representation is like a cartographic map and that the specific cognitive operations of mapping are like those of cartography are reinforced by our continuous exposure to cartography, graphics, geometry, and visualization. Taken together, these provide an apparently "natural" way to think about the spatial environment, to talk about it, and to study it.

The point of this argument is that we must be careful about the insidious interrelations between what we look for, what we expect to find, and how we look for it. For example, if we begin with the idea of a partitioned space (e.g., physically delimited neighborhoods), then we are obviously concerned with the cognitive representation of this space. Are people aware of the names and boundaries, of the differences between a within and a without? Units which are partitioned out of space are readily depicted on cartographic maps. It seems logical to ask people to draw their representations of the units in question. But what can we say about the externalized cognitive maps that are produced? Ignoring differences in graphic ability and comprehension of cartograhic convention, what do the lines mean? Do they mark true discontinuities? Are the areas within homogeneous? Are the areas within continuous spaces? There is an apparent precision, almost decisiveness, to lines on a map. And yet this impression is misleading; even the boundaries represented on a cartographic map are not clear-cut. We have cartographic conventions for establishing the edge of the built-up area or the coastline. Do we have similar cognitive conventions that allow us to delimit our neighborhood?

Bearing these comments in mind, what can we say about the nature of cognitive space? There are four characteristics of cognitive mapping that seem relevant to understanding the relation between neighborhoods and the sense of place. First, cognitive mapping is a purposive, goal-directed activity with two outcomes: plans for solving spatial problems (e.g., wayfinding) and frames of reference for interpreting the spatial environment (e.g., reading a news story about a zoning change). Second, it is an interactive process in which cognitive maps and the developing nature of cognitive mapping emerge from continual interaction with the spatial environment. Spatial behavior, preferably active (e.g., driving a car) rather than passive (e.g., being a car passenger), is essential for linking the external environment with the internal representation. Third, it is a highly selective process. Information becomes part of a cognitive map as a joint result of the functional importance of places and the distinctiveness or imageability of places. Finally, it is an organizational process which makes sense out of the world. Representations are actively constructed, often using an inferential process to go beyond the information given.

The resultant cognitive space is egocentric, parochial, stereotypical, symbolic, and highly subjective. These characteristics lead to the most important question about cognitive maps: What is the agreement between maps of the same spatial environment? The communal sense of place is based on an assumption of generalized agreement; that is, a shared understanding of identity, boundary, attachment, control, and responsibility. To what extent is such agreement possible? An answer to this question falls into two parts, one dealing with the accuracy of cognitive maps, and the other considering the similarity between cognitive maps.

Accuracy is difficult to define. There can be no identity between representation and environment; the mapping process, because of its selectivity, does not generate a one-to-one correspondence. Given a many-to-one mapping between environment and representation, can we speak of correspondence between the two? Even this question is unanswerable. We cannot provide a universally agreed-upon replica of the environment, and even if we could, any person can generate many different external representations of the same spatial environment. The translation from internal-cognitive to external-physical representation affects the form of the representation.

The accuracy question is only meaningful if we ask: Does someone possess the spatial information and problem-solving strategies necessary to live in a particular spatial environment? Are the cognitive maps functional in terms of spatial behavior, interpretation of events, and communication with others? Neither precision nor completeness of representation is important; utility is. The objective is not accuracy, as such; but, given patterns of experience and needs, making the best use of limited information-processing abilities. Since such ends can be attained in many ways, the idea of generalized agreement, as it underpins the communal sense of place, must be questioned.

It must be questioned for a second reason. Can we expect people to produce similar cognitive maps of the same spatial environment? To begin with, age, experience, skill, and training affect the form of the external representation. Even if we control for these effects, a definitive answer is impossible. A best approximation would be: Parts of cognitive maps are common to all members of a large group, parts are common to a sub-group, while still other parts are unique to each person. Variations in similarity result from three factors: the scale of the spatial environment, the sources of information about that environment, and the location of the person doing the representing.

Cognitive maps similar for members of a large group involve representations of large-scale units such as nations. These stereotyped cognitive maps are a *pot pourri* of fact and fiction, truth and distortion, broad-minded views and prejudice. They form the common currency of expression in a culture and lie at the heart of written and verbal communication. Two factors ensure that geocentric views persist: they are rarely based on direct, first-hand experience, and they are self-perpetuating systems.

General consensus breaks down at smaller scales. Residence in an area, shared local interests and communications media, and shared activity patterns result in loyalty, parochialism, and pride of place. The same basic processes operate as in the development of national stereotypes, but the result is a mosaic of different viewpoints. The key is that for each group of people, those *within* the community of interaction, experience of the environment is direct and highly personal. A person subscribes to many viewpoints; a hierarchical sense of place does exist. However, each viewpoint is shared by a decreasing number of people, and disputed by an increasing number.

We must be careful in using this argument to support a communal sense of place. Although parts of cognitive maps are similar, representations reflect personal experiences and meanings. The "fine" structure of organization and detailed contents of cognitive maps are expressions of individuality, superimposed on a common skeleton of similar cognitive mapping capabilities and shared activity patterns. To expect too much similarity is to misunderstand the cognitive mapping function. Environment and behavior offer but the raw material from which we assemble cognitive maps. Cognitive mapping does not duplicate; it selects, constructs, and organizes. Cognitive maps are similar enough to permit us to share and communicate our understanding of the spatial environment. They are personal enough to accommodate unique experiences. They are an outcome of a process that can cope with social requirements and individual needs.

The spatial cognition of urban neighborhoods

As Suzanne Keller (1968) has ably documented, the history of the idea of neighborhood is long, checkered, and confusing. Of the many problems that surround the term, two seem particularly relevant to an understanding of the cognitive representation of neighborhood. The first concerns the character of the debate itself. There is an impassioned, romantic, and often ideological overtone to the literature on neighborhood. One is confronted with powerful images, vividly expressed: Who can read about urban villages (Gans, 1962), about grieving for a lost home (Fried, 1963), without picturing the destruction of "old world," almost "prelapsarian," communities? In a sense, much of the debate depends less on empirical evidence than on emotional predilection, on the writer's image of what a neighborhood is (or was and ought to be). The second problem centers on the range of meanings that are subsumed under the label "neighborhood." Two major dimensions underlie this tangle of meanings: Is neighborhood a unit of physical or social space, and what is the size of this unit? Both problems have bedeviled the literature and must form part of the background to this discussion. Both lead directly to assumptions about the urban resident's knowledge of the city, about the sense of place.

From a cognitive perspective, let me argue that there are two distinct, though overlapping, cognitive systems that are used by urban residents to partition the space in which they live. The precise degree of overlap between these systems is an unanswered empirical question. We must attempt to disentangle the two systems because, although people have access to both, the systems serve very different functions and are generated and maintained by very different processes. The first system is that of *regionalization,* a process whereby the whole city is partitioned and labeled. The second system is that of the *defended neighborhood* (Suttles, 1972), a process whereby the classic community of interaction is generated.

Regionalization is based on a generalized view of the city as a whole and is essentially a process of segmentation for convenience, for coping with the scale and complexity of daily life in the city. There are two facets to the cognitive system of regionalization, one based on a generalized model of a stereotypical city and the other based on a naming system for the parts of a particular city.

The generalized model is a simple schematic which partitions the city into such gross areas as inner city and suburbs. This two-component, "doughnut" model of the city can be further sub-divided, using such categories as downtown, the central business district, the ghetto or the slums, skid row, auto row, the wrong side of the tracks, and generic names for ethnic areas (Little Italy, Chinatown, etc.). The partitions are stereotypes which transcend the particular place. Together, the names provide a descriptive frame of reference, a basis for generating predictions and expectations. The generalized model can be used to cope with new places and fills in where specific information is lacking. It exerts a powerful but hidden shaping influence on what we write (and read), how we think (and interpret), and how we decide (and behave). The stereotypes themselves are generally accepted categories that are part of the common currency of communication and understanding in a society. They are useful tools in the search for an economical means of organizing spatial knowledge. They allow people to cope and to simplify, but at a cost that is unacknowledged. We become, in Stephen Carr's phrase, victims of conventionality.

Interlaced with this generalized model is the naming system that is the essential underpinning for everyday behavior in a particular city. The formation of districts with distinctive names plays a vital role in a range of activities: in the news reports of the mass media, in the operations of the real estate market, in the route planning and use of mass transit systems, in dispatching systems for public services, and in the use of cognitive maps for all forms of personal wayfinding.

What can we say about these two cognitive systems from the perspective of a sense of place? First, both depend upon the process of identification, the attachment of unambiguous labels. Innumerable studies have pointed to the remarkable degree of consensus that exists with respect to the set of district labels in everyday use. For example, Donnelly, Goodey, and Menzies (1973) prepared a pilot study on the utility of perception studies in the local planning process. People in Sunderland (England) were able to demarcate areas in the city and apply a commonly accepted system of place names to these areas. Responses rapidly polarized to one of a very few generally held names; there was more certainty with respect to the "given" name in the outlying districts. However, the genesis of such an informal naming system is not understood and it is not subject to easy manipulation. Willmott (1962) reports that only 30 percent of the interviewed residents were able to name their own neighborhood, despite the longstanding efforts of the Stevenage Development Corporation to stimulate a sense of identity with the neighborhood units that were built into the town. The difference between Sunderland and Stevenage is in the genesis and subsequent history of the naming system. In Sunderland, the neighborhood names were the product of spontaneously generated common usage, whereas in Stevenage the names were imposed by the Development Corporation.

Second, although the sense of identification is crucial to both systems of regionalization, the sense of boundary plays less of a role than might be expected. The whole space of the city is not completely and exhaustively partitioned and labeled. Blank, unnamed areas exist alongside of, and separate, named areas. In this respect, the cognitive space is not a continuous surface. Moreover, for the named areas, the cores seem more significant than the boundaries. Place names seem to "float." In a comparison of residents and city officials in Stamford (Connecticut), Spiro (1974) found that the two groups agreed on the number of areas (6), on their names, but not on the edges of the areas, their respective sizes, and their internal characteristics. Both sets of cognitive regionalizations bore only a slight resemblance to the local planning districts that form the closest level of local governmental space partitions. In a parallel study, Milgram (1972, p. 200) argues that: "New York City, as a psychological space, is very uneven." Given a series of slides taken at various locations throughout the city, Milgram asked respondents to identify the location of the slide. At two distinct spatial scales, New York showed an uneven psychological texture. Boroughs varied widely in terms of "recognizability," while there were also significant differences by borough in the proportion of slides located in the correct neighborhood. Thus, the Borough and neighborhoods of Manhattan are far more distinct and recognizable than the Boroughs and neighborhoods of the Bronx and Staten Island.

One of the best discussions of the nature of regionalization systems is Lynch's *Image of the City* (1960). Districts form a basic component of the resident's image of the city. However, although one can enter "inside of" a district, the acknowledgment of precise boundaries is not crucial to the existence of a district. The set of core characteristics, what Lynch calls the "thematic unit," gives the district a common identifying character. Characteristics such as physical form, street plan, building style, building materials, colors, and social class can all form part of the ensemble of things that typify the district. Boundaries may or may not be significant. A knowledge of where the district begins and ends is less vital than a sense of what it is like. There is one qualification to this statement. As Ross (1962) shows, boundaries are of greater significance to those people who live within the particular district, or more importantly, who would like to live within the district. His work in Boston indicated that people on the edge of two districts generally identified themselves with the name of the "better" neighborhood. This search for the right address is confirmed by Eyles (1968). People living close to the "Village" in Highgate (London) tended, in their cognitive maps, to extend the boundaries of the "Village" towards their own residence. By manipulating boundaries, one can capture the elusive ambience that goes with living in the right place.

The work of Ross and Eyles provides a convenient transition into the second major cognitive system for partitioning urban space: that of the *defended neighborhood*. This epitomizes the classic form of the community of interaction. In this discussion, I would like to bypass the debate over whether such units still exist or whether they are fast disappearing. Instead, I will talk about the defended neighborhood in its idealized form, considering it from a cognitive perspective.

The defended neighborhood provides a view of the larger city from within, from the inside out. In this egocentric cognitive system, the senses of identification, boundary, and attachment are all crucial. Unfortunately, the term "neighborhood" subsumes a series of different, though related, phenomena that we must try to disentangle.

A major controversy exists over whether neighborhood is a spatial, formal unit or a social, functional unit. From the cognitive perspective, it is both simultaneously. As a spatial unit, neighborhood encompasses a continuous area of physical space, giving rise to what Fried (1963) calls a feeling of spatial identity and a sense of attachment to a clearly defined physical space. As a social unit, neighborhood is a functional unit, based upon interactions and links between people and engendering what Fried (1963) calls social identity. The links and interactions are of many forms: kin, friends, neighbors, acquaintances, and faces recognized.

These two units lead to very different forms of cognitive space. Considered alone, the spatial unit is an areal construct with a sense of continuity within and a boundedness defining a without. The social unit is a network of social relationships. The people exist as points in physical space, but the properties of continuity and boundedness do not apply.

Unfortunately, the two units have become identified with two traditions of sociological research, spatial units with the natural area (or ecological) emphasis and social units with the social interaction emphasis. To the extent that these research traditions posit an either/or character to discussions of neighborhood, three problems emerge. First, the situation is not either/or: both types of unit have a cognitive representation. Second, we should not treat physical and social units as though they are incompatible and, therefore, try to reconcile them: they both exist simultaneously for a person. Thus, we can find both in empirical studies. Gans (1962), for example, showed that the West End was distinct in both a physical and a social sense. Third, we should not assume, as do many planners, that there is a simple coincidence between physical and social spaces.

A major difficulty in neighborhood studies relates to the question that is often asked by respondents: What do you mean? The Sunderland study of Donnelly, Goodey, and Menzies (1973) suggests that the definition of neighborhood is neither obvious nor immediately understood. When asked "What do you call this area, which you live in?" people wanted to know whether the researchers meant a street, a group of houses, or a ward. Even the temptation to refer to such obvious physical units as city blocks must be considered carefully. Cybriwsky (1972) discusses the fundamental cognitive distinction between block (as a physical or formal rectangle) and "face-block" (the sociological interaction unit formed from the faces of two blocks.) Face-blocks themselves can be split into still smaller units. In a study of an Italian neighborhood in San Francisco, Appleyard and Lintell (1972) showed that the size and shape of the "home territory" varied with the traffic volume on the streets. As the traffic volume increased, the size of the home territory diminished from the face-block, to one side of the block only, to the apartment or the apartment building.

Different impressions of the shape and size of the neighborhood can be generated, depending on the data collection method that is selected. Abbott and Lee, quoted in Canter (1977), gave the same subjects two separate response formats. In the one, residents were given a list of places and were asked to say which places were in their neighborhood. In the second, they were asked to draw a line on a cartographic map delimiting their neighborhood. Both were expressions of neighborhood. However, on the average, only two-thirds of the named points were contained within the line drawn.

In the face of these problems, it is difficult to

gauge the character of the defended neighborhood. Of the many attempts to do so, Terence Lee (1968) has provided the most convincing framework for addressing the cognitive representation of urban neighborhoods. By adopting a cognitive approach, Lee argues that the spatial-social duality can be understood. From his empirical studies, he identifies three types of neighborhood. The *social acquaintance neighborhood* is a unit of social interaction whose boundaries are set by the feeling of knowing and being acquainted with a limited group of people. It is a small physical area, perhaps ranging from part of a block through to a small group of houses. The binding force is the feeling that everybody knows everybody else, whether or not they have anything to do with each other. The *homogeneous neighborhood* is of different character: social interaction plays only a minor role in the definition of this unit. The boundaries reflect perceived discontinuities in the price, type, and condition of houses. Although there is a social relationship of mutual awareness, a more apt social character is given by the expression "people who live in houses like ours." The *unit neighborhood* is close to the planner's idea of neighborhood unit. Encompassing a wide physical area, a variety of people and housing types, a mixture of friends and acquaintances, the unit neighborhood is a name rather than a clearly bounded unit of space. It lacks either the social identity of the social acquaintance neighborhood or the spatial identity of the homogeneous neighborhood.

Lee's typology raises several important issues about cognitive urban space. First, there is no place name or label for the social acquaintance neighborhood. This is the unit which seems to emerge in most interview situations; it is what people try to draw or to describe. It bears absolutely no resemblance to any formal governmental boundaries, to any communities of interest. (The only possible exception to this statement might be with respect to the boundaries of a census enumeration district.) A Royal Commission on Local Government study (1969) surveyed a large number of people about the area that they belonged to and felt at home in. Not only was this area very small but it bore little or no resemblance to the surrounding local government area. In a study contrasting a middle-income, co-op project and two twenty-story buildings, Cooper (1971) questioned residents about the sense of belonging. In the twenty-story buildings, 60 percent said that they belonged in their apartment and the remainder said "nowhere." For the co-op, 40 percent identified with the development as a whole, one-third with a particular section of the development, and 25 percent with their own building. Gans's (1962) study of the West End reinforces this point about spatial scale. Although the West End formed part of the district-naming system for the city as a whole, the residents identified with and defined street- and block-size units as their neighborhood. Willmott's (1962) findings about neighborhood names in Stevenage must be interpreted in a similar way. The district names were assigned to the planning level of the neighborhood unit. They had little or no salience from a cognitive point of view.

This issue of salience is a second crucial point that we can relate to Lee's typology. The progression from social acquaintance to homogeneous to unit neighborhood reflects a decrease in salience from the point of view of those within. Each level of neighborhood is generated and maintained by a different experiential base and serves a different function. The homogeneous and unit neighborhoods often share the same name as that given to the area in the cognitive regionalization system. But although the defended neighborhood and the regionalization systems intersect, they remain two distinct systems. The character ascribed to a homogeneous neighborhood by the residents is likely to be very different from that accorded it as part of the thematic unit within the regionalization system.

The weight of empirical research also makes two additional telling points with respect to the question of boundaries and cognition. First, urban neighborhoods, particularly the small-scale units, are highly salient, personally meaningful units. They are also highly idiosyncratic units. For example, both Lee (1964) and Ladd (1970) show that even people living in very close spatial proximity generate very different cognitive representations of neighborhood. Second, the boundaries of neighborhoods are only rarely and coincidentally those of the relevant governing unit, the community of interest. Physical barriers, land-use changes, housing-type changes, traffic flows, and a wide variety of phenomena serve as the boundaries of neighborhoods. In some situations, the boundaries are purely cognitive constructs which have no direct, tangible expression in the physical space itself.

Conclusion

Given this brief survey of the links between spatial cognition and urban neighborhoods, we can return to our original question. Is there any psychological validity to the assumption of a sense of place? There are, perhaps, two answers to the question. If one accepts the division between communities of interest and communities of interaction, then it is clear that communities of interest lack any grounding in a sense of place. They have little or no salience, little meaning in terms of name recognition, awareness of boundaries, sense of attachment, etc. If we consider communities of interaction, then there is a strong link with the sense of place. This link, however, is a complex one. The complexity resides in the intersection between spatial scale, cognitive salience, and generating processes. One can have, for example, an intense identification with a defended neighborhood without the existence of any identification in the sense of naming. Similarly, in the case of the district-naming system, it is possible to have identification without a sense of boundedness. Our understanding of the complexity is so weak that we cannot use the sense of place-neighborhood link as a planning tool. Whatever the nature of the generating processes that lead to a sense of place, we cannot yet manipulate them.

References

Appleyard, D., and Lintell, M. "The Environmental Quality of City Streets: The Residents' Viewpoint." *Journal of the American Institute of Planners,* 38 (1972): 84–101.

Canter, D. *The Psychology of Place.* London: The Architectural Press, 1977.

Cooper, C. "St. Francis Square: Attitudes of Its Residents." *Journal of the American Institute of Architects,* 58 (1971): 22–27.

Cybriwsky, R. "Social Relations and the Spatial Order in a Neighborhood in Central Philadelphia." Ph.D. Dissertation, Department of Geography, The Pennsylvania State University, 1972.

Donnelly, D., Goodey, B., and Menzies, M. *Perception Related Survey for Local Authorities: A Pilot Study in Sunderland.* Centre for Urban and Regional Studies, University of Birmingham, Research Memorandum, Number 20, 1973.

Downs, R., and Stea, D. *Maps in Minds.* New York: Harper and Row, 1977.

Eyles, J. "Inhabitants' Images of Highgate Village (London)." Graduate School of Geography, London School of Economics, Discussion Paper, Number 15, 1968.

Fried, M. "Grieving for a Lost Home." In *The Urban Condition,* edited by L. Duhl, pp. 151–71. New York: Basic Books, 1963.

Gans, H. *The Urban Villagers.* New York: The Free Press, 1962.

Keller, S. *The Urban Neighborhood: A Sociological Perspective.* New York: Random House, 1968.

Ladd, F. "Black Youths View Their Environment: Neighborhood Maps." *Environment and Behavior,* 2 (1970): 74–99.

Lee, T. "Psychology and Living Space." *Transactions of the Bartlett Society,* 2 (1964): 9–36.

Lee, T. "Urban Neighborhood as a Socio-Spatial Schema." *Human Relations,* 21 (1968): 241–68.

Lynch, K. *The Image of the City.* Cambridge, Mass.: M.I.T. Press, 1960.

Lynch, K. *Managing the Sense of a Region.* Cambridge, Mass.: M.I.T. Press, 1976.

Milgram, S. "A Psychological Map of New York City." *American Scientist,* 60 (1972): 194–200.

Ross, H. "The Local Community: A Survey Approach," *American Sociological Review,* 27 (1962): 75–84.

Saarinen, T. *Environmental Planning.* Boston: Houghton Mifflin, 1976.

Spiro, L. "A Study of City Structure Through a Comparison of Environmental Imagery and Factorial Ecology." Masters' Thesis, Department of Geography, The Pennsylvania State University, 1974.

Suttles, G. *The Social Construction of Communities.* Chicago: University of Chicago Press, 1972.

Warren, R., and Wechsler, L. "The Multiple Boundaries of Cities: A Framework for the Policy Analysis of Urban Space." In *Problems of Theory in Policy Analysis,* edited by P. Gregg, pp. 103–13. Lexington, Mass.: Lexington Books, 1976.

Willmott, P. "Housing Density and Town Design in a New Town: A Pilot Study at Stevenage." *Town Planning Review,* 33 (1962): 115–27.

Section C Mental Illness

The Ecology of Mental Disorder*

Leo Levy and Louis Rowitz

A review of the literature

In 1939, Robert E. L. Faris and H. Warren Dunham published their pioneering study of mental disorder in Chicago. Their study utilized data on all first admissions to public and private mental hospitals from 1922 through 1934 from the seventy-five community areas of Chicago (Faris and Dunham, 1960). They found that areas of high social disorganization located at or near the city center are the areas in which the highest rates of mental illness occur. Specifically, they concluded:

1. Mental illness, like other social problems, fits into the ecological structure of the city;
2. The highest rates of schizophrenia are to be found in the "disorganized communities at or near the center of the city." The schizophrenia rate tends to decline in every direction the further one gets from the city center;
3. There is a tendency, although not a very strong one, for the manic-depressive cases to come from the higher socio-economic areas. However, ecological patterns of manic-depressive cases tend to be random;

*Reprinted from Leo Levy and Louis Rowitz, *The Ecology of Mental Disorder* (New York: Human Sciences Press, 1973), pp. 1–11, 17–18, 52–55, 59–62, and 153–56, by permission of the authors and Human Sciences Press, 72 Fifth Avenue, New York, New York 10011.

4. Alcoholic psychoses have their highest rates in or near the city center, as do drug addiction and general paresis rates;
5. While the psychoses of old age show a pattern similar to the schizophrenic rates, the rates do not always show a decline farther away from the center of the city.

In brief, these were the main findings of an extremely important study. It was the first systematic foray into a luxuriously rich conceptual area. It had no comparable antecedents and stimulated many studies over the succeeding three decades.

Antecedent studies

Historically, studies of the ecology of mental disorders appeared infrequently in the psychiatric literature. In 1875, P. M. Deas looked at the distribution of insanity in five geographic subsections in Cheshire, England (Deas, 1875). He found that there was a greater development of insanity in some geographic areas than in others. In the high insanity areas, in comparison to low rate areas, Deas noted an increasing ratio of insanity among men as compared to women, a large proportion of cases due to organic disease or degeneration, a small ratio of recoveries, and a large proportion of deaths.

A. O. Wright, writing in 1884, noted the differing insanity rates reported in the 1880 census for different regions in the United States. The rate reported for the state of Massachusetts was the highest for any state and the rate declined roughly in proportion to the distance from that state traveling in any direction (Wright, 1884). The author clearly favors the explanation that the more recently settled areas of the country are inhabited by "a selected population, mostly young and middle aged people of sound minds and bodies. The insane are left behind . . ." (Wright, 1884, p. 232).

The same pattern of geographical distribution of insanity from the 1880 and 1890 censuses, analyzed by W. A. White in 1903, elicits a different interpretation. White's view is that the stresses and strains of civilization encourage the development of mental disorder. He states: "The savage in his simplicity does not know what it is to suffer from the cares and worries which are the daily portion of the European, and it is little wonder that the latter, beset by all manner of disappointments and vexations, should more frequently break down in mind than his less gifted brother" (White, 1903, p. 263). This general view leads one to predict higher rates of mental illness in older (and larger) cities than in the frontiers where life is presumed simpler.

Sutherland (1901), noting both the growth in lunacy ratios and differential distribution among the four main areas of Scotland, sets out five explanatory propositions listed in order of importance:

1. Economic capacity of householders in different counties to maintain the insane without public assistance;
2. Migration of the strong from rural to urban areas leaves the feeble in the rural areas;
3. High infant mortality in the cities eliminates many lunatics that would survive in a rural setting;
4. Conditions of modern life such as competition, abuse of alcohol and tea, poor diet, etc., set up deranged metabolism and disturb mental equilibrium;
5. The definition of lunacy is being widened and hence there is a widened portal of entry to the official registry as a lunatic.

In an interesting letter to the *British Medical Journal* in 1908, W. R. MacDermott cites differences in insanity rates for several counties in Ireland which had stable populations (same families for at least 100 years) (MacDermott, 1908). He found increases in insanity in several areas and concluded that environmental factors rather than inheritance accounted for the increase in insanity rates.

At the seventieth annual meeting of the Medico-Psychological Association in Dublin in 1911, Dr. W. R. Dawson gave a presidential address on the relationship between geographical distribution of insanity and certain social conditions in Ireland (Dawson, 1911). He pointed out that insanity is prevalent in agricultural countries and is closely tied to poverty. He saw some correspondence between insanity rates and emigration rates. Some slight relationship existed between mental illness, the prevalence of criminality, and chronic alcoholism. Dawson saw no appreciable relationship between insanity and density of population, valuation of land, or distribution of drunkenness as opposed to alcoholism.

The contributions of these authors are important in that each documents a differential spatial distribution of mental illness. Sutherland raises issues which are still current concerning socio-economic status, migration, and the problem of defining mental illness. His reflections on the strains of modern life, along with White's similar observations, may be considered romantic and not internally consistent. If, as White states, the more civilized and gifted are more prone to psychological disorder, one must account for the fact that within cities it is the poor and the less-educated who appear to manifest higher rates of mental disorder. It is necessary to note that Dawson found higher rates of insanity in rural areas with a great deal of poverty. Wright's point about the sturdier stock going out to the frontier, leaving the aged, the infirm of body and spirit, and other weaker souls, would sound reasonable if MacDermott had not found that there was an increase in insanity even in areas with very stable populations. The problem is that Wright does not deal well with the low rates in the southern part of the U.S. which was definitely not the frontier, nor with the fact that rates of insanity increase as one goes west from Colorado to California. He explains these inconsistent findings with an assertion that Negroes are mentally healthier than whites (that disposes of the low rates in the south) and that there are a lot of homeless men subjected to hardships of life in mining and lumbering camps in California. If we could discount Dawson's findings of higher prevalence in agricultural areas, it is interesting that none of the other authors appear to entertain the possibility that the differential rates of mental disorder in urban versus rural areas may be explained by the greater accessibility of psychiatric services at or near the larger and older cities and the higher visibility of deviant behavior in densely populated cities.

Studies of ecological patterning of mental illness in urban areas

Following hard on the Faris and Dunham study of Chicago, similarly designed studies were executed in St. Louis, Milwaukee, Omaha, Kansas City, Rockford, Peoria, Cleveland, and Providence (Schroeder, 1942; Queen, 1940). The findings with regard to admissions for all diagnoses generally follow the patterning observed by Faris and Dunham in Chicago. Highest rates tended to be clustered about the city center, and rates tended to lessen as distance from the city center increased. For individual diagnoses, however, the results differed somewhat from the Chicago pattern. In St. Louis, the schizophrenic rates did not follow the pattern in a conclusive way. In Milwaukee, the schizophrenic rates concentrated in the city center, but tended to extend further out than the admission rates for all diagnoses. In Omaha and Kansas City, the schizophrenic rates, with some variation, did follow the pattern for all admissions. The results in Peoria were inconclusive because of the small number of cases studied. Cleveland data lent partial support to the hypothesis that institutionalized mental patients came predominantly from areas of dense population, low economic status, and high rates of delinquency, most of which areas surround the central business district. In Providence, Faris's study of a single public mental hospital yielded results similar to those obtained in Cleveland. Faris and Dunham's findings of a general scattering of admissions for manic-depressive psychosis throughout the city was verified in each of the cities studied. Only the St. Louis study examined admission rates for senile psychosis and found, contrary to the Chicago study, a random pattern of admissions.

Gerard and Houston (1953), studying 305 male first admission schizophrenics to the State Mental Hospital from Worcester, Massachusetts, between 1931 and 1950, found a typical ecological distribution of schizophrenia. However, when analyzed by family setting, it was found that the typical ecological pattern is based on the residential pattern of a minority of patients, i.e., the single men living alone. The majority of patients were living in a family setting at the time of admission. These patients were distributed without central concentration. There was greater residential instability for patients living alone at the time of first admission than for those living in family settings. Stein (1957), in a study of first admissions for several diagnoses from two areas of London for the years 1954–1955, found higher admission rates from the higher social class west boroughs of London than for the lower class east boroughs. This pattern held for schizophrenia as well as for all diagnoses combined. Clausen and Kohn (1959) did not establish spatial patterning in their study of schizophrenics in Hagerstown, Maryland. Specifically, they argue that their data show no apparent relationship between the socio-economic levels of areas of the city and rate of hospitalization for first admissions.

A recent study by Hafner and Reimann (1970), investigating first admissions from the city of Mannheim (Germany), reports topographic distributions of mental disorder similar to that found by Faris and Dunham in Chicago.

What can be gleaned from the multiplicty of studies reviewed to this point? First, it is apparent that the studies differ in quality and design. The area of psychiatric epidemiology is fraught with methodological difficulties. . . . It appears that in large urban areas, patterning of admissions to psychiatric hospitals is a fact. The patterning is clear enough for all diagnoses combined but becomes more disputed as separate diagnostic categories are considered. Further, the spatial arrangement of admissions appears to be subordinate to certain broad socio-economic and demographic variables which create the identity of areas. There is nothing inherent in the spatial location of the central city which produces numerous psychiatric casualties. Rather, it is the fact that large cities generally have their poverty concentrated in the city center and poverty abates as one proceeds outwards to the suburbs. This is certainly not a perfect relationship and all large cities are not similar, but the relationship is good enough, apparently, and cities beyond a certain size are similar enough, apparently, to create a series of effects mainly contingent upon the poverty variable.

Issues raised by studies of ecological distribution of mental disorder in urban areas

Aside from serious problems of methodology . . . , these studies and others which will be dealt with shortly pose a number of fascinating issues:

1. Are schizophrenia and other mental disorders in any important sense caused by the environment in which a person is raised? This has come to be referred to by the shorthand phrase "social causation."

2. If schizophrenia and other mental disorders are not in any important sense caused by factors

definable as environmental, then are there specificable processes by which cases aggregate spatially and demographically? This has come to be referred to by the shorthand phrase "social selection."

3. Do measures of incidence and prevalence produce comparable ecological distributions of schizophrenia and other mental disorders? Incidence is generally defined as first admissions, treated prevalence as unduplicated admissions, and prevalence as complete count of disordered persons for an area.

As one can readily understand, these issues strike at the heart of some vital concerns for a number of scientific disciplines and the Faris and Dunham study of Chicago posed the issues in sharp terms. Myerson (1940), in reviewing the study, immediately raised the issue of the "drift hypothesis" (now subsumed under social selection), an issue much discussed to the present. In general, the drift hypothesis holds that persons with schizophrenia, as a result of the illness, drift downwards in the social structure. Measures of the drifting process can be intergenerational or intragenerational. Operationally, this would mean demonstrating either a status differential between the father of the schizophrenic and the patient or a status differential in the patient at different points in his life (e.g., at the point of first admission and at some later point). The issue of drift has been applied to schizophrenia, but could well be applied to any chronic disabling mental illness which has its onset relatively early in the person's life. It turns out, however, that the other categories of psychiatric diagnosis do not meet these specifications as schizophrenia does or is thought to. Manic-depressive psychosis, for example, while possibly chronic, is not disabling except during relatively brief episodes of illness which are thought to be self-limiting with or without treatment. Psychoneurosis, psychosomatic disorders, and personality disorders, while possibly chronic, are for the most part never as disabling as schizophrenia. Senile dementia is chronic and progressively disabling but its onset is too late in life to make a measure of downward drift meaningful. Perhaps alcoholism and possibly other addictions come closest to the criteria of early onset, chronicity, and serious disability which characterize schizophrenia. Indeed, the rapid downward descent of the alcoholic is much more familiar in the popular literature than that of the schizophrenic.

LaPouse, Monk, and Terris (1956) conclude from their two-year study of first admission schizophrenics to two mental hospitals serving the Buffalo (N.Y.) area that the concentration of schizophrenic admissions from low socio-economic areas is not the result of drift of these patients into these areas. They also conclude that the concentration of these patients in low socio-economic areas is not the result of upward mobility of normal persons and the stagnation of schizophrenic patients. Their study is in agreement with earlier data from a study by Hollingshead and Redlich (1954), which found that parents of schizophrenic patients had not been located in higher social classes than their schizophrenic children.

Turner and Wagenfeld (1967), using data from the Monroe County (N.Y.) psychiatric case register, studied a group of 214 schizophrenic first admissions. They conclude that there is indeed intergenerational drift (which they call social selection) and that this factor substantially explains the heavy loadings of schizophrenic patients in the lower occupational categories. Intragenerational drift (which they call social drift) makes only a minor contribution to this finding. Their conclusions are in agreement with Grunfeld and Salvesen (1968), who found a decline intergenerationally in social status in their sample of 85 schizophrenic patients admitted to Gaustad Hospital (Norway) between 1955 and 1958. A five-year follow-up of these patients compared to a group of 101 reactive psychosis patients showed additional (social) drift.

Dunham has presented arguments in his recent writings to explain the concentration of schizophrenics in the lower socio-economic areas of the city. In *Community and Schizophrenia* (1965), he carefully contrasts two communities in Detroit in which he established the incidence of schizophrenia to differ by a factor of 3. He proceeds to demonstrate that this apparent difference is reduced to insignificance when residential mobility into the areas is taken into account. He states (Dunham, 1965, p. 252): "The evidence pointed very clearly to the fact that the difference in rates was due to the patterns of mobility of those families that produced schizophrenia. . . . The evidence clearly pointed to the operation of certain forces within the social system that apparently selected certain families that produced schizophrenics for certain subcommunities within the city. . . ."

Thus Dunham's position on the concentration of schizophrenics in the lowest socio-economic stratum is clear. It is a function of their disease and its disabling characteristics. He goes on to state (Dunham, 1965, p. 252): "There is no basis for asserting that one social class is likely to produce more schizophrenics than another social class. . . ." On the other hand, Dunham rejects "drift" as a concept, since for him this implies an unconscious process over which the individual exerts no control. The preferable concept is "social selection." In an earlier essay, he describes the process of social selection as follows (Dunham, 1961, p. 247):

> Certain persons because of age, sex, personality traits, intelligence, emotional instability, psychotic proneness, are selected for certain positions in occupational groups, city areas, marital status categories, institutions and the like in contrast to other positions in these structures as the social system moves through time. The process may be either active or passive as far as the person is concerned and through it one can account for significant differences in the rate of mental illness.

Thus there appears to be a shift in point of view from a social causation hypothesis advanced in Dunham's early work to a social selection hypothesis in his later work. However, in a personal communication from Professor Dunham, it was pointed out that social selection is not a true opposite of social causation. Social selection can be used to explain the distribution pattern formed by admission rates, and to provide certain notions as to where to look for those social factors that might induce some psychiatric states. Social causation might be operative even though a process of social selection places more patients in some community areas than others. . . .

The study population

Data in the present study were collected for 10,653 Chicago residents who were admitted to forty-four public and private mental institutions and psychiatric units of general hospitals from July 1, 1960, to June 30, 1961 (13 state hospitals, 18 psychiatric units of general hospitals, and 13 private psychiatric institutions).

The study population accounted for 12,519 mental hospital admissions during the period 1960–61. Specifically, 1,384 people had more than one admission during the study period. Also, 67.8 percent of the study population had stays in mental hospitals prior to the study period. There were 3,431 people admitted to a psychiatric inpatient unit for the first time during 1960–61 (32.2%).

Information was provided for the individual patients on the following variables:

1. Sex
2. Race
3. Marital status
4. Chicago community area of residence
5. Institution assigned to
6. Type of institution (private sanitarium, general hospital, state hospital)
7. Date of birth
8. Place of birth
9. Prior psychiatric history
10. Diagnosis (using standard American Psychiatric Association nomenclature).

No information is available on occupation, educational history, family composition, income, and residential stability. . . .

Major findings

Total unduplicated admissions

The major finding about persons admitted to hospitals (whether first admissions or readmissions) with a psychiatric diagnosis during the study period from Chicago is that the people who demonstrate the highest utilization of mental institutions tend to come from areas with high percentages of poor people ($r = 0.41$) and high percentages of substandard housing (0.57). The high mental hospital utilization areas also tend to be areas which lack residential stability (0.42). The areas also tend to be high delinquency areas (0.46) and areas with high rates of successful suicides (0.54).

It is worthwhile in passing to note that state hospital admission rates in 1967 correlate highly (0.72) with the unduplicated admission rate in 1961. This would seem to indicate that high utilization areas during the study period remained high six years after the study was completed.

Total first admissions

With regard to first admissions as a whole, some interesting variations in pattern are found which contrast sharply with the findings on unduplicated admissions. First admissions tend to come from areas with more economically affluent populations (0.50), with higher median education (0.48), and a larger percentage of white-collar employees (0.67). First admissions also tend to use private sanitaria (0.63) and psychiatric units of general hospitals (0.79) rather than state hospitals (0.35).

Schizophrenia

Many of Faris and Dunham's findings relate to first admission schizophrenia. Very weak relationships were found between the income variables: income under $3000, $r = -0.30$; income over

$10,000, r = 0.26$, which implies that first admission schizophrenics can come from any area of the city. However, with the total unduplicated schizophrenic admissions, the correlations related to income become stronger. Schizophrenics come from areas with poor populations (0.77) and large percentages of substandard housing (0.71). The high utilization areas also correlate highly with the residential mobility variable (0.60) and with high unemployment (0.71). Schizophrenics also tend to come from areas with high male delinquency rates (0.75), percent illegitimate births (0.62), and percent public assistance recipients (0.72). Another interesting finding relates to unduplicated readmissions. A correlation of 0.85 was found between readmitted schizophrenics and income under $3,000, whereas the correlation between readmitted schizophrenics and income over $10,000 is -0.73.

Manic-depressive psychosis

The manic-depressive diagnostic group does not tend to come from areas with a large percentage of blacks (-0.41) but does tend to come from areas with a larger percentage of foreign stock (0.40). Manic-depressive patients tend to come from economically more affluent areas (0.43) with higher median school years completed (0.46). The areas also tend to have large percentages of white-collar employees (0.51). Male delinquency tends to be lower in the areas from which this diagnostic group comes (-0.45).

Diseases of the senium

Those people diagnosed as senile and/or arteriosclerotic tend to come from areas with large black populations (0.67), small aged populations (-0.45), and smaller foreign-born populations (-0.56). These patients also come from relatively poorer areas of the city (0.72) with large percentages of substandard housing (0.70). The neighborhoods tend to be characterized by residential mobility (0.53), with high unemployment rates (0.71). The delinquency rate also tends to be high in the areas from which these people come (0.71), as are the rates on percentage of illegitimate births (0.63) and percentage on public assistance (0.73).

Alcoholism

Alcoholic patients tend to come from low-income areas (0.49) with high percentages of substandard housing (0.63). There is a tendency, although weak, for these patients to come from high unemployment areas (0.41) and also areas with some transiency (0.37). These areas of alcoholic admissions also tend to be high on male delinquency (0.53), the percentage of illegitimate births (0.40), and the percentage on public assistance (0.41). The ecological relationships are on the whole much more difficult to evaluate for the alcoholic group. Not all alcoholics go to mental hospitals. Many go to jail. Further research is needed into the factors which lead the alcoholic into a mental institution rather than a jail.

Psychophysiological, psychoneurotic, and personality disorders

The strongest relationship found for the psychophysiological, psychoneurotic, and personality disorders diagnostic groups is that they tend to come from high-income areas (0.53), with higher median school years completed (0.62), and a high percentage of white-collar employees (0.75). There is also a tendency for these areas to have a small black population (-0.43). Considering the fact that the data in this study exclude outpatient facilities and private psychiatrists, the findings for these diagnoses present some interesting hypotheses as to whether there is a relationship between social class and the diagnoses of psychophysiological, psychoneurotic, and personality disorders similar to the findings reported by Hollingshead and Redlich (1958). Unfortunately, the data in this study do not allow us to directly test these hypotheses since no information is available on the patient population with respect to occupation, education, and income. . . .

Discussion

The evidence for a drift or social selection hypothesis

First admissions for all diagnoses are recorded as higher from areas which may mainly be described as white, nonpoverty areas of the city. These first admissions appear to be routed most frequently to psychiatric units of general hospitals and private psychiatric hospitals. This finding may be contrasted with the observation that unduplicated admissions most frequently originate in areas associated with poverty and social disorganization. The following explanation, while certainly not obligatory, is consonant with the data.

The person having a first episode of severe mental disorder who lives in a well-structured, middle-class community and is in all probability covered by hospitalization insurance will seek treatment in a private rather than a public facility. Chances are that this treatment will be more effective and recovery take place in a relatively briefer period of time than in an overcrowded and understaffed state mental hospital. Also, the prospect of entering a private facility is so much more agreeable than that of entering a state hospital that one may assume persons will seek aid earlier and more willingly, thus again improving their chances for stable recovery. Upon discharge, the patient's chances of avoiding subsequent readmission are held to be much better if his first admission originated in a middle-class area.

Following release from initial hospitalization, it is possible that those patients most devastated by the episode of mental illness and thus the best candidates for subsequent readmissions may actually drift to (or select)[1] other areas of the city to which they will move. These areas will presumably be less socially demanding and more in line with his impaired earning ability.

If the above process in fact occurs, then one has an explanation for the higher first admission rates (with no subsequent readmissions) in the more affluent, better structured middle-class areas and the higher unduplicated admission rates (mainly readmissions) from the less affluent, more fluidly structured lower-class areas.

The most direct evidence to test this hypothesis must be case evidence taken on a person longitudinally over a long period of time. This evidence we do not have but some studies [have been conducted] using this and alternative procedures. . . . The evidence for a drift or social selection hypothesis, while not compelling, is still substantial.

This general line of reasoning is particularly applicable to schizophrenia, where we have noted that first admissions appear to be fairly randomly distributed throughout the city, crossing freely racial, ethnic, and socio-economic boundaries. The drift or social selection hypothesis has most usually been applied to the schizophrenic population, although it may well hold for the senile dementia patients as well. Schizophrenia is generally held to be the most chronic, disabling, and refractory of all mental disorders. Downward drift in the social structure after the onset of this disorder seems very likely in a competitive society which emphasizes superficial sociability, high earning power, and "making it." Our data also supports a drift phenomenon for senile dementia, but not for manic-depressive psychosis, psychoneurotic, psychophysiological, and personality disorders, or alcoholism.

Label versus disease

Consonant with previous studies, psychophysiological, psychoneurotic, and personality disorders appear to be diagnoses associated with the middle class. A similar situation obtains with the diagnosis of manic-depressive illness. The psychiatric epidemiologist is always faced with the problem of determining whether these disorders are really middle-class disorders or whether these labels get assigned to middle-class patients because they are somewhat more hopeful and agreeable than alternative diagnoses such as schizophrenia, senile dementia, and alcoholism. It is obvious that the data to definitively decide this issue is not present in this study. We can simply note that our data does, in fact, break diagnoses along socio-economic lines and express our own preference for a labeling hypothesis. . . .

Diagnosis

Of particular interest is the contrast between our findings with regard to diagnosis of first admissions and those of Faris and Dunham done in Chicago some thirty years earlier. The only area of obvious disagreement is the distribution of schizophrenic first admissions. Our results show schizophrenic first admissions to be distributed in random fashion over the city, while Faris and Dunham found a definite concentration in the city center and, moreover, that the admission rates abated in all directions from the city center to the city perimeter. Later findings by Dunham (1965) based on research in Detroit on schizophrenia did not corroborate his Chicago findings and indeed facilitated an important shift in his thinking from a social causation to a social selection hypothesis. . . . The findings reported in *Community and Schizophrenia* (1965) are compatible with our findings with regard to schizophrenia. Still, the patterning of unduplicated admissions does approximate very well Faris and Dunham's original finding for first admission schizophrenics. This is explained by our study tapping into a "latent" middle-class schizophrenic population because of our coverage of psychiatric units of general hospitals which were largely nonexistent at the time of the Faris and Dunham study. The juxtaposition of patterns between first admission and unduplicated admission (readmission) schizophrenics is our strongest bit of evidence in favor of a drift phenomenon in this disorder.

---

[1] Dunham (1961) makes a case for social selection as a concept superior to and different from drift. However, the end product in any case would appear to be the same, i.e., the person moves downward socially and this downward movement is reflected in his choice of neighborhood in which to reside.

In other respects the two studies do agree. For total first admissions (all disorders), manic-depressive psychosis, alcoholism, and disease of the senium, the distributions are similar. Total first admissions and unduplicated admissions show a concentration in the city center and a tendency to abate as one moves to the outer perimeter. Manic-depressive first admissions and unduplicated admissions tend (weakly) to come from middle-class areas and thus tend to show a fairly random pattern throughout the city. Alcoholism first admissions and unduplicated admissions concentrate in the city center and tend to fall off as one proceeds to the periphery. Senile first and unduplicated admissions cluster in the city center but do not show any uniform decline away from the city center. Hospitalization for this disorder is strongly related to poverty. Additionally, psychophysiologic, psychoneurotic, and personality disorder first and unduplicated admissions show a scattering throughout the city. If any trend is discernible, it is a tendency for such admissions to be correlated with middle-class social credentials at the time of first admission. However, readmission patterns weaken this trend in that admissions come from both lower- and middle-class communities.

Social Change

Much evidence is available to support the contention that social change is correlated with mental hospital admission rates. The literature on migration and mental illness, while not unanimous on this point, does tend to implicate relocation with an elevated risk for developing a mental illness (see, for example, Murphy, 1961; Kantor, 1965). The breakdown of old communities due to urban renewal (Fried, 1963) causes the suspension of established systems of social support; this could be expected to precipitate emotional disorder in persons who might otherwise continue to function well. The risks of rapid social change are particularly great for older persons but are apparently quite general for all ages. In our study it is precisely those areas of the city which are undergoing rapid and drastic social change which appear to create the highest rates of psychiatric casualties.

Racial and Ethnic Composition of Communities

One manifestation of social change which is highly visible in large cities like Chicago is the change in the racial and ethnic composition of a neighborhood. Beyond this, minority status per se in a community area elevates the risk for mental hospital admission. Conversely, it is the all-white areas where rates of unduplicated admission are lowest and the black areas where rates of first admission are lowest.

Poverty

The old villain poverty emerges in our study, once again, as a significant factor related to the prevalence of serious mental disorder. It would appear, however, that the relationship is not causal. Poverty does not relate in any clear fashion to incidence. The poor would appear to develop mental disorders with similar frequency as the well-off, but are less successfully treated and maintained in a functional status. As a side issue, one might recall that on occasion the criterion of social competence is invoked as a measure of mental illness for the poor, but considered as an inadequate criterion for the well-off. One might consider that mental illness results in decreased social competence in the poor because of inadequate treatment, while the ravages of mental illness are better contained in members of the middle class. The phenomenon, instrumentally related to the phenomenon of drift or social selection, causes an accumulation of cases (increased prevalence) of serious mental disorder in the lower classes and thus in lower-class neighborhoods.

Extrusion

Finally, the phenomenon of extrusion bias appears to be critical when one's measure of psychiatric casualties is mental hospital admissions. Communities vary in their propensity to push out or hold on to persons with given kinds and degrees of mental disorder. It may turn out that this property of communities is more significant in determining rates of mental hospital utilization (or other institutional use patterns) than any standard characterization of the properties of a patient ripe for mental hospitalization. It is interesting to reflect parenthetically on the standard criteria for admission to a mental hospital: "dangerous to self . . . dangerous to others . . . incapable of managing his own affairs. . . ." Such criteria, as our friends in the legal profession are bound to point out, are extremely vague and allow for a maximum of interpretation on the part of the person making the judgment. It could be argued that the criteria are vague for the same reason that such vague (and lately thought to be unconstitutional) laws against vagrancy and loitering are put on the books. Vagueness allows selective enforcement of the law at the discretion of the police. Similarly, vague criteria for mental hospitalization allow discretion on the part of communities as to who goes into a hospital and who is handled in the community by other means. Extrusion bias is a characteristic of a community which is determined by the cohesiveness of its social structure, the presence and intactness of interpersonal support systems, and its characteristic attitudes towards mental illness and psychiatry.

References

Clausen, J. A., and Kohn, M. L. "Relation of Schizophrenia to the Social Structure of a Small City." In *Epidemiology of Mental Disorders*, edited by B. Pasamanick. Washington, D.C.: American Association for the Advancement of Science, Publication No. 60, 1959.

Dawson, W. R. "Presidential Address on the Relation Between the Geographical Distribution of Insanity and That of Certain Social and Other Conditions in Ireland." (70th Annual Meeting of the Medico-Psychological Association, Dublin.) *Journal of Mental Science*, 57 (1911): 571-97.

Deas, P. M. "An Illustration of Local Differences in the Distribution of Insanity." *Journal of Mental Science*, 21 (1875): 61-67.

Dunham, H. W. "Social Structures and Mental Disorders: Competing Hypotheses of Explanation." In *Causes of Mental Disorders: A Review of Epidemiological Knowledge*. New York: Milbank Memorial Fund, 1961.

Dunham, H. W. *Community and Schizophrenia*. Detroit: Wayne State University Press, 1965.

Faris, R. E. L., and Dunham, H. W. *Mental Disorders in Urban Areas*. New York: Hafner Publishing Co., 1960.

Fried, M. "Grieving for a Lost Home." In *The Urban Condition*, edited by L. Duhl. New York: Basic Books, 1963.

Gerard, D. L., and Houston, L. G. "Family Setting and the Social Ecology of Schizophrenia." *Psychiatric Quarterly*, 27 (1953): 90-101.

Grunfeld, B., and Salvesen, C. "Functional Psychoses and Social Status." *British Journal of Psychiatry*, 114 (1968): 733-37.

Hafner, H., and Reimann, H. "Spatial Distribution of Mental Disorders in Mannheim, 1965." In *Psychiatric Epidemiology*, edited by E. H. Hare and J. K. Wing. London: Oxford Press, 1970.

Hollingshead, A. B., and Redlich, F. C. "Schizophrenia and Social Structure." *American Journal of Psychiatry*, 110 (1954): 695-701.

Hollingshead, A. B., and Redlich, F. C. *Social Class and Mental Illness*. New York: Wiley and Sons, Science Editions, 1958.

Kantor, M. B., ed. *Mobility and Mental Health*. Springfield, Ill.: Thomas, 1965.

LaPouse, R., Monk, M., and Terris, M. "The Drift Hypothesis and Socioeconomic Differentials in Schizophrenia." *American Journal of Public Health*, 46 (1956): 978-86.

MacDermott, W. R. "The Topographical Distribution of Insanity." *British Medical Journal*, 2 (1908): 950.

Murphy, H. M. B. "Social Change and Mental Health." *Milbank Memorial Fund Quarterly*, 39 (1961): 385-445.

Myerson, A. "Review of 'Mental Disorders in Urban Areas'." *American Journal of Psychiatry*, 96 (1940): 995-97.

Queen, S. A. "The Ecological Study of Mental Disorders." *American Sociological Review*, 5 (1940): 201-9.

Schroeder, C. W. "Mental Disorders in Cities." *American Journal of Sociology*, 48 (1942): 40-47.

Stein, L. "Social Class Gradient in Schizophrenia." *British Journal of Preventive and Social Medicine*, 2 (1957): 181-195.

Sutherland, J. F. *The Growth and Geographical Distribution of Lunacy in Scotland*. Glasgow: British Association for the Advancement of Science, September 1901.

Turner, R., and Wagenfeld, M. "Occupational Mobility and Schizophrenia: An Assessment of the Social Causation and Social Selection Hypotheses." *American Sociological Review*, 32 (1967): 104-13.

White, W. "The Geographical Distribution of Insanity in the United States." *Journal of Nervous and Mental Disorders*, 30 (1903): 257-79.

Wright, A. "The Increase of Insanity." *Conference on Charities and Corrections* (1884), pp. 228-36.

Section D Commuting

Urban Daytime Population: A Field for Demographic–Ecological Analysis*

Donald L. Foley

In large American cities there is a marked discrepancy between the spatial distributions of resident population and daytime population. Viewed dynamically, this involves daily movement from place of residence to various other locations and return. The most typical form of such movement is embraced in the term commuting, referring to the daily trip to and from work, presumably from a residence sufficiently distant so that some form of transportation other than walking is involved.

Three main questions will be treated in this paper: (1) What salient features of urban ecological structure contribute to extensive differences between daytime and nighttime population distributions? (2) What key characteristics of daytime population movement and distribution can we identify, as relating to large American cities? (3) What are the status and prospects of daytime population study as a subfield of demography and ecology?

*Reprinted from *Social Forces*, 32 (May 1954), 323–30, by permission of the author and The University of North Carolina Press.

Urban ecological structure and daytime population

The phenomenon of daytime population has an ecological basis, for it is the fact that large numbers of residents have different *spatial* locations during the day than they have at night that concerns us. But we must not assume that we "explain" daytime population by describing the American city's ecological patterning. We must go beyond the ecological to identify contributing historical, economic, and technological factors.

Two main points deserve stress: (1) In the contemporary large American city, a mosaic of functional areas has evolved seemingly as an inevitable counterpart of the broader fact of economic specialization. Ecologists term this process segregation. So long as a city is characterized by specialization and, specifically, by segregation, we can expect that communication and movement among these divergent functional areas will be necessary if that city is to function as an integrated community. As Hawley has stated:

... from a spatial standpoint, the community may be defined as comprising that area the resident population of which is interrelated and integrated with reference to its daily requirements, whether contacts be direct or indirect.[1]

The development of efficient communication devices, particularly the telephone and postal service, has made it possible for much daily activity to be handled without movement of persons. Nevertheless, as will be further documented below, a vast amount of daily travel *is* necessary. In Liepmann's words, based mainly on British experience but also applicable to our cities:

The functioning of modern conurbations, with dormitories and work places divorced, is absolutely dependent upon an elaborate system of transport services. What has happened is that the technical possibility of carrying masses of people considerable distances to and from work has been utilised, and more or less lengthy journeys have become the routine of millions of workers. . . . [2]

Summarizing, then, we have stressed the general and very fundamental fact that movement of persons in the course of carrying out day-to-day activities provides a dynamic mechanism by which the city's various functional areas are linked. Up to this point, however, we have not indicated anything about the ecological patterning of functional areas as this bears on the daily movement of persons.

(2) A major characteristic of American city ecological structure is that for a variety of reasons centrally located daytime destinations tend to remain fairly strong, while residential areas have for several decades been undergoing considerable dispersal. The full impact of this fact on daytime population will be discussed in the next section of this paper. At this point it seems appropriate to discuss the combination of technological and other factors that have contributed to this centralization of work and other nonresidential functional areas and to the dispersal of residential areas.

The early American factories were generally developed in multi-storied buildings and surrounded by densely packed workers' residences. The particular manner in which steam or water power could be applied to industrial operations and the absence of good transportation discouraged dispersion. Early cities also tended to develop a centrally located commercial district which could conveniently serve the compactly developed community.

In each of our older cities, we have inherited such a densely built-up and functionally important central area, the heart of which we typically term the downtown or central business district. This contains the large department stores, other specialized stores, a welter of offices, and various services and amusements. This district tends to shade off into loft-building types of functions: warehousing and light manufacturing.[3] There may also be other industrial uses but these are by no means restricted to the central area.[4] The surrounding areas that once contained workers' homes have by now typically deteriorated into slums. They have been invaded for various nonresidential uses: parking lots, small repair firms, warehousing, etc.

With the advent of improved transportation, coupled with our particular history of successive waves of immigrants who tended to locate in close-in slum areas, it became possible and socially "desirable" for those who could to move farther out to "better" residential areas. With the exception of certain high- and medium-rent apartment districts in our largest cities, the strong and general trend was toward the development of a veritable sprawl of outlying middle- and upper-class residential districts. The newest post–World War II developments have, of course, pushed far-out suburban "colonies" to new extremes. Some close-in newer apartments have been erected and more may be in store with continued urban redevelopment, but essentially the trend has been for central residential densities to drop off while central daytime destinations—for employment, shopping, business errands, or professional visits—continue in strength.

Manufacturing plants have in recent years been

[1]Amos H. Hawley, *Human Ecology: A Theory of Community Structure* (New York: Ronald Press, 1950), p. 257.

[2]Kate K. Liepmann, *The Journey to Work: Its Significance for Industrial and Community Life* (New York: Oxford University Press, 1944), p. 81.

[3]See Chester Rapkin, "An Approach to the Study of the Movement of Persons and Goods in Urban Areas" (Doctoral dissertation, Columbia University, 1953), especially chap. 6; Robert B. Mitchell and Chester Rapkin, Traffic and Urban Land Use (forthcoming) [Ed. note: Robert B. Mitchell and Chester Rapkin, *Urban Traffic, A Function of Land Use*, Westport, Ct.: Greenwood Press, 1974]; Richard U. Ratcliff, *Urban Land Economics* (New York: McGraw–Hill, 1949), chap. 13; and Gerald W. Breese, *The Daytime Population of the Central Business District of Chicago* (Chicago: University of Chicago Press, 1949), especially chap. 3.

[4]Edgar M. Hoover, *The Location of Economic Activity* (New York: McGraw–Hill, 1948), p. 128.

undergoing very considerable dispersal. In the long run, this outward movement of industry may well revolutionize our cities. But as of the present, two facts need airing: (1) This industrial dispersal is usually not so well coordinated with the development of new satellite communities that we can expect to find most employees of these new plants being accommodated in immediately adjacent communities. Thus, while the direction of home-work movements is being shifted, long trips are by no means eliminated. (2) For a variety of reasons there is still tremendous inertia in industrial location, involving the fact that many well-established, centrally located firms have huge investments in their present plants so that they continue to cling to them. There is evidence, for example, that the net impact of World War II military production, with prime contracts being channeled as they were through the large established firms, was to strengthen the existing pattern. Only currently does the dispersion policy that the federal government has been encouraging for defense reasons seem to be adequately implemented.

It is much more conceivable that suburban residents in the future may be able to shop nearby than to work nearby. The carefully planned outlying shopping center—with one or more branch department stores as a nucleus, with adequate parking, and with convenient shopping hours—has been coming into its own during the past few years. However, there seem to be marked differentials in the extent to which outlying residents do in fact use facilities in their own districts rather than traveling to more distant facilities, particularly those in or near the central business district.[5]

Key characteristics of daytime population movement and distribution

Our next task will be to present something of a profile describing the general character of this travel and the resulting daytime distribution of population for large American cities.

Here we are obviously attempting to cut through the maze of individual trips made in any one city during the course of a day. We shall use Mitchell's and Rapkin's concept, "movement system," which Rapkin defines as

> ... an analytical classification of individual movements within the total structure of movement. It consists of a large group of individual movements which occur within a particular period of time, but which are not necessarily concurrent. ... Systems of movement are specified in such fashion that they relate functionally to organized patterns of social and economic activity.[6]

Two main movement systems are suggested: (1) Movement of persons classified by "purpose" of trip, such as utilized in origin-destination surveys now being conducted in the U.S.: work, shopping, medical-dental care, etc. (2) Movement of persons cross-classified by the general functional areas in which each trip has its origin and destination (hence by such areas as industrial, central business district, residential, etc.). These two movement systems represent alternate ways of classifying total daily trips within a city.

Let us summarize some of what is known about weekday trips, classified by purpose.[7]

(1) The most important single type of trip is the "journey to work." Roughly two out of every five trips from home are to work. Since these trips tend to be concentrated during morning and afternoon peak hours they are a major factor in causing heavy rush-hour traffic. (As will be noted below, the more traditional European concern has been for this work trip only; the British, for example, have defined daytime population as workers at their "work places.")

(2) The next most prevalent trip is for "social-recreational" purposes, accounting for about one-fourth of all trips (as defined for origin-destination survey purposes). During work days, this is typically an evening trip.

(3) "Shopping" and "business" trips are next in importance. Shopping tends to be most prevalent in the afternoon. Thus, the weekday business district density is greatest in the afternoon, and the late-afternoon trips home from the downtown districts directly contribute to the 5:00 p.m. transportation congestion.

The trip to work warrants the closest examination:

(1) Generally, the larger the city, the longer the trip to work.[8] In cities under 100,000 the bulk of these trips takes no more than half an hour.

---

[5]Donald L. Foley, *Neighbors of Urbanites? The Study of a Rochester Residential District* (Rochester: Department of Sociology, University of Rochester, 1952).

[6]Rapkin, op. cit., p. 81.

[7]Based on averages from recent origin-destination surveys (Minneapolis–St. Paul, 1950; Portland, Oregon, 1946; Salt Lake City, 1946; Tacoma, 1947; Harrisburg, 1946; Racine, 1951; and Johnstown, Pennsylvania, 1949).

[8]J. Douglas Carroll, Jr., "Home-Work Relationships of Industrial Employees: An Investigation of Relationships of Living and Working Places for Industrial Employees with Attention to Implications for Industrial

For New York City, on the other hand, a pre-war survey of Manhattan workers showed that two out of three spent at least 40 minutes each way.[9]

(2) According to Carroll, the central district is "generally the greatest single point of employment concentration."[10] Since such a large portion of all employment is centrally located, the farther one lives from the central part of the city, the longer is the average trip to work. A recent Chicago study, for example, found that for a housing project 1.5 miles from the city center, the average distance to work was 1.6 miles; but for a project 16.5 miles from the center, this average increased to 8.6 miles.[11]

(3) Trips to work in the central district are typically longer than trips to work in off-center industrial districts, for any given city.[12] Lower income groups tend to live closer to work, and are more likely to use automobiles.[13]

(4) Carroll has hypothesized that "each worker seeks to minimize distance from home to work...."[14] Ranyak, by way of exploring Carroll's research and striving for a more inclusive theory, has suggested this modification: "People tend to minimize their journey to work, maximize their employment benefits and maximize their residential amenities."[15]

Now let us turn to the second movement system: trips cross-classified by general functional areas of origin and destination. During each weekday there is an exodus from residential areas and an inflow into industrial and commercial areas. The question is: how great are the exodus and the inflow? We shall present certain relevant empirical findings.

The broadest view has been provided by recent special tabulations of origin-destination traffic survey data carried out by a research group at the University of North Carolina.[16] Their findings are summarized as Table 1. For the five cities studied, they found that by early afternoon only about three out of every four residents remain in residential areas. During most of the regular working hours, about 13 percent of all residents are to be found in the city's industrial areas. During the early afternoon a peak of about 16 percent of the residents are in commercial areas. (Which portion is in the central business district and which portion is in outlying shopping districts is not reported.) Streets claim nearly one in ten residents during the late afternoon rush hour and early evening.

We also know that on each weekday there is an overwhelming tendency for population to move in toward the central areas (whether business or industrial) during the morning with a reverse flow back to the outlying residential areas late in the afternoon.

Some general evidence for this inward daily flow is shown in Table 2, providing an interpretive summary of recent population estimates prepared by the U.S. Bureau of the Census for the Federal Civil Defense Administration. Neither the Bureau of the Census nor the F.C.D.A., of course, assumes responsibility for the use or interpretation made of these data. From the 120 U.S. cities included in the project, the five largest cities were selected for which the concentric rings used for estimating purposes appeared to have been centered within their respective central business districts. The daytime population estimates are intended to represent "expected normal maximum" figures in the sense of maximums that might usually be expected to occur under normal daytime conditions. Therefore the overall estimates may be somewhat high for the purposes intended here. It should be further noted that "these estimates can only be considered very rough," because several of the "more important assumptions, while appearing not unreasonable, have been made with little actual supporting evidence." Nevertheless, a general trend is certainly clear. While on the average only 4 percent of the population in these cities resides in a zone approximating a radius of one mile from the central business district, the estimated daytime equivalent amounts to 30 percent. Similarly within a radius of two miles resident population amounts to about 15 percent of the city total, but estimated

---

Siting and City Planning" (Doctoral dissertation, Harvard University, 1950), pp. 34, 40; hereafter referred to as "Home-Work Relationships of Industrial Employees."

[9]Homer Hoyt and L. Durward Badgley, *The Housing Demand of Workers in Manhattan* (New York, 1939), Table 5, p. 29.

[10]J. Douglas Carroll, Jr., "The Relation of Homes to Work Places and the Spatial Pattern of Cities," *Social Forces,* 30 (March 1952), p. 280.

[11]Robert F. Whiting, "Home-to-Work Relationships of Workers Living in Public Housing Projects in Chicago," *Land Economics,* 28 (August 1952), p. 287.

[12]J. Douglas Carroll, Jr., "The Relation of Homes to Work Places and the Spatial Pattern of Cities," pp. 277–79.

[13]J. Douglas Carroll, Jr., "Home-Work Relationships of Industrial Employees," chap. 6.

[14]Ibid., p. 160.

[15]John A. Ranyak, "A Theoretical Approach to the Journey to Work" (Bachelor's thesis, Massachusetts Institute of Technology, 1952), pp. 11–12.

[16]Institute for Research in Social Science, University of North Carolina, Industrial Areas Study, *Population Distribution—Spatial and Temporal: A Study of Daytime–Nighttime Differentials in the Proportionate Distribution of the Total Population of Selected Urban Areas* (September 1952).

Table 1. Mean Percent of Resident Population Present in Functional Areas, Selected Hours, Five U. S. Cities*

Hour of Day	Total Study Area	Commercial Areas	Industrial Areas	Residential Areas	Streets**
3:00 a.m.	101	4	7	89	1
6:00 a.m.	101	3	7	89	2
9:00 a.m.	102	9	16	73	3
12:00 noon	102	13	16	68	4
3:00 p.m.	102	13	16	66	6
6:00 p.m.	101	5	9	78	9
9:00 p.m.	101	5	9	76	9
12:00 midnight	101	3	8	85	5

*These cities with date of survey and population of study area in thousands were: Philadelphia-Camden (1947–2,479)
Minneapolis-St.Paul (1949 – 919)
Grand Rapids (1947 – 221)
Flint (1950 – 206)
Erie (1948 – 135)

**Peak "Street" percentages occurred at 8:00 a.m., with 6 percent, and at 5:00 p.m. with 10 percent, as computed for each hour on the hour.

Source: Adapted from Institute for Research in Social Science, University of North Carolina, Population Distribution, Spatial and Temporal (September 1952), Table 5, p. 76. Figures may not add to totals because of rounding and the omission of a "miscellaneous" category.

daytime population equals half of the city's resident population. And within a radius of a four miles, the resident population totals less than half of the city population, while estimated daytime population slightly exceeds three-fourths of the city's resident population. Clearly, these figures illustrate the definite tendency for daytime population to show concentration within the metropolitan area.

A recent study of travel to the various commercial centers within the Washington D.C. metropolitan area compared trips to the core area of the central business district with trips to 16 outlying shopping centers. It is possible that the central business district is less important in Washington than it is in other large American cities because of Washington's unique economic base. Yet slightly over three-fifths of the total trips were to the core area. Of all trips to commercial districts for "work," 77 percent of the destinations were in the core area. Shopping and recreational trips, on the other hand, were about evenly split between the core area and the outlying centers (as a group).[17]

In a recent study of daytime population entering and accumulating in central business districts of large American cities, it was found that for every 100 metropolitan residents, between 40 and 50 persons enter the business district during each weekday, about 20–25 have destinations there, and from 8–13 are there in the afternoon at the time of maximum accumulation. Since this study has been reported in some detail elsewhere,[18] we shall indicate only two or three main conclusions at this time. In general, the larger the city the lower are these ratios, presumably indicating that in the large cities daily facility uses are more likely to be handled by outlying centers. Insofar as admittedly scanty data permit, these ratios seem to be holding up over time, with post–World War II ratios comparing favorably with those in the late 1920s. This writer is also inclined to support Carroll's contention that the central business district has relatively less drawing power when a city has strongly dispersed industrial employment. Detroit is a case in point.

Daytime population as a subfield for study

The interest that some social scientists are currently showing in daily trips from home to various activities is by no means new. Since summaries by Liepmann and Breese of the earlier European studies are available in published form, we shall limit our remarks to a few comments about the major emphases of the earlier work.[19]

[17]Gordon B. Sharpe, "Travel to Commercial Centers of the Washington, D.C. Metropolitan Area" (Washington: Highway Transport Research Branch, Bureau of Public Roads, January 1953), mimeographed; see, especially, Table 1.

[18]Donald L. Foley, "The Daily Movement of Population Into Central Business Districts," American Sociological Review, 17 (October 1952), pp. 538–43.

[19]Kate K. Liepmann, op. cit., especially pp. 111–30; Gerald W. Breese, op. cit., pp. 6–13.

Table 2. Estimated Normal Maximum Day Population in Relation to Resident Population, by Concentric Zones From City Center; Means for Five of the Largest U.S. Cities, 1950

Concentric Zones*	Percent of Total Resident Population		Cumulative Percent of Total Resident Population		Ratio of Daytime to Resident Population**
	Resident	Daytime	Resident	Daytime	
Zone 1	1.4	11.0	1.4	11.0	23.84
Zone 2	2.9	19.4	4.3	30.4	7.41
Zone 3	5.1	11.0	9.4	41.4	2.33
Zone 4	5.6	7.5	15.0	48.9	1.53
Zone 5	7.1	7.9	22.1	56.8	1.15
Zone 6	7.6	6.8	29.7	63.6	.93
Zone 7	7.3	6.8	37.0	70.4	.95
Zone 8	8.9	8.1	45.9	78.5	.94
Total Zones 1-8	45.8	78.5			1.77
Remainder of City	54.2	45.7			.82
City Total	100.0	124.2			1.24

*Zones are groups of census tracts intended to approximate the populations of concentric half-mile rings and extending out to a distance of four miles or city limits, whichever is closer.
**Each ratio is the unweighted mean of the respective city ratios. It differs from the ratio that would have been obtained by dividing each second-column figure in this table by the corresponding first-column figure.
Source: Computed from U.S. Bureau of the Census special reports.

The bulk of the earlier research in this area was conducted in Great Britain, Germany, and France in the 30-year period up to the outbreak of World War II. In these and other European nations, government statistical agencies were typically called into service, either at the national or municipal levels. The focus was characteristically confined to the relation of workplace to residence, with the flow that in our country we label "commuting" termed *Pendelwanderung* or *migrations alternantes*. Industrial conurbations and the largest metropolitan areas were particularly investigated. For example, the Leipzig area,[20] Paris,[21] and London[22] were selected. In some of the studies, such as those for Hamburg, the city was districted much as we would introduce census tracts, and cross-tabulations by district of residence and district of workplace were made.

European studies since World War II have not been adequately summarized, although Iklé's doctoral dissertation includes a valuable annotated bibliography.[23] Menzler has reported an estimate of London's daytime population based on the London Travel Survey of 1949.[24] According to this report, over 700,000 persons daily enter the central commercial area, swelling the resident population of 170,000 to about 878,000. Caplow's interesting discussion of city structure in France contains references to recent studies and the brief observation that:

> The degree of centralization in France ... is always less marked than would be expected in an American community of comparable size.... Schaeffer and Chaumbart de Lauwe are able to demonstrate the absence of any single point of concentration in Paris, and the absence as well of the radical daily movement of population which characterizes Chicago. Instead there is a sort of general spiral movement, inwardly directed, but without convergence to a central point, and the volume of this spiral movement appears to be diminishing.... The separation of home from workplace is far less complete in France than in the United States.[25]

[20]Statisches Reichsamt, "Die Pendelwanderung im Mitteldeutschen Industriegebiet," *Vierteljahrshefte zur Statistik des Deutschen Reichs,* Heft 1, 40 (Jg. 1931), pp. 132–48.

[21]Henri Bunle, "Migrations alternantes des professionels dans la région parisienne," *Bulletin de la Statistique Générale de France,* Tome 21 (1932), pp. 585–612.

[22]Census of England and Wales, 1921, *General Report,* Part XII, "Workplaces," especially Tables 90 and 91.

[23]Fred Charles Iklé, "The Impact of War Upon the Spacing of Urban Population" (Doctoral dissertation, University of Chicago, 1950).

[24]F. A. A. Menzler, "An Estimate of the Day-time Population of London," *Journal of the Town Planning Institute,* 38 (March 1952), pp. 116–20.

[25]Theodore Caplow, "Urban Structure in France," *American Sociological Review,* 17 (October 1952), pp. 545–49.

Watson has recently reported on some aspects of commuting in New Zealand.[26]

Realistically viewed, there has been no great burst of American activity in this field during recent years. At a theoretical, social science level there has been but a handful of contributors, so far as we are aware. The work by these persons has in general already been referred to during this presentation: Breese, Carroll, Mitchell, Ranyak, and Rapkin.[27] A limited amount of work has been carried out by city or regional planning agencies.[28] The report by F. Stuart Chapin, Jr., and others in the Department of City and Regional Planning, University of North Carolina, should be seen by those interested in this field.[29] The American Society of Planning Officials recently published a report entitled *The Journey to Work*.[30]

The greatest promise of empirical data comes from the highway research field, as carried out by traffic engineers. A bibliography by Barkley lists origin-destination surveys that have been conducted.[31] Two recent papers by Sharpe[32] and Hitchcock[33] provide good examples of traffic survey analyses.

What of the future prospects for daytime population study? We might break this down into the two requisite phases between which there must be effective collaboration: theoretical formulation and data collection.

The crying need is for imaginative conceptualization. The problem here has an interdisciplinary setting, for it can be approached by land economists, geographers, sociologists, and students of transportation. To be maximally useful, such theoretical work should be essentially dynamic in its orientation so as to handle population movement and functional interrelationships. The theoretical formulation by Mitchell and Rapkin, to be forthcoming in monograph form, may well be an important steppingstone.[34]

But for the field of daytime population to prosper, there must be organized and, probably, *official* resources at the data collecting phase. A most logical next step would be some additional Census assistance. In some cases, even one or two additional questions to relate place of residence with workplace or facilities would provide a vast new sphere in which cross-tabulations could be run. Information on travel time, mode, and cost might be considered, although these are secondary to the main need for cross-tabulating the origin and destination of daily trips. There is also considerable room for collaborating with traffic engineers who now conduct large-scale sampling surveys of trip movements. . . .

In conclusion, analyses in the population field already cover a wide range. Demographers have generally displayed a willingness to search broadly for those factors that have a bearing on or are in turn affected by population size and composition and their changes. At least some of us are of the opinion that urban daytime population is a phenomenon of our times that could stand concerted descriptive and analytical research. We have the hopeful idea that the development of conceptual approaches and operational measures for bringing the nature of daytime population movement and distribution out into the open should considerably further our understanding of the city as a dynamic socio-economic complex.

[26] John E. Watson, "Travelling Time to Work: Some Notes from the New Zealand Census of 1945," *Social Forces,* 30 (March 1952), pp. 283–92.

[27] Since this was written, our attention has been directed to the valuable series of research reports prepared by Otis Dudley Duncan and his group at the University of Chicago, 1951–1953, under contract with the Human Resources Research Institute, Maxwell Air Force Base. Report Numbers 5, 11, 13, 17, and 20 are particularly relevant.

[28] See, for example: Seattle City Planning Commission, "Daytime and Night-time Population Distribution in Metropolitan Seattle: April, 1950" (September 17, 1951), mimeographed; New York Regional Plan Association, "Persons and Vehicles Entering Manhattan South of 61st Street, 1924–1948," *Regional Plan Bulletin,* 74 (October 1949); and Port of New York Authority, "Supplement to Regional Plan Association Bulletin No. 74" (December 1949), mimeographed.

[29] Institute for Research in Social Science, University of North Carolina, Industrial Areas Study, op. cit.

[30] American Society of Planning Officials, *The Journey to Work: Relation Between Employment and Residence,* Planning Advisory Service Information Report No. 26 (May 1951).

[31] Robert Emmanuel Barkley, *Origin-Destination Surveys and Traffic Volume Studies* (Washington: Highway Research Board, December 1951).

[32] Gordon B. Sharpe, op. cit.

[33] S. T. Hitchcock, "Influence of Population, Sales, and Employment on Parking" (Washington: Highway Transport Research Branch, Bureau of Public Roads, January 1953), mimeographed.

[34] Mitchell and Rapkin, op. cit.

Nighttime and Daytime Populations of Large American Suburbs*

Avery M. Guest

Most studies of population distribution in American urban communities focus on the nature of the residential, or what might be called the nighttime, population. Very little research exists on the nature of the workforce, or what might be called daytime population. Most of the limited workforce studies (Breese, 1949; Foley, 1952, 1954; Kain, 1965; Liepmann, 1944; Schnore, 1965:346–65) are restricted to such issues as the general location of residences in relationship to workplace, the numbers of workers in various work sites, and the types of transportation used in the journey to work. Little attention has been devoted to the interrelationships of workforce and residential populations in rather delimited parts of the metropolis outside of the central business district (CBD).

In theory, researchers might devote almost as much attention to the distribution of population by workplace as by residence, since the nonsleeping part of the day (16 of 24 hours) is frequently split

*Reprinted from *Urban Affairs Quarterly*, 12 (September 1976), 57–82, by permission of the author and publisher.

This research has been supported by NIH grant/R01 HD08454-01 POP. I am grateful for comments on this paper from Christopher Cluett, William Frey, and Samuel H. Preston. The following provided able assistance in the coding and processing of data: Janice Jones, Mary Leong, Danny Malet, and Joy Wu.

evenly between workplace and residence. In practice, less attention has been devoted to population distribution by workplace because major statistical agencies such as the U. S. Bureau of the Census have traditionally reported most data by place of residence, rather than place of work. In fact, the U. S. Census has asked questions about place of work in only the 1960 and 1970 regular population censuses, and only limited data on place of work and its relationship to place of residence have been reported.

Tabulations in the 1970 U. S. Census on the size and demographic characteristics of both residential and workforce populations in more than 100 suburban communities of U. S. Standard Metropolitan Statistical Areas (SMSAs) provide a rare opportunity for analysis on the interrelationships between nighttime and daytime populations. The data are reported only for suburbs with more than 50,000 residential population, and thus must be used cautiously to generalize about the nature of workforce and residential populations in all American suburbs. Nevertheless, our knowledge of the interrelationships of workforce and residential populations in American suburban communities is so limited that some analysis of the data is valuable. Better or more complete data for a larger and more representative sample of American suburbs are unlikely to be presented in the near future by the U. S. Census or other major statistical agencies. The cost and time is prohibitive for tabulating complete data for all American communities on the character of their workforce; as the coding of workplace location must be done by hand from reports of street address.

Previous research

While popular folklore often perceives suburbs as places of residence and central cities as places of work, the distinction should not be overemphasized. Suburban rings of metropolitan areas show only a slight tendency to be disproportionately places of residence (Guest, 1974), and studies (Harris, 1943; Schnore, 1957) suggest that suburban communities may be distinguished as places of employment or places of residence. One observer of suburban trends, Dobriner (1963:155–56), describes residential suburbs as "suppliers of labor and consumers of commodities," while the employing suburbs are "consumers of labor and suppliers of commodities."

Most previous studies of suburban workforce populations in U. S. metropolitan areas have relied on data drawn from the censuses of manufacturing and business. These data provide total employment figures for urban places in specific industries, but they do not indicate general social and demographic characteristics of workers in the specific industries, much less the total workforce. Studies (Schnore, 1965: 135–83) using these selective industry censuses have focused, first, on the functional differentiation of suburbs, that is, the ability to characterize suburbs as places of residence or places of work. Some research (Harris, 1943) has also attempted to characterize suburbs in terms of the dominance of various industries such as manufacturing or retailing, although more attention has been devoted to the issue of whether suburbs may be characterized as employing or residential. Studies have also determined whether the employment versus residential character of suburbs relates to other social and demographic characteristics such as social status, family and racial-ethnic composition, and housing structure. This paper will not deal directly with the importance of the employment versus residential distinction in predicting other social and demographic characteristics of suburbs; another paper in preparation deals more directly with this issue. Recently, Smith and Manton (1975) have suggested some problems in using the employment versus residential distinction in predicting suburban population and housing characteristics.

Research questions

Four major questions will be analyzed sequentially in this paper. First, to what extent may suburbs be considered places of residence as opposed to places of work? Reiss (1956:575), for instance, has suggested that suburbs may often be "polarized" as residential or employment areas. And in reviewing the literature on suburbs, Dobriner (1958:29) observed that "a strong case has been made in the literature for the recognition of only two major types—industrial and residential." The issue at hand, then, is whether large numbers of residents and workers tend to be found jointly in most suburbs, or whether suburbs tend to be predominantly places of residence or places of work.

This research question has been difficult to answer because complete data on industrial composition of suburban workforces have rarely been available. Schnore (1965: 149), a pioneer in using the residential versus employment distinction in studying suburbs, has cautioned that more research is needed on its basic legitimacy.

Second, to what extent do workers live and work in the same communities? The development of the automobile and high-speed freeways has presumably freed the typical worker from a close relationship between home and workplace, although the looseness of this relationship is not clear. It would be valuable to know not only what proportion of suburban workers also live in the same community but also what features of suburban communities attract workers to "live in." An analysis of such characteristics would indicate how suburban communities might maximize their attractiveness to desired residential groups.

Third, how do social and demographic characteristics of the workforce community correlate with the social and demographic characteristics of the residential community? Industries employ workers of different skills, abilities, and demographic characteristics—a factor which undoubtedly leads to community differences in workforce characteristics. If workers live relatively close to workplaces, one might expect a high correlation between the social status, racial composition, and life-cycle composition of workforce and residential populations in the same community. If the location of workplace has little effect on the location of home, the social and demographic character of community workforces might be somewhat independent.

Fourth, what are the consequences of having large workforce populations on the ways of life in suburban communities? More specifically, we shall investigate two general areas, retail sales and community municipal expenditures. It is rather obvious that the presence of large workforces in a community would increase the sales of businesses and force governments to increase their services; but, it is less obvious how important workforce populations are in relationship to residential populations in affecting these community activities.

The data

The 1970 U. S. Census special report on the journey to work (U. S. Bureau of the Census, 1973a) contains data on the location of home and workplace in parts of the SMSA with 250,000 or more residential population. The parts of the SMSA generally include the central city, suburban ring communities of 50,000 or more, and the balance of the ring area, generally reported by county. Locations of home and workplace are shown for both the total population and important subgroups by such characteristics as race and ethnicity, sex, household relationship, age, income, educational achivement, occupation, and industry of work. From these data it is possible to determine the characteristics of people who work and live in communities and people who work in but do not live in the same community. It is also possible to determine the characteristics of people who live in the community and work in the surrounding SMSA. Unfortunately, specific characteristics of the generally small number of persons who live in each community but work outside the SMSA are not reported in the census special report. It is possible, however, to determine the total number of employed persons who live in each community from the general volumes of the 1970 Census of Population (U. S. Bureau of the Census, 1973b).

Some 35 of the SMSAs have data reported in the journey to work publication on what we consider suburbs. Our sample of 129 suburbs consists of two types of places. A total of 126 are simply incorporated urban places of 50,000 or more residential population which lie in the SMSA but outside the central city. We also included as suburbs the central cities of the Anaheim–Garden Grove–Orange SMSA, which adjoins the Los Angeles–Long Beach SMSA. All these places were part of the Los Angeles SMSA in 1960, and their whole history indicates that they have grown as part of the general development of the Los Angeles area.

We turn now to each of the research questions which were asked in the first part of the paper.

Extent of Polarization

The value of the residence-employment dichotomy in distinguishing American suburbs will be tested in two ways. First, we focus on the degree to which suburbs are places where urbanites work as opposed to reside. This may be determined by taking a ratio of the number of persons, 16 and over, employed in community X to the number of persons, 16 and over, who are employed somewhere else and live in community X. A ratio of 1 would indicate that the number of workers in a community is equal to the number of employed residents, and would indicate little polarization in either direction for a specific suburb. Ratios above 1 would indicate a workplace suburb while ratios below 1 would indicate a residential suburb. Over the 129 communities, we have found an average ratio of .83, indicating that the average suburb in the study could be categorized to a slight extent as a place of residence. The standard deviation of .37 indicated that about two-thirds of the suburbs had ratios between .46 and 1.20 (.38 ± .37). There were more possibly polarized residential than employment suburbs, as 25 places had at least twice as many employed residents as workers while only one suburb had twice as many workers as employed residents. The values of this measure of polarization must be interpreted carefully, because it would probably vary more among a sample of smaller suburbs. In the absence of more complete data on other suburbs and a clear standard for polarization, we cannot conclude whether suburbs are particularly polarized as residential or employment centers. It does seem, though, that a large number of suburbs in our sample are not clearly polarized as either residential- or employment-oriented.

A second, and more important, approach to the question of polarization involves an analysis of whether employment in various industrial activities tends to be closely associated with each other, while employment in various industries does not strongly relate to the residential population of the community. In short, if the employment versus residential distinction has clear validity, we should find that most types of employment vary closely together, while types of employment do not vary closely with the residential population. To determine this, we have computed indices of dissimilarity (Taeuber and Taeuber, 1969:195–245) among workforces in various industries and the total employed residential population. These indices, computed over 129 suburbs, indicate the percent of workers in an industry or in the total residential population who would have to shift communities to be equally distributed. Thus, an index of 0 would indicate that the groups were equally distributed over all communities, while an index of 100 would indicate that communities with persons in one category had absolutely no persons in the other category.

The indices in Table 1 clearly suggest that most groups are not very segregated from each other. In general, less than 20 percent of one group, employment or residential, would have to shift communities to be equally distributed with another group. More importantly, the data suggest that workforces in the various industries are not particularly associated with each other. In fact, the residential population is more closely associated with various workforce populations than some of the workforce populations are associated with each other. The most clear-cut case is manufacturing activity, which tends to be relatively segregated from both residential population and such industries as finance, communications and other public utilities, personal services, professional and related services, and public administration. Many types of manufacturing activity have been unattractive due to their nuisance (pollution, appearance). Activities such as retailing tend to be associated closely with the residential population. Given the dependence of activities such as retailing on proximity to potential customers, this pattern should not be particularly surprising.

We summarize the interrelationships of workforce and residential populations by using Kruskal's (1964, 1968) multi-dimension scaling program, MDSCAL (Versions 4 and 4m). The program sorts the data into major dimensions on the basis of similar or dissimilar location patterns. Groups of activities which "load" together on a dimension may be considered similar in location patterns. A display of the first two, and presumably most important, dimensions obtained from this program is shown in Figure 1. The ability of two dimensions to summarize the relationships among the variables is indicated by a stress figure of .21, which is considered "good" (Kruskal, 1968).

The figure nicely shows that workers in manufacturing, transportation, public administration, and other industries seem to have somewhat unique location patterns in relationship to a large cluster of activities containing both the total residential population and various residential-population-serving industries such as wholesaling and retailing, personal services, and construction. The figure, then, simply tends to reinforce the impressionistic observation from an inspection of the data matrix that various types of workforces are more closely associated with the residential population than with other workforce populations.

In summary, the second set of data raises serious questions about the basic clarity of the employment versus residential distinction. The major types of industries with similar location patterns tend to be service activities for the residential population, and the location of residential population is closely related to these activities. Some industries such as manufacturing and public administration are not closely associated with the residential population, but then they are not closely associated with each other, either.

Living in

The overlap in residential and workforce populations in most suburban communities suggests that many persons may live and work in the same community. In fact, the data do not permit clear-cut conclusions about whether communities are relatively self-contained. Over the 129 communities, we found that the average workforce contained 61.9 percent (roughly 3 in 5) who lived outside the community of work, while, on the average, 38.1 percent of workers (roughly 2 of 5) lived in the same community. At the minimum, we can say that a fairly sizable number of suburbanites in these large communities live most of their lives within segmented parts of the metropolis—namely their suburban community.

There is some variation in this pattern over communities, as the mean percentage living in their workplace community has a standard deviation of 13.2 percent. Four types of factors may be tested for their importance in determining whether workers live in their community of workplace:

(1) *Transportation system.* When workers have available a well-developed automobile-highway transportation system, one would expect them to choose locations outside the workplace community more frequently than when the transportation system was dependent on slower means such as bus or railroad. Using the 1970 special report on the journey to work, it is possible to determine the percentage of all workers in a community using the automobile to reach their workplace.

Table 1. Indices of Dissimilarity Among Number of Workers in Various Industries and Number of Employed Residents, 129 Suburbs, 1970

		(1)	(2)	(3)	(4)	(5)	(6)	(7)	(8)	(9)	(10)	(11)
(1)	Construction	---	32.3	22.1	13.1	22.1	19.4	20.8	20.2	23.8	26.6	16.4
(2)	Manufacturing		---	33.3	30.3	24.8	31.2	38.1	36.5	38.1	39.1	22.3
(3)	Transportation, Communication, and Other Public Utilities			---	21.5	23.8	23.2	26.5	24.9	24.8	33.5	18.6
(4)	Wholesale and Retail Trade				---	17.3	16.5	16.7	16.4	24.1	23.9	11.6
(5)	Finance, Insurance, and Real Estate					---	17.1	18.8	18.7	25.0	26.9	19.4
(6)	Business and Repair Services						---	19.6	18.1	24.9	24.4	14.7
(7)	Personal Services							---	18.6	25.0	24.2	18.9
(8)	Professional and Related Services								---	22.7	28.2	8.0
(9)	Public Administration									---	33.3	21.6
(10)	Other Industries, excluding Armed Forces										---	25.1
(11)	Employed Residents											---

Source: U. S. Bureau of the Census (1973a: Table 2).

(2) *Housing opportunities.* Workers would presumably be more likely to live in their workplace community when alternative housing opportunities outside were relatively few. Thus, one would predict large proportions of workers living in their workplace community when it was large in residential population (providing a large number of nearby opportunities) and when the total metropolitan area (SMSA) was small in residential population (providing few alternative housing opportunities). The population size of the workplace community was measured by the total residential population living in the community, while the SMSA population was operationalized as the total residential population. Furthermore, spatial position in the metropolis might affect the number of housing opportunities. Communities which are located on the outskirts of a metropolitan area, particularly outside the densely settled part, might also have few alternative residential opportunities to those found in the local community. As a crude measure of spatial position, we have determined the location of the suburb in one of the following distance zones from the nearest CBD of a central city: 0–9.9 miles, scored a 1; 10–19.9 miles, scored a 2; 20.0 or more miles, scored a 3. This variable was treated as a continuous variable in the regression analysis.

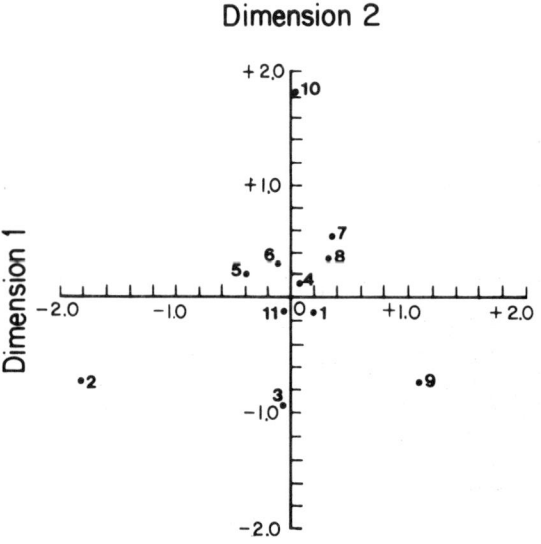

Figure 1—Loadings of workforce and residential groups on two principal dimensions (see Table 1 for number identifying group).

(3) *Attractive features.* Communities may exert a residential pull on workers when they have attractive housing and a desirable environment. We shall operationalize two measures of housing desirability, the single-unit character and the age of the housing. Single-unit housing often provides a large amount of internal building space and external yard space. Older housing often suffers from deterioration and structural obsolescence. Two measures of community attractiveness will be considered, the ratio of manufacturing workers in the community to the total area and the percentage of households headed by a black. Manufacturing activity is often unattractive in physical appearance and characterized by high rates of pollution and noxious fumes. The race variable may have effects due to two possible causes, both related to the fact that the vast majority of the suburban workforce is white. First, whites may flee blacks, or second, and just as likely, whites may live elsewhere when community housing is separated into white and black housing markets, and thus the potential supply of housing becomes somewhat limited.

(4) *Military population.* Unfortunately, the data do not permit the exclusion of armed forces personnel from the total workforce figures for the use of transportation. We have thus included them in the percentage of workers who live in the same community. Military bases often provide local facilities for personnel, or even require that personnel live locally, and it thus seems important to include the proportion of all workers in the armed forces as an independent variable.

Table 2 indicates that most of the hypothesized variables have low zero order correlations with the dependent variable, the proportion of workers in each suburb who also live in the same community. Several of the independent variables are strongly intercorrelated, however, suggesting the wisdom of a multiple-regression analysis to determine the joint effect of the variables and the relative importance of each.

As the table also shows, the eight independent variables jointly explain more than half the variance in "living in." In the multiple-regression equation, two variables appear as powerful predictors when the other variables are controlled: automobile suburbs have low proportions of workers living in, and suburbs with large amounts of manufacturing activity also have low proportions of workers living in.

There may be some problems in interpreting the effects of the automobile variable, since some workers may not use an automobile because they

Table 2. Effects of Community Structural Characteristics on Percentage Who Work and Live in the Same Community

	(1)	(2)	(3)	(4)	(5)	(6)	(7)	(8)	(9)	r	B	Mean	S.D.
(1) Military – Percentage of Workforce in Military	---	.18	-.16	-.07	-.12	-.00	-.01	-.02	.05	.30	.11	9.3	4.30
(2) Community Size – Number of Total Residents in Community		---	.18	.05	.00	-.04	.05	-.01	.00	.21	.25[1]	78,153	28,539
(3) SMSA Size – Total Population Size of SMSA			---	.45	.02	-.02	.11	.07	.07	-.27	-.39[1]	4,066,792	2,798,519
(4) Distance – Linear Distance From CBD				---	.33	.20	-.08	-.18	.10	.07	.38[1]	1.67	.73
(5) Transportation – Percentage of Workers Who Use Automobile to Get to Work					---	.74	-.20	-.68	.42	-.27	-.81	84.91	9.36
(6) Units – Percentage of Single Family Housing Units						---	-.14	-.71	.50	-.06	.08	62.4	20.3
(7) Negro – Percentage of Occupied Households Headed by a Negro							---	.12	-.17	-.08	-.12	4.89	10.92
(8) Manufacturing – ratios of Number of Manufacturing Workers to Land Area in Square Miles								---	-.34	-.18	-.59[1]	292.01	266.11
(9) Age – Year When Community First Reported in Census									---	-.17	-.11	1,898	47
(10) Live – Percentage of Workers Who Live in Same Community												38.06	13.24
							R^2 = .566		D.F. = .119				

Sources: Military, Transportation, Manufacturing, Live—U. S. Bureau of the Census (1973a: Table 2); Community Size, SMSA Size—U. S. Bureau of the Census (1971: Tables 31, 32); Age—U. S. Bureau of the Census (1963: Table 5); Units, Negro—U. S. Bureau of the Census (1971b: Table 1).

r = Zero-order correlation of each independent variable with percentage who work and live in the same community.
B = Standarized partial regression effect of independent variable on percentage who work and live in the same community.
[1]Significant at .001 level, one-tailed F-test; all other coefficients did not reach .05 level of significance.

"live in" the workplace community. In short, the use of transportation and "living in" may be simultaneously determined. When the transportation variable is eliminated from the equation, only 33.6 percent of the variance in the dependent variable is explained. The signs of the other variables remain the same.

Of the other variables, only those indicating housing opportunities have partial regression coefficients above .20. The regression results suggest that large suburban communities have high proportions of workers living in, while a large SMSA and a central spatial location induce a low rate of residence in the workplace community. The importance of community size in the regression suggests that a sample of somewhat smaller suburbs would show lower average rates of living in than we found in this study of relatively large suburbs.

It is possible, even likely, that more of the variance could have been explained by knowing more about the pull characteristics of communities around the workplace. Obviously, community features can become attractive or unattractive only in relationship to the other available communities. Guest and Cluett (1974), for instance, have shown that characteristics of the surrounding communities do exert effects on the tendency of workers to live near the workplace community. These features of other communities are more difficult to operationalize with a national sample of suburbs.

Workforce characteristics

Research (Guest, 1972; Fine, Glenn, and Monts, 1971) on the social and demographic characteristics of suburban residential populations suggests that suburbs vary greatly in their social status and family composition. Most suburban communities in the United States are overwhelmingly white in racial composition, but there are a number of predominantly black suburbs (Hermalin and Farley, 1973). One might expect a close relationship between the character of the workforce and residential populations. Given that large numbers of workers live in the community of their employment, there will almost inevitably be some overlap between the characteristics of the workforce and residential populations. The overlap between the two populations may be enhanced by decisions of contractors and businessmen. Home builders may purposefully construct housing in a community which would appeal to specific types of workers in nearby industries. At the same time, industries may locate in a community because their requirements for labor of certain skill levels may be met by the types of persons already residing in the community.

As we have noted, the 1970 journey to work special report makes possible the comparison of the age, sexual, racial, and status characteristics of all workers, 16 years and over, who live in a community while working somewhere in the SMSA. The residential character as determined by this procedure thus excludes all persons living in the community who work outside the SMSA. Fortunately, in most communities, the number of such persons should be small enough to have little effect on our comparisons. As Table 3 shows, there is generally a high, although not overwhelming, correlation between most community workforce and residential characteristics. The only exception is sexual composition, where the proportion of males in the workforce has only a very slight relationship to the proportion of males in the residential community. In general, then, status, racial-ethnic, and age characteristics are similar or vary together, whether one investigates the workforce or residential populations. One reason for the low correlation of the residential and workforce sexual compositions may be the fact that sexual composition showed little variation in workforce and residential populations across the communities and thus it was difficult for one variable to account for much variation in the other. The mean percentage of males in the workforce was 61.6 percent, with a standard deviation of 3.8 percent while the mean percentage of males in the employed residential population was 61.0 percent, with a standard deviation of 6.4 percent.

The relationship of black residential and workforce populations deserves some further attention since the two variables are highly correlated, suggesting their strong interdependence. This strong interdependence is particularly interesting because few blacks live in most suburban rings (Hermalin and Farley, 1973), and blacks are more represented in suburban workforce populations than suburban residential populations (Guest, 1975). The relationship suggested by our data may be better understood by presenting the unstandardized regression equation of the percentage black in the workforce on the percentage of blacks in the residential population:

$$\text{Percentage black, residential} = -3.86 + 1.62 \; \text{Percentage black, workforce}$$
$$R^2 = .60$$

Table 3. Correlation of Workforce and Residential Population Characteristics, 129 Suburbs, 1970

Characteristic	Total Workforce-Total Employed Residents	Only Work-Only Live	Only Work-Work and Live	Only Live-Work and Live
Occupation-Percentage of civilians in professional and managerial occupations	.68	.40	.45	.78
Income-Proportion of earners with annual incomes of $10,000 or more	.46	.20	.53	.63
Negroes-Percentage of Negroes of all races	.76	.28	.31	.96
Young-Percentage of persons 16-24, of population 16 and over	.55	.05	.52	.32
Child-bearing-Percentage of persons 25-44, of population 16 and over	.75	.54	.61	.85
Sex-Percentage of males, of population 16 and over	.15	-.12	.64	-.14

Source: See Table 1.

The unstandardized regression coefficient of 1.62 indicates that the percentage of blacks in the residential population increases 1.62 percent for every 1 percent increase in the percentage of blacks in the workforce. Yet, the regression equation also suggests that blacks are often underrepresented in the residential population relative to the workforce percentage of blacks. This is due to the fact that the constant in the equation is negative, indicating that almost no blacks will live in a residential community when few blacks are in the workforce. More specifically, the equation suggests that, on the average, no blacks will live in a suburb until 2.4 percent of the workforce is black. The proportion of blacks in the residential population will increase rapidly with increases in the black proportion of the workforce, so, for instance, a community with a workforce which is 20.0 percent black would be predicted to have a residential population which is 28.6 percent black.

This result should not be too surprising. Given a situation in which blacks generally face discrimination in most white housing markets, it would be difficult for black workers to gain entrance in white residential communities when they are few in number, but, once the size of the black workforce grows to a rather sizable number, it becomes easier by sheer force of numbers to open black residential enclaves. Once these are established, the demand for housing by black workers is apt to lead to a relatively rapid increase in the local black residential population.

While the generally high correlations between the character of the workforce and residential populations are important, they result partially from the fact that many persons in the workforce population are also found in the residential population. This may be determined by removing persons who both work and live in a community from the workforce and residential populations, and then determining the correlations between the characteristics of persons who only work or only live in the community. As Table 4 shows, these correlations are often strikingly lower than those for the total residential and workforce populations. For all characteristics except perhaps occupational and child-bearing status, the correlations suggest very little covariation in the nature of the two populations.

The characteristics of persons who both live and work in a community must be determined by some complex interrelationships between the natures of the total workforce and the total residential populations. This is suggested by the fact that the characteristics of persons both living and working in a community tend to be at least moderately correlated with both the characteristics of persons only working and persons only living in the community. Assuming that many persons desire to live near workplaces, one can speculate that the character of the workforce has some effect on the resulting social status, age, and racial character of the surrounding communiy. At the same time, the social status, age, and racial character of the already existing residential population must have some effect on the types of workers who decide to live in the residential community.

Workforce consequences

An analysis of the character of workforce populations may have some interest in itself, but it undoubtedly assumes a more clear-cut value when it can be shown that the character of the workforce has consequences on the way of life in local communities. We pick two areas of suburban life which may be related empirically and unambiguously to the character of the workforce and residential populations: total retail sales by various types of business establishments and levels of municipal government expenditures.

Most analysis of the determinants of retail sales in metropolitan subareas has focused on the size and composition of the local residential population, and there is clear evidence that the character of the residential population is related to the character of retailing (Berry, 1963). There is, however, very little analysis of the relative impact of workforce and residential populations on retailing for a large sample of communities and for various forms of retailing. Hoover and Vernon (1962: 119–20), in their study of 19 counties in the New York Metropolitan Region, show how employment in various retail sales activities may be predicted by both county residential and workforce populations, although the high correlations between the two variables over the counties caused some difficulty in determining the relative importance of residential versus workforce populations. Friedman (1973) has shown how the number of workers in the central business districts of U. S. SMSAs has an effect on retail sales independent of other variables. And Kasarda (1972) has shown in an analysis of SMSAs how central city sales per capita (residential population) in various retail and wholesaling activities are affected by the size of the suburban community, presumably an indicator of the rate of daily commuting to the central city.

Previous analysis of government expenditures also suggests the importance of workforce populations, but once again there is little systematic evidence about the effects of the workforce versus residential populations in determining levels of government services. In his study of municipal expenditures among New Jersey communities, primarily suburbs of the New York Metropolitan Region, Wood (1964) found that the most consistent positive predictor of various types of government expenditures was the level of "industrialization," a measure of the extent of nonresidential activities in the community. In a similar study of suburbs around Philadelphia, Williams, Herman, Liebman, and Dye (1965) found that the percentage of the land in industrial and commercial uses was the strongest and most consistent positive predictor of government expenditure levels. Finally, Kasarda (1972) found that the number of job commuters from suburban rings to central cities of SMSAs had significant positive effects on various types of central city expenditures per capita.

Retail expenditures in various major categories, two-digit codes of the Standard Industrial Classification (SIC), are found in the 1967 Census of Business (U. S. Bureau of the Census, 1969). Selected municipal expenditures are found in the 1967 Census of Governments (U. S. Bureau of the Census, 1970). Municipal expenditures generally include education as a sizable component, and thus it is important to look at specific other types of expenditures which would not be so obviously dependent on the size of the residential population. While data on retail and government expenditures are not available for the same year (1970) as the residential and workforce populations, the two dates are close possible approximations. In some communities, data on selected types of retailing and government expenditures were not available due to disclosure problems, methods of government financing, or lack of incorporation in 1967. The largest possible sample size for any regression was 128.

In Table 4, we have presented the unstandardized regression effects of the two population variables on the various dependent variables. The regression coefficients indicate the increase in total sales or expenditures in dollars for each increase of one person in each category of the independent variables. The two populations include military personnel since they are potential customers and users of services. We do not investigate per capita sales because it is unclear whether the base population should be the number of residents in the community or the number of workers. Over the 128 communities analyzed at least once in Table 4, the sizes of the workforces and residential population correlated at .65.

In general, the explained variances for the various equations are quite high for retail sales and somewhat less impressive for government expenditures. Methods of government financing differ somewhat between states, and this may help to account for the higher error of prediction.

The most significant finding is the greater importance of workforce populations, in contrast to residential populations, in predicting total levels of retail sales. The equation thus suggests that each worker, on the average, makes somewhat more retail purchases near his place of work than his place of residence.

Of the various retail expenditures, only food and general merchandise are more strongly associated with residential than workforce populations, while items such as furniture, automobiles, and apparel are clearly more related to the number of workers. The other retail items seem to be influenced relatively equally by the size of residential and workforce populations. The perishability of much food undoubtedly necessitates purchases close to residence, but it is not entirely clear why general merchandise is closely related to the residential population while furniture, automobiles, and apparel are more closely related to the workforce population. One hypothesis is that furniture and automobile stores are more often located in strings along major routes of transportation, while general merchandise stores seem to be more isolated in shopping centers, and thus perhaps less accessible during the journey to work. It is also possible that zoning regulations differ among suburbs so that residential areas are less hospitable than employment areas to the location of activities such as automobile and furniture dealers.

Total government expenditures are clearly more dependent on the size of the residential population, a finding undoubtedly related to the high proportion of educational expenditures in the total municipal budget of many communities. Educational activities would, of course, primarily serve the residential population. Residential and workforce populations have more equal effects on various types of selected expenditures. For instance, workforce populations have more influence on increasing governmental expenditures on highways, sewer and sanitation services, and parks and recreation. The findings for highways and sewer-sanitation are not surprising, given the dependence of many workplaces on highways for the movement of workers and goods and the need for facilities to remove industrial and human waste from workplace sites. The finding for parks and recreation is less interpretable. While police and fire expenditures are more closely related to the size of the residential population, it is clear the workforce populations are associated with relatively high expenditures in those areas.

Summary and conclusion

We have answered four basic questions about the nature of the daytime and nighttime, or working and residential, populations in a sample of 129 American suburban communities. First, most suburbs in the study are both places of residence and places of work; there is some difficulty in finding suburbs in which various types of employment cluster together, without significant numbers of residents. Second, a relatively large number of persons both live and work in the same suburban community, almost two out of five, suggesting that many suburban residents live their lives within geographically segmented parts of the metropolis. Variations in the tendency for workers to "live in" are strongly related to the nature of the local transportation system, the industrial composition of the workforce, and the number of housing opportunities. Third, workforce and residential populations are generally quite similar in terms of social status, race, and family-life cycle status, while there is less correlation between their sexual compositions. In other words, similar types of persons are found in the workforce and residential populations of individual communities. The character of those who only work and those who only live in a community is less strongly associated. Fourth, and finally, two indicators of suburban life, the sales of retail establishments and the levels of governmental expenditures are clearly related to the sizes of both local workforce and residential populations, and for some types of expenditures,

Table 4. Partial Unstandardized Regression Effects of Workforce and Residential Populations of Government Expenditures and Retail Sales

Government Expenditures	Workforce Population	Residential Population	R^2	Sample Size
All Municipal Expenditures	130.70[2]	265.44[1]	.32	126
Highways	19.19[2]	17.52[3]	.25	126
Police Protection	15.72[1]	27.72[1]	.63	126
Fire Protection	12.37[1]	18.19[1]	.43	118
Sewage and Sanitation	11.72[2]	6.38	.14	122
Parks and Recreation	19.56[1]	6.61	.45	121
Retail Sales				
All Sales	2,687.69[1]	1,735.92[1]	.63	128
Building Materials, Hardware, and Farm Equipment Dealers	88.13[1]	66.20[2]	.42	125
General Merchandise Group Stores	455.81[2]	590.90[2]	.31	117
Food Stores	267.63[1]	581.38[1]	.64	128
Automotive Dealers	536.93[1]	122.39	.27	124
Gasoline Service Stations	148.73[1]	123.20[1]	.69	128
Apparel and Accessory Stores	233.25[1]	-4.32	.23	127
Furniture, Home Furnishings, and Equipment Stores	246.83[1]	39.17	.49	125
Eating and Drinking Places	271.65[1]	101.69[3]	.56	128
Drug Stores and Proprietary Stores	78.57[1]	56.26[2]	.41	124
Miscellaneous Retail Stores	164.18[1]	88.05	.34	128

Source: U.S. Bureau of the Census (1969: Table 3; 1970: Table 38; 1973a: Table 2; 1973b: Table 8b).

[1] Significant at .001 level, one-tailed F-test.

[2] Significant at .01 level, one-tailed F-test.

[3] Significant at .05 level, one-tailed F-test.

the size of the workforce population is more strongly related.

Overall, then, our results indicate the close interdependence between residential and workforce populations. Most suburbs contain a mixture of both groups, and each group in turn has an effect on the other group. Clearly, further insights into the nature of the metropolitan spatial organization may be developed by further study of the interrelationships of workforce and residential populations.

In particular, further research on the nighttime and daytime populations of American suburbs should explore two important areas: the viability of the residential versus employment distinction in predicting other characteristics of suburbs and the consequences on suburban life of different types of workforces.

As we noted in the initial part of the paper, some research effort has been devoted to specifying the consequences of having a residential versus employment suburb on social, demographic, and housing characteristics of communities. Yet, our research suggests that this distinction is not very clear-cut, and, in fact, some industries are more closely associated with the residential population than with other industries. Most of the previous research (Schnore, 1965:169–83) on the employment versus residential distinction has not found clear-cut differences in social and demographic characteristics—a finding which also underlines the need for more research attention to this area.

Researchers on the nature of suburban life may also suggest that the character of the workforce may also affect other social institutions. For instance, Schnore and Alford (1963) argue that the city-manager form of government and nonpartisan elections serve the interests of higher status residents more than other social strata. The status of the workforce may also affect forms of governments, since the workforce could presumably serve as pressure groups. It also seems likely that the sheer existence of large workforce populations in suburbs may increase the general level of political conflict, for there are certain to be numerous conflicts over the nature of land use and the use of community resources.

All the empirical conclusions which we have reached must be accepted with some caution as indicative of the general nature of American suburbs, because the sample of places is relatively small in number and characterized by communities with generally large residential populations. Hopefully, the U. S. Census Bureau will provide better and more complete data on the workforce populations of American suburban communities.

References

Berry, B. J. L. *Commercial Structure and Commercial Blight: Retail Patterns and Processes in the City of Chicago.* Chicago: University of Chicago, Department of Geography, 1963.

Breese, G. W. *The Daytime Population of the Central Business District of Chicago.* Chicago: University of Chicago Press, 1949.

Dobriner, W. *The Suburban Community.* New York: G. P. Putnam, 1958.

Dobriner, W. *Class in Suburbia.* Englewood Cliffs, N.J.: Prenctice–Hall, 1963.

Fine, J.; Glenn, N.; and Monts, J. K. "The Residential Segregation of Occupational Groups in Central Cities and Suburbs." *Demography,* 8 (February 1971):91–101.

Foley, D. F. "The Daily Movement of Population into Central Business Districts." *American Sociological Review,* 17 (October 1952):438–543.

Foley, D. F. "Urban Daytime Population: A Field for Demographic–Ecological Analysis." *Social Forces,* 32 (May 1954):323–30.

Friedman, J. J. "Variations in the Level of Central Business Retail Activity Among Large U. S. Cities: 1954 and 1967." *Land Economics,* 49 (November 1973):326–35.

Guest, A. M. "Retesting the Burgess Zonal Hypothesis: The Location of White Collar Workers." *American Journal of Sociology,* 76 (May 1971):1094–1108.

Guest, A. M. "Patterns of Family Location." *Demography,* 9 (February 1972):159–70.

Guest, A. M. "Functional Differentiation in the Metropolis." Unpublished. Seattle: Center for Studies in Demography and Ecology, University of Washington, 1974.

Guest, A. M. "The Journey to Work: 1960–1970." *Social Forces,* 54 (September 1975): 220–25.

Guest, A. M., and Cluett, C. "Workplace and Residential Location: A Gravity Model." Unpublished. Seattle: Center for Studies in Demography and Ecology, University of Washington, 1974.

Harris, C. D. "Suburbs." *American Journal of Sociology,* 49 (July 1943):1–13.

Hermalin, A. I., and Farley, R. "The Potential for Residential Integration in Cities and Suburbs: Implications for the Busing Controversy." *American Sociological Review,* 38 (October 1973):555–610.

Hoover, E. M., and Vernon, R. *Anatomy of a Metropolis.* Garden City, N.Y.: Doubleday, 1962.

Kain, J. F. "The Commuting and Residential Decisions of Central Business District Workers." In *Trans-*

portation Economics, pp. 245–73. New York: National Bureau of Economic Research, 1965.

Kasarda, J. D. "The Impact of Suburban Population Growth on Central City Service Functions." *American Journal of Sociology*, 77 (May 1972):1111–24.

Kruskal, J. B. "Multidimensional Scaling by Optimizing Goodness of Fit to a Nonmetric Hypothesis." *Psychometrika*, 29 (1964):1–27.

Kruskal, J. B. *How to Use MDSCAL, A Program to Do Multidimensional Scaling and Multidimensional Unfolding*. Murray Hill, N.J.: Bell Telephone Laboratories, 1968.

Liepmann, K. K. *The Journey to Work*. New York: Oxford University Press, 1944.

Reiss, A. J., Jr. "Research Problems in Metropolitan Population Redistribution." *American Sociological Review*, 21 (October 1956):571–77.

Schnore, L. F. "Satellites and Suburbs." *Social Forces*, 36 (December 1957):121–27.

Schnore, L. F. *The Urban Scene*. New York: Free Press, 1965.

Schnore, L. F., and Alford, R. R. "Forms of Government and Socioeconomic Characteristics of Suburbs." *Administrative Science Quarterly*, 8 (June 1963):1–17.

Smith, J., and Manton, K. "The Socioeconomic Differentiation of Suburban Places." Paper presented at the Annual Meeting of the Population Association of America, Seattle, Washington, 1975.

Taeuber, K., and Taeuber, A. *Negroes in Cities*. New York: Atheneum, 1969.

U. S. Bureau of the Census. *U. S. Census of Population: 1960. Vol. 1: Characteristics of the Population, Part for Each State*. Washington, D.C.: Government Printing Office, 1963.

U. S. Bureau of the Census. *Census of Business: 1967. Retail Trade, Report for Each State*. Washington, D.C.: Government Printing Office, 1969.

U. S. Bureau of the Census. *Census of Governments: 1967. Vol. 7: State Reports, Reports for Each State*. Washington, D.C.: Government Printing Office, 1970.

U. S. Bureau of the Census. *U. S. Census of Population: 1970. Number of Inhabitants, United States Summary*. Final Report PC(2)–A1. Washington, D.C.: Government Printing Office, 1971 (a).

U. S. Bureau of the Census. *U. S. Census of Housing: 1970. General Housing Characteristics, Final Report for Each State*. Washington, D.C.: Government Printing Office, 1971 (b).

U. S. Bureau of the Census. *U. S. Census of Population: 1970. Subject Reports: Journey to Work*. Final Report PC(2)–6D. Washington, D.C.: Government Printing Office, 1973 (a).

U. S. Bureau of the Census. *U. S. Census of Population: 1970. Vol. 1: Characteristics of the Population, Part for Each State*. Washington, D.C.: Government Printing Office, 1973 (b).

Williams, O. P.; Herman, H.; Liebman, C. S.; and Dye, T. R. *Suburban Differences and Metropolitan Policies: A Philadelphia Story*. Philadelphia: University of Pennsylvania Press, 1965.

Wood, R. C. *1400 Governments*. Garden City: Anchor Books, 1964.

The Journey to Work in Australia*

Ian Manning

Factors influencing the journey to work

The fact that far fewer people work in their home local government area than could shows that work travel in Sydney is caused by more than the mere geographic separation of workplaces and dwellings. Confronted with a geographic distribution of jobs and houses over which they individually have very little control, the workers of Sydney choose to travel further than they might. This decision will be discussed by taking, firstly, the choice of job starting from a fixed dwelling, and secondly, the choice of a house by people who have a fixed job.

Those who are looking for a job without wanting to change their dwelling include most young people seeking their first job, wives returning to the work force, and indeed anybody changing jobs who is committed to a particular dwelling through home ownership, the convenience of others in the household, or simply an attachment to the home locality. They confront the geographic array of jobs from a fixed point, from which they will have to travel daily to whichever job they choose. The factors which may influence this choice begin with the willingness to travel, which depends mainly

*Reprinted from Ian Manning, *The Journey to Work* (Sydney: George Allen and Unwin, Ltd., and Boston: Allen and Unwin, Inc., 1978), pp. 19–21 and 176–87, by permission of the author and George Allen and Unwin, Ltd.

upon the urgency of other calls on the traveller's time. We may guess that people with domestic responsibilities (housewives, single parents) will be more anxious than others to minimise time spent travelling (Liepmann, 1944, p. 40). Willingness to travel is also a matter of the ability to pay for it, depending in turn on wage rates (economists can enjoy themselves intersecting travel cost gradients with wage gradients, both increasing with distance) and the urgency of other financial commitments. If a person's income is already heavily committed he may seek to economise on travel spending, either by switching to slower but cheaper means of transport (thus trading time for money) or seeking a closer job. Average travel distances will be shorter in kilometres if not in minutes among those who, for lack of a licence or financial stringency, do not own a car: the young just embarking on their first job, the ill-paid perhaps, and those whose incomes are fully committed to maintaining a family or buying a house.

While work journeys will tend to be long if people are willing to put up with the time and cash costs of travel, a wide local choice of jobs can make such travel unnecessary. This choice is not just a matter of number, but of suitability, depending in turn on the characteristics of the job and the worker; his degree of attachment to a particular kind of job, and his willingness to accept inferior local jobs in preference to better work at a greater distance. Again, a worker's chance of acquiring a nearby job depends on the competition for these jobs. If the turnover of suitable local jobs is high, and there is not much competition, a worker who so desires has a good chance of a short journey to work. On the other hand, if the turnover is low, as in a recession (Schaeffer and Sclar, 1975, p. 3), and if the number of locals far exceeds the number of local jobs, he is likely to have to go further afield. The limitations to this are reached when suitable jobs are available only at such a distance that the worker finds it worthwhile to shift his dwelling (this in turn depends on the cost of shifting house, which varies greatly from family to family); or when there is no suitable work, in which case he either revises his idea of what is suitable, withdraws from the labour market (a course forced on many working wives during recessions), or joins the reserve army of the unemployed.

As soon as a change of dwelling is at issue, the factors influencing the journey to work become more varied. Shifting house is usually both costly and troublesome, so it is most frequent among renters (provided they are not locked in by below-market rents or neighborhood associations) and among people buying their first home. People may change their house when they change their workplace, but this would not be common for changes of workplace within the metropolitan area. In addition, there may be owners who sell and buy to upgrade their accommodation. Such changes are probably not frequent in the life of any family, though in the aggregate they may contribute a significant proportion of total moves (Urban Research Unit, 1973, p. 64).

According to the accepted theory of residential location, an individual considering where to live balances his distaste for the journey to work against his demand for a pleasant house and his ability to pay for it (Evans, 1973, chs. 3 to 7). Thus all the factors influencing the choice of work from a fixed dwelling come into play, with some extra as well.

If a worker's job lies in an area of cheap low-density housing—which, according to the theory, means that it is probably in a place where jobs are few—conflict between housing costs and travel costs does not arise. If, however, he works in an area where housing is cramped and expensive (which in all probability is an area where many other people work as well) he can but trade travel costs (in money and time) against housing costs and the quality of the housing on offer. Apart from individual preferences, the demand of space is likely to increase with family size (Evans, 1973, ch. 9). Again, of all people, the man who can leave a housebound wife to perform the domestic chores is most likely to be willing to lengthen the working day with long travel. The theory therefore predicts that those with the longest journeys to work will tend to be the heads of families where only one adult works, while single parent workers and single people will try harder to save travel time. The position is, however, complicated by the money costs of housing and travel. A low income may not be enough both to pay for a long journey to work and to cover the mortgage on even the cheapest house on the spacious fringe. Workers on low incomes may therefore squeeze their perhaps large families into cramped quarters within walking distance of work (Evans, 1973, ch. 8). By contrast, the wealthy family man can afford a large purchase of space (in which he gets best value at a distance) and the speediest form of travel. This may result in a long journey to work. However, at present levels of traffic congestion there is a limit to the extent to which money can buy speed, so some of the wealthy, particularly those with a taste for urbane living, may opt to buy a house near their workplace.

These fine calculations of optimal journey distance become less certain when other factors are taken into account (Wheaton, 1977, p. 360). In many families, for example, the wife works in a different place from her husband, and they may seek a house that is convenient to both. Again, accessibility to work is not the only consideration in choosing a place to live. There is also accessibility to other destinations and the attractions they offer—the schools and universities, the more specialised entertainments of live theatres and orchestras. Further, for some people the immediate surroundings of the house are more important than its location in relation to the services and activities of the rest of the city—topography, neighbours, and the snob rating of the suburb (Evans, 1973, ch. 8). Of course, pleasant topography and prestigious neighbours are not necessarily incompatible with short journeys to work. Wealthy people who want to live close to the employment opportunities and diversions of the central city have established high-class residential suburbs in parts of the inner areas. Again, the surroundings of a centre of employment may or may not be attractive topographically. These forces serve mainly to concentrate demand in particular areas (Davis and Spearritt, 1974, Maps 1 and 48). On the other hand, insofar as a lack of local employment is a necessary attribute of a prestige suburb, the well-paid are likely to have long journeys to work.

In Sydney it is normal for families to strive for home ownership. Given the disruption involved in shifting house, having once achieved home ownership people tend to stay put (Neutze, 1971, p. 14). Housing in the older suburbs thus comes available for resale or redevelopment rather slowly. In a growing city, many first-time buyers will have to go to the fringe—a trend which will be encouraged if finance is more easily arranged on new houses than old. Not for them any fine balancing of the utility of space against the disutility of work travel. At the most, the location of the workplace will influence their choice of the sector of the fringe in which to buy. Further, if this has been going on for some time (as it has in Sydney), the length of a man's journey to work will depend as much on his age as on his family size or income. . . .

Facts about workplaces and work travel in Sydney

In describing work travel in this study, four distinctions were frequently made. Journeys to work can be classified according to destination (workplace), origin (residence), and the sex and occupation of the traveller. These distinctions provide the basis for a summary of the results, which broadly confirm the theoretical predictions [made earlier].

Jobs in Sydney are more centralised than dwellings. A quarter of the jobs are in the City of Sydney, a quarter in three other major employment concentrations (two of which are adjacent to the city centre), and the other half are dispersed. People who work in the city centre have, on average, long journeys to work; those who work in the suburban concentrations have journeys of middle length, and those who work elsewhere have, on average, journeys only half as long as those of city centre workers. Workers come to the city centre from all over the metropolitan area, though particularly from the inner areas; they come to the major suburban concentrations mainly from nearby and from areas radially outwards, while those who work at jobs dispersed in other suburbs by and large live locally. City centre workers travel by public transport (particularly train) more than others, while some local workers walk, but in nearly all employment areas except the city centre the main means of getting to work is the motor car.

The population living in any residential area tends to divide into those who work in the city centre, who except in the inner suburbs have long journeys to work, and those who work locally and have short journeys. In most places except the outer western fringe, these two together comprise at least 40 percent of the resident work force, even when "local" is defined as working within three kilometres of home. The proportion of resident workers who work locally depends mainly on the local availability of jobs, though such factors as the ease of transport to and from other areas, and the extent of competition from job-deficient areas nearby, also affect the proportion.

The distribution of men's and women's jobs is fairly similar, the main difference being that the major manufacturing areas are relatively short of women's jobs. The sexes have roughly the same opportunity to work locally, roughly the same opportunity to work in the city centre. Despite this equality of opportunity, women's journeys are shorter on average than men's. This is mainly due to the job choices of married women, who tend to have work journeys short in distance and brief in time. Single women's journeys are almost as long as men's, and include a relatively high proportion to the city centre. They take more time than men's journeys, because single women tend to travel by public transport rather than by car.

These sex differences are reflected in journey lengths by occupation (women in any occupation tend to have shorter journeys than men) and are reinforced by a tendency for people on lower incomes to have shorter journeys, and to use public transport more often (women's incomes are lower than men's). However, the average distance travelled by workers in any occupation also depends on whether that occupation has a tradition of living locally (like doctors), and, most importantly, on the extent to which jobs in that occupation are dispersed around the suburbs. Where jobs are widespread, journeys are short.

These findings confirm the fundamental hypothesis given [earlier], that people (particularly married women) prefer short journeys to work rather than long. However, journeys to work in Sydney are not as short as they might be, for two reasons. First, by historic standards the present metropolitan area is developed at a low density, which provides spaciousness but demands long journeys. Second, people travel further than apparently they need, even with the present spread of jobs and houses, the benefit being that they can choose from among a wider range of each. It is very difficult to weigh these two benefits

Table 1. Duration of the Journey to Work, Australia 1974.

City	Population '000	Hypothetical radius of the urban area km	Half the journeys were less than mins	Three-quarters of the journeys were less than mins
Sydney	2874	21	30	44
Melbourne	2583	21	24	43
Brisbane	911	15	22	34
Adelaide	868	13.5	21	33
Perth	739	13	21	34
Canberra	185	6.5	18	26
Hobart	158	6	17	24
S.A. (excl. Adelaide)	various		11	17
W.A. (excl. Perth)	"		11	16
Albury/Wodonga	43	3	11	17

Note: Hobart data from before the Tasman bridge fell down.

Sources: ABS, Journey to Work and Journey to School, 1974.
Cities' Commission, <u>Australians' Use of Time</u>.

against the known and calculable costs of the journey to work: the hour a day of an average worker's time; the 5 percent or so of the household budget. However, the evidence of certain groups (married women, or the citizens of Manly/Warringah) is that, where so minded, people can give greater emphasis to short journeys without obvious hardship.

A comparison with the rest of Australia

Though this study has concentrated on Sydney, many of its findings should apply in other parts of Australia. Unfortunately, the 1971 census material has not been processed for any other city, but some comparison is possible using the Survey of the Journey to Work of August 1974 (ABS, Journey to Work, 1974). This survey did not give any indication of geographic distances, and the sample was too small to divide cities into areas of journey origin or destination, but certain broad comparisons are possible as to journey times and means of transport.

Sydney is the largest city in Australia, and one might therefore expect journey distances within it to be longer than in other cities. This is true, at least insofar as distances are reflected in times (Table 1). Median journey times in the cities and towns of Australia are ranked quite strictly according to their population. In country towns and small cities like Albury/Wodonga, half the journeys typically take less than 11 minutes, while in Sydney half take more than half an hour. This gives a difference of three times in the duration of the typical journey, which is quite out of proportion to the population difference of nearly seventy times.

These time differences broadly reflect differences in actual journey lengths, at least for those cities where journey lengths are known. In Albury/Wodonga the median journey to work was 4 actual kilometres, in Melbourne about 12 actual kilometres (Cities' Commission, 1975, p. 67), and in Sydney about 9.4 straight-line kilometres. The difference in geographic distance is once again about three times.

This lack of proportion between city size and journey lengths is partly explained by the facts of geometry. In a circular city, with all the jobs in the centre and the population spread uniformly over the circular area, the length of the average journey to work is proportional to the radius, for all journeys are from points within the circle to the centre. To take examples, neither Sydney nor Albury/Wodonga is exactly circular, but supposing that their existing populations were living in circular cities at the same density as now, the radii would be roughly 21 and 3 kilometres, respectively—a difference of seven times (Table 1). If journeys to work in both cities were arranged in the same orderly pattern from the edge towards the centre, the difference of average length would thus be roughly seven times—much less than the difference in population, though more than the observed difference in journey length. Average journey lengths greater than the radius of the urban area, such as the 4-kilometre average in Albury/Wodonga, come about if people are not concerned about how close their job is to their house. The Albury/Wodonga urban area is sufficiently small for most people to work locally (or at any rate within 5 kilometres of home) without any special effort to select a nearby job. On the other hand, average distances fall substantially below the radius of the urban area when at least some people are trying to get jobs near home, or homes near their job. In a small city people can reconcile their preferred home and job without incurring excessive journey distances or times (provided they travel by car), whereas in Sydney, for some at least, there comes a compromise: a job other than the best available in the urban area, a dwelling other than the most pleasant the family could afford, or perhaps a journey to work that takes more time and money than they would wish.

Given that Sydney and Melbourne are of very nearly equal hypothetical radius, it is of interest that the median duration of the journey to work in Melbourne is less than in Sydney (Table 1). This may mean that distances in Melbourne are a little shorter, perhaps because its city centre is relatively less important in its total metropolitan job market. It may also be due to faster average speeds, for a higher proportion of Melbourne workers travel to work by car (Table 2).

The importance of rail transport for the journey to work varies with the size of the city (Table 2). The importance of cars varies inversely with the importance of trains, which suggests that these are an alternative means of transport—trains being used for city centre traffic which would go by car if there were less congestion. Buses and trams, on the other hand, carry between 9 and 13 percent of all work travellers in all the capital cities except Hobart, where they are more important. This relatively constant share of total travellers is consistent with the finding that buses are used mainly by short-distance travellers who cannot afford a car. Such people are probably a fairly constant proportion of the work force in all cities. Finally, the proportion walking to work varies between 5 percent and 8 percent, again a relatively constant figure, which may indicate that the proportion of the work force working locally is similar in all Australian cities.

Table 3 confirms the association between trip duration and the means of transport. Average journey times in Sydney by any particular means of transport are but little longer than in Australia as a whole, and the difference in overall average journey times arises almost wholly because more people in Sydney travel by train. Journey times by rail are typically long, not so much because trains are slow but because they carry people long distances.

Sydney thus differs from other Australian cities as average journey distances are greater, durations are longer, and more people travel by train. All these differences come about simply because it is larger and has a relatively important city centre. A further difference is that average spending on fares in 1974 by those who used public transport was greater (Table 4). This, however, was as much due to the particular fares in force as to differences in distances travelled.

Despite these effects of city size, the behaviour of Sydney people is typical of other Australian workers: married women have the briefest journeys, men next, and single women the most time consuming (Table 5). Insofar as inferences can be made by comparing journey durations from the 1974 survey with journey lengths from the census, Sydney people are also typical: clerks have the longest journeys to work, and sales, service, and process workers the shortest. In these and presumably other behavioural respects, therefore, the results of this study of Sydney should be directly applicable to other places in Australia.

Implications for public policy

The policy instruments by which governments can influence work travel stop far short of enabling them to dictate where people shall live, where they will work, and how they will get from the one to the other. Even so, town planning regulations can be used to influence the location of jobs and housing areas, as can direct state involvement in land development and redevelopment. And even if governments eschew town planning and land development (as some would do, through devotion to laissez faire or cynicism as to the effectiveness of planning), the state and its municipalities are still responsible for all transport infrastructure in Australian cities, and for the operation of most public transport. Many of the demands for transport investments, and a proportion of the bus and train deficits, are due to work travel. Whether or not governments believe that they can influence urban layout for the better, concern for the use of public

Table 2. Main Mode of Travel for the Journey to Work, Australia 1974 (percent)

Area	Train/Ferry	Bus/Tram	Car	Walk	Other
Sydney	17	13	62	6	2
Melbourne	13	12	66	8	2
Brisbane	9	11	69	6	3
Adelaide	4	13	73	6	4
Perth	3	13	75	5	4
Hobart	–	19	71	8	2
Canberra	–	9	83	6	2
Rest of Australia	1	5	77	12	6
All Australia	8	10	70	8	4

Source: ABS, Journey to Work and Journey to School, 1974.

Table 3. Median Duration of the Journey to Work by Main Travel Mode.

Main Mode of Travel	Sydney 1971 mins	Australia 1974 (ABS survey) mins
Car travellers	20	18
Train travellers	57	57
Bus travellers	37	34
Pedestrians	12	11
TOTAL	30	20

Sources: SATS tapes; and ABS, Journey to Work and Journey to School, 1974.

Table 4. Average Weekly Fares Paid for Work Travel, August 1974.

Main Mode of Travel	N.S.W. $/worker/week	Australia $/worker/week
Train	4.00	3.55
Bus	2.95	2.88
Total Public Transport	3.50	3.24

Note: Though these figures include travel in provincial areas, they are dominated by the respective capital cities.
Source: ABS, Journey to Work and Journey to School, 1974.

Table 5. Median Duration of the Journey to Work by Sex.

Population Group	Sydney 1971 mins	Australia 1974 (ABS Survey) mins
Men	30	21
Single women	34	22
Married women	28	17
Total	30	20

Sources: SATS tapes; and ABS, Journey to Work and Journey to School, 1974.

funds demands that they try to have a coherent policy for the journey to work.

Yet work travel is only a small part of life in cities. Concern that work journeys are too long, or too slow, or too expensive, should not be allowed to dominate policy to the exclusion of all else. Urban policy should be taken as a whole, and is so discussed by Neutze (1978). A study of workplaces and work travel cannot come up with definite policy recommendations, but only with comments on the effects of particular policies. . . .

One of the most consistent aspirations of Australian politicians, and indeed of the people also, is that as many families as possible shall have their own house and garden, set in a residential area from which most kinds of employment-generating activities have been banned (Stretton, 1975, ch. 2). This suburban ideal has been pursued consistently and with some success. On the other hand, there is less of a consensus as to the ideal work environment, which is seen as varying with the kind of job. Some kinds of work are found "naturally" in small enterprises, like a mum-and-dad shop; others are found in large factories. Some are scattered "naturally" around the suburbs—teaching, local retailing jobs—while others "naturally" concentrate in the city centre. However, other things being equal, there is at least a widespread belief that it is better to work in an employment area which is full of things to do, a place vibrant with life, rather than in a neighborhood shop or a factory in a large industrial estate. For many people the city centre is still the archetype of the preferred place to work.

If housing is to be dispersed, and if as many jobs as possible are to be concentrated in the city centre (or that plus a small number of other major employment centres), the necessary result in a city the size of Sydney is journeys to work which are both time-consuming and expensive. The obvious response to this dilemma is to propose that transport be quicker and cheaper. Government responsibility for the transport infrastructure is already established, and there are government engineers anxious to get on with the task of improving it, so administratively this response comes naturally (Sandercock, 1975, p. 192). Further, there have been occasions in the past when happy technical innovations have meant that transport could be speeded and journey times cut at the same time as costs were reduced. An example would be the substitution of electric for steam traction on tramways and railways earlier this century. However, at present extra speed is available only at extra cost, for it is attained by switching from public to private transport, and by building freeways. Car travel is more expensive than train in private costs, while the public costs of building freeways to carry city centre traffic have proved prohibitive. For this reason increased speed cannot reduce the burden of long city centre work journeys. People therefore turn to policies which would reduce the distance to be covered on the typical journey to work, for such policies would seem to have the benefit of reducing both the time and money costs of the journey.

If it be that dispersed housing and concentrated employment generate excessively long work journeys, it may be that the housing pattern is at fault. People therefore propose that residential densities should be increased, even at the sacrifice of the traditional house and garden. They argue that well-designed, medium-density housing is at least as pleasant to live in as the single-family dwelling on its quarter-acre block, and that redevelopment of the inner suburbs in this form would in the long run be cheaper than adding to the urban sprawl. Those against this policy deny both these arguments, and point to the high costs and lack of public acceptance of inner area high-rise flats as substitutes for brick veneer villas on the urban fringe (Halkett, 1976, ch. 1; Jones, 1972, ch. 5). While some degree of flat building and redevelopment may continue, for the moment the fringe-extenders seem to have won the argument.

If the people do not want to be brought to the jobs, then surely the jobs should be brought to the people. Since this would result from a continuation of present trends it seems the most likely way of reducing the burden of the journey to work. The effectiveness of such a policy of job decentralisation depends on public response in two ways. First, how far will people respond by reducing their travel distances, and second, to what extent will they switch to a more expensive means of transport?

So long as all employment in a city is concentrated in one place the question of providing a choice of jobs does not arise. Anybody who can get to the city centre (and with a developed system of suburban railways anybody can) has the choice of all the city's jobs. But what happens when jobs are dispersed? It cannot be arranged that everybody works at home, so there will still be journeys to work. The crucial question, therefore, is how far people will travel in search of a suitable job. It is conceivable, indeed quite likely, that a policy of job decentralisation would result in longer work journeys (Thomson, 1977, ch. 3). This is associated with a third issue. So long as most of the jobs in a city the size of Sydney are at the centre, the main means of transport for the journey to work has to be the railway, with some use of other kinds of public transport. There is little room for motor cars in such high-density city centres, and the distances are too long for people to walk. Once employment is dispersed passenger flows become too diffuse to support public transport, particularly the forms with high capacity and low costs. The resulting switch to private motor transport will mean that journey times are shorter, but will costs be less? And what becomes of those who do not have cars? It could be that a policy of job decentralisation would result in people travelling further, and at greater private and social cost, than they would have if most jobs were in the city centre. The bright lights would have been sacrificed to no avail.

To a considerable extent these questions are being posed after the event. Jobs *have* been decentralised, and journeys to jobs in the suburbs *are* shorter than journeys to the city centre. The motor car *has* become the most common way of getting to work. And yet questions still arise: should the trend to the decentralisation of employment be encouraged or resisted? And should its form be changed, with greater emphasis on regional centres, or on dispersed local employment?

The trend towards the decentralisation of employment has reached the stage where only 14 percent of the jobs in Sydney are in industries classified as highly concentrated, and but 22 percent of all jobs are in highly concentrated occupations. Employment is dispersed in all manufacturing industries, in road transport, in retailing and in most services like health and education. For these, the great majority of industries, the question is not whether the overall degree of decentralisation is too little, but whether the distribution is satisfactory. In manufacturing, for example, the existing pattern of jobs can be criticised in two ways. First, despite the generally high degree of dispersion, a satisfactory local balance between workers and workplaces has yet to be achieved in some suburbs, particularly those on the outer western fringe. Efforts should be made to remedy this, since process workers (particularly women) appreciate local employment opportunities, and tend to work locally when they can. Second, modern manufacturing plants are developed at a low density. Most factory jobs are beyond walking distance from any residential area, and indeed are so spread even within industrial zones that it is difficult to provide them with a good bus service. This dispersion of jobs within manufacturing areas comes about for technical reasons (single-story factory layouts are efficient, and firms like to have spare room for expansion), but there may be scope for improvement of design which would make it easier to provide factory workers with the option of getting to work by public transport.

In retailing and services the problems are often those of excessive scatter:

> In North America and Australasia too many hospitals, universities, technical and teachers' colleges and high schools and trade schools, suburban office developments, supermarkets, filling stations, theatres, cinemas, pool halls, squash courts and other high and low entertainments have been scattered like confetti along main roads through characterless expanses of housing, in the name of decentralisation.... Not houses and gardens, but that formless uncentred litter of facilities on and off public transport routes.... is the true and disastrous meaning of "suburban sprawl." Policies of *re*centralisation should try to gather those activities to support each other.... in strong town centres with their own local transport, local government, day-and-night life, and sense of identity (Stretton, 1976, p. 225).

In other words, dispersion "in the uncoordinated pursuit of cheap sites" can result in a city which is not only visually (and perhaps spiritually?) depressing, but which is expensive to live in, for it depends entirely on motor transport. Is it possible to have decentralisation without these disadvantages?

Much depends on the design strategy adopted. When facilities are most scattered the proportion of workers who can work locally is maximised, journey lengths on average are shortest, and the greatest number can walk to work. However, in most occupations a majority will still choose not to work locally (e.g., the case of schoolteachers) and these people will have to drive, for there is little hope of providing a competitive public transport service in a city where facilities are all over the place. There is, therefore, much to be said for promoting regional commercial centres, situated at intervals along railway lines and at the focus of bus routes for each group of suburbs. Public transport patronage could also be increased by grouping such other facilities as hospitals and tertiary institutions in line nearby, so that they can be served by the bus routes before they diverge. The best new example of such a regional centre in Sydney is that at Mt. Druitt.

If there are to be regional centres, would it be possible for some of them to attract the office industries, and so reduce the concentration of clerical employment in the city centre? While it is true that clerks have a low response rate to local employment and that decentralisation of clerical jobs would invite more long-distance motoring than most kinds of work, it still seems unfair that employment opportunities for clerks should be so heavily concentrated in the city centre.

The most spectacular suggestion is that a "twin and rival" city centre should be built at Parramatta, able to compete with the City of Sydney itself in the brightness of its lights (Stretton, 1975, p. 255). This is but a dream, for if employment in the Parramatta business district were increased much beyond its present 11,000, new roads and rapid transit lines would have to be built to bring these people to work. The insertion of such new transport routes into the existing urban fabric would be too costly. On the other hand, where the planning is done in advance and land for the necessary transport routes is reserved, it may be possible to develop major new centres at a reasonable cost. The new centre at Campbelltown provides an example: even if it achieves its planned total employment of 38,000, it will be no rival for the city centre (employment 207,000) but it will still make a substantial contribution to job opportunities in the outer suburbs (SPA, 1973, p. 88).

In that none of the proposed suburban centres would be very large, the present city centre would still be very attractive to those establishments which require centrality. This might mean that employment in it would rise, though only slowly. While it is true that journeys to work in the city centre are long (on average), they are mostly by train, the form of transport with the lowest operating costs per passenger kilometre. Any additions to the city centre work force would likewise have to travel by public transport, but it should be possible to provide for them by works to increase the capacity of the present suburban railways and by increased use of transit lanes for buses, or even trams.

Though continued decentralisation of employment should reduce the average length of journey to work, there will still be people (mostly men) who insist on driving a long way across town to their jobs. Collectively, these gentry make considerable demands on the road system. Should expensive arterial roads and orbital freeways be built to meet their requirements? The extra journeying range provided by such roads will be something of a luxury once employment has been dispersed to all suburbs. Again, it may not be a wholly bad thing for a district to have its road access restricted. In Sutherland and Manly/Warringah, journeys to work are shorter than elsewhere, without any vociferous complaints that the residents are being denied their due choice of jobs. The example of these two regions would recommend a policy of traffic restraint rather than roadbuilding (Thomson, 1977, ch. 7).

A policy to guide the existing trend towards decentralisation of employment would thus try to constrain the amount of motor traffic by limiting construction of new roads in built-up areas and by maintaining public transport services to the city centre and to regional centres. Some local workers might find their journey lengths so reduced that they can switch back from vehicular transport to walking or cycling. However, most workers in local employment will continue to have the same choice as now: bus or the private car. Given reasonably easy car parking (and the proposed regional employment centres will not be large enough to generate severe parking difficulties), present-day bus speeds and costs are not competitive with the private car except where the bus enables the traveller to dispense with car ownership entirely. It must therefore be expected that most workers outside the city centre will continue to drive to work (Smith, 1976, p. 221). Bus services will have to be maintained for the benefit of those who cannot drive, and by focusing both services and employment on regional centres it should be possible to make bus travel attractive to those who could drive but would rather avoid the expense of car (or second car) ownership.

For the present, therefore, a policy of job decentralisation to suburban centres can be expected to reduce distances travelled to work, without unduly limiting people's choice of jobs, and without depriving them of public transport. In addition to these benefits, job decentralisation would seem a prudent policy in a world in which energy supplies are dwindling. Should a coming energy crisis increase the cost of transport, particularly the energy-intensive (private) forms of it, a policy which enables and encourages short work journeys in the present will have incalculable benefits in the future.

References

ABS (Australian Bureau of Statistics). *Journey to Work and Journey to School Aug. 1974,* Ref. 17.5. Canberra, 1976.

Davis, J. R., and Spearritt, Peter. *Sydney at the Census: 1971, a Social Atlas.* Canberra: Urban Research Unit, 1974.

Evans, A. W. *The Economics of Residential Location.* New York: St. Martin's, 1973.

Halkett, I. P. B. *The Quarter-Acre Block.* Canberra: Australian Institute of Urban Studies, 1976.

Jones, M. A. *Housing and Poverty in Australia.* Melbourne: Melbourne University Press, 1972.

Liepmann, K. K. *The Journey to Work.* London: Kegan Paul, Trench, Turbner & Co., 1944.

Neutze, Max. *People and Property in Bankstown.* Canberra: Urban Research Unit, 1971.

Neutze, Max. *Australian Urban Policy.* Sydney: Geo. Allen & Unwin Australia, 1978.

Sandercock, Leonie. *Cities for Sale.* Melbourne: Melbourne University Press, 1975.

Schaeffer, K. H., and Sclar, Elliot. *Access for All; Transportation and Urban Growth.* Harmondsworth: Penguin, 1975.

Smith, A. B. "A Review of the Factors Affecting the Outlook for Urban Passenger Transport." *Bureau of Transport Economics* (1975): 201–42.

SPA (State Planning Authority, NSW). *The New Cities of Campbelltown Camden Appin Structure Plan.* Sydney, 1973.

Stretton, Hugh. *Ideas for Australian Cities,* 2nd edition. Melbourne: Georgian House, 1975.

Stretton, Hugh. *Capitalism, Socialism and the Environment.* Cambridge: Cambridge University Press, 1976.

Thomson, J. M. *Great Cities and their Traffic.* London: Victor Gollancz, 1977.

Urban Research Unit. *Urban Development in Melbourne.* Canberra: Australian Institute of Urban Studies, 1973.

Wheaton, W. C. "Income and Urban Residence: An Analysis of Consumer Demand for Location." *American Economic Review,* 67 (1977):620–31.

Section E Social Area Analysis (Factorial Ecology) of the City

Residential Area Characteristics: Research Methods for Identifying Urban Sub-areas—Social Area Analysis and Factorial Ecology*

R. J. Johnston

The preceding chapters have discussed the mechanisms by which people and households either choose or are allocated to residential areas within a city, a series of individual and corporate decisions which collectively produce the urban mosaic. If these mechanisms were deterministic, a clearly segmented map would ensue. Unfortunately for the analyst, this simple situation does not exist; although neighbourhood differences can be observed in most cities, so too can overlaps and deviant cases. Thus, the study of social areas, both as descriptive mapping and as preparation for predictive analyses, requires a methodology which will unravel the complexities of spatial differentiation between and within districts. The present chapter reviews those methodologies which have been, and are still being, used.

Many reasons can be proposed for the failure of location/allocation mechanisms to produce patterns of complete separation of various socioeconomic, demographic, and housing groups. The first rests on their lack of deterministic power; because some residents, at least, have a degree of freedom of choice of where to live, there is a possibility either of "irrational" behaviour or of individuals basing their location decisions on attitudes and values which are not consonant with those of the social group(s) to which they belong. Secondly, the various mechanisms are not independent, but they may vary in their relative effect. Nor are the various groupings within society independent; most people are members of several

*Reprinted from D. T. Herbert and R. J. Johnston, *Social Areas in Cities*, Vol. 1 (London: Wiley, 1976), pp. 193–209, 214–20, and 230–35, by permission of the authors and John Wiley and Sons, Ltd.

(socioeconomic, demographic, ethnic, life-style, etc.) and may differ as to which group's norms they rate most highly when choosing where to live. Thirdly, and following from this, the social groups which we identify rarely have precise boundaries. In residential choices, there are some groups with clearly defined limits, as in societies where there is a maximum income above which no householder will be allotted a publicly-owned or -financed home. But most group boundaries are ill-defined, resulting in overlapping residential patterns—as noted in Chicago (Duncan, 1956; Duncan and Duncan, 1955). Finally, individuals, households, areas, cities themselves—all are continually changing, the first two in their characteristics and attitudes, the second two in their attributes and forms. Change may be sudden, but is usually slow, in which case observation at one time may view different areas at different states and rates of change.

All of these factors suggest that the residential mosaic of a city, especially one with considerable freedom of choice for at least some individuals and reliant on market operations for its form, will be far from clear-cut in its patterns. This is why methodologies are needed for its description and analysis and why it is a challenging field for researchers. The problem is invariably compounded, however, by the data which are available for such investigation, and since in this, as in many large-scale research areas, methodology is so dependent on data, discussion of information sources is a prerequisite for this chapter.

The data of urban sub-areal study

The average town, let alone a city, is too large for researchers with relatively limited resources to allow collection of specific data for a study of the residential mosaic. There are, of course, exceptions to this, the classic being Charles Booth's mammoth survey of London at the end of the nineteenth century (Booth, 1889–1902; Pfautz, 1967). Others have sampled households, perhaps as part of a data collection exercise by or for a planning agency (Forrest and Tan, 1971), many of whom have developed massive data banks to support their operations. Data on individual properties may be obtained from such sources as property tax rolls in several countries (Davies, Giggs and Herbert, 1968; Forrest, 1968), but these deal only with "impersonal" measures. If the researcher is unable to design and conduct his own surveys, he must rely on those operated by others.

He then confronts the confidentiality problem; many such sources are made available only when they are irrelevant to living persons, as with the release of British census enumerators' books a century after their collection (Lawton, 1955). Thus it is that research in urban historical geography has been based on some of the most detailed data (Doucet, 1972; Goheen, 1970; Ward, 1969), though the advantages that these bring are often counterbalanced by the tedium involved in their collection.

Most students of the residential mosaic who are interested in its general patterning rely, often exclusively, on decennial and quinquennial censuses for their data. They are thus dependent on the policy of the collecting bodies—invariably government departments—regarding the nature and quality of those data. Two major aspects of this reliance, which very much influence the methodology applied, concern the questions asked, plus the variables tabulated from them, and the areas for which tabulations are produced.

Censuses are objective counts, their origins being in tax-gathering exercises (Glass, 1973). They aim to record the number of people in a place at a given moment, but they also collect information on resident and dwelling characteristics which are relevant to their sponsor, to bodies allied with this sponsor (such as local governments), and perhaps also to other bodies who can influence the census-takers in their choice of questions. A census questionnaire must be simple and easy to answer, as well as easy to administer, requirements which are increasing in importance with the trend towards self-enumeration. Questions must be straightforward, perhaps requiring only a multiple-choice answer. They can record income and occupation, which are "objective indices," but not socioeconomic status, which involves attutudes; age and household structure can be recorded, but not life-style or activity patterns. Thus, the researcher is constrained to variables which can be construed as indices of the constructs he investigates. Often the data provide poor surrogates, because what he considers to be necessary questions are not asked. Many census authorities have been wary of asking for information on "touchy" subjects; income has been one of these and religion another (not asked in Britain since 1851, for example). In addition, information is rarely collected which is relevant to studies of housing markets.

Censuses are confidential documents; their publications refer to population aggregates only. As well as national totals, almost all publish at least some tabulations by areal subdivisions. The constitution of these reflects two constraints: the responsibility of the census to provide information for subnational territories, such as local government areas, and ease of administration. The latter concerns the actual census-taking. Most censuses employ enumerators, who distribute and collect the questionnaires in a defined area, demarcated for logistic convenience. Such enumerators', or collectors', districts form the basic building blocks. Data are too numerous to be published at this level, though they can usually be purchased, often in machine-readable form.

Together, these two constraints produce the basic areal mesh of census districts. This is hierarchical, since the enumerators' districts must have common boundaries with local government areas. (The hierarchical structure may be a severe constraint if there are overlapping independent areas—such as political constituencies and local governments—for which data must be provided.) The value of such data varies from place to place, from census to census. Often the design of enumerators' districts pays no heed to the social pattern on the ground, producing a set of areas whose analysis suggest spatial heterogeneity, perhaps only as a result of boundary locations. Census authorities vary in the extent to which they work with local interests to produce a "relevant" set of areas. The United States censuses are exemplary in this way, for they have designed special areas, at all scales, such as census tracts, urbanized areas, and standard metropolitan statistical areas, to assist users of their material in the portrayal of spatial realities. (Of course, these spatial realities change, and often the areas lag behind in accommodating these changes; see Berry, Goheen, and Goldstein, 1968; and Guest and Zuiches, 1972. They sometimes lag in the information they tabulate; there are several large ethnic groups in Toronto, for example, for which no detailed areal breakdown is provided.) By way of contrast, the British census has traditionally offered much less to the researcher (see Hall and coauthors, 1973). A final parameter of a census of relevance here is its representativeness. Some censuses ask all of their questions of every household and resident (e.g., Australia and New Zealand); others ask a basic set of questions of all, but a more detailed set of only a sample (such as the 10 percent which is the basis of many detailed tabulations in the British census); some collect information from all, but then only tabulate it for a sample. Often such samples are so small that their standard errors are very large for small-area data, with consequent confounding effects on areal analyses.

Most research on urban residential mosaics is, and must be, based on census material, therefore. In this way, its philosophy, its methodology, and its results are all subject to external decisions concerning the nature and the quality of the data. Since such research concerns areal characteristics—amalgamations of individuals—the pattern of such areas, their size, shape, and relevance to other features (and the variance of all of these) all very much affect the input-output matrix of such investigations. It is in this light that their procedures and findings must be assessed.

Social area analysis

Early studies of social patterns within cities were limited by both data paucity and technical ability to handle the available information. Thus Burgess's pioneer studies of Chicago in the 1920s were based on single-variable maps (see Burgess and Bogue, 1967, p. 6) and verbal comparisons of maps and "gradients" (Burgess, 1927). By the 1940s, the more detailed census tract data available in the United States were being employed in relatively sophisticated mapping of "natural areas" (Hatt, 1946), an exercise which was the prelude to the development of a "census-tract-based methodology."

The "theory"

The breakthrough was made by three interrelated groups of researchers at universities in the three main cities—Los Angeles, San Francisco, Seattle—on the United States' West Coast. Characteristics of all three included a concern for theory, rather than description as a prelude to practice which typified the Chicago approach; a focus on *areas* as aggregates of individuals, rather than on individuals within areas; and the development of multivariate analyses.

The initial statement, a monograph (Shevky and Williams, 1949), was an empirical presentation of census tract differences in the metropolitan area of Los Angeles. Three indices of tract characteristics were derived: social rank, urbanization, and segregation. Reviewers pointed out the lack of any theoretical rationale for the selection of these three—the work was judged a novel method for using census tract data to display areal differentiation in selected population and dwelling characteristics.

Following the Los Angeles study, the methods

were refined for a similar investigation of San Francisco (Bell, 1953), after which the Los Angeles data were reworked. In 1955, the basic statement on the method was published (Shevky and Bell, 1955). In this, social area analysis is presented as part of an apparently deductive model of social change, based on a concept of increasing scale or interdependence (Wilson and Wilson, 1945), on Colin Clark's (1940) observations concerning the division of labour in a society, and on Louis Wirth's (1938) classic deductions of the relationships between population concentration and social forms. (Note, however, a later admission by Bell and Moskos [1964] that the theory was an *ex post facto* rationalization of the earlier empirical work.)

Shevky and Bell's model (Figure 1) is based on three postulated major trends of industrial societies: changes in the range and intensity of relations, differentiation of function, and complexity of organization. Operational measures of these trends are then listed—for the first, changing intersectoral division of labour; for the second, the declining role of the household as an economic unit; and for the third, greater population mobility and concentration. From these are derived the three basic constructs of increasing scale: social rank, urbanization, and segregation. (These were Shevky's terms; as made clear in an appendix, Bell preferred economic status, family status, and ethnic status.) A range of possible variables indexing these constructs was provided, from which seven were selected—three each for the first two and one for the third. (The third index for social rank—rent—was later omitted because of the influence of rent controls.) Choice of these variables has been criticized (e.g., by Udry [1964] who claims that the trends Shevky and Bell describe were not typical of the United States over previous decades). Further, as will be suggested in a later section, selection of the variables to represent urbanization provided a set poorly related either to the general trends or to the sample statistics.

Several strong attacks were launched by reviewers against this theory and its application. Shevky and Bell (1955, p. 20) had defined social areas as "containing persons with similar social positions in the larger society," but, as Hawley and Duncan (1957) argued in the most detailed critique, there is no way of relating social differentiation, which is a product of increasing scale, with spatial differentiation at the census tract level. Shevky and Bell (1955, p. 20) may argue that:

The social area . . . is not bounded by the geographical frame of reference as is the natural area, nor by implications concerning the degree of interaction between persons in the local community as is the subculture. We do claim, however, that the social area generally contains persons having the same level of living, the same way of life, and the same ethnic background; and we hypothesize that persons living in a particular type of social area would systematically differ with respect to characteristic attitudes and behaviours from persons living in another type of social area.

Their methods might, within the bounds of census tract data (on which Hawley and Duncan are sceptical, although Duncan—Duncan and Duncan, 1955—was using them for somewhat similar descriptive purposes at the same time), show differences between the residents of different areas, and their field studies may indicate the hypothesized attitude and behaviour differences (Bell, 1969; Bell and Boat, 1957; Bell and Force, 1956a, 1956b, 1957; but see Palm, 1973a), but nowhere do they suggest the mechanisms by which the different groups are propelled to, or choose, their allotted social areas. As Jones (1969, pp. 18–21) has observed, Shevky and Bell's model is better suited to inter-urban studies, of places at different levels of increasing scale, than to intra-urban investigation. It has been left to later theorists . . . to suggest the relevance of Shevky and Bell's indices, if not their theory, to intra-urban areal differentiation.

While other scholars have been reviewing, testing, and sophisticating Shevky and Bell's methods, McElrath (1965, 1968) has revised the theory. He replaces the master trends by two major interrelated processes of social change, industrialization and urbanization (Johnston, 1973a). The former produces the changing intersectoral and interpersonal divisions of labour covered by the social rank and urbanization constructs; the latter produces population aggregation (migrant status) and a greater spatial hinterland for metropolitan growth which brings visible minorities to the cities (ethnic status).

The method

Social areas are defined according to the three-dimensional space of the Shevky–Bell model. For each dimension, a standardized index ranging from zero to one hundred is defined as an unweighted average of similar standardized scores for each of the relevant variables. The standardized score for a tract on a variable is

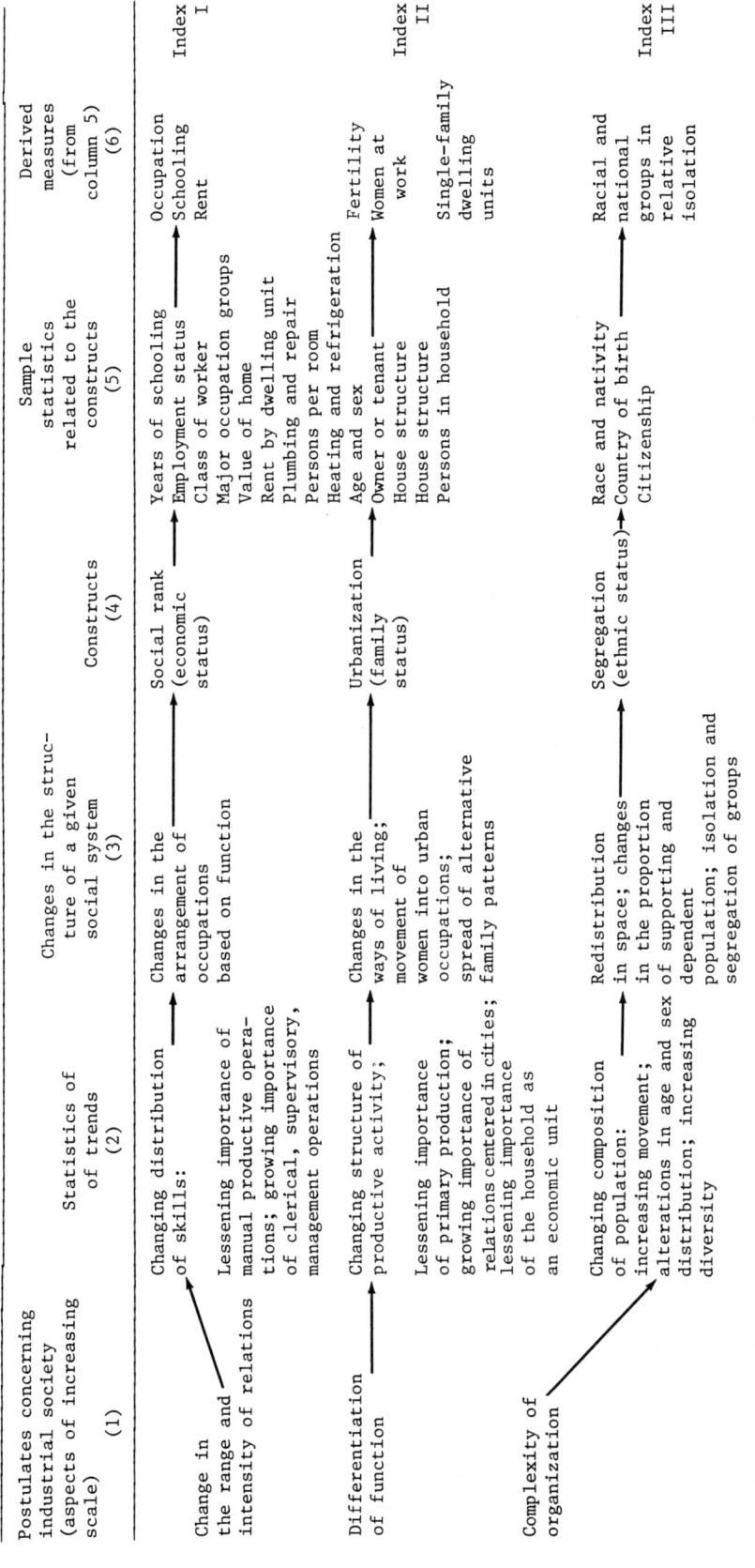

Figure 1—The Shevky-Bell theory of increasing scale and its relationship to index construction (reprinted with permission of Stanford University Press, from *Social Area Analysis: Theory, Illustrative Application and Computational Procedures*, by Eshref Shevky and Wendell Bell, 1955, Figure IV, p. 26).

$$S_t = X(r_t - 0) \tag{1}$$

where
- 0 = the lowest value of the variable over all census tracts
- r_t = the variable for tract t
- X = 100 divided by the range for the variable
- S_t = the score for tract t

Each variable set (in all cases a vector of ratios, e.g., the proportion of the population aged 25 and over who had no more than eight years' schooling) is ordered so that within any one construct, a high value on one variable has the same status connotation as a high variable on another. The final index is simply

$$I_t = \Sigma\, S_t/n \tag{2}$$

where
- S_t = the score as defined in equation (1)
- n = the number of variables forming the index
- I_t = the index for tract t

In the Los Angeles study, eighteen social area types were defined. The social rank and urbanization indices for the universe of tracts were plotted against each other. The social rank indices were then trichotomized, using the thirds of the range of values as the inner boundaries; and the urbanization indices were trichotomized, using the area within ±2 standard errors of the regression line of social rank on urbanization to define the average, with high and low beyond these. The resulting nine-cell social space (Figure 2) was further dichotomized into areas of high and low indices of segregation. For the 1955 monograph, the procedure was simplified, replacing the regression approach—variable between cities in its slope, and also (not noted by Shevky and Bell) in whether urbanization or social rank is used as the independent variable—by a standardized fourfold division of each variable, using index values of 25, 50, and 75 as the boundaries. The resulting sixteen cells (Figure 2) are further subdivided into low and high segregation tracts, with the percentage of the total population of the relevant urban area in the designated ethnic groups as the dividing line: thirty-two cells ensue as the major social areas.

The procedure adopted by Shevky and Bell as a standard, and presumably generally applicable, method is not problem-free. As Bell notes in his factor analyses of the same data sets, the constructs are related, especially social rank and segregation. If these relationships vary, by time

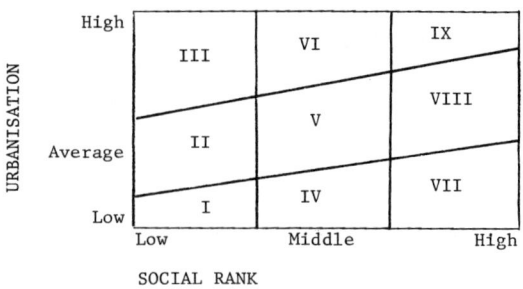

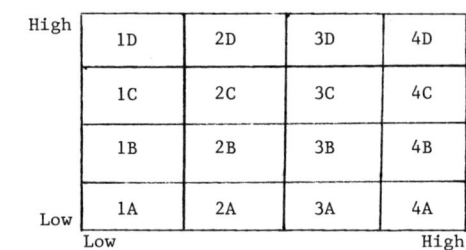

Figure 2—The Social Area Classification Schemes (reprinted with permission of Stanford University Press, from *Social Area Analysis: Theory, Illustrative Application and Computational Procedures*, by Eshref Shevky and Wendell Bell, 1955, Table II.I, p. 4).

and place, comparative study of social areas defined by the standard diagram will be impeded (as it will also by the use of local averages, especially in the segregation index). Such problems are likely to be especially acute in applying the method in other societies, as Herbert (1967) discovered in his work on Newcastle-under-Lyme. (Extra problems may have been introduced there by the use of only one urban unit of a larger conurbation.) Within the monograph, however, the use of only Los Angeles and San Francisco as examples, especially the latter, restricts the recognition of these possible problems; there is no reference, either, to Bell's (1955) factor analyses, which suggests that the manuscript for the monograph was completed several years before its publication. Discussion concentrates on the social space, on its constituent variables, and on relationships with others, notably referring to age and sex composition. The social areas of San Francisco are

Use, developments, critiques

Perhaps because of the initial debate on its value, the Shevky–Bell methodology has not been widely used. In the United States, several "disciples," notably Greer and McElrath, employed the procedures (see the list in Bell, 1969), but Timms' (1971, pp. 150–51) listing shows only seven social area analyses conducted in other countries (a later addition is Parkes, 1972). In part, this represents a time-and-space lag in diffusion, but more importantly it reflects the later availability of census tract or equivalent data in other countries, which availability coincided with the rapid worldwide spread of high-speed computer hardware plus the software for the factorial ecology method discussed in the next section.

As already noted, one of the initial criticisms of Shevky and Bell's methodology was its "theoretical" base, the imposition of the three-dimensional social area framework. Yet parallel to their work, Tryon's (1955) cluster analysis of the San Francisco tract data produced an empirical set of dimensions very similar to Shevky and Bell's. This may be taken as indicative of the validity of the latter's procedure, irrespective of its relationship to their general theory, but it may be interpreted more skeptically as showing that both studies—separately or in consultation—exhausted the main possibilities of the available census data.

The Seattle group of researchers, headed by Schmid, initially followed a different line, being more concerned with the interrelationships among a wider set of variables over the universe of census tracts, and with changes over time, than in the characteristics of areas *per se* (Schmid, 1950; Schmid, MacCannell, and Van Arsdol, 1958). In this, their work was closer to the Michigan–Chicago-based investigations of intra-urban spatial differentiation, although the latter used a simpler, statistically less-demanding index (Duncan and Duncan, 1955). But the Seattle group were attracted to social area analysis, both for the particular purpose of Schmid's (1960) crime area study and "to determine by empirical research whether or not Shevky's scheme might possess distinctive features as well as indicate a compelling persistence and a pragmatic value" (Van Arsdol, Camilleri, and Schmid, 1962, p. 10). This was done by replicating Bell's factor analyses (see below), on a sample of ten medium-sized cities. The aims of these analyses were (a) to see whether the three Shevky–Bell dimensions were separate indices of areal differences, and (b) to see whether the individual variables were closely related to the relevant dimension of the theory. Bell's (1955) work provided strong support for the dimensional model; the Seattle workers discovered some variations which were explicable in terms of the different social environments of the sample cites (Van Arsdol, Camilleri, and Schmid, 1958a, 1958b). However, they then argued (Van Arsdol, Camilleri, and Schmid, 1961, p. 27) that "despite the theoretical difficulties, the case for social area analysis would be strengthened if it could be shown that empirical advantages are obtained by combining the six traditional census tract measures into composite indexes and types," which they investigated by relating both the indices and their constituent variables to measures of spatial stability and age structure. The variables performed better than the indices; the pragmatic value of the latter was disclaimed, despite contrary arguments that high-level generalizations are not meant to predict specific patterns (Van Arsdol, Camilleri, and Schmid, 1962; Bell and Greer, 1962).

Factorial ecology

As a research procedure, social area analysis has virtually disappeared. In large part, it has been made redundant by technological advances, which have brought with them a greater concern for inductive theory, based on empirical results, as opposed to deductive formulations (though the pedigree of social area analysis as a deductive theory remains in doubt). Social area analysis has received a fairly detailed review here, however, both because it has provided the general framework within which most recent studies of intraurban residential mosaics have been set and because the later, factor analysis-based studies of its dimensions heralded the application of that body of techniques to social area investigations.

Factorial ecology described

The term factorial ecology was coined by Sweetser (1965a, 1965b) "as a model for ecological structure, . . . the method par excellence for comparing cross-nationally (and intra-nationally) the ecological differentiation of residential areas in urban and metropolitan communities" (Sweetser, 1965a, p. 219). He was not the first to use this method, Bell (1955) having preceded him by at least a decade, followed by the Van Arsdol and coauthors' studies, by Anderson and Bean's (1961) Toledo investigation, and by Schmid (1960), Gittus (1964), and Jones (1965), among others. Sweetser's term was rapidly adopted by other researchers, most notably at the University of Chicago's Center for Urban Studies, as a collective description of all studies applying the techniques of factor analysis to the study of areal differentiation. It thus can be applied to any spatial scale and any set of characteristics, not exclusively to intra-urban population and housing patterns, but it is commonly applied only to studies of the type discussed here.

A more detailed definition of factorial ecology than given by Sweetser is provided by Berry (1971). The term "ecology," he claims, indicates that the research focus is a system of component parts (census tracts or their equivalents) which interact among themselves and with their environment. Interest is in the variance between the components only; variability within tracts is ignored. Factorial ecology asks (Berry, 1971, p. 215): " 'How does the system cohere and pattern?' The answer is sought by trying to identify repetitive sequences of spatial variation present in many observable attributes of areas." Factor analysis methods are used in the answer.

The methods of factorial ecology have been applied widely in the last decade to cities in every continent (Rees [1972] gives perhaps the most comprehensive, recent catalogue). As with many innovations, it developed something of a bandwagon effect, with many somewhat "unthinking" replications applied to whatever data were available. Not surprisingly, this flood of studies has been followed by retrenchment and rethinking. Nevertheless, factorial ecology demonstrably offers a widely accepted approach to identification of underlying determinants of intra-urban residential patterns, as well as providing series of indices of value for further study of ecologies; hence its preeminence in this chapter.

The procedures of factorial ecology

The set of procedures and operations which comprise a factorial ecology is outlined in Figure 3; Table 1 and Figure 4 provide an example of these, for the Whangarei (population 34,029) urban area, New Zealand, in 1971. Some of the stages are standard parts of the process and are indicated thus in the diagram; others are only included where either necessary or desired.

The first stage, assembly of the data matrix, is straightforward, although, as indicated later, some technical and interpretative problems may arise from the selection of variables. In the Whangarei example, eight variables were selected—four (X_1–X_4) representing the Shevky–Bell family status construct, three (X_5–X_7) representing economic status, and one (X_8) for ethnic status. According to Bell's (1955) hypothesis, these should cluster on three separate dimensions in a factor or component solution. Stage Ia, data transformation, is not always employed. Since principal components and factor analyses usually operate on matrices of product moment correlations, the data should meet the requirements of those correlations, especially those relating to *linearity* of relationships (Poole and O'Farrell [1971] succinctly list these; some are not relevant if description is the aim of the analysis, as against inference to population from a sample). With the Whangarei data, inspection of frequency distributions (Figure 4) indicated positive skewness for variables X_3, X_7, and X_8, and these were transformed to base ten logarithms, giving approximate normal distributions. Each vector of the data matrix was then standardized to Z-deviate form where

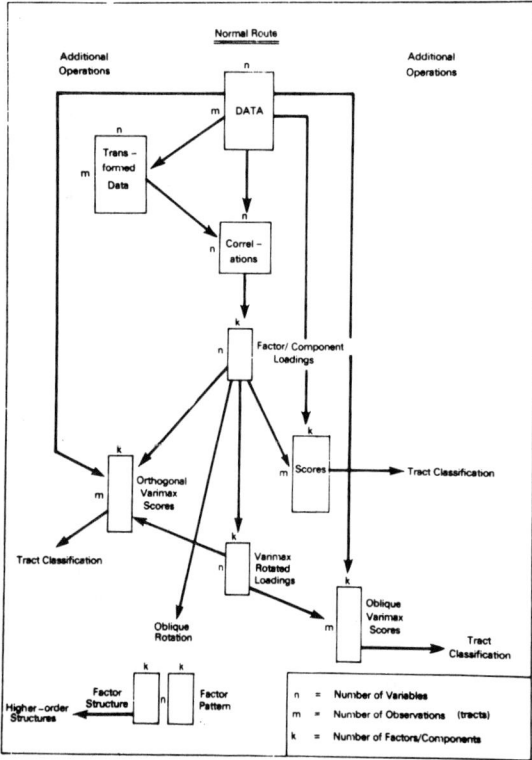

Figure 3—Flow diagram of the procedures involved in the factorial ecology method.

$$Z_{ij} = \frac{X_{ij} - \bar{X}_i}{S_i} \quad (3)$$

where X_{ij} = the value of variable i in district j
$\bar{X}_i$ = the mean for variable i
S_i = the standard deviation for variable i
Z_{ij} = the Z-deviate of variable i in district j

The matrix of intercorrelations gives a general impression of the strength of intervariable relationships and the degree of independence between clusters of variables. For Whangarei, variables X_1–X_4 apparently form one cluster and variables X_5–X_7 another, with X_8 related to variables of both clusters, but especially the latter. Such visual interpretation is difficult, however, especially where many variables are used; hence, the next stage.

There is a great variety of ways to factor a data matrix. The two main alternatives differ in their treatment of the communalities, the entries on the major diagonal of the correlation matrix. Principle components analysis uses unities as communality estimates, making it a closed model which accepts that all of the variance among the variables can be accounted for endogenously. Factor analysis, on the other hand, accepts that there are exogenous variables influencing those in the matrix; it isolates only the common variance. To do this, its communalities are estimates of that common variance.

This is frequently done by entering in the main diagonal the squared multiple correlation between the relevant variable and the others.

Once the communality estimate has been determined, factoring of the correlation matrix proceeds. The common methods produce an iterative set of hybrid, orthogonal variables, with each successive hybrid (factor or component) accounting for a smaller proportion of the variance. In components analysis, the total possible number of hybrids is the number of variables in the original matrix (except in cases where there are fewer observations than variables, in which case the number of observations determines the maximum). Factor analysis will produce fewer factors than variables, because of the reduced communalities. In both cases, however, it is usual to extract fewer components or factors than variables, since the aim is to represent the major common patterns and to ignore the minor elements. Determining the number of components or factors to extract involves subjective judgement, though a variety of rule-of-thumb methods has been advanced (Davies and Barrow, 1973). A common method is to extract only those with eigenvalues, sums of squared loadings, exceeding 1.0, which account for a greater proportion of the variance than does one of the original variables.

Both models are frequently employed in factorial ecologies, though there has been little discussion of the relevance of the various communality estimates (Tarrant, 1974). A components approach was used with the Whangarei data and

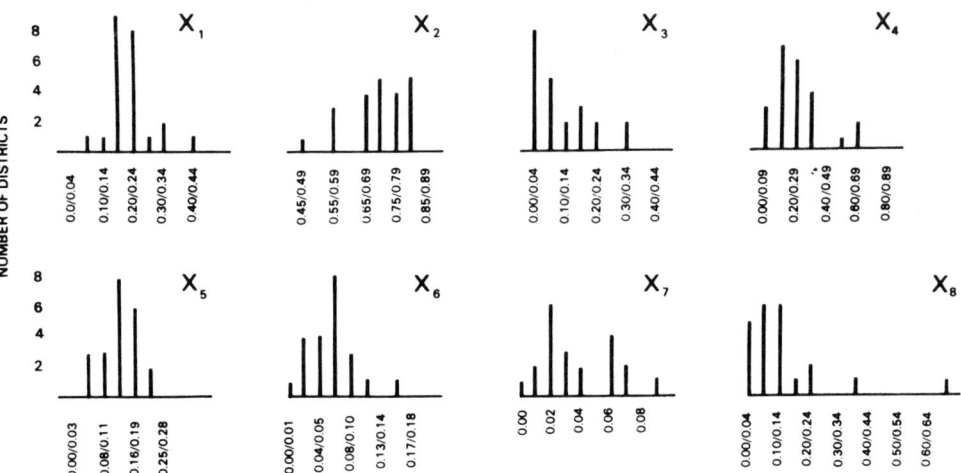

Figure 4—Factorial ecology of Whangarei, 1971: frequency distributions (for key to variables, see Table 1).

Table 1. Whangarei, New Zealand, 1971: A Factorial Ecology

Variables	
X_1	Proportion of those aged 16 + not married
X_2	Proportion of households that are one-family households
X_3	Proportion of dwellings that are flats
X_4	Proportion of dwellings that are rented
X_5	Proportion of male workers in professional/managerial occupations
X_6	Proportion of male workers earning $6,000 + per annum
X_7	Proportion of male workers with a university degree
X_8	Proportion of population who are Maori

A. Original data

District	X_1	X_2	X_3	X_4	X_5	X_6	X_7	X_8
1	0.18	.79	.07	.14	.17	.07	.04	.12
2	0.12	.83	.05	.15	.10	.04	.02	.12
3	.17	.72	.11	.24	.19	.15	.06	.03
4	.23	.67	.13	.54	.08	.02	.02	.39
5	.21	.76	.04	.16	.16	.02	.02	.03
6	.20	.66	.16	.32	.14	.05	.03	.07
7	.20	.59	.16	.26	.14	.06	.03	.03
8	.29	.55	.32	.37	.21	.11	.07	.04
9	.24	.66	.16	.27	.12	.06	.02	.08
10	.33	.59	.21	.35	.21	.04	.02	.12
11	.40	.48	.34	.64	.13	.06	.06	.09
12	.18	.69	.08	.21	.21	.07	.01	.09
13	.31	.70	.23	.35	.16	.08	.09	.15
14	.16	.83	.03	.17	.15	.08	.04	.06
15	.16	.78	.06	.24	.06	.02	.01	.23
16	.20	.73	.08	.21	.08	.04	.01	.11
17	.17	.81	.03	.13	.16	.08	.02	.09
18	.08	.84	.01	.08	.19	.07	.07	.04
19	.16	.74	.00	.67	.06	.00	.00	.68
20	.16	.80	.02	.09	.14	.06	.06	.12
21	.17	.74	.00	.13	.13	.06	.06	.13
22	.20	.77	.00	.19	.08	.03	.03	.20

B. Product moment correlations

	X_1	X_2	X_3	X_4	X_5	X_6	X_7	X_8
X_1	1.00							
X_2	-0.84	1.00						
X_3	0.57	-.61	1.00					
X_4	0.60	-.66	.15	1.00				
X_5	0.01	-.09	.36	-.34	1.00			
X_6	0.04	-.10	.40	-.26	.77	1.00		
X_7	0.29	-.22	.61	-.30	.67	.71	1.00	
X_8	0.02	.15	-.42	.45	-.75	-.64	-.67	1.00

Table 1. Whangarei, New Zealand, 1971: A Factorial Ecology continued

C. Unrotated component solution

Variable	Component I	Component II	Component III	Communality
X_1	.36	.85	-.06	.87
X_2	-.41	.85	-.07	.90
X_3	.73	.44	-.37	.85
X_4	.49	.88	.36	.93
X_5	.82	-.33	.32	.88
X_6	.82	-.26	-.35	.86
X_7	.89	-.10	-.19	.84
X_8	-.81	.33	.06	.77
Eigenvalues	3.67	2.70	0.53	
Cumulative % of variance	45.88	79.66	86.33	

D. Unrotated component scores

District	Component I	Component II	Component III	District	Component I	Component II	Component III
1	0.29	-0.73	-0.14	12	0.39	-0.33	1.03
2	-0.62	-0.79	-1.24	13	0.81	0.70	-0.13
3	1.50	-0.79	1.67	14	0.28	-1.04	0.27
4	-0.88	1.42	-0.36	15	-1.30	0.04	-1.38
5	0.12	-0.49	-0.90	16	-0.59	0.14	-1.40
6	0.30	0.32	0.46	17	0.09	-1.01	0.57
7	0.78	0.28	-0.44	18	0.38	-1.88	0.48
8	1.85	0.72	1.24	19	-2.85	0.89	2.76
9	0.20	0.43	-0.55	20	-0.19	-0.91	-0.39
10	0.22	1.39	-0.85	21	-0.14	-0.57	-0.41
11	0.89	2.61	0.23	22	-1.50	-0.38	0.42

E. Varimax-rotated component solution

Variable	Component I	Component II	Component III
X_1	-0.01	.87	-.33
X_2	-.10	-.91	.24
X_3	.30	.49	-.73
X_4	-.27	.87	.31
X_5	.93	-.04	-.11
X_6	.92	.03	-.09
X_7	.68	.06	-.60
X_8	-.76	.14	.43
Sum of squared component loadings	2.93	2.60	1.37
Trace (%)	34.12	69.12	86.25

Table 1. Whangarei, New Zealand, 1971: A Factorial Ecology continued

F. Varimax-rotated component scores

District	Oblique			Orthogonal		
	I	II	III	I	II	III
1	0.41	-0.60	-0.76	0.42	-0.66	-0.20
2	-0.41	-0.99	1.53	-0.82	-1.18	-0.67
3	1.54	-0.20	-3.41	2.27	0.00	0.74
4	-1.10	1.04	2.04	-1.37	1.04	-0.00
5	0.16	-0.47	-0.93	-0.15	-0.66	-0.78
6	0.14	0.37	-1.60	-0.07	0.25	-0.58
7	0.55	0.48	-3.24	0.35	0.32	-0.80
8	1.44	1.32	-5.80	1.86	1.40	0.05
9	0.03	0.44	-1.40	-0.23	0.31	-0.62
10	-0.21	1.33	-2.21	-0.67	1.14	-0.97
11	0.13	2.74	-4.28	-0.02	2.70	-0.51
12	0.47	-0.14	-0.31	0.91	0.03	0.71
13	0.50	0.90	-3.31	0.38	0.80	-0.59
14	0.50	-0.87	-0.22	0.70	-0.84	0.19
15	-1.18	-0.44	3.37	-1.73	-0.59	0.52
16	-0.62	-0.13	0.81	-1.18	-0.36	-0.91
17	0.36	-0.90	0.70	0.68	-0.79	0.54
18	0.79	-1.63	0.08	1.16	-1.56	-1.40
19	-2.39	0.08	1.91	-1.41	0.88	3.73
20	0.03	-0.93	0.81	-0.03	-0.99	-0.15
21	-0.00	-0.60	0.43	-0.11	-0.67	-0.22
22	-1.12	-0.80	5.81	-0.93	-0.58	1.16

G. Inter-component correlations

	I	II	III		I	II	III
I	1.00	0.03	-0.86	I	1.00	-0.03	0.00
II	0.03	1.00	-0.41	II	-0.03	1.00	0.00
III	-0.86	-0.41	1.00	III	0.00	0.00	1.00

H. Direct oblimin rotation

Variable	Factor structure			Factor pattern		
	I	II	III	I	II	III
X_1	.02	.82	-.60	-.05	.71	-.46
X_2	-.10	-.86	.57	-.10	-.81	.35
X_3	.39	.36	-.91	.03	.17	-.86
X_4	-.35	.93	.04	-.04	.98	.24
X_5	.94	-.20	-.40	.97	.01	.05
X_6	.92	-.12	-.41	.97	.09	.07
X_7	.77	-.10	-.78	.46	-.14	-.60
X_8	-.82	.30	.58	-.60	.26	.36

I. Inter-component correlations

	I	II	III
I	1.00	-.20	-.47
II	-.20	1.00	-.22
III	-.47	-.22	1.00

three dimensions were extracted: the third has an eigenvalue of 0.53 but was used (a) because the Shevky–Bell theory suggested the need for three, and (b) because after rotation (see below) the summed squared factor loadings all exceeded 1.0.

The aim of this part of the analysis is to reduce the overlap among variables, which are different measures of some more general concepts. The hybrid variables are interpreted by their loading vectors, their correlations with the original variables. Thus, the first Whangarei component (Table 1C) suggests, through high positive correlations with variables X_5, X_6, and X_7, that areas with high proportions of professional and managerial male workers also have high proportions of high-income earners and of university graduates. Such areas tend also to have low proportions of Maoris, suggested by the negative correlations with X_8, and high proportions of flats (X_3). (Note that these are ecological correlations—see Alker, 1969; Robinson, 1950—indicating characteristics of areas, but not implying that, for example, university graduates live in flats or that no Maoris earn more than $6,000 annually.) In general, therefore, this first component suggests the validity of Shevky and Bell's economic status dimension. The correlations with the second, notably those with variables X_1, X_2, and X_4, are very suggestive of the family-status dimension, but no such easy interpretation can be made of the third component. Together the three components account for over 86 percent of the variance in the distributions of the eight variables over the twenty-two districts, thereby presenting a parsimonious description of the patterns displayed in the original data matrix.

The usual methods of component and factor extraction are variance-maximizing; they locate the hybrid variables in order to account for as much of the variance as possible. Thus in the n-dimensional variable space, it may be that components/factors bisect two or more separate variable clusters. (This point is best illustrated by geometrical analogies, as in Rummel [1967].) For this reason, the pioneer factor analysts introduced the notion of simple structure, which involved rotation of the factor/component axes to another best-fit position. The most common of these, the Varimax procedure, aims to maximize the fit of hybrid variables to original variable clusters, by producing as many loadings as close to either ±1.0 or 0.0. Application of this procedure to the Whangarei matrix (Table 1E) more clearly identified the first two components as, respectively, economic/ethnic status and family status. The third stands out as intermediate to these two, emphasizing that the areal distribution of flats and university-educated persons tended to be relatively independent of the two main dimensions, with a slight negative relationship between these two distributions and that of Maori persons. (Note that the relative importance of the components, indicated by the eigenvalues and the summed squared loadings, is usually of little relevance to the interpretation, since it is largely a function of the number of variables selected to represent each major social area concept.)

Components analysis may be likened to a process of superimposing sets of maps portraying the distributions of different variables over a common set of areas and attempting to isolate shared patterns. Having identified these shared patterns from the various loading vectors, the computation of component scores produces composite maps, standardized values for the observations on the hybrid variables. In the case of the unrotated solution, these are obtained by summing the products of the original standardized variables and the component loadings (weighted by the reciprocal of the eigenvalue). The result is a vector of scores, which has the attributes of a normal curve, on which the observations are scaled according to their rating on the hybrid variable. In Whangarei, therefore, the district with the highest positive score on component I (8) can be interpreted as that with the highest economic status, whereas District 19, with the largest negative score, has the lowest such status. (Note that, if a factor analysis is used, scores can only be estimated, because the correlation matrix factored did not use all of the information in the original data matrix but only the common variance.)

Component scores may also be estimated after the rotation procedure. Two approaches are available. Using the same formula as in the unrotated case produces score vectors which are non-orthogonal, the scores being crude weighted indices of the original values. The alternative formula retains the othogonal independence of the score vectors (see Table 1G). In many examples in the literature, the approach which has been taken is not stated. Parkes (1973), however, has compared them in a study of Newcastle, N.S.W., arguing that the correlated vectors provide a clearer classification of suburban types. The argument is that, although the "pure" dimensions of the residential mosaic may be independent, additive components, any one set of sub-areas may be far from independent in their scores on these "pure" constructs. Thus, in Whangarei, the oblique or correlated rotated component scores indicate a very strong correlation between the third dimension and the first, and a lesser one between two and three; the latter is clearly a composite map, intermediate to the independent first two components.

The component scores can be used as mappable data, providing representations of areal differences on these components of the residential mosaic. They can also be used as input to classification procedures, producing groups or types of social areas. Such exercises are particularly valuable in analyses using very large numbers of observations and for general multivariate regional delimitations, but they involve many decisions. A wide array of classificatory procedures is available (see Johnston, 1968, 1970; Spence and Taylor, 1970; and chapter 5 in this book by Adams and Gilder), involving considerable analyst choice and judgement. One method—intercolumnar correlation analysis (McQuitty and Clark, 1968)—has been applied to distance matrices derived, using Pythagoras's theorem, from the three matrices of component scores obtained in the Whangarei analyses. . . .

A final stage, or an alternative to the Varimax rotation, is to obtain an oblique rotation of the factor or component loading matrix. This removes the orthogonality constraint and allows the identification of separate yet related clusters of variables which underlie the residential mosaic. As shown in Bell's (1955) original use of the method, related dimensions are very likely and, indeed, arguments are frequently advanced that the orthogonality constraint is unrealistic—that it is unlikely that dimensions of the mosaic are independent. Few oblique rotations have been undertaken, however, largely because of the absence of easily available computer programs—that most generally used (developed by Jenrich and Harman) requires analyst decisions which can only be intuitively made. Where oblique rotations have been conducted, they either have not been fully interpreted (Van Arsdol, Camilleri, and Schmid, 1958a, 1958b) or have been accepted as little different from an orthogonal solution (Carey and Hughes, 1972). Only Timms (1970, 1971) has used multiple-group factor analysis, which is a hypothesis-testing device producing factors which may be either oblique or orthogonal, and only Davies and his associates (Davies and Barrow, 1973; Davies and Lewis, 1973) have used oblique solutions as the input to higher-order factor analyses, which suggest hierarchical or nested factor structures of dimensions and sub-dimensions.

As the orthogonality criterion introduces an unnecessary constraint on the researcher's ability to identify the proper structure of the underlying dimensions of residential differentiation, a strong argument can be made for using oblique rotation procedures, either the inductive ones such as Davies uses or the multiple-group procedures espoused by Timms. An orthogonal result would then be a special case of the more general oblique. Unfortunately, application of oblique methods involves more analyst decision and experimentation.

The Jenrich–Harman method was applied to the Whangarei matrix and, with a delta value of 0.0, produced the results given in Table 1H. The factor structure matrix, depicting the correlations between variables and components, emphasizes, more clearly than the Varimax solution, the interrelationships among the three dimensions. Variables 2, 3, 7, and 8, in particular, have considerable correlations with more than one of the dimensions, clearly stressing that, in this urban area at least, socioeconomic, family, and ethnic status do not produce independent processes leading to spatial differentiation, at the given areal scale, of residences.

The findings

Variants of the factorial ecology method have been applied to many cities since the mid-1960s. The rapidity of antibandwagon developments in modern social science means that criticisms, both general and specific, have appeared already, although the first major geographical studies were not published until 1969. Nevertheless, the method is still being widely applied, being the basis, for example, of major urban studies presently being conducted by the Association of American Geographers. No detailed review of the completed studies is to be presented here, along with a complete bibliography, since such a task has recently been undertaken by Rees (1970, 1971, 1972; see also Berry and Rees, 1969). Instead, the major findings are outlined, as an introduction to a critique of the approach.

By far the major finding, common to a majority of studies, irrespective of the location and cultural context of the relevant city, is the generality of Shevky and Bell's three-dimensional model of the bases to residential area differentiation. This must, in part, reflect the data used, the variables collected by census authorities and made available for small areas, and the inference that Shevky and

Bell derived their theory concurrently with their experimentation with census data. Yet, within this constraint, there can be no doubt that socioeconomic status, family status/life-cycle, and ethnic status are consistently major determinants of where people live, irrespective of the degree of institutional intrusions to the processes of residential location.

The factorial ecology methods, and the computer hardware and software on which they rely, allow researchers to employ a larger number of variables than was available to Shevky and Bell and to Schmid and his coworkers. Not surprisingly, extra variables tend, at least partly, to confuse the basic three-dimensional pattern. Often these do not deny the Shevky–Bell theory, but rather add to it, developing aspects which were either overlooked by those authors in their search for high-level generalizations or were not relevant to their data sets and study areas. Examples of such findings are the following:

1. Identification of dimensions whose main loadings are with variables representing aspects of population mobility. In their model, Shevky and Bell placed demographic urbanization—the spatial concentration of people—within the ethnic status dimension; McElrath (1965), however, recognized the need for a separate construct representing the intra-urban location of recent immigrants, from whatever origin, a construct which has been identified in studies including variables indexing movement into the city during a recent period (e.g., Schmid and Tagashira, 1965). Shevky and Bell also presented only static analyses of individual cities, not identifying the intra-urban mobility characteristics (often related to other population and dwelling variables) of areas of recent residential development and/or redevelopment (e.g., Murdie, 1969; Schmid and Tagashira, 1965).

2. Identification of clusters of dimensions rather than the single dimension postulated by Shevky and Bell. For example, those authors derived a single index of ethnic status composed of the residential distributions of several minority groups. Factorial ecologists have often placed each group as a separate variable, with several ethnic status dimensions emerging rather than one (e.g., Murdie, 1969). Each is a reflection of similar social tendencies, which produce "segregation," but indicates that such groups are often as spatially separate from each other as they are from their host society. (Indeed, if intra-cultural data were available by island of origin among Cook Islanders in Auckland, for example, or village/district of origin among Italians in many cities of the "New World," analyses would undoubtedly derive further "clusters within clusters.")

Similar clusters of dimensions have been obtained for the other Shevky–Bell axes. Indeed the first factorial ecology of a wider range of variables (Anderson and Bean, 1961) isolated two family status dimensions, one with loadings on life-style variables (dwelling type, dwelling tenure, marital status, etc.) and the other on family composition variables (fertility, working females). This indicates that areas of familism are differentiable according to their age structure (often also identified when age variables are employed; see Johnston, 1973b), as well as showing that high fertility—and the associated proportion of women working—is a feature of certain family life-style areas only. The latter feature is becoming more significant with the trend to concentration of child-bearing into but a few years of the average woman's fertile life (Willmott, 1969), perhaps indicating that the fertility variable is no longer particularly valid as a general index of life-style. Finally, one of Parkes's (1973) analyses of data for Newcastle, N.S.W., suggested a division of the socioeconomic status dimension into "wealth" and "occupational status."

3. Identification of specialized areas, usually as a consequence of heteroscedastic correlations (Johnston, 1971). Examples of this include Schmid and Tagashira's (1965) "Skid Row" dimension for Seattle. All Skid Row areas are very low on both socioeconomic and family status; but since the converse does not hold (all low socioeconomic status and all low family status areas are not Skid Rows), a separate dimension representing such specialized areas may emerge from the analysis (see also Johnston, 1973b). Indeed, I have argued that ethnic status may properly be related to this type of sub-dimension (Johnston, 1971), as Berry and Rees (1969, p. 468) also hint. My suggestion is that the larger the minority group relative to a city's population, the more closely socioeconomic and ethnic status dimensions will be related

With the above caveats, which are relatively minor in the total set of factorial ecology results, the basic conclusion to be derived from this burgeoning literature is the general applicability of Shevky and Bell's model, at least within the constraints of most census data sets. (See, however, the criticisms of Meyer [1971] and Palm and Caruso [1972], which are discussed below.) There are patterns to which the model is not directly applicable, in cities with no obvious ethnic minorities, for example, and in cities set in very different socioeconomic/cultural environments (with consequent differences in the census variables provided). Among the latter, the limited range of studies (e.g., Abu-Lughod, 1969; Berry and Spodek, 1971; Curson, 1973; Schwirian and Smith, 1971; Timms, 1970, 1971) suggests a sequence of dimensional states in non-socialist cities whose progress follows that of "modernization":

- (a) Pre-industrial, in which the only dimension is communal, usually reflecting areal and perhaps tribal origin of groups.
- (b) Transitional, in which the communal division becomes associated with occupational divisions and eventually gives way to a socioeconomic status dimension.
- (c) "Modernizing," in which life-style choice becomes increasingly divorced from socioeconomic status.
- (d) Industrial, with the full Shevky–Bell range of dimensions.
- (e) Post-industrial.

Analyses of change

The residential pattern of a city is not a constant, therefore, and the long-term changes in its structure, just identified, are not the only alterations which are in progress. In most cities, certainly in those of the "developed" world, population movement is considerable. The personnel of areas changes—sometimes rapidly, often rather slowly. The fabric changes also; new buildings are added, others are removed, while a third category have their characteristics markedly changed. A factorial ecology of one data set, referring to one census, thus provides but a single snapshot of the ongoing readjustment of urban form, function, and population. Indeed, because areas may be changing at different rates and in different ways, the single snapshot may produce results which are not typical of the system at other times, or in its long-term, almost certainly only partial, equilibrium.

The need is for studies of systems at more than one date and for studies of changes between dates. Unfortunately, these are few and are difficult because of data (both variables and areas) incompatibilities between censuses, which may be taken a decade or more apart—and therefore may miss important, short-term changes. Nevertheless, several attempts at spatiotemporal factorial ecologies have been made, using one of two basic methods. In the first, separate factorial ecologies for the various dates are compared, either by visual interpretation (e.g., Hunter, 1971; Timms, 1971), or by the use of correlation-type methods such as congruence coefficients (Haynes, 1971) and Veldman's (1967) RELATE procedure (Johnston, 1973b). In the other, variables measuring the changes between two dates in each of a set of variables, over a constant set of areas, are constructed and the resulting matrix submitted to factorial ecology procedures (Murdie, 1969; Brown and Horton, 1970; Johnston, 1973c). From both, the findings suggest:

1. Considerable stability in the basic dimensions of the residential structure.
2. Variations which parallel the basic dimensions, in that variables which cluster together on the same component in a "static" analysis tend to cluster similarly in a "dynamic" analysis.
3. Generally little change between two dates in the characteristics of most areas (see also Tryon, 1967), with the main exceptions being areas of rapid change in their ethnic status.

Factorial ecology—a critique

Widespread continuing use of the factorial ecology method indicates its acceptance for studies of intra-urban residential differentiation. At the descriptive level it provides, within the constraints of variables and areas, suggestions of the basic patterns of differentiation and maps of these. Such output can be used to test general hypotheses concerning social areas and to generate further hypotheses concerning the mechanisms producing the patterns, concerning the place-to-place variations in patterns, and concerning the processes of "neighbourhood" change. Further analytical work can make use of various pieces of output, in particular the component or factor score matrices, which can be employed in analyses of spatial form . . . , as the bases for sampling frameworks in the mould of social area analysis (see Herbert and Evans, 1974) and as independent variables for a wide variety of ecological regression analyses. . . .

A methodology of such potential power requires careful appreciation in order that these latent qualities may be realized. The analyst must choose from a number of basic techniques, must select an appropriate path through the methodological procedures, and must make a number of subjective judgements (for example, in the classification procedure; see Johnston, 1968). Interpretation of the output, notably of the "meaning" of the factors/components which is basic to the whole procedure, also involves personal assessments. Development of a detailed appreciation of the methodology, which may lead to its modification or extension, is necessary. . . .

Conclusions

Unravelling the complexity of the urban residential mosaic is clearly a major research task. Its multivariate causes, ongoing processes, and static and dynamic patterns demand comprehensive data, referring to relevant variables and areas, and a technical arsenal for their analysis and interpretation. At present, the factorial ecology method is the most favoured, which accounts for its primacy within the present chapter. As it is presently practised, the method is not ideal for the test set, yet it appears to promise greater sophistication as it embraces a wider range of multidimensional scaling algorithms.

Perhaps more concern should be expressed about the nature of the material fed into factorial ecologies than with the method itself. On the surface, replication to various cities suggests clear basic dimensions to intra-urban spatial differentiation, yet much of this common structure may be an artifact of the data which censuses collect. Palm (1973a), for example, has shown that the social areas of Minneapolis, as defined by a factorial ecology, are not communuties of common interests, as indicated in their reading patterns of journalistic material; nor are they clearly defined functional regions, as indexed by telephone call volumes (Palm, 1973b). Smith (1973), too, has suggested the need for a much broader data base than presently employed if factorial ecologies are to be of value in uncovering spatial patterns of welfare and "illfare" within cities. In part, however, it may be that the variables Smith considers indices of pathologies are, in part, consequences of the social environmental variations defined by the factorial ecologies, and should therefore be analysed in a more obvious causal modelling program. . . . For all its disadvantages of omission and commission, therefore, the factorial ecology method has a clear lead at present in the search for ways to describe intra-urban environments and to uncover resulting man—environment interactions.

References

Abu-Lughod, J. L. "Testing the Theory of Social Area Analysis: The Ecology of Cairo, Egypt." *American Sociological Review,* 34 (1969): 198–212.

Alker, H. R. "A Typology of Ecological Fallacies." In *Quantitative Ecological Analysis in the Social Sciences,* edited by M. Dogan and S. Rokkan, pp. 69–86. Cambridge, Mass.: The M.I.T. Press, 1969.

Anderson, T. W., and Bean, L. L. "The Shevky-Bell Social Areas: Confirmation of Results and a Reinterpretation." *Social Forces,* 40 (1961): 119–24.

Bell, W. "The Social Areas of the San Francisco Bay Region." *American Sociological Review,* 18 (1953):39–47.

Bell, W. "Economic, Family, and Ethnic Status: An Empirical Test." *American Sociological Review,* 20 (1955):45–52.

Bell, W. "Urban Neighborhoods and Individual Behavior." In P. Meadows and E. H. Mizruchi, *Urbanism, Urbanization, and Change: Comparative Perspectives,* pp. 120–46. Reading, Mass., Addison-Wesley, 1969.

Bell, W., and Boat, M. D. "Urban Neighborhoods and Informal Social Relations." *American Journal of Sociology,* 62 (1957):391–98.

Bell, W., and Force, M. T. "Urban Neighborhood Types and Participation in Formal Associations." *American Sociological Review,* 21 (1956a):25–34.

Bell, W., and Force, M. T. "Social Structure and Participation in Different Types of Formal Association." *Social Forces,* 34 (1956b):345–50.

Bell, W., and Force, M. T. "Religious Preference, Familism and the Class Structure." *Midwest Sociologist,* 19 (1957):79–86.

Bell, W., and Greer, S. "Social Area Analysis and Its Critics." *Pacific Sociological Review,* 5 (1962):3–9.

Bell, W., and Moskos, C. C. "A Comment on Udry's 'Increasing Scale and Spatial Differentiation'." *Social Forces,* 42 (1964):414–17.

Berry, B. J. L., ed. "Comparative Factorial Ecology." *Economic Geography,* 47 (1971):Part 2.

Berry, B. J. L.; Goheen, P. G.; and Goldstein, H. *Metropolitan Area Definition.* Washington, D.C.: United States Bureau of the Census, 1968.

Berry, B. J. L., and Rees, P. H. "The Factorial Ecology of Calcutta." *American Journal of Sociology,* 74 (1969):445–91.

Berry, B. J. L., and Spodek, H. "Comparative Ecologies of Large Indian Cities." *Economic Geography,* 47 (1971):266–85.

Booth, C. *Life and Labour of the People of London* (17 volumes). London: Macmillan, 1889–1902.

Brown, L. A., and Horton, F. E. "Social Area Change: An Empirical Study." *Urban Studies,* 7 (1971): 271–88.

Burgess, E. W. "The Determination of Gradients in the Growth of the City." *Publications of the American Sociological Society,* 21 (1927):178–84.

Burgess, E. W., and Bogue, D. J. *Contributions to Urban Sociology.* Chicago: University of Chicago Press, 1967.

Carey, G. W., and Hughes, J. W. "Factorial Ecologies: Oblique and Orthogonal Solutions: A Case Study of the New York SMSA." *Environment and Planning,* 4 (1972):147–62.

Clark, C. *The Conditions of Economic Progress.* London: Macmillan, 1940.

Curson, P. H. "A Factorial Ecology of Avarua, Cook Islands." *Pacific Viewpoint,* 14 (1973):23–37.

Davies, W. K. D., and Barrow, G. "Factorial Ecology of Three Prairie Cities." *The Canadian Geographer,* 17 (1973):327–53.

Davies, W. K. D.; Giggs, J. A.; and Herbert, D. T. "Directories, Rate Books, and the Commercial Structure of Towns." *Geography,* 53 (1968):41–54.

Davies, W. K. D., and Lewis, G. J. "The Urban Dimensions of Leicester." *Institute of British Georgraphers, Special Publication,* 5 (1973):71–85.

Doucet, M. J. *Nineteenth Century Population Mobility: Some Preliminary Comments.* Discussion Paper No. 4. Toronto: Department of Geography, York University, 1972.

Duncan, B. "Factors in Work-Residence Separation: Wage and Salary Workers, Chicago, 1951." *American Sociological Review,* 21 (1956):48–56.

Duncan, O. D., and Duncan, B. "Occupational Stratification and Residential Distribution." *American Journal of Sociology,* 50 (1955):493–503.

Forrest, J. "An Approach to the Analysis of Subareas in Timaru." *New Zealand Geographer,* 24 (1968):195–201.

Forrest, J., and Tan, M. "Residential Location and Place of Work in a New Zealand City." *Proceedings, Sixth New Zealand Geography Conference* (1971):51–57.

Gittus, E. "The Structure of Urban Areas." *Town Planning Review,* 35 (1964):5–20.

Glass, D. V. *Numbering the People.* Farnborough: D. C. Heath, 1973.

Goheen, P. G. *Victorian Toronto, 1850–1900.* Research Paper No. 127. Chicago: Department of Geography, University of Chicago, 1970.

Guest, A., and Zuiches, J. "Another Look at Residential Turnover in Urban Neighborhoods." *American Journal of Sociology,* 77 (1972):457–71.

Hall, P.; Thomas, R.; Gracey, H.; and Drewett, J. R. *The Containment of Urban England,* 2 volumes. London: George Allen and Unwin, 1973.

Hatt, P. K. "The Concept of Natural Area." *American Sociological Review,* 11 (1946):423–27.

Hawley, A. H., and Duncan, O. D. "Social Area Analysis: A Critical Appraisal." *Land Economics,* 33 (1957):227–45.

Haynes, K. E. "Spatial Change in Urban Structure: Alternative Approaches to Ecological Dynamics." *Economic Geography,* 47 (1971):324–35.

Herbert, D. T. "Social Area Analysis: A British Study." *Urban Studies,* 4 (1967):41–60.

Herbert, D. T., and Evans, D. J. "Urban Sub-Areas as Sampling Frameworks for Social Survey." *Town Planning Review,* 45 (1974):171–88.

Hunter, A. A. "The Ecology of Chicago: Persistence and Change, 1930–1960." *American Journal of Sociology,* 77 (1971):425–44.

Johnston, R. J. "Choice in Classification: The Subjectivity of Objective Methods." *Annals, Association of American Geographers,* 58 (1968):575–89.

Johnston, R. J. "Grouping and Regionalizing: Some Methodological and Technical Observations." *Economic Geography,* 46 (1970):293–305.

Johnston, R. J. "Some Limitations of Factorial Ecologies and Social Area Analysis." *Economic Geography,* 47 (1971):314–23.

Johnston, R. J., ed. *Urbanisation in New Zealand: Geographical Essays.* Wellington: Reed, 1973a.

Johnston, R. J. "The Factorial Ecology of Major New Zealand Urban Areas: A Comparative Study." *Institute of British Geographers, Special Publication,* 5 (1973b):143–68.

Johnston, R. J. "Social Area Change in Melbourne, 1961–1966." *Australian Geographical Studies,* 11 (1973c):79–98.

Jones, F. L. "A Social Profile of Canberra, 1961." *Australian and New Zealand Journal of Sociology,* 1 (1965):107–20.

Jones, F. L. *Dimensions of Urban Social Structure.* Canberra: Australian National University Press, 1969.

Lawton, R. "The Population of Liverpool in the Mid-Nineteenth Century." *Transactions, Lancashire and Cheshire Historical Society,* 107 (1955):189–200.

McElrath, D. C. "Urban Differentiation: Problems and Prospects." *Law and Contemporary Problems,* 30 (1965):103–10.

McElrath, D. C. "Societal Scale and Social Differentiation: Accra, Ghana." In *The New Urbanization,* pp. 33–52. New York: St. Martin's Press, 1968.

McQuitty, L. L., and Clark, J. A. "Clusters from Iterative, Intercolumnar Correlational Analysis." *Educational and Psychological Measurement,* 28 (1968):211–38.

Meyer, D. R. "Factor Analysis versus Correlation Analysis: Are Substantive Interpretations Congruent?" *Economic Geography,* 47 (1971):336–43.

Murdie, R. A. *Factorial Ecology of Metropolitan Toronto, 1951–1961.* Research Paper No. 116. Chicago: Department of Geography, University of Chicago, 1969.

Palm, R. "Factorial Ecology and the Community of Outlook." *Annals, Association of American Geographers,* 63 (1973a):341–46.

Palm, R. "The Telephone and the Organization of Urban Space." *Proceedings, Association of American Geographers,* 5 (1973b):207–10.

Palm, R., and Caruso, D. "Labelling in Factorial

Ecology." *Annals, Association of American Geographers,* 62 (1972):122–33.
Parkes, D. N. "A Classical Social Area Analysis: Newcastle and Some Comparisons." *The Australian Geographer,* 11 (1972):555–78.
Parkes, D. N. "Formal Factors in the Social Geography of an Australian Industrial City." *Australian Geographical Studies,* 11 (1973):171–200.
Pfautz, H. W. *Charles Booth on the City.* Chicago: University of Chicago Press, 1967.
Poole, M. A., and O'Farrell, P. N. "The Assumptions of the Linear Regression Model." *Transactions, Institute of British Geographers,* 52 (1971):145–58.
Rees, P. H. "Concepts of Social Space." In B. J. L. Berry and F. E. Horton (eds.), *Geographical Perspectives on Urban Systems,* 306–94. Englewood Cliffs, N.J.: Prentice-Hall, 1970.
Rees, P. H. "Factorial Ecology: An Extended Definition, Survey, and Critique." *Economic Geography,* 47 (1971):220–33.
Rees, P. H. "Problems of Classifying Sub-Areas within Cities." In B. J. L. Berry (ed.), *City Classification Handbook,* pp. 265–330. New York: John Wiley, 1972.
Robinson, W. S. "Ecological Correlations and the Behavior of Individuals." *American Sociological Review,* 15 (1950):351–57.
Rummel, R. J. "Understanding Factor Analysis." *Journal of Conflict Resolution,* 40 (1967):440–80.
Schmid, C. F. "Generalizations Concerning the Ecology of the American City." *American Sociological Review,* 15 (1950):264–81.
Schmid, C. F. "Urban Crime Areas." *American Sociological Review,* 25 (1960):527–42, 655–78.
Schmid, C. F.; MacCannell, E. H.; and Van Arsdol, M. D. "The Ecology of the American City: Further Comparison and Validation of Generalizations." *American Sociological Review,* 23 (1958):392–401.
Schmid, C. F., and Tagashira, K. "Ecological and Demographic Indices: A Methodological Analysis." *Demography,* 1 (1965):194–211.
Schwirian, K. P., and Smith, R. K. "Primacy, Modernization, and Urban Structure: The Ecology of Puerto Rico Cities." Unpublished Paper. Columbus, Ohio: Department of Sociology, Ohio State University, 1971.
Shevky, E., and Bell, W. *Social Area Analysis: Theory, Illustrative Application and Computational Procedure.* Stanford, Calif.: Stanford University Press, 1955.
Shevky, E., and Williams, M. *The Social Areas of Los Angeles.* Los Angeles: University of California Press, 1949.
Smith, D. M. *The Geography of Social Well-Being in the United States.* New York: McGraw-Hill, 1973.

Spence, N. A., and Taylor, P. J. "Quantitative Methods in Regional Taxonomy." *Progress in Geography,* 2 (1970):1–64.
Sweetser, F. L. "Factorial Ecology: Helsinki, 1960." *Demography,* 2 (1965a):372–85.
Sweetser, F. L. "Factor Structure as Ecological Structure in Helsinki and Boston." *Acta Sociologica,* 8 (1965b):205–25.
Tarrant, J. R. *The Identification and Interpretation of Principal Components.* Monash University Publications in Geography No. 5. Clayton, Victoria: Monash University, 1974.
Timms, D. W. G. "Modernisation and the Factorial Ecology of the Cook Islands, Brisbane and Auckland." *Australian and New Zealand Journal of Sociology,* 6 (1970):139–49.
Timms, D. W. G. *The Urban Mosaic.* Cambridge. Cambridge University Press, 1971.
Tryon, R. C. *Identification of Social Areas by Cluster Analysis.* Berkeley, Calif.: University of California Press, 1955.
Tryon, R. C. "Predicting Group Differences in Cluster Analysis: The Social Area Problem." *Multivariate Behavioral Research,* 2 (1967):453–75.
Udry, J. R. "Increasing Scale and Spatial Differentiation: New Tests of Two Theories from Shevky and Bell." *Social Forces,* 42 (1964):404–13.
Van Arsdol, M. D.; Camilleri, S. F.; and Schmid, C. F. "An Application of the Shevky Social Area Indexes to a Model of Urban Society." *Social Forces,* 37 (1958a):26–32.
Van Arsdol, M. D.; Camilleri, S. F.; and Schmid, C. F. "The Generality of Urban Social Area Indexes." *American Sociological Review,* 23(1958b): 277–84.
Van Arsdol, M. D.; Camilleri, S. F.; and Schmid, C. F. "An Investigation of the Utility of Urban Typology." *Pacific Sociological Review,* 4 (1961):26–32.
Van Arsdol, M. D.; Camilleri, S. F.; and Schmid, C. F. "Further Comments on the Utility of Urban Typology." *Pacific Sociological Review,* 5 (1962): 9–12.
Veldman, D. J. *Fortran Programming for the Behavioral Sciences.* New York: Holt, Rinehart and Winston, 1967.
Ward, D. "The Internal Spatial Structure of Immigrant Residential Districts in the Late Nineteenth Century." *Geographical Analysis,* 1 (1969):327–53.
Willmott, P. "Some Social Trends." *Urban Studies,* 6 (1969):286–308.
Wilson, G., and Wilson, M. *The Analysis of Social Change.* Cambridge: Cambridge University Press, 1945.
Wirth, L. "Urbanism as a Way of Life." *American Journal of Sociology,* 44 (1938):1–24.

Social Area Analysis and Factorial Ecology: A Review of Substantive Findings*

Bernd Hamm

Introduction

Classical human ecology, social area analysis, and factorial ecology offer different, yet mutually interdependent, approaches to the analysis of urban socio-spatial structure (Senior, 1973; Timms, 1976; Hamm, 1977c). It is the argument of this paper that only if all three approaches are kept in mind simultaneously can progress in cumulative knowledge and the limitations of each tradition be adequately assessed. In the past, too little attention has been paid to the restrictions involved in isolated research within each of the approaches and too little effort has been dedicated to their integration.

In this paper I intend to specify the relationships between classical human ecology, social area analysis, and factorial ecology. Substantive evidence is my primary object, rather than methodological criticism (see the preceding paper by R. Johnston). Dimensions of urban ecological structure and the socio-spatial pattern of the city are the subjects of interest.

Classical human ecology in the Park, Burgess, and McKenzie tradition attempted to describe and to explain the regularities of spatial patterning and change of urban land uses and population. Its crucial assumption was stated by Park: physical distance is strongly correlated with social distance (Park, 1925); or as we prefer to put it nowadays: the spatial and social organizations of a population are mutually interdependent, reinforcing, and synomorphic. Social as well as spatial organization of a population are answers to the restricted resources of a limited environment, mediated by technological development—a frame of reference which later on was conceptualized as the "ecological complex" or the "ecosystem" (Duncan, 1959, 1961).

---
*This article was written for this volume and has not been published previously.

A first draft of this paper was read by Wendell Bell and George A. Theodorson. While the former's detailed comments were of invaluable help and provided deeper insight into many facets of social area analysis, the latter's press to shorten the earlier version turned out to contribute significantly to further clarification and concentration on the essential arguments. Jutta Dagli, Martin Litsch, and Ingrid Wacht dedicated considerable effort to compile the tables and the bibliography. Some of the material mentioned became available to me as unpublished research reports by the respective authors. The support of all of them is very gratefully acknowledged. Any shortcomings and misinterpretations are, however, my sole responsibility.

Social area analysis (SAA) began with the publication of *The Social Areas of Los Angeles* by E. Shevky and M. Williams (1949). The main interest of the Shevky group of scholars was in social differentiation and inequality. So, the subject of their reasoning was society as a whole. Empirical demonstration, however, concentrated on the city, as was the case in classical human ecology. Severe criticism, and some growing dissatisfaction with SAA, but also the success of early factorial studies (Bell, 1953, 1955b), fostered the proliferation of factorial ecology (FE) since the mid-1960s. Data availability and easy access to appropriate computer software significantly contributed to the change in research strategies.

SAA and FE have, in general, made relatively little use of classical human ecology. Despite this and despite the lack of theoretical clarity objected to in SAA and FE, their empirical research revealed highly consistent evidence on urban socio-spatial differentiation throughout the many cities that have been studied.

Ideally, FE should provide us with the necessary experience to decide what is relevant in our data sets and what is not—it turns out to be a kind of inductive dimensional analysis and part of the operationalization process. If this information were available to us and if relevant dimensions could be condensed to not more than a few composite indexes, the easily manageable technique of SAA could be used more confidently to test substantial hypotheses on urban structure and change included in ecological theory, as will be shown later.

Dimensions of urban differentiation and spatial patterning

A more general look at the Burgess hypothesis

The theory of concentric zones proposed by Burgess (1924; reprinted in this volume) was based on detailed maps showing the spatial distribution of single variables in Chicago. Despite its essentially ideographic character, it was proposed as an ideal-type conception, the generality of which had to be tested. Therefore, it may be stated in a more general form, which can only be attempted here very briefly.

As societies become more complex and more interdependent in the course of social development, they tend to differentiate specific functions which formerly were not separated. This aspect of social development has been widely agreed upon since the days of Durkheim and throughout social ecology (SAA's increasing societal scale is but another term for the same thing.) Different functions (e.g., production, distribution, consumption, communication, etc.) have to be located in space. Space is not an amorphous phenomenon but a product of social action preformed and structured in history. So space provides for a variety of different locational qualities, among them accessibility with respect to the transportation network, topography, micro-climate, and the like.

In the search for appropriate location, every societal function elaborates a more or less clear pattern of locational preferences. The extent to which every function is able to realize its locational preferences depends on the distribution of power in a society. Where such preferences are conflicting between different functions, there will be competition for locations or, more precisely, for the locational qualities concerned. The outcome of this competitive process depends on (1) the resources a function has available to succeed over other functions, and (2) the profits (in a very broad sense, pertaining to the function's goals) expected from realizing the optimal location providing for resources in the future. There will result a power hierarchy among different functions as well as within them—the classical concept of dominance. As easy access to all other located functions figures among the locational qualities competed for, the function with greatest power will succeed in obtaining the most central location. It now depends on the ways and means by which, in a given society, power is exerted: in capitalist societies, the regulative mechanism is land market and price; in socialist societies, it will be public administration and planning.

Dominance means the property of one unit within a system to integrate the functions of other units and to control for the conditions of their development. The dominant unit sets and regulates the prerequisites for development and change of subordinated units (Quinn, 1950). As transportation, communication, and trade are the means to integrate other functions, administrative, service, and retail functions are likely to be located in central areas, pushing away other functions to less central locations according to their position on the dominance scale.

Every function has a specific area which may be world-wide, as in the case of the multi-national corporation, or strictly local, as in the case of a public school. This service area at the same time provides the basis for potential resources. Its extension is determined by the means of transportation and communication available at a given level of technological development and division of labor. If, as demonstrated by Hawley (1950), a certain maximum amount of time is dedicated to daily commuting, the growth potential of an urban area is fixed at a given point in time (but, of course, increases or decreases with changes in the transportation and communication systems). When daily accessibility increases and extends into peripheral areas, the service areas of centrally located functions expand—in a spatial perspective, this means expansion of the central business district (CBD) into the zone of transition, succession, and suburbanization (Kasarda, 1972). So, whatever the incentives for change may be, they will become effective first in the dominant functions or, spatially, the dominant locations.

Housing, in itself, is a consumptive function, not a productive one. So residential land uses, in a very general sense, are expected to range among the most subordinated urban land uses, their location relative to the CBD being determined by the more dominant functions. But within residential areas, too, there is a marked difference with respect to locational qualities, and there is differential power of residents to succeed in obtaining the most advantageous locations. Here, too, we find a pattern of dominance and subordination. The regulating mechanisms, again, are the real estate market and housing prices. So, it is first of all residential segregation according to social class that will be observed (Duncan and Duncan, 1955). As ethnicity or immigrant status tend to be highly correlated with socio-economic status, introducing immigrants and discriminated minorities at the very bottom of the status ladder, these groups will be found to locate at the least advantageous areas of a city. This general pattern of urban differentiation is actually not contradicted by the symbolic land uses emphasized by Firey (1947). Of course, local authorities are among those exerting power to shape the land-use pattern, and public opinion and political participation play their part in the game.

What are the spatial patterns that will result from the processes described? The pattern of specialization of land uses will prove to be concentric in terms of ecological, i.e., time-cost, accessibility. Recent trends for decentralization (Hoyt, 1964) due to industrial relocation, peripheral shopping malls, or the creation of second-order centers seem to change this pattern only very slowly, and only in the largest urban areas. The relationship between specialization of land uses and segregation according to family status and life-cycle has been established by factorial ecology but is more difficult to explain.

Segregation according to social class will find the upper classes residing at the most desirable locations, the lower classes at the less advantageous. As locational qualities in some instances may be distributed sectorally (lake fronts, rivers, hillsides, etc.), social class also may reveal a sectoral pattern (Hoyt, 1939)—some problems of this model will be referred to later. Ethnicity or immigrant status will tend to follow the lines of class segregation. Due to cultural identity, common history, discrimination, and visibility, ethnic and national minority groups will tend to cluster together much more than native groups of the same socio-economic status. Only in the course of their assimilation will the sharpness of residential segregation decrease (Peach, 1975; Esser, 1980). As has been shown by Farley (1977), residential segregation of minority groups is at least as much an effect of discrimination by natives as it is of cultural identification.

As population growth, technological development, the distribution of power, and environmental aspects are included in this frame of reference, it may be systematically connected with the ecosystem model, which, to be sure, operates on a higher level of abstraction. On the other hand, as residential segregation is interpreted as a spatial expression of social inequality, the spatial concentration of social groups may provide one basis for and reinforce the establishment of specific subcultures (Cohen, 1955; Beshers, 1962)—a lower level of abstraction relating to the natural-area concept of classical human ecology. So, it may be restated, the spatial and social organizations of an urban population are mutually interdependent, reinforcing, and synomorphic.

The theory of social area analysis reviewed

In my view, it is not very important whether the concept of increasing societal scale has been introduced into SAA as an *ex post facto* interpretation of empirical results (Hawley and Duncan, 1957) or as an *ex ante* frame of reference, as supposed by Shevky and Bell (1955). I do not intend to trace back to its roots the development of SAA but rather to try to evaluate cumulated evidence. There are three questions to be discussed:

1. How do proponents of SAA link their variables to the concept of increasing scale? Is their research adequate for testing the increasing-scale hypothesis?
2. Has there been any proof of the reliability

and validity of the analytical tool proposed by SAA?
3. What did SAA claim to perform, and what has been the result?

THEORY AND METHODOLOGY. To answer the first question, I start by summarizing the argument by Shevky and Bell, referring for clarification to Wilson and Wilson (1945) and for further extension to McElrath (1965a,b). Shevky and Bell call what they present a "method of urban analysis" (p.v), a statement which receives some ambiguity through the remainder of the book and in the discussions it provoked. The work of urban ecologists is taken as a point of departure. "Beyond this point, however, our concern with problems of social differentiation and stratification has led us to a different kind of analysis, and our attention has been focused on relationships of a different order than those considered by urban ecologists" (p. 1). They "conceive of the city as the product of the complex whole of modern society; thus the social forms of urban life are to be understood within the context of the changing character of the larger containing society" (p. 3). Quite obviously, it is a theory of social change they have in mind. Their concept of increasing scale refers to Wilson and Wilson's investigations in Central Africa: "By the scale of a society we mean the number of people in relation and the intensity of those relations. . . . We measure the intensity of relations within a group, then, by observing the proportion of economic co-operation, of communication of ideas and of feelings within and without the group; together with the relative inclusiveness of value, dogma and of symbolism within and without the group, and the degree of social pressure exerted within and without the group" (Wilson and Wilson, 1945, pp. 25, 29). Shevky and Bell specify three broad sets of interrelated trends as aspects of increasing societal scale: "These are changes in (1) the distribution of skills, (2) the organization of productive activity, and (3) the composition of the population" (pp. 3–4). Arguing on lines proposed, among others, by economist Colin Clark (1951) and sociologist Louis Wirth (1938), Shevky and Bell come to take social rank, urbanization, and segregation as constructs representing the trends mentioned,[1] and they continue to propose occupation, education, and rent (for socio-economic status), fertility, women in the labor force, and single-family dwelling units (for family status), and racial and national groups in relative isolation (for ethnicity) as adequate, as well as available, sample statistics.

If these measures, as Shevky and Bell propose, are used to classify urban sub-areas, then does the change of socio-economic status, family status, or ethnicity give any information about increasing societal scale? This seems somewhat dubious, one reason being the different levels of abstraction. Thus, no matter how these measures are derived, one may hesitate to assess them as valid descriptors of increasing societal scale. One reason for this may be, as is implicit in the contribution of McElrath (1965a,b), that the concept of increasing societal scale has been developed in the study of what Wilson and Wilson (1945) call "primitive" societies. Their increase in scale may be measured by comparing them with "civilized" societies. But comparison by the use of such variables certainly intrudes a severe ethnocentric bias. And, obviously, no further increase in scale can be observed this way for highly developed societies themselves, for which the concept of increasing scale only applies in a very broad and heuristic sense.

SAA received much attention, reflected in several reviews of Shevky and Bell's book in scholarly journals. There is no need to review this discussion in detail, especially since part of it has been superseded by more recent work. There are only a few arguments I would like to add. As Duncan (1955) points out, nothing is contained in the book by Shevky and Bell on the forces and influences shaping urban sub-areas, the notion of "social area" being even less valuable than classical ecologists' "natural area." "Natural areas" were not conceived of as being shaped by nature, but as a result of unplanned competitive forces within the urban society. In our day, however, more or less homogeneous urban sub-areas are to a considerable extent created under the influence of public planning; and so the notion of "natural

---

[1] Obviously, there was some disagreement with respect to the labels of the indicators: Shevky preferred social rank, urbanization, and segregation, while Bell used the terms economic status, family status, and ethnic status. In a recent discussion I had with Bell on this subject, he suggested that today the clearest names of the three constructs would be socio-economic status, family life, and ethnicity, given the results of the research of the last two decades. In my opinion, however, family life will be interpreted by most readers as relating to the single household unit; so family status may be more appropriate with respect to the aggregate measure. Ethnicity, on the other hand, only makes sense in the case of a society with different ethnic groups, which clearly is not the case in most European countries—here, immigrant status would be the more appropriate term. For reasons of terminological clarity, I shall use socio-economic status, family status, and ethnicity or immigrant status throughout the remainder of this chapter.

area" certainly is misleading. But there is another difference Duncan did not mention: a "social area," in the sense of Shevky and Bell, is defined as a class of urban sub-areas which are not necessarily contiguous, while the "natural area" is a contiguous unity, like the Black Belt in Chicago or the ghetto (Wirth, 1938), with its own history. Despite their obvious similarity, "social area" is nominalistic and "natural area" is analytical in character. The analysis, according to Duncan, gives nothing more than an indication of how the indicators used are spatially distributed. Nevertheless, Duncan explicitly agrees with Shevky and Bell's call for comparative analysis and their emphasis on societal influences shaping urban structure. Carpenter (1955) questions the use of census tract data which disguise real heterogeneity—an issue discussed sometimes in ecological research in general (Myers, 1954).

The most detailed and intense discussion of SAA has been given by Hawley and Duncan (1957). Their principal argument may be summarized as follows: classical human ecology has provided much more insight into urban structure and differentiation than is reflected in the Shevky–Bell book. What is treated as homogeneous social areas by no means has to be homogeneous in reality. SAA contains no theoretical argument about why areas are or tend to be homogeneous. Their discussion of social trends is, according to Hawley and Duncan, at no point relevant for our understanding of urban differentiation, and there is no theory relating social organization to spatial patterns.

Other criticisms have been summarized by Timms (1971) and Hamm (1977a) and need not be repeated here, as the central points of discussion have already been mentioned. In summary, the first question may be answered thus: There is no clear and convincing linkage between the concept of societal scale and its operational measures, as proposed by Shevky and Bell. So, the method of urban analysis cannot systematically be related to the theory of increasing societal scale, and it does not provide for a valid device to test any hypothesis which may be derived from this frame of reference. The relationship between urban differentiation, on a census tract base, and social change remains unclear.

VALIDITY OF INDICATORS. Now let me proceed to find an answer to the second question. To do this, I shall refer to empirical studies explicitly conducted to test the validity of the Shevky–Bell model. There were, in the early 1950s, three lines of research aiming in the same direction: Tryon (1955), working at Berkeley, studied social areas in San Francisco by cluster analysis; Schmid (1950) proposed his "Generalizations Concerning the Ecology of the American City," at which he arrived by applying scale analysis to a sample of 23 U. S. Cities; and Bell (1955b) published a paper using factor analysis. As the three research teams were in contact with each other, their work may be assessed as mutually supportive.[2]

Of these devices, scale analysis has never been used again to analyze ecological data. As Schmid concludes, median school-year completed is the most valid index of a population's socio-economic status; in differentiating ecological areas in large cities, one or at most a few selected indexes may be much more valid than a large number of indexes; and relationships between ecological data show a marked tendency to conform to curvilinear patterns (p. 281). The use of a few indexes, including education, occupation, and rent, supports the Shevky–Bell model with respect to their socio-economic status index. In a later study, Schmid, Bowerman, and Shanley (1954) extended this result to include an index of "family life" (reported by Van Arsdol, 1957, p. 9).

In 1955, R. C. Tryon published his *Identification of Social Areas by Cluster Analysis*. Using 33 variables of the 1940 census for each of 243 census tracts of the San Francisco Bay Area, Tryon came to define three clusters to account for nearly the entire variance of the initial attribute space. These clusters were labeled "family life," "assimilation," and "socio-economic independence." The parallels to the Shevky–Bell model are even more obvious than they are in the Schmid studies: "family life" resembles the family status dimension, and, somewhat less obvious, "assimilation" reveals some similarity with ethnicity in that it contains foreign-born minority groups.

Two papers have been published using materials of Bell's doctoral dissertation of 1952 (Bell, 1953,

---

[2]The Social Science Research Council supported a committee called the Pacific Coast Committee on Community and Area studies, among the members of which we find R. Tryon, E. Shevky, L. Broom, W. S. Robinson, C. F. Schmid, and W. Bell. Thus, while there were three research centers and three different approaches, there was communication among them. Bell held, in 1951–52, a Social Science Research Council Pre-doctoral Training Fellowship to support his work comparing the results of the Tryon and Shevky methods. Although he was working at UCLA with Broom and Shevky, he spent one summer working with Tryon in Berkeley and, in his dissertation, he compared the Los Angeles and San Francisco studies.

1955b), the latter being of special interest to the question raised here. Its purpose was "to test the extent to which the three dimensions are necessary to account for social differentiation between urban sub-populations in two metropolitan areas, and to determine whether the indexes selected to measure the three dimensions are unidimensional measuring instruments" (p. 45). While three factors, according to his first hypothesis, are necessary to account for most of the variation between census tracts of Los Angeles and the San Francisco Bay area in 1940, the correlation matrixes between factors for both areas show considerable covariation of socio-economic status and ethnicity, as well as of socio-economic and family status.

The most intense efforts to evaluate SAA are reported from a group of scholars at the University of Washington under the direction of Calvin F. Schmid. Starting with the doctoral dissertations of M. Van Arsdol, Jr., and E. H. MacCannell (both of 1957), many papers were presented to evaluate the validity of SAA measures in a comparative mode of analysis, factoring census tract data of 10 American cities. This investigation indicated considerable support for the stability of SAA dimensions, although there are two deviant cases (Atlanta and Kansas City) where fertility was more closely related to socio-economic status than to family status (Van Arsdol, Camilleri, and Schmid, 1958b).

On the whole, the Washington studies can be assessed as having confirmed socio-economic status, family status, and ethnicity as important dimensions of urban differentiation. The variables used to build each of the Shevky–Bell indexes are highly intercorrelated, the indexes more or less independent aspects of social areas. The connection between theory and indexes, however, continued to be in question.

There was some discussion of SAA outside the Washington group of researchers, too. In these contributions, theoretical problems were much more the focus of interest than methodological aspects. W. Kaufman, a student of Bell's at Northwestern University discussed in his dissertation (1961) the relationships between classical urban ecology and SAA. In his opinion, they have some common elements but are substantially different, in that the first is centered on spatial distribution and the latter on social differentiation (an argument used later by P. Orleans [1966] to compare both approaches). Kaufman was the first to mention possible similarities between the concept of dominance and the family status index (p. 70). He proposed a revised version of social area analysis indicators to be tested by replication of the Bell (1955b) factor analysis with data for the San Francisco Bay Area and a sample of Chicago's census tracts. His results support the Bell hypothesis, with a slightly higher proportion of explained variance.

Anderson and Bean (1961), in a study of Toledo, used an extended 13-variable set, including the Shevky–Bell variables, to test the social area analysis scheme. Four factors were extracted from the correlation matrix to account for 83 percent of total variance. Family status appears to be interpretable as a twofold dimension, one aspect being connected with multiple-family structures and owner-occupied housing (urbanization), the other with family status. Income did not, as was expected, have only one loading on social rank, but also a significant negative loading on urbanization.

In a subsequent study, Anderson and Egeland (1961) used the Shevky–Bell dimensions to test Burgess's concentric zone against Hoyt's sectoral hypotheses in four U. S. cities. The Burgess model was supported with respect to urbanization, but not with respect to socio-economic status, while Hoyt's sectoral scheme received support with respect to socioeconomic status, but not with respect to urbanization.

While these studies carefully avoid the frame of reference represented by the concept of increasing societal scale, this comes to be the focus of Udry's (1964) study. As societies increase in scale, so his argument goes, their sub-areas become more and more functionally specialized discrete units. Udry's was the first attempt to establish this relationship in conceptually clear ways. "What will be the axes of functional differentiation of sub-areas of a society? This is the real question which Shevky and Bell and others working with the Shevky–Bell theory have been trying to answer." Udry concludes:

> But if you are interested in describing axes of differentiation of social areas, ... then why not construct a typology on the basis of independent factors derived from factor analysis of census tract data? ... I suggest that we consider these two separate but co-ordinated theories: one a theory of increasing scale; the other, using the same axes and variables, but not deducible from the first, a theory of sub-area differentiation. The two theories are logically co-ordinated through the proposition that, as a society increases in scale, its sub-areas become functionally differentiated" (pp. 408–9).

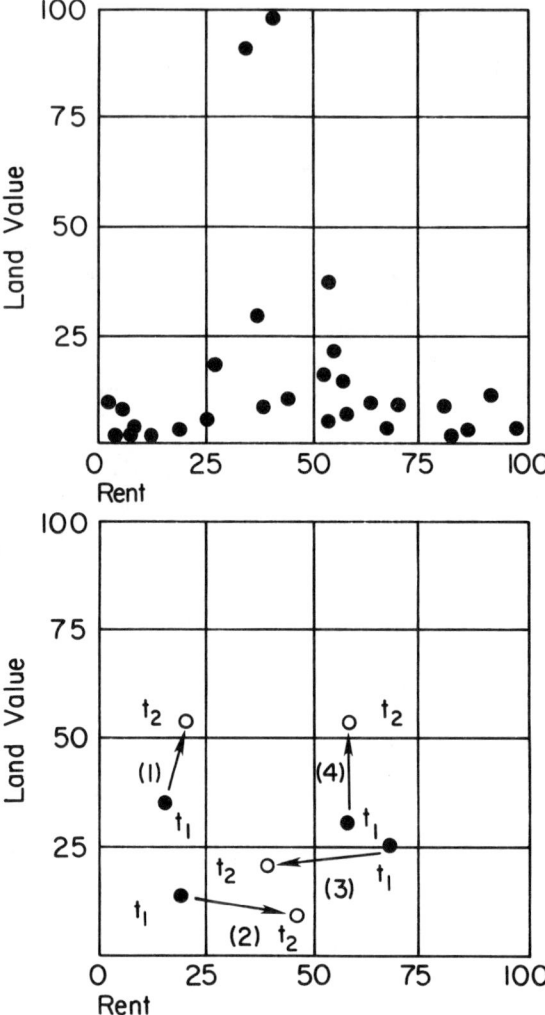

Chart 1—Social area diagram. *Upper chart*: Social areas of Berne, Switzerland, 1970, with land values and median rent substituted for urbanization and social rank. *Lower chart*: Dynamic analysis of single observational units (1) CBD expansion, (2) gentrification, (3) filtering down, and (4) second-order center in suburban residential area.

This statement is cited here at some length because it represents the logical starting point of factorial ecology. But, as far as I know, Udry's suggestion to study within and between variations of census tracts to test the Shevky–Bell model has not been followed.

While this line of research was focussed on increasing-scale theory, a study in Berne, Switzerland, proceeded on a quite different argument (Hamm, 1977a): If, so the reasoning, human ecology delivers, in the form of the generalized Burgess model described above, a theory of urban development, and if SAA is to be conceived of as a tehnique to desribe and analyze urban structure, but not necessarily bound to increasing-scale theory, it might make sense to reformulate the Shevky–Bell indexes in order to fit the ecological model. The socio-economic status index poses no special problem, the segregation of social class being known since the investigation of Duncan and Duncan (1955), and the generality of the socio-economic status dimension is well established by factorial ecology. On the other hand, the family status index has always been under discussion for reasons of theoretical and empirical ambiguity.

Using information from various sources, it was possible to include, in the Berne study, land value, ecological distance from CBD, workplaces per inhabitant, and retail surface per inhabitant in the attribute space. It was hypothesized, then, that the family status dimension would be closely correlated with these variables. Correlation coefficients between the family status factor scores and variables representing specialization of land use proved to be high, accounting for 47 percent to 68 percent of the variation, while the socioeconomic status factor scores contributed only slightly to the explanation of these characteristics. As in U. S. studies, family status and socio-economic status were uncorrelated.

To carry the argument a bit further, some more tests were conducted (Hamm, 1977b) by factor analysis and multiple-correlation procedures. The most important of these tests contrasted only two variables against the indexes: land value and rent, which were considered to represent, at best, the distributing forces as assumed in the generalized Burgess model. Their performance in multiple correlation with a set of 49 variables turned out to be even better than was the performance of the Shevky–Bell indexes. Land values were closely correlated with all the variables included to describe family status, life-cycle, mobility, and, of course, land use. In the author's view, this result was to be interpreted as a strong support for the assumption that SAA might allow substantive interpretation if based on the ecological theory of urban development. So, another technique of SAA could be formulated: rent and land value were substituted for socio-economic status and family status in the social area diagram (Hamm, 1977c), which is considered as a useful aid to illustrate urban structure and change. The scattering points, representing census tracts in their relative position to each other, could then be substantially interpreted as indicating the ecological position of each census tract within the urban network (see Chart 1).

In the context of a dynamic theory of urban ecological development and change, the ecological position (a concept quite unclear in classical human ecology) of a sub-area is assumed to be specifically related to the intensity and pace of change, and processes of invasion, succession, dispersion, concentration, and expansion can be graphically and analytically analyzed more adequately. In its revised version, SAA may be valid to describe the ecological conception of urban structure and change. If so, the missing link between increasing societal scale and sub-area differentiation probably may be found to be urban ecological theory. By this, a tentative answer to the second question might have been found, which, to be sure, needs confirmation and generalization.

PERFORMANCE OF SOCIAL AREA ANALYSIS IN CROSS-LEVEL ANALYSES. The third question to be answered refers to what SAA claimed to perform and what it actually did. As Shevky and Bell suggested (1955, pp. 20–22), SAA might be used
—for the analysis of other areal units such as cities, counties, regions, or even countries.
—for the delineation of sub-areas.
—for comparative studies at one point in time.
—for comparative studies at two or more points in time.
—as a framework for the execution of other types of research in cross-level analysis or as a sampling device.

As far as I can see, SAA in its proper sense has not been applied to areal units other than census tracts. One of the reasons for this might be, as Shevky and Bell suggest, that different specific measures in the indexes might prove necessary. The delineation of sub-areas was sometimes accomplished by mapping social area indexes or factor scores, and some students have used one or the other technique to delineate zones and sectors. Most of these studies, however, were in the factorial ecology tradition. The same holds true for comparative studies, both cross-sectional and longitudinal. So, social area analysis as a sampling technique and as a basis for contextual studies are the things in question here, and both of them usually coincide.

During the fifties, there were numerous studies investigating the contextual properties of social areas. It is to be mentioned, however, that

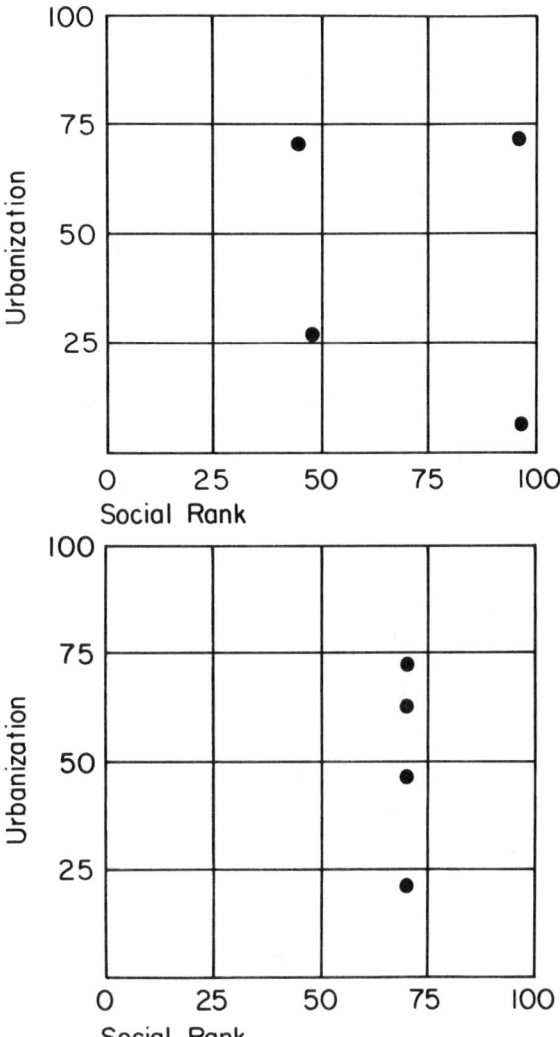

Chart 2—Social area analysis as a sampling device. *Upper chart*: W. Bell, M. D. Boat, and M. T. Force (1954). *Lower chart*: S. Greer and E. Kube (1959).

contextual analysis, as we use the term now, was the exception rather than the rule. Census tracts classified according to the Shevky–Bell typology were used to draw a sample of dichotomized areas as a base for field studies (see Chart 2). In a strictly methodological sense, this kind of procedure does not yet constitute the basis for contextual analyses (Green, 1971) in which individual and context effects have to be separated, which, in fact, has not been done in most of the studies. But these are two logically different questions:

1. Does social area analysis differentiate urban sub-areas so as to indicate specific levels of living, ways of life, characteristic attitudes, and forms of behavior?
2. Does the method of analysis allow the separation of contextual from individual factors of influence?

It is only for the first type of question that we have considerable evidence available.

Thus, the variation produced in independent variables is clearly defined by social area types. Within these sample areas, usually random samples of households were drawn for questionnaire investigation. Issues studied included participation in formal associations (Bell and Force, 1956); informal relations (Bell and Boat, 1957); anomie and isolation (Bell, 1957); religious preference (Bell and Force, 1957); neighboring, participation in cultural events, formal associations, localism, selection of friends (Greer, 1956; Greer and Kube, 1959); individual prestige ratings (McElrath, 1955); marital adjustment (Williamson, 1953, 1954); employability of aging workers (Curtis, 1957); petition for change of name (Broom et al., 1955); minority differentials (Broom and Shevky, 1949); suicide and juvenile delinquency (Wendling, 1954; Polk 1957a,b); political preferences and voting behavior (Tyron, 1955; Kaufman 1956; Kaufman and Greer, 1960), and delinquency rates (Polk, 1957b). The most intensive use of the social area typology has seemingly been made within the context of the Metropolitan St. Louis Survey (for a summary description, see Bollens, 1961) and within investigation of reference groups (Sherif and Sherif, 1964).

It is not possible here to refer to these studies in greater detail. Their cumulative evidence clearly indicates that the social area typology is indeed useful for the study of individual behavior, attitudes, and life style. It may not be surprising, then, that some practioners with various fields of interest also directed attention to this type of methodology—see, for example, Bange et al. (1955), for social welfare; Curtis et al. (1957) and Sullivan (1961), for Catholic parishes; Brindley and Raine (1979), for planning research; Struening (1975) and DeWitt (1978), for evaluation research.

METHODOLOGICAL LIMITATIONS. No doubt, the cumulated experience reported in the preceding section points to considerable empirical evidence on the usefulness of the SAA technique, despite the theoretical and methodological objections raised against it. Some of the concerns relating to the method of urban analysis have already been mentioned; but there are others which are equally important.

The procedure of gaining standard scores for the variables to be included in index construction necessarily produces a certain amount of dissimilarity between the units of observation, no matter how important original variances of the variables are. So, independent of initial variation, statistical procedures produce heterogeneity as an artifact of the standardization formula, if there is no reference city to draw on for comparative purposes. And there is no argument for using the Los Angeles conversion factors for 1940, as reported by Shevky and Bell (1955, p. 67), as the only significant point of reference. The classification of social areas seems to be likewise arbitrary and without clear theoretical meaning. The variables included have already been put to question (see, for this point, Bell [1969]). The simple fact that social area indexes may be computed as long as the necessary data are available does not, per se, constitute any theoretically meaningful device to analyze urban structure.

Factorial ecology

Frank Sweetser (1965a,b,c) has usually been credited with having coined the term "factorial ecology." If, however, we accept as "ecological" any study of areas as observational units (Rees, 1971b), FE as the manipulation of ecological data by factor analysis dates back to the early forties (Hagood et al., 1941; Price, 1942). There are several excellent reviews available summarizing FEs on scales (Russett, 1967; Berry, 1972) other than the city census-tract scale which need not be reported here. And there are summary reports on urban factorial ecology, too (Rees, 1971a,b; Timms, 1971; Janson, 1980). The purpose of this section is not to add another detailed overview to those already published, but rather to condense existing knowledge to a point that may easily be related to the problems raised in the introduction. I will report on studies concerned with the general structure of urban differentiation, but not with specific features, such as crime, voting, religion, or education (see Janson 1980). As a kind of dimensional and classificatory analysis, FE may answer three types of questions: First, what are the dimensions of urban ecological differentiation? To what extent may these dimensions be generalized? How can differences be explained? Second, what

Table 1. Selected Dependent Variables Studied by Social Area Analysis.

City	Dependent Variables	Reference
Buffalo	Employability of aging workers	Curtis, 1957
Los Angeles	Neighboring, participation in cultural events, formal associations, localism, selection of friends	Greer, 1956; Greer and Kube, 1959
Los Angeles	Individual prestige ratings	McElrath, 1955
Los Angeles	Marital adjustment	Williamson, 1953, 1954
Los Angeles	Petition for change of name	Broom, Beem, and Harris, 1955
Los Angeles	Jews	Broom and Shevky, 1949
Portland	Delinquency rates	Polk, 1949
San Diego	Juvenile delinquency, suicide	Polk, 1957a,b
San Francisco	Participation in formal associations	Bell and Force, 1956
San Francisco	Informal relations	Bell and Boat, 1957
San Francisco	Anomie, isolation	Bell, 1957
San Francisco	Religious preference	Bell and Force, 1957
San Francisco	Suicide	Wendling, 1954
San Francisco	Social welfare	Bange et al., 1955
San Francisco	Political preference, voting participations, psychiatric hospitalization, university attendance, rent change	Tryon, 1955
St. Louis	Political preferences, voting behavior, local government, metropolitan problems	Kaufman, 1956
St. Louis	Catholic parishes	Curtis, Avesing, and Klosek, 1957; Sullivan, 1961
St. Louis	Parapolitical structure, direction of vote, local social participation, political participation, political competence	Greer, 1962; Kaufman and Greer, 1960 Greer and Orleans, 1962
St. Louis	Voting behavior	Kaufman and Greer, 1960
Selected Cities	Episcopal, Methodist, Presbyterian, Lutheran Churches	Wilson, 1958
Selected Cities	Goals and values, attitudes and behavior of adolescents, perception, life histories	Sherif and Sherif, 1964

Table 2. Selected Factorial Ecology Studies in the U.S.

City	Time	Method	NVAR	NFA	Dimensions				Reference
					1	2	3	4	
Boston	1960	PC, OR	20	3	SES	FS	URB	–	Sweetser, 1965b
Boston	1960	PC, OR	34	3	SES	FS	ES	–	Sweetser, 1969
Buffalo	1960	n.r.	12	2	SES	FS	–	–	Salins, 1971
	1950		11	2	SES	FS	–	–	
	1940		10	2	SES	FS	–	–	
Chicago	1950	PC, OR	53	3	SES	FS	ES	–	Brown and Horton, 1970
	1960	PC, OR	53	4	SES	FS	ES	FS	
Chicago	1930	VA	9	3	SES	FS	ES	–	Hunter, 1974
	1940	VA	9	3	SES	FS	ES	–	
	1950	VA	9	3	SES	FS	ES	–	
	1960	VA	9	3	SES	FS	ES	–	
Chicago	1940	CEN	7	3	SES	FS	ES	–	Kaufman, 1961
	1950	CEN	7	3	SES	FS	ES	–	
Indianapolis	1960	n.r.	8	2	SES	FS	–	–	Salins, 1971
	1950		7	2	SES	FS	–	–	
	1940		6	2	SES	FS	–	–	
Kansas City	1960		9	2	SES	FS	–	–	Salins, 1971
	1950		8	2	SES	FS	–	–	
	1940		7	2	SES	FS	–	–	
Los Angeles	1940	CEN, OB	7	3	SES	FS	ES	–	Bell, 1955b
Newark	1960	PF, OR	48	6	ES	URB	SES	FS	Janson, 1968
Providence	1950	PC, OR	23	3	SES	FS	?	–	Dent, 1972
	1960	PC, OR	23	3	SES	FS	?	–	
Seattle	1960	PA, OR	42	8	FS	SES	–	MOB	Schmid and Tagashira, 1964
			21	4	FS	SES	–	ES	
			12	4	FS	SES	?	ES	
			10	3	FS	SES	ES	–	
Milwaukee	1940	PC, OR	10	3	SES	FS	ES	–	Ottensmann, 1975
	1950	PC, OR	12	3	SES	FS	ES	–	
	1960	PC, OR	12	3	SES	FS	ES	–	
	1970	PC, OR	12	3	SES	FS	ES	–	
Minneapolis	1970	FA, OR	12	3	FS	SES	ES	–	Nordstrand, 1973
San Francisco	1940	CEN, OB	7	3	SES	FS	ES	–	Bell, 1953
San Francisco	1940	CEN	7	3	SES	FS	ES	–	Kaufman, 1961
	1950	CEN	7	3	SES	FS	ES	–	
Spokane	1960		12	2	SES	FS	–	–	Salins, 1971
	1940		11	2	SES	FS	–	–	
Syracuse	1940	PC, OR	20	3	SES	FS	?	–	Dent, 1972
	1950	PC, OR	23	3	SES	FS	?	–	
	1960	PC, OR	23	3	SES	FS	?	–	

Table 2. Selected Factorial Ecology Studies in the U.S. continued.

City	Time	Method	NVAR	NFA	Dimensions 1	2	3	4	Reference
Toledo	1950	CEN, OR	13	4	URB	FS	SES	ES	Anderson and Bean, 1961
18 U.S. cities	1950	MG, OB	6	3	SES	FS	ES	–	Van Arsdol et al., 1958 a, b
5 U.S. cities	1950	MG	6	3	SES	FS	ES	–	Van Arsdol, 1957

Note: NVAR = Number of Variables NFA = Number of Factors Interpreted

ANALYSES
- AL = Alpha Factor Analysis
- CEN = Centroid Factor Analysis
- FA = Factor Analysis
- IA = Image Analysis
- MG = Multiple Group Factor Analysis
- OB = Oblique Rotation
- OR = Orthogonal Rotation
- PA = Principal Axes Factor Analysis
- PC = Principal Components Analysis
- RAO = Rao's Canonical Factor Analysis
- VA = Various Algorithms
- n.r. = not reported

DIMENSIONS
- AGE = Age
- CLER = Clerks
- DENS = Density
- ES = Ethnic Status
- EUR = Northwestern Europeans
- FS = Family Size
- HOUS = Housing
- INF = Infertility
- INT = Integration
- JEW = Jews
- MOB = Mobility
- POV = Poverty
- PROB = Social Problems
- PROF = Professionals
- REL = Religion
- RES = Residentialism
- SES = Socio-Economic Status
- SEX = Sex
- SUB = Suburban
- WOM = Women
- ZIT = Zone in Transition

are the spatial patterns in the distribution of such dimensions? Do they show regularities that can be related to existing models of urban structure? Third, what kind of change can be observed in dimensions and patterning over time?

DIMENSIONS OF SOCIO-SPATIAL DIFFERENTIATION. Urban factorial ecologies in the United States very regularly have extracted at least three separate dimensions: socio-economic status, family status, and ethnicity (see Table 2). Where oblique rotation of the initial factor pattern was conducted, socio-economic status and ethnicity tend to be significantly correlated. So, the three factors seem to be fairly general for U. S. cities, despite the fact that they are not necessarily completely independent. Australian, Canadian, and European studies (see Table 3) are less clear with respect to ethnicity, owing to different immigration experiences. As minority groups tend to cluster in the core areas of European cities (Burgess's zone in transition), and as they occupy a specific position on the family status dimension due to patterns of selective migration, it is not surprising to see some common variation between foreign immigrants and family status, as well as some interdependence with socio-economic status. What seems to be the most general rule, then, for industrial capitalist societies is the existence of at least two independent dimensions, viz., socio-economic status and family status.

Only one FE has come to my attention from socialist countries: Weclawowicz's study of Warsaw in 1931 and 1970. Its results tend to confirm the two independent dimensions. Studies in non-industrial societies, however (Table 4), suggest that socio-economic status and family status, which also are extracted quite regularly, might be considerably less independent than they are in more advanced societies.

Obviously, the factors to be extracted are completely determined by the variables included in the analysis. If there are no variables relating to specialization of land use, no such factor will appear, no matter how important such a dimension may be in the generalized Burgess model. Moreover, the factor pattern is also strongly influenced by the relative weight of the variables within the attribute space. If there are only 2 variables relating to socio-economic status in a 50-variable set, social rank will contribute only very little to the overall variation and an independent factor may not be extracted at all. If there are variables included which show very little initial variation, then they will be arbitrarily associated with factors, thus disturbing substantive interpretation.

Finally, FE is bound to the implicit premises of correlation; while normality is not a very severe assumption, provided that the interpretation is strictly descriptive and no inferential argument is used, homoscedasticity and linearity of bivariate relationships prove to be more important. Some authors, if they suspect non-linear relationships in

Table 3. Selected Factorial Ecology Studies in Europe, Canada, Australia, and New Zealand.

City	Time	Method	NVAR	NFA	Dimensions 1	2	3	4	Reference
EUROPE									
Geneva	1965	PC	19	4	SES	FS	AGE	–	Bassand, 1974
Payerne	1970	PC, OR	22	9	INT	INT	SES	FS	Campiche and Zimmermann, 1975
Eaux-Vives	1972	PC, OR	22	9	INT	INT	SES	REL	
Esserfines	1970	PC, OR	22	9	INT	INT	SES	PROF	
Lignon	1970	PC, OR	22	9	INT	SES	INT	MOB	
Helsinki	1955	CEN	13	3	URB	SES	FS	–	Grönholm, 1961
Helsinki	1960	PC, OR	20	3	SES	FS	URB	–	Sweetser, 1965a
Helsinki	1960	PC, OR	42	6	SES	FS	URB	RES	Sweetser, 1965b
Helsinki	1960	PC, OR	20	6	SES	FS	URB	RES	
Helsinki	1960	PC, OR	33	6	SES	FS	ES	WOM	Sweetser, 1969
Berne	1970	PA, OR	7	2	SES	FS	–	–	Hamm, 1977 a, b, c
Berne	1970	PC, OR	57	10	FS	SES	URB	MOB	Hamm, 1978
		PC, OB	57	10	FS	SES	?	CLER	
		PA, OR	57	10	FS	SES	URB	MOB	
		PA, OB	57	10	FS	SES	ES	DENS	
		AL, OR	57	10	FS	SES	URB	MOB	
		AL, OB	57	10	FS	SES	ES	DENS	
		RAO, OR	57	10	FS	SES	SES	URB	
		RAO, OB	57	10	FS	SES	SES	MOB	
		RAO, OB	63	7	SES	SES	MOB	FS	
		RAO, OB	53	5	SES	URB	MOB	SES	
Pforzheim	1970	PA, OB	33	7	SES	HOUS	PROB	FS	Mischke, 1976
Vienna	1961	PA, OR	35	7	SES	FS	SES	SUB	Sauberer and Cserjan, 1972
Warsaw	1931	PC, OR	26	3	SES	ES	FS	–	Weclawocicz, 1979
	1970	PC, OR	41	4	SES	HOUS	SES	FS	
Sunderland	1961	PC	30	4	SES	HOUS	HOUS	POV	Robson, 1969
Mannheim	1970	PC, OR	28	6	SES	FS	FS	FS	Bähr, 1978a
Newcastle	1961								Herbert, 1967
Copenhagen	1950 1960		14	3	FS	SES	MOB	–	Pedersen, 1967
Hamburg	1970	OR	16	5	FS	SES	MOB	DENS	Friedrichs, 1977
Trier	1968/70	PC, OR	12	4	SES	FS	URB	–	Jung, 1980

Table 3. Selected Factorial Ecology Studies in Europe, Canada, Australia, and New Zealand continued.

City	Time	Method	NVAR	NFA	Dimensions				Reference
					1	2	3	4	
CANADA									
Toronto	1951	PC, OR	86	6	SES	FS	ES	MOB	Murdie, 1969
	1961	PC, OR	78	6	SES	ES	FS	FS	
Ottawa	1961	FA, OB	6	4	FS	INF	ES	SES	Schwirian and Smith, 1974
Windsor	1961	FA, OB	6	4	HOUS	SES	ES	FS	
Kingston	1961	FA, OB	6	4	FS	SES	INF	ES	
Winnipeg	1961	PC, OR	70	5	SES	FS	ES	ES	Nicholson and Yeates, 1969
Winnipeg	1951	PF, OR	15	3	SES	FS	ES	–	Hunter and Latif, 1973
		PF, OR	15	3	SES	FS	ES	–	
		PC, OR	15	n.r.	SES	FS	ES	–	
		PC, OB	15	n.r.	SES	FS	ES	–	
		AL, OR	15	4	SES	FS	ES	–	
		AL, OB	15	4	SES	FS	ES	–	
		IA, OR	15	n.r.	SES	FS	ES	–	
		IA, OB	15	n.r.	SES	FS	ES	–	
	1961	PF, OR	15	n.r.	FS	SES	ES	–	
		PF, OB	15	n.r.	FS	SES	ES	–	
		PC, OR	15	n.r.	FS	SES	ES	–	
		PC, OB	15	n.r.	FS	SES	ES	–	
		AL, OR	15	n.r.	FS	SES	ES	–	
		AL, OB	15	n.r.	FS	SES	ES	–	
		IA, OR	15	n.r.	FS	SES	ES	–	
		IA, OB	15	n.r.	FS	SES	–	–	
Montreal	1951	PC, OB	90	5	SES	FS	MOB	ES	Haynes, 1971
	1961	PC, OB	90	5	SES	FS	MOB	ES	
AUSTRALIA									
Auckland	1966	MG, OB	11	3	FS	SES	SES	–	Timms, 1971
Brisbane	1961	MG, OB	11	3	FS	SES	ES	–	Timms, 1971
Canberra			24	4	ES	FS	AGE	–	Jones, 1965
Melbourne	1961	PC	24 SES	3	SES	ZIT	SUB	–	Jones, 1967, 1969
			24 FS	3	FS	–	SEX	–	
			22 ES	3	ES	EUR	JEW	–	
			24	3	SES	FS	ES	–	
NEW ZEALAND									
Whangarei	1971	PC, OR, OB	8	3	SES	FS	–	–	Johnston, 1976

Note: For abbreviations, see Table 2.

Table 4. Factorial Ecologies in Non-Industrial Societies.

City	Time	Method	NVAR	NFA	Reference
Abidjan	1963		11	4	Clignet and Sween, 1969
Accra	1960		11	4	Clignet and Sween, 1969
Ahmedabad	1961	PA, OR	29	7	Berry and Spodek, 1971
Alexandria	1947	PC, OR	13	3	Latif, 1974
Alexandria	1960	PC, OR	13	3	Latif, 1974
Bombay	1961	PA, OR	14	4	Berry and Spodek, 1971
Cairo	1947	PC, OR	13	3	Abu-Lughod, 1969
Cairo	1960	PC, OR	13	3	
Calcutta	1961	PA, OR	37	10	Berry and Rees, 1968
Kanpur	1961	PA, OR	9	3	Berry and Spodek, 1971
Madras	1961	PA, OR	15	5	
Mayaguez	1960	FA, OB	6	3	Schwirian and Smith, 1974
Ponce	1960	FA, OB	6	4	
Poona	1822	PA, OR	9	4	Berry and Spodek, 1971
Rio de Janeiro	1960	PA, OR	22	4	Morris and Pyle, 1971
San Juan	1960	FA, OB	6	4	Schwirian and Smith, 1974
Seoul	1960	FA, OR	181	7	Pitts, 1971
Sholapur		PA, OR	33	9	Berry and Spodek, 1971
Poona	1937	PA, OR	30	9	
Poona	1954	PA, OR	19	6	
Santiago de Chile	1970	PC, OR	28	4	Bähr, 1978b
Taegen	1960	FA, OR	59	7	

Note: For abbreviations, see Table 2. Because of extreme heterogeneity of attribute spaces and factors, the factor labels are not given here.

their correlation matrices, recommend log or ln-transformations, without, of course, solving the problem. In this respect, the results of a study using correlation ratios instead of Pearsonian correlation coefficients may be interesting (Hamm, 1978). While the factor pattern remained fairly consistent between the two solutions, the sequence of extraction changed considerably, together with the variance accounted for by each factor: the factor pattern much more clearly supported the hypothesis based on the generalized Burgess model.

FE, as any other statistical device, provides for tautological transformation of the information included. If so, there is no way to avoid theoretical justification of the selection of variables if meaningful results are to be obtained. This is suggested to be the most important contribution of the Shevky-Bell social area analysis model—that it defines its attribute space as a generally usable set founded on certain theoretical propositions.

There obviously is considerable evidence on the generality of socio-economic status and family status as important dimensions of urban residential differentiation. But these may not be the only factors relevant to adequately describe socio-spatial patterns, and their relative importance might be questioned.

SPATIAL PATTERNS. As a generally accepted rule, the family status dimension shows concentric or gradient variation, socio-economic status is sectorally distributed, and ethnicity may cluster in multiple nuclei according to the Harris–Ullman proposition (Salins, 1971). Whenever the question of spatial distribution was under research, the students proceeded in the same manner: the CBD or intersection of highest land values was determined and a series of concentric zones drawn around it with regularly increasing linear distances. To test the sectoral pattern, the city was subdivided by regular angles from the CBD to the urban periphery. What could not be shown to differentiate according to zones or sectors was interpreted as clustering in the multiple nuclei pattern.

From the point of view of classical human ecology, these operationalizations of zones and sectors are questionable. First of all, as McKenzie (1926) and Quinn (1940) have shown, the zonal pattern cannot be adequately evaluated by the use of geometrical distance on a map. In ecology, distance has always been conceptualized as a time-cost measure or functional accessibility relative to the CBD. Wachs and Kumagai (1973) have pointed out some of the difficulties to be solved with the definition of accessibility (see also: Pirie, 1979). So, the question of operationalization is by no means incidental and the easy way to do it may be misleading.

The same holds true with respect to sectors: geometrical angles drawn on a map cannot adequately substitute for real lines of sectoral patterning, for example, as given by the transportation network, rivers, hills, or railway tracks—only these are really relevant for urban differention. If sectors are to be defined, much more careful inspection of maps is needed and simple geometrical regularity has to be avoided as potentially misleading, possibly meaningless.

The generalized Burgess model suggests land use and family status to be distributed zonally, as accessibility is a major factor in the determination of land use and as radial expansion of the housing stock is a major factor influencing the distribution of dwelling types. Socio-economic status, however, tends to be patterned according to locational qualities highly esteemed by residential uses. This may, but does not necessarily have to, lead to sectoral distribution. Where residential locational qualities are almost equally distributed, socio-economic status is likely to show a concentric pattern. With respect to ethnicity, the multiple nuclei or random pattern is not very likely to exist in reality. Where social distance between minority and dominant groups is high, the minorities will tend to cluster in or near the zone of transition if their social rank is low and in suburban areas if their social rank is high. The more they become assimilated to the dominant group, however, the more their pattern of locational choice will tend to converge with that of the dominants.

FACTORIAL ECOLOGY OF CHANGE. Among the first studies investigating the patterns of change by means of factor analysis is one done by Sweetser (1961). But dynamic analyses have remained relatively infrequent; selected examples are listed in Table 5. As the notion of "factorial ecology of change" is restricted to studies applying factor analysis to change scores, comparison of factor patterns or factor scores between two points in time are excluded from consideration here (on this point, see Janson [1978]). Change scores have been differently defined: Hunter (1974) uses simple differences between his variables, subtracting the t_1 variable from the t_2 variable; Murdie (1969) and, similarly, Brown and Horton (1970), calculate change quotients; and Dent (1972) uses residuals from the regression line of t_2 variables on t_1 variables as change scores.

The factor patterns of structural analysis, if compared along different points in time, reveal considerable stability (Hunter, 1974). This seems not to be the case when change scores are used. First, the proportion of total variance explained by change factors significantly drops, only exceptionally reaching 60 percent. This indicates that change patterns are more complex than structural patterns, including more specific variance on the variables used. Second, factor patterns show less stability in the study of change, and commonly some combination of the underlying structural dimensions emerges. Third, the use of change scores omits changes in means from the analysis, i.e., patterns of differential change between units are described rather than patterns of change (Janson, 1978).

The same objections that have been raised against structural FE apply also to dynamic analysis. Probably the most important problem stems from the definition of the attribute space studied. Since sets of variables for the same unit in different points in time tend to be only partly comparable, only the comparable ones are retained for analysis, resulting in a relatively small set. But as variables are excluded on grounds of technical considerations, this may concern those attributes bearing the most important theoretical meaning. Thus, theoretical interpretation of results is more complicated, more remote from empirical results, and this increases arbitrary judgment.

As the few studies available are vulnerable to these criticisms, and as they use different definitions of change scores and different techniques of analysis, the factorial ecology of change remains at the stage of exploratory investigation.

Conclusion and suggestions for further research

The aim of this paper was to reformulate some relationships between classical human ecology,

Table 5. Factorial Ecologies of Change.

City	Period	Method	NVAR	NFA	% VAR[a]	Reference
Chicago	1930-40	PC, OR	9	3	54	Hunter, 1974
	1940-50	PC, OR	9	3	56	
	1950-60	PC, OR	9	3	57	
Chicago	1950-60	PC, OR	53	5		Brown and Horton, 1970
Montreal	1951-61	PC, OB	90	5		Haynes, 1971
		PA, OB				
Providence	1950-60	PC, OR[b]	19	3	51	Dent, 1972
Rome	1951-61					McElrath, 1980
Syracuse	1940-50	PC, OR	16	3	63	Dent, 1972
	1950-60	PC, OR	19	3	46	Dent, 1972
Toronto	1951-61	PC, OR	56	6		Murdie, 1969

[a] % VAR = % variance explained.
[b] PC, OR and MG, OB
Note: For abbreviations, see Table 2.

social area analysis, and factorial ecology. It was maintained that all three approaches, being mutually interdependent, are necessary to gain more cumulative knowledge on the structure and processes of urban socio-spatial differentiation. To substantiate the argument, a generalized model of urban structure and change, relying in a broad sense on the Burgess hypothesis, was formulated and extensive reviews of SAA and FE were given. The main difficulties in connecting the three approaches are due, of course, to theoretical and methodological problems.

The principal difficulty seems to be implied in the selection of meaningful variables. It is true, as Droth and Fischer (1980) suggest, that there is no study properly solving this problem. The generalized Burgess model sketched above is only an initial step in this direction. But the model has to be carefully elaborated, and there is much theoretical and empirical experience to be included from research conducted by geographers, economists, regional scientists, and planners.

An elaborated frame of reference must be able to permit the formulation of testable hypotheses and to define meaningful variables for investigation. It may even be possible to define, as a tentative first step, a standardized set of variables to be used in the study of ecological structure and change at least within one society (Sweetser, 1978), or perhaps in societies of a similar type—capitalist, industrial societies being one example. Certainly, the difficulties in obtaining relevant data are manifold, but easy access is no substitute for theoretical meaning. Census information alone is not enough; it has to be supplemented with data from other sources, such as departments of urban planning and transportation or taxation, possibly even survey research. Areas in question include land use and land values, accessibility, workplaces, as well as residential segregation according to socio-economic status, life cycle, and ethnicity, mobility or migrant status, and density. As FE has demonstrated, there may be, in some cases, suitable substitutes for variables not obtainable, but attention should be paid to content validity.

On the other hand, the choice of units of observation—blocks, census tracts, or another basis for aggregation—seems to be of minor influence, as long as factor patterns are the relevant results of a study (see, for example, Hunter's [1974] study of Chicago local communities as compared with census tract-based studies of the same city). If, however, cross-level analysis is in question, using factors for the definition of contexts of behavior, things are different—homogeneity of the units of observation may then become critical to separate individual from context effects, and a block base of aggregation might be more appropriate. Factor patterns seem to be fairly robust against different algorithms for extraction (Dent, 1972; Hunter, 1972; 1974; Hamm, 1978). It is the author's view that SAA techniques are highly appropriate to test substantive hypotheses on urban differentiation, despite the fact that their potential is far from being exhausted. The critical prerequisite seems to be the choice of indicators. At present, specialization of land use and residential segregation according to social class seem to be of prime importance. This may be, however, questioned by further research. But let me, for a moment, assume that such indicators have been defined: social area analysis may then prove most useful in comparative research, cross-sectional and longitudinal, for describing the whole of the urban network or for research on single units of observation. The typical picture of urban structure as illustrated in the social area diagram is expected to form the shape of a pyramid, as in the case of Berne in 1970 (Chart 2), with land values on the vertical axis and median rent per inhabitant on the horizontal axis. The CBD will be highest on the vertical axis and approximately medium on the horizontal axis, due to its very few and heterogeneous permanent inhabitants (see Chart 2).

A set of issues included in the frame of reference may then be approached more precisely: expansion of the CBD into the zone of transition; filtering down or gentrification of residential areas; degree and pace of residential turnover over time and according to ecological position relative to the CBD; dominance and subdominance; the development of second-order urban centers, and so on. Of course, the technique only provides for a descriptive device; but morphological description is just a first step to substantive explanation (Schnore, 1958). If our concern is with how urbanization affects urban structure, we can define a reference city as proposed by Shevky and Bell (1955) and cross-sectionally describe cities of different size or the same city over different points in time. It might even be possible to contrast two conflicting theories on urban development: one suggesting a universal trend of successive convergence of urban structure to the pattern typical for modern industrial societies; the other (Berry, 1976) inclined to see urban development as determined by specific intra-societal forces without universal trends to convergence. There certainly are numerous further issues that might be investigated by the use of SAA. The social area diagram provides for a useful illustrative device and the indicators are open for analytical manipulation.

FE, then, has its significance for the definition of appropriate indicators (which may, of course, also be defined as factor scores). It helps us to gain more insight into the concomitants of urban differentiation. By this, it sets a broader stage for substantive interpretation than is possible with original social area indicators. But substantive interpretation can only be made relative to the theoretical frame of reference. So, the three approaches to urban structure and development—classical human ecology, SAA, and FE—do provide supplementary tools for gaining cumulative knowledge on the structure and change of the urban network.

Increasing-scale theory cannot be connected with urban differentiation in a definite way. In neo-orthodox human ecology, the ecological complex (Duncan, 1959) has been proposed on a similar level of abstraction. As it is more complex than increasing-scale theory, possible relationships between the two frames of reference might be worthy of further discussion. The essential differences between the two approaches are well-known in the debate within, as well on, human ecology: they pertain to the "biotic" or "subsocial" (which, by no means, was thought of as non-social!) versus the "behavioral" or "social" level of a population's organization.

In any case the adequate theoretical link between highly abstract theories and urban differentiation might turn out to be classical human ecology.

References

Abu-Lughod, J. "Testing the Theory of Social Area Analysis: The Ecology of Cairo, Egypt." *American Sociological Review,* 34 (1969):198–212.

Anderson, T., and Bean, L. L. "The Shevky-Bell Social Areas: Confirmation of Results and a Reinterpretation." *Social Forces,* 40 (1961):119–24.

Anderson, T., and Egeland, J. "Spatial Aspects of Social Area Analysis." *American Sociological Review,* 26 (1961):392–98.

Bähr, J. "Zur Entwicklung der Faktorialökologie mit einem Beispiel einer sozialräumlichen Strukturanalyse der Stadt Mannheim," *Mannheimer Geographische Arbeiten,* 1 (1978a).

Bähr, J. "Santiago de Chile: Eine faktorenanalytische Untersuchung zur inneren Differenzierung einer lateinamerikanischen Millionenstadt," *Mannheimer Geographische Arbeiten,* 4 (1978b).

Bange, E., et al. "A Study of Selected Population Changes and Characteristics with Special Reference to Implications for Social Welfare." A group research project submitted in partial fulfillment of requirements for the Master of Social Welfare Degree, University of California, Berkeley, 1955. (Mimeographed)

Bassand, M. "Développement urbain et Logement: La

situation de Genève." Institut D'Études Sociales, Genève; *Annales du Centre de Recherche Sociale,* 2 (1974).

Bell, W. "The Social Areas of the San Francisco Bay Region." *American Sociological Review,* 18 (1953):39-47.

Bell, W. "Comment on Duncan's Review of *Social Area Analysis,*" *American Journal of Sociology,* 61 (1955a):260-61.

Bell, W. "Economic, Family and Ethnic Status: An Empirical Test." *American Sociological Review,* 20 (1955b) 45-52.

Bell, W. "Anomie, Social Isolation and Class Structure," *Sociometry,* 20 (1957):105-16.

Bell, W. "Urban Neighborhoods and Individual Behavior." In M. Sherif and C. Sherif, eds., *Problems of Youth.* Chicago, 1965.

Bell, W. "The City, the Suburb and the Theory of Choice." In S. Greer et al., eds., *The New Urbanization.* New York, 1968.

Bell, W. "Urban Neighborhoods and Individual Behavior," In P. Meadows and E. H. Mizrichi, eds., *Urbanism, Urbanization and Change.* Reading, Mass., 1969.

Bell, W., and Boat, M. D. "Urban Neighborhoods and Informal Social Relations." *American Journal of Sociology,* 62 (1957):391-98.

Bell, W., Boat, M. D., and Force, M. T. *People of the City.* Stanford, 1954.

Bell, W., and Force, M. T. "Urban Neighborhood Types and Participation in Formal Associations." *American Sociological Review,* 21 (1956):25-34.

Bell, W., and Force, M. T. "Religious Preference, Familism and the Class Structure." *The Midwest Sociologist,* 19 (1957):79-86.

Bell, W., and Greer, S. "Social Area Analysis and its Critics," *Pacific Sociological Review,* 5 (1962):3-9.

Berry, B. L. "Latent Structure of the American Urban System with International Comparison." In B. L. Berry, ed., *City Classification Handbook.* New York, 1972.

Berry, B. L. "Urbanization and Counterurbanization." *Urban Affairs Annual Reviews,* 11 (1976).

Berry, B. L., and Rees, P. H. "The Factorial Ecology of Calcutta." *American Journal of Sociology,* 74 (1968):445-91.

Berry, B. L., and Spodek, H. "Comparative Ecologies of Large Indian Cities." *Economic Geography,* 47 (1971):226-285.

Beshers, J. S. *Urban Social Structure.* New York, 1962.

Bollens, J. C. *Exploring the Metropolitan Community.* Berkeley, 1961.

Borukhov, E.; Ginsberg, J.; and Werczberger, E. "The Social Ecology of Tel-Aviv: A Study in Factor Analysis." *Urban Affairs Quarterly,* 15 (1979): 183-205.

Brindley, T. S., and Raine, J. W. "Social Area Analysis and Planning Research." *Urban Studies,* 16 (1979):272-89.

Broom, L.; Beem, H. P.; and Harris, L. "Characteristics of 1107 Petitioners for Change of Name." *American Sociological Review,* 20 (1955):33-39.

Broom, L., and Shevky, E. "The Differentiation of an Ethnic Group." *American Sociological Review,* 14 (1949):476-81.

Brown, L. A., and Horton, F. E. "Social Area Change: An Empirical Analysis." *Urban Studies,* 7 (1970):271-88.

Burgess, E. W. "The Growth of the City: An Introduction to a Research Project." *Proceedings of the American Sociological Society,* 18 (1924).

Campiche, R., and Zimmermann, E. "Un quartier traditionnel Les Eaux-Vives." In Institut d'ethique sociale de la fédération des églises protestantes de la Suisse, *Changement social et communauté,* Séries Spéciale d'Etudes et Rapports 4.

Carpenter, D. B. "Review of *Social Area Analysis.*" *American Sociological Review,* 20 (1955):497-98.

Clark, C. *The Conditions of Economic Progress.* London, 1951.

Clignet, R., and Sween, J. "Accra and Abidjan: A Comparative Examination of the Theory of Increase in Scale." *Urban Affairs,* 4 (1969):297-324.

Cohen, A. K. *Delinquent Boys.* Glencoe, Ill., 1955.

Curtis, J. H. "The Employability of Aging Workers in Social Areas of High Urbanization and Low Social Rank." Unpublished paper presented at the annual meetings of the American Sociological Society, Washington, D.C., 1957.

Curtis, J. H.; Avesing, F.; and Klosek, J. "Urban Parishes as Social Areas." *American Sociological Review,* 22 (1957):319-25.

Dent, O. F. "Aspects of Change in Urban Social-Spatial Structure." Ph.D. dissertation, University of Michigan, Ann Arbor, 1972.

DeWitt, K. F. "Applications of Social Area Analysis to Program Planning and Evaluation." *Journal of Evaluation and Program Planning,* (1978):65-78.

Droth, W., and Fischer, M. M. "Zur Theoriebildung und Theorietestung. Eine Diskussion von Grundlagenproblemen am Beispiel der Sozialraumanalyse." Hamburg und Wien, 1980. (Unpublished Paper)

Duncan, O. D. "Review of *Social Area Anlaysis.*" *American Journal of Sociology,* 60 (1955):84-85.

Duncan, O. D. "Human Ecology and Population Studies," In P. M. Hauser and O. D. Duncan, eds., *The Study of Population,* Chicago, 1959.

Duncan, O. D. "From Social System to Ecosystem." *Sociological Inquiry,* 31 (1961):140-49.

Duncan, O. D., and Duncan, B. "Residential Distribution and Occupational Stratification." *American Journal of Sociology,* 60 (1955):493-503.

Esser, T. *Aspekte der Wanderungssoziologie,* Darmstadt, 1980.

Farley, R. "Residential Segregation in Urbanized Areas of the United States in 1970: An Analysis of Social

Class and Racial Differences." *Demography*, 14 (1977): 497–518.
Firey, W. *Land Use in Central Boston*. Cambridge, Mass., 1947.
Friedrichs, J. *Stadtanalyse*. Reinbek, 1977.
Green, B. S. R. "Social Area Analysis and Structural Effects." *Sociology*, 5 (1971):1–19.
Greer, S. "Urbanism Reconsidered: A Comparative Study of Local Areas." *American Sociological Review*, 21 (1956):19–25.
Greer, S. "The Social Structure and Political Process of Suburbia: An Empirical Test." *Rural Sociology*, 27 (1962):438–59.
Greer, S., and Kube, E. "Urbanism and Social Structure: A Los Angeles Study." In M. B. Sussman, ed., *Community Structure and Analysis*. New York, 1959.
Greer, S., and Orleans, P. "The Mass Society and the Parapolic Structure." *American Sociological Review*, 27 (1962):634–46.
Grönholm, L. "Ecology of Social Disorganization in Helsinki." *Acta Sociologica*, 5 (1961):31–41.
Hagood, M.; Danilevsky, N.; and Blum, C. "An Examination of the Use of Factor Analysis in the Problem of Subregional Delineation." *Rural Sociology*, 6 (1941):216–33.
Hamm, B. *Die Organisation der Stätischen Umwelt*. Frauenfeld, 1977a.
Hamm, B. "Sozialraumanalyse." *Schweizerische Zeitschrift für Soziologie*, 3 (1977b):97–110.
Hamm, B. "Zur Revision der Sozialraumanalyse." *Zeitschrift für Soziologie*, 6 (1977c):174–88.
Hamm, B. "Segregation and Land Use." Paper presented at the 9th World Congress of Sociology, Uppsala, 1978.
Hawley, A. H. *Human Ecology*. New York, 1950.
Hawley, A. H., and Duncan, O. D. "Social Area Analysis: A Critical Appraisal," *Land Economics*, 33 (1957):337–45.
Haynes, K. E. "Spatial Change in Urban Structure: Alternative Approaches to Ecological Dynamics." *Economic Geography*, 47 (1971):324–25.
Herbert, D. T. "Social Area Analysis: A British Study." *Urban Studies*, 4 (1967):41–60.
Herbert, D. T., and Johnston, R. J., eds. *Social Areas in Cities*. London, 1976.
Hoyt, H. *The Structure and Growth of Residential Neighborhoods in American Cities*. Washington, D.C., 1939.
Hoyt, H. "Recent Distortions of the Classical Models of Urban Structure." *Land Economics*, 40 (1964): 199–212.
Hunter, A. A. "Factorial Ecology: A Critique and Some Suggestions." *Demography*, 9 (1972):107–17.
Hunter, A. A. *Symbolic Communities*. Chicago, 1974.
Hunter, A. A., and Latif, A. H. "Stability and Change in the Ecological Structure of Winnipeg: A Multi-Method Approach." *Canadian Review of Sociology and Anthropology*, 10 (1973):308–33.

Janson, C. G. "The Spatial Structure of Newark, New Jersey, Part I: The Central City." *Acta Sociologica*, 11 (1968):144–69.
Janson, C. G. "The Factorial Study of Socio-Ecological Change." In Z. Milnar and H. Tenne, eds., *The Social Ecology of Change*. London, 1978.
Janson, C. G. "Factorial Social Ecology," *Annual Review of Sociology*, 6 (1980). (Forthcoming)
Johnston, R. J. "Residential Area Characteristics: Research Methods for Identifying Urban Sub-Areas—Social Area Analysis and Factorial Ecology." In D. T. Herbert and R. J. Johnston, eds., *Social Areas in Cities I*. London, 1976.
Jones, F. L. "A Social Profile of Canberra, 1961." *The Australian and New Zealand Journal of Sociology*, 1 (1965).
Jones, F. L. "A Social Ranking of Melbourne's Suburbs." *The Australian and New Zealand Journal of Sociology*, 3 (1967):93–110.
Jones, F. L. *Dimensions of Urban Social Structure: The Social Areas of Melbourne, Australia*. Canberra, 1969.
Jung, R. "Migration und Siedlungsstruktur—Zum Zusammenhang von sozioökonomischen Status, Stellung im Familienzyklus der Migranten und sozialräumlicher Differenzierung. Dargestellt am Beispiel der Stadt Trier." Unpublished diploma thesis, 1980.
Kasarda, J. D. "The Theory of Ecological Expansion." *Social Forces*, 51 (1972).
Kaufman, W. C. "The People of St. Louis." Mimeographed report. Metropolitan St. Louis Survey, 1956.
Kaufman, W. C. "Social Area Analysis: An Application of Theory, Methodology and Techniques, with Statistical Tests of Revised Procedures, San Francisco and Chicago, 1951." Ph.D. dissertation, Evanston, 1961.
Kaufman, W. C., and Greer, S. "Voting in a Metropolitan Community: An Application of Social Area Analysis." *Social Forces*, 38 (1960):196–204.
Latif, A. H. "Factor Structure and Change Analysis of Alexandria, Egypt, 1947 and 1960." In K. P. Schwirian, ed., *Comparative Urban Structure*. Lexington, 1974.
MacCannell, E. H. "An Application of Urban Typology by Cluster Analysis to the Ecology of 10 American Cities." Ph.D. dissertation, Seattle, 1957.
McElrath, D. C. "Prestige and Esteem Identification in Selected Urban Areas." *Research Studies of the State College of Washington*, 23 (1955):130–37.
McElrath, D. C. "The Social Areas of Rome." *American Sociological Review*, 27 (1962):376–91.
McElrath, D. C. "The Social Differentiation of Migrants in Accra, Ghana." Unpublished research report, Evanston, 1965a.
McElrath, D. C. "Urban Differentiation: Problems and Prospects," *Law and Contemporary Problems*, 30 (1965b):103–19.

McElrath, D. C. "Changing Differentiation of Rome: 1951–1971." Unpublished draft, 1980.

McKenzie, R. D. "The Scope of Human Ecology." *Publications of the American Sociological Society*, 20 (1926):141–54.

Mischke, M. *Faktorenanalytische Untersuchung zur räumlichen Ausprägung der Sozialstruktur in Pforzheim*. Karlsruhe, 1976.

Morris, F. B., and Pyle, G. F. "The Social Environment of Rio de Janeiro in 1960." *Economic Geography*, 47 (1971):286–99.

Murdie, R. A. *The Factorial Ecology of Metropolitan Toronto, 1951–1961: An Essay on the Social Geography of the City*. Chicago, 1969.

Myers, J. K. "A Note on the Homogeneity of Census Tracts." *Social Forces*, 32 (1954):364–66.

Nicholson, T. G., and Yeates, M. H. "The Ecological and Spatial Structure of the Socio-Economic Characteristics of Winnipeg, 1961." *Canadian Review of Sociology and Anthropology*, 6 (1969):162–78.

Nordstrand, E. A. "Relationship between Intraurban Migration and Urban Residential Social Structure." M.A. thesis, University of Minnesota, 1973.

Orleans, P. "Robert Park and Social Area Analysis: A Convergence of Traditions in Urban Sociology." *Urban Affairs Quarterly*, 1 (1966):5–19.

Ottensmann, J. R. *The Changing Spatial Structure of American Cities*. Lexington, 1975.

Park, R. E. "The Urban Community as a Spatial Pattern and a Moral Order." *Publications of the American Sociological Society*, 20 (1925):1–14.

Peach, C., ed. *Urban Social Segregation*. London: 1975.

Pedersen, P. O. *Modeller for befolkningsstruktur og befolkningsudrikling; storbyområder—speciell med henblik på Storkøbenhavn*. Copenhagen, 1967.

Pirie, G. H. "Measuring Accessibility: A Review and Proposal." *Environment and Planning*, 11 (1979):299–312.

Pitts, F. R. "Factorial Ecology of Seoul and Taegu, Korea: A Preliminary Report." *Economic Geography*, 47 (1971):2.

Polk, K. "The Social Areas of San Diego." M.A. thesis, 1957a.

Polk, K. "Juvenile Delinquency and Social Areas." *Social Problems*, 5 (1957b):214–17.

Price, D. O. "Factor Analysis in the Study of Metropolitan Centers." *Social Forces*, 20 (1942):449–55.

Quinn, J. A. "Burgess' Zonal Hypothesis and Its Critics." *American Sociological Review*, 5 (1940):713–22.

Quinn, J. A. *Human Ecology*, Englewood Cliffs, N. J., 1950.

Rees, P. H. "Measuring and Modelling the Distribution of Social Groups within Cities." Paper presented at the Regional Science Association Conference, 1971a.

Rees, P. H. "Factorial Ecology: An Extended Definition, Survey and Critique of the Field." *Economic Geography*, 47 (1971b):220–23.

Robson, B. T. *Urban Analysis*. Cambridge, 1969.

Russett, B. M. *International Regions and the International System: A Study in Political Ecology*. Chicago, 1967.

Salins, P. D. "Household Location Patterns in American Metropolitan Areas." *Economic Geography*, 47 (1971):234–48 (Supplement).

Sauberer, M., and Cserjan, K. "Sozialräumliche Gliederung Wien 1961—Ergebnisse einer Faktorenanalyse." *Der Aufbau*, 27 (1972):284–306.

Schmid, C. "Generalizations Concerning the Ecology of the American City." *American Sociological Review*, 15 (1950):264–81.

Schmid, C., and Tagashira, K. "Ecological and Demographic Indices." *Demography*, 1 (1964):194–211.

Schnore, L. F. "Social Morphology and Human Ecology." *American Journal of Sociology*, 63 (1958):620–34.

Schwirian, K. P., and Smith, R. K. "Primacy, Modernization and Urban Structure: The Ecology of Puerto Rican Cities." In K. P. Schwirian, ed., *Comparative Urban Structure*. Lexington, 1974.

Senior, M. L. "Approaches to Residential Location Modelling." *Environment and Planning*, 5 (1973):165–97.

Sherif, M., and Sherif, C. W. *Reference Groups*. New York, 1964.

Shevky, E., and Bell, W. *Social Area Analysis*. Stanford, Conn., 1955.

Shevky, E., and Williams, M. *The Social Areas of Los Angeles*. Berkeley. 1949.

Struening, E. L. "Social Area Analysis as a Method of Evaluation." In E. L. Struening and M. Guttentag, eds., *Handbook of Evaluation Research*. Beverly Hills, 1975.

Sullivan, T. "The Application of Shevky-Bell Indices to Parish Analysis." *American Catholic Sociological Review*, 22 (1961):168–71.

Sweetser, F. L. *Social Ecology of Metropolitan Boston: 1950*. Boston, 1961.

Sweetser, F. L. "Factorial Ecology: Helsinki 1960." *Demography*, 2 (1965a):372–85.

Sweetser, F. L. "Factorial Ecology: Zonal Differentiation in Metropolitan Boston, 1960." Chicago, 1965b. (Mimeographed)

Sweetser, F. L. "Factor Structure as Ecological Structure in Helsinki and Boston." *Acta Sociologica*, 8 (1965c):205–25.

Sweetser, F. L. "Ecological Factors in Metropolitan Zones and Sectors." In M. Dogan and S. Rokkan, eds., *Quantitative Ecological Analysis in the Social Sciences*. Cambridge, 1969.

Sweetser, F. L. "Neighborhood Typologies and Social Ecological Theory." Paper presented to 9th World Congress of Sociology, Uppsala, 1978.

Tiebout, C. M. "Hawley and Duncan on Social Area

Analysis: A Comment." *Land Economics*, 34 (1958):182–84.
Timms, D. W. G. *The Urban Mosaic*, Cambridge, 1971.
Timms, D. W. G. "Social Bases to Social Areas." In D. T. Herbert and R. J. Johnston, eds., *Social Areas in Cities*. London, 1976.
Tryon, R. C. *Identification of Social Areas by Cluster Analysis*, Stanford, 1955.
Udry, R. J. "Increasing Scale and Spatial Differentiation: New Tests of Two Theories from Shevky and Bell." *Social Forces*, 42 (1964):403–13.
Van Arsdol, M. "An Empirical Evaluation of Social Area Analysis in Human Ecology." Ph.D. dissertation, 1957.
Van Arsdol, M.; Camilleri, S. F.; and Schmid, C. F. "An Application of the Shevky Social Area Indexes to a Model of Urban Society." *Social Forces*, 37 (1958a): 26–32.
Van Arsdol, M.; Camilleri, S. F.; and Schmid, C. F. "The Generality of Urban Social Area Indexes." *American Sociological Review*, 23 (1958b):277–84.
Van Arsdol, M.; Camilleri, S. F.; and Schmid, C. F. "An Investigation of the Utility of Urban Typology." *Pacific Sociological Review*, 4 (1961):26–32.
Van Arsdol, M.; Camilleri, S. F.; and Schmid, C. F. "Further Comments on the Utility of Urban Typology." *Pacific Sociological Review*, 5 (1962): 9–13.
Wachs, M., and Kumagai, J. G. "Physical Accessibility as a Social Indicator." *Socio-Economic Planning Science*, 7 (1973):437–56.
Weclawowicz, G. "The Structure of Socio-Economic Space in Warsaw, 1931 and 1970." In R. A. French and F. E. Hamilton, eds., *The Socialist City*. Chichester, 1979.
Wendling, A. "Suicide in the San Francisco Bay Region 1938–1942 and 1948–1952." Ph.D. dissertation, 1954.
Williamson, R. C. "Selected Urban Factors in Marital Adjustment." *Research Studies of the State College of Washington*, 21 (1953):327–41.
Williamson, R. C. "Socio-Economic Factors and Marital Adjustment in an Urban Setting." *American Sociological Review*, 19 (1954):213–16.
Wilson, G., and Wilson, M. *The Analysis of Social Change*. Cambridge, Mass., 1945.
Wilson, R. L. "The Association of Urban Social Areas in Four Cities and the Institutional Characteristics of Local Churches in Five Denominations." Ph.D. dissertation, 1958.
Wirth, L. "Urbanism as a Way of Life." *American Journal of Sociology*, 44 (1938):1–24.

Testing the Theory of Social Area Analysis: The Ecology of Cairo, Egypt*

Janet L. Abu-Lughod

I

Despite the fact that the originators of Social Area Analysis[1] disclaimed interest in traditional ecological preoccupations with urban spatial structure,[2] one unintended but nevertheless highly significant consequence of their efforts has been a revitalization of the latter field, stimulated perhaps by the more workable assumptions they provided. What were some of the older assumptions and how did the approach of social area analysis revise them?

Implicit in earlier ecological research was the idea that *the* spatial pattern of a city was imbedded in objective reality and would be revealed almost automatically once a sufficient number of sensitive indices to it were gathered and processed. By plotting indices against a geographic grid, overlaying and comparing their distributions, one reached a synthesis—a single pattern—corresponding to objective reality. That some indices revealed *the pattern* and others did not (that is, reflected discrepant patterns of distribution) was attributed

---

*Reprinted from the *American Sociological Review*, 34 (April 1969):198–212, by permission of the author and The American Sociological Association.

Research and statistical processing were facilitated by the Social Research Center of the American University at Cairo, MIT-Harvard Joint Center for Urban Studies, National Science Foundation and the University of Massachusetts. Among those who contributed to the statistical methods or read the current manuscript critically are Peter Park, Hilda Golden, T. O. Wilkinson, Brian Berry, Frank Sweetser and colleagues at Northwestern University, to whom I express my gratitude.

[1] Although the name refers to an approach initially reported by Eshref Shevky in 1949, the antecedents and consequences are somewhat broader. In 1947, a Committee on Community Studies was organized on the west coast including, among others, Shevky, Calvin Schmid, Robert Tryon, William Robinson, and Leonard Broom. The cross-fertilization that resulted is reflected in subsequent publications by the participants, although only Shevky was subsequently associated with social area analysis. The approach spread at first largely through personal influence and only now appears to be entering the public domain with widespread but often mechanical applications in various parts of the world (for example, Herbert [1967] and the Hyderabad Study [1966]). A fairly complete bibliography of all but the most recent studies using social area analysis can be found in Wendell Bell (1965).

[2] The classic critique of social area analysis was stated in a review article by Hawley and Duncan in which the authors accused social area analysts of disavowing "a concern with what most students of the urban community think of as area structure" (1957:337).

to the superior power or sensitivity of the former.[3] Furthermore, since ecological forces were thought to operate on the subconsensual level, it was anticipated that fairly universal patterns would be found that transcended cultural variations. (For fuller treatments, see Abu-Lughod, 1966 and 1968.)

These questionable assumptions were already falling into disrepute among urban ecologists when social area analysis entered the field, arguing that the basic problem was taxonomic rather than phenomenological. The goal was to set up categories useful for analyzing and giving order to the world; if patterns were to be found, they might have to be "put into" the city by the investigator. This was similar to the revision that had already been introduced with respect to "natural areas" (Hatt, 1946).

Social area analysts further suggested that there was no reason to assume *a priori* that differentiation within cities would be unidimensional. Shevky and Williams (1949) posited that at least three dimensions of differentiation might be required to comprehend significant variations among census tract populations in contemporary American cities, namely: (1) social rank, (2) urbanization, and (3) ethnicity. Since each of these dimensions would be revealed through a different set of indices, each might have its own unique pattern of distribution. Once this possibility was admitted, it became feasible to resolve several controversies of long standing in the field of urban ecology.[4]

However, Shevky and Bell offered perhaps their most significant revision of the traditional ecological approach in their small monograph of 1955 in which they presented a set of theoretical propositions relating ecological variations to cultural-technological conditions (Shevky and Bell, 1955:3–17). This theory, inadequately explicated as it was and appended uneasily to serve chiefly as an elaborate "rationalization" for Shevky's perspicacious and, as it later developed, "happy hunches" with respect to American urbanism, hinted at the possibility of relating the type and complexity of urban differentiation to the "scale" of the society in which a city was found. The theory suggested that the pattern of social (and physical) differentiation in preindustrial societies (cities) would be relatively simple and perhaps virtually unidimensional; as the scale of society increased, there would be increased complexity of differentiation and a separation of the axes or dimensions of differentiation.[5]

If this hypothesis should prove correct, the theory, duly qualified and developed, might provide a fruitful framework for comparative urban studies which hitherto have been conducted on an *ad hoc* and highly eclectic basis. Eventually, it may even yield what has long been recognized as an urgent need in the field of urban ecology—namely, true theory that both illuminates empirical findings and guides research in meaningful directions.

To say that the theory offers such potential is not to claim that it has already demonstrated its value. In fact, the theory as originally formulated was not operational at all, i.e., the propositions were not in testable form.[6] Furthermore, as we shall argue, the very method of social area analysis obviated the possibility of using it to test the theory from which it had been derived, for it took for granted *a priori* just those propositions most in need of proof.

What we propose to do in this paper, by a somewhat circular route, is to suggest some operational propositions, point to a method better suited than social area analysis to testing them, and present some preliminary evidence in support of the propositions, in the hope that we may contribute to an eventual and fuller formulation of theory.

---

[3]The most detailed and honest account of this process appeared in Murray Leiffer's research paper (1933), which followed the technique first explored by Burgess (1923).

[4]For example, the conflict between the concentric circle "theory" (read "generalization") of Burgess and the Hoyt sector "theory" was easily resolved by a simple but highly significant study by Anderson and Egeland (1961), which suggested that the distribution pattern of socio-economic variables was sectoral (Hoyt had based *his* conclusions on rents and house values), and that of family and social organization was largely a function of distance from the center (Burgess had based *his* generalization primarily on density of settlement and house types). The crucial discovery was that the two patterns were not mutually exclusive but were in fact legitimate descriptions of two different aspects of reality.

[5]Our understanding is only approximate. Actually, the "model" presented by the authors (Shevky and Bell, 1955: Tables II–I), purporting to relate societal scale to urban differentiation, does not spell out increasing scale and its implications but merely postulates certain characteristics of modern western industrial society. One must "read into" the model any dynamic implications.

[6]This soon became obvious, for Udry's (1964) attempt to "test" the theory, using longitudinal United States data, not only fails to confirm it but—a much more serious criticism—strikes the reader as quite irrelevant to the main issue. Another attempt (McElrath, 1962), to test the theory by applying the method to Rome, a city in a society of presumably smaller "scale," was similarly unsatisfying, although we believe that the results, when reinterpreted, are highly suggestive.

II

To accomplish this end, it will be necessary to review briefly the method itself, to trace the application of what we contend to be a superior solution to the basic methodological problem, and to review some of the "deviant" findings (including our own on Cairo) which have led to a reformulation of certain theoretical propositions.

Shevky's early study of Los Angeles (1949) diverged little in orientation or aim from the traditional ecological case study. Its contribution was more methodological than conceptual, although a brief theoretical rationale was included to defend the methodology. Instead of plotting selected indices of social differentiation separately and then combining them into a single geographic pattern, Shevky advanced a set of related hypotheses concerning the ways in which urban subpopulations in America are differentiated, selected a few indices designed to measure each hypothesized dimension of differentiation, combined (averaged) the standardized scores of the indices to obtain census tract scores for each dimension, and then plotted their patterns on base maps of Los Angeles.

The innovations of greatest importance were: (1) a shift from the inductive method characteristically employed in ecological studies to a more deductive one, and (2) a reversal of the procedural order, substituting mapping as a final step *after* index construction for the more conventional approach of mapping as a step *preliminary* to synthesis. These innovations should be clearly distinguished since, in our view, the latter represented a significant advance, whereas the former was an innovation of dubious reliability and limited applicability.

This is a strong indictment and in need of defense. It does not question the value of Shevky's conceptualization of separate dimensions, a conceptualization which many studies of United States cities have tested and found exceedingly useful, but it does question the method used to measure these dimensions. Before proceeding further, then, let us stop to consider whether the deductive approach was an essential or accidental part of social area analysis.[7] Could an alternate approach have achieved the posited separation of axes of differentiation while resolving the measurement problem in more determinant fashion and leaving open the real question posed by the theory of increasing scale: namely, whether social differentiation within cities of smaller-scale societies could be comprehended by a simpler schema than that required to understand cities in large-scale industrialized societies?

We contend that the deductive derivation of indices was *not* essential to social area analysis; it merely represented a crude and approximate solution to a problem for which a superior methodology already existed, even though the latter had not yet been applied in urban ecology. This alternative was factor analysis (a variant of which had actually been employed by Shevky's colleague, Robert Tryon),[8] which was called in only belatedly to buttress the validity of the social area approach.[9] More than a decade earlier, Margaret Hagood had suggested that factor analysis offered an appropriate method for "synthesizing data on characteristics with respect to which subregions are to be homogeneous" (Hagood et al., 1941). The urban ecologist immediately recognizes in this statement his *central problem*. Nevertheless, despite the obvious linkage, it was not until the late 1950s that factor analysis began to be substituted for the less rigorous methods of social area analysis. Led by Calvin Schmid,[10] investigators in the field of urban ecology have subjected batteries

---

[7] I do not wish here to defend some false dichotomy between induction and deduction. Deduction requires some preliminary induction, no matter how unsystematic, and induction, by its very act of selecting *relevant* items, conceals much unacknowledged deduction. The distinction is to be found in that portion of the inevitably circular process of reasoning in which *crucial* research decisions are made.

[8] Tryon (1955) applied cluster analysis (first explicated by him in 1939) to 33 variables in the San Francisco Bay area, confirming that at least three dimensions were necessary to account for the variance between census tracts. Shevky and Bell (1955) considered Tryon's study an empirical verification of their theory but, while Tryon acknowledged that the three most independent clusters were *roughly* equivalent to Shevky's three dimensions, he argued that he could obtain higher validities by including more and better measures than the handful selected *a priori* by Shevky and by manipulating the data according to a more determinant weighting system.

[9] Bell (1955) used Thurstone's centroid method of factor analysis in a hypothesis-testing framework, not to extract factors from a battery of indices, but to prove that the few preselected variables of social area analysis could be factored into three dimensions and that these axes could be rotated to approximate simple structure. This he accomplished by sacrificing orthogonality. The intercorrelations between factors required to approximate simple structure ran as high as $-.73$ between economic and ethnic factors and $-.50$ between economic and familism factors. If the goal was to *separate* the dimensions of differentiation, then it is difficult to defend such close positioning of the axes.

[10] Schmid's original investigation (1950), based upon 23 American cities as of 1940, did not go beyond a correlational analysis. Nevertheless, the similarity of ecological correlations within the cities suggested a

of census-derived indices to correlational and factorial analysis.

We now have results from both social area analysis and factor analysis for many cities in the United States, several in Europe, and some even in non-Western countries. Since no summary list exists in published form, we have enumerated them here.[11]

American cities that have been studied by social area analysis (and/or have had a factor, correlational, or cluster analysis *test* of social area variables) include: *Akron, O.* (Van Arsdol, Jr., 1957, and Van Arsdol, Jr., et al., 1958; MacCannell, 1957; Anderson and Egeland, 1961); *Atlanta, Ga.* (Van Arsdol, 1957, 1958; MacCannell, 1957); *Birmingham, Ala.* (Van Arsdol, 1957, 1958; MacCannell, 1957); *Chicago, Ill.* (Kaufman, 1961); *Cleveland, O.* (Moush et al., 1960); *Dayton, O.* (Anderson and Egeland, 1961); *Indianapolis, Ind.* (Anderson and Egeland, 1961); *Kansas City, Mo.* (Van Arsdol, 1957, 1958; MacCannell, 1957); *Los Angeles, Calif.* (Shevky and Williams, 1949; Bell, 1952, 1955); *Louisville, Ky.* (Van Arsdol, 1957, 1958; MacCannell, 1957); *Minneapolis, Minn.* (Van Arsdol, 1957, 1958; MacCannell, 1957); *New York: Bronx* (Sullivan, 1961); *Portland, Ore.* (Van Arsdol, 1957, 1958; MacCannell, 1957); *Providence, R.I.* (Van Arsdol, 1957, 1958; MacCannell, 1957); *Rochester, N.Y.* (Van Arsdol, 1957, 1958; MacCannell, 1957); *San Diego, Calif.* (Polk, 1957); and *San Francisco, Calif.* (Bell, 1952, 1953, 1955; Shevky and Bell, 1955; Kaufman, 1961).

Among the Western cities outside the United States studied by this method are: *Newcastle, England* (Herbert, 1967); *Quebec, Canada* (Gagnon, 1960); and *Rome, Italy* (McElrath, 1962).

Cities in the non-Western world for which similar studies are available include: *Abidjan, Ivory Coast* (Clignet and Sween, 1969); *Accra, Ghana* (McElrath, 1968); and *Hyderabad, India* (Hyderbad Metropolitan Research Project, 1966).

Increasingly, however, factorial analysis is gaining precedence, although methods for selecting variables still remain relatively unstandardized. The U.S. cities for which factor analysis solutions (whether accompanied or not by geographic mapping) have been performed include: *Boston, Mass.* (Sweetser, 1965a and earlier reports); *Chicago, Ill.* (Rees, 1969); *Cleveland, O.* (Beshers, 1957, Simplex analysis); portions of *New York City* (Carey, 1966); *San Francisco, Calif.* (Tryon, 1955); *Seattle, Wash.* (Schmid and Tagashira, 1964); *Toledo, O.* (Anderson and Bean, 1961); and *Washington, D.C,* (Carey, Greenberg, and Macomber, mimeo, 1968).

In addition, students in the Center for Urban Studies at the University of Chicago, in a project undertaken in 1968, have analyzed the factorial ecology of the following American cities: *Birmingham, Ala.; Canton, O.; Columbus, O; Fort Worth, Tex; Gary, Ind.; Homestead, Pa.; Honolulu, Hawaii; Louisville, Ky.; Miami, Fla.; New Haven, Conn.; Harlem and Yonkers, N.Y.; Omaha, Neb.; Providence, R.I.; Richmond, Va.; Rockford, Ill.; Sacramento, Calif.; St. Louis, Mo.; San Diego, Calif.; San Jose, Calif.; Shreveport, La.; South Bend, Ind.; Syracuse, N.Y.; Tacoma, Wash.; and Worcester, Mass.*

Western cities outside the United States for which factor analyses are available include: *Toronto, Canada* (Berry and Murdie, 1965; Goheen, 1969; Murdie, 1968); *Hampshire, England* (Gittus, 1964); *Liverpool, England* (Gittus, 1965); *Sunderland, England* (Robson, 1969); *Copenhagen, Denmark* (Pedersen, 1967); *Helsinki, Finland* (Sweetser, 1965a, 1965b); and *Brisbane, Australia* (Timms, forthcoming).

To my knowledge, only two factor analyses have been made of non-Western cities: *Cairo, Egypt* (Abu-Lughod, 1966); and *Calcutta, India* (Berry and Rees, mimeo, 1968). What do these studies indicate concerning the nature of social differentiation within cities of various types? And if the results have not always supported the Shevky–Bell typology, have there been any consistent deviant findings that might permit us to reformulate the underlying theory?

---

common underlying factorial pattern. In Schmid et al.'s (1958) replication based upon 1950 census data, the matrices of intercorrelations remained relatively constant, although a few significant deviations did appear. Some of these deviations may have reflected actual changes, whereas others may have been due to a modification of method, i.e., a substitution of one variable for another. The conclusions drawn from this and related analyses have been cogently questioned by Beverly Duncan (1964). Two doctoral dissertations under Schmid's direction attempted further analyses of the correlational matrices of ten of the sampled cities. Earle MacCannell (1957) used Tryon's method of cluster analysis, while Maurice Van Arsdol, Jr. (1957) did a factor analysis, the results of which were reported in Van Arsdol, Jr., et al. (1958). Most recently, Schmid and Tagashira published the results of a factor analysis of 1960 data for Seattle (1964).

[11] I am indebted to a number of persons who assisted in making this list more exhaustive, and especially to Brian Berry, whose bibliography (mimeographed and updated) is the most complete I have seen. Additions to the literature are now being rapidly made, often in the form of inaccessible theses and dissertations. I would greatly appreciate hearing from individuals engaged in factor analysis of specific urban areas.

III

First we might examine the congruencies. In each one of the studies cited, a dimension has been found which discriminates among subareas within the city in terms of the socio-economic rank of residents. While indices of occupation and education are invariably linked to this factor, other measures clearly economic in nature, such as rent, income, housing quality, and density of dwelling-unit occupancy also relate to this dimension. Not only is a socio-economic factor always isolated, but it is frequently the factor that accounts for the largest proportion of the variance in the correlation matrix.[12]

A second uniformity has been the occurrence of correlational linkages among variables measuring family size and age composition. This can be interpreted as indicating the selective attractiveness of various parts of the urban community as residential environments for different kinds of families or families at different stages of their life cycles.[13] Shevky originally termed this "urbanization" (meaning the absence of family orientation) but later investigators have redefined it and indeed measured it in reflected form, labeling it, variously, "familism" (Bell, 1953), "young family cycle stage" (Beshers, 1959; Kaufman, 1961), or "progeniture" (Sweetser, 1965a, b). The fertility ratio usually constitutes a sensitive indicator of this dimension. Less universally reliable or reproducible have been the other two components of Shevky's urbanization score. The proportion of single-family detached homes, for example, although a discriminating index in most large American cities, has had little significance in studies of foreign cities where one notices a greater uniformity of building types.[14] The proportion of females in the labor force does not relate closely to variations in family types in those societies where the over-all proportion is either very high or very low.[15] Substitute measures, such as the dependency ratio, average family size, and the proportion of one- and two-person families, appear better designed for cross-culture application than are the more culture-specific measures used by Shevky.

If these are the congruities, what are the deviations? They appear to be of two kinds: first, those deviations that relate to the appearance of more than these two factors; and second, those deviations that relate to the degree of "independence" between the key dimensions of socio-economic rank and familism. For the first, a relatively simple explanation exists and the issue does not appear to be critical for theory. For the second, however, additional probing is required. We would suggest that a basic question, related in essential fashion to the theory of increasing scale, is involved here. We might then dispense quickly with the former before addressing ourselves to the latter.

It will be recalled that the third dimension posited by Shevky was "ethnicity." Despite the obvious relevance of this dimension to the analysis of U.S. cities, ethnicity was found to be the least independent of the three dimensions included in Bell's factorial test (1955: see note 9 *supra*) and the least clearly paralleled in Tryon's cluster analysis.[16] Nor has this factor ever been replicated outside the United States, even when it has been consciously sought and appropriate modifications in its measurement made. McElrath (1962) eliminated it from his social area analysis of Rome for lack of information, although he suggested that migration

---

[12] Several exceptions should be noted. In Schmid and Tagashira's study of Seattle (1964), a principal axes factor solution yielded first a family status factor and then a socio-economic one. A similar result was obtained in Pedersen's study of Copenhagen (1967). In Berry's study of Calcutta (1968), the principal axes factor analysis yielded socio-economic rank as the second factor, following one chiefly indicative of family-life orientation. It is important to point out that the particular "mix" of initial variables determines the order of factor extraction.

[13] It must be borne in mind that we are not dealing with individual correlations but with ecological ones. The inferences that can be drawn must therefore be related to this constraint.

[14] In studies of Rome (McElrath, 1962), Helsinki (Sweetser, 1965a, b), and Cairo (Abu-Lughod, 1966), for example, this measure was irrelevant or nondiscriminating. Land-use density in residential quarters probably constitutes a more useful measure for cross-cultural research.

[15] In Scandinavian countries female participation in the labor force is higher than in the U.S. There, although family cycle stages are relatively segregated residentially, gainfully employed women appear to be less so (Sweetser, 1965a, b). On the other hand, in certain developing countries such as Egypt and India, where females traditionally have been excluded from the urban (but not the rural) labor force, what little female labor-force participation there is reflects socio-economic status; areas with employed women are those where resident domestic servants abound. African cities present a third variation. Female labor-force participation in urban areas is largely confined to petty marketing, an activity, that in no way conflicts with child-rearing. See studies of McElrath on Accra (1968) and Clignet on Abidjan (1968).

[16] Although Shevky and Bell (1955) claim that their dimension of ethnicity was paralleled by Tryon's Assimilation II cluster, the three variables of this cluster were: foreign-born from Northwestern Europe, females, and females in white collar employment; a measure of non-whites does not even appear in it (Tryon, 1955).

status, had information been available, might have tapped a third dimension of social differentiation within that city. Tribal origins and migration status were analytically separated by him in his study of Accra (1968), but factorial independence was not tested for. Sweetser attempted to isolate the dimension of ethnicity in Helsinki by distinguishing between the Finnish-speaking majority and the Swedish-speaking minority elite, and Abu-Lughod tried a similar approach by including religious and nationality variables for Cairo, but in neither case was a separate "ethnic" factor extracted under conditions of orthogonality. In the two latter cases, ethnic variables loaded chiefly upon the socioeconomic vector.

However, studies have yielded more than the two basic factors of social rank and familism, even though ethnicity has not been among them. Some investigators have isolated factors relating to building type or dominant land use (Anderson and Bean, 1961, for Toledo; Berry, 1968, for Calcutta); others have found factors indicative of mobility and/or age and sex imbalance resulting from selective migration (Berry and Rees in Calcutta; Schmid and Tagashira, 1964, in Seattle; and Abu-Lughod, 1966, in Cairo), and factors of size, density, and even social disorganization have also been found. The appearance of these additional factors is due clearly to the initial "mix" of variables and to cultural differences too complex to be synthesized or explained on the basis of diverse studies undertaken without a common framework of analysis. While it would be important eventually to understand these variations, the state of the art at present precludes this. The deviations are scattered and inconsistent.

The opposite is true of the deviations that have occurred when the vectors of social rank and familism have been distinguished. Here a basic framework of analysis has guided the investigators; roughly the same variables have been included; the number of tests has been relatively more adequate; and the deviations, where found, have been in a consistent direction. We seem to be approaching a situation where theory can be reformulated with respect to these two dimensions, where hypotheses can be framed, and where an experimental design for testing them can at least be suggested. Let us examine the evidence thus far.

IV

It will be recalled that early in the development of social area analysis, Bell (1955) conducted a factorial test, designed to substantiate the hypothesis that the dimensions of social rank and familism measured relatively independent aspects of social differentiation within American cities. This he found to be approximately correct for Los Angeles and San Francisco, even though the correlations between factors were, in our opinion, exceedingly high.[17] A similar test conducted by Van Arsdol, Jr., et al. on a sample of ten other U.S. cities (1957, 1958) substantiated Bell's results in six cases but failed to reproduce factorial separation in the remaining four. On the assumption that deviant cases may be critical for the advancement of theory, let us look more closely at their peculiar characteristics and the ways in which they deviated.

The six cities where factorial separation between social rank and familism variables was achieved were predominantly white, Northern communities. The deviants included three Southern cities (Birmingham, Atlanta, and Louisville) and the single "border city" in the sample, Kansas City. All four cities contained high proportions of non-whites; in fact, when the ten cities were ranked according to percentage of Negroes, the deviants were the top four.

How did their factor structures deviate from the Shevky–Bell model? The fertility variable loaded higher than anticipated on Factor I (Social Rank). On the basis of the theory, fertility should have been disassociated from social rank as in the Northern cities. However, despite tortuous rotations, the best reduction obtained in factor loadings of fertility on social rank was +.492 for Birmingham and +.360 for Louisville, the two partially deviant cases. The clearly deviant cases of Kansas City and Atlanta presented greater problems, since the factor loadings of fertility on social rank (+.613 and +.528, respectively) even *exceeded* the loadings of this variable upon the familism factor (+.523 and +.475, respectively). The authors concluded:

> ... [T]hese four cities include relatively large proportions of Negroes. This fact, combined especially with the unfavorable economic position of the Negroes, may indicate that the *range of family forms* in these cities, as described by the fertility measure, *has not yet become disassociated from social rank* (Van Arsdol, Jr., et al., 1958:282, with emphasis supplied).

[17]This does not appear too critical, since other studies have since confirmed separation of social rank from familism variables, even under the stringent requirements of orthogonality.

In fact, the authors suggest an alternate factor model for these four cities in which occupation, education, and fertility are viewed as components of social rank, a hypothesis which, when tested, yielded a neater and virtually orthogonal factor solution.

McElrath, investigating the relationship in Rome between variables ostensibly measuring social rank and those measuring "urbanization," also concluded that the degree of disassociation there was considerably less than in American cities. Although he did not include a factorial test, given the zero-order correlation coefficients between variables (1962: 383), it appears that fertility would not have factored on urbanization at all but would have loaded on social rank.[18] Reanalysis of the data, however, is necessary before a firm conclusion can be reached.

Deviations noted in the American South and in Rome were even more marked in Cairo, Egypt (Abu-Lughod, 1966). When census tract indices were subjected to principal axes factor analysis and subsequent varimax orthogonal rotations, no factorial separation between indicators of social rank and indicators of family cycle stage could be obtained. Since these data have heretofore not been published, we present them in some detail.[19]

From the Egyptian censuses of 1947 and 1960 for the Governorate (metropolitan province) of Cairo, thirteen indices reflecting variations in demographic structure, family characteristics, socio-economic level, and ethnic composition were computed[20] for each of the 216 census tracts (pl. *shiyākhāt*) into which the city was administratively divided in 1947.[21] Zero-order correlation coefficients, as shown in Table 1, were computed for each of the data years. These matrices were then subjected to principal axes factor analysis (using Hotelling's iterative procedure for estimating communalities). Seven factors, accounting cumulatively for 90 percent of the total variance in each year, were successively extracted, of which the first four (having sums of squared values of unity or higher) were retained for rotation.

Solutions were comparable for both years; in each case the first four factors accounted for more than 75 percent of the total variance. More significantly, in each year the first orthogonal factor explained about half of the total variance in the matrices.[22]

Inspection of the graphs suggested that the factorial solution could be improved and a varimax orthogonal rotation was therefore performed. Factors I and II were significantly clarified, although Factors III and IV remained reversed in the 1947 and 1960 solutions. A further graphic rotation (still orthogonal) was performed on the latter two factors to arrive at a third factor of comparable pattern in the two years. The factor loadings after rotation are presented in Table 2.

Interpretation of the factors was not difficult. Factor I, while clearly representing socio-economic status, also includes many variables indicative of family life, suggesting that it is to be interpreted as a "style of life" vector in which class and family patterns are inextricably linked. Thus, census tracts with high Factor I scores[23] are characterized by commodious housing accommodations, by

---

[18] Fertility correlated −.676 with non-manual occupations; +.744 with the illiteracy rate; and +.783 with dwelling-unit overcrowding. These intercorrelations were virtually as high as those between the social rank indices themselves.

[19] Analysis was begun in 1958 when the author was affiliated with the Social Research Center, American University at Cairo. Preliminary processed data for Cairo census tracts as of 1947 were published in Abu-Lughod and Attiya (1963). Due to distance, however, the author remained ignorant of factor analytic applications being attempted in the States until 1961.

[20] Space does not permit presentation of the operational definitions and computational formulae of the thirteen indices nor of means and standard deviations in each year. (In the initial study, more variables had been included although the list was far from ideal; some were later eliminated because 1947 and 1960 data were not comparable; others which proved relatively nondiscriminating were dropped.) Details can be found in Abu-Lughod (1966) or obtained from the author.

[21] Careful adjustments were made to retain uniform metropolitan boundaries and to assure congruency of census tract boundaries (political redistricting and certain combinations and subdivisions of tracts had been made between census dates). For the final study, 206 congruent replicated subareas were retained. Given the large size of the city—two million in 1947, three and a half million in 1960—and the relatively small number of subdivisions, average size of census tract was large; furthermore, census tracts ranged widely in population, reducing statistical reliability. These defects, while recognized, could not be rectified. Absolute size was not included as a variable and all ratios were converted to standardized scores.

[22] This dominance of the first factor is not uncommon in principal axes factor solutions, but it should be noted that our results seem even more extreme than can be accounted for by the inherent bias of the method. We might note, in passing, that a Thurstone centroid extraction, based upon fifteen variables in 1947, had yielded a similarly dominant centroid factor, suggesting that, at least for the limited variables included in our study, and for a city such as Cairo, marked as it is by gross cultural variations, the major social differentiations reflected in ecological organization are almost *unidimensional*.

[23] The factor-score computational method is explained in Abu-Lughod (1966).

Table 1. Intercorrelation Among 13 Variables, 206 Census Tracts, Cairo

1947 (above the diagonal); 1960 (below the diagonal)

Variables	1	2	3	4	5	6	7	8	9	10	11**	12	13
1 Persons per Room		.22	.05	.60	-.65	-.32	-.42	.11	-.65	-.70	-.51	.47	.44
2 Density	.28		-.10	.24	-.18	.02	-.11	.07	-.09	-.17	-.18	.00	.13
3 Sex Ratio (15-49)	-.03	-.05		-.19	.09	.15	.55	-.10	.04	.18	.13	-.23	-.02
4 Fertility Ratio	.73	.18	.02		-.81	-.48	-.66	.22	-.62	-.82	-.48	.47	.55
5 Females (16+) Never Married	-.74	-.13	-.00	-.84		.42	.75	-.17	.81	.89	.63	-.59	-.51
6 Ever-Married Females (16+) Divorced	-.29	-.09	-.12	-.54	.44		.41	-.05	.36	.40	.23	-.50	.04
7 Males Never Married (16+)	-.52	-.05	.39	-.66	.78	.42		-.17	.68	.73	.55	-.59	-.33
8 Handicapped Rate	.50	.23	.02	.43	-.40	-.15	-.24		-.08	-.14	-.14	-.02	-.03
9 Male Literacy Rate	-.74	-.22	-.04	-.64	.78	.31	.58	-.51		.89	.71	-.70	-.45
10 Female Literacy Rate	-.84	-.23	.11	-.80	.90	.35	.71	-.50	.89		.87	-.67	-.64
11 Females Employed	-.85	-.28	.08	-.85	.81	.45	.60	-.52	.71	.87		-.49	-.35
12 Males Unemployed*	.43	.18	-.03	.39	-.37	-.19	-.23	.32	-.32	-.42	-.46		.24
13 Percentage Muslim	.51	.21	-.07	.47	-.47	.08	-.31	.17	-.37	-.50	-.45	.16	

Note: Correlation coefficients, computed to six decimal places, have been rounded for presentation only to the nearest one-hundredth. Complete figures were retained for computations based upon these matrices.
*The usual manual-nonmanual occupation index could not be computed from census tract cross-tabulations, since members of the labor force were subclassified partly by skill categories and partly by industry groupings.
**Females in school. See note 7 to Table 2.

a highly literate population having low rates of dependency and unemployment, and by the presence of resident domestic servants. These are economic aspects of a life-style which, in today's Cairo, is increasingly finding expression in more "modern" family patterns of female education, delayed age at marriage, and lower fertility. In addition, since foreign and indigenous Christians in Cairo enjoyed economic and social advantages under colonial conditions and, because of their greater "westernization," followed modern family patterns somewhat earlier, tracts in which they are well represented are among the highest scoring. At the opposite extreme are the tracts scoring lowest on Factor I. House overcrowding often reaches astronomical heights, most residents are illiterate, and females seldom attend school or commute to paid employment. Here the traditional patterns of female exclusion from the educational system mesh with marriage at a very young age and a post-nuptial life of high, sustained, and generally uncontrolled fertility.[24]

Factor II was interpreted as representing "male dominance," i.e., selective settlement of unattached migrant males. The spatial patterning of the Factor II scores confirmed this, since high scoring tracts were confined almost exclusively to areas around the central business district and its strip extensions (see Abu-Lughod, 1961), or to institutional quarters of male predominance.

Factor III was interpreted as reflecting "social disorganization" of the urban variety. High scoring tracts were located chiefly in the dense central slum quarters to which the physically handicapped, the unemployed or dubiously employed, and divorced women had gravitated. These forms of social dependency can rarely be tolerated in the more rural social organization found towards the periphery of the metropolitan area, even though the socio-economic level there may be as low as, or lower than, that found in the slums flanking the Western-style central business district.

For present purposes, however, the most significant finding is the close association between certain variables of family status and the variables of social rank. It is this to which we wish to address ourselves in the final section of this article.

V

Under what circumstances would one expect an ecological disassociation between family and socio-economic variables? Under what circumstances would it be reasonable to expect a close association between the two? Is there any relationship between these circumstances and the "scale" of society? Or, if the concept of scale is too broad to make the necessary distinctions, what other concepts relevant to the analysis of social change might more meaningfully be incorporated into a general theory of ecological organization? These critical questions, hitherto raised only in the most cursory fashion by earlier investigators, lie at the crux of a general comparative theory of urban social structure. To answer them one must return to the most elementary level of analysis.

First, under what circumstances would one expect to derive an independent socio-economic rank vector from the specific variables included in ecological studies, namely, occupation, education, etc.? If social rank were determined by criteria untapped by our measures (which are, after all, adapted to the analysis of "modern" social structures), and/or if residential segregation either did not occur at all or occurred according to principles *unrelated* to social rank, the vector would not be identified through the ecological method.[25] Therefore, for our measures to yield a vector of social rank, we require: (a) that the effective ranking system in a city be related to our operational definition; and (b) that the ranking system be manifested in residential segregation of persons of different ranks at a scale large enough to be picked up at our level of analysis (the census tract).

Second, under what circumstances would one expect to find a "familism" vector, either independent of or coalesced with a socio-economic vector? Consider a situation in which the extended family is so pervasive an element of urban life that various stages of the family cycle are played out within

---

[24]A more detailed description, a theoretical framework, and a discussion of the physical ecology of the city, together with a map based upon factor scores, can be found in Abu-Lughod (1969). The ecological correlations between socio-economic variables and the fertility ratio, apparent here on the macro level, have been confirmed on the non-ecological level as well. Age and education of the wife and, to a lesser extent, occupational rank of the husband were variables significantly related to the median number of children ever born within various duration-of-marriage classes in Cairo in 1960 (Abu-Lughod, 1965).

[25]So much for the claim that social area analysts can be concerned with social space *independent of* physical space. While it is true that social rank might still be a relevant category of social space even if it failed to be manifested in physical segregation, *this could not be picked up through an ecological analysis,* i.e., social area analysis, which depends upon spatially-grouped aggregates.

large stable households rather than isolated in sequential residential settings. Consider the theoretically conceivable situation where no differences between social classes, ethnic groups, or other critical subcategories in urban society exist with respect to fertility, family size, or the propensity to maintain extended families. And consider, further, a situation where housing facilities are so distributed (at random) throughout the city that no area is selectively attractive to special family types. In these cases, one would hardly expect to find a family vector in ecological analysis, for family types are relatively undifferentiated in social space, and urban subareas are relatively undifferentiated in physical space. Therefore, for our measures to identify a familism vector, we require: (a) that family types vary, either due to "natural" causes, such as those associated with sequential stages in the family cycle, or to "social" causes such as those associated with other divisions in society, whether ethnic, socio-economic, or other; and (b) that subareas within the city be differentiated in their attractiveness to families of different types.

Now let us look at the more complex problem. Granted a situation in which both family and social rank variables are reflected in the ecological pattern, under what circumstances would one expect these variables to lie on the same vector or to lie upon somewhat separate vectors? A neces-

Table 2. Rotated Factor Loadings, Cairo Census Tracts, 1947 and 1960*

Variables	Factor I 1947	Factor I 1960	Factor II 1947	Factor II 1960	Factor III 1947	Factor III 1960
Persons Per Room	-.80*	-.81*	.15	.01	.10	+.32*
Density (persons/sq.km.)	-.18	-.03	-.16	-.10	+.62*	+.72*
Sex Ratio[1]	-.00	-.01	+.96*	+.97*	-.05	-.03
Fertility Ratio[2]	-.81*	-.89*	-.20	-.01	+.20	-.00
Never-Married Females[3]	+.91*	+.95*	.12	.04	-.07	-.03
Females Divorced[4]	+.42*	+.52*	.19	.03	+.31*	+.41*
Never-Married Males[5]	+.67*	+.76*	+.62*	+.49*	.01	.11
Handicapped Rate[6]	-.08	-.40*	-.04	.13	+.68*	+.32*
Male Literacy Rate	+.90*	+.81*	.06	-.07	.12	-.20
Female Literacy Rate	+.95*	+.92*	.17	.10	-.03	-.19
Female Sch/Employed[7]	+.72*	+.85*	.13	.06	-.07	-.15
Males Unemployed[8]	-.68*	-.31*	-.28*	-.03	-.38*	.18
Percent Muslims	-.65*	-.56*	.02	-.06	.16	+.55*
Sum of the Squares	6.26	6.70	1.40	1.26	1.20	1.13
Variance Explained by Factor (%)	48	52	11	10	9	9

*Values ± .20 are considered significant.
[1] Males per 100 females in the migration-prone ages between 15-49.
[2] Children under 5 per 1000 women 15-49 (median age at marriage for Cairo females is under 20).
[3] With marriage almost universal, this measure of females 16 years of age and over who have not yet married is a sensitive indicator of the "usual" age at marriage in a census tract; the higher the rate, the older the age of the typical bride.
[4] As expressed: Currently divorced women as a percent of ever-married women 16 years and older. Remarriage for females is uncommon.
[5] This ratio reflects usual age at marriage, as well as selective single status of migrant groups; hence, it appears on both Factors I and II.
[6] This variable was poorly replicated in the two censuses. Reporting procedure was changed, a fact that did not become evident until the analysis was completed.
[7] In 1947 this measure was estimated from the labor-force category, "unproductively employed," and measured female school enrollment. No labor-force participation figure for females could be computed since, in the 1947 census, all housewives were tabulated as employed in domestic service, undifferentiated from domestic servants working for wages. The 1947 measure correlated almost perfectly with the 1960 measure of employed females (which fortunately no longer included housewives). Fuller defense of what appears to be a cavalier treatment of data appears in Abu-Lughod (1966).
[8] An inexplicably large drop in rates of unemployment between 1947 and 1960 creates suspicion here.

sary but not sufficient condition of the former is that social rank and family type be correlated, that the "typical" family of one social class be significantly different from the "typical" family of another class.[26] Conversely, a necessary but not sufficient condition of the latter is that there be no major overpowering linkages between class on the one hand, and such familism variables as fertility, completed family size, and tendency to remain within extended households on the other hand. These conditions exist within the realm of social space. They are necessary but not sufficient because they fail to take into account the character of physical space. Thus, even given some relationship between class and family type, it would still be possible to have "familism" variables somewhat disassociated from socio-economic rank variables if two further conditions were present: (1) if stages in the family cycle were clearly distinguished from one another, each stage being associated with a change in residence; and (2) if sufficiently large subareas within the city offered, at all economic levels, highly specialized housing accommodations suited to families at particular points in their natural cycle of growth and decline. A third condition is also assumed, namely, cultural values permitting and favoring mobility to maximize housing efficiency, unencumbered by the "unnatural" frictions of sentiment, local attachments, or restrictive regulations.

Taking the above propositions as our frame of reference, we can now return to the substantive findings reported in earlier sections of this article. The disassociation between social rank and familism variables found in contemporary Western cities in societies at the terminal stages of the demographic transition can be attributed to the reinforcing and cumulative effects of several conditions that "define" the nature of urban organization in such cities: (1) residential segregation according to modern ranking systems, (2) relatively low correlations between social rank and differences in fertility and family styles, (3) high differentiation of residential subareas by housing types, (4) mobility, and (5) predominance of independent households. To the extent that these conditions are not perfectly fulfilled, the vectors will not be totally disassociated. For example, fertility differentials by social class have not completely disappeared, even in cities of highly industrialized Western countries; special house types are available, but not on a basis of equality for different economic groups; residential segregation by class is not total, despite zoning attempts to achieve homogeneity; and even for the most "rational industrialized man," sentiments and inertia, as well as law, inhibit frictionless mobility. These prevent a total disassociation between socio-economic and familism vectors.

The same set of variables accounts for the close association between familism and socio-economic rank in Cairo. While there is a marked degree of residential segregation by rank in that city, there is also an increasingly pronounced relationship between rank and family style (patterns of fertility, completed family size, and differential tendency to form nuclear units). But thus far no clear differentiation of neighborhoods with respect to housing accommodations is noticeable. Suburbanization is minimal; most dwellings are in multi-family structures and dwelling-unit density is tied to economic constraints more than to distance from the center. Therefore, segregation by family type in the city occurs largely as a by-product of economic segregation.

Between these two extremes is the city of Rome, characterized by residential specialization by social rank with some association (albeit of declining significance) between class and fertility, and very little differentiation of residential areas by housing types. Whatever minor significance family differentiation has in shaping the urban pattern, then, is attributable more to social rank than to the independent force generated by family-cycle-linked housing specialization. The two vectors, then, are naturally in closer position (the variables more correlated), although one can predict that they will push farther apart as fertility differentials by social class continue to narrow and residential neighborhoods become more specialized by housing accommodations.

Another intermediary type which can be explained by the same set of factors is the American Southern city with a large non-white minority. The peculiar location in social space of the non-white subgroup (lower in socio-economic rank but higher in fertility) creates an association between social rank and fertility possibly similar to that apparent in the early stages of the demographic transition. This, coupled with lesser segregation by family-cycle stage, especially for residentially restricted

[26] We are not prejudging the causal sequence underlying the correlation. In a rigidly age-graded society the association between "class" and "family type" might be due to parallel status and family-stage variables linked to advancing age. Conversely, in a society deep within the demographic transition, the association would clearly be attributable to class-linked fertility and family-style variations.

non-whites, accounts for the strong association between rank and family vectors.

This association between ethnicity and fertility, hinted at but obscured in Southern cities, is more clearly apparent in the African cities of Accra and Abidjan, which have been subjected to social area analysis. Several salient findings need to be explained by our theory. First, although four dimensions were identified, namely, family, rank, ethnicity, and migration status, only the last three were considered to be relatively independent. Family variables were slightly related to rank and migration status but much more strongly related to ethnicity. Second, the zero-order intercorrelations of variables were lower than have been found elsewhere. What accounts for these findings?

We would suggest that the African cities studied by McElrath (1968) and Clignet (1969) are still ecologically structured according to the pre-industrial principle of ethnicity, linked by endogamy to extended kinship networks. This, combined with the yet weak influence of "modern" ranking criteria, makes the usual measures of rank (occupation and literacy) somewhat irrelevant. Low ecological correlations are the result. Furthermore, the societies are still too early in the demographic transition for clear fertility differentials by class to have become pervasive enough to be picked up by the gross method of ecological analysis. (Individual correlations are undoubtedly higher.) Family differences are still linked to cultural variations between ethnic or tribal groups which, in the premodern period, control family type variations not through conception or gestation variables but through the intercourse variables of mating. (The distinctions are Davis and Blake's, 1956). When these factors are combined with the low level of housing-type differentiation also found throughout these cities and with the residual elements of extended households also present, the result is weak vectors of residential differentiation, as measured by the indices that have proven powerful discriminators in Western industrial societies. Factor analysis using a larger battery of variables should permit the development of an analytical schemata better suited to these cities than the conventional methods of social area analysis have provided.

In the preceding discussion we have not used the concept of increasing scale directly for our explanations. This does not necessarily mean that the concept is irrelevant. We suggest that scale represents a shorthand compression of several factors we find valuable to separate. In that sense it has the same assets and defects as the concept of "modernization," which more and more is being broken down into its constituent elements when detailed comparisons between "modernizing countries" are made. At some future time we may have sufficient knowledge to move to the next level of abstraction represented by the theory of increasing scale, but I would suggest that we are not yet ready to do so. Working with the three variables of (1) degree of residential segregation by "modern" criteria of social rank; (2) the degree of correlation between rank and family variables with relation to the demographic transition; and (3) the extent of residential and family type specialization, as related to housing-land use on the one hand and the isolation of discrete stages in the family cycle on the other, appears to be a necessary preliminary. Using these variables in controlled combinations, it would be possible to design an experiment, to be tested by the method of factor analysis, to measure the influence of each upon the ecological structures of a widely diverse sample of cities. Herein may lie the hope of developing a detailed theory of comparative urban ecology.

References

Abu-Lughod, Janet. "Migrant Adjustment to City Life: The Egyptian Case." *American Journal of Sociology,* 67 (July 1961):22–32.

―――― "The Emergence of Differential Fertility in Urban Egypt." *Milbank Memorial Fund Quarterly,* 43 (April 1965):235–53.

―――― "The Ecology of Cairo, Egypt: A Comparative Study Using Factor Analysis." Ph.D. dissertation, University of Massachusetts, 1966.

―――― "The City Is Dead—Long Live the City: Some Thoughts on Urbanity." *CPDR Monograph Series, No. 12.* Berkeley: University of California Center for Planning and Development Research, 1968.

―――― "Varieties of Urban Experience: Contrast, Coexistence and Coalescence in Cairo." In Ira Lapidus (ed.), *Middle Eastern Cities,* Berkeley and Los Angeles: University of California Press, 1969.

Abu-Lughod, Janet, and Attiya, Ezz El-Din. *Cairo Fact Book (Arabic and English Maps and Tables).* Cairo: Social Research Center, American University at Cairo, 1963

Anderson, Theodore, and Bean, Lee. "The Shevky–Bell Social Areas: Confirmation of Results and a Reinterpretation." *Social Forces,* 40 (December, 1961):119–24.

Anderson, Theodore, and Egeland, Janice. "Spatial Aspects of Social Area Analysis." *American Sociological Review,* 26 (June 1961):392–98.

Bell, Wendell. "A Comparative Study in the Methodology of Urban Analysis." Ph.D. dissertation: University of California at Los Angeles, 1952.

———. "The Social Areas of the San Francisco Bay Region." *American Sociology Review*, 18 (February 1953):39–47.

———. "Economic, Family and Ethnic Status: An Empirical Test." *American Sociological Review*, 20 (February, 1955):45–52.

———. "Social Areas: Typology of Urban Neighborhoods." In Marvin Sussman (ed.), *Community Structure and Analysis*, p. 61–91. New York: Thomas Y. Crowell, 1959.

———. "Urban Neighborhoods and Informal Social Relations." In Muzafer and Caroline Sherif (eds.), *Problems of Youth*, p. 235–64. Chicago: Aldine Press, 1965.

Berry, Brian, and Murdie, Robert. *Socioeconomic Correlates of Housing Condition*. Toronto: Metropolitan Planning Board, 1965.

Berry, Brian, and Rees, Phillip. "The Factorial Ecology of Calcutta." Unpublished mimeographed paper. Chicago: Department of Geography, University of Chicago, 1968.

Beshers, James. "Census Tract Data and Social Structure: A Methodological Analysis." Ph.D. dissertation, University of North Carolina, 1957.

———. "The Construction of 'Social Area' Indices: An Evaluation of Procedures." *Proceedings of the Social Statistics Section, American Statistical Association*, pp.65–70. Washington, D.C.: 1959.

Burgess, Ernest. "The Growth of a City: An Introduction to a Research Project." *Proceedings of the American Sociological Society*, 18 (1923):85–97.

Clignet, Remi, and Sween, Joyce. "Accra and Abidjan: A Comparative Examination of the Theory of Increase in Scale." *Urban Affairs Quarterly*, 4 (1969):297–324.

Davis, Kingsley, and Blake, Judith. "Social Structure and Fertility: An Analytical Framework." *Economic Development and Cultural Change*, 4 (April 1956):211–35.

Duncan, Beverly. "Devolution of an Empirical Generalization." *American Sociological Review*, 29 (December 1964):855–62.

Gagnon, Gabriel. "Les Zones Sociales de l'agglomeration de Quebec." *Recherches Sociographiques*, 1 (July–September 1960).

Gittus, Elizabeth. "The Structure of Urban Areas: A New Approach." *Town Planning Review*, 35 (April 1964):5–20.

———. "An Experiment in the Definition of Urban Sub-Areas." *Transactions of the Bartlett Society*, 2 (1965):109–35.

Goheen, Peter. "The North American Industrial City in the Late Nineteenth Century: The Case of Toronto." Chicago: University of Chicago Department of Geography Research Paper, 1969.

Hagood, Margaret; Danilevsky, Nadia; and Beum, Corlin. "An Examination of the Use of Factor Analysis in the Problem of Subregional Delineation." Presented at Conference on the Analysis of Social and Economic Data, University of North Carolina, June 1941. *Rural Sociology*, 6 (September 1941):216–33.

Hatt, Paul. "The Concept of Natural Area." *American Sociolgoical Review*, 2 (August 1946):423–27.

Hawley, Amos, and Duncan, Otis Dudley. "Social Area Analysis: A Critical Appraisal." *Land Economics*, 33 (November 1957):337–45.

Herbert, D. T. "Social Area Analysis: A British Study." *Urban Studies*, 4 (February 1967):41–60.

Hyderabad Metropolitan Research Project. *Social Area Analysis of Metropolitan Hyderabad*. Hyderabad: Osmania University, 1966.

Kaufman, Walter. "Social Area Analysis: An Explication of Theory, Methodology and Techniques, with Statistical Tests of Revised Procedures, San Francisco and Chicago, 1950." Ph.D. dissertation, Northwestern University, 1961.

Lieffer, Murray. "A Method for Determining Local Urban Community Boundaries." *Proceedings of the American Sociological Society, 26 (1933):137–43*.

MacCannell, Earle. "An Application of Urban Typology by Cluster Analysis to the Ecology of Ten Cities." Ph.D. dissertation, University of Washington, 1957.

McElrath, Dennis. "The Social Areas of Rome: A Comparative Analysis." *American Sociological Review*, 27 (June 1962):376–91.

———. "Social Differentiation and Societal Scale." in Scott Greer et al. (eds.), *The New Urbanization*, pp. 39–51. New York: St. Martin's Press, 1968.

Moush, Edward; Scrivens; and Aresing. *Social Area Analysis of Cleveland and Metropolitan Area, 1950*. Cleveland: Le Play Research, Inc., 1960.

Murdie, Robert A. *The Factorial Ecology of Metropolitan Toronto, 1951–1961: An Essay on the Social Geography of the City*. Department of Geography Research Paper No. 116. Chicago: University of Chicago, 1968.

Pedersen, Poul O. *Modeller for befolkningsstruktur og befolkningsudvikling i storbyområderspecielt med henblik på Storkøbenhavn*. Danish text with English summary. Copenhagen: State Planning Institute, 1967.

Polk, Kenneth. "The Social Areas of San Diego." Master's thesis, Northwestern University, 1957.

Rees, Philip. "The Factorial Ecology of Metropolitan Chicago." In Berry and Horton (eds.), *Geographic Perspectives on Urban Systems*. Englewood Cliffs, N.J.: Prentice-Hall, 1969.

Robson, Brian. *Urban Analysis: The Social Ecology of Sunderland*. Cambridge: Cambridge University Press, 1969.

Schmid, Calvin. "Generalizations Concerning the Ecology of the American City." *American Sociological Review*, 15 (April 1950):264–81.

Schmid, Calvin; MacCannell, Earle; and Van Arsdol, Maurice, Jr. "The Ecology of the American City: Further Comparisons and Validation of Generalizations." *American Sociological Review,* 23 (August 1958):391–401.

Schmid, Calvin, and Tagashira, K. "Ecological and Demographic Indices: A Methodological Analysis." *Demography,* 1 (1964):194–211.

Shevky, Eshref, and Bell, Wendell. *Social Area Analysis: Theory, Illustrative Application and Computational Procedures.* Stanford University Series in Sociology, No. 1. Stanford: Stanford University Press, 1955.

Shevky, Eshref, and Williams, Marilyn. *The Social Areas of Los Angeles: Analysis and Typology.* Los Angeles and Berkeley: University of California Press, 1949.

Sullivan, Terence. "The Application of Shevky–Bell Indices To Parish Analysis." *American Catholic Sociological Review,* 12 (Summer 1961): 168–71.

Sweetser, Frank. "Factor Structure as Ecological Structure in Helsinki and Boston." *Acta Sociologica,* 8 (1965a):205–25.

———. "Factorial Ecology: Helsinki, 1960." *Demography, 2 (1965b):372–86.*

Timms, Duncan W. G. *Social Area Analysis in Brisbane.* Brisbane: University of Queensland Press, forthcoming.

Tryon, Robert C. *Cluster Analysis: Correlation Profile and Orthometric (Factor) Analysis for the Isolation of Unities in Mind and Personality.* Ann Arbor: Edwards Brothers Press, 1939.

———. *Identification of Social Areas by Cluster Analysis: A General Method with an Application to the San Francisco Bay Area.* University of California Publications in Psychology, 8 (1). Berkeley and Los Angeles: University of California Press, 1955.

Udry, J. Richard. "Increasing Scale and Spatial Differentiation: New Tests of Two Theories from Shevky and Bell." *Social Forces,* 42 (May 1964): 403–13.

Van Arsdol, Maurice, Jr. "An Empirical Evaluation of Social Area Analysis in Human Ecology." Ph.D. dissertation, University of Washington, 1957.

Van Arsdol, Maurice, Jr.; Carmilleri, Santo; and Schmid, Calvin. "The Generality of Urban Social Area Indexes." *American Sociological Review,* 23 (June 1958):277–84.

Part III Urban Patterns in Different Cultural Settings

Introduction

Studies of urban ecological structure in cultural settings other than that of the United States enable us to test the generalizability of principles derived in studies made in this country. Further, these studies may permit us to determine, in cases where these principles are found not to hold universally, the conditions under which they are applicable. From the large number of studies of cities around the world that are now available, we have attempted to include here a selection reflecting a variety of cultural settings, research techniques and orientations, and dealing with issues of significance in urban ecology.

A number of studies have dealt with the ecological structure of Latin–American cities and have found a traditional Spanish pattern very different from that of cities in the United States. As early as 1934, Hansen studied the ecology of Merida in Yucatan. He found a traditional pattern which had existed from the sixteenth to the late nineteenth century, in which the plaza was the center of the city and the center of the area of highest prestige. The cathedral and government buildings adjoined the plaza. Nearby was the retail business district, and in an adjacent area were the homes of the upper class. As one proceeded from the center of the city outward, socioeconomic level declined, the reverse of the classical ecological pattern of large modern cities in the United States. The area surrounding the center of the city was divided into five semiautonomous districts, called barrios, each with considerable community cohesion and organization. Each barrio had its own plaza. This pattern has been changing in Merida, and it has been becoming more like the typical pattern found in the United States. This change has accompanied growing outside contact and industrialization.[1] A similar pattern in Mexico City was reported by Hayner in 1945. For centuries Mexico City grew slowly, and therefore a zone of transition near the central business district never developed. As in the case of Merida, residential desirability declined with distance from the central plaza. In recent years with the growth of population, industry, and transportation, the pattern has been shifting to one like that found in the United States. Hayner reported, however, that in the town of Oaxaca, which had been fairly inaccessible until recently, the traditional Spanish–American pattern still prevailed.[2] Essentially the same situation was reported by Hawthorn and Hawthorn in 1948 for Sucre, Bolivia. The traditional pattern placed the upper class around a central plaza, but there, too, recent changes have been taking place. However, in Sucre the new pattern had not yet completely prevailed, and a mixed pattern was found with lower-class families residing on almost every block in the city.[3]

In the first paper presented in Part III, Caplow reviews in some detail these and other studies of the ecology of Spanish-American communities. He then presents a detailed ecological study of Guatemala City, discussing first its historical development. He shows that Guatemala City did not sim-

---

[1]Asael T. Hansen, "The Ecology of a Latin American City," in E. B. Reuter (ed.), *Race and Culture Contacts* (New York: McGraw–Hill Book Co., 1934), pp. 124–42.

[2]Norman S. Hayner, "Mexico City: Its Growth and Configuration," *The American Journal of Sociology*, 50 (January 1945): 295–304.

[3]Harry B. Hawthorn and Audrey E. Hawthorn, "The Shape of a City: Some Observations on Sucre, Bolivia," *Sociology and Social Research*, 33 (November–December 1948): 87–91.

ply grow on its own through the operation of "natural" forces, but from an early date was planned in accordance with the traditional Spanish–American pattern described above. Thus in the ecological structure of Guatemala City were found the central plaza surrounded by important buildings, the desirability of residence near this plaza, the location of slums on the outskirts of the city, and the division of the city into barrios, each with its own smaller plaza. Caplow also reports certain changes that have occurred, including a decline in the autonomy of the barrios. However, other traditional patterns were found to have a strong tenacity. Despite the growth of certain foreign-type wealthy suburbs, many wealthy families, especially the older upper-class families, maintained a central location. Crime was found to increase with distance from the center of the city; and no process of invasion and succession has existed in Guatemala City, due to a marked tendency to maintain traditional land-use patterns. Finally, Caplow discusses the factors responsible for the maintenance of an ecological pattern in Spanish–America different from that which is characteristic of the United States.

London reviews the literature on Latin–American cities that has been published since Caplow's article on Guatemala City appeared thirty years ago, and reports that in the major cities the traditional Spanish pattern no longer prevails in its original form. However, rather than a prototype of the Burgess pattern, typically a mixed pattern has emerged. Emphasizing the processual nature of urban ecology, London concludes that it is not possible at this time to determine whether Latin–American cities are evolving toward the Burgess model or whether a new ecological form is emerging.

Thus throughout Spanish–America there existed a traditional ecological pattern very different from that found in the United States. That this pattern was due to a definite Spanish colonial policy, may be demonstrated by the differences between Spanish and Indian towns. McBryde, in a study of Guatemala, found that when a sizable number of Ladinos (of mixed Indian and Spanish descent, but considering themselves Spanish) is present, a settlement will take the form of a centralized town, while without many Ladinos an even larger Indian population aggregate will form only an agricultural settlement.[4] The next paper

[4]Felix W. McBryde, *Cultural and Historical Geography of Southwest Guatemala,* Smithsonian Institution, Institute of Social Anthropology, Publication No. 4 (Washington, D.C.: United States Government Printing Office, 1947), p. 85.

consists of a selection from a study made by Stanislawski of eleven towns in the state of Michoacán in Mexico. The selection includes a detailed description of an Indian town, Pichátaro, and then presents the conclusions of the study. Stanislawski found a striking difference between Spanish and Indian towns. The Spanish towns follow the pattern described above. Stanislawski reports an additional interesting spatial distribution in these towns based on occupational prestige. High prestige occupations, such as retail merchants, mule drivers, and leatherworkers, tend toward central locations on main arteries, while low prestige occupations, such as nonleather crafts, are in peripheral locations or in blind alleys. In Indian towns, in contrast, the plaza is unimportant and no spatial distribution of prestige is found.

While the spatial distribution of prestige may not be characteristic of American–Indian towns, it is found in cities throughout the world, and traditionally many cities have followed the Spanish colonial pattern. In the selection by Sjoberg we find that the basic elements of the traditional Spanish-American pattern are typical of preindustrial cities in many periods of history and parts of the world. Sjoberg describes the preindustrial socioeconomic distribution as the inverse of the Burgess model, with the dwellings of the wealthy centrally located and the poorest elements on the city's fringe. The center of the preindustrial city usually houses important political and religious edifices which dominate the horizon and overshadow surrounding commercial activities. Frequently, too, the city is divided into quarters inhabited by occupational and ethnic groups. Sjoberg's analysis of the ecology of the preindustrial city has been very influential, and has come to be seen by many writers as a polar type in contrast to the Burgess model.

Having considered the implications of lack of industrialization for urban ecology, in the next piece Bulliet calls our attention to the impact on urban patterns of the absence of the wheel. In a short but very interesting selection, we read of the ecological consequences in the cities of North Africa and the Middle East that followed from the use of pack camels rather than wheeled vehicles for transportation.

Returning to the ecology of the preindustrial city in general, Abbott provides an illustration of a test case of the Sjoberg model. Using census data from 1897, when it was still essentially preindustrial, Abbott analyzes the residential patterns of Moscow. He finds that the socioeconomic status of residents tends to decline with distance from the city center, as do literacy and educational level. Measures of disorganization, such as rates of physical defects, emotional disorders, divorce, and infant mortality, increase from the center of the city to the periphery. Thus, Moscow in 1897 fits the Sjoberg model of a preindustrial city.

In the following article, Morris and Pyle present a picture of a city transitional between a traditional and industrial ecological structure. Rio de Janeiro in 1960 still showed a persistence of the traditional pattern in which centrality is associated with higher socioeconomic status, but signs of change in this pattern are also evident. Morris and Pyle's study of Rio de Janeiro, however, goes beyond consideration of this question. Using factorial ecology, they give us a detailed look at the social areas of the city. Particularly interesting is the comparison of the characteristics of the zonas (established wards) and the favelas (shantytowns).

Eyre focuses our full attention on shantytowns in the next article. He defines the shantytown as a specific type of urban settlement, widespread in tropical and subtropical lands, and not simply synonymous with slums. Their chief characteristic is the absence of an official subdivision or plan for housing in the area before the dwellings are constructed and usually the absence of any legal title to their dwellings by the occupants. They are associated with rapid urban growth in which the demand for cheap housing outstrips the supply, a common condition in developing nations. Eyre analyzes shantytowns, specifically those of Montego Bay, Jamaica, in terms of their spatial organization, developmental stages, demographic structure, economic activity, use of savings and investments, inadequacy of public amenities, degree of population control, and need for security. He suggests there is a short-sightedness and a lack of understanding on the part of those who out-of-hand condemn all aspects of shantytown living and seek to eradicate this phenomenon with a bulldozer.

Next, we turn our attention to western Europe, considering first Caplow's study of urban structure in France. Caplow emphasizes the importance of quartiers—local communities with strong identification and distinct customs—in both large and small cities. He also found a number of other ways in which the French city differs from the typical pattern of the United States, including: less centralization, no increase in centralization with increased city size, a different pattern of daily population movement—no convergence at a central point, no conspicuous pattern of invasion and succession, no regular association of social class with spatial distribution, and no uniform tendency for density of population to decrease toward the periphery of the city. Caplow also found, as Firey did in Boston, that land values were not all-important in determining urban patterns, but that historic parks, palaces, etc., were able to resist commercial pressures. Caplow's study illustrates that in France strong traditional values may interfere with the operation of ecological processes based on a primacy of economic values. It should also be noted, however, that France has had a low population growth, and her cities have not been rapidly expanding.

To provide a brief overview of trends in Europe today, London surveys recent ecological research on European cities, particularly focusing on the contrasting Sjoberg and Burgess models. He notes evidence of what may be a called a "post-Burgess" pattern in some western European cities, characterized by the upgrading ("gentrification") of some inner city areas; while eastern European cities seem to be evolving from a Sjoberg to a Burgess pattern.

Mukerjee's description of traditional communities in India again shows the important part values may play in determining spatial structure. In the development of the compact village as contrasted with the development of the scattered farmstead type of settlement, Mukerjee shows the interplay of geographic factors and cultural values. The reader will undoubtedly notice that in this article Mukerjee gives considerable emphasis to the positive functions of the caste system (although in other writings he has recognized its negative aspects). However, from the ecological point of view Mukerjee's discussion of caste is of interest primarily in showing the importance of religious values in determining village structure. Mukerjee describes how villages are divided into distinct caste areas, each with a sacred grove, a shrine, and a place for the residence of the headman. Moreover, where caste structure is strict the highest castes tend to be located near the village temple and ceremonial bathing tank and the lowest castes on the outskirts of the village. In India, religion has affected ecological structure not only through the impact of caste, but also through the use of mystical symbols in town planning. Mukerjee presents a most interesting description of the ancient Indian tradition of planning villages and towns according to the form of supernatural figures and mystical symbols. Finally, Mukerjee tells us of the breakdown of traditional patterns in the rapidly growing, Western-type industrial cities. There have been a few industrial cities in which there has been an attempt to maintain some of the traditional patterns.

Berry and Spodek provide another persepective on ecological structure in India. While briefly surveying ancient literature on the spatial structure of communities and noting residence patterns in villages and small towns, their primary concern is the ecology of the large Indian city of today. Six large Indian cities are studied by means of factorial ecology, using data from 1961, and in addition one of the cities is surveyed historically (from 1822 to 1954). The major change found by Berry and Spodek is that socioeconomic status has become the dominant factor in the ecological structure of Indian cities, outweighing caste in importance. However, centrality still tends to be associated with higher status, and traditional elements show strong persistence. The authors analyze the complex interplay of traditional and modern influences on the structure of urban ecology in India today.

Returning to the United States—Lind's study of Honolulu finds a basic ecological pattern of community disorganization similar to that which characterizes other large American cities. Highest disorganization is found in areas of transition, adjacent to the central business district. However, the pattern of suicides shows certain interesting deviations from this typical distribution. Slum areas inhabited by Portuguese and Hawaiians have a very low suicide rate, while areas inhabited by Japanese, Chinese, and Koreans have disproportionately high rates, showing the effect of different cultural attitudes toward suicide. Lind also shows how cultural influences affect the distribution of juvenile delinquency and vice, altering to some extent the typical pattern. Thus we find that Honolulu follows the general American ecological pattern except where specific cultural characteristics intervene.

Since Lind's original study of Honolulu was basically analytic rather than merely descriptive, it remains significant today despite the passage of many years. Nevertheless, it is of interest to know what changes have occurred in the ecological structure of Honolulu in the past fifty years and whether the patterns of disorganization found earlier still persist. In the final paper in Part III, an addendum to his earlier study, Lind gives us an overview of the situation today. The biggest change, he reports, has been a tremendous population explosion, both in permanent residents and visitors. A second major development has been the assimilation of ethnic groups. Lind discusses how these two developments have contributed to changes in the ecology of community disorganization in Honolulu.

From the studies we have considered of communities in various parts of the world, two opposing models appear to emerge: the Sjoberg model of the preindustrial city, and the Burgess model of the large, rapidly growing industrial city. It should be kept in mind that these are theoretic models (in Weber's sense, "ideal types") from which concrete cities will necessarily deviate. Many cities today are in an in-between stage, with old patterns still strong while new ones are emerging. Moreover, as we have seen, even in the most industrialized setting, strongly held values and traditions may interfere with the operation of economically based principles, and divergences from the general Burgess pattern may be found. Furthermore, it cannot be assumed that the Burgess pattern is necessarily the final form toward which industrialized cities will evolve. The "gentrification" of central areas of Western cities has been noted—a process encouraged by factors such as urban renewal programs, the increased cost of commuting, and the cultural and educational attractions of the central cities. It is not possible to predict how far this process will go or what new patterns may emerge.

Finally, it is important to reemphasize the point made in the first edition of this book that a specific pattern of spatial distribution should not be confused with the principles of urban ecology. Human ecology is not merely a descriptive, but is an analytic, discipline. Urban spatial arrangements undoubtedly will continue to change as they have changed in the past. It is the task of urban ecologists, using all the theoretic and research tools available, to define the significance of these changing urban spatial patterns in understanding the social, cultural, and economic organization of communities.

The Social Ecology of Guatemala City*[1]

Theodore Caplow

I

The sociological literature on Latin-American cities is still very scanty. Both the urban ecologists of the United States and the urban geographers who pursue their parallel efforts in Europe have until very recently confined their attention to the more industrialized parts of the world. There has been, of course, enormous progress in these fields within the last few decades. Considerable contributions have been made to "political arithmetic," and there has developed an impressive body of theory concerned with both social behavior and spatial patterns. Nevertheless, urban theory has often confused local peculiarities with universal factors. The realization of this has led to increasing interest in the urban patterns of areas outside the immediate orbit of nineteenth century industrialism. For the North American student, the most important of these—from the standpoint of intrinsic interest and comparability—are the urban communities of Latin America.

A special advantage of Guatemala as a field for this type of research is that the culture pattern and social organization of its rural communities—the necessary background for urban research—have been brilliantly documented in recent years by Redfield, Gillin, Tax, Tumin, Lincoln, Davidson, McBryde, and others.[2] Among their specific emphases has been the impact of urban traits on the indigenous culture; but the urban milieu in which they arise has not been described in detail.

The first ecological study in Middle America was Hansen's study of Merida in Yucatan.[3] Some

---

*Reprinted from *Social Forces*, 28 (December 1949), 113–35, by permission of the author and The University of North Carolina Press. (Sections of pages 113, 118, 119, 121 and 134, and page 135 were omitted.)

[1] The preliminary field study, June-September 1948, was supported by a grant from the research funds of the Graduate School, University of Minnesota.

[2] See, for example: Robert Redfield, "Ethnographic materials on Agua Escondida," *Microfilm Collection of Manuscripts in Middle American Cultural Anthropology* (MCMMACA), University of Chicago Libraries, 1945, #3; John Gillin, "Race Relations Without Conflict: A Guatemalan Town," *American Journal of Sociology*, 53 (1948), 337–43; "Houses, Food and the Contact of Cultures in a Guatemalan Town," *Acta Americana*, 1 (1943), 344–59; "Parallel Cultures and the Inhibitions to Acculturation in a Guatemalan Community," *Social Forces*, 24 (1945), 1–14; Sol Tax, "The Municipios of the Midwestern Highlands of Guatemala," *American Anthropologist*, 39, #3, 423–44; Melvin M. Tumin, "San Luis Jilotepeque: A Guatemalan Pueblo," MCMMACA, #2; J. L. Lincoln, "An Ethnological Study of the Ixil Indians of the Guatemalan Highlands," MCMMACA, #1; William Davidson, "Rural Latin American Culture," *Social Forces*, 25 (1947), 249–52; Felix Webster McBryde, *Cultural and Historical Geography of Southwest Guatemala*, Smithsonian Institution, Institute of Social Anthropology, Publication #4, 1948.

[3] The complete study has not yet been published. It is found in part in Asael T. Hansen, "The Ecology of a Latin American City," Chapter VIII of *Race and Culture*

of the salient points in his analysis should be noted here. A distinction is made between the traditional ecological pattern which persisted with relatively little change from early in the sixteenth to late in the nineteenth century and an urban structure approaching that of northern cities to which the traditional pattern is now giving way.

This newer pattern is marked by the growth of upper-class residential suburbs and the displacement of upper-class families away from the urban center by increasing intra-urban mobility; the lessening of locality identifications; the increasing concentration of business and, to a lesser extent, of industry; the development of sub-communities based on land prices and common employment rather than upon the traditional status arrangements. Hansen sees the disintegration of the stable societal arrangements which underlay the earlier pattern as the basis for these changes. Redfield comments that "It [Merida] stands out among the communities of Yucatan as the place where the old culture has suffered the greatest amount of disorganization and where new ways of life, borrowed from other urban societies or developed under the stimulus of its own urban conditions, are most in evidence."[4]

The older pattern involved what has come to be known as "the plaza plan." The geographic and social center of the community was an open square, surrounded by the city hall, the state government building, and the cathedral. The municipal market and such stores and offices as existed at that time were concentrated in an adjacent area. In every other direction to the depth of a few blocks were the dwellings of the upper class. Blocks were laid out on a uniform gridiron.

In the conventional pattern, the gradients of status typically found in North American cities were literally reversed. In general, the social and economic status of the population declined from the center toward the periphery.

At points less than a mile from the center were five other squares, each the social hub of a lower-class community (barrio) which retained considerable community organization and isolation in the earlier period. The functional separation of the center from the barrios was a basic feature of the traditional pattern. The center was characterized by smooth streets, the concentration of municipal and cultural institutions, and the large one-story patio houses of the wealthy. In the barrios, the typical dwelling was an oval thatch-roofed hut, of a type dating back to the days when the barrios were close-knit Indian settlements clustered about the area of Spanish residence. The life of barrio-dwellers tended to be concentrated within the district, local identifications were taken seriously, and each barrio was somewhat hostile to the others.

Currently, as well as in past generations, the city is marked by an extreme concentration of political power, literacy, educational institutions, formal art, and prestige. The wealthy, the powerful, and the educated all tend to remain in the city. The few families of entirely European ancestry, including both the old aristocracy and more recent immigrants, are heavily concentrated in the city. A much wider range of occupations is found in the city than in the rest of the Yucatan, together with the marked concentration of skills.

In Mexico City, where commercialization and industrialization have proceeded very much further, Hayner finds the process of pattern change advanced to the point where "One wonders whether under the influence of increasing population and modern means of communication and transportation, all other large Latin-American cities are assuming an ecological pattern similar to that of cities in the United States."[5] The original configuration of Mexico appears to have been quite similar to that of Merida. Hayner notes the limitation of the central area to the radius of easy walking distance from the plaza and its relation to the formerly universal practice of returning home during the siesta period. He makes the interesting comment that, in a city which is not growing, there is no zone in transition and points out that the development of the vecindad—the one- or two-story patio-centered tenement so characteristic of Mexico City—was associated with the rapid decline of upper-class residential neighborhoods in the central district.

The shift toward a highly dynamic structure based upon economic centralization is far from complete, however. Although the land value gradients are quite North American in general trend and the patio-centered single-family residence is almost obsolete, the four worst slums in

---

Contacts, ed. E. B. Reuter (New York: McGraw-Hill, 1934), and in Robert Redfield, *The Folk Culture of Yucatan* (Chicago: University of Chicago Press, 1941), especially pp. 19–35.

[4]Ibid., p. 35.

[5]Norman S. Hayner, "Mexico City: Its Growth and Configuration," *American Journal of Sociology,* 50 (January 1945), 295–304.

1941 were all peripheral and brand new.⁶ Thus, the old public market, still centrally located, expanded side by side with the one-price retail shops stocked with imported merchandise. While the upper class has been completely displaced from the center, the slums may still be found in areas that have been characterized by lower-class housing for centuries. The growth of the city in 1930–1940 was more rapid in the central areas than on the outskirts. The concentration of skills, literacy, and leadership in the capital is still extraordinarily high. The extreme sex ratio (83.8 in 1940) is also suggestive of the older pattern. Thus, Mexico City illustrates the persistence of many elements of the stable ecological structure of an earlier day even under the impact of rapid and unregulated growth.

Hayner has also analyzed the ecology of Oaxaca, an isolated provincial city of 30,000 in southern Mexico.⁷ Distinguishing between privileged and underprivileged families, he found that the privileged families are still located in the center within walking distance of the Zocalo while the underprivileged are concentrated on the periphery. The situation in Oaxaca is complicated, however, by the special circumstances of the Revolution, the mass emigration of the upper class, the presence of a foreign colony, and shifts in the exterior road net. In this connection, it is interesting to note that, "Social change in Oaxaca has varied markedly in the different social classes. To some extent the physical pattern has outlasted the set of values from which it arose."

Even in very small communities, the basic forms of the pattern may be found. Gillin has shown, for San Luis Jilotopeque, the tendency for Ladino families to concentrate toward the plaza and, what is equally important in Guatemala, the preference shown for corner locations.⁸ The latter preference in some cases leads to marked diversity in the populations of small areas, as in Quezaltenango where very rich and very poor families sometimes occupy the same block.

McBryde has described the settlement patterns of the western highland region of Lake Atitlán in great detail.⁹ He points out that, "With a fairly large Ladino nucleus, a sizable center of population generally assumes the character of a town, even though Indians may greatly predominate. Without many Ladinos, a much larger center of population may be a big agricultural village." Among the points of major interest is the system of a municipal ownership which still prevails in the villages around Lake Atitlán. Although land may be held privately, and taxed, the more common arrangement is for the local government to grant life-tenancy of tillage land to residents when they reach the age of eighteen. He also comments on the conversion of the *Plaza* into the *Parque* or *Parque Central* when the market is removed to another location in the interest of beautification, the centralization of major institutions, and the early attempt at true north-south orientation of streets.

Tax has distinguished between the nuclear Guatemalan town, conforming in its essential features to the basic Spanish colonial model, and the vacant-town type of settlement, characterized by the "geographic duality" of its partly resident Indian population.¹⁰ He also describes certain intermediate varieties of settlement pattern quite in keeping with the ethnological complexity generally characteristic of the highlands.

The development of Sucre in Bolivia is conceived by Hawthorn as a modification of the standard design laid down in the Laws of the Indies by the following factors: the convenience of having the service trades close at hand, the impoverishment of the formerly wealthy, the development of the type of dwelling known as the *casa quinta* (a garden retreat within the city limits), the modern upper-class suburban home, recent commercial and railroad projects, and the irregular topography of the site.¹¹

An important element which none of these studies has stressed is the role of systematic and deliberate planning in the early period of urban growth. The rectangular street plan, the rational siting of churches and open spaces, the systematic subdivision of blocks, the sumptuary and technical regulation of construction, and the control of marketing, water supply, and commercial location are all elements that loom large in the early history of Guatemala and were presumably operative in Mexican communities as well.

For a general survey of urban demographic

---

⁶Norman S. Hayner, "Criminogenic Zones in Mexico City," *American Sociological Review*, 11 (August 1946), 428–38.

⁷Norman S. Hayner, "Differential Social Change in a Mexican Town," *Social Forces*, 26 (May 1948), 381–90.

⁸John Gillin, "Parallel Cultures and the Inhibitions to Acculturation in a Guatemalan Community," *Social Forces*, 24 (October 1945), 1–14.

⁹McBryde, op. cit., pp. 85–100.

¹⁰Tax, op. cit.

¹¹Harry B. and Audrey E. Hawthorn, "The Shape of a City: Some Observations on Sucre, Bolivia," *Sociology and Social Research*, 33 (November–December 1948), 87–91.

trends and rural-urban population differentials in Latin America, we are still restricted to a single source,[12] although a considerable amount of local material, of varying quality, exists for each major city. Davis and Casis find that in view of its relatively small industrial development and the low average density of its population, Latin America is urbanized to a surprising degree. Further, it may be shown that the rate of urban population growth exceeds that of the rural population and is being supported by heavy migration from rural areas. The rate of growth is, in general, directly correlated with the size of the city, but suburbanization is less marked than in more industrialized areas.

There appears to be a consistent rural advantage in fertility, and the fertility ratio of large cities for those areas where data are available is lower than that for small cities. Mortality, on the other hand, shows no consistent urban-rural differences. The natural increase of the rural population is "far superior."

Both internal migration and foreign immigration contribute heavily to city growth in Latin America. "There is a tendency for the cities to attract both men and women in the vigorous period of life, and the larger cities exert a greater pull than the smaller ones."

At all ages above 15, there is a marked concentration of women in the city. With one or two exceptions, the sex ratio of small cities is higher than that of large cities. Rural migrants are mostly female, although foreign immigration may—as in the case of Panama—consist mostly of males.

The typical urban-rural differentiation found elsewhere with respect to marriage and legitimacy is reversed in Latin America, the proportions of married persons and legitimate births being much higher for the cities. The explanation, however, is held to lie in the term "married"; it is believed that the proportion of persons living together in consensual union is much greater for rural areas than for urban.

City populations are far more literate than country populations but there is evidence that the literacy rate in rural areas is increasing faster. Literacy and the proportion of the population speaking an indigenous language are found to be inversely associated.

[12]Kingsley Davis and Ana Casis, "Urbanization in Latin America," *Milbank Memorial Fund Quarterly,* 24, No. 2 (April 1946), and 24, No. 3 (July 1946), Pts. I & II.

II

Guatemala City (La Nueva Guatemala de la Asunción) is located in the highland area of south-central Guatemala at an altitude of 1750 meters. Its population is estimated at 230,000 in 1948—less than 8 percent of the national population—but it exceeds in size the ten next largest urban places combined. The site is a fertile, level valley surrounded by volcanic peaks, and crisscrossed by deep ravines. Like other highland capitals in Middle America, it is fairly remote from water transportation, lying about two hundred miles from the Atlantic Coast and seventy-five from the Pacific.

Unlike most communities of comparable size in Spanish America, it has a rather short history. Throughout most of the Colonial period the capital of the Kingdom was that other Guatemala City now called Antigua (Santiago de los Caballeros de Guatemala), which was in size and wealth the third city of the New World after Mexico and Lima, and whose ruins in the Panchoy Valley, about twenty-eight kilometres from the present capital, still preserve more architectural monuments than can be found in the newer city. Repeatedly damaged by earthquake, Antigua was finally abandoned between 1773 and 1776, after a bitter struggle between the secular powers and the Church. In the meantime (1773), a provisional capital had been established in the village of La Ermita, in the Valle de las Vacas, while a technical commission investigated a number of sites. The chief considerations were water supply, salubrity, and earthquake safety. A site very close to La Ermita was finally chosen among other reasons, because it was believed that the surrounding ravines would absorb the shock of earthquakes. It is a passing irony that the abandoned capital has had only minor tremors since 1773, whereas the new Guatemala City has been severely damaged several times, most recently in 1917, when a substantial portion of the central area was destroyed.

At the time of settlement, it was necessary to purchase or acquire by transfer, 13 large holdings (ejidos) comprising a total of 204 caballerías.[13] (A caballería is 640,000 square yards or 33.33 acres. This area exceeds even the present extension of the city.) It is not known in detail how these lands,

[13]The discussion which follows is based largely upon a collection of old plans of Guatemala, many of which were copied from the files of the Seccion de Urbanismo, Oficina de Obras Publicas. I am greatly indebted to the brilliant and energetic chief of that section, Sr. Ing. Amilcar Gomez Robelo, for his advice and cooperation.

which were inalienable, came into private possession, but there appear to have been several methods. Grants were made at first, on the basis of holdings in Antigua, to individuals and institutions that had abandoned land in the older city. Later, land was sold to provide operating revenues, and rentals under leaseholds came to be regarded as taxes. In the outlying areas, the boundaries of the original purchase were soon forgotten, and land ownership came to be based either upon the old village holdings, for which deeds still exist, or upon a sort of small-scale enclosure movement, which created suburban plantations late in the eighteenth century.

Included in the tract were a number of small settlements, the largest of which was that at the base of the small hill of Cerro del Carmen, on which a shrine had been erected as early as 1620. One source gives the population of the village as more than 1600 in 1773.[14] The largest of the ejidos, the Hacienda de la Culebra y Lexarcia, included no less than 11 clustered villages, which soon merged together to form Ciudad Vieja. The process of urbanization by merger, which later absorbed Ciudad Vieja into Guatemala City, was thus operative at a very early date, since a map dated 15 years later shows it as a single settlement. Other settlements in the neighborhood of the new site were Jocotenango, an Indian village important in the later development of the city, and San Felipe, whose name is still preserved in an outlying section.

A rather rough plan dated 1782 shows the central part of the city to consist of an oblong 15 blocks long (north to south) and 13 blocks wide. Some of the blocks were subdivided with short streets running between the major streets and avenues, but, with the exception of the irregular section on the northeast that marks the Cerro del Carmen, all blocks are perfectly rectangular, and all major streets are of the same width—about 12 yards, located 280 feet apart.[15] The central square occupies the same position as today and is surrounded by four slightly smaller squares, one to each cardinal point, each distant four blocks. Jocotenango to the north and Parroquia Vieja to the southwest are shown separated by open spaces from the center. There appear to be four large parks or fields on the four outer edges.

[14]Vera Kelsey and Lilly deJongh Osborne, *Four Keys to Guatemala* (New York: Funk & Wagnalls, 1933), p. 233. Original source not cited.

[15]Francis Violich, *Cities of Latin America* (New York: Reinhold, 1944).

The city had a population of perhaps 13,000 in 1782, but the central district was not noticeably smaller than it is today. It had been laid out with regard to future population and must have been very thinly populated at the edges.

A very poor map of 1800 shows the city and at some distance from it, quite separate, the towns of Jocotenango, Ciudad Vieja, the Villa of Guadelupe, and the villages of Carmen, La Libertad, San Gaspar, Santa Ana, and San Pedro. The growth of these outlying Indian communities seems unquestionably to have been sharply stimulated by the growth of the central city, although some of them, like Jocotenango, were transferred more or less intact from Antigua. Not until well into the twentieth century are they completely absorbed into the city proper, and even now, Guadalupe and Ciudad Vieja preserve something of an independent character.

. . . . .

A plan drawn in 1821 shows the city as having the same extension, with the concentration of colleges, churches, and offices around the major plaza slightly reduced in favor of the four minor plazas. The entire outside area to a distance of 12 or 15 miles from the city is shown divided into large-scale land holdings which include all of the territory but the five major suburban communities and the ravines which surround the city to the north, south, and southwest. Except for the irregular section of Cerro del Carmen, the city is almost perfectly square. There has been some peripheral growth. Foreign travelers had commented on the absence of walls in Antigua Guatemala as far back as the sixteenth century. There is no trace of fortification in the new city at this date.

. . . . .

An excellent map of 1842 shows the division of each block by lots and discloses the preference for corner locations previously mentioned. In almost every case, the corner lots are much larger than the interior on the same block. In some cases, they are ten or twenty times as large. In general, the larger blocks have fewer lots, but these large lots are not concentrated in any section of the city. Nor is there any close relationship between density, as suggested by lot size, and proximity to the central plaza except that the two conspicuously dense areas are each located close to a secondary plaza. Although the list of institutions and churches suggests a considerable growth in density, the central area still appears to be uncrowded, and this is borne out by contemporary observations. Of peripheral growth, there is very little. The fortress of San Jose, commanding the roads to the south, appears for the first time, as does the fifth minor plaza, the Plaza de Toros.

. . . . .

A map dated 1870 and showing the illuminated area of the city is especially interesting for two points. In the first place, the area having street lighting is almost exactly identical to that shown in the plan of 1782. Thus it appears that, although peripheral areas such as Jocotenango were being absorbed into the city, they still retained their indigenous character and were not allotted urban services. Secondly, the light posts are only slightly centered in the neighborhood of the Plaza Mayor, almost all of the blocks within half a mile of the center being adequately illuminated. Sometime in this period, a new administrative organization of 13 cantons had been set up, replacing the old division into four wards. A few years later, in 1882, this had again been replaced by a division into four numbered cantons. It is curious that the 1882 plan omits the suburban areas and shows almost the same limits as those of a hundred years before. However, the era of great stability was coming to a gradual end.

The introduction of modern urban facilities coincided with the 10-year administration of J. Rufino Barrios (1874–1884). A gas street-lighting system was completed in 1879 and converted to electricity in 1885. The horse-drawn streetcar system was put into service in 1882, and the installation of telephones began in 1884. During the same period, the railroad reached the capital, the police force was reorganized several times, and a public school system was founded. The relatively slight time-lag between the introduction of these municipal measures in western Europe and the United States and their introduction in the isolated and conservative Central American city is in many ways comparable with the "modern" appearance upon which tourists commented in the 1940s. It should also be noted in passing that the extension of new functions to the municipality was markedly facilitated by the tradition of almost unlimited regulative and sumptuary authority which persisted from the colonial cabildo. Despite the prominence of foreign speculative capital on the scene, the majority of the technical improvements remained under close local control.

The earliest directory I have been able to find for the city is part of a tourist guide published by Ovalle in 1889.

Ovalle's plan of the city shows, for the first time, well-developed areas outside of the old Centro. The most important of these are Jocotenango, still partly separated from the city, and, to the south, three cantons whose combined area about equals that of the center. These were the cantons of Barillas, LaPaz, and San Pedro Las Huertas.

The directory for 1889 shows 2 banks, 3 hotels, 8 inns, 8 pensions, 5 printing establishments, 3 book and stationery shops, 4 lithographers, 4 photographers, 6 book-binders, 17 pharmacies, 3 oil-cloth factories, 3 rendering works, a match factory, 11 commercial baths, 4 money changers, an antique shop, 11 pawnshops, 2 breweries, 3 charged-water bottling plants, 2 ice-plants, 10 jewelers, 10 silversmiths, 11 restaurants and canteens, 6 coppersmiths, 14 carriage-makers, 4 machine shops, 6 engravers, a flour mill, a paint factory, a riding school, 32 musical composers (!), 3 bull fighters, and a teacher of deaf mutes, among others.

It is a moot question whether the community was not then more urbanized in proportion to population than it is today. At any rate, one can not but be struck by the similarity of the directory to that published in Guatemala City in 1948. The list of professionals included a dozen language teachers, and many commercial establishments bear non-Spanish names which witness the substantial English, American, Jewish, and German immigration which had already taken place. There was a telephone system, although it did not extend beyond the central area, and a telegraph network which already connected the capital with places as remote as San Tomas Chichicastenango. In this era, steamship communication with Panama and San Francisco was established from Champerico and San Jose on the Pacific; and the development of Atlantic ship routes on the East coast of the isthmus was well under way. The question as to how so rapid a degree of urbanization could have overtaken the sleepy isolated capital in so short a period without marked disorganization will be reserved for later consideration. At the moment, it should be noted only that it was the dominant group of land-holding "white"[16] families who, together with the immigrant businessmen, had taken over the new commercial functions, and that the addition of new lower-class areas had only strengthened the dominance of the old central area. In an 18-page list of "proprietors and

---

[16]The term *ladino* was in this era still roughly equivalent to *mestizo*.

negotiators" less than a dozen addresses outside the 1783 boundaries of the city are given.

An elaborately drawn plan of 1894 illustrates a further stage of this explosive growth. Although it does not distinguish between existing and projected areas, it is noteworthy in that it gives the city approximately the boundaries reached in 1936, and shows in detail the suburban development of Tivoli and Santa Clara (not then named). These two areas now represent the closest approach to North American suburban architecture, consisting chiefly of two-story, detached residences, without patios. Included in the 1894 plan is the large boulevard developed in imitation of Mexico City's Reforma, and Aurora Park with its athletic and recreational facilities. The sudden transition from the tightly centered square of three decades before to the extensive suburbanization of the 1894 plan and its imitation of the street plans of crescive industrialized cities is striking; but it should be noted, on the other hand, that this projected growth was to occur—and actually did occur—within the framework of the same sort of systematic planning which created the older Guatemala.

The plan of 1900 is the last in which a separation still exists between the three ancient suburbs of Ciudad Vieja, Guadalupe, and Jocotenango. Thereafter, they appear as parts of the city. Growth by districts proceeded at a very rapid rate throughout this period, although Tivoli was still largely cultivated farmland as late as 1917. At the periphery of the expensive residential suburbs, which were growing slowly, there sprung up a number of middle- and lower-class subdivisions not contemplated in the 1894 plan. The ultimate development toward which this tended was a narrow strip of expensive housing and landscaping extending out from the southern extremity of the old center and terminating in a park. Along both sides of this belt grew areas of mass housing, which eventually enveloped the old independent communities in their path. Further out still, there grew up after the earthquake of 1917, a number of modern slums of which El Gallito at the eastern extremity and Colonia Abril at the western were the worst, being, in quality of street-layout and construction, probably the worst sections that had developed in the city since its founding. Thus, continuity between the old upper-class residential district and the new ones with their pattern of suburban living was essentially maintained, and the reversed gradient persisted even in the face of rapid and commercialized growth. A similar process took place to the north, but was sharply limited by the ravines, which prevent extensive growth in that direction, and by the very old communities of Ermita and Jocotenango.

Nevertheless, a visitor in 1909 comments upon the remoteness and isolation of the city: "Guatemala City is a perfect place to play with life, cloistered away from the active world, and yet so near to its bustling stir. The real world and its manners are here, but here are none of its problems. . . . People who have money have inherited it or made it easy; those who have it not, never expect it. There is no hustling, ambitious middle class to stir up rivalry and discontent. . . ."[17]

Again there is a description of the one-story patio-centered houses, built of adobe, or stone covered with stucco, with their red-tiled roofs and massive doors, and of the roughly paved streets with sidewalks of flagstone. The visitor is struck, like his compatriots of a century before, by the great churches towering over the even rows of low houses. There is a sense of identity with the older configuration which is unmistakable. Yet even this rather hasty account itemizes a number of newly imported institutions: the small formal parks, the Jewish synagogue, the Chinese colony, the fine modern homes along the Reforma, the American Club, the Presbyterian mission, the new railroad station, and the two large markets. The principal market has finally been moved from the plaza to the site at the rear of the Cathedral which it still occupies, and a second market had been added in the western part of the city. Their internal arrangements were almost precisely as they are today. The population was between 75,000 and 100,000.

The series of earthquakes in 1917–1918, although they occasioned considerable destruction, had no substantial effect upon the plan of the city. Rebuilding was on an ambitious scale, but, with few exceptions, the new structures were substituted literally for the old. The brief interregnum between the dictatorships of Cabrera and Ubico was marked by extensive building, and a considerable number of public buildings were erected by Ubico, so that the city acquired a certain architectural elegance well suited to its favorable setting.

From the ecological standpoint, the main outlines of the new development show no real variation from the cartographical prediction of 1894. Plans of 1927 and 1936 show some growth of lower-class colonias on the eastern and western

[17]Nevin O. Winter, *Guatemala and Her People of Today* (Boston: L. C. Page, 1909), p. 80.

outskirts, the blocking in of suburban areas in Santa Clara and Tivoli, the development of slum populations inside and across the ravines to the north and west, and the improvement of the exterior road net. The street-railway was destroyed in 1917 and never replaced: a system of small busses contributed to the gradual obliteration of boundaries between districts, so that, except for the poorest outlying sections, there has come to be no sharp visible differentiation between one barrio and the next. The city has occupied almost all of the available land within the ring of ravines; except for the filling in of a few vacant sites on the south, future physical growth must take place across the ravines, and experiments in leveling and draining the ravines are under way. Just as the city remained for more than hundred years with the general boundary of the central area planned by Navarro and Sierra in 1776, so the development of the last half century has followed the plan of an expanded center trailing suburbs in one direction which was sketched by Urrutia and Gomez in 1894. The dependence upon a pre-formulated plan and the steady growth through increasing density of built-up areas continue to characterize the development of Guatemala.

The elements of the ecological pattern which exists today will be suggested in the following section.

III

Guatemala City, according to Tax, is more like Detroit than like any other place in Guatemala.[18] Certainly, the appearance of Guatemala is strikingly "modern." All of the streets in the central and many of those in the peripheral districts are well-paved or cobbled and have adequate sidewalks. Drainage is good, and the streets and facades are far cleaner than in most American or European cities. A well-equipped modern airport ties the city into the network of world transportation; its streets are crowded with late-model cars, the yellow school busses used for public transportation, and motorcycles—as well as with oxcarts, pack-mules, burro carts, horse-drawn vehicles, pack-horses, and a variety of wagons. (The cargador from the country with his heavy load of pottery, wood, vegetables, or blankets is no more a rarity in this environment than the large new convertibles of the wealthy.) Typical urban services, from soda fountains to the repair of calculating machines, are readily available. There are well-equipped theatres showing the latest Hollywood attractions, and English is widely spoken in the shops and hotels. The villas in the Santa Clara district typically have suburban architecture, picture windows, and distinguished landscaping.

Nevertheless, the sparkling modernity of the city is in many ways deceptive. Part of it is due to an integrated network of urban services which goes back to the eighteenth century and the centralized control and wide powers of the colonial Ayuntamiento; in part it arises from the naturally functional house plans suggested by earthquake safety and the patio-centered home. These chance to be aesthetically pleasing now; earlier travelers complained of their monotony and lack of decoration. In part too, the impression which the casual traveler derives is based upon the circumstance that the center of the city is an area subject to little blight, and that slums and blighted areas are in almost inaccessible locations on the urban periphery. The extent to which Guatemala departs from the distributional pattern of Detroit may be suggested by the consideration of several related phenomena: the relative weakness of commercial concentration, the peripheral location of disorganized areas, the persistence of upper-class residential areas close to the commercial center, the stability of commercial enterprises, the absence of blight in connection with succession, and the stability of certain characteristics which may be derived from census figures.

1. The relative weakness of commercial concentration is reflected in a variety of phenomena. The central business district covers an enormous area extending about 15 blocks from north to south and about nine from east to west. Within this area there is no clearly defined central point; the maximum traffic concentration is at 6th Avenue and 11th Street, the four corners of which are occupied by a theatre, a bar, a vacant lot, and a small drygoods store. Despite some clustering of similar functions—three of the six theatres are close together, and lawyers may be found in great numbers in the Pasaje Aycinena—the leading stores are found in various parts of the area, factories are encountered at widely-separated points, and doctors' offices show an almost random distribution. There is no markedly deteriorated area in the entire central district, nor is there any area without substantial traffic. Even where some local concentration exists, as in the row of tourist shops opposite the Palace Hotel, the effect

---

[18]Sol Tax, "Culture and Civilization in Guatemala Societies," *Scientific Monthly,* 48 (1939), 467–71.

upon neighboring blocks is negligible. There tend to be several types of shopping districts at the edge of the central area which are intermediate between the expensive shops found near the plaza and the native markets. These represent some clustering by economic level, but in no case is location the emphatic factor which it tends to be within the more dynamic and centralized pattern of the crescive North American city.

Equally significant is the fact that this commercial area tapers off gradually and never gives way to a completely residential area except in the foreign-type suburbs to the south. The writer and two assistants mapped a sample of 100 linear blocks drawn at random from the 500 square blocks (2,000 linear blocks) surrounding the point of traffic concentration. This area includes almost half the municipality and the bulk of the urban population. Only 11 (11 percent) of these blocks were purely residential in character, and none was entirely non-residential. Stores, ranging from small well-stocked department stores down to the one-room *tienda* with its open doorway on the street, are scattered throughout the urban area proper, as are schools, textile factories, printing plants, churches, government buildings, and motorcycle agencies. A fairly typical block more than a mile from the center includes: a barber shop, a dressmaker's shop, the office of an accessories firm, five extremely modern duplexes occupied by middle-class families, a 22-room private residence about forty years old, and two slightly smaller houses occupied by upper-class families. Across the street is a row of old subdivided dwellings rented one room at a time to impoverished families, a plant nursery, a small woodworking plant, a large grocery store, a doctor's office, and several homes of moderate size.

2. The peripheral location of disorganized and criminogenic areas has been noted by Hayner in Mexico City as a current phenomenon, even though Mexico in other respects conforms rather closely to the pattern of large industrial cities. The phenomenon is so well marked in Guatemala as to require no detailed demonstration. To the north, east, and west, the poorest and the least prosperous segments of the population are located peripherally. Furthermore, there is only one area of markedly poor housing within two kilometres of the commercial center. Many foreign and some upper-class residents are not aware of the existence of really impoverished and deteriorated areas in the city, so thoroughly are these isolated from the main currents of urban traffic.

That this is expressible in terms of social disorganization is illustrated by a chart prepared by Cesar Meza to show the location of areas which represent a problem from the public health point of view.[19] These areas form almost a continuous border around the city, broken only at Jocotenango, and to the south.

Additional evidence of the same sort appears from a study of school attendance based upon the Censo Escolar of 1946. The percentage of non-attendance among children of school age rises consistently as one moves out toward the urban periphery.[20]

Thus, the gradient which shows a decline in all disorganization phenomena as one moves from the area in transition to the suburban fringe in industrial cities is literally reversed in Guatemala. Correspondingly, the variation in density which we find as we move from central to suburban areas in the crescive city is not encountered here to any great extent. Considered in relation to total area, of course, the urban periphery is not densely populated. However, in terms of persons per room, or houses per block, it is precisely in the colonias at the very edge of open country that the highest degree of crowding may be found.

3. Both Hayner and Hansen called attention to the displacement of the upper class from its traditional station near the plaza. In Guatemala, this displacement involves an additional shift from the patio-centered house of colonial architecture to the walled house with its non-characteristic suburban style. These latter are concentrated in the three southern suburbs of Santa Clara, Tivoli, and Guadalupe, all the fruition of plans made in the 1890s and modeled rather closely after the development of the Paseo Reforma and Las Lomas in Mexico City. It may even be suggested that the development of suburbanization, when it did occur in Guatemala, was as much a response to foreign influence and a direct imitation of foreign patterns as a development process arising out of local conditions. Although the oldest part of this expensive residential suburb has been settled for more than a generation, displacement of upper-status groups from the center is far from complete. On the contrary there has been the same tendency for the more traditionally-minded groups to cling to a central location within two or three blocks of

---

[19] Cesar Meza, *Guatemala y el Seguro Social Obligatorio: Estudio Medicosocial* (Tipografia Nacional, Guatemala, C. A., 1944), p. 60.

[20] Unpublished data, Dirección General de Estadística.

the Parque Central as Hansen observed in Merida. This can be rather strikingly illustrated:

In an appendix, Thompson gives a list of the 30 principal (richest) families of 1826.[21] Of these there are 13 known to the writer to have retained both wealth and status as of 1948. Residential telephone listings were used to check the location of this group; since there is less than one telephone to every 20 inhabitants, and most of these are commercial, the use of residential telephones has the effect of excluding poor relations. There were, in all, 39 residential telephones listed for these 13 families in 1948. Of these, 28 were within six blocks of the Parque Central; the remainder were outside the Central district. Thus more than 70 percent of these oldest upper-class sub-families remained in their traditional location.

Another group of 13 families was selected, all of which are listed by Ovalle as among the *Negociantes* of 1889 and are not found in the 1826 list. This is a group which has enjoyed continuous wealth for at least two generations; it is probably somewhat wealthier than the older group, taken en masse. There were a total of 48 telephone listings; of which only 22, or 46 percent, were located in the center; the remainder were suburban.

A still more recent group was obtained by selecting 25 family names from a membership list of the German Club in 1923.[22] Besides being more mobile, this group was somewhat affected by the expropriation of enemy-alien property in 1942. A total of 14 listings was found divided among 10 families. Only two of these, or 14 percent, were in the central district; the remainder were suburban.

The existence of several distinct patterns of housing should be noted. Unlike many other urban areas in Latin America, Guatemala retains the patio-centered house as the dominant form of housing (although modern architecture has reduced the number and size of patios). Houses built to this pattern tend to preserve certain physical and social features which are distinctive. The large gate of which the ordinary door forms a part, is almost invariably retained. The entrance arch in some new projects serves as a garage. The dining-room forms a separate structure on the patio, there is an open arcade connecting the rooms, distinctive and carefully tended plants are found in the patio, bedrooms are dark, and the living-room tends to be a formal parlor rather than a center of family activity. At the rear of the house there is found the service patio, the pila—which is a combination fountain, washtub, and kitchen sink—and the rooms for servants.

The only other type found in the conventional housing pattern is the *barraca,* a one- or two-room dwelling with a doorway directly on the street, and no interior corridors, stairwells, or open air spaces. These are often combined into long, shallow "row" buildings housing a number of families. The only data on the distribution of this type are given by the census of 1921 which classifies 43 percent of the dwelling units in the municipio as *barracas.* It should be noted, however, that the barraca, too, is a distinctly urban form of housing. More than 90 percent of these dwellings had tin, tile, or composition roofs in 1921, and that proportion appears to have increased since.

The living-room-centered house, built to conventional foreign models and often executed with architectural elegance and vigor, is characteristic only of the southern suburbs and of isolated structures within the city proper. Even apartment buildings tend to preserve the basic patio form in certain of its essential elements.

It would appear that the retention of the traditional housing pattern and the form of inwardly focussed family living which it implies has been possible to a great extent; the development of new patterns, as in Santa Clara and Tivoli, appears to arise from the imitation of foreign patterns (and also as a direct accomodation to resident foreigners) rather than from the collapse of traditional habits.

4. The stability of commercial enterprises is another factor which invites attention. Jones[23] has demonstrated the essential instability of the Guatemalan urban economy, dependent as it is upon world coffee and banana prices and fluctuations in the tourist trade, all of which are largely independent of local control. In addition, the drastic alternations characteristic of Central American politics have often involved the financial standing of participating pressure groups. In view of these conditions, it is surprising to note that the average life of a business in Guatemala tends to be long. About half of the major retail enterprises in Guatemala City are operated by second-, third-, or fourth-generation owners.

The best explanation appears to be that which is almost uniformly put forward by local business-

---

[21]G. A. Thompson, Esq., *Narrative of an Official Visit to Guatemala from Mexico* (London: John Murray, 1829), Appendix.

[22]From Jose A. Quinonez, *Directorio General de la República de Guatemala,* Tipografia Nacional, undated.

[23]Chester Lloyd Jones, *Guatemala Past and Present* (Minneapolis: University of Minnesota Press, 1940).

men, namely, the informal limitation of competition. Although the agricultural life of the country is largely dominated by the enormous holdings of a few families and corporations, the commercial atmosphere of the city is marked by the relative absence of concentrated economic power. This appears to have been achieved in part by direct action taken by the business group to limit the expansion of nascent tycoons, and in part by the absence of a well-developed credit system. The role of bank credit in Guatemala's urban economy has, until recently, been almost negligible. This accords rather neatly with the implicit limitation of competition for location which can be seen as the salient feature of the commercial district. Two factors which lie somewhat further in the background of this pattern of attitudes are the importance of non-monetary channels of status achievement and the fundamentally limited and inflexible market situation.

5. Studies of Chicago and other crescive cities of North America and Europe have familiarized us with a syndrome of changes in land use which follows almost uniformly the sequence of homogeneity: invasion by a new function—blight—succession—homogeneity—invasion, etc. It is this process which produces the concentric growth rings of the ideal-typical industrial community and is responsible for the disorganized transitional area which is found in the area of maximum change at the edge of the central business district and, on a smaller scale, at other points in the urban configuration where a change in the fundamental land use is under way. This syndrome does not exist in Guatemala, although there are distinct indications that it may develop in the future. The rate of succession is relatively low, No process that can properly be described as invasion can be discerned, since homogeneity of function exists only at the periphery, and changes in land use do not ordinarily appear to produce blight or neighborhood disorganization.

In the absence of acute competition for favorable sites, succession proceeds at a very moderate pace indeed. Considerable portions of the central district—including the Parque Central and the areas to the west and north of it—are still devoted to the same uses which characterized them in the eighteenth century. The extension of the commercial district has been gradual and marked by a distinct continuity of land ownership. (At least one site in the commercial center is still in the hands of the original grantees.) The central market is still only one block from the plaza in which it was originally located. Even the southern suburbs, as we have seen, follow an allocation of land made more than fifty years ago.

There are several reasons why large-scale invasion cannot take place; the most obvious of these is the absence of concentration and functional homogeneity. In addition, the extended housing pattern and the existence of a city plan which has consistently expanded faster than the actual use of land have allowed the density of population to increase by concentration in already built-up areas without much change in structures and without violent alteration in the use of existing structures. Finally, the basic relationship of high land values and low rents which characterizes the blighted area cannot develop in an area with little variation in land value, few speculative opportunities in the anticipation of land price increases, and a pattern of housing which imposes relatively low density.

Another way of expressing this generalization is to point to the absence of land or site obsolescene. The social, historical, economic, and physiographic factors which have already been mentioned sufficiently account for the tendency of areas to satisfy the same functional demands over long periods of time. The demands of an impoverished in-migrant population for housing thus tend to be met by the utilization of marginal peripheral land, rather than by the increase of density on obsolescent sites and in obsolete structures. The worst slums of Guatemala are comparatively new: El Gallito, where personal and social disorganization is at a maximum, is a district which grew up after the earthquake of 1918, and it is possible today, in areas on the opposite edge of the city, to see new slums under construction. Even in Mexico City, where the pressure of tremendous commercial expansion did overtake the central district and create tremendous transitional areas, there is still a tendency for disorganized areas to be newly constructed and peripheral.

There is a tremendous amount of misery, poverty, and hunger among the lower-class half-urbanized population of Guatemala City, and the worst Guatemalan housing is as spectacularly inadequate in some respects as the worst in New York or Washington, although overcrowding on the North American scale is unknown. What is striking, however, is the fact that urban growth as such cannot be held directly responsible, and the tendency to interior breakdown of the urban configuration which so concerns contemporary North American planners is not a problem in Guatemala.

For related evidence of urban stability, we may turn to the limited numerical data available. The

Table 1. Population of Guatemala and Guatemala City.

From The Census Of	Guatemala City	Total Republic	Percent In City
1880	58,000	1,225,000	4.7
1893	72,000	1,365,000	5.3
1921	121,000	2,005,000	6.0
1940	186,000	3,283,000	5.7

Table 2. Concentration of Foreigners, Professionals, and Clergy in the Department of Guatemala. (Percent of each group in Department of Guatemala.)

	1893	1921	1940
Foreigners	23	29	31
Doctors	56	43	62
Lawyers	63	61	65
Clergymen	39	43	40

Table 3. Literacy in Guatemala and Guatemala City. (Percent literate—reading and writing.)

	City	Total Republic
1893	41	11
1921	55	13
1940	95	36

records of the Republic consist of the published Censuses of 1880, 1893, 1921, and 1940, and scattered references to the unpublished Censuses of 1902 and 1930. With the nominal exception of the 1940 Census, none of these was taken by direct enumeration. Even the base population figures are subject to marked internal inconsistencies. Little use can therefore be made of the series presented by these sources, and the data which follow are presented only as approximations. Their use in any connection is justified only by two considerations: the probability that figures for the capital are in each case somewhat more accurate than those for the rest of the country, and the circumstance that the phenomena described are of such an order as to permit very large errors without impairing the conclusions.

In the first place, it would appear that the population of the capital has grown at a rate closely proportionate to the rest of the country.

The proportion of Indians in the capital and in the Republic is another measure of differentiation: In 1893, 8.5 percent of the city's population were enumerated as Indians, in 1921, 7.0 percent; in 1940, 6.5 percent. The corresponding figures for the Republic were 64.8, 64.8 (!), and 55.4.

The concentration of foreigners, professional men, and clergy, in a Latin-American capital is an interesting phase of its relation to the hinterland. The degree of concentration has apparently not varied greatly in the last half century.

A related measure is the differential in literacy. While literacy has increased sharply in the same period, the ratio between the city and the Republic has not greatly altered.

Information on population structure scarcely deserves tabulation. In general, it would appear that the median age of the city's population has exceeded the median age of the Republic's population by two to four years, and that the sex ratio of the capital has been consistently lower than that of the remainder of the population. It also appears probable, from scattered evidence, that the city has consistently shown some excess of births over deaths.

What these data suggest in general is that the role of the capital in relation to the country has not been drastically changed over a considerable period of time, a supposition which indirectly confirms the hypothesis of regulated growth previously presented.

IV

The basic question raised by all ecological research, namely, what relation exists between the spatial community structure and the social organization associated with it, assumes additional interest in this case as soon as it can be demonstrated that growth under certain cultural conditions did not involve the ecological processes which the student of urban life in the United States takes for granted. In this section some attempt will be made to relate the topographical and economic aspects of the ecological pattern to an embarrassing variety of background factors which seem to underlie the observed differentiation between Guatemala City and St. Paul or Dubuque.

Hoyt and others have substantiated the hypothesis advanced by Park (and earlier by Hurd) that the competition for land and the resulting steep gradient of land prices exerted a determinative influence upon the centralized concentric segregated city pattern.[24] If we take as our starting point, the necessity for explanation raised by the lack of violent variation in land price in Guatemala City, we see at once a number of major factors which seem relevant.

1. In Spanish America, the existence of the city, in colonial times and later as well, depended to a great extent upon the centralization of political administration and upon a system of absentee land ownership. Public functionaries and rentiers, plus their servants, dependents, and the personnel of services which depended upon their patronage, have always comprised a substantial proportion of the population of Guatemala City. Neither of the two capitals was located with primary emphasis upon commercial routes, and neither witnessed any significant development of industry. Although the capital could not fail to attract a good deal of commerce, whose function went beyond that of merely supplying the needs of residents, its growth was not initiated nor limited by the intersection of commercial routes, a break in means of transportation, nor even by the resources of the immediate hinterland.

[24]Homer Hoyt, *One Hundred Years of Land Values in Chicago* (Chicago: University of Chicago Press, 1933); R. E. Park and E. W. Burgess, *The City* (Chicago: University of Chicago Press, 1925); Richard M. Hurd, "Principles of City Land Values" (New York: *The Record and Guide,* 1903).

2. Attention has been called to the obscure and dubious origin of urban land tenure. Land passed into private ownership under several distinct sets of restrictions. The first and least important of these was the nominal ownership by the municipality of all residential land—an ownership which appears never to have been formally abrogated. More significant was the tradition of urban planning and the control of growth, which allowed the government to retain powers of regulation that have not yet been approached in North American cities and implicitly subjected all private land use to official aprproval. Closely supporting this tradition were the legal and later customary restrictions on the residence of the indigenous population. Their location on the periphery of the city or even in communities apart from the city was gradually transformed from a strategic administrative policy to a time-honored custom. Similarly, the attachment of the upper-class population to the center of the community arose from the planned location of the ruling group in colonial times, and persisted in terms of both the symbols of status represented by central location and the social habits which became associated with the palacio, the cathedral, and the plaza.

In terms of the developing situation, the municipality seems to have maintained a comfortable margin between the extension of the street-plan and the growth of the population, so that increases of density imposed no great strain upon the community. A degree of local centralization around the "satellite" plazas also contributed to the systematic distribution of new population.

3. The development of the land market was further inhibited by the tendency of land to be a value in itself, rather than a means to consumption income. On the one hand, the Spain which set the pattern of settlement in Central America was still close enough to feudalism to attach extraordinary status values to the possession of land. On the other, the lack of investment opportunities and the perennial instability of the currency in early times, and the weakness of the credit system in later times, forced a certain permanence upon land investments. Land tended to be leased or converted in use rather than sold. Conversion in use was vastly facilitated by the fact that the agricultural and political upper class was not restricted from trade. Even small retail trade carries no stigma for members of families whose value assumptions and pattern of living are essentially those of land-holding aristocracy. Thus the development of the business district, as distinguished from the central

market, can be clearly traced in the conversion of portions of large family dwellings in the central district to shops and offices, typically occupied at the outset by members of the owning family.

4. The function of earthquakes in preventing blight by periodically removing structures fit for demolition should not be overlooked. Whimsical though this point of view may appear, the writer is convinced that it has been a major factor in preventing land obsolescence. There can be no serious question that the action of earthquakes is selective, and they have occurred with sufficient frequency to insure a constant check upon structural soundness. Unlike the gradual decay of an urban neighborhood in the United States, this type of land clearance raises no economic or social obstacles to rebuilding since the quality of the neighborhood as a whole tends to be unaffected. The partial demolition of houses may even have led to a higher level of maintenance than would otherwise have been achieved. A subsidiary effect was the virtual limitation of vertical growth in the interest of earthquake safety. By expediency, as well as statute, buildings were until recently limited to one story, a limitation which automatically set strict restrictions on any rise in land values.

5. While location near the center remained an important element of status, it is striking that the use of housing itself as a form of conspicuous display or as a means of social mobility was inhibited by a number of characteristics in the Spanish colonial housing pattern. Both the climate and the culture helped to maintain the interior privacy of the dwelling which turned a blank wall or barred windows toward the street. This, added to the one-story limitation, accounts for the curious fact that even today it is sometimes impossible to distinguish between the four-room marginal slum dwelling in a built-up area and the twenty-room palace which may be next to it, by their external appearance. The necessity for housing the numerous members, dependents, and servants of the extended collective family also contributed to the maintenance of the patio-centered house with its isolation from the street. In turn, this interior reference permitted activities of a different nature to exist side by side with little mutual interference. The amount of passing traffic or the proximity of a workshop matter very little to the inhabitants of a patio-centered house, most of whose activities are completely isolated from the street outside.

6. Although there are many American communities whose rate of growth over long periods of time has been no greater than that of Guatemala City, it is notable that short-run absolute rates of population increase have been held down to a comparatively low level by historical and ecological circumstances. Thus, Guatemala City has never passed through a period of short-run growth so rapid as to impair the capacity of the existing pattern to absorb new population and to make adjustments. The present era represents perhaps the closest approach to that condition. Despite the violence of the local business cycle, the building cycle is dampened by the fact that there is seldom a need to absorb any tremendous sudden influx of population, a phenomenon which characterizes North American cities at periodic intervals.

This brief and incomplete consideration of the factors governing ecological distribution immediately suggests that there may also be differences in societal organization which depart from the urban norms typical of an industrial society in a parallel direction. So long as the urban milieu is viewed from the point of view of the rural culture, as in Redfield's classic study of Yucatan, the city appears unmistakably as an area in which conventional social patterns are transmuted and disorganized. The basic function of the city as a center for the two-way diffusion of culture traits implies the destruction of the culture complexes and the related social forms carried by incoming rural migrants as well as the disintegrating influence of foreign traits transmitted by the city to rural communities within its area of influence. If, however, we compare the Central American city with urban settlements of the same size elsewhere, we may note a stability and continuity of pattern which is equally striking. Without offering any detailed proof for the hypothesis that Guatemala shows a continuity in social structure which is analogous to its continuity in physical pattern, it may be pertinent to suggest certain conservative factors which arise from the total social situation:

1. The continuity of the upper-class group is apparently great. Of Thompson's 30 principal families of 1826, at least 13 would have to be included in any such list today. A similar continuity may be noted for the leading commercial enterprises, a large proportion of which date back seventy or eighty years to the beginnings of formal commerce in the modern sense. Such superficial factors as the building of villas for temporary residence at Amatitlán, the service functions performed by the Indian residents of the neighboring (eight kilometres) town of Mixco, the persistence of the siesta and the noon-time family association

it permits, the organization of the University and of the private secondary schools, the peculiarities of local accent and usage, the organization of the native markets, the practices associated with shopping, the differential religious participation of men and women, illustrate a continuity of specific local habits for more than a century. Descriptions of the urban diet written in the 1820s are, except for imported canned goods, quite applicable to the present day. The network of communication and transportation centralized in the capital is organized almost precisely as it was in 1889 although the specific means of transportation have changed.

2. The curious phenomenon associated with the maintenance of status by these upper-status groups has been the success of the city as an instrument of upward social mobility. It is the more plausible opinion that the growth of population has been achieved by the ladinoization of Indians rather than by the in-migration of Ladinos. Numerous investigators have called attention to the largely sociological character of ethnic identification throughout this area; it is often impossible to distinguish between Ladinos and Indios on the basis of physical characteristics alone. Unlike Merida, the indigenous element in Guatemala City's population has always been small (6.5 percent in 1940 compared to 55.4 percent for the country as a whole), and maintenance of the Indio identification is normally dependent upon remaining in an Indio community. Thus, Indio migrants rise by ladinoization. There is some evidence that Ladino migrants also rise by the mere acquisition of urban status and literacy. The efficacy of social reward is even more striking in the case of immigrant foreigners. There is a distinct tendency for immigrants to rise one major step in the social scale. And this is facilitated by the great amount of personal service which is still part of the pattern of living. (Lower-middle-class Americans, joining the American colony and acquiring three servants and caste privileges, have in a definite sense been promoted.)

3. Finally, it should be noted that this urban society has developed, in its own manner, a characteristic noted by many observers in the surrounding rural environment, to wit, a basic traditionalism which permits the absorption of many alien traits and practices with comparatively little effect upon the underlying patterns which shape behavior. The comparison may be slightly strained, but the all-pervading authority ascribed to "custom" by the rural Indian seems to be matched in a sense by the influence of "the cultural tradition" on the urban intellectual, or "the national character" used by businessmen to explain the more uneconomic aspects of Guatemalan trade practices. Guatemala is an area of only partly explicable survivals: in costume, ritual, community organization, and group behavior generally. Although Guatemala City has developed several substantial foreign colonies, whose members play a large part in local commercial activity, and is subject to the influence of foreign films, tourists, automobiles, furniture, refrigerators, books, sports, and styles, it is very doubtful whether the typical life-cycle of the workman or servant clearly manifests a marked degree of cultural inconsistency. It is equally striking that the expansion of the upper class both in numbers and diversity has been marked by relatively slight displacement of the older dominant group. Although this tendency may easily be exaggerated, it should certainly not be overlooked.

. . . . .

The Social Ecology of Latin American Cities: Recent Evidence*

Bruce London

The literature on the ecological structure, the internal spatial structure, or the distribution of demographic and ecological factors within Latin American cities has grown considerably since Caplow published his article on Guatemala City in 1949. This body of work has tended to focus on a small number of key themes or issues. In general, researchers in the post-World War II era were beginning to recognize the ethnocentrism of earlier writings in human ecology and were, as a result, beginning to conduct cross-cultural comparisons of urban structure. This issue was clearly addressed by Caplow; his paper attempted to outline the spatial development of Guatemala City in order "to account for the absence or attenuation of ecological processes which have come to be taken for granted in the analysis of urban growth elsewhere" (Caplow, 1949:113).

To be more specific, the emergent comparative literature constitutes a critique of aspects of Burgess's (1925) concentric zone hypothesis. Burgess was attempting to answer the question of whether or not cities, despite obvious variations, have an underlying "ideal typical" form by which they may be described. With his series of five concentrically arrayed zones, he implied a direct relationship between socioeconomic status and distance from the city center. We will focus our attention on this gradient hypothesis.

Empirical studies of both American and foreign cities have demonstrated that the Burgess construct was both time-bound and culture-specific. Researchers like Caplow were beginning to discover that Burgess's theoretical construct does not have the degree of universal applicability which urban sociologists of an earlier day (and perhaps Burgess himself) attributed to it (cf. Dotson and Dotson, 1957:1–5). In many Latin American cities, "the gradients of status typically found in North American cities were literally reversed" (Caplow, 1949:114).

The search for ideal types or cross-cultural universals ultimately reduces to a question in the realm of the sociology of knowledge—is cross-cultural generalization possible? Or, are all such attempts at generalization subject to the pitfalls of historicism or cultural relativism? Criticisms of Burgess's attempt to form an inclusive generalization have thus evolved into a debate over the relative import of unique cultural variables (indigenous to a given society) on the spatial structure of a given city in that society. For analytical purposes, we can distinguish two divergent

---

*This article was written especially for this volume, although much of the material presented is adapted from a paper by London and Flanagan (1976).

schools of thought. On the one hand, researchers such as Dotson and Dotson (1954, 1956, 1957) and Caplow (1949)—generally writing prior to 1960—emphasize the empirical refutation of Burgess's hypothesis in cities in Latin American countries, and cite certain culturally "unique" variables such as land-use planning and value orientations as the ecologically relevant forces which determine either a non-zonal or an inversion of the zonal pattern posited by Burgess. On the other hand, researchers such as Schnore (1965; 1972) and Hawley (1971)—writing more recently—recognize the import of the culturally unique, but emphasize the need to search for more inclusive generalizations in spite of the difficulties inherent in such an endeavor.

In order to explore the possibility of formulating qualified generalizations, it would be advantageous to have an ideal type with which to contrast the Burgess construct. As a quasi-theoretical framework, we would then have two distinct ideal types, each standing at opposite ends of an hypothetical continuum. Empirical cases would be assumed to fall somewhere between the extremes.

Sjoberg (1960) has provided just such a device in his description of the spatial ecology of the pre-industrial city. Typically, this city was walled, with its central area containing the prominent governmental and religious structures and the main marketplace. The city also tended to be sectioned off along ethnic and occupational lines (with the two often coinciding). However, of central concern here, there was an indirect relationship between power and wealth, and distance from the center of the city—exactly the reverse of the socio-economic status gradient posited by Burgess.

Approaching cross-cultural generalization within a historical perspective, the empirical question becomes not one of determining whether Burgess or Sjoberg is "right" or "wrong." Rather, it becomes a matter of determining just where on our continuum a particular city (in a particular society at a particular time) lies. In effect, was are attempting to reconcile the two opposing types within a single framework.

Schnore's "evolutionary sequence hypothesis" (1972:17–21) attempts just such an analysis. After reviewing the literature on Latin American cities, and noting that their spatial structures conform to neither Burgess's nor Sjoberg's types, Schnore (1972:21) advances the possibility that these two types "are special cases more adequately subsumed under a more general theory of residential land uses in urban areas." More specifically, he feels that it may be possible to demonstrate that, starting with a pattern similar to that of the pre-industrial city, the residential structure of the city evolves in a predictable direction (towards the structure posited by Burgess) as city and nation experience the process of development.

The strategy employed in this paper will be to review the literature in order to see if certain evolutionary processes are evident in the spatial structure of cities in Latin American countries. Do cities, early in their histories, display a spatial structure similar to Sjoberg's pre-industrial city type in which centralization of upper-class residence, and a corresponding decentralization of lower-class residence, is the general rule? Do these same cities, later in their histories, display a spatial structure similar to Burgess's concentric zone type in which centralization of the lower classes, and decentralization of the upper classes, is the rule?

If any initial support for an evolutionary sequence hypothesis is to be gleaned from the literature, we must observe the changes that particular cities undergo over time in order to discover if, or to what extent, the assumption of elements of the Burgess pattern has taken place. If the cities examined were initially pre-industrial in form (at least in terms of the status gradient which is our central focus), then our focus should be on whether or not upper-class "suburbs" and centralized "slums" developed over time.

Latin American cities—the evidence

Even those Latin American studies which—by contrasting Anglo and Latin "cultures"—emphasize the importance of culturally unique factors in determining a city's internal spatial structure provide "evidence" which supports an evolutionary-sequence hypothesis. After reviewing the literature on the spatial structure of Latin American cities, Schnore (1965:358) noted that in *all* the studies reviewed the traditional or colonial pattern (characterized by a "pre-industrial" status gradient) was "reported to be in one or another stage of breakdown," with an apparent tendency to shift in the direction of the North American pattern.

This observation was reported as early as 1934 in Hansen's study of Merida. Hansen felt that the traditional relationship between status and residential location with reference to the center was reversing over time largely because of population growth and concomitant organizational change. Caplow (1949:132)—after reviewing much the same literature as Schnore—augments Hansen's initial insight by pointing out that, after arranging all the Middle American cities upon which some ecological data are available in order of size, "it is at once apparent that the larger the community the further it has departed from the traditional colonial pattern."

This is not to imply that Caplow is optimistic about the possibility of cross-cultural generalization. He feels that the reverse gradient persists in Guatemala City despite rapid growth (1949:123) and that the syndrome of changes in land use common to the North American "crescive city" does not exist in Guatemala City (1949:127)—largely because of a tradition of land-use planning and an elite value system which leads them to "cling to central location." Nevertheless, Caplow cannot avoid pointing out that there are distinct indications that the North American pattern may develop in the future for there is already some displacement of upper classes from their traditional central location (1949:125).

Not unlike Caplow, the Dotsons (1956, 1957) attribute causal primacy to culturally unique variables in determing the spatial structure of Mexican cities. They, too, however, cannot avoid citing descriptive evidence which lends support to an evolutionary sequence hypothesis. They cite the importance of the upper-class view that the urban life is "the good life," claiming that such a value system is one key to the persistence of a reverse gradient.

The Dotsons themselves, however, also point out how the elite urbanite is, with increasing frequency, "forced" from his traditional location by the expansion of the central business district. There comes a point at which culturally defined ideals are no longer able to withstand the economic pressures of growth. Again in agreement with Caplow, the Dotsons recognize "a correlation between the degree of deviation from the colonial pattern and the size and rate of growth of the city" (1954:367). In smaller cities such as Puebla, we do not find some of the "new ecological tendencies." Guadalajara, however, has a mixed pattern in which most of the housing near the center is middle or upper class; and the periphery is the locus of the poorest housing; *but,* the very best housing is partly located on the edge or outskirts of the city. Finally, Mexico City displays the most marked shift towards the North American pattern, including much new middle-income housing on the fringes, suburbs for the wealthy ten miles from the central business district (Hayner, 1968:166), and, most significantly, a considerable deterioration of parts of the central business district which is very reminiscent of Burgess's zone of transition. Amato's (1969a, 1969b, 1970) studies of the South American cities of Bogota, Quito, Lima, and Santiago generally reinforce the emerging image of a clear shift away from traditional elite centralization towards a rather mixed pattern marked by the persistence of some elite centralization and low-status decentralization, a frequently sectoral decentralization of elite groups to environmentally desirable locations, and a delayed but increasing centralization of low-status groups.

Rather than interpret this emerging pattern in terms of convergence with the North American structure, the Dotsons view it in terms of the emergence of "an essentially new ecological form." Nonetheless, this new ecological form may just as easily be interpreted as a shift *towards* the North American pattern, especially in light of the fact that this form, although new to Mexico, is probably very similar (in terms of its mixed residential distribution) to American industrial cities of the nineteenth or early twentieth centuries. Indeed, recent studies by Schwirian and Rico–Velasco (1971) and Schwirian and Smith (1974) of the three largest cities in Puerto Rico make just this point by linking the centralization patterns of a city's status groups to, first, a society's position in the economic development process, and, second, a city's position in that society's urban hierarchy. Just as Caplow and the Dotsons found larger communities departing further and further from the Sjoberg pattern, the Puerto Rican studies reveal that "Mayaguez, . . . the smallest of Puerto Rico's metropolitan centers . . . , shows an ecological pattern characteristic of the traditional city"; while the somewhat larger city of "Ponce is in the midst of a spatial shift in population distribution" such that "the traditional spatial pattern is disrupted while the more modern has yet to emerge"; but that "the ecological patterning of San Juan—the primate city—is very similar to that of cities in highly developed societies" (Schwirian and Rico–Velasco, 1971:334). Indeed, "in San Juan the most centralized groups are generally those of lowest status" (Schwirian and Smith, 1974:420). (For a detailed description of San Juan's changing ecological structure, see Caplow et al., 1964:1–63.)

It should be noted that there is at least one criticism in the literature of this frequently cited relationship between city size and the evolution of city structure. Yujnovsky (1976:24–25) notes that

Buenos Aires, one of the largest and most industrialized cities of Latin America, exhibits suburbanization and central district deterioration to a relatively small degree. He concludes "that a typical pattern of internal urban development in Latin America does not exist" (1976:25). In a related paper, however, the same author (Yujnovsky, 1975) separates cities by "population rank." He asserts that, in metropolitan areas (including Buenos Aires), there is on the one hand an "upper-class axis" or ecologic segregation of upper strata in sectors extending into the periphery, and, on the other hand, "a trend toward low-income groups from the center to the periphery, contrary to the ecology of U. S. cities. This is due to the small size of the central slum area as compared with the number of low-income persons living on small peripheral lots . . . or in squatter settlements" (Yujnovsky, 1975:211–12). In intermediate-size and small cities, however, he confirms "the general position of the upper stratum at the city center, surrounded by middle-income groups and finally lower-income strata in a peripheral location" (Yujnovsky, 1975:212), despite some incipient suburbanization.

In one of the most balanced reviews of the findings, Portes and Walton (1976) attempt to summarize the mixed patterns emerging in Latin American cities. Elite displacement from central locations began to occur as early as the latter half of the nineteenth century in some Latin American cities. In any given city, movement tended to be in a single direction, towards "the most desirable areas of the urban periphery" (Portes and Walton, 1976:22). At the same time, the majority of the urban poor have tended to settle in the less desirable portions of the urban periphery in squatter settlements, although in a number of cities (e.g., Santiago, Mexico City, Lima, and others) slum-like residences of the poor have developed. For the most part, however, the central area has come to be the residence of "the growing middle and lower-middle strata" (Portes and Walton, 1976:67). Whether these changes are viewed as the emergence of a new pattern or as an evolution toward the Burgess pattern, the fact remains that the traditional, colonial, Sjoberg pattern has been superseded in major Latin American cities.

By the same token, the Burgess pattern no longer adequately describes North American cities as gentrification, the revitalization of inner-city areas, the decentralization of industry, and increases in the numbers of lower-income groups in the suburbs have altered the status gradients typical of an earlier day (Theodorson, 1961:330). One might conclude that there is no uniform pattern to be observed everywhere, and, as far as it goes, this would be correct. Such a conclusion, however, misses a very basic point: "It is important not to confuse a specific pattern of spatial distribution with the general principles of . . . ecology" (Theodorson, 1961:330). Ecological principles must reflect the fact that structure arises out of *process*. The city is a dynamic emergent—growing and changing. This processual nature has meant that the spatial relationships among groups within the city are constantly in flux. Any static, cross-sectional description of what is the most recently valid pattern of residential land use will soon be outmoded. Just as cities in the Third World are moving away from the Sjoberg pattern, cities in North America are transcending the Burgess pattern. Even though both are evolving under the impetus of different causes, each is incorporating what were once solely elements of the other's ideal status-gradient pattern, and are thus becoming structurally more and more similar. Thus, the hypothesis that some sort of overall convergence towards similarity is taking place, or that the residential structure of the city is evolving in a predictable direction, should be neither constrained by the reification of a given "ideal type," nor precluded by the fact that there is now no uniform pattern observable everywhere.

References

Amato, P. W. "Population Densities, Land Values, and Socioeconomic Class in Bogota, Columbia." *Land Economics,* 40 (February 1969a):66–73.

Amato, P.W. "Environmental Quality and Locational Behavior in a Latin American City." *Urban Affairs Quarterly,* 5 (September 1969b):83–101.

Amato, P.W. "A Comparison: Population Densities, Land Values, and Socioeconomic Class in Four Latin American Cities." *Land Economics,* 41 (November 1970):447–55.

Burgess, E. W. "The Growth of the City: An Introduction to a Research Project." In R. E. Park, E.W. Burgess, and R. D. McKenzie (eds.), *The City,* pp. 47–62. Chicago: University of Chicago Press, 1925.

Caplow, T. "The Social Ecology of Guatemala City." *Social Forces,* 28 (December 1949):113–33.

Caplow, T. *Ecologia de la Cuidad Guatemala.* Guatemala, C.A.: Centro Nacional de Investigaciones Sociales, 1966.

Caplow, T.; Stryker, S.; and Wallace, S. E. *The Urban*

Ambience: A Study of San Juan, Puerto Rico. New York: Bedminster Press, 1964.

Dotson, F., and Dotson, L. O. "Ecological Trends in the City of Guadalajara, Mexico." *Social Forces,* 32 (May 1954):367-74.

Dotson, F., and Dotson, L. O. "Urban Centralization and Decentralization in Mexico." *Rural Sociology,* 21 (March 1956):41-49.

Dotson, F., and Dotson, L. O. "The Ecological Structure of Mexican Cities." *Revista Mexicana de Sociologia,* 19 (1957): mimeographed translation.

Hansen, A. T. "The Ecology of a Latin American City." In E.B. Reuter (ed.), *Race and Culture Contacts,* pp. 124-42. New York: McGraw-Hill, 1934.

Hawley, A. H. *Urban Society: An Ecological Approach.* New York: Ronald Press, 1971.

Hayner, N. W. "Oaxaca: City of Old Mexico." *Sociology and Social Research,* 29 (November-December 1944):87-95.

Hayner, N. W. "Mexico City: Its Growth and Configuration, 1345-1960." In S.F. Fava (ed.), *Urbanism in World Perspective: A Reader,* pp. 166-77. New York: Crowell, 1968.

London, B., and Flanagan, W. G. "Comparative Urban Ecology: A Summary of the Field." In J. Walton and L. H. Masotti (eds.), *The City in Comparative Perspective,* pp.41-66. New York: Saga-Halstead, 1976.

Portes, A., and Walton, J. *Urban Latin America.* Austin: University of Texas Press, 1976.

Schwirian, K. P., and Rico-Velasco, J. L. "The Residential Distribution of Status Groups in Puerto Rico's Metropolitan Areas." *Demography,* 8 (February 1971):81-90.

Schwirian, K. P., and Smith, R. K. "Primacy, Modernization, and Urban Structure: The Ecology of Puerto Rican Cities." In K. P. Schwirian (ed.), *Comparative Urban Structure: Studies in the Ecology of Cities,* pp. 324-38. Lexington: D.C. Heath, 1974.

Schnore, L. F. "On the Spatial Structure of Cities in the Two Americas." In P.M. Hauser and L.F. Schnore (eds.), *The Study of Urbanization,* pp. 347-98. New York: Wiley, 1965.

Schnore, L.F. *Class and Race in Cities and Suburbs.* Chicago: Markham, 1972.

Sjoberg, G. *The Preindustrial City: Past and Present.* New York: Free Press, 1960.

Theodorson, G. A., ed. *Studies in Human Ecology.* Evanston: Row, Peterson, 1961.

Yujnovsky, O. "Urban Spatial Structure in Latin America." In J. E. Hardoy (ed.), *Urbanization in Latin America,* pp. 191-220. New York: Anchor, 1975.

Yujnovsky, O. "Urban Spatial Configuration and Land Use Policies in Latin America." In A. Portes and H. L. Browning (eds.), *Current Perspectives in Latin American Urban Research,* pp. 17-42. Austin: University of Texas Press, 1976.

The Anatomy of Eleven Towns in Michoacán*

Dan Stanislawski

The towns of Michoacán have distinct personalities. Of this I was sure after several field trips into the state. But "personality differences," although readily felt, are not easily expressed. A method was needed to determine these differences, one that would indicate them in a manner useful to geographers.

In the small town of Latin America the home is also the center of most economic activities. The major exceptions to this are trading in the open market and farming. This being the case, it seemed probable that if a record were made of all dwellings and the economic activity of each, the information, when placed upon a map, would show distributional aspects that might indicate differences between towns. Through this, the elusive quality of these differing towns might be understood.

To procure such information I had to canvass all dwellings or get the information from informants. In the latter case I used always two and sometimes three informants for the same area to be certain of accuracy.

For mapping, all activities were arranged in four major categories: (1) stores, (2) crafts, (3) administrative offices, and (4) services. In addition to

*Reprinted from *Latin-American Studies X* (Austin: University of Texas Press, 1950), pp. 1–3, 40–48, and 71–75, by permission of the author and the publisher. (Original article, pages 1–75.)

this, a subjective estimate was made of the quality of each house.

I chose eleven towns to study. One is located in the coastal mountains of the southern part of the state; three are in the low Balsas valley; and two are in the contact zone between the hot Balsas valley and the temperate slopes to the north. Four are within the mountain valleys of the volcanic ranges, and one, the highest, is on the cool slope at about 8,000 feet elevation.

These towns were chosen originally because of an assumption that each would be strongly influenced by its geographical area. From this assumption, it seemed probable that a town of one geographical area would exhibit intrinsic differences from a town of another area. From this, it was inferred that the four towns in the mountain valleys would show similarities due to their environment, and that the Balsas valley towns would show similarities to each other but differ from those of the mountain valleys. These assumptions were quickly shattered by the analytic maps. It became obvious that the geographical region could not explain the differences between the towns except in part.

It has long been recognized, not only for Spain and the Spanish New World but for other parts of Europe and North America as well, that importance and prestige is associated with the central square, plaza, commons—whatever it is called. In Latin America, this tendency seems to be accentuated.

In studying and comparing the maps made of these selected towns, the overwhelming importance of the plaza in some towns and its neglect in others was obvious. That this is directly correlated with racial and cultural backgrounds makes it fundamental to the interpretation of the results of this study.

The position relative to the plaza in Hispanic towns is also a rough index as to the value of property. This is also true, although less obviously, in relation to the main arterials. It is not possible to place exact money values on property, either for business or residence, but Hispanic informants never hesitated to express the opinion that the best property was that on or near the plaza or on the arterials.

Among the eleven towns studied, some are of basically Iberian character, others are of the New World with sixteenth-century innovations but not in sufficient numbers or strength to nullify the essential Indian character. A third group must be noted. It cannot be accurately called mestizo or blended because it possesses both pure Indian and Hispanic qualities rather unmixed.

It seems logical to classify all of these towns in terms of two basic culture groups, Hispanic and Indian, with a third category, the dual-character town.

Under these major categories a further breakdown must be made in terms of function, but function altered by geographical position and time of settlement.

The Hispanic towns are: Pátzcuaro, Ário de Rosales, Tacámbaro, Purépero, Apatzingán, Buena Vista, Churumuco, and Arteaga. Erongarícuaro is a dual-character town, rather more Hispanic than Indian. Chilchota falls within the same category but is more Indian than Hispanic. Pichátaro is Indian with its character essentially unaltered by the fact that the stores of the town are largely mestizo-owned.

. . . . .

Pichátaro

The village of Pichátaro is on a low knoll looking out over a beautiful and fertile mountain valley of approximately 8,000 feet elevation. It lies about 1,000 feet higher than Erongarícuaro and is three hours distance from it by muleback. It is high enough to lie on the border of the Cwb and Cwc climates, but the growing season is long enough for a prosperous agriculture.

There are ample lands to feed the villagers well and to produce a surplus sufficient to support a number of craftsmen who are, partly at least, producers of luxury goods.

Pichátaro has never been a large settlement. It was in existence before the arrival of the Spaniards. They altered the form of the unplanned Indian village by imposing the grid. They also added the church. But seemingly the life of the Tarascans was otherwise changed but little.

Near the end of the sixteenth century, Ponce[1] found it to be a small village with water brought in to irrigate its crops. There were many fruit trees. It has grown since then, but part of his description still holds. The good springs nearby are still furnishing the water for irrigation of their lands. There are still many fruit trees. In fact, Pichátaro has started, in a small way, the commercial production of apples.[2]

We have another brief report on Pichátaro in 1788 when it was described as having well-placed houses on beautiful little streets.[3] It was completely Indian. The orchards were still notable, as also were fine harvests of wheat and maize. Only one craft, carpentry, was mentioned.

Pichátaro has had good luck. It has always had good lands and sufficient acreage for its population. It was sufficiently removed from contact with the Spanish landholders around Lake Pátzcuaro so that the village lands were not appropriated to be included in haciendas as was the case near the lake. Yet being situated on the route between the lake and other mountain villages farther to the west, it has not stagnated by isolation.

This explains a good deal of the quality of present Pichátaro. Its inhabitants have come in contact with other places and people. A few "foreigners" settled there, and it is no longer 100 percent Tarascan in speech but it was never made a stopping place for mule-drivers. Its location only three hours away from Erongarícuaro precluded this.

[1] Fray Alonso Ponce, *Relación breve y verdadera de algunas cosas de las muchas que sucedieron al padre Fray Alonso Ponce en las provincias de la Nueva Espana . . .* , II, 6.

[2] Trees are notable in all of the mountain villages of these Tarascan Indians. The appearance of the villages is often that of houses in the midst of orchards (although not commercial orchards with the trees in seried rows). This is all the more noticeable inasmuch as Hispanic towns in this area are almost treeless when viewed from the street. This may involve a difference of cultural attitude between the Spaniard and the Indian.

[3] Archivo General de la Nación (AGN), *Ramo de História*, Vol. 73, Expediente 18, p. 100.

The effect of the few mestizos who have settled in town is chiefly seen in the stores. Of eleven in the village, six are owned by mestizos. This does not properly indicate their importance, however, for the mestizo stores do probably 95 percent of the business. They are nearest to the plaza. The Tarascan stores are farther out and with one exception have nothing more in stock than a few bottles of soft drinks, a few cakes of soap, and perhaps a few candles. They are hardly businesses. Hour by hour the owner may be found sleeping, chatting, or working at odd chores without a glimpse of a customer. Toward the end of the afternoon and in the early evening when business is brisk, one or two of the mestizo stores are crowded and do a bustling business while the little Tarascan stores have only an occasional customer.

The idea of the store is somewhat alien to these Indians. They transact their business and exchange their surplus goods between one another or in the market places of Erongarícuaro and Pátzcuaro. Each Sunday several groups jog downhill to Erongarícuaro to the market. Their chief product for exchange is pine wood that they cut in the surrounding hills. In the Erongarícuaro market they meet the lake fishers, and there they barter their wood for fish. When there is a surplus of wheat or maize, and fruits in season, they also are taken to market.

All stores, mestizo and Tarascan, are located on the main street, which is the "through route." This is in keeping with Hispanic practice and is due to the influence of the mestizos who have come into the town.

The anatomy of the town indicates its difference from Hispanic settlements. There is far less concentration of activities. That which does exist is due to mestizo store influence on the main street. As for Tarascan crafts, they are widely scattered and show no tendency to be near to the plaza or in any other particular part of town. As far as one can determine, there is no preferred location. There are carpenters near the plaza, on the main street as well as in the outlying blocks. The same is true of hatmakers and weavers. Position carries no prestige.

The same is true of houses. There is little difference in quality between a house on or near the plaza and a house at the outskirts. In fact, the two chief officials of town at the time that this inquiry was made lived at one extreme corner of the village. They both agreed that one place was as good as another for one's home. They had no idea of the value of the lot. They said that sales were so infrequent that it was hard to state a price. It all depended on "how much you wanted to pay."

The function of crafts within the village and the attitude of the Indians toward crafts is quite different from that in Hispanic settlements. It is non-commercial insofar as trade outside of the village is concerned. The crafts are almost all Indian and the exceptions have been fitted into an Indian pattern of use and wont. This is in spite of mestizos and their control of the main stores. It is so because crafts are partly "luxury." Tarascans of Pichátaro are fundamentally farmers who make or support the making of things for their greater comfort and satisfaction. Now crafts are not taken up commercially but to supply things that can be used in town. Only occasionally are the embroidered blouses or belts of woven wool taken into the markets. Actually, the only craft that is pursued for commercial purposes is the most ancient of their crafts, that of woodworking. But this is only one specialized part of their woodworking—that of making canoe paddles. Pichátaro, three hours away from Lake Pátzcuaro, makes most of the paddles used by the boatmen of the lake. Any Sunday these paddle-makers can be found in the market in the city of Pátzcuaro. (One man from Pichátaro also takes wooden chairs to the Pátzcuaro market.)

In the matter of woodworking the Tarascans were noted at the time of the Conquest. It amounted almost to a cult at that time and has remained a strong part of their culture pattern. In a document of 1788 it was noted as being the only outstanding craft endeavor in Pichátaro.[4] Today it is the most important and most spectacular of the skills. There are nearly a dozen men in Pichátro now known to the villagers as "maestros" in woodworking, fine craftsmen working with an aesthetic sense. They are not mere repairmen as is the case with the carpenters in Hispanic towns.

Most of the houses of Pichátaro, as in many towns of the mountain Tarascans, are built of sawed logs. Formerly these logs were hewn. The carving of the pillars, capitals, lintels, and door frames as well as other parts of the house is done with the fine precision of a craftsman. The lich gate with its two-slope roof may have been introduced into this woodworking country by churchmen or it may have been developed with some idea of utility. Indians now justify them by saying that they protect the gate against the rain. In view of

[4] AGN, *Ramo de História,* Vol. 73, Expediente 18, p. 101.

their elaboration and the comparative simplicity of the gates this is hardly reasonable. It seems to be just another opportunity to exercise their skill in carving and to satisfy their aesthetic desires and love of the craft. Woodworkers also occupy themselves in making chairs, tables, and other household furniture.

There are many good weavers. Some of the handsomest of the native woolen blankets are made by the men of Pichátaro. Women weave belts in both men's and women's types and napkins of cotton. There is one weaver of women's shawls.

Sweaters are "knitted" (with a buttonhook) on a wooden frame with a row of nails on either side of the elongated middle opening.

There are several hatmakers who weave the strips of palm fibre, purchased in the market at Erongarícuaro, and stitch them together on a sewing machine. At the end of the nineteenth century Pichátaro was the center for the manufacture of hats. Early in this century many of the hatmakers moved, due to chaotic political conditions, to a place of refuge on the island of Xarácuaro in Lake Pátzcuaro. The hat industry of Xarácuaro has become dominant in the area and only a remnant is left in Pichátaro. A few brooms are made from the palm fibre left over from the hatmaking process.

There are numerous masons, a craft obviously from Spain. The masons of Pichátaro may hire out by the day if they feel so inclined, but a proletarian class of masons has not been formed. With stone carving, as with wood, much of the effort expended cannot be explained in terms of cash value. Many of the streets are lined with walls of chiseled stone. The foundations of many of the log houses are made of shaped stone. In this there is clearly a pride of craftsmanship that is beyond mere utility.

Leatherworking is weakly represented in Pichátaro. There are enough shoemakers to supply the needs of the village but that is all. There is no attempt to make anything else nor to elaborate on their shoemaking process as they have with their Indian craft of woodworking. Tarascans, of course, worked leather and made a sort of shoe before the advent of the Spanish. There are no tanners in town. Presumably this activity has been appropriated by Hispanic towns and was never a strong enough part of the Tarascan village to be maintained.

There are no metal workers in Pichátaro.

. . . . .

General conclusions

1. Great persistence of culture groups.
2. The character of towns is largely based on the cultural background of the dominant group and this cannot be recognized casually. Probing into the anatomy of towns brings to light important survivals that will not otherwise be suspected.
3. The original function and purpose of a firmly established town tends to continue. For example, Ário de Rosales was a commercial town from its inception. It is now, in spite of changes in the economy and economic geography of Mexico, in spite of changes in transport, a commercial town. There is no question that its geographical position was excellent for precisely this function but there are other locations that might have assumed its function. They did not because Ário had supplied the service. Tacámbaro and Pátzcuaro continue, in spite of major changes in Mexico, to function much in the way that they have done. The same can be said of most of the other towns.
4. The presence or absence of certain traits in a town may be a clue to its period of settlement and its type of settlers. And this may be reversible, for granting the serviceability of the conclusions, if the type of settler and the time of settlement is known, one may have a fair indication of the character of the settlement, although obviously, the variations within the town and, indeed the activities, are altered by physical conditions.

The qualities of the Hispanic town

A. Historical-Cultural Aspects

1. Prestige of the plaza and arterials

This is true for homes as well as for all activities. Usually it includes all sides of the blocks contiguous to the plaza. The land is of higher price here and values diminish in all directions out from them with the exception of the properties along the arterials. Positional importance decreases less rapidly from the plaza toward the peripheries along these streets leading to the main cross-country routes.

2. Social position of trade

The retail merchants have, in general, choice positions. Invariably many of the best locations on the plaza are theirs. Most of the street corners along the arterials are occupied by retail stores.

These are middlemen, and there is seldom any manufacturing connected with their activities. There was and is a New World middle class, Spanish or mestizo, not Indian. It has never been large but always, presumably, vitally important.

Judging by their geographical position in towns, the individuals have long enjoyed prestige and respect.

In Pátzcuaro one of the merchants on the corner of the main plaza is considered to represent the finest graces of the disappearing colonial nobility. His family has been involved in trade through all known generations. Usually the merchant lives with his store. There is no detachment of living from business. There is no desire to finish the disagreeable work of the day so that one can remove himself to his home for his hours of enjoyment.

   3. Economic prestige of mule-driving

The positions of the *mesones* is almost always favorable. This is surprising to one of the Anglo–American point of view. The stopping places of mules and their drivers are breeders of odors, flies, and presumably disease. In the sixteenth century the Spanish king issued explicit instructions to the effect that noisome and objectionable activities should be relegated to beyond the limits on the low side of town. This obviously wise regulation is still largely honored. But in the minds of the Spaniards there could have been no idea that there was anything distressing about the *meson*. It has a location side by side with that of the most properous merchant or the grandest "grandee" of town. They are intimately associated with the plazas or are among the better locations of the chief arterials.

There is, in most towns, a clear positional relation between the *meson* and the church—note especially Pátzcuaro, one of the oldest and certainly the most aristocratic of the towns. It points certainly to one conclusion, that mule-driving was an ancient and honored profession. Whether it was equally so in Iberia would be interesting to know. But why is it associated with the church? Did churchmen desire the association? Did they perhaps use mule-drivers in furthering their work? Or was it, a more probable explanation, a simple relation between two honored institutions of Hispanic life? The *mesones* seem to have been among the most stable things in Pátzcuaro. Of those recorded in this study most appear on the map of 1895. It is entirely reasonable to assume that prior to 1895 they had been there for long generations.

   4. Prestige of leatherworking

The anatomy maps show that towns that are clearly Hispanic in background have a variety of craftsmen working in leather, making huaraches, shoes, gloves, vaqueros necessities, bridles and saddles. Their position in town is good. They do not have the preferred position of the storekeeper, but they are usually close to the plaza or on the arterials.

This again raises the question as to the type of persons who followed the professions. In Spain shoemaking and training were officially designated as "oficios viles y bajos."

Were there a group of depressed laborers in Spain who found new and better lives for themselves in the New World? Or did the necessities of the situation offer opportunities and prestige denied in Spain? In any event the record of this study seems clear beyond cavil.

   5. The weakness of all crafts except leatherwork

Whatever other crafts exist are found located in inferior positions toward the periphery or in blind alley locations between the arterials. This is notably the case with regard to woolworking. Weavers are nearly always in the worst possible locations and never in the best. It is not true that Hispanic peoples and towns avoided textile-making. In some of the towns considered in this paper cotton-weaving was important during colonial times and was in the hands of mestizos. It is probably true that they were never strongly interested in the weaving of wool. It seems that Indians at the cooler elevations were glad enough to relinquish cotton to the Hispanic peoples in exchange for wool.

Woodworking is absent or, if present, unimportant. This is probably a universal trait of Hispanic settlements. The lack of trees in such settlements probably reflects an age-old Iberian disinterest in trees and their products. Metalworking is important nowhere.

Spain was a raw-material-producing country with income derived from her raw materials. The Moors had introduced crafts but presumably the Vizigoths after their reconquest looked with disfavor upon these crafts that had been in the hands of Moors and the Jews. The Spanish kings, at various times, had seen the advantage of craft development and had tried to promote them but with small success. If the evidence of town anatomy in Michoacán may be accepted, the basic Iberian attitude was transferred here.

   6. Conservatism

After four centuries many of the traits shown in the anatomy of these towns are similar to those demanded in the sixteenth century. For example, the position of the slaughter houses, probably that of the *mesones,* and the churches.

B. Distinction between Towns on the Basis of Function and Period of Settlement
  1. Landholding (hacienda) towns, e.g., Pátzcuaro, Tacámbaro
  Greatest prestige of plaza
  Best homes
  Services in greater number
  2. Early commercial trade route towns, e.g., Ário de Rosales and Tacámbaro
  Activities more evenly distributed
  3. Mule-drivers towns, e.g., Purépero, also Ário de Rosales and Tacámbaro
  Importance of leatherwork
  Importance of the arterials

In the above three groups there is one town in each group that stands out above the others as distinct in its class. For example, Pátzcuaro was the aristocratic seat of the great hacendados of the region. Tacámbaro was the town center for some haciendas but was clearly in second place to Pátzcuaro. Tacámbaro functioned also as a mule-drivers center. In its anatomy it shows the fine homes and their concentration on the plaza that would be expected from an hacienda center. It also shows the importance of the arterials and the distributional factors associated with mule traffic and trade routes. Ário has never been associated with the aristocracy. In times past there have been haciendas in greater importance than today. This may be reflected in its anatomy. But dominantly it is commercial. Purépero, designed almost completely by mule-drivers, shows this in the distribution along its arterials as well as in its leather crafts.

The qualities of the Indian town

  1. Plaza unimportant
  2. No positional value or prestige
  3. Strength of Indian crafts
  Woodworking, weaving (although wool has supplanted the cotton of aboriginal Michoacán)
  4. Weakness or absence of Hispanic activities
  No mule-driving. However, this may be true only of this area. In the colonial records there is evidence of Indian mule-drivers elsewhere. It seems that the Tarascans had little interest or that the particular Hispanic peoples of the area preempted the business.
  Almost no commerce in stores
  Leatherwork of minor importance
  5. Town loosely constructed
  In the center of Pichátaro buildings are closer together than at the outskirts, which is probably the effect of the few mestizos who are concentrated on the main street. In fact, they undoubtedly have made it the "main street." The Indian towns of the Tarascan mountain region all suggest a transitional stage between the *rancheria* and the agglomerated town proper.

The *solares* or town lots are large and serve as house gardens. The houses themselves may or may not face directly upon the street. There are large numbers of trees. In fact the towns often look more like wooded areas with houses scattered among trees than like towns with trees in them.

The dual-character town

There is not much to be said about the anatomy of the "dual-character" town except that it shows traits of both Hispanic and Indian towns.

It certainly lacks strong character. Mule-driving is usually present but weakly represented. No craft is strong, although both Hispanic and Indian crafts may be found. In terms of the total number of activities, the position is midway between Hispanic and Indian towns.

The anatomy map in such a settlement probably shows the degree of dominance of either Indian or Hispanic character (not the number of people, for the Spaniard will dominate with smaller numbers, especially in commerce).

Conclusions

In reviewing the work done and the results of it, one pronounced weakness is at once apparent. It is the need for more examples of Indian towns. One almost pure product and two with traits that are presumed to be Indian in the light of evidence from the one good example is hardly sufficient evidence. For a person who is acquainted with the area it seems tentatively acceptable, however, as there is no apparent discrepancy in terms of other Indian towns of the region even though they have not been dissected anatomically.

Obviously the ancient Indian craft of pottery-making should not have been neglected. An inquiry should be made into the anatomy of a pottery-making village to compare it with Pichátaro. It is equally obvious that samples should be taken from other areas and other native culture groups as well as of Hispanic settlements in other environments.

This neglect was a product of an original assumption that anatomical differences would reflect regional difference. On the basis of that belief, the towns chosen for study had very little relation—only accidental—to racial and cultural groups.

That this assumption was erroneous was not apparent while the field work was being done or a different tack could have been taken. Unfortunately it did not come out clearly until all of the information had been mapped, studied, and reduced to generalizations. By this time the field had been left behind.

The Preindustrial City*

Gideon Sjoberg

Spatial arrangements

If industrial man could remove himself temporarily from his milieu, with all its comforts and conveniences, and set himself down within a non-industrial city, he would immediately perceive the dramatic contrast between its physical complexion and that of the modern industrial metropolis. We first discuss some salient features of the cityscape that would strike his eye, then consider specific land-use patterns.

A general overview

Typically, all or most of the city is girdled by a wall. Inside, various sections of the city are sealed off from one another by walls, leaving little cells, or subcommunities, as worlds unto themselves. Walled cities have been the generalized pattern throughout the Middle East from North Africa to Central Asia, and in India and China during much of their history. Even certain pre-Columbian cities of Meso–America conformed to this pattern.[1]

Walls, moats, and similar devices have been common as defensive bulwarks, though only partially effective, as evidenced by the numerous once-thriving walled cities now lying in crumbled ruins. And with the advent of modern weapons, they have become obsolete. But the city's ramparts serve other functions, like regulating the activities of merchants and other visitors to the city: watchmen and other agents can readily be stationed at the few points of entry to collect tolls or ward off social "undesirables." And the circumvallated districts or quarters reinforce the segmentation of social groups that obtains in many cities. Minority ethnic groups, especially, may be so sealed off in quarters of their own—e.g., as in Fez—that they have only tenuous relations with the general populace. As with the gates to the city proper, the entrances to the various walled quarters are secured at night, bringing communication among the units to a halt. With the ever-present threat of thieves and other marauders, locking the interior gates provides an added measure of protection for the inhabitants.

Within the walled precincts congestion is the order of the day. Although crowding in the center is alleviated somewhat by the spacious dwelling units of the wealthy, the urban poor, except those on the outskirts, beyond the walls, live closely packed. Given the scanty transport.media, people reside and work where they have access to the city's special facilities, and because the technology does not allow many multistoried structures, buildings are set closely together, often immediately juxtaposed, to permit a maximum number of people to partake of the advantages of life within the city walls.

The clumping of buildings is intensified by the narrowness of the streets, mere passageways for

---

*Reprinted from Gideon Sjoberg, *The Preindustrial City* (Glencoe, Ill.: Free Press, 1960), pp. 91–107, by permission of the author and Macmillan Publishing Co., Inc.

[1] See Notes section at the end of this article.

humans and animals, though a few permit the circulation of small wheeled vehicles. Two-way auto traffic in most traditional cities is out of the question. Cities in India, the Near East, Tibet, China, and parts of Europe and Latin America still contain streets wherein the pedestrian can touch buildings on either side simultaneously. The "Street of the Kiss" in Guanajuato, Mexico, derives its name from the fact that at least part of it is so constricted that lovers in houses on opposite sides of the street could kiss from the overhanging balconies.[2]

The usual street, as opposed to the few main thoroughfares, is narrow, winding, unpaved, poorly drained, and apt to turn to mud during periods of snow and rain, making transportation slow and uncomfortable. Medieval Paris was notorious in this respect.

The typical street is greatly congested during the day. Here ambulatory vendors hawk their wares, and numerous small shops and stalls front on the street with little or no sidewalk intervening. Indeed, much business is transacted in the street itself. Combine this with the din of children playing, adults gossiping or bargaining, and animals being led to market, and we find life in feudal cities far from placid and uneventful.

Consider the discomforts and dangers of unpaved, congested, poorly lighted, and poorly drained streets in a city wherein public services are almost totally wanting. Garbage collectors and street sweepers are sparse relative to those in industrial centers. The limited technology of the preindustrial city and the dearth of scientific knowledge about sanitation procedures or the very need for these have meant that waste materials are apt to litter the community. Salusbury[3] describes street life in late medieval England:

> ... butchers and poulterers were by no means alone in their careless disposal of animal refuse; fishmongers and cooks and the ordinary households were all guilty. In Chester (1475) ... women carrying entrails of animals from the butchers carried them uncovered and threw them out near the gates, to the public nuisance.... The private citizen was only too ready to dispose of dead dogs and cats by dropping them into the river or just over the town wall, or even by placing them in any open space....

Furthermore:[4]

> The final disposal of filth and rubbish collected from the streets presented a grave problem. The town ditch and the river were used very frequently for this, but in the fifteenth century a greater population made these places less desirable for the purpose, and indiscriminate showering of garbage over the walls and just outside the gates became intolerable.

These patterns, quite generalized in preindustrial cities, have also been vividly described for Lhasa[5] and for Rabat, Morocco, at about the turn of the present century. Concerning Rabat, Caillé remarks as follows (in translation from the French):[6]

> The removal of waste is effected in a very primitive manner by donkey-drivers who load it in the *chouaris* their animals carry. But this service includes only five or six donkeys for the whole city. Besides, it is suspended on holidays and when it rains; also, in winter, the streets of the city often show a layer of liquid mire more than ten centimeters deep. When the waste-matter has been removed it is thrown into the sea; or often it is simply heaped up at the gates to the city, where it forms a veritable cesspool.

And an Indian comments concerning present-day, pre-industrial-like Hyderabad city:[7]

> A high majority of the citizens commit nuisance promiscuously in open spaces.... Public latrines are few and far between ... [and] are not kept clean by the scavengers, and it is an annoying sight to see many a scavenger emptying his bucket full at some street corner or under a culvert....

Of course in many cities human waste is collected by scavengers in baskets or other receptacles to be distributed to farmers in the immediate countryside.

In addition to humans, dogs, pigs, birds, or other animals roam about serving as scavengers of the offal indiscriminately thrown into the streets. Though frequently provided with a depression in the center for drainage purposes, streets are more often than not choked with refuse. This explains in part why preindustrial cities have risen continually throughout the ages—some of the ancient ones uncovered by archeologists reveal many layers of habitation, the newer buildings having been constructed upon the refuse of the past.

The problem of sanitation is compounded by the difficulty of obtaining an adequate water supply; little can be spared for keeping streets and houses clean. The aqueducts of Rome were marvelous inventions for their day, but they did not serve the city's tenement areas. In non-industrial cities many households are dependent for their drinking water upon water-carriers. Or a few wells, apt to be contaminated, may serve a great number of homes. Inhabitants of Rabat have used the water conduits for bathing, washing their clothes, and watering their herds. Andkhui, a small urban community in northern Afghanistan, has long depended, for both drinking and cleansing purposes, upon a large water tank teeming with finger-length worms and other wriggling creatures. Facilities in Indian cities are still notoriously inadequate by industrial standards, a protected water supply being unavailable to a large portion of the citizenry.[8]

Under these conditions, and in the absence of a scientific understanding of the causes of communicable disease, the prevalence of epidemics in preindustrial cities, both past and present, should elicit no surprise.[9] Once underway, these sweep through a city like a fire fanned by violent winds, leaving thousands of dead in their wake. In the face of such devastation the inhabitants have little more than their religious and magical beliefs to cling to.

Preindustrial urbanites, at the mercy of the periodic epidemics and subject to the general malaise resulting from the non-salubrious conditions, are also prey to destructive fires. Having at their disposal only the meagerest of fire-fighting equipment, cities like those in Japan, with their closely-bunched dwellings built flimsily of wood, have been ripe for annihilating conflagrations.[10] Even cities with many fewer wooden structures—e.g., late medieval London or mid-nineteenth-century Cairo—have existed under the perpetual threat of fire because of the excessive crowding and the paltry technological means for coping with such eventualities.

The foregoing discussion, though not of central concern to urban ecology, nonetheless serves to capture some of the flavor of preindustrial city life. But let us turn our attention to patterns of greater theoretical interest to ecologists.

Specific land use patterns.

This section takes up three patterns of land use wherein the non-industrial city contrasts sharply with the industrial type: 1) the pre-eminence of the "central" area over the periphery, especially as portrayed in the distribution of social classes; 2) certain finer spatial differences according to ethnic, occupational, and family ties; and 3) the low incidence of functional differentiation in other land-use patterns.

As to the first, concentrated in the city's "central" area (often coterminous with the physical center, but not necessarily so) are the most prominent governmental and religious edifices and usually the main market. The chief public buildings either crowd around an open square, or plaza, onto which converge a number of streets—as in numerous cities of the Near East, southern Europe, and Latin America—or stand along, or at the end of, a broad, straight thoroughfare—as in Thebes (in ancient Egypt), Vijayanagar, and Peking. Considering the narrow lanes that wind through much of the city, the main plaza or street can be nothing but imposing.

The plazas or main streets serve as meeting places and ceremonial sites for the populace. Into these open spaces flow the citizenry to hear public pronouncements, mass communication media being non-existent, or there they engage in the elaborate processions and pageantry that mark the ceremonial life of these cities.

Both physically and symbolically, the central governmental and religious structures dominate the urban horizon. The Acropolis (or citadel) at Athens contained the chief religious buildings and in its early days served as the royal headquarters. So did the Temenos at Ur. Peking has had as its focus the "Forbidden City," site of the residences of rulers and of the leading temples. But often the religious edifices at the center dwarf the nearby governmental structures. This was the case in pre-Columbian American and ancient Cambodian and Ceylonese cities. The public buildings in many a medieval European city were overshadowed by the main cathedral, as in cities today in southern Europe and Latin America. So, too, the typical Muslim city has the chief mosque as its hub, while traditional Buddhist Lhasa radiated outward from its own "Cathedral," the focus of most community activity.

An understanding of these patterns is dependent upon knowledge of the technology and the total social structure. Any explanation of the ecology through a biotic, i.e., non-social, orientation collapses when applied to the preindustrial city, for an exceedingly high correlation exists between the technology, the social structure, and the spatial distribution of the city's inhabitants, and between all these and the urban center's physical appearance. Because we have yet to analyze the social structure, only some of the more pertinent relationships are suggested here.

Because political and religious activities in feudal cities have far more status than the economic, the main market, though often set up in the central sector, is subsidiary to the religious and political structures there. Interestingly, the chief market is apt to be located next to, or nearby, the dominant religious edifice. (In fact, markets and ambulatory vendors tend to cluster about religious buildings throughout the city, apparently to take advantage of the considerable pedestrian traffic these attract.) Nevertheless, the commercial structures in no way rival the religious and political in symbolic eminence; typically these tower above all others and are the most resplendent—the resulting skyline is far different from that of industrial cities where commercial structures tend to loom over all others. These land-use patterns refute the still widely accepted proposition of the Chicago school that the "central business district" is the hub of urban living, a generalization fulfilled only in industrial cities, where commercial activities are necessarily more prominent, supporting as they do the complex industrial system.

The preindustrial city's central area is notable also as the chief residence of the elite. Here are the luxurious dwellings, though these often face inward, presenting a blank wall to the street—a reflection of the demand for privacy and the need to minimize ostentation in a city teeming with "the underprivileged."

The disadvantaged members of the city fan out toward the periphery, with the very poorest and the outcastes living in the suburbs, the farthest removed from the center. Houses toward the city's fringes are small, flimsily constructed, often one-room, hovels into which whole families crowd. (Still farther out, well beyond the city limits, are the summer homes or ancillary dwellings of the elite.)

This general elite and non-elite ecological arrangement can be adduced from archeological finds at Ur, Knossos, and numerous other ancient urban sites. We have historical evidence of it for medieval Europe, and it persists today in parts of Italy, France, England, and other European countries, although modifications wrought by industrialization have for some time been underway. Descriptions of Cairo, Timbuctoo, Aleppo, Fez, Lhasa, Calcutta, and cities in Japan and Indonesia and Latin America all confirm the universality of this land-use pattern in the non-industrial civilized world.[11]

Urban sociologists themselves have documented the tendency of the elite to cling to the city's center, the poor being concentrated on the outskirts. Gist[12] observes this for Bangalore, India. The investigations of the Dotsons[13] in Guadalajara, Mexico, reveal that the upper class, traditionally clustered about the central plaza, has only recently been attracted to the suburbs. And Rosenmayr[14] stresses that the traditionally high prestige accorded residence in or near Vienna's historic nucleus is responsible for the low incidence of suburbanization, although he falsely considers this a peculiarly Viennese phenomenon. Like most urban sociologists, these writers have not perceived the generality of their findings, each being content to immerse himself in the culture or society that is his specialty.

How do we explain this distribution of classes as between the central sector and the periphery? Throughout feudal cities, values operate defining residence in the historic center as most prestigeful, location on the periphery as least so. But reference to values alone cannot account for the ecological differences between the traditional and the modern urban community. Far more pertinent is technology. For one thing, the feudal society's technology permits relatively little spatial mobility, thereby setting limits to the kinds of ecological arrangements that can obtain. People travel mostly on foot, occasionally on animal-back; only the privileged ride in the human- or animal-drawn vehicles, slow and uncomfortable though these be by industrial standards.

Assuming that upper-class persons strive to maintain their prerogatives in the community and society (here social power enters as a factor in ecology), they must isolate themselves from the non-elite and be centrally located to ensure ready access to the headquarters of the governmental, religious, and educational organizations. The highly valued residence, then, is where fullest advantage may be taken of the city's strategic facilities; in turn these latter have come to be tightly bunched for the convenience of the elite—patterns that are readily revised with the introduction of rapid transit, telephones, and so on. Residence in or near this high-status area reinforces one's social position. This locale is, moreover, the best-protected sector of the city, often enclosed by a wall of its own, whereas residence on the urban periphery is hazardous in time of war or in the face of the recurrent banditry.

The poor, in conformance with the rigid class structure, are kept toward or upon the outskirts and must accept all the disabilities of this location. It is they who travel the farthest to gain access to the city's facilities. Even in this community where time is not highly valued, those in the suburbs, though least able to afford it, often must traverse the greatest distance to reach their place of work. With heavy rain or snow the environs may be cut off completely from the central district as the streets become veritable seas of mud and slush.

Various low-status groups are relegated to the city's outskirts through efforts of the elite to minimize social contact with them. This is most apparent for those workers in malodorous occupations—tanning, butchering, and the like—mainly the province of outcaste groups.[15]

Some lower-class groups settle on the urban extremities so as to supplement their meager incomes and food supply through the cultivation of crops, impossible in the heart of the city. These patterns persist today in Mexico, India, and the Near East, to mention just a few sectors of the world. In turn these farmers at the city's edge are functional to the maintenance of the urban populace, for they make available perishable commodities that the urban community with its poor storage and transportation facilities would be unable to import from remote regions.

Having considered the gross disposition of the elite and lower-class (and outcaste) groupings in the central city and the periphery, respectively, we can examine some of the finer ecological demarcations according to ethnic, occupational, and kinship affiliation that are generally subsumed under the broader class categories.

Subdivisions along ethnic and/or occupational lines are mainfested in the preindustrial city in the numerous wards or quarters, well-defined neighborhoods with relatively homogeneous populations that develop special forms of social organization. Bokhara at the turn of the present century allegedly included 217 different residential quarters, many with their own leaders and special paraphernalia for their periodic ceremonial activities.[16]

Segregation by ethnic groups, which in turn are associated with specific occupations, occurs widely in preindustrial cities. The Jews in Europe have had their well-defined ghettos, persisting in some locales well into the twentieth century, and Jewish quarters have long been part of the urban scene throughout the Middle East. Ethnic quarters tend to be self-sufficient entities to the extent that urban living allows, physically and socially separated from the rest of the community. Often they have their own unique social structure—political leaders, schools, etc.—and even some ecological differentiation according to "class" within the quarters. It must not be thought that the Jews are alone in carrying out this role; there have been Muslim quarters in Chinese cities, Christian quarters in Near Eastern cities, and Hindu quarters in the urban areas of Central Asia, to name a few.[17]

Differentiation of land use according to occupation is usual. A special quarter, district, or street is allotted to a particular economic pursuit. Here are grouped merchants dealing in certain kinds of produce or handicraftsmen plying a specific trade. In Fez craftsmen have had shops strung along special streets, whereas a few groups, like potters, have had quarters all to themselves. Gray[18] in his description of Canton in the nineteenth century lists dozens of streets, each restricted to the shops of artisans and/or merchants dedicated to making and/or selling a specific product—paper objects, gongs and bells, pork fat, or betel and cocoa nuts. In a host of preindustrial cities streets commonly bear the names of their occupational groups—the street of the goldsmiths, the street of the glass workers, and so on. Sometimes just a few blocks may be devoted to one occupation, while another group takes over farther along, with resultant name-changes several times along the length of the street.

This localization of particular crafts and merchant activities in segregated quarters or streets is intimately linked to the society's technological base. The rudimentary transport and communication media demand some concentration if the market is to operate: in this way producers, middlemen, retailers, and consumers alike can more readily interact. How much business could a seller of hides transact in a day if his prospective customers, the leather workers, had their shops scattered helter-skelter about the city? Moreover, the social organization, especially the guild system, itself largely interwoven with the technology, encourages the propinquity of members of an occupation, which in turn fosters group cohesion.

Other dimensions to occupational differentiation cut across the previously discussed elite-nonelite residential gradient. Domestic servants are one lower-class, or outcaste, element, that live in upper-class districts, in quarters provided by their employers, the better to serve their masters. As a further complication, which finds its counterpart in the social structure, merchant groups occupy a wide range of ecological positions. A few well-established businessmen are situated close to the principal markets at the city's core, whereas the least prosperous, including the outcaste foreign merchants, huddle on the urban fringe.[19]

Land-use differentiation, aside from that related to class, ethnic, and/or occupational groupings, occurs along family lines. A particular family, or an extended kinship unit, may control a given street, forming a well-defined subsystem. If the family is prominent the street may bear its name, one apt to persist long after all remnants of the kinship group have died or moved away, as evidenced in some European and Latin American cities today.

Although sharp ecological differentiation along certain lines is characteristic of the feudal city, absent are a number of forms of land-use specialization so typical of industrial cities. Frequently a single plot of land serves multiple functions in the non-industrial-urban milieu. Religious edifices functioning concurrently as schools are a not-uncommon sight. And markets are likely to be set up on the grounds adjoining the church, mosque, or temple. In medieval Europe a favorite spot for fairs was in or near the city's principal church. Or, we can turn to a description of Bokhara for a colorful protrayal of the multiple uses to which the zone around the religious structure may be put.[20]

> Meschidi Namaziya, or Namazi-gah, is a great mosque, with an immmense platform before it.... Prayers are read in it during the Ramazan and Kúrban; at which periods the public also resort there for amusement.
> The whole square ... is covered, on such occasions, with temporary booths, in which confectioners, vendors of dried fruit, ... exhibit their tempting merchandize [sic] to the gaze of the crowds which rush to and fro; ... Behind the tents and the booths, wrestlers show off their feats; races, also, are set on foot; and camels are made to fight.

The dearth of land-use specialization is also attested to by the fact that the residential units of artisans and merchants often serve simultaneously as their places of work, the living quarters being behind or just above the shop. The ecological situation wherein a person may reside, produce, store, and sell his wares within the confines of the same structure has been a feature of preindustrial-urban life from the earliest cities in Mesopotamia down to the present day.[21]

To reiterate: the feudal city's land-use configuration is in many ways the reverse of that in the highly industrialized communities. The latter's advanced technology fosters, and is in turn furthered by, a high degree of social and spatial mobility that is inimical to any rigid social structure assigning persons, socially and ecologically, to special niches. Moreover, the improved technology and the greater rationality of the economic organization in industrial cities both require and permit considerable specialization in land use. For example, rapid transit enables the urbanite to live far removed from his place of work.

Temporal patterning

Implicit in the discussion of the spatial ordering of social relationships have been certain temporal patterns. With the poor technology, communication is necessarily slow and movement from one part of the city to another, relative to that in the industrial city, is most time consuming. Travel between cities or between rural and urban communities is a major undertaking, one limited to specialized groups—merchants, administrators, and the like. The preindustrial order requires a certain amount of intercity and intracity mobility, but nothing on a scale comparable to that in industrial cities where the technology demands and makes possible spatial movement in a grand manner. The labor force must be mobile, and scientists and other experts who man the key positions in the system must maintain a great deal of social contact if they are to keep it operating.

The temporal ordering of life within the city merits special comment. Activity proceeds at a slow pace, and the preindustrial urbanite, compared to industrial man, does not think of time as a "scarce commodity," except within broad limits such as days or weeks. The shopkeeper opens and closes his establishment at varying times as the spirit moves him. And keeping an appointment at a stated time is more the exception than the rule. Even businessmen may spend several hours in pleasant conversation before getting down to the matter at hand. All this is quite in keeping with the technology of the city, which typically lacks any precise measuring instruments. Clocks, a recent phenomenon, are mostly confined to the small urban upper class. The average person must observe the position of the sun or gauge time by the pealing of church bells and the like: in Muslim cities the periodic calls to prayer from the mosque orient the actor to the daily tempo of city life. One illustration of the crude time-reckoning technology, from about the turn of the present century, makes these generalizations more empirically meaningful.[22]

> In Kabul and the principal cities time is kept by means of a sundial, but though there are tables

printed in Persian of the daily difference between solar and mean time, the time given by them is only approximate, for the dials have been constructed for other latitudes, and they are fixed in the direction of the magnetic north instead of the true one.

Under these circumstances, people can hardly maintain rigid, time-bound schedules, even if such were their goal. Admittedly, chronological measurement over days, months, and years is much more precise. All preindustrial cities have had a calendric system, in large part used to standardize the routine of religious ceremonial activity.

Quite a different picture emerges in the industrial city. It demands of its populace extreme consciousness of time and provides the means for such. In industry the wheels must turn at just the right moment; the workers on the assembly line must synchronize their efforts with the tasks of others or costly bottlenecks will ensue. And science, the basis of modern technology, could never proceed without precise measurement of the "time factor."

Another distinguishing feature of the temporal distribution of activity in the non-industrial city is the restriction of so many activities to the daytime. The typical preindustrial city is without permanent artificial street lighting, making passage through the city difficult and dangerous at night. Handicraft activities, for example, have to be carried on during the day, and the work-day varies from season to season depending upon the number of hours of daylight. Of course, on ceremonial occasions torches, or similar devices, may be set up in the main squares, but urban night life on a grand scale is definitely a post-industrial development.

Notes

1. E.g., see Wolfram Eberhard, "Data on the Structure of the Chinese City in the Pre-Industrial Period," *Economic Development and Cultural Change,* 4 (April 1956): 253–68; J. Sauvaget, *Alep* (Paris: Librairie Orientaliste Paul Geuthner, 1941); Robert E. Dickinson, *The West European City* (London: Routledge and Kegan Paul, 1951); Pedro Armillas, "Mesoamerican Fortifications," *Antiquity,* 98 (June 1951): 77–86.

2. George Woodcock, "Guanajuato," *Mexican Life,* 34 (December 1958): 15.

3. G.T. Salusbury, *Street Life in Medieval England,* 2d ed. (Oxford: Pen-in-Hand, 1948), p. 77.

4. Ibid., pp. 90–91.

5. Austine Waddell, *Lhasa and its Mysteries* (New York: E. P. Dutton, 1906), pp. 340, 352.

6. Jacques Caillé, *La Ville de Rabat jusqu'au Protectorat Français,* I (Paris: Vanoest, 1949), p. 558.

7. S. Kesava Iyengar, *A Socio-Economic Survey of Hyderabad-Secunderabad City Area* (Hyderabad: Government Press, 1957), p. 353.

8. E.g., K. N. Venkatarayappa, *Bangalore: A Socio-Ecological Study* (Bombay: University of Bombay, 1957); Sevak Samaj Bharat, *Slums of Old Delhi* (Delhi: Atma Ram and Sons, 1958), pp. 28ff. Also, recent studies of Baroda, Bombay, Lucknow, Gorakhpur, and Calcutta, mimeographed summaries of which were examined by the author, lend dramatic support to this conclusion. For medieval Europe, see R. J. Forbes, *Studies in Ancient Technology,* I (Leiden: E. J. Brill, 1955), pp. 175–76.

9. E.g., Roger Mols, *Introduction à la Démographie Historique des Villes d'Europe du XIVe au XVIIIe Siècle,* II (Louvain: Université de Louvain, 1955); David Herlihy, *Pisa in the Early Renaissance* (New Haven: Yale University Press, 1958), pp. 47ff.

10. R. A. B. Ponsonby–Fane, *Kyoto* (Kyoto: Ponsonby Memorial Society, 1956), pp. 403ff.

11. E.g., Marcel Clerget, *Le Caire: Étude de Géographie Urbaine et d'Histoire Économique,* I (Cairo: E. and R. Schindler, 1934), p. 265; Horace Miner, *The Primitive City of Timbuctoo* (Princeton: Princeton University Press, 1953); F. Spencer Chapman, *Lhasa: The Holy City* (London: Chatto and Windus, 1938), pp. 167–68; Thomas O. Wilkinson, "Tokyo: A Demographic Study" (Ph.D. dissertation, Columbia University, 1957), p. 46; Justus M. van der Kroef, "The Indonesian City: Its Culture and Evolution," *Asia,* 2 (March 1953): 27–28; Mols, op. cit., p. 37.

12. Noel P. Gist, "The Ecology of Bangalore, India: An East–West Comparison," *Social Forces,* 35 (May 1957): 356–65.

13. Floyd Dotson and Lillian Ota Dotson, "Ecological Trends in the City of Guadalajara," *Social Forces,* 32 (May 1954): 367–74. Also see Pedro Yescas Peralta, "Estructura Social de la Ciudad de Oaxaca," *Revista Mexicana de Sociología,* 20 (Septiembre–Diciembre 1958): 769.

14. Leopold Rosenmayr, "Anotaciones sobre el Fenómeno de la 'Urbanización Allende de la Ciudad,' " *Revista Mexicana de Sociología,* 20 (Septiembre–Diciembre 1958): 737–38.

15. E.g., Chapman, op. cit.

16. O. A. Sukhareva, "Byt Zhilogo Kvartala Goroda Bukhary v Kontse XIX–Nachale XX Veka," *Akademiya Nauk Soyuza SSR, Institut Etnografii, Kratkie Soobshcheniya,* 28 (1957): 35–38.

17. E.g., J. Weulersse, "Antioche: Essai de géographie urbaine," *Bulletin d'Études Orientales,* 4 (1934): 27–79; O. Olufsen, *The Emir of Bokhara and His Country* (London: William Heinemann, 1911).

18. John H. Gray, *Walks in the City of Canton* (Hongkong: De Souza, 1875).

19. van der Kroef, op. cit.; Sauvaget, op. cit.

20. Khanikoff, *Bokhara: Its Amir and Its People,*

trans. Clement A. De Bode (London: James Madden, 1845), p. 120.

21. Dickinson, op. cit.; Mary L. Shine, "Urban Land in the Middle Ages," in Richard T. Ely et al., *Urban Land Economics* (Ann Arbor: Edwards Bros., 1922), p. 79; Clerget, op. cit., II, 140; Olufsen, op. cit., A. Bopegamage, *Delhi* (Bombay: University of Bombay, 1957), pp. 173ff.

22. Frank A. Martin, *Under the Absolute Amir* (London: Harper and Bros., 1907), pp. 91–92.

A Society Without Wheels*

Richard W. Bulliet

Whoever has attempted to characterize medieval Middle Eastern and North African cities has sooner or later commented upon the narrow streets, the blind corners, the encroachment of buildings upon the public way, and in general upon the labyrinthine quality that strikes so forcibly the Western visitor. Many scholars have attributed this quality in some way to the Islamic religion and have implied that it is a universal feature of Islamic cities.[1] None has seen it as a characteristic of a society without wheels.

A rectilinear layout with streets of uniform width lined with buildings of similar height and design has until recently represented in Western thinking good order and intelligence of design. Even individuals not normally given to admiring this type of design are apt to feel the adjective "oriental" come to mind when confronted with a city composed of winding streets and narrow alleys.[2] Dirt, darkness, and crowding are thought of as inevitable and evil conditions of this "oriental" type of city, while parallel Western conditions such as motorized danger and isolating quality of broad avenues are taken in stride. Islamic society is often described as turning its back on the street, hiding inside walled courtyards and behind windowless walls to shut out the cheerlessness and contagion of the public way. Private life is supposedly given precedence over communal life, which is seen by many to be deficient in the Islamic religion.[3] To an earlier generation such streets might be regarded as visible evidence of the supposed inadequacy of Islam.

This entire conception of the "oriental" city plan being generated by Islamic social principles runs counter both to logic and to fact. A great many factors, some deliberately planned and others unconscious and incremental, come into play in the development of a particular urban environment. Religious principles undoubtedly have their place among them. When it comes to the layout of streets, however, what cries out most for explanation is the rigid application of abstract geometrical forms. A particular shape or compass orientation might be dictated by religious belief, astrology, legal principle, or the caprice of a ruler; but whatever the motive, it is reasonable to expect it to be ascertainable.[4] The same cannot be said of a city whose streets are not laid out in a formal pattern but according to the lay of the land and the inclination of the builder. Disorder requires explanation only if order is taken to be normative.

Since from the Western viewpoint regular patterns in urban topography are generally considered

---

*Reprinted by permission of the author and the publishers from *The Camel and the Wheel* by Richard Bulliet, pp. 224–28 (Cambridge, Mass.: Harvard University Press, Copyright © 1975 by the President and Fellows of Harvard College).

[1]See Notes section at the end of this article.

to be good and the absence of such patterns bad, it is not difficult to understand why the feeling arose that the "oriental" city plan had to have an explanation, particularly in view of the fact that Roman cities throughout the Middle East and North Africa are known to have exhibited a uniform rectilinearity.[5] But if narrow, winding streets are not inherently bad, the rationale for seeking an explanation for this alleged falling away from perfection evaporates. And, in fact, narrow, winding streets have much to recommend them. They easily follow the lay of the land; in hot countries they provide shade; they diminish winds; they permit a higher density of habitation which in turn makes a sizable city accessible to pedestrians; they facilitate social relationships; and they are easily defensible. As for enclosed courtyards with windowless exteriors, they provide secluded open spaces where many household tasks are carried out, as is desirable in a warm climate; and they allow for careful regulation of water consumption, as is desirable in an arid climate. There is privacy, too, of course, but that much familial privacy is available in the large tenements built around courtyards that are characteristic of North African cities is open to doubt. A final observation on the sociability of streets of this character may be gleaned from Roger Le Tourneau's monumental study of the city of Fez in Morocco. He observes that the street is the center of a quarter or neighborhood, while the demarcation lines between neighborhoods follows the abutted backs of houses.[6] By way of contrast, American residential neighborhoods often divide along street boundaries with residents on the same side of a street knowing each other better than they know their neighbors facing.

Since, then, the transition from Roman rectilinearity to medieval disorder was not necessarily inherently bad, the need to explain it in moral or ideological terms is greatly diminished, and the way is open for a more prosaic physical explanation. Wheeled vehicles—and this can come as no surprise to today's city dwellers—are inflexible in the restraints they put on city life. Streets must be flat, without stairsteps or precipitous grades, and, if possible, paved. Moreover, they must be maintained in this state if circulation is not to be interrupted. They must always be as wide as a single axle—as wide as two if the citizens are to be spared immoderate language. Corners must not be too sharp or narrow to be maneuvered; dead ends must be eschewed. Encroachments on the public way either by buildings or by merchants displaying goods cannot be tolerated. And on top of all of these burdens is the fact that wheeled vehicles are noisy and dangerous.

Freed from this vehicular straitjacket by the disappearance of the wheel, it is scarcely a matter for wonder that Middle Eastern and North African cities gradually evolved types and arrangements of streets suited more closely to human needs. With only pedestrians and pack animals to accommodate, the street could become an open market or a narrow cul-de-sac giving access to residences. In the absence of any ideological sanction of constant widths and right angled turns, only enough legislation was needed to keep the streets passable. This is very different, needless to say, from having an ideological sanction for disorder. Indeed, on occasion Islamic rulers decreed explicit plans for cities, plans that on one occasion might call for broad thoroughfares and on another for great plazas,[7] but planning was not the rule because without wheeled vehicles the necessity for plans was negligible. As late as 1845 the width of a major new street in Cairo was determined by measuring the combined width of two loaded camels.[8]

The advent of Islam did not destroy previous city plans. Antioch and Herat are two examples of cities that to this day preserve long, straight thoroughfares traceable to pre-Islamic incarnations.[9] Alexandria was still described as being laid out like a chessboard in the thirteenth century.[10] But Islam came into being in a society that had recently abandoned the wheel, and hence it incorporated in its growth no ideological bias in favor of vehicular traffic. The evolution from a geometric to an organic urban design within the zone of the wheel's disappearance followed as a natural consequence. Outside that zone other, equally Islamic—or, rather, equally non-Islamic—urban patterns arose. The dispersed cities of Indonesia made up of discrete, villagelike *kampongs;* the precise rectilinear design of Jaipur in India;[11] and the striking "skyscraper" cities of southern Arabia illustrate the variety of urban design to be found in Islamic lands and testify to the irrelevancy of Islamic religious principles in this domain. It is the absence of the wheel that goes furthest toward explaining this characteristic feature of Middle Eastern and North African urban environments.

Roads are the second area in which the camel's impact on a wheeled economy can be observed in concrete form. What is important here is not which routes were traversed. In the Middle East desert caravan tracks were in use for centuries before the camel replaced the wheel. In North Africa the camel would have become the common carrier on Saharan trade routes regardless of its effect on wheeled transport. And in Central Asia the camel dominated the caravan trade without eradicating the wheel as it was used in agricultural districts or for moving the belongings and portable homes of migrating nomads. What is important is not the choice of routes; it is their actual physical state.

Camels, donkeys, and pedestrians do not need paved roads. Given that throughout the zone of the wheel's disappearance the climate is dry during most of the year, it is more comfortable to walk on dirt. Furthermore, natural obstacles, such as boulders, do not have to be removed to provide for a constant minimum width, nor do ruts have to be filled in. Cost of maintenance is as negligible as cost of construction. In a nonvehicular economy the most important physical features of a road are its bridges. One bridge in place of a ford or ferry can make an enormous difference in the ease and cost of transportation. After bridges, the most important features are accommodations for travelers. A regular daily stage of travel for a caravan does not exceed twenty miles, and a good road will afford a stopping place at the end of every stage, whether it be a town, a village, or a caravanserai. Beyond these two things, bridges and caravanserais, the physical upkeep of roads is insignificant; but bridges and caravanserais themselves can be very costly.

The reflection of this state of affairs is everywhere apparent in the history of the Islamic Middle East. References to the upkeep of roads are almost nonexistent, but powerful dynasties frequently show their interest in promoting trade by building bridges and caravanserais.[12] Investment in these two things is functionally equivalent to roadbuilding in a wheelless society. There is no need to search for an ideological explanation for a nonexistent neglect of public ways. Middle Eastern governments acted with complete rationality in investing in bridges and caravanserais instead of in useless grading and paving.

In the case of roads, as in the case of medieval urban topography faced by modern automobile traffic, what was rational and desirable in a nonvehicular society has proved to be highly undesirable in the wheeled economy of modern times. In Europe, road improvement and advances in vehicular design went hand in hand. Heavy vehicles drawn by several animals meant that load size could be greatly increased over the quarter-ton limit imposed by the pack camel, but their efficiency could only be fully realized on roads that were straight, level, and paved. Consequently, the infrastructure of carriageable roads that Europe took into the period of the industrial revolution far outstripped what the Middle East had going into the twentieth century.[13] Almost all non-Western countries, of course, have been faced with the need to build a network of motorable roads as a prerequisite for modernization, and many areas are endowed with much greater physical obstacles than is the Middle East, which is dry, devoid of forests, and relatively free of rivers. Strictly from the Middle Eastern perspective, however, it is evident that the area would have entered the period of modernization with a much better road system had it not been for the dominance of the pack camel and the absence of wheeled vehicles. Given the vital role of transportation in the industrialization process, both for centralizing manufacturing and distributing manufactured goods, it is possible that this deficiency was crucial.

Notes

1. For example, Xavier de Planhol, *The World of Islam* (Ithaca: Cornell University Press, 1959), pp. 14–22. He is forced to observe (p. 22) that Mecca itself does not look like an Islamic city since it has straight streets.

2. Lewis Mumford so describes fifth century B.C. Athens in *The City in History* (Harmondsworth: Penguin Books, 1966), p. 192. One nineteenth-century British lady traveler wrote of Cairo: "Only a few narrow streets and old houses are left . . . where you can yet dream that the 'Arabian Nights' are true." Norman Daniel, *Islam, Europe and Empire* (Edinburgh: Edinburgh University Press, 1966), p. 50.

3. Planhol, *The World of Islam*, pp. 7–8.

4. For an extensively illustrated analysis of the rationale for specific urban designs, see Sibyl Moholy-Nagy, *Matrix of Man* (New York: Praeger, 1968).

5. See, for example, the plans of Roman Timgad and Cuicul in Julien, *Histoire de l'Afrique du Nord*, I, pp. 168–69.

6. Roger Le Tourneau, *Fès avant le Protectorat* (Casablanca, 1949), p. 229.

7. For Baghdad, a round city, see K. A. C. Creswell, *A Short Account of Early Muslim Architecture* (Harmondsworth: Penguin Books, 1958), pp. 164–70; for Samarra, a city centered on a long, wide axial street, see J. M. Rogers, "Sāmarrā: A Study in Medieval Town-planning" in *The Islamic City,* eds. A. H. Hourani and S. M. Stern (Oxford: Bruno Cassirer, 1970), pp. 119–55.

8. Marcel Clerget, *Le Caire: étude de géographie urbaine et d'histoire économique* (Paris: Paul Geuthner 1934), I, p. 289.

9. For comparative maps of Roman and Islamic Anti-

och, see Glanville Downey, *Ancient Antioch* (Princeton: Princeton University Press, 1963), pls. 4–5; for a map of Herat showing its crossed north-south, east-west main streets, see Alexandre Lézine, "Hérat: Notes de Voyage," *Bulletin d'Études Orientales*, 18 (1963–64), fig. 1.

10. Yāqūt, *Muʿjam al-Buldān*, I, 186 (Beirut: Dar Sader and Dar Beirut, 1955–57).

11. Moholy–Nagy, *Matrix of Man*, fig. 170.

12. One of the few individuals who ever displayed an interest in the upkeep of roads was the Persian heretic Bihāfrīd who was executed in northeastern Iran in 749. The fact that upkeep of roads and bridges was remembered as one of his prescriptions would seem to testify to its being an unusual concern at that time. E. G. Browne, *Literary History*, I, 308–10.

13. The mileage of carriageable roads in Iran in 1914 was still extremely limited. Issawi, ed., *The Economic History of Iran, 1800–1914*, pp. 203–4.

Moscow in 1897 as a Preindustrial City: A Test of the Inverse Burgess Zonal Hypothesis*

Walter F. Abbott

"The typical process of the expansion of a city can best be illustrated, *perhaps,* by a series of concentric circles, which may be numbered to designate both the successive zones of urban extension and the types of areas differentiated in the process of expansion" (Burgess, 1925:50; italics added). In its origins, the well-known Burgess zonation thesis was thus tentatively proposed as a pattern of spatial organization expected to emerge in communities in the United States experiencing population growth and geographic expansion. Burgess clearly recognized in his original paper (1925:51–52) that there would be many deviations from this model and specified later (1929:113), in a more systematic exposition of the thesis, that this pattern may be modified by topographical features and the street plan of the city. Despite these theoretical qualifications to the zonal hypothesis, substantial debate has ensued among students of communities in the United States and elsewhere whether it is an adequate model for analyzing community spatial organization and expansion.

Larger cities in the United States have been researched extensively since Burgess formulated his thesis in the twenties. Studies of Long Beach (Longmoor and Young, 1936), Rochester (Bowers, 1939), and Philadelphia (Blumenthal, 1949) indicate that status tends to be decentralized. Queen and Carpenter (1953:99–104) found support for the Burgess thesis in their study of St. Louis. The results of Duncan and Duncan's (1955) research on Chicago using 1950 census data also indicated decentralization of occupational status. Green's (1932) analysis of Cleveland indicated that social status was decentralized, although not necessarily in a radial pattern. In a later study based on Duncan and Duncan's methodology, Uyeki (1964) also found a decentralized status pattern for Cleveland. Evidence presented by Schnore (1965b:207) from the 1960

---

*Reprinted from the *American Sociological Review,* 39 (August 1974):542–50, by permission of the author and The American Sociological Association.

Author's note: I should like to acknowledge Calvin F. Schmid for initial interest in this form of sociological research and the assistance of Edward Raymaker in the details of preparing this paper. Two anonymous reviewers provided useful suggestions for the revision of a previous draft.

United States census for two hundred urbanized areas indicates that status differences exist between central cities and suburbs for income, education, and occupation in the larger but not necessarily smaller urbanized areas. Although research does not tend to confirm Burgess's model taken in its most literal and concrete form, and the model has also received fundamental criticism by Davie (1938) and Hoyt (1939:72–78, 112–22), there is substantial support for the general thesis that status tends to become decentralized in the process of community growth and expansion in the United States.

Burgess (1925:47) appears to have specifically held the American system as his main referent in the analysis of urban life: "All the manifestations of modern life which are peculiarly urban . . . are characteristically American." Following World War II, however, American sociologists gave particular attention to the comparative method. Therefore, whereas the zonal hypothesis had reached its peak of interest in research on American cities by World War II, the comparative approach brought about its application in a radically different form to the study of preindustrial communities. Sjoberg (1960:97–98) thus provides this ideal-type description of the spatial structure of preindustrial cities: "The preindustrial city's central area is notable . . . as the chief residence of the elite. . . . The disadvantaged members of the city fan out toward the periphery, with the very poorest and the outcastes living in the suburbs, the farthest removed from the center." Whereas social class is positively related to distance from the city center in the Burgess model, social class is inversely related to distance in the preindustrial model. There is thus an inverse Burgess zonal pattern in the preindustrial city model that has been set forth by Sjoberg.

The inverse zonal hypothesis is supported in many detailed studies of contemporary cities with a preindustrial heritage in Europe, Latin America, Asia, Africa, and North America in the nineteenth century, but is not supported in other studies. In Francis Hauser's (1951) comparative study of London, Paris, and Vienna, a zonal model does not appear to be useful. Castle and Gittus (1957) claim support for the Burgess thesis in their study of social disorganization in Liverpool. Nevertheless, the inverse model has been found to apply to such European cities as Oxford (Collison and Mogey, 1959), Rome (McElrath, 1962), Stockholm (Olsson, 1940; Hauser, 1951), Prague (Moscheles, 1937) and Budapest (Beynon, 1943). The urban structure of larger cities in Latin America has been well researched. The pioneering study of the ecology of Latin American cities was conducted by Hansen (1934) in the early thirties of Mérida in the Yucatan. Hansen found that the Spanish or higher status residential district was clustered around the plaza and that the Indians resided in the barrio, which is farther out. Similar status patterns have been reported for Mexico City (Hayner, 1945), Guatemala City (Caplow, 1949), Sucre, Bolivia (Hawthorn and Hawthorn, 1948), and Guadalajara (Dotson and Dotson, 1954). Although the preindustrial pattern appears evident, Latin American urban ecological research has been criticized by Schnore (1965a) and Stanislawski (1961). Studies of Asian and African communities also support the preindustrial model. Indian communities are especially consistent with the preindustrial pattern. Bangalore (Gist, 1957), Poona (Mehta, 1968), and Calcutta (Ghosh, 1950) have a decentralized status system. Based on a study of eighty cities in India, Brush (1962:60) thus generalized that in the indigenous sections "Brahmans and other castes are usually in the best-built residential areas in or near the center of the old cities . . . the laboring castes and menial outcastes of lowest socioeconomic status occupy the poorest houses and tend to be located in the outskirts rather than the center." Although research on the ecological patterns of preindustrial communities of North America in the nineteenth century and Africa is not as extensive as it is for Europe, Latin America, and Asia, limited findings for Timbuctoo (Miner, 1953:44), Khartoum (Hamden, 1960: 23,28), Detroit (Wilson, 1938), and Philadelphia (Johnston, 1966) are consistent with Sjoberg's preindustrial community model.

Moscow is a major world city that has not been researched for its implications on the theory of urban structure. This paper presents evidence on the urban structure of prerevolutionary Moscow using Sjoberg's inverse zonal model of a preindustrial community as the orienting framework. The date of 1897, the year in which the first All-Russian census was conducted, precedes the revolution of 1917 by only a few years, and provides baseline data for analyzing ecological developments in Moscow in the Soviet period.

Moscow in 1897—the census data

The 1897 All-Russian census (Russia, 1899–1905) is the most detailed yet published of Russian society, before and after the revolutions of 1917.

Census data were also collected and published by police zones within the boundaries of Moscow. Figure 1 is a reproduction of the official census map of Moscow published in the census volume on the province of Moscow. Gradients were established on the basis of distances from the Kremlin, the historic city center. The city divisions identified by name (except the Kremlin) are the police areas. Although Lyall (1823:106) reported that twenty police zones existed in 1792, only seventeen were reported within the limits of the city in 1897. The data on suburbs constitute the eighteenth statistical unit.

Table 1 reports the spatial structure of Moscow as a test of the inverse zonal hypothesis. The areal characteristics are organized into the infra-ecological, socioeconomic and disorganizational dimensions. Two types of statistical methods are used to indicate distribution patterns. The first is regression analysis in which the characteristics of the police areas are treated as functions of distance from the Kremlin. The Pearsonian r, constant, and slope are reported. Zone I (the Gorodskaia district) is excluded from the regression analysis since it includes the Kremlin. Table 1 also reports the mean values of the characteristics of the police areas of each distance zone as a form of gradient analysis discussed in a technical paper by Burgess (1927). Whereas the regression analysis provides an abstract conception of spatial structure, the gradient analysis indicates distributional patterns more concretely.

Ecological infra-structure

Infra-structure refers to exogenously explained phenomena that are basic to interpreting other phenomena. The infra-structure consists here of distance from the city center, population density, and the percent change of areal population from 1871 to 1897. The distance gradients are the mean distances in miles of the zonal subareas from the Kremlin as indicated in Figure 1.

Density is a particularly strategic variable in community analysis. Wirth (1938) traced many of the dysfunctional features of city life to density, although Hawley (1972) more recently has proposed that we reconsider commonplace sentiments about the dysfunctions of density in city life. In the Burgess model, density is inversely related with distance from the city center, and the higher socioeconomic classes thus reside in the low density areas. (Burgess did not indicate this precisely in his original paper, but it is reasonable to make this inference.) It has been implicit throughout the present paper that density thus decreases with distance from the center of Moscow. If this were not the case, any reverse pattern

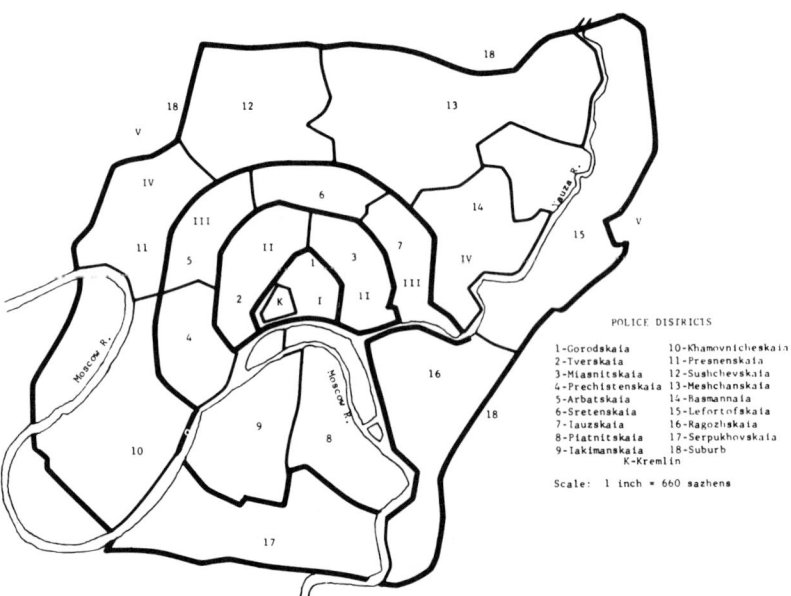

Figure 1—Police districts and distance gradients in Moscow: 1897.

Table 1. Distance from City Center and Ecological, Socioeconomic, and Disorganizational Characteristics of the Areas of Moscow in 1897: Regression Analysis and Gradient Values.

Characteristics of Areas	Mod-el*	Regression Analysis: Distance and Areal Characteristics			Gradient Values					Data Fit+	Data Fit++
		r**	a**	b**	I	II	III	IV	V		
Ecological Infra-structure											
Distance in miles from Kremlin	--	--	--	--	--	.9	1.4	2.6	4.4	--	--
Density (population per 1000 sq. sazhen)***	Neg.	-.72	133.19	-29.65	93.9	116.7	90.8	54.0	--	+/-	+
% change in population 1871-97***	--	.17	48.00	13.21	-7.9	29.5	71.4	87.2	--	--	--
Socioeconomic Status											
% nobility, clergy, merchants and petty bourgeoisie	Neg.	-.74	41.67	-4.50	21.9	33.7	36.6	30.4	19.1	-	+/+
% literate	Neg.	-.85	67.62	-5.22	64.9	64.2	60.3	53.9	43.5	+	+/+
% 17 years and above with middle and higher education	Neg.	-.69	20.86	-4.41	6.2	17.7	15.3	8.6	2.5	+/+	+
% mfg. occupations	Pos.	.68	24.26	6.83	33.7	28.2	34.9	41.8	54.6	+/+	+
Disorganization											
Sex ratio	Pos.	.64	102.52	13.51	237.7	129.1	116.0	137.8	164.8	-	+/-
% with defects (blind, deaf, mental disorders	Pos.	.63	.03	.15	.2	.1	.3	.4	.8	+/+	+
% infant survival****	Neg.	-.59	87.93	-8.21	66.7	78.0	77.0	67.2	46.7	+/+	+
% divorced (17+ years)	Pos.	.28	.17	.02	.1	.2	.2	.2	.3	+	+

*Association with distance in preindustrial model.
**Gorodskaia division (Gradient I) excluded. Distance is treated as the independent variable.
***Suburbs excluded.
****Miasnitskaia district excluded.
+Data fit including Zone I (A "+" indicates that the pattern is consistent with the preindustrial model, a "_" indicates that the pattern is inconsistent with the preindustrial model, and "+" indicates that there is an essential fit with minor deviations.
++Data fit excluding Zone I.

in the location of socioeconomic status might simply be a function of a peculiar distribution of the population in Moscow, and would thus be consistent with the Burgess model of urban structure. Table 1 indicates that the density pattern for Moscow was generally consistent with the pattern expected in any large community. Since Zone I included the Kremlin, the gradient values indicate it had a lower density than Zone II. However, the remaining zones indicate a declining density with distance from the city center. The regression analysis thus indicates a general inverse distance and density pattern. The correlation coefficient between distance and density is $-.72$ and the slope is -29.65.

The Burgess zonation thesis is a model of the expansion of cities in response to increases in population. The data (Table 1) indicate that not only was Moscow in a state of change, but the population change was in the outlying areas. The city center (Zone I), for example, lost population over the 1871–1897 period; whereas the mean areal increase in population in Zone IV was 87.2 percent. The ecological structure of Moscow was thus quite consistent with the premise of the Burgess model that population density decreases and yet population growth increases with distance from the city center. Yet the sociological importance of the Burgess model is not to be found in such variables as density and population change. The question is: Did the spatial distribution of socioeconomic status fit the predictions of the Burgess model?

Distance and socioeconomic status

Data on estates, literacy, educational achievement, and occupation are used to study the distributional pattern of socioeconomic status. Russia had a stratification system based on estates until the October revolution. The estates reported in the 1897 census included the following: the nobility, clergy, merchants, petty bourgeoisie, peasants, foreigners, and a residual category for all other estates. The peasants comprised the predominant share of the population. The upper and middle estates (nobility, clergy, merchants, and petty bourgeoisie) as a proportion of the total population is the status index reported here. The inverse zonal hypothesis predicts that status will be inversely related with distance from the city center. In regression analysis, taking status as a dependent variable and distance as independent, the sign of the correlation coefficient and the slope should thus be negative if the inverse zonal hypothesis holds. As Table 1 reports, the correlation coefficient between distance and the proportion of the population in the upper and middle estates is $-.74$ and the regression coefficient is -4.50, indicating findings consistent with the preindustrial or inverse zonal model.

The mean values of the status characteristics indicate that the zones or gradients were far from homogeneous. The zones were not "pure" upper, middle, or lower estate zones. Zone III, the zone with the highest proportion of upper and middle estate populations, had 36.6 percent in that category. Furthermore, the interpretations of the distance pattern depends on whether Zone I is included in the analysis. The fit of the data with the preindustrial model (excluding and including Zone I) is indicated in the last two columns of Table 1. In the case of social status, for example, whether Zone I is included greatly affects the fit of the data. If Zone I is included, for example, the proportion in the middle and upper estates first increases with distance from the Kremlin (reaching a peak in Zones II and III) and then returns to a lower level in the suburbs. The data do not fit the model. However, if Zone I is excluded, and since Zones II and III are observed to be quite similar, the status trend is downward with distance and may be considered to be consistent with the preindustrial model. Literacy is a particularly sensitive indicator of socioeconomic status in preindustrial societies. Closely allied with literacy, of course, is educational achievement. Table 1 thus indicates the zonal pattern of literacy and the proportion of the population seventeen years and above with either a higher, technical, middle, or intermediate education. In the preindustrial model, the proportion of the population that is literate or has some form of advanced education decreases with distance from the city center. The correlation coefficients between distance and literacy and percent with advanced education are $-.85$ and $-.69$, respectively, and the regression coefficients are thus also negative. The literacy rates for the gradients consistently decline with distance from the city center whether or not Zone I is included. The pattern of educational achievement is affected by including Zone I in the analysis, however. Nevertheless, the evidence from both the regression analysis and the gradients support the preindustrial hypothesis.

Evidence on the distribution of the economically active population engaged in manufacturing is also presented in Table 1. In the Burgess model the working class population tends to reside toward the center of the city. Consequently, in a preindustrial city the working classes may be expected to reside toward the outskirts and suburbs. A positive relation between distance and working class status is thus expected in a preindustrial community. As Table 1 reports, the correlation and regression coefficients for distance and percent in manufacturing are positive (.68 and 6.83), and the mean percentages beginning with Zone II consistently increased from 28.2 percent to 54.6 percent for Zone V. The data on manufacturing occupations also indicate that Moscow approximated a preindustrial community.

Distance and social disorganization

Early studies of communities conducted in the Chicago ecological tradition often dealt with social disorganization. Although this concept is subject to various interpretations and criticisms, it has nonetheless served as a guide for ecological investigation by various members of the Chicago School (e.g., Faris, 1955:3–25; Schmid, 1933). For present purposes, four ecological measures are considered which indicate patterns of social disorganization in Moscow: the sex ratio, percent with certain defects (deafness, blindness, and mental disorders), the percent of population seventeen years and above that is divorced, and the percent infant survival. The task is to ascertain, using these as indexes of disorganization, whether the pattern is consistent with the essential idea of the preindustrial model: that the incidence of these factors (with the exception of infant survival) progressively increases with distance from the city center.

The sex ratio is an index of community organization or disorganization because substantial deviations from a balanced sex distribution have implications for other social patterns, such as family structures and work-force patterns. Burgess (1925:54) used a biological analogy to interpret variations in sex patterns in community areas: "Marked variations, as any great excess of males over females, or of females over males . . . are symptomatic of abnormalities in social metabolism." Newcomb (1951) indicates how the sex ratio for Chicago varied in 1920, and Schmid (1964) reports age and sex patterns in Seattle in 1960. In the United States, there is thus evidence that the inner city is characterized by a higher sex ratio, decreasing outward as socioeconomic status increases. In 1897 Moscow this pattern was reversed: the correlation coefficient between distance from the inner city and the sex ratio is .64. The gradients indicate the suburbs had a sex ratio of 164.8 and the second and third gradients had mean sex ratios of 129.1 and 116.0, respectively. The inner gradient had the highest sex ratio of all, however.

Data on the population with physical defects and emotional disorders are available in the 1897 census. On the basis of the preindustrial model, the incidence of physical defects and emotional disorders may be expected to increase with distance from the Kremlin. The defects included are blindness, deafness, and mental disorders, which have been aggregated into a total index of defects and disorders. As Table 1 indicates, the proportion with physical defects and emotional disorders increased with distance from the Kremlin. The correlation between distance and the percent of the total population with physical defects and emotional disorders is .63. The gradient data, however, do not show that a high proportion of the population was reported to have had these defects: .1 percent of the population had defects in Zone II, progressively increasing to .8 percent in the suburban areas.

The rate of infant survival may be estimated from age data stated in months of age for the first year. Assuming that the data are accurate, mobility is not substantial, and seasonal variation in fertility is not substantial, the ratio of the population in the latter months in the first year to the first months is an index of survival; the higher the ratio the greater the survival rate. The ratio of infants ten to twelve months to zero to two months has been constructed and stated as a percentage for this purpose. We find that the percent of infant survival decreased with distance from the Kremlin. The correlation between distance and survival is $-.59$. The gradients indicate that this varied from 78 percent in Zone II (the Miasnitskia district was excluded because of apparent inaccuracies in the data) to an incredible 47 percent in the suburbs. The overall pattern of infant survival is thus consistent with the preindustrial community model.

We expected that divorce rates should be inversely related to the pattern found in American cities. Although the data on divorce rates for Chicago from 1919 to 1935 reported by Mowrer (1938:348–49) are impressionistic, it was concluded that higher rates tend to have occurred in the Wilson Avenue districts, and the Loop and Ashburn districts, which consisted of rooming houses and lower middle class and high mobility areas, with some higher rates in high class areas.

An inverse distance pattern of divorce rates is thus the general picture that emerges. The concentration of divorced women in downtown areas in Philadelphia in 1930 was also found by Bossard and Dillon (1935). The expected pattern for 1897 Moscow, then, would be for the incidence of divorce to increase with distance from city center. Table 1 indicates that this was the general pattern, although the pattern was not strong. The divorce rate in Moscow is defined as the percent of the population seventeen years and above (both male and female) that was divorced in 1897. The correlation between distance and the divorce rate is .28, and the regression coefficient is .02. The gradient data indicate that the incidence of divorce was low in this preindustrial city, however. In the inner gradient .1 percent of the population was divorced which increased to .2 percent for the next three gradients, reaching .3 in the suburban areas.

Conclusion

The purpose of this paper has been to add to the literature on the comparative study of urban structure by analyzing prerevolutionary residential patterns in Moscow. The data are from the 1897 All-Russian census. The date of 1897, a year which represents virtually the end of Romanov rule in Russia and a feudal system of social stratification based on estates, thus makes the findings usable as baseline data for historical studies of Moscow in the Soviet period under a socialist economic system, a totalitarian political system, and an open stratification system.

The preindustrial model of urban structure has been the orienting framework of the study. In this model, formulated by Sjoberg in *The Preindustrial City*, social status is centralized. The higher class thus tends to reside near the inner core of the city, and social class progressively decreases with distance from the city center. The preindustrial model is thus the inverse of the well-known Burgess model of city structure in which social status tends to become radially decentralized in the process of population growth and geographic expansion. The preindustrial urban structure has been found in such European cities as Budapest, Oxford, Rome, and Stockholm, as well as leading cities in Latin America, Asia, Africa, and possibly the United States in the nineteenth century.

The study finds that the urban structure of Moscow in 1897 clearly resembled the Sjoberg model of the preindustrial community. Four indices of socioeconomic status were used. Social status was based on a system of estates in prerevolutionary Russia. The percent of the population in the upper and middle estates (nobility, clergy, merchants, petty bourgeoisie) generally declined with distance from the Kremlin, the historic city center. A clear inverse distance pattern is found for literacy and education. The proportion of the working population in manufacturing increased with distance from the city center, a finding which is also consistent with the preindustrial model. Indexes of disorganization were also considered to test for the relevance of the distribution of socioeconomic status. Assuming that the pattern of disorganization in a preindustrial city is the inverse of that postulated by Burgess for a modern commuinity, we thus expected, and found, that the sex ratio, rate of physical defects and emotional disorders, and the rate of divorce increased with distance from the Kremlin. The estimated percent of infant survival decreased with distance from the city center, a pattern also consistent with the preindustrial model. Although the ecology of Moscow may have been in the process of changing towards a modern structure in 1897, the traditional form clearly appears in this study conducted for a single period of time.

References

Beynon, E. "Budapest: An Ecological Study." *Geographical Review,* 33 (April 1943):256–75.

Blumenthal, H. "On the Concentric-Circle Theory of Urban Growth." *Land Economics,* 25 (May 1949):209–12.

Bossard, J., and Dillon, T. "The Spatial Distribution of Divorced Women: A Philadelphia Study." *American Journal of Sociology,* 40 (January 1935):503–7.

Bowers, R. "The Ecological Patterning of Rochester, New York." *American Sociological Review,* 4 (April 1939):180–89.

Brush, J. "The Morphology of Indian Cities." In Roy Turner (ed.), *India's Urban Future,* pp. 57–70. Berkeley: University of California Press, 1962.

Burgess, E. W. "The Growth of the City: An Introduction to a Research Project." In R. E. Park, E. W. Burgess, R. D. McKenzie (eds.), *The City,* pp. 47–62. Chicago: University of Chicago Press, 1925.

——— "The Determination of Gradients in the Growth of the City." *Proceedings of the American Sociological Society,* 21 (December 1927):178–84.

——— "Urban Areas." In T. V. Smith and L. D. White (eds.), *An Experiment in Social Science Research,* pp. 114–23. Chicago: University of Chicago Press, 1929.

Caplow, T. "The Social Ecology of Guatemala City." *Social Forces,* 28 (December 1949):113–33.

Castle, I., and Gittus, E. "The Distribution of Social Defects in Liverpool." *Sociological Review,* 5 (July 1957):43–64.

Collison, C., and Mogey, J. "Residence and Social Class in Oxford." *American Journal of Sociology,* 64 (May 1959):599–605.

Davie, M. "The Pattern of Urban Growth." In G. Murdock (ed.), *Studies in the Science of Society,* pp. 133–61. New Haven: Yale University Press, 1938.

Dotson, F., and Dotson, L. "Ecological Trends in the City of Guadalajara, Mexico." *Social Forces,* 32 (May 1954):367–74.

Duncan, Otis Dudley, and Duncan, B. "Residential Distribution and Occupational Stratification." *American Journal of Sociology,* 60 (March 1955):493–503.

Faris, R.E.L. *Social Disorganization,* 2nd ed. New York: The Ronald Press, 1955.

Ghosh, S. "The Urban Pattern of Calcutta, India." *Economic Geography,* 26 (January 1950):51–58.

Gist, N.P. "The Ecology of Bangalore, India: An East-West Comparison." *Social Forces,* 35 (May 1957):356–65.

Green, H.W. "Cultural Areas in the City of Cleveland." *American Journal of Sociology,* 38 (November 1932):356–67.

Hamden, G. "The Growth and Functional Structure of Khartoum." *Geographical Review,* 50 (January 1960):21–40.

Hansen, A.T. "The Ecology of a Latin American City." In E.B. Reuter (ed.), *Race and Culture Contacts,* pp. 124–42. New York: McGraw–Hill, 1934.

Hauser, F. "Ecological Patterns of European Cities." In T. Smith and C. McMahan (eds.), *The Sociology of Urban Life: A Textbook with Readings,* pp. 370–88. New York: Dryden Press, 1951.

Hawley, A. "Population Density and the City." *Demography,* 9 (November 1972):521–29.

Hawthorn, H., and Hawthorn, A. "The Shape of a City: Some Observations on Sucre, Bolivia." *Sociology and Social Research,* 33 (November–December 1948):87–91.

Hayner, N. "Mexico City: Its Growth and Configuration." *American Journal of Sociology,* 50 (January 1945):295–304.

Hoyt, H. *The Structure and Growth of Residential Neighborhoods in American Cities.* Washington, D.C.: U.S. Federal Housing Administration, 1939.

Johnston, N. "The Caste and Class of the Urban Form of Historic Philadelphia." *Journal of the American Institute of Planners,* 32 (November 1966):334–50.

Longmoor, E., and Young, E. "Ecological Interrelationships of Juvenile Delinquency, Dependency, and Popluation Mobility: A Cartographic Analysis of Data from Long Beach, California." *American Journal of Sociology,* 41 (March 1936):598–610.

Lyall, R. *The Character of the Russians and a Detailed History of Moscow.* London: T. Cadell, 1823.

McElrath, D. "The Social Areas of Rome: A Comparative Analysis." *American Sociological Review,* 27 (June 1962):376–91.

Mehta, S. "Patterns of Residence in Poona (India) by Income, Education, and Occupation (1937–65)." *American Journal of Sociology,* 73 (March 1968):496–508.

Miner, H. *The Primitive City of Timbuctoo.* Princeton: Princeton University Press, 1953.

Moscheles, J. "The Demographic, Social and Economic Regions of Greater Prague." *Geographical Review,* 27 (July 1937):414–29.

Mowrer, E. "The Trend and Ecology of Family Disintegration in Chicago." *American Sociological Review,* 3 (June 1938):344–53.

Newcomb, C. "Graphic Presentation of Age and Sex Distribution of Population in the City." In P. K. Hatt and A. J. Reiss, Jr. (eds.), *Reader in Urban Sociology,* pp. 287–97. Glencoe, Illinois: Free Press, 1951.

Olsson, W. W. "Stockholm: Its Strucutre and Development." *Geographical Review,* 30 (July 1940):420–28.

Queen, S., and Carpenter, D. *The American City.* New York: McGraw–Hill, 1953.

Russia. Tsentral'nyi statisticheskii komitet. *Pervaia vseobshchaia perepis' naseleniia Rossiskoi Imperii 1897,* g. 89 vols. 1899–1905.

Schmid, C. F. "Criteria for Judging Community Organization and Disorganization." *Publications of the American Sociological Society,* 27 (May 1933): 116–22.

——— "Age and Sex Composition of Urban Subareas." In R. W. O'Brien, C. Schrag, and W. Martin (eds.), *Readings in General Sociology* (3rd edition), pp. 158–60. Boston: Houghton–Mifflin Co, 1964.

Schnore, L. "On the Spatial Structure of Cities in the Two Americas." In P. Hauser and L. F. Schnore (eds.), *The Study of Urbanization,* pp. 347–98. New York: John Wiley and Co, 1965a.

——— *The Urban Scene.* New York: The Free Press, 1965b.

Sjoberg, G. *The Preindustrial City.* New York: Free Press, 1960.

Stanislawski, D. "The Anatomy of Eleven Towns in Michoacán." In G. Theodorson (ed.), *Studies in Human Ecology,* pp. 348–55. Evanston: Row, Peterson, 1961.

Uyeki, E. S. "Residential Distribution and Stratification, 1950–60." *American Journal of Sociology,* 69 (March 1964):491–98.

Wilson, L. S. "Functional Areas in Detroit, 1890–1933." *Papers of the Michigan Academy of Science, Arts and Letters,* 22 (1938):397–409.

Wirth, L. "Urbanism as a Way of Life." *American Journal of Sociology,* 44 (July 1938):1–24.

The Social Environment of Rio de Janeiro in 1960*

Fred B. Morris
and Gerald F. Pyle

Urban social geography has been a neglected area and one of general misunderstanding on the part of geographers from the United States working in Latin America. When Preston James studied and described Rio de Janeiro and São Paulo in the 1930s, it was possible to cover both cities on foot and compile land use maps. It was also characteristic to relate transportation change and urban growth to the physical environment. Some geographers writing in both Portuguese and English have never broken from this tradition, and even now they simply do not perceive the fact that in addition to these traditional aspects of study much can be contributed through the study of the social environment of Latin American cities. This exploration of Rio de Janeiro's social geography was undertaken to suggest the remedial potentialities of factorial urban ecology, within the particular context of Brazil and physical setting of Rio.

The growth of Rio de Janeiro

The more striking features of Rio de Janeiro include the enormity of Guanabara Bay and drastic differences in elevation due to mountains, hills, and an escarpment. Although there is plenty of level and slightly sloping land within the city, the *Tijuca-Carioca* (Andarai Range) extends from east to west dividing the city into northern and southern zones. A second range (*Pedra Branca*, or *Rural*) forms a western rural boundary, and a third (*Rural Marapicu-Gericinó*) forms the northern boundary. Unlike the *Planalto* of São Paulo, the areas interior to Rio, the rugged *Serra do Mar,* provided an effective barrier to hinterland growth for several centuries.

From 1700 to 1821, Rio grew rapidly. The city spread outward from the older *castelo* and bay shore centers to fill much of the level and slightly hilly land. The steeper hills and slopes were shunned as poor residential sites by the Portuguese–Brazilian culture. It was only after the movement of the capital to Rio at the beginning of the nineteenth century that some of the higher areas were settled: this settlement consisted of European "colonies" in Gloria, Flamengo, Botafogo, Floresta da Tijuca, and Santa Teresa.

Rio continued to grow after independence (1822), and nineteenth century innovations in transportation and communications helped to maintain the city as a major seaport, a function

*Reprinted from *Economic Geography,* 47 (June 1971), 286–99, by permission of the authors and the publisher.

which outlasted the mineral boom of the interior. Near the end of the nineteenth century, shantytowns started to develop on the steeper hillsides. These less desirable residential locations became filled with former slaves and soldiers returning from interior campaigns. In remembrance of a strategic battle hill, Antonio Conselheiro's victorious foes named their settlement (Morro da Providencia) *favela*. Hence a generic name was derived for the shantytowns which now appear as areal pockmarks across the face of Rio.

During the twentieth century, Rio has taken up the appearance of a modern cosmopolitan metropolis. Many small hills have simply been removed. Wide avenues have been extended through the older heart of the city and around the shores of Guanabara Bay and the Atlantic Ocean. New areas for growth have been opened by the tunneling of former mountain barriers. A fashionable beach area, Copacabana, lined with 12- to 13-story (legal limits) apartment buildings has become the center of Rio's social life. Many modern shops and new office areas have done much to shift the center of the city's commercial emphasis.

At the same time there has been growth outward along rail lines and on the urban fringes. Many of these areas, however, do not yet share some of the social amenities of urban life, and are made up of predominantly middle- and lower-middle-class residents. Although many of these areas have been urbanized in the physical sense for many years, a *suburbano* in the popular *carioca* mind is one from the countryside who lacks sophistication and culture. Yet the favela paradox represents a transfer of Brazilian rural culture to the very heart of Rio.

Four factor analyses

The factor analytic methods applied in previous studies were also used in studying the resultant social geography of Rio de Janeiro. The units of observation for the study are the 74 municipal or census enumeration *zonas* of the city. This does not include all of the metropolitan area of Rio, as several large suburbs are across the state line in the State of Rio de Janeiro, but the census for 1960 did not include data for comparable units from that state. However, the 1960 census did provide separate coverage of 48 favelas that are found within the 74 wards.[1] The wards have little, if any, homogeneity socially, economically, or politically, and vary greatly in size. Gross population was included as a variable in the factor analysis to account at least for the size differences. The other 22 variables were converted to ratios or percentages.

Some limitations of the analysis are inherent in the data. For example, no data are available in a useable form for analyzing the racial or ethnic distribution of the population. In addition, the data on the favelas are certainly limited and somewhat controversial. In the census of 1950, the *Instituto Brasileiro de Geografia e Estatístico* (IBGE) adopted the following criteria of classification of a community as a favela:

1. Groups of buildings or residences in units of 50 or more.
2. Predominance of shanties or rustic structures, constructed mainly of tin plate, boards, and so on.
3. Unlicensed constructions not regulated by zoning laws, built on land owned by others, or whose ownership is undetermined.
4. Absence of public facilities, including sewers, water, telephone, electricity, etc.
5. "Non-urbanized" area with no regular streets, no numbering system, and so on.

The same criteria were used in the 1960 census, except for the first one. This was not followed because it was felt that by 1960 the favelas, even the newer ones, had become easily identifiable, making it a simple matter to locate them for the purposes of the census.

Many observers, however, including official agencies, would put the favela population at a figure two or three times higher than that revealed by the IBGE census. For example, in 1960 the official census of the IBGE counted 147 separate favelas in 48 of the wards of the city, with a total population of 335,063. But in the same year the *Serviço Nacional de Febre Amarela* (National Yellow Fever Service) estimated that there were 850,000 persons in Rio's favelas. The *Serviço Federal de Habitação* (Federal Habitation Service) counted 183 favelas with 174,000 dwellings and 900,000 occupants. A police report in 1961 estimated nearly one million persons lived in the favelas (2). Thus, this study is made on the assumption that the census data used are representative of the reality of the favelas, though they may be challenged as to their completeness.

Because of the very general distribution of the favelas among the zonas (wards), four separate analyses were made, each with 22 variables. One is of the 74 zonas as presented in the census data. These data include the favela population indis-

---

[1] The IBGE reported on a total of 147 separate favelas located within 48 of the 74 wards. All of the favelas in each ward were considered as one, statistically.

criminately as part of the zonas. Because of what is known of the sharp contrasts that usually exist between the favelas and their larger environment, it was felt that this study alone would not give an accurate picutre of the ecology of the city. A second analysis was made of the 74 zonas, after subtracting the respective favelas, to see what the city would look like without the favelas. A third study was made of the 48 favelas alone, and a fourth dealt with the data of the second and third analyses simultaneously, treating a 121 × 22 data matrix.[2] Principal axis factor analysis with normal varimax rotation of all factors with eigenvalues exceeding unity was used.

The descriptive statistics prepared for the factoring process showed some striking differences between the favelas and the zonas. Before taking a detailed look at the four analyses, it is instructive to look at some of these broad differences. The first and perhaps most striking contrast is in the age structure of the favelas as compared to the zonas. The favelas are markedly younger, having a mean age of 18.7 years versus 24.6 years in the zonas. In addition, 43.1 percent are under 15 in the favelas as opposed to 30.8 percent in the zonas, and only 3.3 percent are over 60 as opposed to 6.9 percent in the zonas.

A second strong contrast is that there are fewer women in the favelas; the means are 49.8 percent female versus 51.5 percent in the zonas. This is somewhat surprising in the light of Medina's survey work which showed that the social unit of the favelas is the family (3). But this is probably the result of the fact that so many women and girls from the favela are employed as "live-in" maids, and thus appear as residents of the zona though they be of the favelas.

The average family size is remarkably similar, the favela having a mean of 4.64 compared to 4.45 in the zonas. This is to be understood in the light of the traditional Roman Catholic influence on the size of families in all classes of Brazilian society (6). Foreign observers are universally struck by the genuine love of children shown by Brazilians of all classes. And, though no conclusive data are available, it may reasonably be presumed that the favelas would experience a higher infant mortality rate, so that a somewhat greater tendency to reproduce would be offset.

Common-law marriages are more frequent in the favelas than in the zonas—27.5 percent of all recognized unions in contrast to 8.7 percent. Emilio Willems (7) offers the following explanation:

> During the Empire (1822–1889) the religious marriage ceremony performed in a Catholic church by an ordained priest was not only legally recognized but also the only form of marriage with religious *and* legal effects. Separation of church and state led to a distinct civil marriage procedure. Since then, the religious marriage ceremony has been regarded as the act by which social and supernatural sanctions may be obtained, while the civil procedure has been accepted as indispensable legal formality, chiefly by the middle and upper classes. To lower-class rural people, however, who do not have to worry about property and inheritance, the civil procedure adds nothing to the religious sanction, except fees.

Willems further observes that in areas that are not visited regularly by Catholic priests common-law marriages are accepted freely. Thus the high rate of common-law marriages in the favela can be understood as a reflection of the lower-class, propertyless state of the inhabitants.

Strangely enough, employment figures are higher for the favelas, having a mean of 34.0 percent employed in industry, compared to only 21.1 percent in the zonas, and an unemployment rate of 21.8 percent against 28.1 percent. This is understood in the light of the fact that the construction trades are included in the category of industry. Also it must be recognized that there is a great deal of disguised unemployment in the favelas, with many workers involved in *biscate* or temporary, part-time, marginal work, which was not discriminated in the census data.

Literacy is 58.8 percent for the favelas and 84.3 percent for the zonas. Though the difference is great, it should be noted that the rate for the favelas of Rio is higher than that for a majority of the states of the Republic. The percentage of students is not too different between the favela and the zona, a mean of 18.1 percent versus 21.3 percent. When seen, however, in the light of the much larger percentage under age 15 (43.1 percent vs. 30.8 percent), this means that many school-age children in the favela are not in school.

The favelas show a much higher percentage of owner-occupied housing—68.8 percent as compared to 32.0 percent in the zonas. As observed elsewhere, many favelados build their own shacks, thereby being owners; though they often do not own the land on which the dwelling stands.

---

[2] One of the favelas has achieved the status of a zona, or ward, and thus appears in the 74 zonas and in the 48 favelas.

The zonas factored globally

The first of the analyses was that of the zonas including the favelas. Table 1 shows the factor loadings. The results are a clear and rather well-defined three-factor solution for the city (the wood stoves and the gross population load together as Factor IV, having meaning only as a curiosity). The first or strongest factor gives a good indication of socioeconomic status in the city. Factor II can best be described as an in-migration factor, and Factor III is clearly a stage-in-life-cycle factor.

The scores for Factor I when mapped (Figure 1) show a trend for higher status to cluster in the *zona sul*, though there are "pockets" of high status on the map in the northern area. At least three of these, however (Meier, Guadalupe, Praça Sêca), are definitely not considered high status areas in the popular mind of the carioca, thus indicating that the inclusion of the favelas distorts the description of the zonas. Guadalupe is on the border of the state of Rio and if it is known by the carioca at all, it is considered *subúrbio* and definitely low status. Meier and Praça Sêca are generally regarded as areas of good housing and as being adequately urbanized, but they are definitely not prestige areas sought out for reasons of status.[3]

Factor II, the migration factor, when mapped shows (Figure 2) that the areas of highest in-migration are the *zona sul*, the central areas, and the outer fringe areas. The *zona norte* is easily the most stable part of the city.

Factor III, the life-cycle factor, mapped in

[3] Meier is particularly interesting to note at this point. It is one of the few zonas that has no favela. It is one of the more homogenous *barrios* of the city, being nearly solidly middle class, predominantly one- or two-family dwellings with relatively few high-rises. It has a high percentage of literacy, electrification, water, and so on. It should be noted that data on rentals and incomes were not available and thus did not permit the analysis to discriminate more precisely on the matter of prestige. But, as we shall see later, it is clearly the inclusion of the favelas into the zonas statistically that give a false picture of the status areas of Rio in this analysis. Niemeyer, for example, which is generally regarded as a fairly high status area, has one of the largest favela populations, being nearly three-quarters of the zona in 1960. Without question this accounts for its extremely low score here (-3.09) and the much higher scoring (1.06) on the fourth analysis. Other zonas react similarly.

Table 1. Factor Loadings: Zonas

Variables	I	II	III	IV
Common law (%)	-.92	-.05	.19	-.12
With radio (%)	.89	.20	-.21	.24
With refrigerators (%)	.84	-.31	-.37	.04
Rooms per dwelling	.84	-.42	.15	-.13
Literate (%)	.84	.04	-.48	.09
With water (%)	.84	.14	-.24	.31
With electricity (%)	.72	.17	-.13	.36
Female (%)	.69	-.50	.17	.15
In industry (%)	-.59	.48	.37	.22
Alien to Guanabara (%)	-.04	-.93	-.21	.15
In Guanabara 1-5 years (%)	-.15	-.92	-.08	-.02
In Guanabara 1 year (%)	.07	-.87	-.01	-.03
Unemployed (%)	-.14	.80	.21	-.13
Over 15 years married (%)	-.59	.59	.34	-.13
Average family size	-.11	.29	.87	-.17
Median age	.47	-.30	-.76	.25
Under 15 (%)	-.50	.35	.73	-.26
Owner-occupied (%)	-.21	-.35	.72	.37
Age 60 years (%)	.55	-.16	-.68	.29
Student (%)	.22	.34	.64	-.56
With wood stove (%)	-.45	-.11	.08	-.75
Gross population	.18	-.16	-.14	.55
Percent of Common Variance	39.6	66.1	89.1	100.0

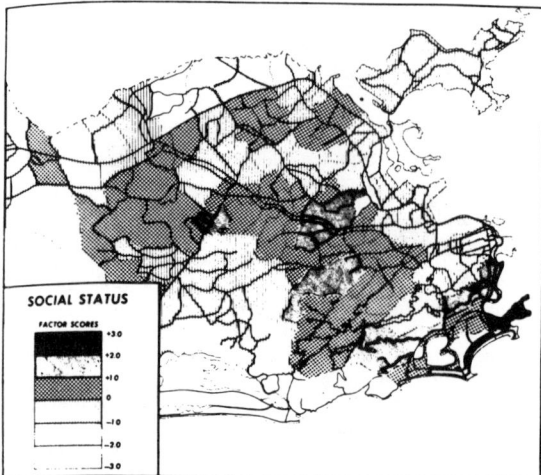

Figure 1—Social status for zonas including favelas.

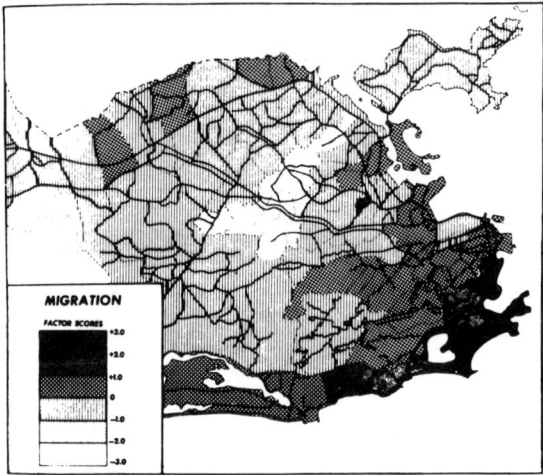

Figure 2—Migration factor: zonas including favelas.

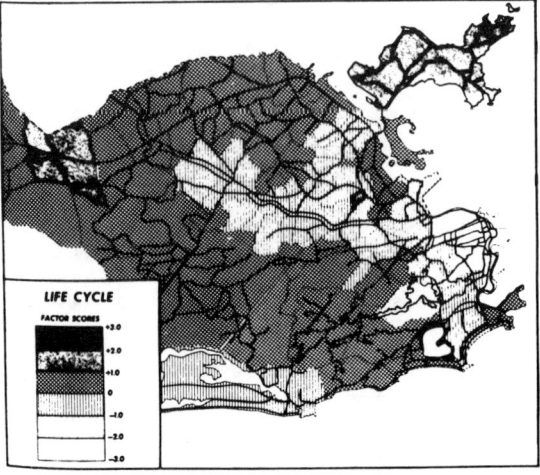

Figure 3—Life-cycle factor: zonas including favelas.

Figure 3, reveals a remarkable degree of homogeneity for the city as a whole. The range of factor scores for nearly the whole area is from 1.00 to −1.00. The notable exception is the CBD, which has a markedly older population.

Zonas without favelas

The second analysis, that of the zonas after the favelas have been statistically removed, gives a much less clear factor solution, but, perhaps, a better picture of the social space of the city (see Table 2).

Factor I in this case is made up of 12 variables. But here it is a mix of status and life-cycle elements that load on this factor. In addition to the same nine variables that loaded on this factor in the first analysis, additional variables are included: the percent over age 60, percent over age 15 married, median age, and percent under age 15 (negative). It can be seen that for the zonas considered alone, status is related to age: the higher status areas tending to be older as well.

An interesting sidelight here is that Jacarézinho, the favela that has become a zona, actually ranks higher in this analysis than two other regular zonas, Barros Filho and Barra da Tijuca, thus showing the difficulty in defining or distinguishing the favelas.

Factor II is still the migration factor and the scores show little difference from the first analysis. The same five zonas with the highest scores appear in the same order in both analyses.

Factor III is another life-cycle dimension. The percent over age 60 had a score of 0.62 on Factor III. Median age also loaded on both factors, with 0.60 here. The percent student scored 0.62 on this factor and also had 0.63 on Factor IV. Again, there are few significant changes in the two analyses for this factor. Jacarézinho ranks second in both cases after Guadalupe, both zonas having very large, young families. And both have much higher scores than the third-place zona, indicating that they are somewhat atypical. Factor IV again put wood stoves and gross population together and added the percent student.

Favelas alone

The third analysis was of the favelas by themselves (Table 3). In this case a six-factor solution was produced, confirming observations (4) about the great variation to be found in and among the favelas. The apparent homogeneity of the favelas is on the surface only.

Factor I is identified by five variables and is

clearly the migration factor; it also includes percent owner-occupied housing.[4]

Factor II is basically a life-cycle one, though two of the variables usually associated with status (percent with water, percent refrigerators) loaded here, too. Factor III is a family-size factor, loading family size, rooms per dwelling, and percent student. Factor IV is difficult to label, including the three variables of percent common-law marriages, percent in industry, and percent female. Factor V groups radios and electricity. Factor VI loads the percent literate negatively with the percent unemployed and can be called a literacy factor. Few generalizations can be made about the spatial distribution of the favelas and their social structure from the results here.

Zonas and favelas separated

The fourth analysis was of the zonas and favelas considered as mutually exclusive units, thus regarding the city as composed of 121 subareas, some favelas and some zonas (Table 4).

Factor I grouped fourteen variables. This factor is virtually identical with Factor I of the analysis of the zonas minus the favelas, the only difference in variables being that gross population loaded on Factor I here rather than on the fourth factor. This again is a socioeconomic status and life-cycle mix (Figure 4).

The second factor is again an in-migration one (Figure 5). The third contains one variable (percent wood stoves loading independently) and the fourth has the variables of family size and percent student, being basically an index of family age (Figure 6).

On looking at the three analyses of the zonas, taken globally, alone, or with the favelas mutually exclusive, several things become clear about the social environment of Rio. First, blurring the favelas into the zonas as though they were not separate units seriously distorts the picture of the human ecology of the city. This distortion occurs, in part, because of the varying size of the favelas in relation to their respective zonas. In some cases the zona is so much larger than the favela that its inclusion has no weight at all statistically. In other cases, the favela is so large that it virtually outweighs the zona.

[4]The percentage of owner-occupied housing loading with in-migration in the favelas is not difficult to understand. Basically, there is no housing market as such in the favelas. The citizens must provide their own shacks. However, in the older favelas, that is, in those whose occupants have been there for a longer period of time, there would logically be more turnover of housing with some people leaving and renting their shacks to newcomers.

Table 2. Factor Loadings: Zonas without Favelas

Variables	FACTORS			
	I	II	III	IV
Common law (%)	-.94	-.11	.11	-.06
With radio (%)	.93	-.09	-.16	.22
With water (%)	.88	-.07	-.15	.32
Literate (%)	.85	.13	-.39	.17
With refrigerators (%)	.81	.38	-.25	-.02
With electricity (%)	.76	-.12	-.06	.34
Rooms per dwelling	.70	.53	.22	-.25
Female (%)	.68	.48	.22	.09
Age 60 years (%)	.63	.12	-.62	.27
Over 15 married (%)	-.62	-.57	.28	-.07
Median age	.62	.34	-.60	.27
In industry (%)	-.60	-.51	.34	.20
In Guanabara 1-5 years (%)	-.05	.96	-.06	-.03
Alien to Guanabara (%)	.07	.94	-.18	.15
In Guanabara 1 year (%)	.06	.89	.03	-.01
Unemployed (%)	-.18	-.79	.12	-.21
Owner-occupied (%)	-.10	.26	.84	.23
Average family size	-.15	-.27	.82	-.32
Under 15 (%)	-.58	-.39	.63	-.27
With wood stove (%)	-.48	.08	-.04	-.73
Student (%)	.03	-.26	.62	-.63
Gross population	.14	.11	-.11	.61
Percent of Common Variance	41.5	69.0	88.2	100.0

Table 3. Factor Loadings: Favelas

Variables	Factors					
	I	II	III	IV	V	VI
Alien to Guanabara (%)	-.87	-.21	.01	-.04	-.09	.06
In Guanabara 1 year (%)	-.81	-.03	-.14	.02	-.30	-.03
In Guanabara 1-5 year (%)	-.79	-.12	-.19	-.17	-.12	-.04
Owner-occupied (%)	-.71	-.03	.14	.33	-.12	.09
With wood stove (%)	.54	.18	-.21	.50	-.14	.22
Under 15 (%)	-.10	-.91	.13	.02	-.12	-.04
Median age	.06	.78	-.23	-.07	-.10	.13
With water (%)	-.07	.70	.26	.03	.31	-.06
Over 15 years married (%)	-.36	-.62	-.18	-.30	.06	-.10
Age 60 years (%)	.44	.61	-.11	-.18	.18	.40
With refrigerators (%)	.31	.53	.28	-.20	.47	-.14
Average family size	.04	-.02	.84	-.01	.07	-.13
Rooms per dwelling	-.06	.01	.83	-.27	.27	.14
Student (%)	.44	-.09	.57	-.36	.05	-.27
Gross population	-.24	.30	.35	.27	.32	.18
Common law (%)	-.19	-.03	-.31	.82	-.00	-.08
In industry (%)	-.06	-.02	.04	-.63	.01	.23
Female (%)	.19	-.09	.49	.60	-.07	.22
With radio (%)	.15	.13	.29	.01	.88	-.06
With electricity (%)	.35	.07	-.04	-.06	-.88	.09
Unemployed (%)	.04	.20	.02	-.18	-.02	.77
Literate (%)	.89	.27	.27	-.48	-.10	-.53
Percent of Common Variance	24.2	44.7	61.6	77.2	91.3	100.0

Table 4. Factor Loadings: Zonas Plus Favelas

Variables	Factors			
	I	II	III	IV
With refrigerators (%)	-.92	-.12	.26	-.01
Age 15 years (%)	-.88	.08	-.22	.36
Median age	.88	-.07	.22	-.35
Rooms per dwelling	.87	-.05	.25	.32
Over 15 years married (%)	-.86	-.16	.11	.13
Age 60 years (%)	.83	-.27	.20	-.33
Literate (%)	.82	-.33	.37	.01
In industry (%)	-.80	-.02	.02	.03
With water (%)	.80	-.35	.38	.01
Common law (%)	-.74	.33	-.44	-.12
Female (%)	.74	.32	-.11	.22
With radio (%)	.72	-.40	.34	.10
With electricity (%)	.59	-.45	.17	.00
Gross population	.56	-.17	.40	-.18
Alien to Guanabara (%)	-.09	.90	-.10	-.19
In Guanabara 1 year (%)	-.24	.84	.09	-.10
In Guanabara 1-5 years (%)	.13	.85	.15	-.03
Unemployed (%)	-.03	-.62	.23	.07
Owner-occupied (%)	-.53	.60	-.07	.14
With wood stove (%)	-.28	-.06	-.87	.01
Average family size	-.36	-.07	-.05	.83
Student (%)	.23	-.42	.07	.76
Percent of Common Variance	54.1	77.8	88.9	100.0

A comparison of Figures 1 and 4 shows how this works with regard to status areas. In general the trend toward higher status in the *zona sul* is the same, but some zonas show up as high status on Figure 1 that are not as high on Figure 4. Figure 4 gives a much more accurate picture of the status areas of the city.

Figures 2 and 5 show the same general pattern of in-migration with one important difference. In-migration, like status, in general is related to centrality, with the exception of the fringes of the city, and the very important exception of the favelas. As noted elsewhere, the favelas show a much higher index of recent in-migration further out from the center. This can be easily understood in terms of land availability. The land useable for the growth of the favelas has long been occupied in the central areas, so that recent arrivals must find space further out, whereas the zonas, especially in the *zona sul*, have experienced vast reconstruction with tremendous vertical expansion providing space for in-migrants who can afford it.[5]

Figures 3 and 6 are not quite comparable, as Figure 3 shows a general life-cycle factor, while Figure 6 is a more restricted family-size factor. Both maps are notable in the striking homogeneity of the city as a whole. The CBD is the main exception. Figure 6 shows a remarkable similarity of family size between most of the zonas and their respective favelas.

Additional aspects of social space

As previous factorial ecology studies have shown, it is often extremely useful to plot factor scores of one factor against another in an effort to learn more about social space. This has been done for the two factor analyses of Rio de Janeiro which covered the entire population of the city, i.e., the first and fourth analyses above. The three major dimensions of the city's social space, status, migration, and family size, are shown to form definite patterns in various sections of the city. However, it is important to note that it takes both of the analyses to identify the major social areas of the city, and the main function of separating the favelas in the mutually exclusive analysis is to show

[5] The recent expansion westward along the beach areas referred to here is an indication of how near the saturation point the other areas have come, as are the data on population density. In 1960 Copacabana had 41,329 persons per sq. km.; Ipanema, 29,977; Leblon, 18,550; Leme, 15,612; Flamengo, 26,982.

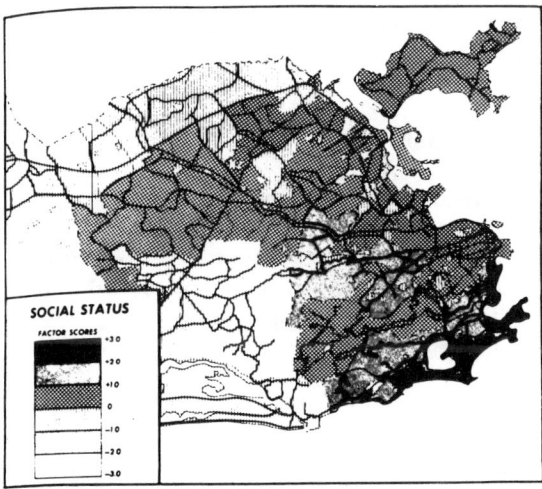

Figure 4—Social status: zonas and favelas separate observational units.

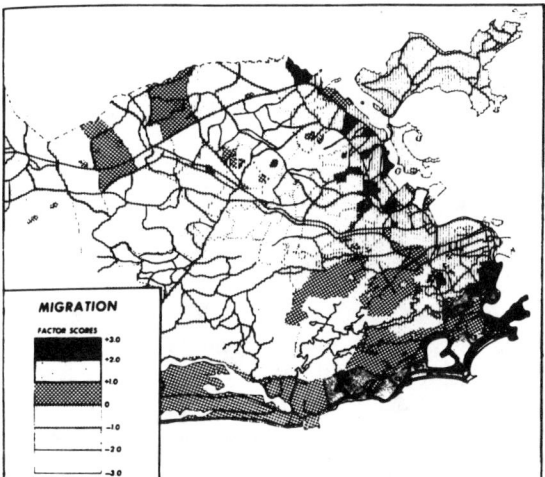

Figure 5—Migration factor: zonas and favelas separate.

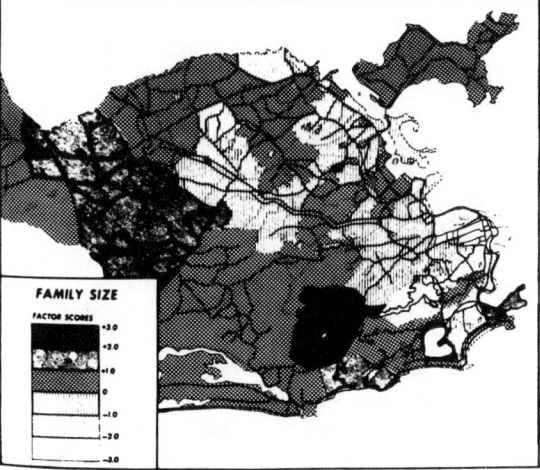

Figure 6—Family-size factor: zonas and favelas separate.

where they differ from the zonas and where they do not.

Conclusions

It would appear that the city of Rio de Janeiro displays distinctive patterns of social geography, to be seen in the mosaic of the zonas most notably in the way centrality and status are related. On the other hand, the city is quite homogeneous regarding the stage in the life-cycle, with the exception of the old central area. The favelas form a pattern different from the basic one of the city in that they are scattered "on top of" the city itself, imposed but not integrated into any other patterns of the city's social space. These ingredients clearly indicate a traditional urban ecology under the impact of modernization.

In reference to Spanish colonial planning policies, the term *Hispanic* is often used in haste as a convenient synonym for describing most urban forms in Latin America. However, the early Spanish and Portuguese colonial urban models were in fact very different, and unfortunately the few geographic studies of cities of Portuguese origin have failed to make this point clear. Latin American cities show substantial similarities in their forms because the Spanish did not waste much time in patterning cities after the edicts of Philip II; the Portuguese simply let things "grow" as they would without reference to the colonial planning ideal. This growth means that Gideon Sjoberg's models of preindustrial cities with later colonial modifications (5) still serve as an excellent point of departure for understanding the sort of Portuguese city transferred to the New World, and hence the early social ordering of Rio de Janeiro.

In the early Portuguese colonial city, the elite commanded proximity to the core. There may have been several elites in the early continental versions, including conquering knights from a half-dozen European kingdoms. Those who could compete the most successfully for space adjacent to the elite were the middle classes, often sharply segregated into quarters. The two more central areas were surrounded by increasingly lower orders of urbanites. On the city fringes were to be found the most recent of migrants from the rural hinterland.

Though Brazilian cities in general did not follow the rigid grid pattern of their Spanish counterparts, the evidence indicates that centrality was primal to the colonial elites of Rio de Janeiro. The desire and need to be near the sources of power, both secular and ecclesiastical, resulted in a pattern of residential location that placed the elites in the center of the city. And although the period of prosperity brought by gold in the eighteenth century led to some movement of the upper classes toward the more open areas of the *várgea* and Tijuca, with the resultant construction of the *chácaras* of that period, centrality was still important and most important families appear to have maintained a residence at the center even while building a more comfortable home a bit farther out. And, of course, these chácaras were not really very far from the center.

Even cities of today reflect many of the traits described by Sjoberg as typical of the ideal-type of the preindustrial city. The dominant pattern of the Brazilian families today, both upper and lower classes, is strikingly similar to that of Sjoberg's hypothesis. Kinship ties remain crucial in the bureaucracy, the government, in employment and so on. Craftsmen are still to be found (shoemakers, furniture makers) though their numbers are rapidly diminishing. The traditional Roman Catholic religion still finds its most faithful adherents among the upper class (women), and magic plays a vital role in the life of the masses.

At the same time, Rio is obviously showing signs of the impact of industrialization upon the social order. The tremendous influx of persons from the rural areas is clear evidence of the economic, educational, and social advantages Rio offers as a center of modernization. Though many of these in-migrants find themselves in situations of dreadful squalor in the city, the fact is that for the great majority, even of the most miserable of the favelados, things are better in the city than in the place they left.

Sjoberg notes that during the period of transition a number of different processes occur more or less simultaneously: (1) the persistence of traditional forms, (2) revisions or modifications of traditional forms, (3) disappearance of traditional forms, and (4) emergence of new structures. This is certainly to be seen in Rio de Janeiro today.

The favelas are providing an arena for these processes of transition and adaptation. This also is in accordance with Sjoberg's construct, in the formation of subsystems and enclaves through which the recent arrival is oriented to the new world of the city, and by which the urbanite maintains his ties with the traditions of his rural past.

Though many would see Rio de Janeiro as an

industrial city today, in terms of its social organization it clearly maintains many of the features of the preindustrial city and of the transitional city, and relatively few of the patterns that are expected of the industrial "type," though these may be emerging.

The Burgess theory of expanding city growth in which there is a direct relationship between socioeconomic status and distance from the center clearly does not apply to the Rio de Janeiro of today, as it has been shown to be inapplicable elsewhere in the socioeconomic case. In Rio, however, it must be noted that the center has effectively shifted since 1900, (or, at the very least a major second center has grown up), with the opening of the tunnels to Copacabana and the beach areas of the *zona sul.* It might be hypothesized: since the mountains in the heart of the city effectively isolated these areas from serious urban development until the beginning of the twentieth century, these areas were preserved for the "suburban" expansion of an industrial era. That is, the fact that considerable land became available (by the opening of the tunnels) close to the center caused what might have been a suburban expansion by the elite à la Burgess. The additional fact that this land has very high amenity value (the beaches and a cooler microclimate) allowed the carioca elite to have it both ways: prestige and centrality.

Today, however, this area is rapidly filling up and a process of succession is under way, with middle- and upper-lower-class groups already occupying many of the apartments in Copacabana. The area of high prestige is progressing further down the beach, forming the "Gold Coast" sector à la Hoyt. As noted earlier, a new tunnel will soon provide easy access to the next beach area, Barra da Tijuca, and every indication is that it will be the next prestige area.

This pattern of shifting elite areas coincides with Amato's observations (1) about elite settlement patterns in Bogatá, Quito, Lima, and Santiago. And the expansion of social services to these areas, such as roads, tunnels, water, and sewers even while multitudes in other areas of the city (e.g., the favelas) are destitute, would lend support to his thesis that the self-interest of the upper classes is a dominating factor in urban development in Latin American cities.

References

1. Amato, P. W. "Elitism and Settlement Patterns in the Latin American City." *Journal of the American Institute of Planners,* 36 (March 1970):96–105.
2. *Correio da Manha.* July 1, 1961.
3. Medina, C. A. *A Favela e o Demagogo.* São Paulo: Livraria Maitius, 1964.
4. Morris, F. B. "A Social Geography of Rio de Janeiro in 1960." M.A. Thesis, University of Chicago, 1970.
5. Sjoberg, G. *The Preindustrial City.* Glencoe, Ill.: Free Press, 1960.
6. Smith, T. L. *Brazil: People and Institutions.* Baton Rouge: Louisiana State University Press, 1963.
7. Willems, E. "The Structure of the Brazilian Family." Paper presented at Fifteenth Annual Meeting of the Southern Sociological Society, Atlanta, Georgia, March 28, 1952.

The Shantytowns of Montego Bay, Jamaica*

L. Alan Eyre

The existence of peri-urban shantytowns, an increasingly visible and spectacularly expanding urban phenomenon in tropical and subtropical lands, has not escaped the notice of geographers. But for the most part their attention is characterized by brief and casual references rather than by detailed information and analysis. Also, superficiality based on academic middle-class prejudices and, worse, misunderstanding based on the classic danger of a little learning are features of many comments on these settlements in geographical texts. Evidence of conceptual confusion between urban slums and truly peri-urban shantytowns can often be found. There is also an unfortunate tendency to invoke stigmatizing value judgments that stereotype these settlements as social aberrations,[1] cancers overwhelming a supposedly otherwise healthy economic body. A number of social scientists have directed attention to various economic and social aspects of life in squatter settlements, but few studies have appeared that analyze the geographical dimensions of such settlements and their role in the spatial web of human and economic relations. Only one geographer, Lucien Parisse, has made shantytowns a specific field of published research.[2] The present paper aims to identify, through a variety of research approaches, a number of diagnostic characteristics by which shantytowns can be identified as a geographical phenomenon.

The term "shantytown" is one of many used to identify taxonomically a widely distributed type of urban settlement; others are uncontrolled settlement, squatter settlement, *barrio*, *barriada*, *favela*, and *bidonville*. The "bustee" of urban India differs in its demographic structure and, therefore, despite a superficial similarity in appearance, must be considered a different type. Despite the variety in nomenclature and, as will appear later, at the expense of some oversimplification, a shantytown is considered here to be any peri-urban collection of dwellings erected without an official subdivision plan on land to which the occupants, or at least most of them, hold no title and which, at least initially, is not subdivided for housing purposes.

Shantytowns, however, reflect more of urban vigor than of urban breakdown. During a period of development or when development is accompanied by rapid population growth, as in Latin America generally, the focuses of development, usually major cities, are unable to allocate scarce

---

*Reprinted from *Geographical Review*, 62 (1972), 394–413, with the permission of the author and the American Geographical Society.

[1] For example, Berckholtz Salinas, *Barrios marginales: aberración social* (Lima, 1963).

[2] Lucien Parisse, "Les favelas dans la ville: le cas de Rio de Janeiro," *Revista Geográfica*, 70 (1969):109–30; see also Robert S. Harrison, "Migrants in the City of Tripoli, Libya," *Geogr. Rev.*, 57 (1967):397–423.

economic resources to the housing sector. The financial institutions are unable to fund directly either the kind or the quantity of housing that would satisfy needs according to generally acceptable planning standards, and the uncertain nature of the shantytown residents' employment prohibits access to such institutions anyway. The shantytown is a way around such barriers. It is financed by small savings and petty loans from small businesses. It emerges as a relatively successful adaptation to the problem of providing cheap housing with fairly secure tenure for a segment of the population that as yet lacks a stable economic role in a developing society.

The city of Montego Bay, Jamaica (1972 estimated population approximately 50,000, boundary as in Figure 1) was chosen for the present study for several reasons. Although it has experienced dynamic growth (71 percent in the decade of the 1960s) owing to tourism, commerce, and light industry, its modest size permitted a comprehensive survey. Differentials of income and wealth are particularly wide, and both the forces operating on urban form and their results are expressed with clarity.

There are no municipal boundaries in Jamaica, and the lowest echelon of administrative units is the parishes. The mayoralty of the city of Montego Bay and the chairmanship of the council of St. James parish in which it is located are therefore one and the same office. Boundaries for the city are set somewhat arbitrarily by the administrators of building codes, market limits, and censuses, acting independently. I have been equally arbitrary, and the choice of the number and distribution of the statistical divisions to be called "Montego Bay" is an empirical one of my own.

The city can be broadly zoned into the central business district, the tourist sector, the commercial and industrial area of the free port, the inner-city slums, an expanding arc of middle- and upper-class residential property, and ten shantytowns in various stages of development (Figure 1). Of the shantytowns, North Gully and Mount Salem are within 2.5 kilometers of the (burned-out) City Hall, and Albion and Glendevon are at an intermediate distance, 3 to 4 kilometers. The remaining six, Flanker, Rosemount, Tucker, Granville, Bogue, and Green Pond,[3] are 5 to 6 kilometers distant. Squatting began in North Gully and Mount Salem in the 1930s or even before, and their combined population is 9600 (1971 estimate).

[3] Not all of Green Pond is shantytown; the squatters adjoin an ordinary rural farming district.

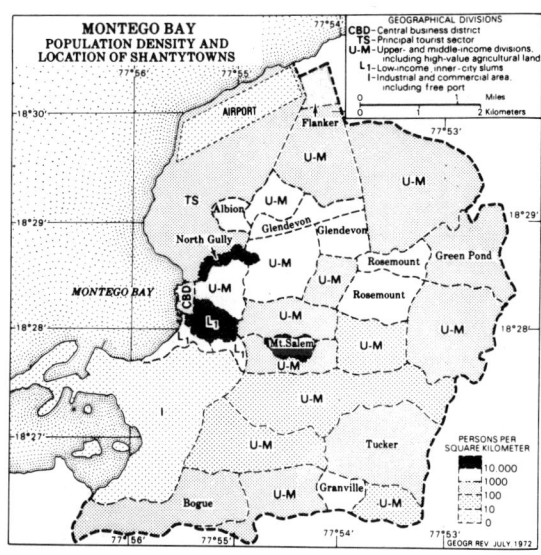

Figure 1

Granville, Glendevon, Albion, and Tucker have grown from squatter nuclei in the 1940s to a total population of 7800. Flanker and Bogue date from the 1950s (combined population 1350), and pressures in the 1960s have led to the occupance of Rosemount and Green Pond (combined population 1350). Thus the total population of all ten shantytowns is estimated to be approximately 20,100 or 40 percent of that of the city as a whole.

It is necessary to visualize the spatial and dynamic urban network into which the shantytowns fit (Figure 2). Although the theoretical aspect of this schema is based on the results of the entire study, it is presented first so that the analysis that follows can be more easily integrated into the whole. Simplified to bare essentials, there are five flows of population. (1) Migrants from rural areas to the city proper form a large flow. Most are single, poor, and often looking for their first jobs. They lack funds and so gravitate to the crowded inner-city slums. (2) A second flow, from these city slums to the shantytowns, consists largely of people who desire to escape from many aspects of life in the inner city (crowding, high rents, crime) and wish to find alternative accommodation. This flow consists in large measure of persons in the stage of family formation who have acquired petty capital while living in the city but who have either no permanent employment or low-paying jobs. (3) A flow from the inner city to the middle- and upper-income ring represents the familiar suburbanization process of the modern sector. Also

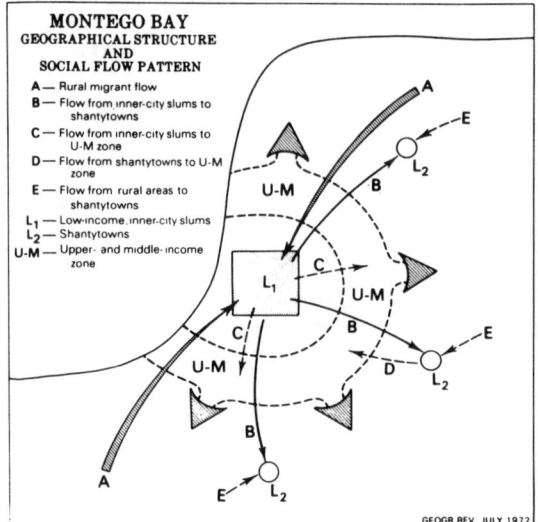

Figure 2

included is a relatively minor flow to government low-income housing. The ring is rapidly expanding outward, usually avoiding the wedges in which the shantytowns are located but sometimes exerting pressure on them. (4) A small flow of upwardly mobile migrants from the shantytowns to the middle ring consists of families who have acquired the permanent and assured income required by mortgage companies. (5) Finally, there is a flow from rural areas directly to the shantytowns. However, contrary to information that appears in references to shantytowns both in the geographical literature and in the popular press, this is not the main source of recruitment for the shantytowns nor even an important one. The flow is restricted almost exclusively to younger relatives of persons who already reside there.

Within this overall framework, then, the specific diagnostic features of the shantytown can be classified under seven broad aspects of their total geography: spatial organization, developmental stages, demographic structure, economic activity, the role of savings and investment, the problem of public amenities, and population control.

Spatial organization

Three diagnostic features may be subsumed under spatial organization. In the first place, shantytowns are usually on land of marginal utility, for land values are often a critical factor in their location. In Montego Bay these settlements have come into being on land that is steep, rough, or subject to flash floods. In North Gully homesites are either on precipitous slopes or are directly on the banks of a torrent. Except for Flanker, which grew up on a part of the reserved airport lands, all the rest are on rough limestone terrain that had previously gone out of agricultural use. Some gully sites pose daunting problems of construction and access, with buildings jutting out from near-vertical cliffs.

Secondly, the distribution of homes is random, a feature which enables shantytowns to be identified clearly from the air. This characteristic reflects the relatively casual manner in which settlement has proceeded and the absence of any coordination in occupancy. Although shacks are removed, the pattern created by the earliest squatters frequently governs the form of subsequent infilling. This pattern is noticeable even in Flanker, the most impermanent of the squatter settlements. Even when a shantytown has been absorbed by the municipality, its original features can be discerned despite a more formal street layout superimposed on it.

Finally, since shantytowns are essentially the product of personal initiative, construction is highly individualized and incorporates anything ready to hand—scrap metal, old packing cases, and junked car bodies, as well as ordinary building materials. Shipping crates are popular, and one home in Rosemount has "made in Czechoslovakia" prominently stenciled across the veranda. However, the use of bright paint can quickly obscure the origins of many homes, and the appearance of the shantytown matures with time and industry. In Flanker, where 77 percent of the homes have been built since 1960, most now have a quasipermanent character. Of the 245 homes in this shantytown, 46 percent are built of lumber and are painted; 29 percent are of lumber, unpainted; 20 percent are of miscellaneous materials; 3 percent are of wattle; and 2 percent are of cement block, plastered.

Almost without exception, homes begin as one-room shacks. Additions are made slowly, as funds become available or are accumulated: a bedroom for a growing family here, a veranda there. Latrines and kitchens are accommodated in flimsy outbuildings, and any kind of bathroom is a rarity in all the shantytowns except Glendevon. Although the exteriors of many homes may indicate both the piecemeal construction of the dwellings and the lowly economic status of the builders, virtually every home is clean and tidy within, often immaculate. The determined polishing of steps and similar rituals say much for the spirit of those who reside in these do-it-yourself projects. I know dozens of shantytowns intimately, and I consider that the value judgment "squalid," so frequently applied to these settlements by superficial observers, does their inhabitants a serious injustice. In the great majority of cases meager resources have been strained to the limit to obtain louvered windows, polished flooring, furniture, and decorative articles.

The four phases of shantytown development

Every shantytown passes through four stages in its secular development, but this development can be halted at any stage or even destroyed completely by government action.

The initial occupancy stage. During this stage unused land becomes available by one means or another, and the poineers move in. They may rent "spots"[4] or may simply squat. Initially, the land area available to each family unit may be quite large, and the time and effort devoted to part-time cultivation or livestock rearing may be considerable. At this stage the feeling of insecurity is high, and consequently most homes are simple if not crude. As other families move in, the area available for farming steadily diminishes, and the population inevitably depends more on the city for subsistence.

The transitional stage. The continuing influx poses an inherent dilemma. A sparse distribution of shacks offers space for cultivation but invites their removal by the authorities. On the other hand, each additional squatter's hut reduces such space but increases the sense or illusion of security. In this stage investment begins to be made in the homes, fences are grown or erected, and small businesses come into being.

The stage of attaining secure tenure. Final attainment of secure tenure is by no means inevitable and can be aborted by any of several factors. Where renting from a multiplicity of small landowners is characteristic, as in Granville, the purchase of house spots may be slow, difficult, or even impossible. There are several reasons for this. Some landowners are disinclined to alienate their land and prefer to multiply their mini-tenancies until saturation is reached, thus maximizing their rental income. In other instances the land is "family land," owned by many kin in common,

---

[4]As an example, for a house spot in Green Pond, the newest settlement, each family pays approximately $45 a year.

and complete agreement is difficult to achieve. Another important reason is that many "owners" have no valid title or have lost the unregistered common-law titles which they may have had originally. Some landowners evict tenants deliberately to avoid any possibility of permanent occupancy, fearing that such occupancy will jeopardize ownership rights. In other cases the transition has been relatively smooth, as in Glendevon; here most of the squatters have attained reasonably secure tenure, since the central government has intervened to purchase certain tracts on their behalf.

The stage of absorption. Only when tenure is reasonably secure does established authority, municipal or national, move in. The effect is to absorb or incorporate the former shantytown into the urban body. Streets may be laid out according to some plan, services and utilities are provided, schools are built, and public transportation is provided, usually by the private sector under license. Glendevon is a typical case of a shantytown in the fourth phase of development. However, this development may be aborted and absorption may be only partly achieved. The land may be too steep to be suitable for a normal street plan, and other physical disadvantages may exist, as at North Gully. Land may be difficult or expensive for the government to acquire or it may be required for other purposes, as was the case in Flanker, where the extension of the airport runway necessitated the eviction of some squatters. The government may prefer total redevelopment as part of a comprehensive plan rather than acceptance or perpetuation of the form that has evolved.

Demographic structure

The role of shantytowns in the urban milieu of demographically dynamic developing countries has been briefly mentioned earlier. They provide an escape from slum life for those who, owing to excessive crowding and other pressures, prefer to be independent. In most cases the reason for moving to a shantytown is to establish a home for a growing family. It is not that accommodation in the inner city is not available, but that many families prefer living in the shantytowns to living elsewhere, their economic circumstances being what they are. A distinctive demographic pattern is recognizable in shantytowns, where the proportion of young children is large. The number of children less than 4 years of age per 100 women 15 to 44 years of age in the shantytowns averages 72, as compared with 53 in the rest of the city. Also, 19 percent of the population of the shantytowns is under five years old, whereas the national average is 16 percent. On the other hand, only 11 percent of the population is 45 years and older, compared with 14 percent in the city proper. Thus a demographically significant diagnostic characteristic may be noted: family formation is relatively more important in shantytowns than elsewhere in the urban nexus.

The detailed census tables for 1960 add a number of other features to the demographic picture. The sex ratio was more balanced in the shantytowns than in the urban area as a whole, where (as is normal in Jamaica) females are overwhelmingly in the majority. Many urban census tracts reported more than 150 females for every 100 males; the shantytowns varied within the 113 to 120 range. More children in the 10 to 20 years age bracket attended secondary school (24 percent) than in the inner-city slums (10 percent), but many fewer than in the middle- and upper-class ring (55 percent). The percentage males gainfully employed at the time of the census was 61 percent in the inner-city slum areas, 67 percent in the shantytowns, and 88 percent in the middle- and upper-class zone.

An additional demographic feature is important: the majority of shantytown dwellers are urbanites of long standing, not rootless rural migrants who have drifted cityward. This fact is not generally appreciated, but in every one of dozens of shantytowns known to me the statement is true. The average household head had lived eleven years in Montego Bay, including his residence in the shantytown. More than three-quarters of the population of all ten shantytowns had been born within the city limits as defined. The rural migrant stream is directed primarily toward the inner city, where rows of slum tenements specifically cater for such transients. A central location is more convenient for job seeking, since a peri-urban location incurs travel expenses. Moreover, many rural migrants may have no choice, since in the main shantytown homes are constructed by and for families.

Once established in the city and in the course of family formation, some slum dwellers may contemplate squatting or renting a spot for a shack on the urban fringe. In Granville, for example, the great majority of early householders have moved out from inner Montego Bay, particularly from such infamous addresses as Railway Lane and Love Lane. Most older adults were born in rural parishes but nearly all of them had lived for a time in the inner city before moving to Granville. This particular shantytown is of long standing, and a considerable proportion of the present new-home builders are children of the first settlers. They build in close proximity to the parental home, and often an informal "compound" results.

Direction of economic activity

The gainful effort and the time of shantytown dwellers is characteristically channeled in three directions—intermittent employment, subsistence agriculture, and petty commerce—all of a casual nature.

The labor force is primarily engaged in intermittent employment, particularly in construction. Shantytowns provide simple housing for certain categories of workers essential to the economy of an originally poor but developing urban region. These workers are essentially service personnel, whose labor is needed in periods of vigorous activity but who may not be able to qualify for a secure and permanent role in the city's economic life. It is not unusual for 20 percent to 50 percent of household heads in shantytowns to be technically unemployed during some part of the year. They may "scuffle" and "ping-pong around" (to use two graphic local idioms) until better luck turns up. But a multitude of small, insecure, impermanent gainful tasks is more characteristic than permanent employment.

The disproportionate importance of construction work as a means of livelihood in shantytowns is shown in Figure 3. In Mount Salem, in the southern part of Glendevon, and in North Gully, more than a third of all gainfully employed males work in construction, and in the eastern part of North Gully, the proportion is almost two-thirds. Flanker, curiously, is in the lowest category with less than 10 percent, but this is because its role is mainly to provide casual service employment for the tourist industry; the famous Rose Hall golf course is nearby and caddying is a seasonal occupation.

A great deal of effort is expended on agriculture to provide as much subsistence as the limited land area will permit and as a hedge against hard times. Many early residents stressed that a major reason for squatting was the desire to cultivate because of temporary unemployment or to eke out a low income with homegrown food. Some have moved outward more than once in an effort to achieve this goal. In Granville, for example, goats, pigs, and fowl can all be raised, and in most yards the people grow maize, bananas, coconuts, yams, peppers, pumpkins, a few sticks of sugarcane, cocoyam, dasheen, and one or more fruit-bearing trees such as mangoes, naseberries, akees, and breadfruit. There are few households in which some member does not engage in petty agriculture at some time during the year. Often cultivation has to be conducted on precipitous slopes and on rough, unsuitable terrain.

Soon after establishment, shantytowns acquire commercial characteristics best described as minitrading. Overheads are low, since much of the activity is carried on outdoors or under the simplest of shelters. Bars are often the earliest forms of business enterprise, but small general stores soon proliferate, and frequently an open-air market selling basic foodstuffs arises. It is usually not long before a "central business district" develops, with a few trees, a small open space for petty trade, social activities, or simply lounging, and with one or two stores and a bar or club nearby. An occasional political meeting may be held there, though career politicians are not always unequivocally welcome in shantytowns!

In the more mature settlements, a few additional quasi-industrial operations can be found: for example, most shantytowns have a mechanic's shop and one or more automobile body-repair shops, which are sometimes simply a collection of old tools under a tree or a lean-to. Table 1 indicates the commercial activity in Glendevon (1971 population about 3800) and in Flanker (about 1000). Only premises with a visual indication of the activity are included. For example, dozens of dressmakers live in Glendevon, but only one has special premises with a sign indicating the nature of the business. The rest ply their trade irregularly in their homes.

The role of savings and investment

Direct private investment in construction and improvement is remarkably high and takes a far greater proportion of the total income of shantytown residents than of those in the middle- and upper-income ring. An important part of the present study was the valuation of residential buildings in shantytowns and a comparison with homes elsewhere in the city. A $50,000 home cannot be valued separately from the lot on which it stands, whereas a $1000 home is movable, and it has therefore been necessary in Figure 4 to

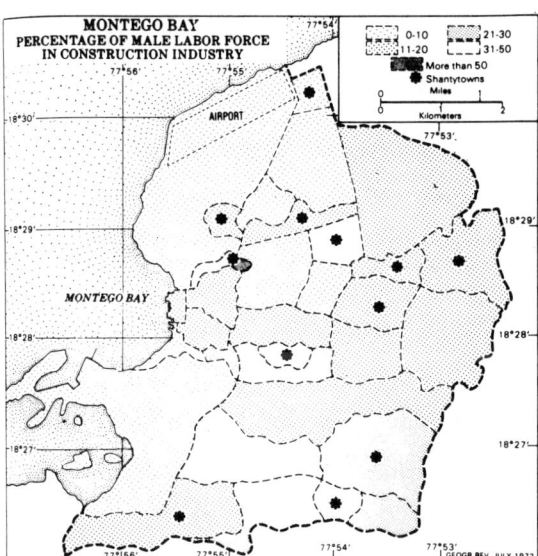

Figure 3

eliminate the landprice factor.[5] Since no security of tenure exists in most shantytowns, homes in them if sold can be moved to another spot.

The valuation of homes in the shantytowns was based on what owners expected to realize if they had to sell, and in most cases their figures agreed well with my own knowledge of the replacement cost in materials and labor. Of the 245 dwellings in Flanker, 12 percent were valued at $2000 or more, 68 percent between $1000 and $2000, 9 percent between $500 and $1000, and 11 percent at less than $500.[6] The last category included some bamboo and tar-paper shacks and several unfinished huts that were occupied by their male builders.

The homes in the newer areas of the more recent shantytowns usually average less than $1000 per unit, but in parts of Glendevon many squatter families now own $10,000 homes. Naturally, in these cases the householders have a strong desire

[5] Figure 4 indicates that valuations in some shantytowns are as high as or higher than in adjoining divisions identified as upper- and middle-income on Figure 1. This anomaly is attributable to the fact that several of the latter have been allocated to middle-income housing by being formally subdivided but are in an early stage of development and still contain poor rural housing.

[6] Valuations are in United States dollars. Prices of homes in the major urban areas of Jamaica are comparable with those in the United States, but it should be remembered that construction is much more labor intensive in Jamaica.

to obtain secure title from the government, and when roads and other amenities have been put in, this may be possible, since the surveying of roads and lots is usually done at the same time. But it is a goal that may take years to achieve.

Inadequacy of public amenities

Official bodies are notoriously ambivalent to shantytowns. The growth of these settlements reduces the demand for public housing, yet they are rarely accepted for what they are, valiant attempts at self-help (which government so often exhorts its people to apply). Through a fear of implying permanency and acceptance of the situation that has arisen, government agencies frequently provide too little too late. In Montego Bay electricity is provided by a private company, and only a few areas lack power. Sometimes in shantytowns several homes may "pirate" a line and so share a meter, and I even know of one or two unauthorized extensions.

Water is another story. In one part of Flanker, five hundred persons have to depend on a single half-inch standpipe that issues water for only two hours a day, between 3:00 a.m. and 5:00 a.m.. In all shantytowns the number of standpipes is woefully inadequate. Transportation is another major problem, since the city bus service does not serve all shantytowns and in any case has had financial trouble and is unreliable. Minibuses and taxis provide more frequent service, but at prices that the average resident can ill afford. This problem affects shantytowns the world over; because they are peri-urban, low income, and have population densities relatively lower than the inner city, they do not attract mass transportation.

The most critical problem, however, is education. The more developed communities, such as Glendevon and Granville, have schools, but they are unbelievably inadequate because the dimensions of growth are unknown to the authorities and the needs of these communities have a low priority. Albion School is dark, dirty, drab, and utterly inadequate for the purpose of education. In Granville whole classes of children wait outside the school for their turn to get in; when they do, it is only to stand huddled against walls while other classes also crowd the room. Amazingly, a few students have eventually progressed to the university.

In 1968 Flanker, which had a population of almost a thousand, of whom 220 were of primary school age, had no school of any kind. All children traveled a difficult three miles to the city schools or to rural schools even more distant. Some families paid as much as a tenth of their weekly income for taxis in order to keep their children in school. This drain on resources was then the largest single problem encountered by the residents of the community. Finally in 1969, after much lobbying and political pressure, and more than twelve years after the first squatters moved to Flanker, a new school was opened.

One division of the public sector about which shantytowns are particularly sensitive is law enforcement. In the absence of unambiguous government policy toward such settlements, the police are viewed with ambivalence. In most cases routine police protection is very limited, and the publicity given to the occasional raids tends to be resented. However, the subjective opinion of the vast majority of residents who have previously lived in crowded urban tenements is that there appears to be less crime and antisocial behavior in the shantytowns than in the inner-city slums. A large number of persons gave as the reason for their greater satisfaction with life in the shantytown the fact that it was less troublesome and that people were less quarrelsome and more law-abiding. "People live good around here, except for a few young fellows" was a typical comment. Certainly I have found this to be generally the case; for despite a difference in life style, I have always received every cooperation and courtesy in all

Table 1. Commercial Premises in Glendevon and Flanker

Type of Enterprise	Number Glendevon	Flanker	Type of Enterprise	Number Glendevon	Flanker
Groceries and soft drinks	24	9	Mortician	1	
Bars	5	1	Dry cleaner	1	
Auto-body shops	3	2	Dressmaker	1	
Clubs (serving liquor)	2	2	Sign maker		1
Ice vending	1	2	Jeweler	1	1
Tailor	1	1	Open-air market	1	1

those settlements that I have come to know intimately.

Nevertheless, a current trend is causing some anxiety to both the residents and the authorities. It is likely that certain shantytowns are increasingly providing a haven for young criminal elements for whom the slums have become too hot. The larger the city, the greater the pressures in this direction. Some of the much larger shantytowns on the fringes of Kingston and St. Andrew are more affected by this trend than are those in Montego Bay.

Population control

Any tropical country with an exploding population that has a proven reservoir of good will toward family planning is fortunate. Positive attitudes toward population control definitely exist in Jamaica, though they are strongly expressed only by a few. Despite an extremely difficult familial milieu from the point of view of the usual family-planning program (only a minority of children are born and raised in a nuclear family with married parents), the shantytowns form a reservoir of potential participation.

The St. James Family Planning Association runs a free clinic in downtown Montego Bay.[7] In 1970, almost a fifth of the women in Granville and in Mount Salem were classified as active participants in the clinic, but only 4 percent in Flanker and 9 percent in North Gully took part. It should be clearly understood that, although women in many middle- and upper-income areas do not participate in the clinic, this does not imply a lack of family planning on their part. Residents in these areas get their information mainly from private doctors and pharmacies.

Despite a massive publicity campaign from 1969 to the present, the precentage of women in the shantytown areas who participate has not shown a spectacular increase. Nonetheless there are definite indications that women in low-income areas are motivated to control the number of births. In the rural areas grannies or some other kin were always ready to assist with child rearing; in the shantytowns the nuclear family may live far from kin, urban pressures are being felt to a considerable degree, and child care has to be paid for if both parents are employed. In communities such as Flanker, North Gully, and Rosemount, a limited

[7]The association is affiliated with the national organization, which in turn is assisted financially by the World Bank.

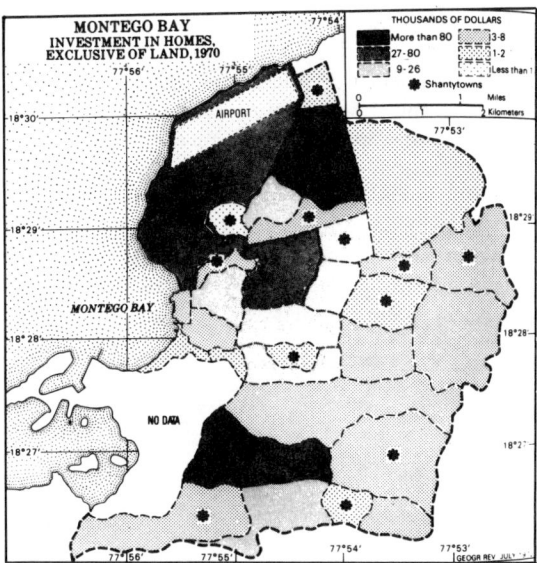

Figure 4—Value of investment in existing homes at 1970 prices, per residential unit.

increase in participation should not be difficult to achieve if this economic motivation can be exploited.

The need for security

The shantytown dweller is the poor suburbanite of the developing world. He is usually upwardly mobile. He is generally industrious. He is a saver, not a prodigal spender as the slum dweller often is; in every shantytown there is evidence of almost incredible efforts at investment and thrift. Although employment may be casual and irregular, I found households where as many as seven different gainful activities were being carried on by the members. On the other hand, the shantytown resident suffers deprivations owing to distance from work, schools, medical care, shops, and government services. However, because of the availability of a few fruit-bearing trees and a yard for livestock and food plants, he may have a better diet than the slum dweller and may avoid the stresses and degradation of slum-tenement life.

Above all, the shantytown resident longs for security of tenure, a spot to call his own. Unlike the stereotype, he is far more often a conservative than a radical. He does not beg government for subsidized or low-cost housing; he gets on with that job himself. What he needs is the security of knowing that he is not doing this in vain, that all his efforts and savings will not subsequently be bulldozed away. He does not resent guidance, but he is impatient with the ambivalence of authority and with the slowness in receiving the few amenities he cannot provide for himself: power, water, transportation, education.

Before anyone castigates shantytowns as undesirable urban outgrowths and laments that they are not removed, this fact should be borne in mind: in Montego Bay between 1960 and 1970 more than three thousand families who moved into the ten shantytowns during the decade invested at least two million hard-earned dollars in materials and labor. Like other members of society, they want their investments protected. In urban areas all over the tropics and subtropics the shantytown dweller and his fellows are a sizable proportion of the total population, and both his better-off fellow citizens and his reluctant governments—local and national—must learn to live with him and to utilize his energies.

Urban Structure in France*

Theodore Caplow

Until quite recently, urban sociology had only a shadowy existence on the Continent. In part this indifference seems to have stemmed from the identification of sociology as a branch of speculative philosophy, which still inhibits empirical research everywhere in Europe. Then, too, the scarcity of funds and facilities for empirical work has scarcely been favorable for large-scale team projects.

Despite this persisting handicap there has been an impressive development of urban research in France (and elsewhere in Europe) in the few years since the war.[1] Among recent publications are the survey of Auxerre by Bettelheim, Frère, and their associates,[2] the work of George, Agulhon, Lavandeyra, Elhai, and Schaeffer[3] on the suburbs of Paris, the migration analyses of Chevalier,[4] and the charting of ecological distributions by Bardet.[5] Among the significant studies in progress should be cited the interesting work of Jean LaPierre in the Marseilles region, and the monumental study of the "social ethnography" of Paris by a team headed by P. H. Chaumbart de Lauwe.[6] Meanwhile, and beginning somewhat earlier, the human geographers have developed an impressive literature on dwelling types, the distribution of economic functions, and the evolution of street

---

*Reprinted from the *American Sociological Review*, 17 (October 1952), 544–49, by permission of the author and The American Sociological Association. (Sections of pages 544 and 548 and page 549 were omitted.)

[1] A useful summary of the present state of knowledge and research in this area will soon be available in the published proceedings of the Deuxième Semaine Sociologique held in Paris in March, 1951. Georges Friedmann (ed.), *Villes et Campagnes: Civilisation urbaine et civilisation rurale en France* (Bibliothèque Générale de l'Ecole Practique des Hautes Etudes, 1952).

[2] Charles Bettelheim and Suzanne Frère, *Une ville Française moyenne: Auxerre en 1950: Etude de structure sociale et urbaine* (Paris: Armand Colin, 1950).

[3] Pierre George, "La banlieue: une forme moderne de développement urbain"; Maurice Agulhon, "L'opinion politique dans une commune de banlieu sous la Troisième République. Bobigny de 1850 a 1914"; L. A. Lavandeyra, "Saint-Maur-des-Fossés"; H. D. Elhai, "Les H. B. M. de la Porte j'Aubervilliers"; R. Schaeffer, "La répartition géographique du personnel d'une grand usine de la banlieue parisienne." Published together as *Etudes sur la Banlieue de Paris* (Paris: Armand Colin, 1950).

[4] Louis Chevalier, *La formation de la population Parisienne au XIX siècle* (Presses Universitaires de France, 1950).

[5] Summarized in Gaston Bardet, *Le Nouvel Urbanisme* (Paris: Vincent Fréal, 1948).

[6] Organized under the general heading of "Ecologie de l'Agglomeration Parisienne," the project is the work of a social ethnography team housed at the Musée de l'Homme (Centre d'Etudes sur la Région Parisienne) under the general sponsorship of the Centre d'Etudes Sociologiques. The methods used in the construction of base maps, and the recording of long-term and diurnal population movements represent, in the opinion of the writer, a distinct refinement of current American practice.

plans—including the methodologically valuable work of Tricart and his students in Strasbourg.[7]

These data are still limited and somewhat disconnected, but taken together, they permit certain tentative statements about urban structure in France:

(1) The familiar ecological principles which explain urban structure in the United States do not adequately account for urban structure in France.

The degree of centralization in France appears to depend upon a variety of local circumstances, but is always less marked than would be expected in an American community of comparable size. It is not a function of increasing community size in any regular relationship and, in the major metropolitan areas, there is some evidence that centralization has declined since the Middle Ages. Schaeffer and Chaumbart de Lauwe are able to demonstrate the absence of any single point of concentration in Paris,[8] and the absence as well of the radical daily movement of population which characterizes Chicago.[9] Instead there is a sort of general spiral movement, inwardly directed, but without convergence to a central point, and the volume of this spiral movement appears to be diminishing.

The maximum height of buildings in French cities has been limited at least since the fourteenth century, and the only skyscrapers in the country are in semi-suburban housing projects. When to this factor are added the extreme durability of construction and the historical autonomy of specialized urban districts, it is easy to understand why invasion and succession are not conspicuous aspects of the French pattern.

Indeed, without the influence of uncontrolled land values, spatial distributions in France are often singularly inexpressive of social structure. The writer has studied a community of moderate size in Southern France which, almost at first glance, shows an unusually distinct social cleavage and an equally marked physical division. Yet, upon careful examination, the association of the social structure with the spatial pattern is found to be quite irregular. Neither the distribution of the population by social class, nor the distribution of institutions by class affiliation, conforms to the fundamental spatial division of the community or to any other precise demarcation.[10]

Dispersion of population is similarly inconspicuous in the French city. There is no uniform tendency for density to decrease toward the periphery, even in Paris with its long history of expansion. Finally, there is no reason to believe that as the urban population increases, the average density of settlement within the urban community decreases, as happens in Minneapolis or Dubuque. The point has not been made more often only because dispersion of population is readily confused with suburbanization. Many French suburbs are of ancient origin: Nîmes, for example, must have been extensively suburbanized in Roman times. These suburbs, however, do not seem to be associated with a continuously declining density in the central city.

A tentative explanation can, perhaps, be approached by a return to the topic from which urban sociology originally sprung—a consideration of land values. Differential land values are sharply limited in France by the limitation of building heights, which has been continuously enforced for centuries. In many cities variations of land values are further restricted by the presence of historical monuments, palaces, churches, and parks, which are almost completely immovable, in what might otherwise be a central commercial district.

[7] J. Tricart, *L'Habitat Urbain,* Volume 2 of his *Cours de Géographie Humaine* (Paris: Centre de Documentation Universitaire, 1951). This volume, strongly Marxist in its interpretation, contains an excellent general bibliography of urban geography.

[8] Op. cit., p. 7.

[9] Gerald Breese, *The Daytime Population of the Central Business District in Chicago* (Chicago: University of Chicago Press, 1951).

[10] Aix-en-Provence. The city, which has a population of 27,000, is characterized by extreme social differentiation. The number of *rentiers,* professionals, and white-collar persons is high, and they are, as a class, extraordinarily secure, while at the same time the working class is extremely solidary. The community possesses a small university, a court of appeals, limited health resort facilities, some local industry, a medical center, some suburban functions, and serves as a market center for a very large rural area. About 40 percent of the electorate are communist; nearly the same number belong to parties of the extreme right, including a substantial royalist faction. Public opinion studies show an almost complete dichotomization of opinion. The real separation of these two halves of the community is illustrated by the tendency to double all institutions. To match the middle-class university, there is a strong "popular" extension university; the theatres, the bus-lines, and the restaurants are as unequivocally segregated as they might be under a caste system. The religious activities of the Church are matched down to such details as lotteries for school children by the parallel activity of the Communist Party.

The community is topographically divided, as well, into an Old Town and a New Town, separated by the principal street. The Old Town conserves the medieval street plan and has many 14th-century houses, while the New Town, designed under Cardinal Mazarin, is an early example of "modern" city planning, with broad rectangular streets, architectural unity, and a partial greenbelt.

Without the sorting out of land uses and population types under the influence of differential land values, questions of ecological location take on an entirely different significance. The siting of slums, factories, ethnic colonies, or apartment houses cannot be explained in simple monetary terms. The real estate market can no longer be regarded as the principal agency regulating urban growth and no equally simple mechanism takes its place.

The variety of structure found in European cities is not fundamentally explicable in the elementary terms of zonal analysis and density gradients. Any serious attempt at comparative ecology would require the substitution of a whole range of determining factors for the monistic land-value device which works so well on the American scene.

It should be noted in passing that the dominance of land values over site-determination is no longer what it used to be even in the United States. The constriction of central business districts, the deconcentration of industry, the suburbanization of the lower income groups, and the vast expansion of urbanized areas in the last twenty years, all tend together to increase the number of alternate locations for any particular group or function, and therefore compel us to pay increasing attention to more subtle factors than the straightforward compulsions of land value.

(2) Rural-urban migration in France has not depended upon the growth of the total population, and mobility of the urban population has not been a function of expansion of the urban population.

The inadequacy of the blanket concept which equates varieties of behavior ranging from status achievement to migratory farm work under the general heading of social mobility is nowhere better illustrated than in the social history of France since the Revolution.

As everyone knows, the population surge which has since overtaken the rest of the world came to France and ran its course early, so that even without the catastrophic losses of World War I, it would have been the first Western country to show an absolute decline of population. The demographic history has actually been somewhat more complicated than this suggests, but it remains true that no civilized area had so limited a growth of population after 1800. (The relative stability of population and a certain stability of resources account perhaps for the great rigidity of social classes, and for the preservation of rural and village sub-cultures throughout the country.) What emerges surprisingly in the studies of Bettelheim and Chevalier is that none of this interfered with a very high rate of internal migration. For centuries the Himalayan isolation of the provincial town has been a literary and popular theme in France. However, the survey of Auxerre, a provincial city of 30,000 selected on a series of neutral criteria similar to those used in the selection of Middletown, "revealed the falsity of the stereotype of the provincial city, isolated, closed in, and maintaining a minimum contact with the exterior."[11] They found instead that three-fourths of the population were not native to the town, that 92 percent of all married couples included at least one non-native. It should be remembered that this refers to a community whose population has not increased sharply since 1801.

Similarly, Chevalier is able to show, using documentary sources which any American investigator would envy, that throughout the nineteenth century the proportion of Parisians born in Paris remained at about one-third, that in-migration (in concordance with Stouffer's intervening opportunity hypothesis) was based on real differences in wages and working conditions, was specialized by occupation by region, and was directed from each province toward particular sections of the metropolis.

(3) A third finding which may be drawn from the French data is that France—and perhaps Europe generally—presents more alternative types of urban settlement than can be found in the United States; more varieties of internal adaptation, of spatial patterning, of architectural structure, and of change. This may be only an elaborate statement of the obvious, but some of its implications are interesting.

It is difficult to estimate the extent to which the physical uniformity of our own cities has colored our general view of the urban environment. It is probable, for example, that many of the features of the conventional ecological diagram depend in large measure upon an undifferentiated gridiron plan, with almost uniform street design and essentially temporary buildings.

French cities show a much greater variety of forms. Of the types of street plans and block arrangements described by Lavedan,[12] only a very few are commonly utilized in this country. Thus, only one of our large cities has a radial plan; none,

---

[11]Bettelheim and Frère, op. cit., p. 258 (above translation by the present author).

[12]Pierre Lavedan, *Geographie des Villes* (Paris: Gallinard, 1936).

to the writer's knowledge, shows the separated sector plan, the fish-bone plan, a system of related terraces, a regular elliptical plan, or even a ribbon pattern. (The case of Duluth, Minnesota, where a gridiron plan has been imposed on a steep precipice, is fairly representative.) In addition to the variety of formalized plans, European cities are conspicuous for the variations occasioned by widely-separated periods of growth.

Even apart from the walls and fortifications, the historical squares and formal gardens, the canals, and other specific antiquities, the sheer diversity of spatial patterns in French cities points up, by contrast, the startling uniformity to which we are accustomed from Maine to Texas. Thus, the *modern* streets in Paris include some which are much wider and others which are much narrower than any streets in New York. We are unfamiliar, in the United States, with the double street, the covered street, the pedestrian street, or the stairway street. Such a simple device as the arcading of a shopping area is uncommon in this country, although arcades would be far more functional in Minneapolis than in Toulouse. Much the same thing may be said of the numerous forms of the public place, the park, the promenade, the patio, the riverfront quai, the bridge, the hillside terrace, and, most important, the family dwelling unit.

(4) The fourth finding from these studies has to do with the continued social viability of the *quartier,* in both large and small cities, although not in all of them.

All of the investigators of the Latin American city have been struck by the former significance of the *barrio* as a community of identification and as a functional unit, and have documented the disappearance of its autonomy within the last generation or so. As the writer has noted in a report on Guatemala City,[13] if we arrange the Middle American cities on which ecological information is available by size, the rank order of size will correspond to the rank order of disappearance of *barrio* autonomy and differentiation.

The French situation is far more complex. Anciently, all cities of any consequence were divided into quarters, and these survive in the largest cities (Paris, Marseilles, Lyons), in many of the middle-sized cities (Nice, Poitiers), and in some of the smaller ones. Throughout France and the Low Countries, differentiation by *quartier* seems to be even more conspicuous than in modern Latin America.

The *quartier* is in a real sense a community. Fully developed, it has its distinct appearance, its local industry and commerce, its housing forms, family types, collective attitudes, formal associations, and often its own dialect and folklore. The *quartier* tends to be clearly demarcated even where it does not correspond to an administrative division. In Paris, where the *arrondissement* is both an administrative and an electoral unit, some *arrondissements* like the XVIth and the Xth correspond to *quartiers,* others do not.

It appears that *quartiers,* once fully developed, tend to be exceedingly stable, perhaps more so than rural settlements. Thus the Faubourg St. Antoine, the Latin Quarter, or the Temple, retain local traditions which have been continuous since very early times. Whether such a tradition can survive in a peripheral district is less certain. One of the two suburban studies previously cited traces the development of a recognizable social type in a low-rent public housing project,[14] another documents the disappearance of community identification in a dormitory suburb.[15] Even in the latter case, however, what remains is a greater intensity of interaction than we should expect to find anywhere in an American city. This is likely to remain so as long as the separation of home from workplace is far less complete in France than in the United States.

[13]Theodore Caplow, "The Social Ecology of Guatemala City," *Social Forces,* 28 (December 1949).

[14]Elhai, op.cit.
[15]Lavandeyra, op. cit.

Urban Structure in Europe: Recent Evidence*

Bruce London

The literature on comparative urban ecology, especially that work which compares the ecological structure of Third World, pre-industrial cities with the ecological structure of North American industrial cities, has yielded some widely examined generalizations. For example, there is evidence of an "evolutionary sequence" from the pre-industrial status gradient described by Sjoberg (1960)—i.e., centralization of elite residential areas and decentralization of lower-class residential areas—towards the North American industrial status gradient described by Burgess (1925)—i.e., elite decentralization and lower-class centralization (see especially Schnore, 1965, 1972; London and Flanagan, 1976; London, this volume). An underlying assumption of this argument implies in very general terms a correlation between levels of national and urban development and urban spatial structure. The Burgess pattern is most likely to be found (a) in the most highly developed nations, and (b) in any nation's largest, most industrialized cities (Schwirian and Rico–Velasco, 1971; Schwirian and Smith, 1974).

A consideration of the literature on European cities, however, casts some doubts on such generalizations. No general urban pattern can be described within or between nations, and this is the case regardless of whether we group cities by age, size, function, or political system. Today, some European cities have extremely mixed ecological structures with high- and low-status groups inhabiting both the center and the periphery. Other cities seem to have retained a pre-industrial status gradient. Still others have manifested a movement away from the Sjoberg pattern to closely approximate the pattern described by Burgess.

A number of factors unique to Europe must be taken into account in any attempt to understand this extreme ecological variability. European cities are much older than American cities. Built on a long-established pre-industrial base, and often having their change guided by a combination of government urban planning as well as the operation of market forces, many of these cities are likely to locate new industries on their peripheries. This both reduces the tendency towards change in the core areas and encourages workers to reside in peripheral locations. In addition, in many European cities the persistence of traditional status considerations contributes to a tendency for high-status residents to cling to inner-city locations rather than moving to the suburbs (Hauser, 1968; Thomlinson, 1969; Timms, 1971).

---
*This article was written for this volume and has not been published previously.

Those European cities such as Oxford (Collison and Mogey, 1959), Rome (McElrath, 1962), Prague (Moscheles, 1937; Musil, 1968), Madrid (Abrahamson and Johnson, 1974), and, in general, French cities (Caplow, 1952) and cities of the Alfold (Beynon, 1961b), which had not undergone particularly rapid growth, retained a pre-industrial gradient. However, even those large cities which grew rapidly, and which thus might be expected to conform to the pattern described by Burgess because of a need to "create" locations for an expanding middle class and for industry—cities such as London, Paris, Vienna, Stockholm (Hauser, 1968), and Budapest (Beynon, 1961a)—exhibit extremely mixed patterns.

London, for example, has long been a city of ecological contrasts. Centrally located slums and upper-status areas stand in close juxtaposition, while sectors of both high- and low-status residences are found to extend from the center into the periphery (Johnston, 1971; Wohl, 1975). Edinburgh and Sunderland appear, at first, to approximate the Burgess pattern, but Edinburgh displays the continued prestige of many inner areas along with much recent upgrading in the center (Johnston, 1971), while Sunderland's Burgess-like status gradient describes only the distribution of private residences. When public "council housing" for lower-status groups (located predominantly in the peripheral ring) is taken into consideration, a highly mixed situation becomes evident (Johnston, 1971; Robson, 1969).

Still other European cities seem to generally conform to the Burgess pattern. Upper-class movement from center to periphery, leaving the inner areas the locus of "inferior dwellings," has been documented for Manchester (Johnston, 1971). Belfast, too, seems to display a positive relationship between distance from its center and socioeconomic status, although superimposed on this general pattern is also a pattern of sectors (Boal, 1975). Finally, Rex (1968) describes Birmingham as having an inner zone of transition whose residents are clearly disadvantaged relative to suburbanites.

Even in those cases where some inner-city deterioration and suburbanization seem to have occurred, a new ecological trend is taking place that makes it even more difficult to formulate generalizations. This trend is called "gentrification." The term was apparently coined by Glass (1964) to describe changes in London neighborhoods in the early 1960s.

One by one, many of the working class quarters of London have been invaded by the middle classes—upper and lower. Shabby, modest mews and cottages . . . have become elegant, expensive residences. Larger Victorian houses, downgraded in an earlier . . . period . . . have been upgraded once again. . . . Once this process of "gentrification" starts in a district, it goes on rapidly until all or most of the original working class occupiers are displaced. (Glass, 1964:xviii)

By 1964, gentrification was well underway in the central London areas of Hampstead, Chelsea, Islington, Paddington, North Kensington, Notting Hill, Battersea, and others, and it persists today (Hamnett, 1973; Williams, 1976; Williams and Hamnett, 1980). In addition, there is evidence of gentrification in other British cities, on the European continent, and in Scandinavian cities, as well as in cities in the United States, Canada, and Australia (cf. London et.al., 1980).

This trend flies in the face of the most widely held theories of neighborhood change, which suggest that inner-city residential areas pass through a series of stages which lead to the gradual *decline* of the area and eventual conversion from residential to commercial or industrial land use (cf. Faris, 1967; Vernon, 1960; Hoover and Vernon, 1962). It also represents a clear reversal of patterns predicted by any evolutionary-sequence hypothesis of ecological structure and change. No gradient generalization is possible as gentrification contributes to the increased mixing of socioeconomic groups throughout the city's space.

Confounding attempts at generalization even further is the case of Eastern European socialist cities. To begin with, are we even able to discuss ecological segregation by social class in ostensibly classless societies? The answer to this question is clearly yes. Wage and social-status differences do exist in East European societies and "those with higher wages and qualifications live in better housing and what is even more important, in more subsidized housing types" (Szelenyi, 1977:19).

In a paper reviewing the literature on recent ecological patterns in Czechoslovakian, Polish, Hungarian, and Yugoslavian cities, Szelenyi (1977) presents evidence which suggests that Eastern European cities may well be undergoing an evolutionary sequence much like that described by Schnore (1965) for Latin American and North American cities. In other words, before World

War II these cities displayed a strong negative correlation between social status and distance from the center. More recently, however, Eastern European cities seem to be evolving from this Sjoberg-like configuration towards the pattern described by Burgess. "The status of the housing estates which are built far from the center is higher than average," while, simultaneously, "the social status of the zone of transition declines" (Szelenyi, 1977:17).

Significantly, however, these similar patterns are clearly caused by different factors. East European cities operate under a socialist mode of production. Hence, the land market so central to the North American situation is not relevant here. "Most of the urban housing is built by the State and allocated under highly subsidized rents, thus replacing the housing market by a redistribution of national income through the housing system" (Szelenyi, 1977:12). Why, then, do we find an apparent convergence on the Burgess pattern in both contexts?

Szelenyi (1977) suggests that the planning priorities of the State are the key determinants:

> Urban planning which concentrates only or mainly on the construction of new estates on the outskirts of the cities might start a process of "slumming" in the so-called area of transition. The physical breakdown of the buildings in the transition area is accompanied by the concentration of the people with smaller incomes, as the families with higher statuses are "siphoned off" from these neighborhoods by the new developments. (Szelenyi, 1977:17)

In sum, tentative support for an evolutionary-sequence hypothesis is gained from an examination of North American, Latin American, and Eastern European cities. However, the extreme ecological variability of Western European cities presents us with a serious exception to the hypothesis. Is such a "general theory of residential land uses in urban areas" (Schnore, 1972:21) possible? At the moment, we can only conclude that no uniform pattern is to be observed everywhere. Such a conclusion, however, ignores a very basic point. "It is important not to confuse a specific pattern of spatial distribution with the general principles of . . . ecology" (Theodorson, 1961:330). Structures change over time. We must remember that our ultimate goal is to understand the process of ecological change; models of ecological structure are simply a means to that end.

References

Abrahamson, M., and Johnson, P. "The Social Ecology of Madrid: Stratification in Comparative Perspective." *Demography,* 11 (August 1974):521–32.

Beynon, E. "Budapest: An Ecological Study." In G. A. Theodorson (ed.), *Studies in Human Ecology,* pp. 357–70. Evanston: Row, Peterson, 1961a.

———. "The Morphology of the Cities of the Alfold." In G. A. Theodorson (ed.), *Studies in Human Ecology,* pp. 355–56. Evanston: Row, Peterson, 1961b.

Boal, F. "Social Space in the Belfast Urban Area." In C. Peach (ed.), *Urban Social Segregation,* pp. 245–65. New York: Longman, 1975.

Burgess, E. W. "The Growth of the City: An Introduction to a Research Project." In R. E. Park, E. W. Burgess, and R. D. McKenzie (eds.), *The City,* pp. 47–62. Chicago: University of Chicago Press, 1925.

Caplow, T. "Urban Structure in France." *American Sociological Review,* 17 (October 1952):544–50.

Collison, C., and Mogey, J. "Residence and Social Class in Oxford." *American Journal of Sociology,* 64 (May 1959):599–605.

Faris, R. E. L. *Chicago Sociology: 1920–1932.* Chicago: University of Chicago Press, 1967.

Glass, R. "Aspects of Change." In Centre for Urban Studies, *London: Aspects of Change,* pp. xii–xlii. London: MacGibbon and Kee, 1964.

Hamnett, C. R. "Improvement Grants as an Indicator of Gentrification in Inner London." *Area,* 5, No. 4 (1973):252–61.

Hauser, F. L. "Ecological Patterns of European Cities." In S. F. Fava (ed.), *Urbanism in World Perspective: A Reader,* pp. 193–216. New York: Crowell, 1968.

Hoover, E. M., and Vernon, R. *Anatomy of a Metropolis.* Garden City: Doubleday, 1962.

Johnston, R. J. *Urban Residential Patterns.* London: G. Bell, 1971.

London, B. "The Social Ecology of Latin American Cities: Recent Evidence." In G. Theodorson (ed.), *Urban Patterns: Studies in Human Ecology,* pp. 374–78. University Park: The Pennsylvania State University Press, 1982.

London, B., and Flanagan, W. G. "Comparative Urban Ecology: A Summary of the Field." In J. Walton and L. Masotti (eds.), *The City in Comparative Perspective,* pp. 41–66. New York: Sage-Halstead, 1976.

London, B.; Hudson, J. R.; and Bradley, D., eds. "Special Issue: The Revitalization of Inner City Neighborhoods." *Urban Affairs Quarterly,* forthcoming.

McElrath, D. C. "The Social Areas of Rome: A Comparative Analysis." *American Sociological Review,* 27 (June 1962):376–91.

Moscheles, J. "The Demographic, Social, and Economic Regions of Greater Prague: A Contribution to

Urban Geography." *Geographical Review,* 27 (July 1937):414–29.

Musil, J. "The Development of Prague's Ecological Structure." In R. E. Pahl (ed.), *Readings in Urban Sociology,* pp. 232–59. Oxford: Pergamon Press, 1968.

Rex, J. A. "The Sociology of a Zone in Transition." In R. E. Pahl (ed.), *Readings in Urban Sociology,* pp. 211–31. Oxford: Pergamon Press, 1968.

Robson, B. T. *Urban Analysis.* Cambridge: Cambridge University Press, 1969.

Schnore, L. F. "On the Spatial Structure of Cities in the Two Americas." In P. M. Hauser and L. F. Schnore (eds.), *The Study of Urbanization,* pp. 347–98. New York: Wiley, 1965.

———. *Class and Race in Cities and Suburbs.* Chicago: Markham, 1972.

Schwirian, K. P., and Rico-Velasco, J. L. "The Residential Distribution of Status Groups in Puerto Rico's Metropolitan Areas." *Demography,* 8 (February 1971):81–90.

Schwirian, K. P., and Smith, R. K. "Primacy, Modernization, and Urban Structure: The Ecology of Puerto Rican Cities." In K. P. Schwirian (ed.), *Comparative Urban Structure: Studies in the Ecology of Cities,* pp. 324–38. Lexington: D. C. Heath, 1974.

Sjoberg, G. *The Preindustrial City: Past and Present.* New York: Free Press, 1960.

Szelenyi, I. "Urban Sociology and Community Studies in Eastern Europe: Reflections and Comparisons with American Approaches." *Comparative Urban Research,* 4, Nos. 2, 3 (1977):11–20.

Theodorson, G. A., ed. *Studies in Human Ecology.* Evanston: Row, Peterson, 1961.

Thomlinson, R. *Urban Structure.* New York: Random House, 1969.

Timms, D. W. G. *The Urban Mosaic: Towards a Theory of Residential Differentiation.* Cambridge: Cambridge University Press, 1971.

Vernon, R. *Metropolis 1985.* New York: Doubleday, 1960.

Williams, P. "The Role of Institutions in the Inner London Housing Market: The Case of Islington." *Transactions,* 1, No. 1 (1976):72–82.

Williams, P., and Hamnett, C. R. "Social Change in London: A Study of Gentrification." *Urban Affairs Quarterly,* forthcoming.

Wohl, A. S. "The Housing of the Working Class in London, 1815–1914." In C. Lambert and D. Weir (eds.), *Cities in Modern Britain,* pp. 109–13. Glasgow: Fontana/Collins, 1975.

Ways of Dwelling in the Communities of India*

Radhakamal Mukerjee

The village outlives all

More than five-sixths of the world's population still find in the regular village their permanent abode and means of living. Rice-culture in particular demands cooperation in the use of land and water resources, and is especially conducive to the compact village form. Wherever mountains, forests, or marshes break the monotony of continuous cultivation, however, we find the country broken up into hamlets of scattered houses and isolated farmsteads, perched on the crests and recesses of mountains, in the midst of forest clearings, or on elevated sites surrounded by floods. Also, most new countries, where the combination of a large-scale agriculture and animal industry is facilitated by the use of power-driven machinery, have adopted the dispersed type of rural habitation and settlement. Nowhere are these general principles more evident than in the Ganges Plain of India, where ancient occupation and a heavy burden of population have established an adjustment of ecological factors, agricultural practices, and rural settlement forms unparalleled for its closeness in any other part of the world.

Wherever a group of mud huts is built, the excavation of earth in the neighborhood during their construction gives rise to a pond or tank in which rain water is made to accumulate for facilities of irrigation and drinking water supply for the rural community. Where there is neither any stream nor pond nor swamp in the open plain, the village pond used for irrigation becomes the binder, and accordingly we find many compact villages in North India situated near or around big tanks or possessing their own ponds, the cultivators having their rights of irrigation from such tanks and ponds and of pasturage in the adjoining meadows. The larger the groups of dwellings concentrated in the village site, the greater becomes the need of centralizing and coordinating agricultural tasks adopted for the advantage of all. Fields are open and distributed in diverse parcels throughout the entire area of settlement. All paths and tracks across the fields, which for each cultivator lie scattered like autumn leaves, lead to the village that is the meeting-ground of all. Wells, temples, panchayat-houses, threshing-floors, and social facilities which are subject to community use are all concentrated here. The scattered field system, which in the history of agriculture equalizes opportunities for all in an agricultural community, cannot work without the centripetal force of the village type of settlement.

*Reprinted from *Asia*, 40 (July–August 1940), 287–90, 375–78, and 439–42, by permission of the author.

In India, the characteristic regions of dense compact villages are the United Provinces, Bihar, West Bengal, Orissa, and the river valleys in Madras. The fringes of deserts, where agriculture exists on sufferance, the mountain sides where drainage and terracing problems long defy human engineering, the moving waters of the active rivers whose storms and waves challenge the stability of human settlement, the stagnant waters and bogs where malaria is a devastating scourge—these seem the natural habitat of scattered settlement. The regions of hamlets and isolated farmsteads are the submontane areas along the entire stretch of the Himalaya and the Vindhya Hills, and East Bengal and Malabar, in the extreme southwest. The areas of dry crop cultivation in central and western India exhibit a disseminated pattern of rural distribution, with a marked tendency towards agglomeration wherever conditions are more favorable.

As early as in Buddha's time we find regular villages established throughout North India. The houses were all together in a group separated only by narrow lanes. The village proper was enclosed by a wall or stockade with gates. Immediately adjoining were the grazing ground for the village cattle and the sacred grove of trees from the primeval forest, left standing when the forest clearing had been made. The fields were all cultivated at the same time, the irrigation channels being laid by the community, and the supply of water regulated by rule, under the supervision of the headman. No individual or corporate needed to fence his portion of fields. There was a common fence. The arable land was made up of individual holdings separated from one another by channels dug for cooperative irrigation. These dividing ditches, rectangular and curvilinear, which the Buddha saw among the cultivated fields of Magadha, suggested to his mind the pattern of his monk's robe, a patchwork of torn pieces of cast-off clothing. Of their own accord the villages combined to build the assembly-halls and resthouses and reservoirs, to mend the roads between their own and adjacent villages, and even to lay out parks.

Throughout North India, the permanency of village sites is proverbial. The communal organization of the village largely contributes to maintain its original boundaries and vital economic unity. Almost everywhere the village in India is divided into wards, *parhas, pattis, cheris,* or *desams,* each with its central site, the residence of the headman, the sacred grove, and the shrine of the Village God or Mother Goddess. These hamlets, or wards, inhabited by different castes, are divided from one another by streets and lanes which usually run from east to west or north to south. The houses cluster as far as possible near the waterside, whether river, tank, or *jhil* (marsh), with every facility at hand for bathing and drinking and for washing clothes. Where caste segregation is rigid, as on the Malabar coast, the houses of the well-to-do castes, like the Brahmans and Nayars, stand apart, each in its own compound, loosely grouped around the village temple and ceremonial bathing-tank. The huts of the low-caste artisans, potters, carpenters, blacksmiths, oil-pressers, irrigation men, barbers, and so on are grouped together in one or more separate wards, or hamlets, usually with their own tanks or wells. But the vagrant and forest tribes, hunters, fishers, and food-gatherers, who are on their way to assimilation with the Hindus through adoption of such crafts as basket-making and leather-work, and who are usually employed as hired laborers for whom there is a very pronounced seasonal demand in the field, as servants, as watch and ward, and even as drummers and sweepers of the village shrine, are regarded as untouchable. They live outside the settlement in hamlets of their own. In western India, the huts of the untouchable castes are close to the village gates and often outside them, and near them are the monumental stones, an image of Hanuman or a small Devi.

In the compact village, comprising several castes, the hierarchy of Indian caste gradation stereotypes economic activities and relations, and subordinates competition to social peace. Agricultural laborers, artisans, peasants, traders, and priests represent the order of social stratification, which in large measure approximates to economic status and scale of living. There is neither inelasticity in the caste structure, however, nor a rigid fixation of rural wages and services. Artisan groups which show an upward economic movement and improve their income, art, and skill, obtain social recognition by gradually crystallizing themselves into new castes. Wages given to artisans and village servants are measured in shares of grain, which also change with the nature of the harvest, and which guarantee a minimum living standard of the family. The cooperation between the castes is backed up by cooperation within the joint family of the peasant.

It is this cooperation which is the essence of village life in India. The village is the spontaneous form and natural symbol of stable, integrated social relationships. And this not only for the present. For the village holds in its bosom the menhirs, dolmens, cists, and circles of stones, the cemetery and the ancestral temple, which pledge continuity of ideas and feelings, goods and relations, from generation to generation. The peasant is firmly rooted to the soil; he is a part of the permanent fields and the monuments of his ancestors.

Throughout India, the peasant's hut is distinctly an implement of agriculture and varies in form and structure with the amount of rainfall and luxuriance or sparseness of the vegetation, which supplies the materials of construction. It is this which is responsible for the difference between the dusty, sun-baked huts, with mud walls and mud or tiled roofs or reed thatches, of the United Provinces and Bihar, and the better built cottages of Bengal. These are usually constructed of bamboo framework with conical roofs of straw thatch well adjusted to the heavy seasonal downpours, walls made of wattle, split bamboo or reed plastered with mud, and courtyards which obtain every morning a thin plaster of cow-dung interspersed with floral drawings. In the treeless new land of the lower delta, however, the kerosene-tin and the corrugated-iron roof are much in evidence.

The gradual reclamation of the *chars,* or new land of the delta, forms an interesting and unique chapter in agricultural colonization and enterprise. Due to the meandering of the mighty rivers, which deposit a stupendous amount of silt, the active delta is intersected by narrow ridges and islands which rear their heads even when the entire area is submerged. It is here that hamlets establish themselves, consisting of little groups of huts, with intervening stretches of rice and jute fields.

A network of water courses and canals forms the only means of communication from July to December. Canoes of all kinds, earthen tubs, plantain and bamboo rafts are all mobilized, during the rains, for both marketing and agricultural operations such as weeding and harvesting. At important confluences of the rivers or in the strips between the ridges, markets or bazars emerge— sometimes growing into considerable villages but more often not.

In the stir and bustle of wholesale exodus when the mighty river swallows up a vast area dotted with habitations and the peasants have to desert their homes and fields in order to carve out new holdings in new chars, ancient social bonds no longer endure. Migration to new reclamations is at the beginning intermittent and tentative. The Muslim population is dominant in this area. As a Muslim youth leaves his family home for an uncertain career in a new char, he marries a new wife, even though he may have children at home. The Islamic customs, widow remarriage and the taking of more than one wife, contribute towards the success of agricultural colonization in virgin islands and swamps, affording an interesting instance of the correspondence between marital custom and economic advantage. Another striking social deviation arising out of the changing social scene is furnished by the prevalence of concubinage and the institution of marriage with boat-girls, a much less formal alliance than is prevalent in the plains villages. In this case, it is the Hindu community which has tolerated the novel rite, born of the hurry which the mighty rivers and sequence of floods import into the landscape.

Throughout the marsh and delta region the peasant lives a life of less economic uniformity than in the upper area. Land transfers are frequent. Habitations leap suddenly into prosperity and decay as suddenly. Individual fortunes are made and dissipated quickly. Even peasants are infected with the gambling and speculative spirit, though this coexists with a courage, enterprise, and initiative unknown among the farmers of the North. In some of the deltaic districts caste has lost its rigidity, and well-to-do members of lower castes can obtain admission into a caste of higher rank; not all at once, but by change of surname, association with members of the better caste, and judicious expenditure. Where society moves at a rapid pace, individual misfits are also more common, and crime, vice, and other anti-social behavior more frequent.

In those communities where village life is decaying, crime is equally evident. In North India, as in the West, the spread of commercial farming and capitalistic landlordism in modern times, with the inclusion by the landlords of communal meadows, pastures, and irrigation tanks in their private estates, has marked the beginning of a process which, in many parts, has led to the virtual extinction of village communities. Rural exodus is proceeding in India, due to the decline of peasant farming, the reduction of small proprietors into tenants and of tenants into agricultural laborers— all portending an agrarian crisis, which is aggravated by increase of population and fractionalization of holdings. But, at the same time, the relative absence of factories has prevented wholesale desertion and the substitution of dispersed for clustered dwellings. Moreover, the seasonal idleness of the peasants has contributed to the development of a large variety of cottage and village industries, of which the wares—whether silks or metal utensils, artistic pottery or embroidery—sometimes acquire wide celebrity.

Clustered villages thronged with population have also brought into existence an array of itinerant small traders, who periodically collect the surplus village produce from the fields or from the homes of cultivators, as well as the products of cottage industries, and supply cotton piece goods, kerosene, salt, and other commodities, including yarn for the hand-loom weavers and metal for the brass-workers. Not less important in this connection is the part played by the weekly markets, where a large number of intermediaries flock together. The chain of peripatetic primary collectors, middlemen, and retail dealers is a dangerous necessity for the village, threatening its collectivism and self-sufficiency. It is they, as well as the local grocers and money-lenders, often the agents of wholesale traders and financiers of the cities, who maintain and expand economic contacts between clustered villages, scattered hamlets, and isolated homesteads and the industrial and commercial world.

On the whole, however, the peasantry in India is too much accustomed to the ancient collectivism to permit economic forces to get the better of the habits and traditions of the old order. True, agricultural life shows high mobility in some of its phases, as transplantation of vast stretches of rice fields by meticulous hand labor, and hectic buying and selling in a big agricultural fair. But such mobility is a temporary though recurrent episode in rural economy. There is no psychological mobility among farmers, who often remain mentally stagnant in the midst of vast social and technical changes.

The Indian peasant's main object in life is to transmit the family land and homestead, faith and tradition, intact to the next generation. The values of his life are centered round the expansion of his heritage of the soil. Thus frugality, endurance, love of routine and stubbornness, conservatism, lack of initiative, and strict adherence to the established social structure are the prized virtues in the countryside. In the world of agriculture it is not the individual that counts, but team work; rural persons prefer collective mores to individual initiative, and the wisdom of the aged to versatility. Social groupings may be stereotyped, but the farm pays little heed to the distinctions of rank and position. The hired farm hands, men and women and even children, may help and direct a young owner's field operations without any risk of being reproved. Collective living and working mitigate the distinctions of proprietor and tenant, of employer and hired labor, and service, however insignificant, is raised to the dignity of a profession.

In the rural-agricultural world, the tempo of life is adjusted to the balance and rhythm of Nature's processes, the sequence of agricultural seasons with their distribution of sunshine and rainfall, and the cycle of soil recuperation and vegetative growth. Man himself is as peaceful here as the tempo of his work, which is guided by the slow ripening of crops in his field, the slow movement of his bullock cart. Nor does he strive materially and spiritually for comforts and satisfactions much beyond what his activity on the farm and spare time in the home supply. The weeks of suspense and anxiety during which he watches for rainfall at certain critical periods in agriculture are devoted to religious contemplation, fasts, and offerings to gods or ancestors. Even the tools and implements of agriculture and handicrafts receive offerings, along with the cattle and buffalo. The Indian villager, however, finds ample opportunities not only for dream and reverie, myth-making and communion with self and nature, but also for vital kinship and neighborhood activity and collective recreation and diversion. The peasant women sing and dance to tide over the crisis of anxious expectancy, and no villager would regard such time divorced from toil as waste. The villager must have his idle time and rest every evening and night, in the middays in hot summer, and in the fast and festival days of the year; for without these, he knows, he cannot show his marvelous and powerful achievements. He cherishes the idle weeks when there is little work to be done in the fields, even when culture does not utilize them. Economic transactions are placed in the background of the cosmic forces, the setting of the larger artistic and religious life of the community. The villager believes in a long scheme of things, and it is he who has given to the world visions of eternity, cosmic justice, and human brotherhood. The natural and the social slip easily into the supernatural. All toil, all art, all morality spring from his ardent personal religion, which, indeed, is the Indian peasant's greatest strength. His moderate speed of living, his equanimity in the midst of calamities, his capacity for living and working with others, and his deep concentration of purpose are all grounded in his belief in a personal God who takes charge of his rainfall and frost, insects and parasites.

Every big village in India is distinguished from a

distance by the tapering spire of the village temple, which rises elegantly against the silhouettes of the conical thatched roofs of the peasant's huts. Its architectural form symbolizes one of India's profound spiritual convictions. In the Indian mind, nothing is truer in the search for the deity than the discipline of gradual affirmation and negation of all forms and expression, until the mind can reach the Beyond which is like empty space. The temple architecture incarnates this dialectic of the human spirit by its arrangement of conical towers that gradually diminish and taper off into the spire, and then into the trident or disc. Many village temples shelter totem and animal guardian spirits, earth goddesses, and disease divinities. In others are installed images of deities which symbolize profound truths of human destiny in cosmic life. The progress from the worship of fetishes and totems to metaphysical abstractions is, indeed, a marvelous record for the rural mind. It is largely the outcome of the strong rural predilection of the intellectual classes in India. The ancient tradition was that the elite should seek the grove and the forest in old age for meditative and intellectual life. "It is impossible for one to attain salvation who lives in a town covered with dust," declares one of the ancient Indian *rishis*. Village life underlies all the social and economic changes which civilization has passed through and it outlives them all. Its settled existence, its security, its certainty of employment, its face-to-fact contacts, its intimate relationships, its essential self-government, its solidarity, and its spontaneous social restraints still remain permanent cultural values, since it is these which have helped mankind to wrest its living from the unyielding earth.

Traditional civic patterns

Little towns thrive in the East, not merely because their role is much simpler than in the complex industrial civilization of the West, but also because over large areas communications are difficult or have not yet developed beyond the dusty road and bullock-cart stage. But neither the little towns nor the great cities grow by themselves. It is the villages in the countryside that set them to undertaking the tasks of coordinating markets and fabricating certain goods which are beyond village resources. Moreover, as the small village or hamlet grows into the larger village, the larger village into the town, or the town into the city, the village still supplies not only crops of each type of soil and men of each type of stock but also elemental concepts and faiths, the physical and spiritual raw materials which go to replenish the culture of cities and will doubtless continue to supply them for ages to come.

Throughout rural India the distance which the cultivator will travel to buy his goods directly is limited by the time he can take from his agricultural work. The primary marketing area is eight to ten miles in diameter in the Ganges Plain, and somewhat larger in a region like the Bengal delta, with its netlike distribution of streams and canals, permitting boat traffic, which is quicker than the bullock cart. Distance, however, is not a deterrent for the litigant set of the rural public, who frequently resort to the collector's court in the district town, or the union *panchayat* at a far-off village. Nor does the villager mind either distance or time in the slack agricultural season. At this period, the temple of a certain god or goddess famed for the prevention or cure of some special disease acts like a magnet, drawing people from long distances.

In the larger village or small town are found the smithy, where plowshares are repaired, the carpenter's shop, which supplies the doors and windows of village huts, the silversmith, the goldsmith, and the utensil-dealer. Such a village is the secondary market supplying the retail dealers in the primary markets with most of their goods at wholesale prices. Here, again, biweekly, weekly, or fortnightly markets are held, and also, at certain seasons of the year, fairs, which are usually associated with some religious celebration in the temple. The itinerant grain-dealers, who collect grain from the homes of the cultivators in distant villages, obtain the bulk of the produce of the region at these periodical markets and fairs, which attract swarms of villagers. But the little town is not only the collecting center for agricultural produce from the surrounding villages; it is also a distributing center for manufactured goods imported from the still more distant town or city. It has its group of wholesale dealers and money-lenders, who largely finance agriculture and trade, its absentee landlords, its shop district for the display of manufactured goods.

In such a town, the temple, the court, the money-lender's business platform, the retail shop, and the rich man's mansion may jostle one another in the same district, usually at the center where communications intersect and focus. Residential buildings are as a rule, however, distributed according to religion, kinship, occupation, or caste groups; it is cultural rather than economic factors which govern the spatial distribution of the various communities and services. In recent decades the new palaeotechnic industrialism has been responsible for the development of a rapidly growing number of towns and cities whose overcrowding squalor and insanitation, with accompanying social and moral degradation, far eclipse the misbuilding and malformation in the West. Yet the traditions of orderly communal living in village and city are still strong and persistent.

The transformation of the village into the small town takes place only as the village adds to its farming routine other interests and occupations. A village on the cross roads, on the river at a point where the agricultural produce has to be taken over on a ferry, on the edge of a plateau or desert where the products of the mountain-dwellers and nomads are exchanged for the merchandise of the plains, develops into a small country town. Again, where a few cottage arts and crafts establish their reputation, the village in the open plain may furnish the background of an industrial workshop. In India, however, there are still many villages of specialized arts and crafts which have not as yet grown into towns. The small town itself is hardly industrial in the modern sense. It is a center of local distribution rather than of manufacturing enterprise, though it gathers up workshops, where the artisans of the region ply their specialized crafts. It has its temples, schools, and bathing-ghats for the organization of leisure, its dispensaries, resthouses, and almshouses for the relief of suffering and distress, even its high school. With well-known temples and educational institutions, it becomes a place of pilgrimage and retirement. Each small town develops a personality of its own; no two are alike as regards their historic traditions and cumulative allurements. In the East, thousands of towns fall into this category. They will never expand into cities, because their economic base cannot be broadened.

The ancient classification of Indian towns and cities clearly indicates their diverse origins. There is the capital city *(Rajadhani)* surrounded by ramparts and moats and containing wards, or other divisions; there is the commercial emporium on the sea, frequented by traders from foreign countries *(Pattana)*; there is the fortress town, with its unfailing provision of food and water for the army *(Durga)*; there is the little town surrounded by villages and inhabited by cultivators, with facilities of communication by land or water *(Kheta)*; there is the market town in the countryside, from which the villages obtain their daily supply of foodstuffs and other articles of consumption, and which is slightly bigger than the little town *(Kharvata)*; there is the bigger market town on a river or on the sea, where local products are offered for sale, frequented by traders *(Dronamukha, Viramba)*; there is the frontier town situated in a forest or hilly region *(Kotmakolaka)*; there is the trading-outpost, a meeting-place for artisans, traders, and caravans *(Nigama)*; and there is also the university town, with its monasteries and residential quarters for students and teachers from different lands, well-defended by soldiers *(Matha,* or *Vihara)*.

Villages and towns were often planned according to a religious symbolism, which governed the layout of wards and streets, the location of temples, monasteries, and village halls, as well as of open spaces, tanks, and gates. In the *Silpa Sastra,* the ancient manual of Indian arts and crafts, various plans are distinguished, which follow mystic figures, symbols, and images. There is, for instance, the village or city laid out in the form of the body of Vishnu, or of the sacred bird *Garuda,* or of the mystic figure *Nandayavarta* (a flower), the latter with broad streets running from east to west and from north to south. Where these intersect in the middle, there is erected a temple of Brahma or Vishnu, or the village assembly-house. This plan is intended for a mixed population of different cultural grades, the Brahmans occupying the central four streets of the village, the Vaisyas the next, and the Sudras the last. There is also the village or town built in the form of the eight-petaled lotus *(Padmaka)*; here also the temple or the council-house is in the center, and the gates are placed at the four cardinal directions. Then there is the habitation with its streets laid out like the branches of the mystic *Swastika,* or the cosmic cross. There is again, the semi-circular village or town, in the shape of a bow *(Karmuka)*, generally suited for the river bank or coast, with its radial streets converging towards the center, where a temple of Vishnu or Siva is located. River ports and harbors are often planned in this fashion. Bow-shaped are the holy cities of Benares, Ayodhya, and Madura. There is also the habitation shaped like a conch *(Pastara)*, or like the auspicious pitcher *(Kumbhaka)*, with its streets following the contours. Finally, the hermitage or *asrama,* where the Brahmans traditionally lived and taught, was planned in the form of the ascetic's staff (hence called *Dandaka)*, with five long parallel streets running east and west, and three shorter intersecting ones, in the middle and at the two ends.

In the Indian conception, the site on which the city is built is as sacred as the altar in the Indo-Aryan family home where is lit the sacred Vedic fire. In ancient village planning, in the center of the village at the intersection of the cross-way there is located a pipal, mango, or margosa tree, or a pavilion of wood or stone, which is the seat of the village assembly. Here also is the village shrine, with associated bathing-tank. Where temples are not built, the gods are placed in the open air under the shadow of trees associated with them, and sometimes there is no god, but the sacred tree receives all acts of devotion intended for the deity. Each city has also its tutelary god or goddess installed where once was lighted the never-failing divine fire. Even the modern city of Calcutta has its Nakuleswar, guardian of the pilgrims that flock to the temple of Kali, as Bombay has also her Siva and Mumbai Devi. Benares, the eternal city of India, loved by the old and the devotee for ages, is protected by Kala–Bhairava, the sentinel of eternity. Throughout India, from end to end, every organized village or city has its gods and temples, sacred bathing-ghats and groves, watching over the vicissitudes of human life, and private affairs and civic festivals mingle in a manner unknown in the West.

In the *Silpa Sastra* texts, we read that the temple of Vishnu should be erected at the center of the town and that Brahma should be placed at the crossing of the two principal highways with his four faces turning towards the four cardinal directions, as if looking down the streets and watching all quarters of the city. There are also deities placed at the gates, as divine sentinels. In the modern village or city, their role is left to Siva, Jagaddhatri, Ganesh, or Mahavira. Aboriginal and Aryan deities now vie with one another in protecting a town from disease, pestilence, and anarchy.

Nearly every Indian town is divided into a number of natural areas, called *muhallas, paras,* or *pattis,* on the basis of clan or kindred, occupation or caste, community or race, rather than of economic status. In the villages and cities of South India, which is the least influenced by foreign social influences, each ward is still generally inhabited by a different caste. The houses of the Brahmans stand in one block, arranged as a rule in double rows facing one another across the principal street, which often encircles the central temple and tank. Like the caste panchayat in each ward of the village, there is a panchayat or some kind of local shadowy organization for each caste or house group in the separate wards of the city, which appoints its own watchmen, has its own sanitary arrangement, and manages its own school, temple, almshouse, or orphanage. As a result of this clustering of similar occupational groups, artisans such as goldsmiths, stone-carvers, furniture-makers, and leather dealers thrive, since customers must go to the special areas, as they resort to the streets of tailors, lawyers, physicians, and motor-car dealers, jostling one another in the same street or block in the modern western city. As in the West, the *bazar* or general market is often near the center of the Indian town or city, close to its main gates, and it is here that the merchants and wholesale dealers have their business platforms, or *gadis,* for banking and financial transactions that regulate the business of the region, the hinterland of the town.

Indian town-planning solves the problem of maintaining social amity among heterogeneous races and peoples and safeguarding their communal autonomy through the separation of house-groups, or wards, for each caste or community, each with its communal centers such as temple, square, and council-tree. At the same time, for the entire town or village, the central shrine, tank, and quadrangle inculcate humanizing sentiments which bring together different ethnic elements for the common civic tasks and rejoicings, radiating unifying influences that have fought the forces of caste separation. A festival in the central shrine and a procession of the town deity, which periodically passes through the different wards of the town, still safeguard the civic life and consciousness in South India, engendering pride and enthusiasm. The temples, guild-halls, and resthouses, noble monuments of the Middle Ages, still keep alive cooperative living on the basis of essential cultural values which competitive industrialism has disregarded. Even the casteless have some role to play in the temple organization of labor, as drummers, sweepers, bat-beaters, and the like, while they may bear some days' expenses of civic festivals when the gods are taken out of the temple in cars in grand procession. The car and the vehicles are periodically overhauled, giving an opportunity to the village or city artisans to show their skill and delicacy of workmanship and to all the villages and cities in the neighborhood to vie with one another in display. To the car procession we also owe not merely the fine layout of the main quadrangle of the streets of a temple city, but also a high standard for the other streets of the city as well, even as the floating boat and water festival assures respect to and periodical purifying of the central tanks of the city.

E. B. Havell, the well-known critic of Indian art, has observed, "The most advanced science of Europe has not yet improved upon the planning of the garden cities of India, based upon the Indian village plans as a unit." In a temple village or city, almost every house has a garden. Usually every street is lined with shady trees, and its width is guaranteed by the periodical car procession of the tutelary deity of the city. The main streets run east and west, or north and south, in order to allow the free entry of the sun from morning till evening and to insure a perfect circulation of air. The diagonal plan of streets is definitely prohibited, because in that case the street runs in the wrong direction for the sun's sacred course through the sky and because it tends to the congestion of traffic and admits of an uncomfortable plan of house and garden. This anticipates modern town-planning developments, where the dwellings are constructed on streets at right angles to the line of traffic, spacious and verdure-lined. At the intersection of the principal streets are the central park, tank, and temple. The tanks are sacred, the trees are sacred, and the temple, which perhaps give its name and distinction to the village or city, is sacred. Upon it were lavished the devotion of the Brahmans, the munificence of the merchants, and the skill of architects, sculptors, and painters, supported by rich and powerful guilds and by the entire brotherhood of artisans. It is from the temple that radiates the impulse which uplifts every house.

Nor is the temple merely a place of the gods. In its assembly-hall, meetings of the royal court and the town or village assembly, were, or are, still held. Stories from the great epics are still recited, temple theatricals, and religious and philosophical discussions still attract the townsmen. In the dance-hall, the crowds still watch religious dancing. Many temples have their guest-houses, where pilgrims stay for a day or two, and there is also a concourse of merchants in the temple avenues and quadrangles. The outer enclosure of the temple of Sri Rangam, near Trichinopoly, in South India, is practically a central market filled with shops, where pilgrims are also lodged. The temple tank with its surrounding colonnades, and the "hall of the thousand pillars," discernible in almost all important temples, offer pilgrims facilities for their ablution, dressing, performance of morning, noon, and evening prayers, and day or night stops.

No age has been characterized by such lack of cultural integrity, such complete dissocation of the esthetic impulse from practical activity, as the present. The degradation of art has attended the spread of competitive industrialism everywhere. But that art was the vehicle for inner poise and mass vision—man's inner serenity and his outer sense of oneness with man and the universe—is fully revealed by the glorious architecture of the early temples and public buildings in India and the bas-reliefs, statues, and paintings which decorate them. These Indian temple frescoes and images of divinities and of the deified heroes and heroines of the epics and the Puranas, skillfully woven into pillars and architectural designs, most fittingly and enduringly express the things communally known and communally loved. For in India there is never one kind of art for the few and another for the many.

The cultural tradition of Indo–Aryan town-planning is today best discernible in the sacred cities of the Jains, such as Palitana and Girnar, in Gujarat, or Parasnath, in Bihar, and in the villages and temple cities of South India, which have escaped Muslim domination, on the one hand, and the tragic incursions of modern competitive industrialism, on the other. Many new towns and captial cities, however, were established under the Muslim rule. Several of them followed the traditions of Indo–Aryan town-planning without, of course, the detailed religious guidance in the allocation of wards and the location of streets, gates, and public buildings. Experts have suggested that the design of the Taj Mahal is a replica of the traditional Indo–Aryan garden plan, which the Buddhists taught the Moguls. Mogul capital cities like Agra, Fatehpur–Sikri, Bijapur, and Lucknow all showed a magnificent layout. Roads, tanks, public baths, wells, and canals were constructed to provide amenities for the different quarters of the city, while beautiful gardens associated with mosques, temples, and colleges extorted the admiration of foreign travelers.

As late as 1728, a Bengali Brahman, Vidyadhar Bhattacharya, followed the *Silpa Sastra* in planning the town of Jaipur, in Rajputana, when the capital was transferred from the old hill fort palace at Amber. The town plan, which is an example of "conch" type, is characterized by the great circumambulatory road just inside the wall, and by the royal road running across the town from east to west, with the royal palace area facing the center. The town is divided into seven nearly square and equal wards by main cross streets. These wards are again subdivided into either four, nine, sixteen, or sometimes sixty-four plots, by other and narrower streets and lanes, according to the particular caste for which accommodation is required. Height, fenestration, and even color treatment were all regulated, to safeguard the beauty of the city.

But the traditional unity of Indian civic life, centered round the temple and its ceremonies and processions, where man's imagination soared to its highest levels, is now broken. The city has become the battle-ground for conflicting cultures, the scene of devasting encroachment of shop and slum on temple and square.

From the village to the city

. . . . .

In India, where the rural pattern of life has remained intact for nearly ninety percent of the population, the country town and the small city exhibit, on the whole, the ecological and social organization of an overgrown rural community. With increase of population, however, the small country town shows its areas of deterioration. There is, first, congestion in the retail shopping area, which expands on all sides. This encroaches upon the central square of the town with its temple, tank, and garden, its cluster of trees offering shade for the cattle, a playground for the children, and a place of community rest and recreation. With the attenuation and denudation of these, the community life of the city suffers, and social distance widens between the various population groups.

In Calcutta and Bombay economic segregation based on the division between the slum area and the rich man's ward, and districts graded according to the economic status or position of the inhabitants in the industrial civic complex, have asserted themselves. Between the slum area fringing Burra Bazar, or the market district in Calcutta, which has specialized in trade in clothing, food, and other neessities, and Chowringhi, the European residential and shopping district, between Malabar Hill, the residential and shopping ward of the Indian rich of Bombay, and Kamatipura, the area of single-room tenements for industrial workers, the social distance is as great as between the "East Side" and the "West End" of the western metropolitan communities. As the city in Asia approximates more and more to the methods of industry, trade, and commerce of Europe and America, an economic segregation superimposes itself upon the ancient division into homogeneous areas based on factors of kinship and caste, ethnic and cultural traditions.

In Madura, some of the most sordid slums of India have risen in temple lands not far from the great temples of Minakshi and Sundareswar, which hide in their bosom a few of the most magnificent creations of human imagination. There is an all-round deterioration of life among the people working in the cotton mills of the city which is in strange contrast with the mode of living and rejoicing of the unending procession of pilgrims who spend their day in the temple guest-houses. The introduction of machine industry not merely revolutionizes the personal and social habits, hygiene, recreation, and diet of city-dwellers, but also brings about serious over-crowding in tenements, barracks, and *bustis,* involving disease and family breakdown, physical and moral instability. The large factory replaces handicraft shops, everywhere disestablishing a mass of workers and converting them into a mobile and floating crowd that eats, works, moves and drinks together.

Calcutta, like Greater New York or Greater London, has become a "super-city," drawing within its orbit a string of settlements on both banks of the Hooghly, from Chinsura to Budge-Budge, Howrah being the largest with a population of 224,873, as compared with Calcutta's 1,196,734. With its wide constituency of subordinate centers, Calcutta not only has become the headquarters of most important mercantile firms, exchange banks, and financial houses in India, but also is linked up with London in an interlaced system of world-wide economic organization. Calcutta is also the hub of western fashion, the focus of western education and new standards of living, which are bringing within its orbit remote villages and towns. Its industry and trade give employment to many thousands of laborers. Its banks and financial agencies, British and indigenous, spread out the net of financial suzerainty covering distant upper reaches of the Ganges. Its schools, colleges, museums, theaters, and cinema-houses act as magnets for those who seek the new learning and ways of living. Thus the Calcutta region is altogether a new pattern of circulation of men, commodities, and services, which has revolutionized traditional rural-urban relations in India. Calcutta—as also the foreign-born city of Bombay, on the other side of India, and fast-growing Karachi further north—has become an enormous octopus, sucking out the life-blood from the older towns and villages of the immediate hinterland.

Under favorable conditions a city and its hinterland are mutually dependent. The ruin of the one involves the ruin of the other. But this is not always the case. Sometimes the modern industrial or commercial town thrives as a "tentacular city" on the basis of a one-sided exploitation of the hinterlands. In certain highly and quickly industrialized regions along a river or railway track, we come across those hybrid industrial towns, which may be suitably designated as "liminal towns."

The transition from the village to an industrial town is nowhere quicker, sharper, and more socially disastrous than in the East. On a narrow ecological foundation the liminal town rears itself, shoots up in size, and becomes feverishly active, attracting migrant workers from distant provinces. In Bengal, for instance, the jute mill town has come into existence as a result of the province's monopoly of jute, and has thriven mightily on this slender economic base. Such a town is superimposed from without on the region—in this case by foreign capital and enterprise—and lives apart from the traditions and culture of the rural community. It neither derives its nourishment from its own culture, nor can it be assimilated into any new cultural pattern. Like the oil or mining town, hectic and expansive so long as coal, oil, or any other mineral resources are not exhausted in the seams, the jute mill town does not stop growing as long as jute remains a prized monopoly and the industry thrives, a cancerous growth on the body politic.

Economic instability is increased by the specialization of the city in one industry and social instability by the influx of a heterogeneous working class population. A haphazard, pell-mell growth does not permit even that customary division of the liminal town into *muhallas,* or natural areas, inhabited by separate castes and communities, as in the small, conservative country towns. Physical and social mobility here is phenomenal; it proves disastrous to the family, and disintegrates the traditional means of social control, resulting in the most profound personal and social deterioration.

Between 1911 and 1921, the mill town of Bhadreswar showed a population increase of approximately 100 percent, Tittagarh, 200 percent, and Kharagpur, 400 percent; between 1921 and 1931, Kharagpur showed a further increase of 130 percent, Bansberia, 123, Halisahar, 129, Kanchrapara, 45. The smaller increases during the past decade were the result of the acute and prolonged depression which almost overwhelmed the jute industry. Only about one-fifth of the total population of these liminal Bengal mill towns was born in the area surrounding the towns. In Tittagarh, a compact manufacturing area, no less than 90 percent of the inhabitants were born outside Bengal. In Bhatpara, 79 percent were born outside Bengal. In Champdani and Bhadreswar, 70 percent, in Kharagpur, 50 percent, and in Bally, Baranagar, Budge-Budge, Howrah, and Serampore, more than 33 percent. The proportion of outsiders in Calcutta falls just short of one-third, being 31.9 percent. Although a few migrant workers have children with them who were born since they moved in, practically none of them have settled down. The majority, estimated at 75 to 85 percent of the workers, are of the floating type, revisiting their village homes once a year or once in eight months and absenting themselves from the factories for a couple of months or so. The labor turnover at the Angus Jute Works, an important representative concern, has been estimated at about 12½ percent per month, which means that the average duration of employment is eight months.

A gradual decline has been observed in the number of females in the rural areas of Bengal, but the greatest disparity is in the industrial areas, where the ratio has been markedly declining since 1882. The ration of females to males in the 20 to 40 age group in Calcutta is only 172 to 500. Here, to the persistent demand for only male labor and the chronic housing shortage, are added two other factors, the large student population and the tendency of traders, merchants, and professional men to visit the city only for a few days, while newcomers never bring their families at once.

When laborers flock to the bustis and *bazars* honey-combed with single huts in which there is little privacy, and no opportunity to live with their families, immorality naturally becomes common. Some who have left their families behind in village homes marry again; others become addicted to vice and settle down in the busti. Most of the industrial workers of India are recruited from the lower castes. In the industrial towns in Bengal nearly half of the women workers belong to the depressed castes, and they do not maintain high standards of morality. In one of the mill towns of Hooghly district, out of the 300 migrant woman workers, one in three admits to being a prostitute; among the people born in Hooghly, of whom one-third of the families work in the mills, one woman in every four professes to be a prostitute. Such is the bitter loneliness and coarse relaxation of an alien urban life. The entire trend of social and economic forces in India is thus bringing about both a social and an individual crisis in our new cities and industrial towns, where the simultaneous disintegration of family, caste, and community habits has made it more and more difficult for many individuals to construct a new life-organization out of the remnants of the old. These become the unfortunate victims of social disorganization: the thieves, the paupers, the vagabonds, the prostitutes, who have lost their vital contact with society.

Perhaps the present transition from village to city life for large numbers of people is too brief a period to justify the condemnation of all city norms and attitudes, which are not as yet settled and have not passed through a sieve of selection of vital import. A correct understanding of the settled social attitudes, habits, or institutions which the city-dweller still cherishes, and their conscious nurture and adaptation to the new milieu and to its wider range of needs and interests, seems to hold out the best means of social integration in the new environment. All attempts at "urbanization" are bound to fail if we do not take due cognizance of the psychological roots of the more naturally adapted, the more humanized, and the more firmly settled rural patterns of living that one or two generations of industrial development and urban growth have not been able to obliterate completely.

Amidst the social deviations of the new urban-industrial environment, man often seeks to restore the elementary occupations as well as human relations and social attitudes of the rural-agricultural regime. In the slums and tenements of every modern city, whether in the East or in the West, angling, hunting, and the cultivation of plants and flowers in pots or cans, or the care of pets such as birds, cats, dogs, rabbits, and fish, or again, organized hiking and camping trips on holidays, represent a return to aboriginal patterns, to which is added some esthetic and romantic fervor.

In a few industrial towns in India we find the migrant workers living with their caste-men in separate muhallas and hamlets, importing into the slums the caste *panchayat,* which acts as a stable, socially defining organization. It is the panchayat which warns a girl of loose morals, brings to book a man of evil reputation, settles a dispute with a usurious money-lender, and organizes religious story-telling, song, and music in many a moonlit night of well-earned rest. The caste-men also bring their gods into slumdom: Ramji, Mahavira, Ganesh, and the rest. Thus the validity of old norms is reëstablished in an environment of social unsettlement and deviation. The rehabilitation of caste brotherhood and panchayat government, binding together workers hailing from different regions, is a deeper and more vital social adjustment than the kind of "cultural homesickness" illustrated in the stray, vestigial remains of rural attitudes referred to above. Similarly, in the markets and slum areas of an Asiatic city, the recital of religious prayers, ballads, and epics on evenings and holidays in ramshackle, improvised temples and mosques revives deep, rural-bound natural feelings, which once created folk art and folk religion during long hours of leisure in the old rural communities.

It may be that some of the rural attitudes, manners, and ways of living, based as these are on primary face-to-face relationships, will be largely modified as being too archaic for the urban-industrial milieu which calls for standardized and impersonal patterns of action. Yet by piecing together the broken fragments of community life and inspiring community cooperation through organizaton into neighborhoods, castes, and functional groups, the urban area may secure physical and social stability and unity, which were once thought obtainable only in simple, self-sufficient rural society. It is at least incumbent on us in India to derive our warning lessons from the nineteenth-century colossal disorderliness and disorganization of the beehive city in Europe, and to plan and build our new towns with the coherent social experience and the confident esthetic command which in the past produced some of the best-planned villages and cities in the world.

Comparative Ecologies of Large Indian Cities*

Brian J. L. Berry
and Howard Spodek

Factorial interpretations of the ecology of Indian cities cannot be cast into the narrow conceptual mould of studies of American cities, for they must contend with a differing form of social differentiation, complex intermixture of occupation and ethnic background, substantially less family-type specialization, differing prevailing modes of urban technology, differing historial attitudes to the city, and contrasting perceptions of the amenities of sites within it. An initial attempt to come to terms with the complexities of these differences has been provided by Berry and Rees (4). This paper seeks to extend the exploration by examining a group of cities (Ahmedabad, Bombay, Kanpur, Madras, Poona, Sholapur) at a point in time (1961)—one of these cities (Poona) over a considerable time-span (1822–1954), and one (Bombay) at a range of observational scales. A primary concern is to penetrate more deeply into traditional styles of Indian culture as they affect residence patterns.

The caveats are predictable. Data are very uneven. The materials on Poona and Sholapur were produced in social surveys under the direction of D. R. Gadgil at the Gokhale Institute of Politics and Economics in Poona and are quite

*Reprinted from *Economic Geography*, 47 (June 1971), 266–85 (tables omitted), by permission of the authors and the publisher.

good (10, 11). Data for the other cities are derived from the Census of India, 1961, which has a somewhat higher degree of error. The Census operation was largely in the hands of state government bureaus responsible to the Central Government Census Office, and these state governments collected different varieties of data. Thus the data for Ahmedabad are richer than those for Madras, which are in turn somewhat richer than the materials available for Bombay and Kanpur. In addition, the census tracts within and among cities differ widely in size, from a few hundred residents to tens of thousands, and very obviously, also in internal homogeneity.[1]

Data over time for all cities except Poona are unavailable since the Indian census only began reporting data for cities by ward in 1961. Also, the number of cities covered is severely limited. The 1961 census promised sections from each state census office treating the large cities of the state. For reasons of non-compliance with this directive,

[1] In the case of Ahmedabad, we undertook tests, using analysis of variance, to assay the relative variability between and within the census tracts. Lacking variance estimates for individuals, internal variability was measured by block-to-block variations within wards, with an average of 20 blocks per ward. As a case in point, female literacy produced an F of 26.7 (3); the greatest source of variability remained that *between* the observational units.

or more simply of delay, most of these city volumes have yet to appear. Some data on a ward basis are available in the census volumes on the districts of India, but even where such data were available, as for Poona in 1961, there were no accompanying maps to indicate which ward was which. Finally, much data on occupation is gathered on a vertical industrial basis, i.e., the classification of workers by industry included everyone from owners of large factories to machine operators, to factory scavengers, to apprentices in three-man workshops. All these obstacles contribute to making this study still a preliminary one. But it was felt that it would be worthwhile to proceed, despite the difficulties, since these were all the systematic data on Indian cities likely to be available for some years to come.

Traditional Indian residence patterns

The empirical part of this study uses factorial methods, results of which have no ready interpretations without an adequate contextual base. Fortunately, there are at least two sources that give a picture of normative patterns of residence in India. The first is ancient religious and policy-making texts, the second is residence patterns in village India.

The ancient texts

The texts, dating from perhaps 300 A.D. back to 500 B.C. and covering both the northern Aryan and southern Dravidian areas of India, represent both texts on political management like the *Arthasastra* and texts on architecture and design such as the *Agni Purana,* the *Manasara,* and the *Sukranitisara.* These texts indicate the antiquity of urbanization of India and the types of towns which characterized the subcontinent in ancient times. They included princely cities, temple centers, market places and trading posts, university towns, industrial centers, and nucleated towns formed by the unions of two or more villages. But the essential features of all these types of cities included a fortress, religious institutions, and socioeconomic separation of the people into distinct neighborhoods or "natural areas" (12:9–10). In the Vedic literature the word *pura,* the modern Sanskrit synonym of town, was used in the sense of a fort or stronghold (7:70–71). The major towns were associated with capitals of local rulers. "Because India was not brought under the suzerainty of one single emperor, but on the contrary, was a medley of small principalities fighting with one another for overlordship, the military camps were turned into royal capitals. Indeed, each prince had to build a new city in order to demonstrate his grandeur and not to be ensnared in the many byways in the former capital which were better known to the former rulers than to himself" (7:38–39). In the Dravidian south, the Tamil word *nakar* carried at least five different meanings: house, temple, palace, castle, city (2:18–19).

The physical structure of the ancient town reflected its fortress qualities in walls and moats, its royal and administrative buildings either in the center of the town or in the northern quadrant, and temples of the city's tutelary deities, again either in the city center or near the corners of the walls. Most striking was the division of people into distinctly separate quadrants and neighborhoods on the basis of caste or occupational status. For example, the *Arthasastra* describes the kind of frontier town which the ruler should build; the varna-castes are clearly divided into separate parts of town. Brahmins, the priests and teachers, are grouped in the northern section, around the royal palace; Kshatriyas, the military, are in the east along with some traders; the trading and landholding castes of Vaishyas are in the south; and the Sudras or working classes are in the west. Outcastes are totally outside the city walls, near the cremation grounds.

The *Agni Purana* presents a very similar picture of the division of peoples, although there are significant differences. The temples are not at the center of the town, but at the corners of the walls, serving to protect the town spiritually. The shops do not flank the central shrines, but are located near, but not at, the four corners of the town. The main four caste-varna groupings are still divided in the four directions as above, though the agricultural traders are now to be grouped with Brahmins; many professionals are grouped with the Sudras; and dancers, musicians, and prostitutes have been joined with the trading castes in the south. This last combination is not so unusual since each of these occupations involves contact with diverse peoples and occupations and the provision of services which the dominant culture often regards with mixed emotions.

Ayyar's representation of the form of Madura in South India indicates that this town had similar features in ancient times, but the bulk of the population is left unrepresented. Thus the palace and court retinue are located in the northeast; the religious and education establishment has its own quarter, though it is now to the east; and the marginal people such as musicians, artists, and dancers are grouped with prostitutes at the other side of the city to the west. The bazaar area and its craftsmen and merchants are to the south just inside the main gate to the fortified city. No mention is made of the other inhabitants of the city, nor of the outcaste areas, though presumably the latter, as in other ancient cities, are outside the gates.

Was such a rigid form of residential patterns ever employed? Probably not. The projected plans actually indicate not what actual city planning as a socioeconomic phenomenon was like in contemporary times, but more a kind of abstraction following the mechanical setup of different occupational groups, castes, and classes in the city with gods, kings, and priests as the center of the whole scheme. And the plans for different cities, or by different authors, indicate differential groupings, thereby revealing the extent to which values and perceptions can and did vary. The most pronounced characteristic, however, the separation of the court, the Brahmins, and the outcastes into distinct areas, separate from each other and from other elements of the population, seems common throughout. The *Sukranitisara,* one of the architectural treatises, thus specifically groups only the royal palace, court, council buildings, museum, officials, and clerks at the center of the city and the untouchables alongside the cremation grounds on the outskirts.

Caste: classical and modern

Clearly, a fundamental determinant of the locational prescriptions in the texts is caste. The classical textual division of people was that of *varna-ashrama dharma;* the India-wide separation of people into four groupings: Brahmin, or teacher-priest; Kshatriya, or warrior-prince; Vaishya, or wealthy landholder and trader; Sudras, or workers; plus outcastes. It is quite unclear that such a four-fold division of society ever existed in social fact; indeed, in modern functional terms, caste is very different from the system presented in the classical texts. The modern view of the Indian caste system which has been developed largely by anthropological work in India over the past twenty years, sees caste groups—*jatis*—as limited in territory and membership, so that far from there being only four all-India castes, there are instead thousands of castes, of which anywhere from two or three to forty or fifty may be represented in any one area. Membership in a caste comes with birth, but does not necessarily determine occupation. It will most likely determine the limit of eligible people for marriage, and it will significantly affect the scope of social and commensal activities. Caste is hierarchical, but not rigidly so. Thus in any one area it is likely that the top and bottom castes will be distinguishable with relatively little difficulty, but the ranking of intermediate castes may be debatable.

Caste and class are antithetical in that the latter system is said to have a rather high degree of potential socioeconomic mobility while the former has relatively little. Caste also is more a village phenomenon than an urban one since in village societies the essential base of wealth and power is in only one sphere, possession of land, while in cities there are many areas in which a man can achieve wealth and power. Polymorphous urban society shakes loose the rigidities of the caste system. One of the most significant variables which the urban ecologist studies everywhere in the world, the social basis for residential patterns, therefore takes on special interest in India for it relates closely to the power of the caste system in determining residential patterns.

Residence patterns in village India

Since India is an urbanizing society in which the early socialization of many urban dwellers still takes place in villages, the role of caste in residence patterns in these village settings is of some importance. Mayer's perceptive study of one village (15:56) is a useful starting place:

> The village does not have streets whose entry is restricted, even in the wards occupied by the Harijans (outcastes). In general there is a tendency for castes of roughly equal status to inhabit the same locality of the village. All over there is a certain amount of intermingling though it is clear which caste provides the nucleus in any ward. But in the Harijan wards there is clear separation both from other castes and between Harijan castes themselves. The sweepers are to one side of the village and the tanners to the other. The weavers form a more dispersed pattern, ranging behind the houses of the Rajputs who were, and still to some extent are, their masters.

From further south, in the State of Mysore, Epstein (8) presents maps of two villages. In both villages, the groups who are considered outsiders are clearly located on the fringe in their own groupings: untouchables, Muslims, washermen, and even public works department personnel not native to the village. The other castes and lineages are somewhat more mingled, but the leading lineages, including that of the headman, tend to have a central location on the most solid and respectable street. Still, among the Hindu castes there is somewhat more mingling in the residence patterns, so that while richer peasants dominate the central area, the artisan groups for the most part live mingled among the other peasant lineages.

Another village of Mysore State, Aminbhavi, is described by Spate and Learmonth (20:201): "caste and community largely govern the layout.... Each caste tends to occupy a solid block of contiguous houses in a lane named from the caste. Low castes and outcastes live on the fringes of the village and even beyond the old moat." The authors add a note about the high correlation between caste status and occupational status. "Occupations likewise are still mainly on a caste basis: the Lingayats provide the bulk of the tenant-farmers, Talwars and Harijans landless agricultural labor; carpenters, smiths, cobblers, washermen, barbers are all separate castes."

From these examples, a few general patterns are observable. The center of the village is usually preempted by the wealthiest and most prestigious caste groupings, as denoted by the lineage of the headman. Outcastes, untouchables, non-Hindus, and non-natives of the village are located on the periphery of the village, or even outside of it. Other castes often form separate neighborhoods of greater or lesser exclusivity and physical separation from one another, but castes which have only one or two representative families usually locate among more numerous castes of approximately equal ritual, social, and economic status. Both caste membership and class status are thus ingredients in residence patterns, particularly at the extreme ends of the hierarchy.

Residence patterns in a small town

An intermediate case between the large cities analyzed below and the village studies is provided by Fox's monograph on Tezibazar, a town of some 7,000 inhabitants in eastern U.P. He found that "no necessary caste or communal patterning to residential areas exists except for untouchables. Muslims and Hindu, Brahmin and Baniya castes all live intermixed or interspersed" (9:35). In the past, however, neighborhoods may have been built along a caste basis. What has caused the change? Fox argues that it is the development of business and a political change which has led to the passing of control from the hands of the hereditary Zamindar or landed proprietor to freer and more open competition for economic power in the commercial sphere. Now, especially in the newer residential quarters, economic class characteristics predominate. "Even in those relatively noncommercial and peripheral sections of the town where some caste-neighborhood congruence exists, it is a residue of the past rather than a presently significant social patterning of town society. This situation is true for all castes and communities other than untouchables" (9:38). In this town, then, the expected passing of caste-based neighborhoods to those based on economic class is taking place as commerce becomes more lively. Families of various castes come together to live in neighborhoods of a common economic status. Again, however, as in village India, the untouchables remain unintegrated at least in terms of residence into this new class structure.

Ecology of the large Indian cities

Armed with the results of the earlier Berry–Rees study (4), which showed interpenetration of classical and modern elements in the social geography of Calcutta, and a broader contextual base, our analysis of six major cities was designed to explore further the intricacies of India's urban ecology.

Five major categories were chosen as spanning the variables most likely to provide answers relating to these questions concerning the social and physical patterns of India's urban residents. Data were gathered from available sources on: density-distance from center of town; demographic variables; caste, religion, and ethnicity (i.e., ascriptive status); socioeconomic status including education, occupation, and income; and housing data. All cities had at least some data in each category, though the types of data and the amounts varied from city to city.

Since, as in the Berry–Rees study, the primary purpose of the data analysis was an exploratory one of pattern recognition, no elaborate examination of alternative factoring procedures was undertaken. Instead, all the results are based upon principal axis factor analyses with normal varimax rotation and pseudo-inverse orthonormal factor scores. Data were transformed where necessary to satisfy linearity assumptions. A cascading sequence of rotations was undertaken in each case from initial principal axes, beginning with the first pair and then by steps for the first three axes, the first four, etc., until the final rotated solution included all principal axes with eigenvalues exceeding unity. . . .

Status differentiation

The cluster of variables defining high socioeconomic status (SES) in the analyses comprised high rates of male and female literacy, a high percentage of males engaged in occupations of trade and commerce, and, conversely, low percentages of scheduled caste members, and of males in manufacturing. Of these variables, nine cities had available data (Poona in 1822 did not), and eight of them showed this factor as significant in their residential patterns.

The additional variables which correlate with high socioeconomic status in the cities where the variables are available, for the most part, are quite expectable: high proportions of white-collar workers, high proportions of non-workers and of women who do not work, high incomes per capita, and low proportions of huts.

Other parts of the output point to the strong link between caste status and economic status. Wards which had high SES were the same wards which had high proportions of high-caste people, Brahmins, and relatively low proportions of Muslims and migrants (non-Marathi in Poona). This may or may not indicate an identity between high caste status and high economic status, but it does indicate that many wards which are high on one index are high on the other. Other loadings link SES with land-use patterns. High-status areas are located in the center of town rather than in the suburbs, substantiating the ideas of Sjoberg in *The Preindustrial City* (18).

Maps of the factor scores (not reproduced here because of space limitations) show spatial distributions of SES readily explainable by the historical and economic development of the cities. Poona, for example, developed early around a fort center on the Mutha River. To the east and south of the fort was the commercial center of town, and this area became the chief location for Muslims, non-Marathis, and the commerical life of Poona. To the west of the fort, filling the area between the Mutha River and a vanished river which followed the course of today's north-south Katraj Aqueduct, lived the Brahmins of the town. Poona has had for centuries a most distinguished and large group of Brahmins, running to twenty-five percent of the city's population. During the days of the Maratha Empire, the rulers of the region instituted policies of great liberality towards Brahmins both as religious leaders and as members of government. Even after the fall of the Maratha Empire to the British, the city continued as a center of Brahmin strength and a major educational center with a number of all-India institutions. The map of Poona in 1822 shows the east-west split largely in communal terms, but by 1937 the split between the two sides of town was definitely one of socioeconomic status. Rosenthal (17) reports a scheduled caste member of Congress as saying, "There are two parts to Poona: East and West. The West is highly developed. People there are educated, rich, and have high posts. . . . The East is very undeveloped. Most of the people are laborers and there are many poor persons."

The pattern of Madras is similar, though it was a city built from a mere cluster of villages by the British into a major metropolis of some two million inhabitants. The area around Fort St. George, the center of British trade and administration from the seventeenth century onward, has become the hub around which are found the high SES wards. Another center built centuries ago is Santhome and Mylapore, the areas where the Portuguese founded their fort in the late sixteenth century, and it, to this day, remains a high SES area. Finally, a high-status zone in the "new residential" area indicates the beginning of suburbanization in Madras city (6:437). In Madras, the areas of low SES correspond largely to the factory areas in the north, which abut the railroad track, and in the harbor area. The correlation of factory area and low SES residence is obvious and indeed forms a later factor in the analysis.

Ahmedabad, Kanpur, and Bombay also show higher socioeconomic status toward the center of the city, with lower status areas farther out, usually associated with the presence of large industrial areas. In Ahmedabad, the high-status areas are all, except one, located in the old city which was surrounded by a wall until the 1920s. The low-status areas were, it seems, always outside the old city and, indeed, the wealthier citizens wanted it that way. "In a list of municipal wants put forward in the annual report of the Municipal Commission for 1887 to 1888 there was included: The removal of low caste and other such people for the reduction of overcrowdedness" (13:124). The banishment of low-caste people from the downtown was a norm for urban India well before large-scale industrialization. Yet suburbanization has also come to Ahmedabad. The statistics in the fourteenth ward, the large sprawling ward across the Sabarmati River from the old city, do not reflect the coming of increasing numbers of upper middle-class, white-collar workers, or the establishment of the Gujarat University and of the state government secretariat in this area; this is because the ward is so large that it includes diverse elements of population in addition to the upper middle-class suburbanites, averaging them out in the census reports.

Kanpur, similarly, has its high SES areas toward the center of the old city. In addition, the residential section of the cantonment where the British established their homes apart from the Indians, and which has been taken over since independence as a suburban development of the wealthy Indians, is a second, high-status community. Some outlying areas (in ward 84) are also being developed as a suburban, upper middle-class settlement. The low-status neighborhoods tend, again, to be located around the great industrial complexes which are the lifeblood of Kanpur's economy.

Bombay, too, has its areas of highest SES near the center of town, adjacent to the commercial hub of the city. Conversely, low SES areas are near the ports, the industrial complex, and in outlying areas. For Bombay, we have data on increasingly fine-grained geographic units of observation. The largest units, 15 wards, have populations from 58,000 (Ward T) to 660,000 (Ward G), and are as large as whole cities. It was with these units that the first generalization was made. The next scale of analysis, 88 *sections,* reveals additional patterns; for example, a suburbanized area of high status appears along Mahim Bay and Matunga. The most detailed study, of 437 units called *circles,* indicates that the high-status areas downtown are commercial. A few high SES areas are shown scattered even further towards the edge of the city as small suburbanized colonies, but the very lowest status areas dominate the outermost fringes of the city. The industrial and port areas, established for a long time, do not have such low status ratings as the more distant areas, revealing that long-time industrial urbanization has its socioeconomic advantages.

Familism: male migrant areas

Data relating to family structure were available for only four cities (Bombay at each scale of analysis), but on each a "familism" index appeared. However, the factor structure reveals that this element of India's urban ecology is not a measurement of degree of family specialization according to stage in life cycle, as in the United States, but one which picks out areas in which vast numbers of men who are working in the industries of the city come to live. They live without women, though they may be married to women they have left in the village, at least temporarily. A similar factor was noted for Cairo, Egypt by Abu–Lughod (1). Statistically independent of SES, some of these neighborhoods of factory workers may approach middle-class status, or else the factorial output indicates that many of the migrants live in neighborhoods which are predominantly middle class rather than lower class.

The maps of the factor scores indicate distinct land-use associations. In Bombay, the fifteen-ward study shows houseless male workers living without their families to be concentrated in the dock areas, the industrial areas, and downtown. Much as one would expect, the houseless workers gravitate to the areas where jobs for unskilled workers are available. Conversely, the most marked positive area of family life is on the fringe of the region where, one presumes, land is available for home building, and, indeed, the area may not yet be seen as fully urbanized.

The 88-ward study picks up the same areas again in finer detail, revealing additional areas of family dwelling without houseless street-sleepers. Unlike the downtown areas, these areas do not attract transient laborers for, despite their affluence, they do not have many jobs to offer. The fine-grained 437-circle analysis comes closer to pinpointing the areas of job availability for transient male labor, separating them from areas of stable residence without commercial land use. Thus, harbor and textile mill areas are strongly marked as low on the familism index; suburban family areas rank high.

Kanpur exhibits a similar separation of industrial areas from residential areas on this index. Interestingly, Kanpur is a city surrounded to the northwest and the southwest by industry. The northwestern quadrant, however, was pioneered more by the British and adjoined the cantonment; no exaggerated familism index, high or low, is in evidence. This may result from British policies in "their" areas of the city, or from an overlap within the same census ward of residential and factory areas so that one balances out the other. The area to the south of the city includes a number of new industries, as well as older ones, and the main railway depot for industry. It has a decidedly low familism index (a factor score of -5.92). The areas especially high on familism seem to be agricultural areas on the fringe of Kanpur, areas quite removed from industrialization.

The familism index in Ahmedabad appears in two sets of factor loadings. As Factor VI it picks out the railway depot ("Railwaypura" ward; factor score -4.53) and the job-rich, highly transient area around it. The workers here are not in industry, but rather in the commercial and transport occupations connected with the depot. As

Factor VII, it distinctly picks out areas of males in household industry and in construction, particularly the new area developing across the Sabarmati River. Similarly, it picks out zones of household industry for men, where hand weaving and some handicrafts are still practiced in the southern areas of the city both within and without the walls. The areas highest on the familism index are: Ranip, a rural area recently annexed to Ahmedabad and so far distant that is not yet mapped; the military cantonment, which houses an elite; and ward 22, Gomptipur, which was for some time a separate town, suburban to Ahmedabad.

Finally, the Madras familism map picks out most clearly the harbor area, the natural destination of men travelling alone and seeking work.

Communal factors

The most significant aspect of the communal element of India's urban ecology may well be the fact that the Government of India refuses to collect data on caste and communal factors in the official census. The only data available on communal affiliation, therefore, are in the surveys of Poona and Sholapur conducted independently by the staff of the Gokhale Institute of Politics and Economics in Poona.

In the study of Poona, a clear pattern of communal residence appears. Muslims are not prevalent in neighborhoods where Brahmins predominate, and vice versa. Similarly, Brahmins and untouchables or scheduled caste members do not in general share the same localities. Indeed, in Poona, the Brahmins are composed of several groups, most prominently of Konkani Brahmins originally from the Konkan coast of western India, in the Ratnagiri District of modern Maharashtra State, and of Deshasta Brahmins who come from the *Desh* or the "country," the internal plateau regions of Maharashtra (14:5). In 1937 the Konkani Brahmins were heavily over-represented in Sadashiv ward, while the Deshastas were more prominent in Budhwar, Shukrawar, and Shanwar. Konkanis are also more highly represented in the highest income brackets and in the highest levels of occupational types. Within an already elite group, the Konkanis (or Chitpavans, as they are often called) are the most elite.[2] A similar separation between the two principal Brahmin groups appears for 1954 (19).

The Sholapur survey shows that wards with high proportions of Brahmins are also high in Marathas, the dominant indigenous caste of the region in terms of numbers. While the Maratha caste is not high in socioeconomic status, working essentially in farming in the rural areas and in craft and industrial jobs in the cities, it is quite respectable in ritual status. The Marathas are the traditional rulers who employed the Brahmins as the administrators of the empire. Conversely, Lingayats and non-Marathis tend to be disproportionately high in the same wards, again no surprise on ritual lines since neither has a niche in the local Hindu caste structure, non-Marathis since they are immigrants and Lingayats because they follow a deviant form of Hinduism. Occupationally, however, they bridge a wide spectrum of activities from large-scale trade to factory workers and private artisans.

The Sholapur study is unique in that it shows communal status rather than socioeconomic status as the most significant variable in residence patterns. Sholapur did not show the clear presence of a socioeconomic status variable. And though Factor I, the communal status factor, here is linked with socioeconomic status, this factor is recessive while the caste factor is dominant. The analysis of Sholapur broke into nine factors, highest of any city, and it had 33 variables for consideration. This finer-grained information base may well be responsible for the greater weighting of the communal factors as compared with the SES variables, and clearly the two are interrelated. Sholapur remains as testimony of the power of caste and communal factors in the formation of neighborhood in a modern Indian industrial city.

Clustering of communal residential areas is, of course, predictable, and in Poona reveals remarkable long-term stability. Over a period of 150 years, from the ruler's social survey of 1822 to the surveys of the Gokhale Institute in 1937 and 1954, the center of the old city retained its social pattern quite apart from an almost ten-fold gain in population. The Brahmin high-caste, high SES group dominated the western region of the city between the Mutha River and the present-day Katraj Aqueduct. The center of the town, along a north-south axis, was and is dominated by the market functions of the town and houses the traders, the Muslims, and the non-local groups of the society. The eastern and northeastern quadrants of the city house the untouchables and scheduled caste members for the most part. Thus

---

[2]Simply in terms of residence choices, this division in the Brahmin community reinforces the model of different groups sorting themselves out residentially according to SES. Dynamically, the preeminence of the Chitpavans may well owe to the fact that the rulers of Poona in its first heyday, the second half of the eighteenth century to 1818, were Chitpavan Brahmins. Doubtless they gave special encouragement to fellow caste members.

the old central city retained a clear caste-class correlation over time, and residential patterns clearly separate out along this line. However, Poona has recently experienced suburbanization, and for the first time large numbers of people are moving away from the center of town. In the suburbs, class and occupation are bringing together groups from diverse castes. In addition, newer migrants have tended to cloud the older outlines. Rosenthal (17) quotes a Brahmin leader as saying, "Formerly there was a communal aspect to the controversy between East and West, . . . but now there is little because scheduled caste people have settled in large numbers in the West. Very few differences remain."

Variations in time and area

The Poona and Bombay materials permit us to add further insights into variations of the results in time and area.

Data on Poona are unusually rich, both in quantity and time, thanks to the work of Gadgil and Gokhale Institute of Politics and Economics in Poona. Materials are available from the records of the Maratha administrators which date back to 1822, and from Gadgil's social surveys of 1937 and 1954. Therefore, we can examine materials on the size of the city, its density, the external connections with other cities, the economic functions of Poona, and the social areas of the city over time. This provides strong indication of the effects of various processes over time, and thus brings us back to the original idea of ecology as expressed by the early Chicago school, a concern with process over time.

Poona began as a temple and fort complex, located slightly away from the river bank, and it had in the early seventeenth century many of the features of the classical Indian city. The priests and the administrators grouped themselves around the temple and fort-administrative area. Off to the east, on marshy land undesired by other groups, the servants and untouchables resided.

Rivers and streams flowing from south to north divided the city into three major sectors. In the west between the Mutha River and a now-dry stream which followed the course of the contemporary Katraj Aqueduct was the Brahmin sector of the city. In the central city, between the old river and the Nagzari stream was the bazaar and shop area of the city, inhabited by commercial and artisan castes and by non-indigenous people. Initially the untouchables lived in the eastern sector.

Poona took on great importance in the eighteenth century as the location of the court of the local ruler and also of the *Peshwa* or prime minister and de facto ruler of the spreading Maratha Confederacy. The city expanded during this time, essentially in a north-south direction, following the earlier three divisions and natural topographical features. Successive areas were added to the city by the Peshwa, who turned each new ward over to a nobleman for management and ordered him to bring a full population to it. Sometimes people were already living there, sometimes they came later; the dates given to wards indicate the year of assignment to the nobleman. However, as wards were added, their residents took on the same character as earlier ones in their section, and so the 1822 analysis revealed clear communal distinctions as the prevailing element in the town's social geography.

Expansion of the town beyond the triangular area bounded by the hilly ridges to the south, the river to the west, and the marshy land to the east took place only in the nineteenth century, and more prominently in the twentieth. The British, after conquering Poona in 1818, built their cantonments to the east and the north, Poona Cantonment (the civilian area) and Kirkee Cantonment (the miliary area), respectively. Later in the nineteenth century, Indian settlement spread west of the river, establishing there an outpost of the old Brahmin sector in a highly intellectual area of colleges and schools. Finally, the twentieth century, especially after independence, has seen the development of Poona as an industrial area with the spread of factories along the railway lines, especially to the west, toward Bombay. Both the 1937 and 1954 analyses reveal the additional British and industrial ingredients in the town's social map.

The physical expansion of the city, especially in recent years, has not quite led to suburbanization. The population has continued to pour into the old core city, boosting the population densities there. Maps from Gadgil's study (10) indicate the increasing density of the central area. A later compilation by Brush (5) indicates that densities are still increasing in the central areas, and the relative sparseness of population in the surrounding area. Brush notes "in 1881 Poona's population was recorded as 99,421; fifty years later it had increased to 213,680, with little change of area. By 1961 the old city had been enlarged more than ten times, from 2,390 acres to 27,190 acres, and the population had grown to 597,419, but without any large shift of people into areas incorporated into Greater Poona" (5:379). Brush correctly suggests that if Poona follows patterns of western cities, with the spread of cheap and rapid public transportation in the city, the suburban exodus will soon come. At present workers travel by bike from the central city to jobs on the periphery. The city now seems ready for the jump to a reverse situation.

The size and density changes have come about largely by a change in the economic, political, and social functions which Poona has played throughout the historical period covered. It changed from a governmental center under the Marathas to a British governmental and military center, the headquarters of the southern command of the Indian army, to an industrialized metropolis to some extent before independence and much more rapidly since then.

What ecological change has been associated with these functional and demographic changes? At least until the last decade, they did not upset the basic structure of the core city. But the creation of the cantonment areas by the British did lend to these new areas quite different characteristics from the older parts. There was greater diversity, with more foreigners and fewer Hindus. Further change was introduced by industrialization and by the consequences of partition. Each factory attracted its cluster of migrant workers, and the general acceleration of rural-to-urban migration impelled many lower-caste Indians to the cities, where they filled most of the empty spaces, clouding the earlier social differentiation. When more data from the 1961 census become available, they will therefore show substantial smoothing of the social lines that remained essentially stable in the three analyses, 1822–1954.

Bombay's three-level analyses provided the opportunity for exploring the effects of differing geographical units of observation on the results of factorial ecologies of Indian cities. By and large, the factorial results were remarkably stable, which is pleasing. However, as the finer-grained analyses were undertaken, much greater insight was provided into the city's social map. With 437 units of observation, for example, docks were clearly separated from wealthier, higher socioeconomic areas behind them, a feature clouded by the 88 sections and 15 wards. What the results therefore indicate is a clear need in factorial ecologies for some notion of the size and character of neighborhood groupings in Indian cities; hopefully, this will serve as a guide to appropriate observational units that will provide a proper degree of geographic detail to general ecological themes repeating themselves through a variety of scales of analysis.

Conclusions

To the extent that the evidence analyzed reveals, socioeconomic status appears to be the dominant of these overarching themes in the residential geography of the Indian city today. The communal and caste status of the classical texts and of village India, and of Poona in 1822, is being transformed into class status as an outgrowth of city life (16). While this is taking place, residual caste status and class status reinforce each other, so that the dominant spatial pattern remains that of high-status neighborhoods in central areas and low-status neighborhoods at the periphery.

If ecology is to be understood as process in time, much more historical study of urban ecology is needed in India where traditional patterns have great longevity and where social areas display great resistance to change. Among the active forces which have introduced new elements into the social geography of the Indian city are: the British cantonments, built on different social bases than old Indian areas, often creating a dual city structure and a second set of higher socioeconomic status central and lower socioeconomic status peripheral neighborhoods; newer institutional areas devoted to government or education, although the latter tend to be Brahmin and the former (even in Chandigarh) tend to repeat the traditional core-to-periphery distribution by SES; and industry. It is this latter element that created a second factor in India's urban ecology by 1961: the family-type distinction between familial areas and zones housing new male migrants, clustered around the commercial core of the city, docks, mills, and new factory estates. Outlying industry, rather than improved transportation, has also promoted limited "suburbanization" and is one of the contributing causes of increasingly specialized neighborhoods in all cities.

What seems clear is that the increasingly diverse bases of social and economic power, which city life in modern India is generating, are transforming the urban structure. No longer are land or rulership or priesthood the bases of status or neighborhood and community, although, their effects are still marked in India's urban landscapes. The issue is whether the emerging forms are converging on the model of the industrial metropolis, as suggested by Berry and Rees in Calcutta (4), or whether some new synthesis of traditional and modern will emerge.

References

1. Abu-Lughod, J. "A Critical Test for the Theory of Social Area Analysis: The Factorial Ecology of Cairo, Egypt." Unpublished paper, Department of Sociology, Northwestern University, 1968.
2. Ayyar, C. P. *Venkatrama: Town Planning in Ancient Dekkan*. Madras: Law Publishing House, 1916.
3. Berry, B. J. L. "A Method for Deriving Multi-Factor Uniform Regions." *Przeglad Geograficzny*, 33 (1961):263–82.
4. Berry, B. J. L., and Rees, P. H. "The Factorial Ecology of Calcutta." *American Journal of Sociology*, 74 (1969):445–91.
5. Brush, J. E. "Spatial Patterns of Population in Indian Cities." *Geographical Review*, 58 (July 1968):362–91.
6. Dupuis, J. *Madras et Le Nord du Coromandel*. Paris: Librarie d'Amerique et d'Orient, 1960.
7. Dutt, B. B. *Town Planning in Ancient India*. Calcutta: Thacker, Pink and Co., 1925.
8. Epstein, T. S. *Economic Development and Social Change in South India*. Manchester: Manchester University Press, 1962.
9. Fox, R. G. *Tezibazar*.
10. Gadgil, D. R. *Poona: A Socioeconomic Survey*. 2 vols. Poona: Gokhale Institute of Politics and Economics, 1945, 1952.
11. Gadgil, D. R. *Sholapur City: Socioeconomic Studies*. Poona: Gokhale Institute of Politics and Economics, 1965.
12. Ghurye, G. S. *Caste, Class, and Occupation*. Bombay: Popular Book Depot, 1961.
13. Gillion, K. L. *Ahmedabad*. Berkeley: University of California Press, 1969.
14. Karve, D. D. *The New Brahmans: Five Maharashtrian Families*. Berkeley: University of California Press, 1963.
15. Mayer, A. C. *Caste and Kinship in Central India: A Village and its Region*. Berkeley: University of California Press, 1960.
16. Rosen, G. *Democracy and Industrial Change in Modern India*. Berkeley: University of California Press, 1965.
17. Rosenthal, D. B. "The Politics and Government of Two Indian Cities." Unpublished monograph, State University of New York at Buffalo, Department of Political Science.
18. Sjoberg, G. *The Preindustrial City*. Glencoe: Free Press, 1960.
19. Sovani, N. V.; Apte, D. P., and Pendse, R. G. *Poona: A Re-Survey*. Poona: Gokhale Institute of Politics and Economics, 1956.
20. Spate, O. H. K., and Learmonth, A. T. A. *India and Pakistan*. Methuen and Co., Ltd., 1967.

Some Ecological Patterns of Community Disorganization in Honolulu*

Andrew W. Lind

American communities, to judge by newspaper accounts and popular lectures, are continually on the crest of an ever-mounting crime wave. Nor is this alarm over the rising tide of delinquency and general community disorganization confined to the chautauqua speakers and ministers. W. I. Thomas prefaces his recent volume, *The Child in America,* with the following words:

> As a result of rapid communication in space, movements of population (concentration in cities, immigration), changes in the industrial order, the decline of community and family life, the weakening of religion, the universality of reading, the commercialization of pleasure, and for whatever other reasons there may be, we are now witnessing a far-reaching modification of the moral norms and behavior practices of all classes of society. Activities have evolved more rapidly than social norms. . . . At present, however, it is widely felt that the demoralization of young persons, the prevalence of delinquency, crime, and profound mental disturbances are very serious problems,

and that the situation is growing worse instead of better.

Hawaii, no less than the mainland of the United States, has been affected by the "decreasing influence of existing social rules of behavior" and the increasing ordering of the individual's behavior on the basis of personal desires and fancies. The community of Honolulu, including its resort population, has recently been greatly exercised over a series of crimes, each of which has been embellished and emphasized by the local press. The Fukunaga kidnapping and murder case, a group of subsequent crimes involving members of one of the racial groups least given to delinquent behavior, and a number of recent sex offenses have served to focus general attention upon the alleged Hawaiian "crime wave."

Without in any sense assuming to deal exhaustively with the general problem of the trends and conditions of disorganization in Honolulu, the writer has sought to shed some much-needed light upon the cultural factors affecting local delinquency and dependency. Of parallel interest has been an attempt to test the validity of a sociological hypothesis held by many but most clearly formulated by Dr. Robert E. Park:

*Reprinted from *The American Journal of Sociology,* 36 (September 1930), 206–20, by permission of the author and the University of Chicago Press.

It is the immigrants who have maintained in this country their simple village religions and mutual aid organizations who have been most able to withstand the shock of the new environment. . . . In some sense these communities in which our immigrants live their smaller lives may be regarded as models for our own. . . . Our problem is to encourage men to seek God in their own village and to see the social problem in their own neighborhood. These immigrant communities deserve further study.[1]

We wish to know how the racial colony influences the stability and social health of its constituency in Honolulu. This two-fold task has been conceived largely within an ecological frame of reference. An effort has been made to isolate and measure the role of position and movement in space in the disorganizing processes of the various racial communities of Honolulu.

I. Indices of disorganization

The problem of the determination of satisfactory criteria of disorganization in an immigrant community is complicated by the fact of the presence of two or more competing standards of life-organization. For example, in Hawaii we are confronted not only with the waning effectiveness of "the moral norms and behavior practices of all classes of society," a process typical of all mainland communities, but we also encounter an additional demoralizing factor in the counteraction of cultural patterns as diverse as the Hawaiian, Japanese, Chinese, Filipino, and Portuguese. The mere weight of numbers of the population of each of the foregoing cultural groups in a region of such limited size gives to each an added strength in the competition of the social and moral standards of the respective groups. The individual is subject thus not only to the disorganizing influence inherent in the mere contacts outside his own group and acquaintance with competing standards and traditions, but also to the positive claims imposed upon him by cultural systems generally recognized by large groups of the population. In the case of the smaller groups, such as the Puerto Rican, Spanish, and Korean, their numerical weakness in the territory as a whole and in almost any community where they may be located, imposes a well-nigh unsurmountable difficulty in the maintenance of their old-country patterns.

A considerable proportion of the criminal acts as defined by law in the territory prove to be quite normal and desirable forms of behavior as defined by the given cultural code, e.g., suicide among the Japanese, certain types of extortion or graft among the Chinese, and cock-fighting among the Filipinos. These cultural patterns, so well established and recognized among the first-generation immigrants, are not infrequently accepted by the second generation of the same and other cultural groups as possessing unquestioned validity.

Gambling, for example, is an offense for which youngsters of all nationalities are frequently brought into the juvenile court, the pattern being taken over from the first generation with whom this pastime is thoroughly accepted. Day after day schoolboys observe their parents in the home and adults in the public parks under police observation engaging in the conventional "craps," *chee-fa, hana,* dominoes, bridge, and other games of chance. To be haled into court and sentenced to the reform school for engaging in the same practices constitutes one of the many unsolvable mysteries with which the second generation in Hawaii is confronted.

The very composition of our population consisting in 1920 of 42 percent who were foreign-born and less than 10 per cent who had been situated within a social milieu comparable to that of the average American community, has delayed the emergence of a universally recognized standard of behavior. When one considers, likewise, the extreme contrast of the cultural patterns involved, the failure of the territory to effect a unitary and compelling mode of behavior is not surprising. Statistical measures of disorganization, which presuppose a uniform cultural pattern, may, therefore, prove quite misleading.

In spite of the difficulties involved, an effort has been made to devise an index of disorganization based upon the following criteria: the frequency of cases appearing before the juvenile court; the rates of dependency, as measured by the cases receiving assistance from the largest social welfare agency in the city; the distribution of cases of suicides; and the rates of vice, as measured by police arrests. One index serves as a check upon, and supplement to, the others.

The juvenile court cases are, of course, of widely divergent character and when segregated according to types permit of some illuminating comparisons and contrasts. Taken in the mass, these cases tabulated and analyzed over a three-year period probably provide the best available index of the waning influence of old-country controls and the progressive individualization of conduct among the Americanized second genera-

---

[1] R. E. Park and E. W. Burgess, *The City*, pp. 121–22.

tion. Occasionally, the juvenile delinquent has merely substituted for the accepted behavior patterns of his parental group a mode of conduct common in another immigrant group but defined as criminal by law. This is undoubtedly true in many of the cases of gambling.

The dependency cases are in the first instance examples of temporary or permanent personal disorganization, but they also usually represent a similar ineffectiveness of the conventional primary group controls of the family, neighborhood, or racial colony. Undoubtedly, a large proportion of the personal maladjustment in Hawaii is provided for by the informal devices and mutual aid practices of the primary group. The Social Service Bureau cases indicate fairly accurately the disorganization of the primary group controls in the community.

Suicide has a rather ambiguous value as an index of social disorganization. For certain of the racial groups represented in the territory, suicide is still partly within the mores and therefore does not provide an entirely satisfactory test of general community disorganization. The first-generation Japanese in particular regard the taking of one's life as not only permissible but commendable and even obligatory under some circumstances. The institutionalized suicide of feudal Japan, while diminishing in its prevalence as the older generation passes on, still serves as a convenient device for avenging insults, avoiding a shameful situation, proving loyalty, or expiating a crime. In China and Korea, too, suicide has been institutionalized, and the rate of suicide among the Chinese and Koreans greatly exceeds the rate of suicide for the territory. This is due in part to the fact that it is socially sanctioned.

Granting, however, due allowance for the approval of suicide by the foregoing groups under certain circumstances, the fact still remains that the taking of one's life commonly reflects a maladjustment bordering on disorganization. The oriental immigrant does not ordinarily commit violence upon his own person except as a last resort in a crisis situation when other devices have failed. Certainly suicide represents for those elements of the population responsive to the American and European public opinion a marked deviation from the accepted and normal patterns of behavior, an evidence of personal disorganization. The correlation of this form of personal with social disorganization is apparent in Honolulu,[2] although certain variant factors affect the local situation.

[2]Cf. R. S. Cavan, *Suicide,* pp. 77–105.

Accurate information as to the location of commercialized vice is difficult to secure, although something of a check of the more flagrant cases has been possible through reports of arrests for vagrancy and prostitution, gambling, and sale of narcotics. Although the total number of such arrests is only 108, this evidence is substantiated by the statements of social workers and by field observation. The most superficial observer is aware that vice, however defined, is by no means confined to the areas or cases designated in this way, but it is still true that the individuals who most openly flaunt their wares before the public eventually appear on the police blotter whether they are brought to trial and convicted or not.

II. Areas of disorganization

Almost as apparent to the intelligent citizen as the fact of the segregation of the retail business houses within certain sections of the city is the tendency of vice, crime, and dependency to concentrate within other as highly specialized areas. The area and location of the old segregated vice district was as much the consequence of natural selective forces operating within the community as of legislative enactment, and although the latter has been removed the operation of these natural forces is still apparent. The abolition of the segregated area in Iwilei was instrumental in shifting slightly the location and the range of movement of the practitioners, but the area remains. So, with regard to juvenile delinquency, suicide, and other forms of social disorganization, one can easily discover centers of high concentration, shading off into areas of comparative freedom from cases of such phenomena.

In a city of the size and population composition of Honolulu, where the processes of specialization are less advanced and under the influence of such peculiar topographical[3] condition . . . , the precision of measurement of gradients of social disorganizations cannot approximate that possible in such centers as Chicago or New York. One may observe, however, much the same tendencies and processes modified by the peculiar local conditions.

A comparison of five maps showing the distribution of cases of juvenile delinquency, dependency, suicides, and common vice reveals a uniform concentration in two areas, Palama and Kakaako,

[3]The large ridges extending down toward the ocean serve most effectively to break up the city into a number of irregular natural areas.

both located immediately outside the central business district of the city. These are the areas of transition between residence and business, of high value and low residential rents, characteristic of all cities. The rates of dependency and deliquency fall off, although not uniformly, as one moves outward from this zone along the three main gradients. A rough correlation with land-value gradients is likewise observable.

Kalihi, an area of early residential settlement but of increasing industrial invasion and decreasing residential value, shows a higher rate of disorganization than a corresponding area on the other two sides of the city. The rise of business and recreational subcenters in Kaimuki and Waikiki has been attended by the development of peripheral zones of disorganization in the less desirable residential sections on the slopes of Kapahulu and in Waikiki-ewa. Areas of second-immigrant settlement in Makiki, Bingham Tract, Upper Nuuanu, and Kaimuki (the "rayon stocking districts") represent higher stages of economic and social adjustment. The "silk stocking" districts of Manoa, Nuuanu Valley, and Kahala rarely figure in the official statistics of disorganization, although juvenile delinquency, suicides, and dependency are not entirely absent.

That these indices of disorganization are not entirely interchangeable is evident from the most casual study of Table 1. Juvenile delinquency and vice cases, both dependent upon the vigilance and initiative of the public officials, show a decided tendency to concentrate in certain sections of the city or perhaps rather to be lacking in other favored sections. Kakaako, Iwilei, Palane, Central, and Kalihi sections have over a considerable period of time developed a reputation for disorder and crime. It is only natural, therefore, that public officials should expect them to justify the reputation. The distribution of dependency and suicide cases being less subject to the discretionary powers and initiative of public officials shows less tendency toward segregation, although the slum areas rank well toward the top. Isolated cases of dependency and suicide appear in practically all sections of the city.

III. Patterns of suicide

Suicide conforms in general to the other ecological patterns of disorganization, bulking heavily in the amorphous slum and lodging-house areas and practically disappearing in the aristocratic residential sections. Certain marked deviations from the norm, due to cultural traits of the population elements involved, are to be noted in certain sections of the city. The Punchbowl area located just back of the central business section and inhabited so largely by Portuguese is entirely devoid of any cases of suicide during the three-year period. Similarly, the Palama and Kalihi sections, housing such a large proportion of the non-suicidal Hawaiians and part-Hawaiians, present a much more favorable picture than one might expect from such transitional areas.

The areas of Chinese, Japanese, and Korean settlement loom higher in suicide than in the other forms of disorganization. The central oriental district provided one-fifth of all the suicide cases in this city during the sample period and all but two of the cases were Chinese, Japanese, or Korean. Unlike delinquency and vice, suicide appears frequently within the Japanese camps of the city. Suicide is still within the mores of the oriental groups.[4]

Perhaps most striking in the ecology of suicide in Honolulu is the noticeable concentration in the "rayon" as distinguished from the "silk" and the "cotton or no stocking" districts. In Honolulu this form of disorganization moves up and out of the slum sections into the areas of middle and professional classes, probably foreshadowing the direction of future community disorganization and individualizaton of behavior. Particularly in Makiki, Waikiki, and Kaimuki, the strongholds of the white middle class, do the rates of suicide mount higher than normal. This is but one evidence of the demoralization which has set in among the *haole*[5] population. The extremely high rate of suicide in the Middle Street area is occasioned by the heavy concentration of white military population within the section.

IV. Community patterns of delinquency and vice

As in the case of suicide, one cannot adequately comprehend the spatial patterns of delinquency and vice without taking some account of the cultural traits of our population. It is true that the transitional zone of the city bulk largest in the police and court records, but equally significant is the fact that large sections within the slum are

[4] The Japanese (27.0), Chinese (31.7), and Korean (75.1) suicide rates are noticeably above the average rate of 22.5 per hundred thousand for the territory.

[5] A Hawaiian term used to designate the white population from continental United States and Northern Europe and their children.

Table 1. Indices of Dependency, Juvenile Delinquency, Vice, and Suicide, According to Census Enumeration Tracts, Honolulu, 1928*

Name of District	Index of Dependency	Rank	Index of Juven. Del.	Rank	Index of Vice	Rank	Index of Suicide	Rank	Composite Index of Disorgan.	Rank
Hell's Half-Acre	2.33	2	2.24	2	1.09	1	0.48	1	6.14	1
Central Lodging-House	2.20	4	1.60	8	.70	2	.40	3	4.90	2
Kakaako-kai	1.81	5	2.80	1	...	...	.13	17	4.74	3
Industrial	2.63	1	1.45	9	.09	14	.18	11	4.35	4
Liliha Street-ewa	2.25	3	1.44	10	.10	12	.15	15	3.95	5
Kakaako	1.44	11	2.19	3	.19	9	.12	18	3.94	6
Buckle Lane	1.72	6	1.60	7	.28	6	.06	28	3.66	7
Palama	1.71	8	1.83	5	...	...	.06	27	3.60	8
Fort Street	1.66	9	1.23	12	.36	4	.20	8	3.35	9
Kalihi-kai	1.17	18	1.92	4	...	...	.15	13	3.24	10
Iwilei	1.71	7	.57	28	.57	3	.29	5	3.14	11
Palama-kai	1.56	10	1.14	15	.10	13	.17	12	2.97	12
Lanakila Tract	.96	20	1.75	6	.07	15	.11	21	2.89	13
Kapahulu	1.02	19	1.21	13	.06	16	.10	23	2.39	14
Cunha Lane	1.30	13	.76	19	.22	8	.08	26	2.35	15
Central	1.22	16	.33	34	.33	5	.44	2	2.33	16
Central Nuuanu	1.24	15	.72	21	.16	10	.04	32	2.16	17
Lower Central	1.29	14	.43	32	.14	11	.29	6	2.15	18
Kalihi-waena	.91	22	1.16	14	.03	18	.09	24	2.12	19
Palolo Valley	1.37	12	.67	24	...	...	.04	31	2.08	20
Miller Street	1.17	17	.53	30	...	...	.14	16	1.84	21
Lower Pauoa	.75	24	.75	20	.26	7	.06	29	1.80	22
Makiki-kai	.81	23	.59	26	.04	17	.26	7	1.70	23
Upper Makiki	.31	33	1.31	11	...	...	.08	25	1.70	24
Waialae	.60	30	.90	18	...	...	.20	10	1.70	25
Punchbowl-uka	.55	31	1.05	16	...	...	...	...	1.60	26
Kalihi-uka	.93	21	.67	23	...	...	...	34	1.60	27
Upper Nuuanu	.67	26	.92	17	.02	21	...	37	1.59	28
Punchbowl	.72	25	.67	22	.02	20	...	36	1.39	29
Punahou	.60	29	.55	29	...	...	.20	9	1.35	30
Middle Street	.63	27	.57	27	...	...	.30	4	1.33	31
Moiliili and Bingham	.60	28	.65	25	...	...	.05	30	1.30	32
Kaimuki	.27	36	.46	31	.02	21	.10	22	.85	33
Moiliili	.30	34	.35	33	.02	20	.12	20	.80	34
Makiki	.52	32	.11	36	...	...	.15	14	.78	35
Manoa	.28	35	.17	35	.03	19	.03	33	.51	36
Waikiki	0.19	37	.08	37	...	...	.12	19	0.39	37
Average	1.09	...	1.00	...	0.11	...	0.13	...	2.33	...

*Indices represent the ratios per thousand of the total population in each of the census enumeration tracts and in Honolulu as a whole in 1926. These data were secured in a house-to-house canvass through the cooperation of public utility and public agencies.

Table 2. Racial Distribution of Juvenile Delinquency

	Number of Juvenile Court Cases	Ratio of Delinquency per Thousand of Public and Private School Pop.
Hawaiian	208	169.7
Part-Hawaiian	176	46.9
Portuguese	173	65.1
Puerto Rican	51	167.1
Spanish	7	46.3
Other Caucasian	26	10.7
Chinese	114	26.8
Japanese	109	12.1
Korean	43	72.8
Filipino	49	108.1
Total	960*	38.3

*The total number of cases includes four cases classified as "All Others," which do not figure in the rates of the racial groups mentioned above.

quite devoid of these forms of disorganization.[6] Likewise the disparity in the rates of delinquency in certain outlying sections of the city must be interpreted largely in terms of the racio-cultural makeup of the population.

Tabulations and rates of juvenile delinquency for the entire city during the three-year period 1926–28, inclusive, reveal a racial distribution which must inevitably color the situation in the various areas of this city [see Table 2].

No attempt will be made at this time to consider the various factors which contribute to the delinquency of the groups represented here.

In the light of the foregoing data, the abnormally high rates of juvenile delinquency in the Kalihi-waena and Kalihi-kai sections, occupied so largely by Hawaiians and Portuguese, become somewhat more intelligible. Likewise the Kakaako, Upper Makiki, and Kapahulu ratios of delinquency are aggravated by the presence of racio-cultural groups which are so largely maladjusted. In brief, the marked deviations from the normal spatial patterns of delinquency in the city, as outlined by Burgess, Shaw, and others, are found in part in the character of our racial patterns. Delinquency is not arranged in the city according to a strict economic and ecological ordering. The Hawaiians, for instance, who rate well above the Japanese in the economic scale and therefore occupy better residential areas, are far less effectively organized to maintain morale.

Vice likewise deviates somewhat from the normal ecological pattern. The narcotic cases appear

[6]The maintenance of well-integrated forms of behavior within the racial colony of the slum is discussed in a separate article entitled, "The Ghetto and the Slum."

chiefly in the Chinese sections, while a disproportionate amount of gambling comes from among the Filipinos. An interesting variation appears with regard to prostitution. The area of heavy concentration of vice extends a considerable distance into the middle-class oriental section along the Nuuanu gradient. The *haole* proprietors of these houses may violate with impunity the taboos of the oriental residents since their disapprobation figures so slightly with the police or with the *haole* patrons. The high concentration of prostitution in the Palama and Central slum areas represents the lower forms such as street-walking and brothels.

V. Delinquency triangles

Following the suggestions of Dr. Ernest W. Burgess, a series of maps were constructed showing the residence location of various types of juvenile delinquents and the place of the offense. An analysis of these maps reveals a number of significant spatial and racial patterns.

Larceny bulks largest among the various types of offenses for which juveniles are haled before the court, and it is likewise a form of crime in which characteristically more than one individual is involved. Eighty-two percent of all juveniles charged with larceny in 1928 in Honolulu were associated with one or more other persons in the commission of the offense.

Interestingly enough, these delinquents show only a slight disposition to select others of their own racial group when engaging in this unsocial behavior. This also tends to confirm the hypothesis that it is the individual who is maladjusted to his own cultural group who likewise fails to accommodate himself properly to the conventional practices of the larger community. Apparently one of the most effective melting pots for the races is the crucible of crime, for once having been purified of the restraining dross of a distinguishing culture and tradition, the individual mingles freely with other of his emancipated kind. As to the contact agencies which initiate this process of *deculturization,* more intensive case study is necessary.

Of all the groups, the Chinese manifested the greatest disposition to associate in crime with members of his own racial group; three times the expected number of contacts being with the in-group. The Japanese, popularly described as clannish, reveal a lower percentage of participation in crime with the in-group, being as often associated with the Chinese. Indeed, the most striking fact with regard to all the groups in this connection is the high ratios of association in crime of the more serious and demoralizing types with the out-group. The racial composition of the neighborhood gangs, which periodically break into public print because of their criminal behavior, is markedly mixed in character. Crime, as well as politics, makes strange bed-fellows.

Neighborhood organization and crime

One additional index of the effectiveness of local community standards of behavior may be found in the frequency of crime within the neighborhood of the delinquents' residence. An area capable of maintaining the strength of its prohibitions is likely also to discourage its wayward residents from attempting the violation of the taboos within the boundaries of the district, although it may not succeed in entirely repressing the behavior. The ease of movement within the modern city and the anonymity of its life enable youth to satisfy its wayward impulses by merely shifting the scene of its activity from the disapproving scrutiny of the neighborhood preceptors. What the gossips don't know won't hurt them; but where there are no effective gossips and self-appointed guardians of morals, one may indulge his vagrant impulses without seeking shelter in the anonymity of a strange community.

The neighborhood triangle of delinquency represents the situation in which the homes of the two or more delinquents and the place of the offense are found within the same neighborhood.[7] This may be nothing more than the innocent display of youthful energy, which the legal public, however, defines as criminal. The chance gathering of two or

[7]Park and Burgess, op. cit., p. 52.

three neighborhood cronies resulting in a window shattered by a baseball, or a more serious foray on the milk supply on the back steps, suggests a pattern of delinquency fairly common in all sections of the city.

The neighborhood triangle of delinquency, particularly as it involves sex offenses, occurs most often, however, in the slum sections where neighborhood standards are at their lowest ebb. The slum area provides not only the few cases of juveniles engaged in prostitution but also most of the instances of the immoral act occurring within the home of one of the participants. The other participant may reside in the same area or may come from an area of higher social status.

The mobility triangle of delinquency, in which the homes of the two or more delinquents lie within the same local community while the place of the offense is situated outside, is likely to have its base situated in an area of somewhat greater stability and more effective social restraints. A boy and a girl of the neighborhood wishing clandestine sex satisfactions find it necessary to seek a rendezvous outside the range of neighborhood scrutiny. The cheap hotel in the lodging-house area or the automobile parked in another section of the city or on a country road provides the necessary protection from the neighborhood gossip.

Kalihi,[8] for example, furnishes a disproportionate share of the cases of juvenile immorality, but this section still has sufficient morale to prevent the more flagrant abuse of its moral sensibilities within its own borders. A comparison of records over a two-year period reveals, however, a lowering of resistance in this regard and an increase of the "neighborhood triangle" or the two-point pattern.[9] In 1926, only 20 percent of the cases in Kalihi were of the neighborhood triangle type; by 1928, this percentage had increased to 38 percent.

The "promiscuity triangle"[10] very often has its residential bases in well-organized communities such as Makiki, Kaimuki, Nuuanu, or Palolo. The basis of contact varies greatly, of course, but the public dance halls figure frequently enough to be significant.

The various participants in larceny cases are

[8]The Hawaiians and part-Hawaiians who constitute approximately 50 percent of the population in Kalihi and who furnish 75 percent of the cases of juvenile immorality are responding in this behavior to folk ways which are deeply ingrained into the Polynesian cultural system.

[9]When the offense occurs within the home of one of the participants.

[10]When all points of the triangle lie in different local communities.

more likely to come from the same district, regardless of its economic and social status, especially if the delinquents are in the lower age levels and are somewhat inexperienced in crime. Petty thievery of milk from the back porch or a chicken from the backyard coops of the neighborhood sets the patterns of delinquency which later develop into burglary in homes in another part of the city.

The neighborhood pattern represents a prior stage to the mobility or promiscuity patterns. In 70 percent of the 1926 cases involving the neighborhood pattern of juvenile larceny, one or more of the participants later in the same year was involved in a mobility pattern of larceny of a more serious nature. A more exhaustive study than the one here attempted would be necessary to determine the causes of the shift from the neighborhood to the mobility pattern of delinquency.

VI. Summary

The polyglot and polychrome population of Hawaii is responding to the forces of the plantation and the urban environment in a fashion which is fairly regular and predictable. Social disorganization proceeds according to laws which may now at least be hypothetically stated. Certain of these laws or principles lend themselves to most effective statement in ecological terms, i.e., position and movement.

Juvenile delinquency, vice, dependency, and suicide are all territorially distributed in Honolulu after much the same pattern of concentric circles. Public opinion is largely instrumental in affecting the rates of various districts with regard to the first two indices. Dependency and suicide seem to respond more accurately to the natural forces of the community. Suicide is less highly correlated with economic status, and its distribution undoubtedly marks the direction of future disorganization.

The breakdown of personal and cultural insularity and the rise of promiscuous and uncontrolled contacts between groups incident to modern mobile life under the conditions outlined serve to transform the melting pot of the races into a crucible of crime and dependency. Born out of economic necessity and the propulsion of school and street, assimilation and amalgamation proceed with irresistible force. The problem of avoiding the most serious social costs in delinquency and dependency arising out of a too-rapid or misdirected fusion of the various cultural elements constitutes a supreme challenge to socal technology both in Hawaii and the world at large.

Honolulu—Fifty Years Later*

Andrew W. Lind

Dramatic though the transformations in the external appearance of Honolulu have been in the half century since the original study was undertaken, the underlying ecological structure of the city in terms of spatial distributions and relations has continued much as it was, but over a wider territory and with greater intensity. What was in 1930 a moderately-sized urban center of less than 150,000 persons, the outgrowth of a sugar- and pineapple-plantation economy, had expanded nearly six times in population by 1980 to a metropolis dependent chiefly on tourist and military enterprises and comprising over half of the land area of the island of Oahu. Despite this startling development in size and economic character, the essential patterns of allocation of the land have not been modified to any comparable degree. Sections of the island, previously devoted to the cultivation of crops or relatively unoccupied on steep slopes or inaccessible ridges, have been converted to residential use, but the varied social strata and behavioral categories have tended to remain much the same with relationship to geographical position and distribution.

Alterations in administrative procedures, especially at the level of federal census definitions and policies, but also in local cultural conceptions and practices, preclude the possibility of strict statisti-

*This article was written for this volume and has not been published previously.

cal comparison between data for 1930 and 1980. In 1970, for example, there were already 96 separate census tracts within the area which in 1930 had constituted the entire city, but no combination of those tracts would correspond at all adequately with the boundaries of the 37 enumeration tracts designated in 1930. Increasing discrepancies between the definitions of ethnic categories recognized in the federal censuses and those utilized by local residents limit still further the possibility of accurate comparison between situations fifty years ago and today, especially with reference to many forms of community disorganization. Under the circumstances, the comments in this epilogue are chiefly confined to the more obvious modifications which have occurred in the ecological structure and their impact on community disorganization.

The population within the official boundaries (from Maunalua to Moanalua) increased approximately two and a half times between 1930 and 1980, with people of moderate means establishing homes further up the slopes of the Koolau mountains or farther distant from the center at lower levels, while people less fortunate were forced into quarters of increasing density along the low-lying plains. Most visually spectacular in this phase of the urban expansion has been the development along the four miles of ocean front between the original trading center at Honolulu harbor and Diamond Head Crater—two miles of an almost solid wall of multi-storied hotels along the Waikiki peninsula and of commercial buildings on most of the intervening distance.

It was, however, in the broader expanses of the island, outside of the city proper, where Honolulu experienced its highest population upsurge—4.7 times the 1930 figure within the fifty years—in a vast "urban sprawl," comparable to what occurred in so many mainland cities at about the same time. Native Hawaiian villages and former plantation camps have burst into suburban satellites, bound to the central city by high-speed highways, instant communication, and with both the stimulation and the disorganizing forces of the metropolis.

The one form of community disorganization which lends itself most readily to objective analysis within an ecological framework, both in 1928 and today, is the degree to which people in different areas of the city have turned to governmental or other external sources for their economic maintenance. The necessity or disposition among families to depend on public or other outside assistance has not only intensified in the "transitional areas" adjacent to the central business district, but it has expanded considerably in districts within the city which were previously relatively free of such dependence. The proportion of all families deriving part or all of their income from "public assistance or public welfare" in 1969, as reported in the 1970 census, ranged from a maximum of 50.7 percent in the tract just outside the central business district, to a low of zero or less than 1 percent in eleven tracts scattered over the more desirable areas on the heights overlooking the city or along its preferred valleys or beaches. The seven other city tracts with high dependency ratios (15.0 to 39.4 percent) were all from within the expanded transitional zone.

Although the over-all dependency rate was actually higher in the sections outside of Honlulu proper than in the central city, the differences in the degree of dependency between the areas of poverty and of affluence were significantly less in the surburban portions of Oahu than in the central city. The six suburban census tracts with the highest ratios of dependency, ranging from only 12.1 to 21.1 percent, were all situated in the dry northwest sector of the island: and, of these, three of the most remote and least economically productive were disproportionately populated by native Hawaiians. At the opposite end of the island, although surburban and most accessible to the city proper, the Kailua district had the highest ratio of Caucasians (68 percent) in its total population of nearly 36,000 persons, but the lowest ratio of dependency (1.4 percent).

Conscious efforts to rehabilitate the inner city have recently brought about, as in several mainland cities, a minor reversal of the dominant centrifugal trend in the ecological spread of dependency. Thus, one census tract on land once known as "Tin Can Alley," and including part of Chinatown, has been largely "gentrified" by the demolition of the old tenements and the erection in their place of attractive condominium and apartment buildings. None of the 340 families within this ghetto of upper middle-class residents derived any income from "public assistance or public welfare" in 1969, while half of the 363 families in another tract, only a half-mile distant, were "on welfare."

None of the other forms of community disorganization seem to fall as neatly into an ecological pattern or illustrate as clearly the accommodation to this structure which extensive social change over extended time may necessitate. In the case of both crime and suicide, probably the most formidable shift in the social environment, serving to obscure the operation of ecological forces, has been the slow but irresistible assimilation of Honolulu's ethnic groups and the corresponding absorption of the population within cosmopolitan neighborhoods, with little regard for ethnic considerations. Except for relatively small enclaves of recent-immigrant Samoans and other Pacific Islanders, and Southeast Asian refugees seeking the comfort and assistance of their fellow ethnics, the earlier racial ghettos with their impact on suicide, gambling, and prostitution have largely disappeared. Neither is it now possible to identify the belligerence of specific ethnic gangs with particular neighborhoods, such as Palama, Kalihi, or Kakaako, which was so notable in the 1920s and 1930s.

Although the small, dilapidated district situated close to the original center of Honolulu and still known as Chinatown has frequently been associated with organized crime, especially in narcotics and gambling, the central figures have rarely been residents of the area or proportionately members of this ethnic group. The low-income areas surrounding the central business district on three sides have in 1980, as in 1930, reported higher-than-average rates of juvenile and adult delinquency, but the spatial distribution of crime is far wider than in 1930, and no section of the city, including the suburbs, is completely immune. The degree of juvenile delinquency in the area of its highest incidence in 1928, Kakaako, has significantly declined, owing to the conversion of the area from low-income residence to warehouses and industry. Waikiki, on the other hand, with the lowest rate of juvenile delinquency and a very low incidence of adult crime in 1928, is now credited with one of the island's highest, as a consequence of the huge tourist invasion since 1959 and the accompanying traffic in drugs and prostitution.

The supposedly wealthy and unwary visitors to Hawaii, as well as the islanders themselves, are vulnerable to criminal depredations, not only where they sleep, but especially where they look for entertainment and diversion. Thus, the districts of rural Oahu with inviting beaches and scenic facilities, to which unsuspecting residents and visitors are naturally attracted, become also a magnet for persons seeking easy victims to exploit. Formerly peaceful subsistence or plantation communities, such as Nanakuli, Waianae, and Waimanalo, and several other rural recreational areas, have figured prominently in reports over the past several years of theft, sexual exploitation, illicit drug traffic, and physical confrontations in which both "locals" and visitors are involved as aggressors. The disproportionate concentration of underprivileged native Hawaiians living in these portions of Oahu is a factor also in their over-representation in the population of reform and penal institutions.

On the basis of recent analyses of suicide on the island during the four-year period (1974–1977), it appears that this form of disorganization has become no less widely diffused spatially throughout the community than has crime. Although the incidence of suicide in proportion to the total population of Honolulu has unquestionably declined somewhat since the 1930s, the secularizing forces of the community have resulted, as in the case of crime, in dispersion over virtually all parts of the island. According to the recent survey, only 9 of 108 civilian tracts over the entire island were without any instances of suicide during the four-year-period, and only one of the nine was in the upper quartile on the basis of income. The trend already noted in 1930 of the movement of suicide "up and out" of low-income areas has clearly continued, but the earlier comment that suicide was still "within the mores of oriental groups," has obviously lost credibility.

Suicide, although not within today's mores, is practiced at all levels in Honolulu society from the highest to the lowest economically, from teenagers to octogenarians, in both sexes, and in all ethnic groups. The ecology of suicide in contemporary Honolulu requires for its interpretation some elements chiefly introduced since 1930, including the rise of tourism and a heightened secularization of life in general.

Index of Persons Cited and Topics

Abbott, W.F., 356, 398–405
Abrahamson, M., 430
Abu-Lughod, J.L., 160–61, 178, 312, 338–51
Agulhon, M., 425
Aldrich, H., 242, 246
Alexander the Great, 144, 145, 147
Alford, R.R., 285
Alihan, M., 4–5, 73–77, 81, 82, 83, 124, 165, 169, 194
Alker, H.R., 309
Allee, W.C., 124
Allen, W.R., 175, 188–93
Allison, A., 12
Amato, P.W., 376, 415
Anderson, N., 59, 73–74, 222
Anderson, T.W., 303, 311, 321, 341, 343
Animal ecology, 3, 21, 25, 26, 74, 76, 84, 108, 109, 124
Appleyard, D., 254
Axial type of development, 46–47

Backeland, L.L., 214, 216
Bailey, K.D., 7, 165–72
Bange, E., 324
Bardet, G., 425
Barker, R.G., 169
Barkley, R.E., 272
Barnes, H.E., 17
Barrios, J.R., 364
Barrow, G., 305, 310
Barrows, H.H., 25
Bean, L.L., 303, 311, 321, 341, 343
Bell, W., 177, 210, 211, 300, 301, 302, 303, 304, 309, 310, 311, 312, 317, 318, 319–24, 330, 333, 339, 341, 342, 343
Berry, B.J.L., 282, 299, 304, 310, 311, 312, 324, 333, 341, 343, 358, 444–53
Beshers, J.S., 318, 341, 342
Bettelheim, C., 425, 427
Beynon, E., 399, 430
Biehl, K., 5
Biotic balance, 22–23, 24, 27, 108
Biotic level, 3–4, 5, 21, 23, 24, 26, 74, 75, 76, 169, 194, 195, 333, 388
Blake, J., 349
Blumenthal, H., 398

Boal, F., 430
Boat, M.D., 300, 324
Bogue, D.J., 5, 88–103, 299
Bogue, E.J., 5, 88–103
Bollens, J.C., 324
Booth, C., 298
Bossard, J.H.S., 86, 404
Boudon, R., 101
Boulding, K.E., 125
Bowers, R., 398
Bowring, J., 9
Bradley, D.S., 241
Braun-Blanquet, J., 106–7
Breese, G.W., 176, 270, 272, 273
Breton, R., 211, 212–13, 217
Brindley, T.S., 324
Broom, L., 324
Brown, L.A., 312, 331
Brush, J.E., 399, 451–52
Buchanan, W., 11–12
Bulliet, R.W., 356, 394–97
Bunnell, G., 241
Burgess, E.W., 3, 4, 35–41, 56, 77, 177, 189, 207, 216, 224, 226, 229, 230, 241, 242, 247, 299, 316, 317, 321, 322, 327, 330, 331, 332, 356, 357, 358, 374, 375, 376, 377, 398–405, 415, 429, 430, 431, 459
Burke, K., 128
Buttimer, A., 158

Caillé, J., 387
Camilleri, S.F., 303, 310, 321
Canter, D., 254
Capers, G., 219
Caplow, T., 355–56, 357, 359–73, 374, 375, 376, 425–28, 430
Carey, G.W., 310, 341
Carpenter, D.B., 320, 398
Carr, S., 253
Carroll, J.D., Jr., 269, 270, 272
Caruso, D., 311
Casis, A., 362
Castle, I., 399
Caton, H., 224
Central business district, 4, 6, 23, 36, 37, 38, 51, 56–

57, 59, 133, 162, 176, 179, 198, 252, 267, 268, 269, 270, 271, 318, 330, 331, 333, 346, 355, 358, 366–67, 369, 376, 389, 417, 427, 463
Centralization, 4, 30, 31–33, 37, 157, 267, 271, 357, 360, 361, 371, 375, 376, 426, 429
Chabral, le Comte de, 9
Chapin, F.S., Jr., 272
Chaumbart de Lauwe, P.H., 271, 425, 426
Chevalier, L., 425, 427
Church, G., 211, 212, 213
Clark, C., 300, 319
Clark, J.A., 310
Classical position, 3–5, 20–72, 177, 194, 195, 316, 317, 320, 331–32, 333; criticisms of, 73–87, 198, 201, 389
Clausen, J.A., 259
Clements, F.E., 125
Clignet, R., 341, 349
Climate, 9, 372, 380, 395, 396, 415, 435
Climax phase, 24
Cluett, C., 280
Cognitive map, 176, 228, 250–52, 253
Cohen, A.K., 318
Cohen, E., 153, 158
Cohen, L.E., 222, 223
Collison, C., 399, 430
Communication, 3, 26, 31, 74, 76, 83, 84, 113, 118, 120, 143, 152, 198, 208, 252, 266, 267, 317, 318, 360, 386, 390, 437, 454
Commuting, 157, 176–77, 266–96, 318, 358
Competition, 3, 4, 21, 23, 24, 26, 31, 51, 52, 53, 74, 75, 76, 77, 83, 84, 86, 87, 104, 105, 106–7, 118, 120, 129, 198, 201, 288, 317, 369, 371, 447, 455
Competitive cooperation, 3, 21
Comte, A., 8, 17, 116
Concentration, 4, 30–31, 34, 37, 41, 59, 113, 203, 209–10, 216, 247, 269, 270, 293, 294, 295, 300, 323, 360, 361, 366, 369, 370, 388, 390, 426, 433, 454, 456
Concentric zone theory, 4, 36–39, 51, 56, 175, 177, 179–87, 189–93, 317, 321, 330–31, 369, 374, 375, 398, 399, 402, 461
Conurbations, 36, 267
Cooley, C.H., 77
Cooper, C., 255
Cooperation, 3, 23, 26, 77, 107, 231, 433, 434, 439
Cortés, Hernán, 145
Creel, H.G., 144
Cressey, P.G., 74, 246
Crime, 3, 8–10, 11–18, 38, 39, 40, 57, 58, 108, 175, 185–86, 191, 193, 203, 204, 227, 356, 422–23, 435, 454, 455, 456, 457–61, 463–64
Cruse, H., 189
Cultural level, 3–4, 5, 21, 25, 26
Curson, P.H., 312
Curtis, J.H., 324
Cybriwsky, R., 254

Darling, F.F., 125
Darwin, C., 3, 20, 21, 107, 118
Davidson, W., 359
Davie, M.R., 83, 230, 399
Davies, W.K.D., 298, 305, 310
Davis, B. *See* Duncan, B.D.
Davis, J.R., 288
Davis, K., 349, 362
Dawson, W.R., 258
Deas, P.M., 257
Decentralization, 28, 32, 37, 177, 293–94, 295, 318, 375, 376, 377, 399, 429
Density, 14, 30, 35, 59, 117–18, 242, 258, 278, 288, 293, 294, 332, 343, 357, 362, 363, 367, 369, 371, 400, 402, 426, 427, 434, 447, 451, 452, 462
Dent, O.F., 331, 332
DeWitt, K.F., 324
Díaz del Castillo, B., 145
Dice, L.R., 124, 125
Dieterlen, G., 159
Dillon, T., 86, 404
Dispersion, 31, 175, 203, 204, 205, 231, 242, 268, 293, 294, 295, 323, 426, 464
Distance, ecological, 29–30, 32, 112–13, 129, 316, 331, 437, 447
Division of labor, 3, 5, 23, 24, 25, 27, 39, 74, 76, 77, 84, 108, 111, 113, 116–18, 120, 121, 122, 128, 300, 318
Dobriner, W., 274
Dominance, 4, 23–24, 106, 125, 157, 221, 274, 317, 318, 333, 364, 427
Donnelly, D., 253, 254
Dotson, F., 374, 375, 376, 389, 399
Dotson, L.O., 374, 375, 376, 389, 399
Doucet, M.J., 298
Downs, R.M., 176, 249–56
Drake, S.C., 224
Driedger, L., 175, 207–17
Drift hypothesis, 260–63
Droth, W., 332
Duncan, B.D., 88, 89, 298, 300, 303, 318, 322, 398
Duncan, O.D., 6, 88, 119, 123–28, 166, 167, 168, 298, 300, 303, 316, 318, 319–20, 322, 333, 398
Dunham, H.W., 176, 257, 259, 260, 261, 263
Dunn, L.C., 160
Dunn, S.P., 160
Durkheim, E., 6, 115–22, 128, 158, 168, 317
Dutt, A.K., 155
Dye, T.R., 282

Ecological complex, 6, 119, 126–27, 128, 167–72, 333
Ecological constellation, 29
Ecological correlation, 5, 88–103
Ecological fallacy, 5, 88–103
Ecological unit, 29
Ecosystem, 124–28, 316, 318
Egeland, J., 321, 341
Elhai, H.D., 425
Elmer, M.C., 3, 8–10
Elton, C., 22, 109, 125
Engel-Frisch, G., 222
Epstein, T.S., 446
Esser, T., 318
Evans, A.W., 288
Evans, D.J., 312
Evolutionary sequence hypothesis, 375, 429, 430–31
Eyles, J., 254
Eyre, L.A., 357, 416–24

Factorial ecology, 177–78, 297, 303–13, 316–17, 318, 324–33, 340–51, 357, 406, 407–14, 444, 447–50, 452
Faris, R.E.L., 176, 257, 259, 260, 261, 263, 403, 430
Farley, R., 280, 318
Farrell, W.C., Jr., 175, 188–93
Fawcett, C.B., 36
Feagin, J.R., 157
Felson, M., 222, 223
Fertility, 183–84, 311, 319, 342, 343–44, 346, 348, 349, 362
Fine, J., 280
Firey, W., 6, 7, 82, 129–36, 150–64, 165, 167, 318, 357
Fischer, M.M., 332

Fisher, C., 46
Fisher, R., 186
Flanagan, W.G., 429
Fletcher, J., 13, 16–17
Fluidity, 29, 31, 32, 60, 263
Foley, D.L., 176–77, 266–72, 273
Forbes, S.A., 125
Force, M.T., 300, 324
Forrest, J., 298
Fox, R.G., 447
Frazier, E.F., 175, 179–87, 188–93
French, R.A., 155
Frère, S., 425
Fried, M., 161, 252, 264
Friedman, J.J., 282
Functional analysis, 6, 153, 154, 156–57, 162, 266, 267, 317, 318, 321, 372, 384, 386, 390, 400, 452

Gadgil, D.R., 444, 451
Gagnon, G., 341
Gans, H.J., 162, 239, 252, 254, 255
Gardner, P., 245
Gehlke, C.E., 5
Gentrification, 157, 241, 322, 333, 357, 358, 377, 430, 463
George, P., 425
Gerard, D.L., 259
Gettys, W.E., 82, 83, 85, 165, 169
Ghosh, S., 399
Gibbs, J.P., 151, 168, 170
Giggs, J.A., 298
Gillin, J., 359, 361
Gist, N.P., 389, 399
Gittus, E., 303–04, 341, 399
Glass, D.V., 298
Glass, R., 430
Glenn, N., 280
Goheen, P.G., 298, 299, 341
Goldstein, H., 299
Goodey, B., 253, 254
Goodman, L., 88, 89–92, 93, 94, 98
Gradients, 78, 79, 80, 81, 182–86, 191–93, 287, 331, 360, 367, 371, 374, 375, 376, 377, 390, 400, 402, 403–04, 427, 429, 430, 456, 457
Gras, N.S.B., 52
Gray, J.H., 390
Green, H.W., 230, 398
Greer, S., 162, 230, 303, 324
Griaule, M., 159
Grid pattern, 6–7, 142–49, 154, 228, 380, 414, 427, 428
Grunfeld, B., 260
Guerry de Champneuf, A.M. de, 8–9, 17
Guest, A.M., 177, 273–86, 299

Habitat, 5, 21, 22, 24, 25, 26, 108, 109, 110, 124–25, 198
Haeckel, E., 3, 21
Hafner, H., 259
Hagood, M.J., 79, 324, 340
Halkett, I.P.B., 293
Hall, P., 299
Hamden, G., 399
Hamilton, F.E.I., 155
Hamm, B., 177, 316–37
Hamnett, C.R., 430
Handlin, O., 235
Hansen, A.T., 355, 359, 360, 367, 368, 375, 376, 399
Harris, C.D., 274
Harvey, D., 161

Hatt, P., 5, 78–81, 82, 83, 299
Hauser, F.L., 399, 429, 430
Hauser, P.M., 183
Havell, E.B., 440
Hawley, A.H., 5–6, 81, 84, 85, 104–14, 118–19, 150, 151, 222, 241, 266–67, 300, 318, 320, 375, 400
Hawthorn, A.E., 355, 361, 399
Hawthorn, H.B., 355, 361, 399
Hayner, N.S., 355, 360–61, 367, 376, 399
Haynes, K.E., 312
Herbert, D.T., 298, 302, 312, 341
Hermalin, A.I., 280
Herman, H., 282
Herzfeld, E.E., 145
Hill, M.D., 13
Hillery, G.A., 230
Hinterland, 31, 56, 300, 370, 371, 414, 439, 441–42
Hippodamus, 144, 145, 146, 147
Hitchcock, S.T., 272
Hollingshead, A.B., 5, 82–87, 260, 262
Homans, G.C., 166
Hoover, E.M., 282, 430
Hoover, K.E., 175, 202–06
Horowitz, C.M., 235
Horton, F.E., 312, 331
Houston, L.G., 259
Hoyt, H., 4, 42–49, 318, 321, 371, 399, 415
Hudson, J.R., 176, 241–48
Hughes, E.C., 85, 86
Hughes, J.W., 310
Human geography, 4
Hunter, A., 161, 212, 227, 245, 312, 331, 332
Hurd, R.M., 51, 371
Huxley, J., 24, 25

Iklé, F.C., 271
Industrialism, 12, 30, 160, 197, 282, 300, 339–40, 348, 355, 356, 358, 359, 360, 362, 389, 392, 396, 414–15, 429, 438, 439, 440, 441–43, 449, 451, 452
Invasion, 2, 22, 30, 33, 37, 39, 51, 130, 175, 176, 196, 198, 212, 216, 217, 218, 241–48, 267, 323, 356, 357, 369, 426, 457

James, P.E., 4, 61–66, 406
Janowitz, M., 150, 230
Janson, C.G., 324, 331
Johnson, P., 430
Johnston, N., 399
Johnston, R.J., 177, 297–315, 316, 430
Jonassen, C.T., 175, 194–201
Jones, C.L., 368
Jones, E., 161
Jones, F.L., 300, 304
Jones, M.A., 293
Joy, R.J., 211–12, 213

Kain, J.F., 273
Kantor, M.B., 264
Kaplan, L., 235
Kasarda, J.D., 282, 318
Kaufman, W.C., 321, 324, 341, 342
Keller, S., 252
King, M., 137
Kirchwey, G.W., 17
Kiser, C.V., 183
Koch, S., 243, 244, 245
Kohn, M.L., 259
Konrad, G., 159
Kroeber, A.L., 123

Kruskal, J.B., 276
Kube, E., 324
Kumagai, J.G., 331

Ladd, F., 255
Land values, 40–41, 51, 53, 57, 60, 70, 131–32, 133, 158, 317, 318, 322, 330, 332, 333, 357, 360, 369, 372, 380, 418, 426–27, 457
Landes, R., 234
Lansing, J.B., 170
LaPierre, J., 425
LaPouse, R., 260
Latif, A.H., 212
Lavandeyra, L.A., 425
Lavedan, P., 427
Lawton, R., 298
Lazerwitz, B., 214
Learmonth, A.T.A., 447
Lee, T., 254, 255
Leiffer, M.H., 79
Lemon, A., 155
Le Tourneau, R., 395
Levels of analysis, 7, 26, 73–77, 81, 83, 84, 123, 124, 165, 169, 170
Levin, Y., 3, 11–18
Lévi-Strauss, C., 158–59, 160
Levy, L., 176, 257–65
Lewin, K., 214
Lewis, G.J., 310
Lieberson, S., 212
Liebman, C.S., 282
Liepmann, K.K., 267, 270, 273, 287
Lincoln, J.L., 359
Lind, A.W., 358, 454–64
Lindesmith, A., 3, 11–18
Lintell, M., 254
Lombroso, C., 16, 17
London, B., 356, 357, 374–78, 429–32
Longmoor, E., 398
Lyall, R., 400
Lynch, K., 253
Lynd, H., 223
Lynd, R., 222, 223

MacCannell, E.H., 303, 321, 341
MacDermott, W.R., 258
MacIver, R.M., 123
Macro-level, 6, 7, 111, 112, 165–66, 167–68, 169, 170–72
Maine, H., 116
Manning, I., 177, 287–96
Manton, K., 274
Martin, W.T., 168, 170
Masterman, M., 151
Mauss, M., 158
Mayer, A.C., 446
Mayer, P., 155
Mayhew, H., 13–16, 17–18
McAdams, D.C., 157
McBryde, F.W., 356, 359, 361
McElrath, D.C., 300, 303, 311, 319, 324, 341, 342, 344, 349, 399, 430
McGee, T.G., 162
McKenzie, R.D., 4, 28–34, 74, 81, 120, 222, 316, 331
McQuitty, L.L., 310
Mechanical solidarity, 116, 117, 119
Medical ecology, 3, 19, 127
Medina, C.A., 408
Mehta, S., 399

Melbin, M., 222, 223
Mental illness, 18, 39, 108, 176, 257–65, 403
Menzel, H., 88
Menzies, M., 253, 254
Menzler, F.A.A., 271
Merrick, G., 46
Merton, R.K., 162, 238
Meyer, D.R., 311
Michelson, W., 165, 169
Micro-level, 7, 111–12, 165–66, 167, 170–72
Migration, 85, 110, 202–05, 219, 220, 258, 311, 327, 332, 342–43, 361, 362, 409, 410, 411, 413, 414, 417, 418, 420, 425, 427, 435, 442, 449, 452, 454
Milgram, S., 253
Millstein, G., 244, 245, 246
Miner, H., 399
Mitchell, R.B., 268
Mobility, 4, 29, 36, 39–41, 60, 85, 191, 207, 208, 209, 211–12, 214, 216, 311, 332, 343, 348, 360, 372, 391, 427, 436, 442, 446
Mogey, J., 399, 430
Molotch, H., 158
Monk, M., 260
Monts, J.K., 280
Moore, R., 157
Morris, F.B., 357, 406–15
Moscheles, J., 399, 430
Moskos, C.C., 300
Mott, P.E., 168
Moush, E., 341
Mowrer, E., 403
Mueller, E., 170
Mukerjee, R., 357, 433–43
Mulcahy, P., 7, 165–72
Murdie, R.A., 311, 312, 331, 341
Murphy, H.M.B., 264
Murray, G., 26
Musil, J., 430
Myers, J.K., 320
Myerson, A., 260

Natural areas, 4, 5, 23, 51–54, 56–57, 78–81, 158, 207, 233, 254, 299, 318, 319–20, 339, 439, 442, 445
Neoclassical ecology. *See* Neo-orthodox approach
Neo-orthodox approach, 5–6, 7, 104–28, 151–52, 165–72, 202, 333
Neutze, M., 289, 293
Newcomb, C., 403
Niche, 21, 24, 106, 391
Nicholson, T.G., 212
Noble, A.G., 155

O'Brien, R.M., 176, 221–23
O'Brien, R.W., 175–76, 218–23
Odum, E.P., 124
O'Farrell, P.N., 304
Ogburn, W.F., 167, 168
Olsson, W.W., 399
Organic solidarity, 116, 117, 119
Orleans, P., 321

Pahl, R.E., 160, 161
Palm, R., 300, 311, 313
Paradigms, 7, 150–52, 163
Parisse, L., 416
Park, R.E., 3, 4, 20–27, 52, 74, 76, 77, 81, 83, 85, 86, 120, 121, 124, 167, 207, 216, 224, 225, 226, 229, 316, 371, 454–55
Parkes, D.N., 303, 309, 311

INDEX

Parsons, T., 153
Peach, C., 318
Pedersen, P.O., 341
Perin, C., 159
Perlman, J.E., 156
Pfautz, H.W., 298
Pickvance, C.G., 161
Pirie, G.H., 331
Planning, city, 35, 36, 50, 51, 52, 53, 54, 138, 140, 143–49, 151, 155–56, 162, 225–27, 228, 243, 245, 249, 254, 291–93, 295, 298, 317, 319, 356, 357, 361, 366, 369, 371, 395, 414, 429, 431, 439–41, 443, 445–46
Plant ecology, 3, 21, 23, 25, 26, 37, 52, 74, 75–76, 84, 106–7, 108, 109, 124
Plazas, 143, 154, 355, 356, 360, 361, 363, 365, 367, 371, 379–80, 381, 382, 383, 384, 388, 389, 395, 399
P,O,E,T, 6, 7, 126–27, 168
Polk, K., 324, 341
Poll, S., 234
Ponce, Fray A., 380
Poole, M.A., 304
Portes, A., 377
Power, 7, 24, 114, 144, 146–49, 153–58, 189, 317, 318, 360, 371, 389, 414, 452
Preindustrial city, 356, 357, 375, 386–93, 398–405, 414–15, 429, 439–41, 448
Price, D.O., 324
Prostitution. *See* Vice
Pyle, G.F., 357, 406–15

Queen, S.A., 259, 398
Quetelet, A., 8, 9, 17
Quinn, J.A., 5, 81, 82, 83–84, 119, 317, 331

Radial plan, 143, 427, 438
Raine, J.W., 324
Ranyak, J.A., 269, 272
Rapkin, C., 243–44, 247, 268, 272
Ratzel, F., 158
Reckless, W.C., 4, 55–60
Redfield, R., 122, 359, 360, 372
Redlich, F.C., 260, 262
Rees, P.H., 304, 310, 311, 324, 341, 343, 444, 447, 452
Reimann, H., 259
Reiss, A.J., Jr., 274
Rex, J.A., 157, 430
Rico-Velasco, J.L., 376, 429
Riis, J., 50
Ritzer, G., 150, 151, 163
Roberts, B.R., 160
Robinson, W.S., 5, 88, 89–91, 92, 93, 94, 95, 98, 309
Robson, B.T., 341, 430
Root, E., 51
Rosen, L.S., 176, 230–40
Rosenmayr, L., 389
Rosenthal, D.B., 448, 451
Ross, H., 253–54
Rowitz, L., 176, 257–65
Rummel, R.J., 309
Russett, B.M., 324

Saarinen, T., 250
Sacred sites, 130, 133–34, 217, 439, 440
Salins, P.D., 331
Salusbury, G.T., 387
Salvesen, C., 260
Sandercock, L., 293
Schaeffer, K.H., 288

Schaeffer, R., 271, 425, 426
Schmid, C.F., 303, 310, 311, 320, 321, 340, 341, 343, 403
Schnore, L., 6, 115–22, 166, 167, 273, 274, 285, 333, 375, 376, 398, 399, 429, 430, 431
Schroeder, C.W., 259
Schutz, A., 152
Schwirian, K.P., 312, 376, 429
Sclar, E., 288
Sears, P.B., 125
Sector theory, 4, 42–49, 159, 161, 318, 321, 330–31, 377, 430
Seeman, A.L., 6, 137–41
Segregation, 4, 30, 33, 39, 51–52, 53, 55, 56, 60, 156, 161, 162, 175, 177, 179, 182, 187, 189, 193, 207, 208, 209, 211, 216, 219, 220, 266, 299, 300, 302, 311, 318, 319, 346, 348, 377, 390, 434, 441, 456
Senior, M.L., 316
Sharpe, G.B., 272
Shaw, C.R., 74, 459
Shaw, D.J.B., 155
Shelford, V.E., 125
Sherif, C., 324
Sherif, M., 324
Shevky, E., 177, 210, 211, 299, 300, 301, 302, 303, 304, 309, 310, 311, 312, 317, 318, 319–24, 330, 333, 339, 340, 341, 342
Shively, W.P., 88
Simmel, G., 25, 116
Sinclair, U., 50
Sjoberg, G., 7, 150–64, 165, 169, 356, 357, 358, 375, 376, 377, 386–93, 399, 404, 414, 429, 431, 448
Sklare, M., 234
Slums, 23, 33, 38, 50, 56, 57, 58, 59, 60, 67, 68–70, 136, 156, 162, 176, 198, 224, 230, 243, 252, 267, 346, 356, 357, 358, 360–61, 365, 366, 369, 375, 377, 416, 417, 419, 420, 422, 430, 441, 443, 457, 460
Smith, A., 116, 117
Smith, A.B., 295
Smith, D.M., 313
Smith, J., 274
Smith, P., 231
Smith, R.K., 312, 376, 429
Snow, D.A., 157
Snow, J., 19
Social area analysis, 177–78, 210, 297–303, 312, 316–24, 332–33, 338–51
Social morphology, 6, 115, 119–22, 333
Sociocultural approach, 5, 6–7, 84–87, 129–72, 194, 195, 198–99, 201, 202, 206
Sopher, D.E., 162
Sorokin, P.A., 151, 161
Spate, O.H.K., 447
Spearritt, P., 288
Specialization, 31, 32, 113, 266, 311, 318, 322, 327, 333, 348, 391, 426, 439, 441, 442, 444, 456
Spence, N.A., 310
Spencer, H., 17, 21, 116, 117
Spiro, L., 253
Spodek, H., 312, 358, 444–53
Stamp, L.D., 3, 19
Stanislawski, D., 6–7, 142–49, 356, 379–85, 399
Stea, D., 250
Steffens, L., 50–51
Stein, L., 259
Steiner, J.F., 222, 223
Steward, J., 119, 122
Stinchcombe, A.L., 152, 160, 161, 162
Stone, G.P., 161

Stouffer, S.A., 84–85, 427
Street, D., 150
Stretton, H., 293, 295
Struening, E.L., 324
Struggle for existence, 3, 20, 21, 28, 52, 75, 77, 83, 106, 107, 118
Subsocial level, 3, 5, 73, 83, 85, 86, 87, 104, 194, 195, 198, 333
Suburbs, 23, 36, 37, 47, 130, 156, 157, 159, 175, 177, 203, 204–05, 206, 209, 212, 216, 230, 231, 241, 252, 267, 268, 273–86, 288–89, 293, 294, 295, 318, 331, 360, 362, 365, 366, 367, 368, 369, 375, 377, 389, 399, 403, 415, 417, 425, 426, 428, 448–49, 451, 463
Succession, 4, 23, 24, 30, 33, 37, 41, 53, 54, 86, 106, 161, 175, 176, 196, 212, 216, 217, 218–20, 224, 242–47, 318, 323, 356, 357, 366, 369, 415, 426
Suicide, 3, 8, 18, 69, 72, 121, 358, 455, 456, 457, 461, 463, 464
Sullivan, T., 324, 341
Survivals, 160, 161, 373, 382
Sustenance activities, 5–6, 28, 110, 201
Sutherland, J.F., 258
Suttles, G.D., 161, 176, 207, 216, 224–29, 230, 250, 252
Sween, J., 341
Sweetser, F.L., 303, 304, 324, 331, 332, 341, 342, 343
Symbiotic relationships, 3, 21, 25, 26, 74, 78, 80, 81, 106, 107, 108, 110
Symbolism, 6, 7, 129–36, 151, 152, 153, 154, 157, 161, 162, 214, 239, 318, 357, 389, 438
Szelenyi, I., 160, 430–31

Taeuber, A., 276
Taeuber, K., 276
Tagashira, K., 311, 341, 343
Tan, M., 298
Tansley, A.G., 124
Tarbell, I.M., 50
Tarde, G., 77
Tarrant, J.R., 305
Tax, S., 359, 361, 366
Taylor, P.G., 310
Technology, 113, 118, 120, 126–27, 153, 157, 169, 170–71, 176, 197, 242, 267, 316, 318, 386, 387, 388, 389, 390, 391, 392, 444
Temporal ecology, 175–76, 218–20, 222–23, 266–72, 312, 391–92
Terris, M., 260
Theodorson, G.A., 377, 431
Thomas, W.I., 454
Thomlinson, R., 429
Thompson, G.A., 368, 372
Thompson, J.A., 20, 21
Thomson, J.M., 294, 295
Thrasher, F., 74
Timms, D.W.G., 303, 310, 312, 316, 320, 324, 341, 429
Tönnies, F., 116, 122
Tricart, J., 426
Tryon, R.C., 303, 312, 320, 324, 340, 341, 342
Tumin, M.M., 359
Turner, R., 260
Turner, S., 13

Udry, J.R., 300, 321–22
Urban morphology, 4, 109
Urban renewal, 162, 176, 224–27, 358
Uyeki, E.S., 398

Vallee, F.G., 213
Van Arsdol, M.D., Jr., 303, 310, 320, 321, 341, 343
Veldman, D.J., 312
Vernon, R., 282, 430
Vice, 4, 38, 39, 40, 50, 55–60, 69, 358, 435, 442–43, 455, 456, 457, 459, 461, 463, 464
Von Gerkan, A., 143, 145

Wachs, M., 331
Wagenfeld, M., 260
Walton, J., 377
Ward, D., 298
Warner, W.L., 223
Warren, R., 250
Watson, J.E., 272
Web of life, 20–22, 25, 82, 83, 108
Wechsler, L., 250
Weclawowicz, G., 327
Weinstein, J., 161
Wells, G.P., 24, 25
Wells, H.G., 24, 25
Wendling, A., 324
Wheatley, P., 152
Wheaton, W.C., 288
Wheeler, W.M., 25
White, W.A., 258
Whitman, C.O., 75
Whyte, W.H., 239
Wiegand, T., 147
Willems, E., 408
Willhelm, S.M., 165, 166–67, 168, 169, 171
Williams, M., 177, 299, 317, 339, 341
Williams, O.P., 282
Williams, P., 430
Williamson, R.C., 324
Willmott, P., 253, 255, 311
Wilson, G., 300, 319
Wilson, L.S., 399
Wilson, M., 300, 319
Wilson, W.H., 31
Wirth, L., 73, 224, 300, 319, 320, 400
Wissink, G.A., 160
Wohl, A.S., 430
Wood, R.C., 282
Wright, A.F., 152
Wright, A.O., 257–58

Yeates, M.H., 212
Young, B., 137, 139
Young, E., 398
Yujnovsky, O., 376–77

Zborowski, M., 234
Zone of transition, 36, 51, 56, 59–60, 78, 318, 327, 331, 333, 355, 360, 367, 369, 376, 431, 457, 463
Zorbaugh, H.W., 4, 50–54, 67–72, 74, 230, 233
Zuiches, J., 299